Imre Baski

Crimean Tatar Folktales

Studien zur Sprache, Geschichte und Kultur der Turkvölker

Volume 38

Imre Baski

Crimean Tatar Folktales

As Collected by Ignác Kúnos (1860–1945)

DE GRUYTER

ISBN 978-3-11-144197-9
e-ISBN (PDF) 978-3-11-144289-1

Library of Congress Control Number: 2024937892

Bibliographic information published by the Deutsche Nationalbibliothek
The Deutsche Nationalbibliothek lists this publication in the Deutsche Nationalbibliografie; detailed bibliographic data are available on the internet at http://dnb.dnb.de.

Cover image: Marco Polo's caravan on the silk road. Painting by Abraham Cresques, 1375, Catalan Atlas. Source: Gallica. Digital Library ID btv1b55
Typesetting: Integra Software Services Pvt.

www.degruyter.com

Contents

Introduction

In this volume we publish a new part of the collection of Crimean Tatar folklore texts by Ignác Kúnos (1860–1945), collected under extraordinary circumstances more than a century ago.

Ignác Kúnos was born Ignác Lusztig into a Jewish merchant family in Debrecen on 22 September 1860. He attended the famous Reformed College in Debrecen, Hungary, and then studied linguistics at the University of Budapest between 1879 and 1882. As a pupil of Ármin Vámbéry (1832–1913), his interests turned to the Turkish language and philology.

With financial support from the Hungarian Academy of Sciences and the Jewish community of Budapest, he spent five years in Constantinople studying the Turkish language and culture. Between 1885 and 1890 he lived in Constantinople, travelling through Asia Minor and the Middle East (what was then Turkey) as far as Egypt. During his travels he collected Turkish folklore texts. His collection of Turkish folktales and anecdotes is still an irreplaceable source. While working as a collector, he also noted the peculiarities of Turkish dialects. His achievements in documenting the traditional Turkish shadow play (*Karagöz oyunu*) and the Turkish open-air theatre play (*ortaoyunu*) are also significant. In Turkey he is rightly considered to be the pioneer of Turkish folklore research (Szilágyi 2007: 348).

In 1890 he was appointed professor of Turkish Philology at the University of Budapest. In the summers of 1925 and 1926, at the invitation of the Turkish government, he was a visiting professor at the universities of Ankara and Istanbul, and in 1925 he organised the folklore department at Istanbul University.

He was one of the most respected scholars of his time in Turkish folk literature and Turkish dialectology. In 1889 he was an external member of the Société Asiatique in Paris. In 1893 he was elected as a correspondent member of the Hungarian Academy of Sciences. He was also admitted as a member of the Deutsche Morgenländische Gesellschaft, the Istanbul Scholars' Society and the Finno-Ugric Society in Helsinki. He also held the honorary title of Vice President of the International Society for Central and East Asia.

As is seen, Ignác Kúnos was already a respected scholar with great collecting experience at the time of the camp work in question; since he had already made significant folklore and dialect collections, he entered the camp of the "Tatar prisoners" with considerable (Ottoman) Turkish folklore knowledge and collecting experience.

Kúnos calls his Crimean Tatar collection *Tatár népköltési gyűjtemény* [A collection of Tatar folk poetry]. Then, on another manuscript page, he gives the whole material the title *Krími tatár népdalok és népmesék. Az égeri és kenyérmezei tatár foglyok táborában (1915–18 év nyarán) gyűjtötte és feljegyezte Kúnos Ignácz* [Crimean Tatar folk songs and

https://doi.org/10.1515/9783111442891-001

folk tales. Collected and written down by Ignácz Kúnos in the camp of Tatar prisoners in Éger and Kenyérmező (summer 1915–18)].[1]

Notes on life in prisoners' camps

Ignác Kúnos described life in the camps and the circumstances of the collection in his report at a meeting of the I. Section of Linguistics and Literary Scholarship of the Hungarian Academy of Sciences on 3 January 1916. His lecture (1916) was later published in the *Budapesti Szemle* [Budapest Review] and, two years afterwards, a new publication appeared in the *Magyar Figyelő* [Hungarian Observer] entitled *Tatár foglyaink táborában* [In the camp of our Tatar prisoners] (Kúnos 1918). The collection was made known to Hungarian and foreign scholars through the detailed descriptions of Zsuzsa Kakuk (1963, 1978, 1987). Later, István Seres explored the topic and, with the help of new archival sources, he managed to draw a much more detailed picture of the conditions in the camps and the collection work carried out there (2003, 2006, 2010). Most recently, two studies in Turkish (Szilágyi 2007, Çolak 2019) have focused on the work of Ignác *Kúnos* and Turkological research in Tatar prison camps.

Forty-two prisoner-of-war camps were set up in the territory of the Monarchy, and the scholars soon noticed that many of the Russian prisoners of war came from the Finno-Ugric and Turkic peoples of the Russian Empire. In January 1915 the joint Department of War permitted the Hungarian Academy of Sciences members to enter the nine camps set up in Hungary. Among the researchers who visited those camps, Ignác Kúnos was the only Turkologist who had gained international fame by collecting Ottoman Turkish folk songs. By this time he had been studying the Kipchak language for several years. In the years preceding World War I he had visited Dobrudja near the Lower Danube, where he had studied the Nogai (or, as he called it, the "Danubian Tatar" language (Seres 2010: 139). The Russian Muslim subjects who had been taken prisoner by the Austro-Hungarians were gathered in two places. One camp, with nearly 8,000 prisoners, was located in Kenyérmező near Esztergom, Hungary. Much more populous was the other, a prison camp near Éger (now Cheb, Czech Republic), which had up to 30,000 prisoners. The Kenyérmező camp held mainly Crimean and Kazan Tatars. In contrast, the Éger camp held Kazan and Misher Tatars from the Volga region, Bashkirs, Kumyks, Nogais, Turkmen and many other ethnic groups (Kúnos 1916: 209).

Each barracks lining the camp's wide streets housed soldiers from a different Turkish nation. Thousands of Mongoloid-looking people lived together comfortably within the framework of Muslim customs. In the middle of the street, where most of the

1 Found in Magyar Tudományos Akadémia Könyvtára, Keleti Gyűjtemény [Library of the Hungarian Academy of Sciences, Oriental Collection]: Török Qu. 110/II.

barracks were located, stood a prominent building, the great bathhouse. The camp had a bakery and shops for prisoners who wanted to improve their diets.

The prisoners' barracks were furnished in a cosy style. Each prisoner had a separate bed and a wooden shelf above. On this shelf were family photographs and letters that reminded them of their homeland.

The Muslim prisoners were not forced to work too hard. They were only responsible for keeping the large camp area tidy and clean (Kúnos 1918: 143–144). In the camp the prisoners could pursue their professions. There was a master for each craft. The workshop had the necessary tools for each skill so the artisans could engage the arts they used in their home countries (Kúnos 1918: 147).

According to Kúnos (1918: 147), the authorities met all the needs of the prisoners, made their long years of statelessness bearable and set up medical institutions for their physical health.

By the requirements of their religion, their cooks cooked their food and their bakers baked their bread (Kúnos 1918: 147). Kúnos (1918: 151), whose relationship with the prisoners was friendly and who tried to remedy their problems, describes in detail how the prisoners spent their time in the camps playing various parlour games, such as cards or puppet shows mocking the Russians.

In the middle of the prisoners' barracks, there was a fenced-off area where the prisoners could perform their daily prayers undisturbed on prayer mats of their own making. Religious services for the community were held only on Fridays in the large mosque. When the imam of the camp called them to pray, all distinction between them ceased and they gathered together (Kúnos 1916: 213; 1918: 144).

The hodja, who came straight from Istanbul, often visited the camps to inform and cheer up the prisoners, and he went on the same mission on the feast days. A theatre and a reading room were even built for the prisoners in the camp. The playwrights of the theatre and the actors were Crimean and Kazan Tatars. Ten plays were performed in a season (Kúnos 1918: 152–153).

On the collection of Crimean Tatar tales

Ignác Kúnos proudly states in his camp report (1916: 212), which has been quoted several times, "I also managed to record the best of their folk tales, the so-called *ekejets*".[2] We can only regret, however, that Ignác Kúnos did not record, or at least did not leave us, details of whom he collected which tales from, and from which regions or settlements the storytellers came.[3] This omission is inexplicable because, while Kúnos

2 Translated from Hungarian.

3 Even during his collections in Turkey, Kúnos did not record the names or other details of the data providers (Szilágyi 2007:162).

was collecting, Austrian researchers made phonogram recordings with Crimean Tatar informants.[4] For each recording, a datasheet (protocol) was drawn up, which contained all the essential data, such as the name, age, occupation, place of origin of the respondent, the time of the recording and the name of the person who made the recording. The recorded texts, including the lyrics of Crimean Tatar songs, were recorded in Arabic script by the Crimean Tatar Üseyn Asan, a vineyard owner from Kapsihor[5] (Lach–Jansky 1952: 72). Kúnos also contributed to the recording. He was responsible for transcribing the texts of the recordings into Latin script letters for the protocols.[6]

The phonogram recordings include only one tale, but a version of it, *Bir noγaj masalı* [The tale of a Nogai], is also in the collection of Kúnos. The story that survived on the phonogram record was told by a 29-year-old farmer from Akmesdzhit (now Simferopol), Fazil Edilerski, on 8 August 1915 in the prison camp in Éger (Liebl, Lechleitner and Remmer 2018, CD 1: 19). However, there are significant differences between the two tales, so it is unlikely that the same person told them both.

Of all the tales of Kúnos, only the teller of *Kičkene ajdamak masalı* [The tale of the little bandit] is likely to be identified since the title of the tale is preceded by the inscription *Krími mese Mehemmed (akmeǯid)* [Crimean tale Mehemmed (Akmeǯid)]. It is quite evident that, in this case, Kúnos was referring to the informant and his residence.

The contributors to the previous story and the songs in the Phonogramm–Archiv were all from the southern dialect area. Five came from the Simferopol (Akmesdzhit) district, two from different villages in the Yalta district, one from Bakhchisaray and one from Kapsihor (now Morskoye) in the Feodosiya district. Some of the tales may have come from the same person.

Even though, among the data providers known from the Phonogramm–Archiv, we see farmers, vine-growers, gardeners and merchants, we can reasonably assume that there were also educated people (teachers, mullahs) among the storytellers. In the context of his collecting work among Kazan Tatars, Kúnos (1916: 215) mentions that he

4 The recordings were made for the Phonogramm-Archiv of the Imperial Academy in Vienna by Rudolf Pöch and Hans Pollak in August 1915–1916 (Kúnos 1916: 224; Liebl, Lechleitner and Remmer 2018, CD 1: 19–28). The recordings in question were made for Robert Lach, who used them for his research on music (see Lach–Jansky 1930, 1952).

5 He (1930: 88) mentioned *Feodosiya* earlier in this context.

6 Nevertheless, the transcription in the appendix to the CD release, cited above (Liebl, Lechleitner and Remmer 2018), is not his. Perhaps this is because Herbert Jansky (1952: 73), in his foreword to the Kazan Tatar folk song collection, severely criticised Kúnos' transcription system. Among other things, he wrote: "Das angewendete Transkriptionssystem ist mehr als primitiv und so inzulänglich, daß es gelegentlich Zweifel offenläßt; ..." ['The transcription system used is more than primitive and so inadequate that it occasionally leaves doubts]. Ildar Kharissov, a native speaker of Uzbek who produced the new transcription, has transcribed the text of the recordings with numerous errors, and in many places the transcription reflects the sounding of his own mother tongue.

visited the "most valuable elements of the Kazan Tatar prisoners" and had contacts with their "leading figures". The intellectuals and leading stratum of the Muslim prisoners of war knew Ottoman Turkish and could be relied on to help him in case of interpretation difficulties.

While collecting work, Ignác Kúnos used notebooks, from which he later copied the material into half-sheets (15×21 cm) of paper and made a fair copy. The Kúnos legacy, therefore, consists of fair-copied texts written on sheets of paper.

During the collection or subsequent verification, Kúnos would try to find the meaning of unknown words and insert them in brackets in the appropriate place in the text. These words in brackets are usually Ottoman Turkish and, only rarely, Hungarian. This leads us to conclude that someone who knew Turkish was there to help him check the text. Kúnos also retained the bracketed words in his explanatory notes. Still, in most cases, he used strikethroughs to indicate that they were not an integral part of the text and were only necessary for the Hungarian translation he would prepare later.

However, copying the folklore texts written down in the camp for clarification became a source of further errors. The text, already inaccurately and incorrectly transcribed in the draft, deteriorated even further during the copying process. The fact that Ignác Kúnos did not know well enough the Turkic languages in which the folklore texts he had written down were spoken could have played a significant role in this.[7]

According to Kúnos' incomplete table of contents, the collection of Crimean Tatar tales and the Hungarian translation originally consisted of 444 numbered half-sheets (A5) in the manuscript. Two stories (*Muradına irmegen dilber* [Beauty with an unfulfilled wish], and *Bir baj kiši masalı* [A rich man's tale]) were omitted from the 36-title table of contents, and *Čora-batır ekjeti*[8] [*Chora-batir's tale*] is listed at position 21 in the table of contents. Still, in reality, the manuscript pages (648–689) containing the tale are at the end of the collection, so Chora-batir's tale, by the continuous manuscript numbering, is listed at position 38 in the corrected table of contents. The text of Chora-batir's tale is incomplete and its pages have been mixed up when copied, as seen in the text that Kúnos has clarified. The last (i.e. 39th) tale, *Kičkene ajdamak masalı* [The tale of the little bandit], occupies 19 unnumbered half-sheets and was found later, without a Hungarian translation, in a notebook in another location of the inheritance.

István Seres (2003, 2010) published the full text of the heroic epic of Chora-batir, which is part of the collection, in Hungarian and in English. The story of Chora-batir was not included in the collection of tales by chance. The predominantly prosaic text is

7 Contrary to our claim, however, the fact that shortly after the outbreak of the "Great War" he wrote a letter to the War Supervisory Committee, applying for a position as a Tatar interpreter, is a cause for consideration (Seres 2006: 169, 174–175).

8 The real title of the tale is *Čora batırnın ekejesi* [The tale of Chora batir].

interrupted only here and there by passages in verse, which are more characteristic of the epics (*destans*).

From a folkloristic point of view, Seres (2003, 2010) considers the text to be part of the Chora-batir epic cycle. He places it accordingly in the series of corresponding epics of the various Turkic peoples (Kazakh, Nogai, Crimean Tatar, Dobrudja Tatar, Kazan Tatar). By comparing the different language versions of the epic, the author has come to several valuable linguistic and folkloristic conclusions.

Despite the recurrence of specific motifs in the texts published here, the Crimean Tatar tales are very varied in content. The stories of padishahs, padishahs' sons and poor, sometimes simple-minded little boys, full of magical elements and enchanting motifs, reveal the richness of the Crimean Tatar folk imagination.

The tales' protagonists are simple people or padishas or sons of padishas – usually three brothers – the youngest of whom is always on the beam. The elder brothers usually deceive, betray or even try to kill their youngest brother, who, in the end, rescues them or kills them for that very reason. During his struggles he is aided by magical old men, old women or demons while, at the same time, giant dragons and five-, seven-, forty-, eighty- and one hundred-headed devils try to destroy him. In most of the tales, we can meet talking animals with magical powers (snakes, dogs, cats, mice, foxes, horses) and a variety of magical tools, including a magic ring, a mirror, a stick, a whip, a sword (e.g. the Zilfikar[9]), etc. Of these animals, the most common is a talking steed, capable of supernatural feats, which plays the hero's faithful companion.

Language and dialect features of the texts in the collection

Kúnos has studied the dialects of the Crimean Tatars with particular attention. The volume of his collected Crimean folklore material bears testimony to his efforts. In his collection, he bequeathed posterity the texts of several hundred folk songs (*mani, ǯır, čın*),[10] one and a half hundred riddles (*tapmaǯa*)[11] and thirty-nine folktales (*ekjet, masal*).[12] During his collection work he had already recognised the difference between the two main dialects of the Crimean Turkic [Oghuz and Kipchak, I. B.], which he believed to consist mainly of phonetic differences (Kúnos 1916: 210).

9 Zilfiqar (< Ar. Zulfiqar), the legendary double-bladed, magical sword of Caliph Ali, given to him as a gift by the Prophet Muhammad.

10 Published in Turkish by Zsuzsa Kakuk (1993a).

11 Published in Hungarian and German by Zsuzsa Kakuk (1993b, 1994).

12 The current state of publication of the collection is presented by Imre Baski (2019).

Transcription system

At the turn of the century, most of the letters of the transcription system used by Ignác Kúnos were commonly used in Turkology and Finno-Ugric studies. Heikki Paasonen's (1865–1919) system includes (χ), (j), (š) and (ž). In the works of Wilhelm Pröhle (1871–1946), for example, we see the letters (ı with ring above), (č), (ǯ), (ү), (χ), (j) and (š). Gyula Németh (1890–1976) also used the letters (ı with ring above), (č), (ǯ), (ү), (χ), (j) and (š) in his (1911) Kumuk and Balkar glossary. Németh started to denote the velar *i* sound without a dot (ı) relatively late. In contrast to his contemporaries, Kúnos did not use the letters (q), (ŋ), (w), (y), and (ä) in his transcriptions.

Table 1: Comparative table. A comparison of Kúnos' transcription with later official writing systems.

Alphabet of Kúnos (1915)	Cyrillic alphabet (1938–)	Latin alphabet (1997–)	System of IPA[13]
a, ā	а	a, â	a
b	б	b	b
ǯ	дж	c	ʤ
č	ч	ç	ʧ
d	д	d	d
e	е, э	e	ɛ
ė	е, э	e, ı	e, ɘ
f	ф	f	f
g	г	g	g
ɣ (ġ[1])	гъ	ğ	ɣ
h			h
χ	х	h	x
ı[2]	ы	ı	ɨ, ɯ
i	и	i	i
j	й	y	j
k	к	k	k
k	къ	q	q
l	л	l	l (ɫ)
m	м	m	m
n	н	n	n
ñ	нъ	ñ	ŋ
o	о	o	o
ȯ	о	ö	ɵ
ö	ё	ӧ̆	ø
p	п	p	p
r	р	r	r
s	с	s	s
š	ш	ş	ʃ

13 International Phonetic Alphabet.

Table 1 (continued)

Alphabet of Kúnos (1915)	Cyrillic alphabet (1938–)	Latin alphabet (1997–)	System of IPA
t	т	t	t
u	у	u	u, w
u̇	у	u	ʉ
ü	ю	ü	y
v	в	v	v
(ž)	ж	j	ʒ
z	з	z	z

Notes: 1 – It occurs only once in the manuscript (Ms. page 246). 2 – In the 39th tale the close back–central illabial *i (ɨ)* is marked with the letter ı ring above instead of <ı>.

Transcription of vowels

Below, in angle brackets <>, are the letters that Kúnos used to write down the Crimean Tatar texts, followed by the signs of the sounds they denote, first in the IPA system, then in round brackets according to the simplified Turkic transcription.[14] However, examples illustrating the use of each sound are given in the transcription used by Kúnos.

<a> *a (a)* open central illabial; e.g. *aš* 'food', *at* 'horse', *ata* 'father', *bar* 'is', *baš* 'head', *kajta* 'coming back', *kızlar* 'girls', *čatal* 'fork', *tavuk* 'hen', *taš* 'stone', etc.

<e> 1. *ɛ (e)* open-mid illabial front vowel; e.g. *de-* 'to say', *edė* 'was', *čitke* 'ashore', *itler* 'dogs', *keteǯek (=kėteǯek)* 'will go', *jerde (=jėrde)* 'on the ground', etc.
2. in some palatal words, it may denote the same close-mid sound as <ė> (see below); e.g. *itep* (10) ~ *itėp* (1) 'having done, having made', *jiget (*=jigėt*), ketep* (3) ~ *ketėp* (2) 'went away', *kette* (1) ~ *kettė* (6) 'went', *kirde* (12) ~ *kirdė* (2) 'went in', *kitep* (27) ~ *kitėp* (1) 'having gone (away)', *kitte* (41) *(=kittė)* 'went', *kürep* (39) *(=kürėp)* 'having seen', etc.
W. Radloff (1837–1918) marked both variants of the Crimean Tatar */e/* phoneme with <ä> (Işıktaş Sava 2008: 507, 510).

<ė> 1. *ɩ ~ ɘ (ẹ̇ ~ ə)* close-mid front–central illabial, pronounced short, sometimes reduced; a vowel between close front illabial *i* and close-mid front illabial *e*; it appears in palatal words; e.g. *aligė* 'former, preceding', *bilėp* 'knowing', *berirėm* 'I give', *dedė* 'said', ėken 'was', *edė (=ėdė)* 'was', *indė* 'now', *itėp* (1) 'having done, having made', *jigėt* ~ *ǯigėt* 'lad', *keldė* (1) 'came', *özėm* 'myself'.

14 In this work, single sounds are traditionally indicated in italics.

2. ɤ ~ ə *(ê)* close-mid back–central illabial vowel pronounced very short (sometimes reduced); a reduced vowel pronounced between velar *ï (ɨ)* and close-mid front illabial *e*, occurs in velar words; e.g. *krk(!) = kėrk* ‘forty’, *ǯojėldė* ‘lost’, *kıdırėp* ‘searching, seeking’, *barėp* ‘going’, *alėp* ‘taken, (taken)’, *kajtėp* ‘returning’, *kėz ~ kz!* ‘girl’, *kza! = kėza* ‘for a girl’, *kaldė* ‘left’, *krsız =*kėrsız (~ kırsız)* ‘thief’, etc.

Letter <ė> is used in the dictionary of the Crimean Karaim of Kúnos, as well as in his Kazan and Misher Tatar collections to denote the velar *i (ɨ)* and the velar and palatal reduced vowels (*ê, ə*).[15] In the Crimean Tatar folksongs of the musicologist Robert Lach, recorded by ear, the letter <ė> indicates a vowel in a similar phonetic position, e.g. *kirdė* (p. 18) ‘went in’, *kittė* (p. 23) ‘went (away)’, *jeldė* (p. 21) ~ *geldė* (p. 25) ‘he came’, *sewdėm* (p. 19) ‘I loved’, *isterėm* (p. 24) ‘I want (to)’, *kozlärė* (p. 31) ‘their eyes’ etc. (Lach–Jansky 1930).

<i> 1. *i (i)* close front illabial; e.g. *bir* ‘one’, *biš* ‘five’, *bitir-* ‘to finish’, *bùri* ‘wolf’, *ǯüzùkni* ‘ring (Accusative)’, *dij* ‘is saying’, etc.

It may denote the palatalised form of velar *i (ɨ)*, which occurs in specific phonetic positions (s. *Notes on the palatalisation of the back ı (ı→i)*.

2. *ı (ịͅ)* velarised *i*; a closed vowel in the upper position of the tongue between the palatal *i* and the velar *ï (ɨ)*; occurs in all phonetic positions;[16] e.g. *ketti (=kettı)* ‘(went) away’, *šėbirtkı (=šėbırtkı)* ‘whip’, *utėrdilar (=utėrdılar)* ‘they sat (down)’, *ačulanip (=ačulanıp)* ‘getting angry’, *aldi (=aldı)* ‘took’, *aligi (= aligı)* ‘the former, the previous one’, *findik (= findik ~ findık)* ‘peanut’, *jetišti (= jetištı)* ‘reached, attained’, *kaldi (= kaldı)* ‘stayed’, *kopir (= kopır)* ‘bridge’, etc.

The significant number of *i ~ ı* variations may be due to the velarised *i* (s. also *Notes on the palatal-velar harmony*).

According to Zühal Yüksel (1989: 21), almost all the palatal *i*’s in the dialects are represented by their velarised versions. Selimova (2006: 95, 96–97) also mentions a backwards-formed, i.e. velarised, shortened pronunciation of *i*, observed in all dialects.[17]

<ı> 1. *ɨ (ï)* close back–central illabial; it occurs most often in the stressed syllable of velar words; e.g. *ačık* ‘open’, *kırsız* ‘thief’, *kız* ‘girl’, *kızını* ‘his/her daughter (Accusative)’, *ıryat* ‘servant’, *jılan* ‘snake’, *kırk* ‘forty’, *mınau* ‘this (one)’, *nasıl* ‘how’, *sıčan* ‘mouse’, *zıpt* ‘pitch’, etc.

---

15 Zsuzsa Kakuk (1991), however, in her publication of the Crimean Karaim glossary, as well as the Crimean Tatar songs (1993a) and the Misher Tatar material (1982, 1996), completely neglected the original letter <ė>, which, however, remained unchanged in the Kazan Tatar editions (1980, Kakuk–Baski 1989).

16 Kurtbilal (2007: 507) describes the similar short, closed, velarised *ĭ* vowel alternating with *i* from Sudak, belonging to the southern dialect area. See also Memetov (2012: 10).

17 See also Kurtbilal 2007: 507.

On the first several pages of the songs, and throughout the 39th tale – collected on a different occasion and from another person – letter <ı> above ring marks the close back–central illabial *i (ɨ)*. Kúnos also used the same letter to denote the *ɨ* sound in his bracketed Ottoman Turkish explanations.[18]

2. ə *(ê)* and *ı ~ ɘ (ę̂ ~ ə)* close-mid central illabial reduced sound; as vowels marked with <ı>, they can occur in both velar and palatal words in all phonetic positions.[19] It may also be dropped in unstressed syllables, affixes and certain types of syllables (e.g. adjacent to *r, l*); e.g. *bız ~ bz [bəz]* 'we', *asıl ~ asl [asêl]* 'base', *bırak ~ brak [bêrak]* 'let', *bır ~ br [bər]* 'one', *bız ~ bz [bəz]* 'we', *Kırımdan ~ Krımdan [Kêrımdan]* 'from the Crimea', *kıza ~ kza [kêza]* 'to the girl', *kırsız ~ krsız [=kêrsïz]* 'thief', *nasıl ~ nasl [nasêl]* 'how', *sızin ~ szin [=səzin] ~ sizın [=sizən]* 'of you', *slaučan [= sêlaučan]* 'earthworm', etc.

<o> *o (o)* close-mid back (velar) labial; e.g. *bo* 'this', *čoban* 'shepherd', *doru* 'true', *jok* 'is not, there is not', *koba* 'cave', *otur-* 'to sit, to dwell', *sokor* 'blind', *toj* 'feast', *zor* 'difficult, hard', etc.

In standard language it can only be in the first syllable. In some dialects it appears in the second and subsequent syllables due to labial harmony (e.g. *boldo* 'became, was', *boldom* 'I was, I became', *bolson* 'let it be'). In the northern dialect the word, which initially had a palatal sound order, has *ɵ (ȯ)* replaced by *o* after the palatovelar *k* (Jankowski 1992: 12). In the Kúnos collection this tendency can be observed in the following examples: *kȯbete → kobete* 'meat pie', *kȯbük → kobuk* 'foam', *kȯj → koj* 'village', *[kȯpir] → kopir ~ kopır* 'bridge', *kȯp → kop* 'many', etc.

<ȯ> *ɵ (ȯ),* close-mid central labial vowel; transitional sound between the vowels o and *ö*.

1. can also be in the first syllable, e.g. *čȯlde* 'in the desert', *dȯrteu* 'four', *ȯgüzım* 'my ox', *tȯšekte* 'on a mattress/bed', etc.

2. can be after *b, g, k, š*, e.g. *bȯjle* 'so', *gȯt* 'ars', *gȯzgü* 'mirror', *kȯp* 'many', *kȯj* 'village', *sȯz* 'word', *šȯl* 'desert', etc.

In the northern dialect, instead of *ɵ*, it is *o* pronounced after *k* (Jankowski 1992: 12, 40). For examples, see above under the letter <o>.

<ö> *ø (ö)* close-mid front (palatal) labial; e.g. *börek* 'flaky pastry with thin layers of cheese or other filling', *čöple-* 'to pick/scratch about (for food)', *kör-* 'see', *köse* 'bare-faced', *öbür* 'other', *ömer* 'life', *öttör-* 'to kill', *ösken* 'grown', *tök-* 'to pour', *töle-* 'to pay', *töre* 'host, farmer', *tošek* 'mattress, bed', etc.

18 In tale no. 39 every letter <ı> ring above has been changed to <ı> (dotless <i>).

19 A similar observation was made by Jankowski (1992: 13).

In the standard Crimean Tatar language and its dialects, the use of the sound *ö* and the letter <ö> is governed by special rules. According to these rules, the sound *ö* can only occur in the first syllable next to *t, d, s, č, j, k'* and rarely after the consonant *ǯ* (Jankowski 1992: 12). Unfortunately, as the examples cited above show, this phonetic principle is only partially implemented in Kúnos' texts.

<u> 1. *u (u)* close back labial; e.g. *bojun* 'neck', *bu* 'this', *buj* 'stature', *čubuk* 'stick', *ǯurt* 'yurt', *duvar* 'wall', *juk* 'there is not, no', *kutu* 'box', *munda* 'here', *su* 'water', *tuj-* 'to feel', *un* 'ten', *vur-* 'to beat, strike', etc.

1.1. In some dialects it appears in the third and subsequent syllables due to labial vowel harmony (Jankowski 1992: 11). Examples of this in Kúnos' folklore texts are: *arnavuttur* 'he is Albanian', *basturuk* 'dungeon', *boldurdum* 'I got tired', *kavuštur(sun)* 'to meet/let him meet', *oturursuñuz* 'you sit, you lodge', etc.

1.2. In the northern dialect the palatovelar *k* and *g* are followed by *u* instead of *u̇* in the originally palatal words, e.g. *buku̇n* → *bukun* 'today', *guna* 'sin', *gu̇zel* → *guzel* 'beautiful', *ku̇lu̇k* → *kuluk* 'hen', *ku̇n* → *kun* 'sun', *ku̇pır* → *kupır* 'bridge', *kursi* 'small table, throne', *ȯkuzlerge* 'oxen', *tu̇gu̇l* → *tugul* '(final) not, (isolated) no', etc.

2. *ʊ*[20] *(u̯)* close/close-mid back labial; it is a more open phonemic variant of */u/* tending towards *o*, which Kúnos presumably heard as *o* and sometimes marked with <o> accordingly. E.g. *bu [bu̯] ~ bo*[21] 'this', *bujur- [bu̯ju̯r-] ~ bojor-* 'to command', *boldu [boldu̯] ~ boldo* 'became, was', *bolup [bolȁp] ~ bolop, ulup [u̯lu̯p] ~ olop* 'being',[22] *bulut [bu̯lu̯t] ~ bolot* 'cloud', *bolsun [bolsu̯n] ~ bolson* 'let it be', *kujruk [ku̯jru̯k] ~ kojrok* 'tail', etc.

3. *ɰ (u̯, w)* voiced palatovelar approximant (semivowel); e.g. *kijeu* 'bridegroom', *deu* 'giant, devil', *dȯrteu* 'four, foursome'; *ajtıuya* 'to say', *aldau* 'deceive', *anau* 'yonder; that one', *ašau* 'food', *au* 'hunt', *auǯı* 'hunter', *aula-* 'to hunt', *euel* 'first', *kauya* 'quarrel', *sau* 'healthy', *tauk* 'hen', *zeuk* 'pleasure, enjoyment', etc. See also <v> 2.

<u̇> *(ʉ (u̇)* close front–central; typical cases of use are in Crimean Tatar folklore texts of Kúnos: *bu̇ber* 'pepper', *bu̇ju̇k* 'big', *bu̇ku̇n* 'today', *ču̇l* 'desert', *ču̇p* 'splinter of wood', *ču̇ru̇k* 'rotten', *ǯu̇bbe* 'cloak', *ǯu̇gen* 'reins', *ǯu̇k* 'burden', *ǯu̇rek* 'heart', *du̇gme* 'button', *du̇gu̇n* 'feast', *ju̇r-* 'to walk', to go', *ju̇rek* 'heart', *mu̇ftu̇* 'mufti', *mu̇šku̇l* 'difficult, hard', u̇gu̇z ~ *ȯgu̇z* 'ox', *pu̇sku̇l* 'tassel', *pu̇tu̇n* 'whole, all', *su̇j-* 'to love', *su̇t* 'milk', *šu̇jle* 'so, this way', *šu̇ku̇r* 'thanks', *tu̇fek* 'rifle', etc.

---

20 Jankowski (1992: 13, 41) denoted this sound by <ɷ> or <ʉ>.

21 See Jankowski (1992: 13): *bo, şo*; Izidinova (1997: 302): *bo*.

22 The example of the *olop* shows Kúnos' hesitation. The word *ulup*, originally in the manuscript of the songs (Ms 1/4), was subsequently changed to *olop*.

<ü> *y (ü)* close front labial; in Crimean Tatar standard language, pronounced only in specific phonetic positions. The sound *ü* can occur next to *t, d, s, č* and *j* in the first syllable but may be present without restriction in other syllables (Jankowski 1992: 10). The cases at Kúnos are illustrated by the following examples: *ȯbür* 'other', *üč* 'three' *šonın‿čün* 'therefore', *ǯük* 'burden', *ǯür-* 'to walk, move', *ǯüz* 'hundred', *dülber* 'beauty', *dürt* 'four', *gül* 'flower', *güzel* 'beautiful', *jüret-* 'to teach', *jüz* 'hundred', *jüzük* 'ring', *kümüš* 'silver', *kün* 'sun', *mülk* 'wealth', *mürat* 'intention', *püsür* 'dirty', *süpür-* 'to sweep', *šü‿jerde* 'in this place'.

Correlation between vowel phonemes and vowels

/a/ = {a}[23] /o/ = {o}
/e/ = {e, ɛ} /ø/ = {ø, ɵ}
/i/ = {i, ɪ, ɘ} /u/ = {u, ʊ}
/ɨ/ = {ɨ, ə} /y/ = {y, ʉ,}

Transcription of consonants

<č> *tʃ (č)* voiceless, alveopalatal affricate; e.g. *ač* 'hungry', *čal* 'grey', *čeček* 'flower', *čizme* 'boots', *čık-* 'to go out', *coban* 'shepherd', *čȯplük* 'rubbish', *čubuk* 'stick', *čüj* 'nail', *ič* 'inside', *kač* 'how much', *uč* 'end', etc.

<ǯ> *ʤ (j)* voiced alveopalatal affricate; e.g. *aǯı* 'bitter', *ayaǯ* 'tree', *ǯadı* 'witch', *ǯer* 'place, land', *ǯıl* 'year', *ǯol* 'road', *ǯut-* 'to swallow', *ǯümle* 'all', *keǯe* 'night', *u̇čünǯi* 'third', etc.

<f> *f (f)* labiodental voiceless spirant (fricative); a foreign sound in Crimean Tatar, primarily found in loan words; e.g. *aff* (Ar.) 'pardon', *afser* (R.) 'officer', *čift* (P.) 'couple', *efendi* (Gr.) 'lord', *fajda* (Ar.) 'benefit', *fil* (Ar.) 'elephant', *sofra* (Ar.) 'table', *tu̇fek* (P.) 'rifle', etc.
In loan words, it is often replaced by *p* (a bilabial voiceless plosive sound); e.g. *čipt (čift)* (P.) 'couple', *čiput ~ čuput (čufut)* (Ar.) 'Jew', *ependi (efendi)* (Gr.) 'lord', *erip (erif)* (Ar.) 'fellow', *karanpil (karanfil)* (Ar.) 'carnation'.

<g> 1. *g (g)* palatovelar voiced plosive consonant; e.g. *begendi* 'he/she liked', *bergen* 'he/she gave', *du̇gu̇n* 'feast', *geče* 'night', *gȯzgu̇* 'mirror', *gu̇l* 'flower', *guzel ~ güzel* 'beautiful', *nege* 'what for', *u̇gi* 'front or beginning of sg', etc.

23 Phonemes are given between slashes and phoneme variants (allophones) are presented in braces, based on the IPA transcription system.

In the transcription of Kúnos, we found no trace of the strongly palatalised g^j (*g'*), typical of southern dialects, which usually occurs inside of words next to palatal vowels but can also be in the word-initial position (e.g. *g'it*! 'go!').

2. *G (ġ)* postvelar (or uvular)[24] voiced plosive consonant; occurs in velar words; e.g. *agaj (=aġaj)* 'brother', *ajgır (=ajġır)* 'stallion', *bajguš (bajġuš)* 'owl', *bargan (barġan)* 'went', *bolgan (bolġan)* 'became', *gajrı (ġajrı)* 'apart (from), outside (postposition)', *gam (ġam)* 'sorrow', *tugul (=tuġul ~ tüġül?)* '(final) not, (isolated) no; negative copula', etc.

<ġ>[25] *G (ġ)* postvelar (or uvular) voiced plosive consonant (see above); e.g. *baġalar* 'frogs'.

<γ> 1. γ ~ ʁ (γ) postvelar (or uvular) voiced fricative consonant; e.g. *aγaj* 'uncle', *ajγır* 'stallion', *baγla-* 'to tie, fasten', *ǯanγız* 'alone', *daγγa* 'uphill; to forest', *γajb* 'loss', *γajrı* 'apart (from), outside (postposition)', *γam* 'sorrow', *padišaγa* 'to padishah', *Orγa* 'to the city of Or', *tuγul* '(final) not, (isolated) no; negative copula', etc.

We have left <γ> in words of Arabic and Persian origin with palatal vowels because they can also have a postvelar spirant *γ*, e.g. *šaγird* 'pupil, servant'.

Taking into account the cases of regular and irregular marking of *g* sounds, we have adopted the version that conforms to the law of phonetic harmony, meaning that in a palatal word only <g>, and in a velar word usually <γ> and occasionally <g>, can be used.

Analysing the <γ>-letter palatal words in Kúnos' texts, we found that the letter <γ> could also denote the palatovelar plosive consonant *g* or the highly palatalised *j*-like consonant g^j (*g'*) of the southern dialect. For the examples, see *Incorrect alternations* on p. 27.

2. *G (ġ)* post-velar (uvular) voiced plosive consonant[26]; e.g. *agaj (=aġaj)* 'uncle', *ajgır (=ajġır)* 'stallion', *ǯangız(=ǯanġız)* 'alone', *dayga (=daġġa)* 'uphill; to forest', *gajb (=ġajb)* 'loss', *gajrı (=ġajrı)* 'apart (from), outside (Postp.)', *gam (=ġam)* 'sorrow', *padišaga (=padišaġa)* (3) 'to padishah', *Orga (=Orġa)* 'to the city of Or', *tugul (=tuġul ~ tüġül?)* '(final) not, (isolated) no; negative copula', etc.

---

24 Jankowski (1992: 16 21, 2006: 74) defines the *G* sound as an uvular voiced consonant with weak plosive element and lists the following examples: *dałGa* 'wave', *damGa* 'onto the roof', *Gajridan* 'anew', *Garp* 'west'.

25 Letter <ġ> occurs only once in the texts of the tales (Ms. p. 246).

26 Selimova (2006: 70) has detected a final *ġ* in the very mixed dialects of Akmesdjit, Bakhchesaraj and the southern coastal area after the return from the migration. A similar observation is made by Chatskaya and Dmitriev (Chatskaya 1926: 341). The *ġ* of velar-voiced words is considered to be spirant by Samoylovich (1916: 15), Jankowski (1993: 18, 21), Izidinova (1997: 301), Useinov–Ganieva–Seydametova (2011: 46, 49), Memetov (2012: 16, 27), Aydın–Adzhumerova (2014: 212), Medzhitova (2017a: 50).

The few examples of the alternate use of the letters <γ> ~ <g> in velar words demonstrated above are sufficient evidence that, in some instances, Kúnos used <γ> to denote a *G (ġ)*, i.e. a post-velar (sometimes perhaps uvular) voiced plosive consonant (see <g>2. and <γ>2.).

Jankowski (2006: 74–75) explains that two uvular consonants, [ʁ] and [G], are represented in writing by the letter <ğ> in the Latin-based Crimean Tatar alphabet and by the letter <гъ> in Cyrillic. The function of these two consonants is shared: the fricative [ʁ] appears between or before vowels, such as in *sağın-* 'to long for', *quşağı* 'his belt', while the weak plosive [G] appears after nasal occlusives and approximants, such as in *yanğan* 'burning, burnt', *yañğız* 'alone; the only one'.

h *h (h)* voiceless laryngeal/glottal spirant; mainly occurs in words of foreign (Arabic, Persian) origin; e.g. *Ahmad* 'Ahmed' (Ar.), *anahtar* (Gr.) 'key', *bahar* (P.) 'spring', *bahča* (P.) 'garden', *ǯāhil* (Ar.) 'fool', *daha* 'still, too', *eždeṛha* (P.) 'dragon', *haǯi* 'hadji' (Ar.), *hatın* (P.) 'woman', *hem* (P.) 'and', *hič* (P.) 'neither; not at all', *horaz* (P.) 'cock', *hüküm* (Ar.) 'rule, dominion', *šeher* (P.) 'city', *zeher* (P.) 'poison', etc.
See also *Disappearance of consonant h (>0)*, p. 22.

<χ> *x (χ)* voiceless velar spirant; e.g. *aχ* 'ah!', *aχaj* 'it is flowing', *aχır* (Ar.) 'final', *aχšam* (Ir.) 'evening', *baχ-* 'to look', *čoχ* 'many', *faχir* (Ar.) 'poor', *χabar* (Ar.) 'news', *χoš* (P.) 'pleasant', *jaχši* 'good', *joχ* 'is not, there is not', *juχu* 'sleep, dream', *kırχ* 'forty', *ruχsat* 'permit', *suχta* (Ar.) 'pupil, disciple', *vaχit* (Ar.) 'time', etc.
For more examples of the *h*-type spirants (*h* and *χ*), see *Alterning of h ~ χ*, on p. 23 *Alterning of k ~ χ*, on p. 24.

<k> 1. *k (k)* voiceless velar (palatovelar) plosive; e.g. *bek* 'lord', *bike* 'mistress', *bukün* 'today', *kelgen* 'came', *kibik* 'like, as', *kȯk* 'sky', *kürek* 'shovel', *tülki* 'fox', etc.
2. *q (q)* voiceless post-velar (uvular) plosive; e.g. *ačık* 'open, *ak* 'white', *balık* 'fish', *bok* 'excrement', *jok* 'is not, there is not', *kapı* 'gate', *kız* 'girl', *konuk* 'guest', *kolaj ~ kulaj* 'light', *kurku* 'fear', *ufak* 'small', etc.

<l> 1. *l (l')* palatalised alveolar/postalveolar lateral approximant; pronounced in words with front (palatal) phonetic articulation order; e.g. *bel* 'waist', *bil-* 'can', *bülbül* 'nightingale', *čelek* 'bucket', *lākin* 'but', *malim* 'my wealth', *müškül* 'sad'.
2. *ł (ł)* velarised alveolar lateral approximant; occurs in words with the back (velar) vowels; e.g. *ačılıj* '(to) open', *altın* 'gold', *bol* 'abundant', *čıl* 'spotted, *molla* 'mullah', *sular* 'waters', *ulan* 'boy, child', etc.

<ñ> *ŋ (ŋ ~ ñ)* velar nasal occlusive; e.g. *dıñ* 'your name', *ananıñ* 'your mother's', *boldoñ* 'you have become', *buña* 'to this', *kiteñ* 'go! (Plur2 Imp)', *soñ* 'the last', *tañ* 'dawn', *teñiz* 'sea', *vezirniñ* 'the vizier's', etc.
In the manuscript, we witness a significant frequency of <ñ> ~ <n> alternation (s. p. 24).

<v> 1. *v (v)* voiced labiodental fricative (spirant); e.g. *hava ~ ava ~ ave* 'air, weather', *avuz* 'mouth; basin', *havuz* 'basin', *kevde* 'body', *tavan* 'ceiling', *vaχit* 'time', *var-* 'to go, arrive', *ver-* 'to give', *vuǯud* 'body', *vur-* 'to beat', etc.

2. *ɰ (w, ư)* voiced palatovelar approximant (semivowel); e.g. *anavı [anawï]* ~ *anau [anaw]* 'behold, the one over there', *av* ~ *au* 'hunt', *bızau* 'calf', *dev* ~ *deu* 'giant, devil beast', *duva* ~ *dua* 'prayer', *evel* ~ *euel* 'before', *jaulık [jawlık]* ~ *jauluk [jawluk]* 'napkin', *javun [jawun]* 'rain', *jau- [jaw-]* 'to rain', *juv-* ~ *ju- [=ju^w-]* 'to wash', *judu [=ju^{w}du]* 'he/she washed', *juyan [=ju^{w}yan]* 'he/she washed', *kuv-* ~ *ku- [=ku^w -]* 'to chase', *savrul-* ~ *saurıl-* 'to be ventilated', *savul-* 'to be healed' ~ *sau* 'healthy', *sev-* ~ *seu- [=sew-]* 'to love', *sevgili [=sewgili]* 'kind', *tavuk* ~ *tauk* 'hen' etc. Cf. also <u> 3.

Consonants not discussed above are given in *Table 1*, p. 7

Correlation between consonant phonemes and consonants

/b/ = {b}[27]	/χ/ = {χ, *h*}/j/ = {j}	/p/ = {p}
/ʤ/ = {ʤ}	/k/ = {k, k', q}	/r/ = {r, r'}
/tʃ/ = {tʃ}	/q/ = {q}	/s/ = {s, s'}
/d/ = {d, d'}	/l/ = {l, l'}	/ʃ/ = {ʃ}
/f/ = {f}	/m/ = {m}	/t/ = {t, t'}
/g/ = {g, g'}	/n/ = {n, n'}	/v/ = {v, ɰ}
/ɣ/ = {ɣ, G, ʁ}	/ñ/ = {ñ}	/z/ = {z, z'}

Note: The palatalised pronunciation of consonants is marked by an apostrophe (').

Notes on the sound alternations

Vowel alternations

Notes on progressive assimilation

In Crimean Tatar, progressive assimilation is very frequent and vowel harmony is, in fact, the result of it (Izidinova 1997: 301). Most examples are provided by the various conjugated forms of the verb *bol-* ~ *ol-* 'to be', which is very common in the texts.

u → o

**bolur → bolor*[28] (57) ~ *bulur* (1) ~ *bolır* (6) 'will be',
**bolup → bolop* (66) ~ *bolıp* (33) 'being',
**bolsun → bolson* (40) ~ *bolsın* (30) 'let it be',

27 Consonant phonemes are given between slashes and phoneme variants (allophones) are presented in braces, based on the IPA transcription system.

28 The texts of the songs also include the *bolor* (7) ~ *olor* (1) variants.

**boldu* → *boldo(lar)* (66) 'became, was',
**boldum* → *boldom* (1) ~ * *oldum* → *oldom* (1)[29] 'I became, was',
kozu (2) → *kozo* (4) 'lamb',
olsun (3) → *olson* (32),[30]
sokır (3) → *sokur → *sokor* (17) 'blind'.

The following cases of *progressive assimilation* occur in the lyrics:

tolu (7) → *tolo* (16) 'full';
toldur- (1) → *toldor-* (4) 'to fill';
**otuz* → *otoz* (16) 'thirty';

According to Henryk Jankowski (1992: 12), such changes occur in the so-called *labialising dialects*. The predominance of more open, *labialised* form variants is striking in the Crimean Tatar folklore collection of Kúnos. This phenomenon has also been noticed by Turkish Turkologist Zühal Yüksel (1994: 19), who has noted it in her review of Zsuzsa Kakuk's (1993a) *Kırım Tatar Şarkıları* [Crimean Tatar songs] published in Turkey. She supposes that the editor may have misinterpreted Kúnos' transcription system.[31] In our view, the phonetic change should rather be regarded as the result of progressive assimilation.

Notes on the closing of open vowels

Vowels of the Crimean Tatar dialects generally tend to become more closed. Aydın–Adzhumerova (2014: 202–203) gives examples of *e* > *i*, *o* > *u* becoming more closed, and Medzhitova (2017a: 52, 53) mentions the same tendency among the phonetic features of the northern dialect. Several examples of this alternation exist in the Crimean Tatar texts of Kúnos.

o* → *u

o (39) → *u* (207) 'he, it'; The pronouns *o* ~ *u* and their various declined forms occur in large numbers in the texts.
boj (3) → *buj(ı)* (31) 'height, stature; length, size; edge (of a road), bank (of a river)';
dost (2) → *dust* (29) 'friend';
jol (18)[32] → *jul*[33] (167) 'way';

29 This variant can be seen in the songs.
30 It occurs in the songs.
31 Yüksel (1994: 19) writes: "Kunos'un transkripsiyon alfabesi yanlış değerlendirilmiş olmalı ki bazı kelime köklerinde ve eklerde u'lu şekiller o'lu verilmiştir" ['Kúnos's transcription alphabet must have been misinterpreted, as in some syllables and affixes the sound *o* was marked instead of *u*].
32 The *jol* version only occurs in tales.
33 It is the dominant version of the lyrics.

kol (20) → *kul* (90) 'arm';
jok (173) → *juk* (23) ~ *juχ* (3) 'is not, there is not';
**ojna-* → *ujna-* (35) 'to play, dance';
toj (30) → *tuj* (23) '(wedding) feast';
toj- (13) → *tuj-*(10) 'to fill up; to become filled and satiated';
zor (1) → *zur* (8) 'heavy; strong, powerful';
zorla- (2) → *zurla-* (6) 'to force', *zurlan-* (1) 'to be forced'.

ȯ → u̇

bȯjle (5) → *bu̇jle* (58) ~ *bujle* (1) 'so';
čȯl (9) → *ču̇l* (10) 'wilderness, field';
dȯrt (3) → *du̇rt* (25) *dürt* (1) 'four';
kȯč- (3) → *ku̇č-* (10) 'to wander';
kȯj (9) → *ku̇j* (75) 'village';
kȯk (5) → *ku̇k* (38) 'sky; blue';
kȯp (10) → *ku̇p* (25) 'many';
kȯr- (27) → *ku̇r-* (250) 'to see';
kȯz (6) → *ku̇z* (135), *küz* (1), *köz* (1) 'eye';
**sȯjle-* → *su̇jle-* (15) 'to say';
**sȯn-* → *su̇n-* (7) 'to go out (fire);

e → i

de- (680) → *di-* (1024) 'to say';
jer (209) → *jir* (14) ~ *ǯir* (13) 'land, place';
ket- (522) → *kit-* (362) 'to go'.

Zühal Yüksel (1994: 19), in her review mentioned above, found it strange that the texts of the songs show a strikingly high proportion of "Kazan Tatar" language features, i.e. the high ratio of words containing the close vowels *u, ü, i* instead of the standard Crimean Tatar vowels *o, ȯ, e*. This may be due to the migration or borrowing from the north of specific folklore genres.

In this context let us recall Herbert Lach's statement that, in the bound text of folk songs, no one speaks in the usual way of everyday life but adapts to the melody and rhythm. Therefore, only after a long time do the lyrics heard from others begin to be adapted to the dialect of the singer (performer). Consequently, folk songs are unsuitable for dialect research (Jansky 1930: 77). Moreover, since the lyrics follow a fixed pattern, they may retain a higher proportion of foreign linguistic (in this case, even Kazan Tatar) language features. To a certain extent this observation may also apply to tales.

It is also not impossible that a Kazan Tatar scribe helped Kúnos interpret Crimean Tatar texts written in Arabic letters, vocalising the Arabic words in his mother tongue. Kúnos (1978: 147, 1999: ch. V) reports on this subject:

> I found a young, educated Kazan soldier named Velijev [Veliyef] – to him I explained my purpose in the most detail. Having learned Turkish, he understood what I wanted. This Velijev became my main support. Thanks to his help, I was able to speak Tatar better and better and to start talking to the prisoners. *When we collected songs and tales from them, my friend from Kazan and I wrote everything down* [!], so my collection progressed well.

We do not think that Kúnos would have taken notes in Arabic script since he argues at length (1978: 141–142, 146) and convincingly in the quoted passages for the transcription of dialect and folkloric texts into Latin letters.

Notes on the palatalisation of the back ı (ı→i)

The palatalisation of the back *ı*, as can be seen in the examples below, usually occurs in the vicinity of the consonants č, ǯ, j, s, š, and z, presumably as a result of their influence.[34]

– e.g. *čık-* ~ *čıχ-* (320) ~ *čik-* / *čiχ-* (14) 'to go out, to come out', *kamčı* (ca. 30) ~ *kamči* (4) 'whip', *ǯılkı* (36) ~ *ǯilki* (2) ~ *ǯılki* (7) 'horse, stud', *aγajım* (1) ~ *aγajim* (1) 'my brother', *kıjik* (1) ~ *kijik* (3) 'deer', *sarajı* (29) ~ *saraji* (3) 'his palace', *aγası* (36) ~ *aγasi* (3) 'his brother', *anası* (114) ~ *anasi* (7) 'his mother', *babası* (88) ~ *babasi* (8) 'his father', *kapısı* (20) ~ *kapısi* (2) 'his door, his gate', *apšık* (2) ~ *apšik* (2) 'woman', *ašı* (9) ~ *aši* (1) 'his food', *bašı* (269) ~ *baši* (21) 'his head'; head', *dašı* (6) ~ *daši* (1) 'stone (Accusative)', *jaχšı* (42) ~ *jaχši* (23) 'good', *kardašı* (22) ~ *kardaši* (5) 'brother (Accusative), his brother', *tarašı* (2) ~ *taraši* (1) 'his hound', *alıñiz* (1), *alıñız* (4) ~ *alıñiz* (1) 'take it! (plur.)', *bašımız* (10) ~ *bašımiz* (1) 'our head; our leader', *jazı* (6) ~ *jazi* (2) 'writing', *kırmızı* (6) ~ *kırmızi* (2) 'red', *kırsız* (55) ~ *kirsiz* (2) 'thief', *kız* (500) ~ *kiz* (80), *tarzı* (36) ~ *tarzi* (1) 'its way', *tazı* (8) ~ *tazi* (1) 'greyhound', etc.

Present tense verbs ending in *-ıj (<-a-j)* often appear in the palatal *-ij* form, e.g. .**bajlıj* ~ *bajlij* (1) 'ties', *bakıj* (2) ~ *bakij* (6) 'he is looking, he looks', *barıj* (6) ~ *barij* (7) 'he is going', etc. The negative *-mıj (<-ma-j)* present tense forms are often replaced by their palatal *-mij* variant, e.g. *ašarmij* (1) 'does not eat', *jakıšmij* (1) 'is not fitting, does not fit', *bolmij* (5) 'is not', *bulunmıj* (1) ~ *bulunmij* (3) 'cannot be found, is not, there is not', etc.

The outcome of the palatalisation seen above is only vaguely present in the Crimean Tatar texts recorded by Kúnos. As the figures in brackets show, the palatalisation in question was, in most cases, not a normal process but the result of misinterpretation and miswriting of an individual pronunciation. Some instances of palatalisation of the back *ı* sound were also recorded by Robert Lach (Lach–Jansky 1930) in the same camps in 1916, e.g. *baši* (74) 'his/her head', *bašima* (29) 'to my head', *jolği* [read *jolǯi*] (61) 'trav-

34 In his preface to Illésházy's 17th-century (Ottoman) Turkish-Latin dictionary, Gyula Németh (1970: 66) discusses the palatalising effect of the consonants *y (j), ş (š), c" (ǯ), ç' (č)* and *s*.

eller', *ajir* (61) 'no', *sujin* (61) 'its water', *čajiryna* [read *čajirına*] (75) 'to its meadow', *čikajim* (68, 69) 'I'll go out, let me out', etc.

Consonant alternations

The typical consonant alternations are discussed in the section on acceptable consonant changes (s. p. 20).

Types of sound alternations and their interpretation/handling in the text edition

It must be said in advance that, in many cases, it is almost impossible to distinguish between the linguistic (dialectal) phenomena in the texts of the collection and the variants resulting from a transcription error, miswriting or misinterpretation. It is still being determined whether the linguistic phenomenon existed in the dialect of the data contributors or was merely a one-off mistake or a slip of the tongue. Taking the above into account, we have distinguished three types of vowel and consonant alternations:

1. *Acceptable (regular) alternations*
2. *Questionable alternations*
3. *Incorrect alternations*

1. Acceptable (regular) alternations

This group also includes the sound alternations described above in the chapter on *Sound alternations* (p. 15). In addition, all those alternations considered linguistically acceptable can be included in the final edition of the text in their original form.

Acceptable vowel alternations

a ~ e -*ajlar* (91) ~ -*ajler* (50) present tense plural 3rd person verbal suffix, e.g. *bašlajlar* (2) ~ *bašlajler* (7) 'they are beginning', **tašlajlar* ~ *tašlajler* (3) 'they are leaving'; affixes forming plural nouns -*lar* ~ -*ler*: *akajlar* (10) ~ *akajler* (2) 'brothers, uncles', *sarajlar* (38) ~ *sarajler* (4) 'palaces', etc.

ı ~ i in certain phonetic positions this alternation is regular (see paragraph on *palatalisation*), but in other cases there is no explanation for the alternation, which cannot be considered a proper (accurate) alternation – even based on the number of occurrences presented by the figures in brackets. Such so-called "irregular" word forms have been corrected without any notes (see ch. *Incorrect alternations*).

ı ~ ė	very short, reduced sounds, hardly different from each other, which can be used in both palatal and velar words, and often alternate, thus the law of palatal-velar harmony is not always followed: e.g. *adı* (11) ~ *adė(n)* (1) 'name (Accusative)', *batır* (63) ~ *batėr* (1) 'hero, champion', *ǯatıp* (3) ~ *ǯatėp* (2) 'lying', *endı* (2) ~ *endė* (4) 'now', *kutulıp* (5) ~ *kutulėp* (1) 'getting rid of', *tarzı* (36) ~ *tarzė* (1) 'in a manner, by way of (Postp)', etc.
i ~ ė *(~ e)* *~ ı*	e.g. *bekim* (7) ~ *bekėm* (2) ~ *bekım* (2) 'perhaps', *brinǯi* (2) ~ *brinǯė(lerė)* (1) ~ *brėnǯė* (1) 'first', *bzni* (2) ~ *bznė* (1) 'us', *ǯigit* (6) ~ *ǯigėt* (55) ~ *ǯigıt* (2) ~ *ǯiget* (114) 'lad, fellow', *dedi* (61) ~ *dedė* (1) ~ *dedı* (2) 'said', *edi* (24) ~ *edı* (1) ~ *edė* (26) ~ ėdė (2) 'was', *eki* (138) ~ ėki (1) ~ ėkė (1) ~ *ekė* (2) 'two', *elli* (14) ~ *ellė* (2) ~ *ellı* (2) 'fifty', *endi* (17) ~ *endı* (2) ~ *endė* (4) ~ *ende* (13) 'now', *iči* (120) ~ *ičı* (7) 'inside of sg', *idi* (20) ~ *ide* (13) ~ *idė* (1) ~ *idı(m)* (2) 'was', *indi* (2) ~ *indė (inde)* (9) 'now', *jigėt* (9) ~ *jiget* (162) ~ *jıget* (3) 'lad, fellow', *keldi* (119) ~ *keldı* (12) ~ *keldė* (4) 'came', *šindi* (14) ~ *šindė* (2) ~ *šindı* (1) 'now'; – often alternates as the linking vowel of the suffix of *-p* gerund, e.g. *itėp* (10) ~ *itip* (1) ~ *etip* (77) ~ *etėp* (17) ~ *etıp* (6) 'having acted, done', *kelip* (141) ~ *kelėp* (14) 'coming, having come', *ketip* (6) ~ *ketėp* (3) 'having gone', *kirėp* (9) ~ *kirıp* (1) 'going in, having entered'.

Acceptable consonant alternations

j- ~ ǯ-	this alternation is dialectally significant, because the word-initial *ǯ-* is a distinctive feature of the northern (Kipchak) dialect, while the word-initial *j-(y) is* that of the southern (Oghuz) dialect. In the central dialect, these features can even co-exist.

Pairs of words alternating of word-initial *j- ~ ǯ-*

jak (88)	*ǯak* (3) 'side, edge'
jak- (30)	*ǯak-* (4) 'to light, to burn'
jakın / jakınla(š)- (18)	*ǯakın* (2) 'close, to close'
jala- (7)	*ǯala-* (16) 'to lick'
jalvar- (11)	*ǯalvar-* (1) 'to beg'
jaman (5)	*ǯaman* (1) 'bad, lean'
jaša- (32)	*ǯaša-* (1) 'to live'
jat- (122)	*ǯat-* (18) 'to lie'
jedi (30)	*ǯedi* (1) 'seven'
jer (173)	*ǯer* (37) 'place, land'
jet- / yetiš- (18)	*ǯet-* (12) 'to reach, to arrive'
jiber- (1)	*ǯiber-* (58) 'to send'
jiget (139) ~ *jiket* (12)	*ǯiget* (67) 'lad, fellow'
jigil- (1) ~ *jıkıl-* (6)	*ǯıyıl-* (5) ~ *ǯıkıl-* (7) 'to fall down, prostrate'
jıl (8)	*ǯıl* (8) ~ ǯil (1) 'year'

jılan (5)	*ǯılan* (18) ~ *ǯilan* (8) 'snake'
jılkı (1) ~ *jılχı* (1)	*ǯılkı* (37) ~ *ǯilkı* (1) 'horse, stallion'
jılkıǯı (13)	*ǯılkıǯı* (3) 'horseman, horseherd'
jok (118) ~ *joχ* (41) ~ *juk* (18)	*ǯok* (14) 'is not, there is not'
jol (19) ~ *jul* (99)	*ǯol* (11) ~ ǯul (3) 'road, way'
jukla- (32) ~ *juχla-* (2)	*ǯukla-*(12) 'to sleep'
juku (14)	*ǯuku* (2) 'sleep, dream'
juv- (6)	*ǯuv-* (4) 'to wash'
jür (33)	*ǯür-* (43) 'to walk'
jüz (18)	*ǯüz* (1) 'face'
jüzük (68)	*ǯüzük* 'ring' (6).

Pairs of words exhibiting alternation of word-initial *d- ~ t-*

daγ (6)	*tau* (94) 'mountain, forest'
daha (8)	*taha* (58) 'still, too'
dajın (30)	*tajın* (4) 'as, like'
dal (4)	*tal* (7) 'branch'
dal- (1)	*tal-* (1) 'to plunge'
dane (8)	*tane* (14) 'piece'
dart- (1)	*tart-* (13) 'to pull'
daš (31)	*taš* (19) 'stone'
degil (2)	*tegil* (17) 'no, not'
demir (2)	*temir* (12) 'iron'
dene (2)	*tene* (21) 'piece'
deñiz (1)	*teñiz* (34) 'sea'
derd (24) ~ *dert* (9)	*terd* (2) 'sorrow, trouble'
dere (24)	*tere(-tüpe)* (3) 'stream, valley'
dora (6)	*tora* (15) 'straight'
dugıl (1)	*tugıl* (3) 'no, not'
duj- (6)	*tuj-* (52) 'feels'
tujdur- (6) 'to make it feel, hear'	*dujdur-* (2)
tulaš- (2) ~ *tolaš-* (4) 'to walk'	*dulaš-* (37) ~ *dolaš-* (6)
tuman (12) 'fog, smoke'	*duman* (6)
tunja (6) 'world'	*dunja* (7)
turba (2) 'bag'	*durba* (1)
tut- (84) 'to take, to hold'	*dut-* (4)
tügme (5) 'button'	*dügme* (1)
tügün (15) 'feast'	*dügün* (15)
tůkan (4) ~ *tůken* (3) 'shop'	*dükan* (1)
tünja (37) 'world	*dünja* (11)

Alterning of word-initial *g-* ~ *k-*

gajb (5)	*kajb* (1) 'loss'
geč- (7)	*keč-* (115) 'to pass away, pass over'
geče (2)	*keče* (50) 'night'
geǯe (4)	*keǯe* (89) 'night'
gel- (8)	*kel- (1056)* 'to come'
gemi (6)	*kemi* (9) 'ship'
gene (21)	*kene* (41) 'again'
gerek (2)	*kerek* (85) 'necessary'
gez- (14)	*kez-* (23) 'to walk'
gibik (2)	*kibik* (70) 'like, as'
git- (3)	*kit-* (452) 'to go'
gȯt (2)	*kȯt* (1) 'ass'
gu̇tir- (1)	*kȯtu̇r-* (1) 'to take (away)'
gu̇čču̇k (2)	*ku̇ču̇k* (19) 'small'
gu̇gen (1)	*ku̇gen* (2) 'blackthorn'
gu̇l (63)	*ku̇l* (13) 'flower'
gu̇n (3)	*ku̇n* (348) 'day, sun'
gu̇neš (1)	*ku̇neš* (3) 'sun'
gȯzgu̇ (1)	*ku̇zgu̇* (7) 'mirror'.

Disappearance of consonant *h* (>0)

It mainly happens in Arabic-Persian loanwords, e.g.:

ah (22) ~ *aχ* (46)	*a* (24), ā (2) 'ah! oh!'
Ahmat (44)	*Amat* (9) 'Ahmad, Ahmed'
Allah (34)	*Alla* (10) 'Allah'
hal (12)	*al* (18) 'status, round session'
halak (1)	*alak* (2), *elak* (2) 'destruction'
Hasan (5)	*Asan* (14) 'Hassan'
hazır (4)	*azır* (1), 'ready'
hazırla- (4)	*azırla-* (12) 'to prepare'
hebet (2)	*ebet* (2) 'yes, of course'
hekjet (1)	*ekjet* (8) '(fairy) tale'
hem (19)	*em* (40) 'and, also'
hemen (8)	*emen* (9) 'immediately'
hep (6)	*ep* (6) 'all, the whole'
her (18)	*er* (28) 'every, each'
herkes (3)	*erkes* (7) 'everyone'
hič (6)	*ič* (20) 'neither, nothing (in negation)
hoǯa (1) ~ *χoǯa* (1)	*oǯa* (5) 'teacher, husband'
nikah (1)	*nikā* ~ *nika* (4) 'marriage ceremony'
sabah (7)	*saba* (5), *sabā* (60) 'morning'

sahat (9) ~ *saχat* (1)	*saat* (9), *sāt* (19) 'hour, clock'
sahib (4) ~ *sāhib* (4)	*sāb(i)* (10) 'landlord, owner; its owner)'
sarhoš (4) ~ *sarχoš* (2)	*saroš* (3) 'drunk'
**talih*	*talı* (6) 'fate, destiny; luck'

**Note:* Cf. osm. *tali'* ~ *talih* (Ar.) 'good fortune, luck'; Crm. Tat. *talij* 'sud'ba, rok' (Useinov 1994: 272).

Alterning of *h* ~ *χ*

This alterning mostly happens in Arabic-Persian loanwords, e.g.:

ah (22)	*aχ* (46) 'ah, oh!'
anahtar (6)	*anaχtar* (6) 'key'
bahča (43)	*baχča* (5) 'garden'
bahšıš (1)	*baχšıš* (1) 'gift'
habar (2) ~ *haber (1)*	*χabar* (53) ~ *χaber* (1) 'news'
hajır (1)	*χajır* (2) 'goodness, blessing'
halk (8)	*χalk* (5) 'people'
hanım (15)	*χanım* (18) 'woman, lady'
has (1)	*χas* (9) 'own'
hasir (1)	*χasır* (3) 'reedmace'
hasta (3)	*χasta* (6) 'sick'
hatın (24) ~ *hatin* (4)	*χatın (47)* ~ *χatin* (11) 'woman'
hatun (13)	*χatun* (11) 'woman'
hizmet (1)	*χizmet* (2) 'service'
hınzır (1)	*χınzır* (1) 'pig'
hırsız (3)	*χırsız* (5) 'thief'
hoǯa (15)	*χoǯa* (1) 'master, husband'
horaz (5)	*χoraz* (1) 'cock'
hoš (2)	*χoš* (8) 'pleasant'
jahšı (1)	*jaχšı* (41) ~ *jaχši* (33) 'good, excellent'
oh (2)	*oχ* (6) 'oh!'
ruhset (1) ~ *rühset* (1)	*ruχset* (3) ~ *rüχset* (1) 'permission'
sahat (9)	*saχat* (1) 'clock'
sarhoš (3)	*sarχoš* (2) 'drunk'
tahta (4)	*taχta* (10) 'board, plank'
vah (7)	*vaχ* (1) 'oh! ouch!'

Alterning of *k* ~ *χ*

Appearing of consonant uvular *k* as consonant *χ* is a peculiar feature of the southern (Oghuz) dialect, e.g.: *ačkan* (8) ~ *ačχan* (1) 'opened', *ajrılık* (2) ~ *ajrılıχ* (1) 'parting', **ajtajık* ~ *ajtajıχ* (1) 'let's say (it)', *anačık* (1) ~ *anačıχ* (1) 'mommy', *baktık* (2) ~ *baχtıχ*

(1) '(we) looked at', *balık* (25) ~ *balıχ* (6) 'fish', *bardık* (1) ~ *bardıχ* (3) 'we went (there)', *čık-* (237) ~ *čıχ-* (96) 'to go out', *čıkar-* (57) ~ *čıχar-* (65) 'to take out, to pull out', *kart* (149) ~ *χart* (1) 'oldman', *katı* (2) ~ *χatı* (4) 'hard', *katın* (28) ~ *χatın* (100) 'woman', *kadar* (79) ~ *χadar* (5) 'till, up to (Postp)', *kajt-* (169) ~ *χajt-* (3) 'to return', *barmak* (5) ~ *barmaχ* (3) 'finger', *kırmızı* (6) ~ *χırmızı* (1) 'red', *kizmet* (8) ~ *χizmet* (2) 'service, work', *kıdır-* (26) ~ *χıdır-* (11) 'to search', *kız* (667) ~ *χız* (13) 'girl, daughter', *koj-* (6) ~ *χoj-* (4) 'to put (down)', *kork-* (13) ~ *χorχ-* (2) 'to fear', *kuda* (2) ~ *χuda* (5) 'matchmaker', *mıšık* (29) ~ *mıšıχ* (5) 'cat', *sandık* (23) ~ *sandıχ* (11) 'box', *ulanǯık* (1) ~ *ulanǯıχ* (2) 'little boy';

Changing of *χ* to *k*

The velar *χ* consonant of Arabic-Persian loanwords in the northern dialect is usually changed to velar *k*,[35] e.g. *χabar* (53) → *kabar* (6) 'news', *χalk* (3) → *kalk* (31) 'people', *χatır* (3) → *katır* (6) 'memory', *χajır* (2) → *kajır(lı)* (2) 'blessing(s), goodness', *χasta* (6) → *kasta* (6) 'sick', *χızmet* (1) → *kızmet* (1) 'service, work', *χoš* (8) → *koš* (10) 'pleasant, good', etc.

Alterning of *ñ* ~ *n*

This phenomenon is probably due to dialectal differences. The *ñ* → *n* change is expected in southern dialects, but some idioms retain *ñ*.[36] According to some researchers, there is no *ñ* sound at all in the southern dialect.[37] However, N. K. Dmitriev (1898–1954), describing the dialect of the village of Tuvak, which belongs to the southern dialect group, argues that there are "stronger" and "weaker" pronunciations of the consonant velar *ñ*. The "stronger" version is pronounced, for example, in *baña* 'for me' and *saña* 'for you', and the "weaker" one in the possessive case ending, e.g. *naneniñ* 'of (the) bread', and in Sing2 person verb and noun endings, e.g. *ȯldüñ* 'you died', *χojnuña* 'into your lap' (Chatskaya–Dmitriev 1926: 344–45).

In Ignác Kúnos' Crimean Tatar texts, variants containing velar nasal *ñ* predominate as the following comparison shows:

añla- (8)	*anla-* (3)[38] 'understand'
baña (58)	*bana* (4)[39] '(to/for) me'
biñ (25)	*bin* (1) 'thousand'
bunıñ (21)	*bunın* (8) 'of this'
-diñ (77) ~ *-dıñ* (38)	*-din* (5) ~ *-dın* (18) past definite, Sing2 ending
-ıñız (62)	*-ınız* (12) Plur2 verb and possessive ending

35 In words of foreign origin, the same change is reflected in the dialects of the villages Suvuk-Su, Taraktaš, Koz, Tokluk and Otuz in the southern dialect area of Sudak (Kurtbilal 2007: 505).
36 Selimova 2006, 35.
37 Aydın–Adzhumerova 2014, 105.
38 Data from southern dialect (Akmesdjit).
39 It occurs only in the lyrics.

misiñ (4) ~ *mısıñ* (2)	*misin* (14) ~ *mısın* (4) Sing2 interrogative particle
-(n)iñ (258)	*-(n)in* (68) Genitive case and Plur2 imperative ending
-sañ (111) ~ *-señ* (116)	*-san* (5) ~ *-sen* (3) Sing2 conditional suffix
saña (138)	*sana* (4) ‘for you’
-sıñ (122) ~ *-siñ* (47)	*-sın* (330) ~ *-sin* (310)[40] Sing2. verbal suffix
soñ (120)	*son* (170) ‘last, later than; after (Postp)’
soñra (6)	*sonra* (14) ‘(here)after, then; later than (Postp)’

2. Questionable alternations

In questionable cases, we have corrected the word forms that were found to be doubtful and have given the versions from the manuscript in the footnotes.

Numerous data show that Kúnos also marked sounds with the letter <e>, which could be closed or more closed *e (ẹ)*, palatal reduced *(ə)* and, in several cases, even velar reduced *(ê)*. The degree of closure of the *e*-sound pronounced by Kúnos’ informants cannot be determined afterwards. Assuming that there is a real linguistic reason for the sound marking described, the words containing the problematic <e> are left in their original form.

This solution is applied to the gerund forms <ep> and the definite past tense forms <de> ~ <te> of verbs with palatal vowels, which occur massively, almost uniformly.

The above procedure is most often used for the following verbs:

– *bil-* ‘know’, *bir-* ‘give’, *bit-* ‘end’, *ǯiber-* ‘send’, *et-* ‘do’, *ič-* ‘drink’, *it-* ‘do’, *ket-* ‘go’, *ketir-* ‘bring’, *kij-* ‘put on, wear’, *kil-* ‘come’, *kir-* ‘go in, enter’, *kit-* ‘go’, *kür-* ‘see’, *min-* ‘get up on, mount’, *pišir-* ‘cook’, *tüš-* ‘fall’, *ütür-* ‘kill’ etc.

In many cases, we also find gerund forms of velar verbs marked with <ep> instead of the standard <ıp> ~ <ip> gerund forms. The following are examples of this:

– *alep* (< *al-* ‘take’), *aldatep* (< *aldat-* ‘cheat’), *bolep* (< *bol-* ‘to be’), *čakirep* (< *čakır-* ‘call’), *χıdırep* (<*χıdır-* ‘seek, search’), *kıčırep* (<*kıčır-* ‘howl, shout’), *kısılep* (<*kısıl-* ‘be sqeezed’), *kušanep* (< *kušan-* ‘gird oneself, dress up’), *kutulep* (< *kutul-* ‘escape, be saved’), *sukep* (< *suk-* ‘thrust into’), *tartep* (< *tart-* ‘pull, draw’), *urep* (< *ur-* ‘hit, shoot’), *uterep* (< *utır-* ‘sit (down), to live’) etc.

40 In this group there are also several endings which are not only variants of the Sing2 person verbal suffix *-siñ (~-sıñ)* but also that of a suffix of Sing3 imperative (*-sın* ~ *-sin*) and the compound affix *-sın* ~ *-sin* composed of Sing3 possessive suffix (*-sı* ~ *-si*) and accusative case ending *-n*.

Assuming that, for velar verbs, the connective sound of gerund on -p is velar reduced (*ê*), the suffix <ep> has been replaced by <ėp> in the text, and the standard version is given in footnotes after the abbreviation "R" as follows: e.g. *alep* → *alėp* → f.n. *alep*; R *alıp*; *uterep* → *utėrėp* → f.n. *uterep*; R *utırıp*, etc.

It should be noted in connection with this phenomenon that, in the text of tale 39 (which shows middle and mixed dialect features), the connective sound of the gerund on *-p* in the phonetic position in question is always marked by letter <ė> denoting close-mid *e (ẹ)* and reduced sounds *(ə, ê)* (e.g. *bilėp* (1) 'knowing', *češinėp* (1) 'undressing oneself', *etėp* (14) 'doing', *kelėp* (13) 'coming', *ketėp* (3) 'going', *ketirėp* (9) 'bringing', *kijėp* (3) 'putting on (clothes)', *kirėp* (8) 'going in', *koterėp* ~ *kȯterėp* (2) 'lifting', *minėp* (3) 'getting up on, mounting', *tüšürėp* (1) 'dropping, throwing off', etc.).

We face a similar phonetic or transcription issue for some verbs' definite past tense forms. For example, it is a typical case that the definite past tense of the same verb has three different states, even in the same tale, e.g. *ide* (10) ~ *idė* (1) ~ *idi* (19) 'was', *kirde* (12) 'went in', *kitte* (32) ~ *kittė(ler)* (1) ~ *kitti* (30) 'gone', *kette* (1) ~ *kettė* (6) ~ *ketti* (28) 'gone', *minde* (2) ~ *mindi* (2) 'mounted, got up on', etc.

In the case of the following alternations, we decided on a case-by-case basis which version would remain unchanged in the edition and which would be footnoted. Examples showing the different frequency (occurrence) data that helped to interpret each word form:

- *aligi* (6) ~ *alıgı* (12) 'former', *alıñiz* (1) ~ *alıñız* (4) 'take, seize', *bašımiz* (1) ~ *bašımız* (10) 'our head; our leader', *bilėp* (2) ~ *bilep* (8) 'knowing', *birep* (6) ~ *birėp* (2) 'giving', *etep* (2) ~*etėp* (14) ~ *etip* (79) 'doing', *ičep* (12) ~ *ičėp* (1) 'drinking', *ide* (10) ~ *idė* (1) 'was', *irgat* (15) ~ *ırgat* (14) ~ *ıγat* (10) 'worker, servant', *itep* (9) ~ **itėp* 'doing', *jazi* (2) ~ *jazı* (6) 'writing', *karañik* (6) ~ *karañık* (9) 'darkness', *kelėp* (13) 'coming', *ketirėp* (9) ~ *ketirep* (23) 'bringing', *kette* (1) ~ *kettė* (6) 'went (away)', *kilep* (4) ~ **kilėp* 'coming', *kirde* (12) 'went in', *kirėp* (8) ~ *kirep* (22) 'going in', *kitep* (24) ~ *kitėp* (2) 'going (away)', *kitte* (32) ~ *kittė(ler)* (1) 'went (away)', *kırmızi* (2) ~ *kırmızı* (6) 'red', *kürep* (38) ~ **kürėp* 'seeing', *minde* (2) ~ **mindė* ~ *mindı* (2) ~ *mindi* (10) 'got up on, mounted', *minėp* (3) ~ *minep* (40) ~ *minip* (4) ~ *minıp* (1) 'getting up on, mounting', *nasil* (20) ~ *nasıl* (62) 'how, what kind of', *piširep* (8) ~ **pisirėp* ~ *piširıp* 'cooking', *tüšep* (7) ~ **tüšėp* ~ *tüšip* 'falling down', *ütürep* (3) ~ **ütürėp* 'killing'.

The dominant original form is retained, and the less frequent or questionable variants are corrected and footnoted. In this respect the word pairs with *i* ~ *ı* alternation caused the most problems.[41] Therefore, we refer to the more likely variant of the word in question in footnotes. Alternatively, specific data were classified not as questionable but as *incorrect (wrong) alternations* (s. below).

41 For example, Kakuk (1993a) transcribed all *i*'s of velar sounding words in the lyrics – regardless of the phonetic context – into back *i*'s, i.e. *ı*'s.

3. Incorrect alternations

Incorrect, atypical, one-off word forms resulting from mishearing, mispronunciation or misspelling have been corrected to the linguistically correct (and/or dominant) word form without any comments (footnote references). The prevalent variants were determined based on their number of occurrences in the whole Crimean Tatar collection (texts of songs, riddles and tales).

The arrow sign "→" shows the direction of change. For example, in case of presumably incorrect variants *kapi* (7) ~ *kapı* (120) 'door', *kapi* is substituted for *kapı* (*kapi* → *kapı*) in the main text without footnote reference.

Replacing letter <i> of the incorrect data (<i> → <ı>)

This altering has been fulfilled in the following and other similar instances:

– *ačip* (4) → *ačıp* (30) 'opening', *ačti* (5) → *ačtı* (5) 'opened', *ajtip* (1) → *ajtıp* (8) 'saying, speaking', *ajtti(lar)* (9) → *ajttı(lar)* (97) 'said, spoke', *aldi* (8) → *aldı* (25) '(in) front of sg', *aldi* (4) → *aldı* (24) 'taken', *alip* (20) → *alıp* (291) 'taking', *alir* (4) → *alır* (34) 'takes', *arti* (9) → *artı* (72) 'back, behind', *ašadi* (5) ~ *ašadı* (16) 'ate', *atti* (5) → *attı* (27) 'threw', *baχti* (6) → *baχtı* (33) 'looked', *bardi* (15) → *bardı* (57) 'gone', *bašladi* (10) ~ *bašladı* (42) 'began', *batir* (11) → *batır* (66) 'warrior, hero', *boldi* (13) → *boldı* (35) 'became, was', *bolip* (5) → *bolıp* (34) 'being', *boni* (2) → *bonı* (12) 'this', *buni*[42] (23) → *bunı* (138) 'this', *čayir-* (3) → *čayır-* (35) 'to call', *čakir-* (6) → *čakır-* (30) 'to call', *ǯanim* (2) → ǯanım (36) 'my soul, my beloved', *hatin* (4) → *hatın* (24), 'woman, wife', *χatin* (24) → *χatın* (75) 'woman, wife', *jari* (8) → *jarı* (17) 'half', *kajtip* (3) → *kajtıp* (50) 'coming back', *kajtti* (5) → *kajttı* (20) 'came back', *kaldi* (12) ~ *kaldı* (28) 'stayed', *kapi* (7) → *kapı* (120) 'door', *katin* (5) → *katın* (25) 'woman, wife', *kidir-* (4) ~ *kidır-* (5) ~ *kıdir-* (4) → *kıdır-* (26) 'to search, to look for', *kirk* (9) → *kırk* (100) 'forty', *kirsiz* (2) → *kırsız* (55) 'thief', *kiz* (90) → *kız* (650) 'girl, daughter of sb', *kuli* (10) → *kulı* (73) 'his/her arm', *-lari* (20) → *-ları* (606) 'plural possessive ending with all its declined forms', *-larni* (8) → *-larnı* (122) 'ending of plural Accusative (or Genitive) case, *-larniñ* (2) → *-larnıñ* (36) 'ending of plural Genitive case', *minda* (4) → *mında* (28) 'here', *oni* (6) → *onı* (21) 'him, her, it', *padišani(ñ)* (10) → *padišanı(ñ)* (60) '(of) the padishah', *sandiy* (5) ~ *sandiχ* (3) ~ *sandik* (1) → *sandıy* (5) ~ *sandıχ* (10) ~ *sandık* (22) 'box', *surdi* (7) → *surdı* (23) 'asked', *tarzi* (1) → *tarzı* (36) 'way of, as', *tazi* (1) ~ *tazı* (8) 'greyhound', *uli* (29) → *ulı* (71) 'son of someone', *uni* (30) → *unı* (94) 'him, that', *utir-* (6) → *utır-* (75) 'to sit, to dwell'.

42 In the texts of the tales we see several instances of the accusative (and the shortened genitive) case ending *-ni* in velar words (e.g. *boni, uni, buni, oni, padišani* etc.).

Replacing <ɣ> in palatal words with <g>

In our edition of the text, we have taken into account the palatal-velar opposition by *replacing the sporadic <ɣ> in palatal words with <g> (<ɣ> → <g>).* In the texts of tales, it has been done in the following cases:

- *ajɣir* (4) → *ajgir* (1) 'stallion', *büjüɣen* → *büjügen* 'grew', *büjüɣi* ~ *büjüɣı* → *büjügi* ~ *büjügı* 'his grown-up; the old', *bürlenmeɣe* → *bürlenmege* 'to open (from bud)', *ǯiɣit* → *ǯigit* 'lad, bachelor', *deɣen* → *degen* 'said', *gezmeɣe* → *gezmege* 'to walk', *jerɣe* → *jerge* 'to place, into the ground', *jıgeremɣe* → *jıgeremge* 'for twenty', *jüzüɣın* → *jüzügın* 'his ring (Accusative)', *katmeɣe* → *katmege* 'to do what', *kelmeɣe* → *kelmege* 'to come', *kelmijǯeɣin* → *kelmijǯegin* 'his not coming', *kereɣıme* → *keregıme (keregime)* 'I may need it', *kimseǯiɣimiz* → *kimseǯigimiz* 'someone/anyone of us', *kördöɣınız* → *kördögınız* 'whom/what you saw', *köseɣe* → *kösege* 'to/for bare-faced', *küpeɣe* → *küpege* 'to the dog', *kürdüɣinden* → *kürdüginden* 'from what he saw', *kümseɣe* → *kümsege* 'for someone / for nobody', *medreseɣe* → *medresege* 'to school', *olenliɣin* → *olenligin*, 'his/her grass', *öɣine* → *ögine* 'in front of him', *öttörmeɣe* → *öttörmege* 'to kill', *ptirmeɣe* → *ptirmege* 'to finish / to end', *süɣıp* → *sügıp* (442) 'scolding', *süjdiɣım* → *süjdigım* 'whom I love', *tökmeɣe* → *tökmege* 'to pour out', *tüɣerek* → *tügerek* 'round', *tüpeɣe* → *tüpege* 'to hill', *tüɣül* → *tügül* '(final) not, (isolated) no', *ülčmeɣe* → *ülčmege* 'to measure', *ülkeɣe* → *ülkege* 'to the country, to the homeland', üɣret- → *ügret-* 'to teach'.

Vowel harmony

There are two types of vowel harmony. One of these regulates the palatal-velar harmony similarly, i.e. the first syllable containing a palatal vowel can only be followed by a palatal vowel, and the first syllable containing a velar vowel can only be followed by a velar vowel. The other type of harmony is the law of labial harmony, according to which the first syllable containing a labial vowel can only be followed by syllables with labial vowels, and the first syllable containing an illabial vowel can only be followed by syllables with illabial vowels. Exceptions in both cases are words of foreign origin.

As is known, in some Turkic languages and dialects, vowel harmony is not fully applied. O. I. Chatskaya (Chatskaya–Dmitriev: 1926, 343) observed varying degrees of weakening of vowel harmony in the Crimean Tatar dialects of the 1920s. Based on her recent collections, Leniara Selimova also notes that the law of palatal-velar harmony is sometimes not applied in the northern (Kipchak) and central (mixed) dialects, but preserving it is much more vital in the southern dialect. Most of all, the use of reduced *ı* hinders harmony. This is because the reduced vowel in question can be used either in velar or palatal words (Selimova 2006: 143–144). In this respect, the Crimean Tatar dialects, from whose speakers Ignác Kúnos recorded the Crimean Tatar texts, do belong to them. Regarding the Crimean Tatar literary language, researchers generally claim a consistent prevalence of vowel harmony.[43]

Notes on the palatal-velar harmony

Based on the whole Crimean Tatar collection, we can state that the palatal-velar harmony is generally valid, with a number of exceptions.[44] Some of the exceptions are due to phonetic reasons (e.g. *palatalisation*), and some of them are due to *incorrect word forms* (see above).

Cases of palatalisation are those in which the final palatalised consonant of a velar word is accompanied by a palatal suffix (Jankowski 1992: 63, 67). Examples of this can be seen below:

1. palatalised final *-l'* + *-di* past tense suffix, e.g. *ačıldi* (1) 'opened', *ajrıldi* (1) 'parted', *aldi* (4) 'took, got, accepted', *anlašıldi* (1) 'understood', *boldi* (10) 'became', *ǯojıldi* (1) 'lost', *kaldi* (10) 'remained', *oldi* (1) 'became', *saldi* (1) 'laid down, built', *toldi* (1) 'filled'.
   In most cases, however, the *l* sound is not palatalised since the regular past tense *-dı* is used. For more examples, see the *Incorrect alternations, 27.*
2. *-j* (suffix of the present tense)/*-maj* (suffix of the negative present tense) *+ler* Plur3. verbal suffix, e.g. *abdırajler* (1) 'they wonder', *ajtmajler* (1) 'they aren't saying', *bašlajler* (7) ~ *bašlajlar* (2) 'they are starting, they are beginning', *bol‿almajler* (3) 'they cannot be', *jasajler* (4) ~ *jasajlar* (3) 'they are making sg', *jašajler* (3) 'they are living', *kapajler* (2) 'they are closing', *kurkajler* 'they fear, they are afraid of sb/sg', *tašlajler* (3) 'they are leaving', *toktajler* (2) 'they are stopping', etc.
3. word final *-j* of velar nouns sometimes affixed by palatal endings, e.g. *akajler* (2) 'brothers', *sarajler* (3) 'palaces', etc.

Depalatalisation or velarisation of *i* → *ı*, *ȯ* → *o*, *u̇* → *u* can also cause damage to harmony, which is a distinctive feature of the northern dialect (Jankowski 1992: 40; Medzhitova 2017a: 52); e.g. *kupır* (3) 'bridge', *kursi* (4) 'table; throne', *kobete* (1) 'kind of pirogue', *kobuk* (1) 'foam', *koč* (2) 'migration', *kojlerinden* (1) 'from their village', *kolmek* (2) 'shirt', *kop* (2) 'many', *kopır* 'bridge', *kopırden* (1) 'from the bridge', *kopırge* (1) ~ *kopirge* (1) ~ *kopırya* (1) 'to the bridge', *korep* (1) 'seeing', *korgende* (1) 'when seeing', *korseñiz* (1) 'if you saw', *kosterdiler* (2) 'they showed', etc.

The violation of the law of palatal-velar harmony can be caused by several other factors. Some of them may be extrinsic to the language, i.e. human error or individ-

43 Jankowski (1992: 61), Savran (2003: 8), Useinov–Ganieva–Seydametova (2011: 61–63), Aydın–Adzhumerova (2014: 201) and Medzhitova (2017a: 51) observe that vowel harmony is very stable in the northern dialect.

44 Sevortyan (1966: 236–237) and Izidinova (1997: 301) call it progressive palatal harmony. According to Izidinova (1997: 301), this fit extends to almost all syllables and word forms.

ual pronunciation. In many cases, a comparison of the number of occurrences of each variant reveals the one-off, non-typical character of the variant under study.

Notes on the labial-illabial harmony

In Crimean Tatar, labial-illabial harmony is limited. In the steppe (Kipchak) dialect, for example, labial vowels occur only in the first syllable (Jankowski 1992: 64; Savran 2003: 9–10). Medzhitova, on the other hand, strongly states that there is no labial harmony in the steppe dialect (Medzhitova 2017a: 51). The validity of labial harmony in the middle and southern (Oghuz) dialects cannot extend beyond the third syllable, e.g. *kutluyından* 'in his/her happiness', *kutuldım* 'I was rescued', *kuturtup* 'maddening sb' *gürüldisinden* 'because of its buzzing notes', *üldürirler* 'they kill sb', but there are exceptions, e.g. *oturursuñuz* 'you sit/dwell'. In the southern dialects, labial harmony is damaged by northern influence. Since the majority of affixes and different endings are only biform – containing either velar or palatal illabial vowels –, e.g. *–nan* ~ *–nen* (*naznan* 'flirting', *akčanen* 'for money', *julnan* 'on/by road', *ojnen* 'with thought/intention', *sokornan* 'with a blind man', *jumruknen* 'with fist'), or one-form, e.g. *–daj* (*kojdaj* 'like a sheep', *kümesdaj* 'like a henhouse', *sendaj* 'like you'), *–man* (*apakajımman* 'with my wife'), we cannot speak of complete labial harmony in Crimean Tatar.

The Crimean Tatar texts of Ignác Kúnos also support this statement. In the text of tales there is a significant quantitative predominance of words with illabial suffixes typical of the middle and northern dialects. Some of the small numbers of data showing a labial harmony can probably be traced back to a southern dialect, i.e. an Oghuz dialect. We try to show the quantitative differences by the number of occurrences of each word form:

- *bolip* (5) ~ *bolıp* (34) ~ *bolop* (66) 'being (gerundium)', *boldi* (13) ~ *boldı* (40) ~ *boldo* (35) 'became, was', *bolsın* (29) ~ *bolsun ~ *bolson* (41!) 'let it be', *budır* (1) ~ *budur* (2) 'this is', *buyıl-* (2) ~ *buyul-* (2) 'to be drowned', *bunı* (138) ~ *bunu* (7) 'this', *buzıl-* (1) ~ *buzul-* (2) 'to go bad, to sour', *čıkıp* (33) ~ *čıkup* (4) ~ *čıkop* (1) 'going out', *ǯujıl-* (3) ~ *ǯujul-* (1) 'to be lost', *ǯurtı* (9) ~ *ǯurtu* (3) 'his house, home', *dustlık* (1) ~ *dustluk* (2) 'friendship', *julı* (35) 'his way' ~ *joluna* (1) 'to his/its way' ~ *juluñ* (1) 'your way', *kulı* (72) ~ *kolu* (8) 'his arm', *kutulıp* (6) ~ *kutulup* (1) 'getting rid of', *kuturtup* (1) 'maddening sb', *olıp* (2) ~ *olup* (1) 'being', *onı(n)* (16) ~ *onı(ñ)* (6) ~ *onu(n)* (4) ~ *unu* 'him / his (Accusative / Genitive)', *oturop* (1) 'sitting, dwelling', *oturtop (oturtȯp)* (1) 'to seat, to place', *tugel* (3) ~ *tugul* (2) '(final) not, (isolated) no', *tujıp* (8) ~ *tujup* (2) 'feeling, hearing', *turıp* (40) ~ *turup* (7) 'standing', *tutıp* (12) ~ *tutup* (2) 'holding, seizing', *učıp* (5) ~ *učup* (1)[45] 'flying', *unı* (94) ~ *uni* (30) 'him, it', *ulı* (70) ~ *uli* (33) ~ *ulu* (94) 'his son', *utır-* (70) ~ *utur-* (8) 'to sit, to dwell', *üli* (4) ~ ülü (21) 'dead', etc.

45 In southern (Oghuz type) dialect.

There are a significant number of gerund forms with *-ip* (ca. 450) ~ *-ıp* (ca. 1,500) compared to *-up* (80) ~ *-op (-ȯp)* (81) ~ *-üp* (27). The *-up* ending is also seen repeatedly as an archaic gerund form of verb stems with illabial sounds, e.g. *akup* (1) 'flowing', *alup* (6) 'taking', *čavup* (9) 'cutting', *čıkup* (4) 'going out', *javup* (1) 'raining (rain)', *jıkılup* (1) 'falling', *kızdırup* (1) 'burning, kindling', *sarılup* (1) 'embracing', *tapup* (2) ~ *tavup* (2) 'finding', *turup* (7) 'standing', etc. In these instances, the labialising effect of the *p* and *v (w)* bilabial consonants can be assumed.

Dialectal features of the Crimean Tatar folklore collection of Ignác Kúnos

A detailed linguistic and dialectal analysis of Ignác Kúnos' folklore collection has not been attempted in the works published. Unfortunately, the same can be said about the Crimean Tatar language material. Apart from a few sentences of a general commentary on the language of the current collection, there is nothing else to be found in the studies and editions already published (cf. Kakuk 1963: 93, 1993a: VI). An exception is the *Krimtatarisches Wörterverzeichnis*, whose preface (Kakuk 2012: 15–20) devotes five and a half pages to a linguistic analysis of the Crimean Tatar songs that provide the basis for the dictionary.

To be able to complete the backlog of research, it was necessary to create an appropriate database. As a first step, the complete Crimean Tatar collection of Kúnos (tales, songs, riddles) was recorded in document files using a word processing program. From the text files created, the texts were segmented into words, and the words thus isolated were analysed using a database management program. By doing so, around 57,000 text words of the entire data set became easily usable for linguistic, dialectological, lexicological and lexicographic analyses. Our database was also of great use in preparing the tales for publication. The complete material can now be easily and quickly consulted in the database, and the data in question can be checked and corrected by taking into account the analogies. After further development and completion, the database available to us enabled the compilation and publication of a glossary of Kúnos' complete collection of Crimean Tatar folklore in the book.[46]

Crimean Tatar dialects

Turkic literature (Sevortyan 1966: 257; Jankowski 1992: 2–3, 2010: 37–48; İzidinova 1997: 299) traditionally divides the Crimean Tatar language into three dialects.[47] Accordingly,

46 For more on this, see Baski 2019: 116–118, 120.

47 Radloff (1887: 263), Samojlovich (1916: 6–7) and Baskakov (1960: 154) speak of only two basic Crimean Tatar dialects.

we can speak of (1) northern (Kipchak or Nogai),[48] (2) central (or Tatar) and (3) southern (Turkish or Crimean Ottoman)[49] dialects. In addition to the Kipchak tribal dialects, the Oghuz tribal dialects played a role in developing the Crimean Tatar dialects, albeit to a lesser extent, from the earliest times. This may explain the fact that the Crimean Tatar dialects are somewhat mixed. For example, N. K. Dmitriev (1898–1954) states that there are no "pure" dialects of Crimean Tatar (1936: 163). Each dialect differs in the proportion of Kipchak and Oghuz features.[50]

Even in the text of the same tale or song, there are sharply different dialectal features. The reason for this can only be conjectured at the moment. With the help of a computerised database of the text, we have detected the most characteristic dialectal features. We have tried to determine the dialect of each tale and the entire Crimean Tatar folklore material.

Northern (steppe or Kipchak) dialect

In this dialect, the Oghuz (southern) characteristics are overshadowed. Its sound system is "kipchakised", and its grammatical system is Kipchak, but some Oghuz forms are also present. Some subdialects differ mainly in their vocabulary. The southern influence is significant in all of them.

In the early 1900s, the language of the Crimean Tatar intelligentsia was predominantly "kipchakised". However, it also had a significant Oghuz (southern) influence (Radloff 1887: 268; Samojlovich 1916: 7; Baskakov 1960: 154).

The northern dialect is spoken in the regions of Perekop (Or Kapy), Dzhankoj, Yevpatoriya (Kezlev), Kerch Peninsula, Chongar (Dzhangara), and in the northern parts of Belogorsk (Karasubazar) and Feodosiya (Kefe) districts (Medzhitova 1998; Mazinov 2012: 33; Selimova 2006: 42).

Phonetic features

u̇	frequent after *ǯ*, e.g. *ǯu̇r-* 'to go, walk', *ǯu̇k* 'load', *ǯu̇gen* 'halter (of a horse)', *ǯu̇jru̇k* 'swift-footed (horse)', *ǯu̇rek* 'heart', *ǯu̇z* 'face', *ǯu̇zu̇k* 'ring';
u̇ → *u*	after palatovelar *k*; e.g. *kun* 'day', *kupır* 'bridge', *kursi* 'table; throne'; there are also word-forms containing the syllable *-ku̇-*: e.g. *ku̇ču̇ǯu̇k* (small), *ku̇j* 'village', *ku̇n* 'day', *ku̇p* 'many', *ku̇r-* 'to see', etc.

48 The term „Kipchak-Nogai" is also used; see Useinov–Ganieva–Seydametova 2011: 9.

49 The term "Kipchak-Oghuz" is also used; see Useinov–Ganieva–Seydametova 2011: 9.

50 In a recent edition of Jankowski's grammar, the dialects are described in more detail (2010: 37–48)

ȯ → *o*	after palatovelar *k*, e.g. *kobete* 'kind of pirogue', *kobuk* 'foam', *koč* 'migration', *koj* 'village', *kolmek* 'shirt', *kop* 'many', *kopır* 'bridge', *kor-* 'to see', *koster-* 'to show', *koter-* 'to lift', etc. We also see examples with syllable *-kȯ-* in several cases: e.g. *kȯb* 'many', *kȯj* 'village', *kȯk* 'blue; sky', *kȯse* 'bare-faced', *kȯter-* 'to lift', *kȯz* 'eye', etc.
u → *o*	a more open version of *u*, tending to *o* (see also Medzhitova 2017a: 52), e.g. *bolor* 'will be', *bolop* 'being', *bolot* 'cloud', *joko* 'dream, sleep', *kozo* 'lamb', *sokor* 'blind', etc.
e → *i*	a more closed *e*, tending to *i* is typical here (see Jankowski 1992: 65, 115); Medzhitova (2017a: 52) marks the short, palatal vowel between *ï* and *i* with <ь> soft sign (Medzhitova 2017a, 50). At Kúnos, this vowel may also be indicated by <i> or <ė>, e.g. *bermij* 'does not give', *ǯer* ~ *ǯir* 'place, land', *de-* → *di-* 'to say', *eki* ~ *iki* ~ *ėki* 'two', *ende* ~ *endė* ~ *endi* 'now', *et-* ~ *it-* 'to do', *ket-* ~ *kėt-* ~ *kit-* 'to go';
ı	(reduced vowel); the close central *ï* is reduced in unstressed positions; Kúnos marks the reduced vowels with both <ı> and <ė>, e.g. *ǯojıldı* ~ *ǯojėldė* 'lost', *kız* ~ *kėz* ~ *kz (=kız)* 'girl' ~ *kza (= kıza)* 'for a girl', *nasıl* ~ *nasl* 'how, what', *alıgı* ~ *algı* 'the previous', etc. Medzhitova notes that the velar *ï* (<ы>) is a weak vowel, and the reduced velar *ə* (<ъ>) is barely audible. The high voiced *i, ı*, are significantly reduced after consonants *p, ǯ, č, k* and *t*, and between the sonorant consonants *j, l, n* and *ñ* (Medzhitova 2017a, 50, 51).
b-	(word-initial); e.g. *bar* (65) 'is', *bar-* 'to go', *ber-* (393) 'to give', *bol-* 'to be, exist';
	Different conjugated forms of the substantive verb *bol-* are used instead of *ol-* in precisely 980 cases, e.g. *bola* 'is, will be', *bolaǯaksın* 'you will be, become', *boldı* ~ *boldo* 'became', *bolγan* 'became (indefinite past)', *bolor* 'will be', *bolsın* ~ *bolson* 'let him/it be', etc.
ǯ-	(in word-initial position); At the beginning of a word, it is typical of the Kipchak dialect (the examples of *j* ~ *ǯ* alternation see on p. 20).
t-	(word-initial); e.g. *tau* (50) 'mountain; forest', *taš* (19) 'stone', *taha* (58) 'still, too', *tajan-* 'to lean on' etc.
k-	(palatoalveolar word-initial *k*); e.g. *kel-* 'to come', *kit-* 'to go', *kůč* 'power', *kůz* (10) 'eye' etc. For examples of *k*~ *g* alternation, see p. 22.
-γ- → *-j-*	(word-internal *j* instead of *-γ-*); e.g. *bajla-* ~ *bajlan-* (29) '(to bind, to be bound'
χ- → *k-*	(word-initial uvular *k*); the velar *χ* of Arabic-Persian loan words is replaced with postvelar (uvular) *k*, e.g. *χabar* → *kabar* 'news', *χalk* → *kalk* (31) 'people', *χatır* → *katır* 'memory', *χajır* (2) → *kajır* 'blessing, goodness', *χasta* → *kasta* 'sick', *χızmet* → *kızmet* 'service, work', *χoš* → *koš* 'pleasant, good', etc.
χ-	preserved in some cases, e.g. *χoš* 'pleasant', *χabar* 'news';

h- → Ø — (not pronounced *h* at the beginning of a word), e.g. *hajvan* → *ajvan* (5) 'animal', *hal* → *al* 'condition, fate', *hava* → *ava* 'air', *hazırla* → *azırla-* 'to prepare', *her* → *er* 'each', *hem* → *em* 'and, also', *herif* → *erif* (11) 'lad, fellow', etc. However, in the texts, the consonant *h* also occurs several times. The examples of the *h* ~ *0* alternation, s. on p. 22.

-k- → *-γ-* — (post-velar voiced consonant γ in intervocalic position in place of *k*), e.g. *čıkar-* → *čıγar-* 'pulls out, takes out', *ǯakın* → *ǯaγın* 'close', *ǯıkıl-* → *ǯıγıl-* (5) 'to fall, to be collapsed';

-k- → *-g-* — (mediopalatal voiced consonant *g* in intervocalic position in place of *γ*), e.g. *ȯküz* → *ȯgüz* 'ox', *sekiz* → *segiz* 'eight';

m- — (word-initial), e.g. *mojın* (9) ~ *mujın* (2) 'neck', *murın* (2) 'nose';

p- — (word-initial), e.g. *parmak* (3) 'finger', *pıč-* 'to cut', *pıčak* (4) ~ *pıčaχ* 'knife', *pıčkı* 'saw', *pit-* 'to end', *pitir-* 'to finish', *pütin* ~ *pütün* 'whole; all', but in some tales the consonant *b-* may be retained;

f → *p* — (mainly in words of foreign origin), e.g. *čipt* (P.) (2) 'couple', *čuput* ~ *čiput* ~ *čıpıt* (Ar.) etc. 'Jew', *pajton* (Fr.) 'carriage, carriage', *pelaket* (Ar.) 'tragedy, disaster', *pikir* (Ar.) 'idea, thought', *pilan* ~ *pılan* (Ar.) '(a) certain', *pilǯan* ~ *pilčan* (Ar.<P.) 'cup', *pırılda-* ~ *pirılda-* 'to shine', *purun* (Gr.) 'furnace';

ñ — (regular use of it where other dialects have *n*), e.g. *saña* (101) 'for you', *oña* 'for him' etc. Examples of *ñ* ~ *n* alternation can be seen on p. 24.

-u — (voiced palatovelar semivowel (*w* ~ *u̯*) in word-final position), e.g. *anau* 'yonder', *ašau* 'food, eating', *kijeu* 'bridegroom', *tau* (50) 'mountain; forest' etc.

-ıp ~ *-ip* ~ *-ėp* — (the linking vowel of the suffix of *-p* gerund is always illabial, close high (reduced) vowel), e.g. *alıp* ~ *alėp* 'taking', *birėp (~ birep)* 'giving', *etip* ~ *etėp* ~ *etıp* 'doing', *ičėp* 'drinking', *kajtıp* ~ *kajtėp (~ kajtip ~ kajtep)* 'returning', *kelip* ~ *kelėp* 'coming', *kujıp* 'placing, putting', *minip* ~ *minėp (~ minep)* ~ *minüp* 'getting on, sitting on', *suγıp* 'hitting', *turıp* 'standing', *tüšürıp* ~ *tüšürėp* 'dropped', *urıp* 'struck, beaten', etc.

Morphological features

-γa ~ *-ge, -kA* — (Dative case ending), e.g. *amamγa* 'to bath', *ǯigitke* 'for a lad, young man', *karγaγa* 'to crow', *kartka* 'to old man', *mollaγa* 'to molla', *sarajγa* 'to palace', *teñizge* 'to sea', *utka* 'to grass', etc.

-nI — (Accusative case ending); e.g. *adını* 'his/her name', *balanı* 'chld', *ǯigetni* 'lad', *sıčkannı* 'mouse' etc.

-n — (Accusative case ending associated with with the 3rd person possessive suffix *-I* ~ *sI*), e.g. *adın* 'his name', *babasın* 'his father', *bašların* 'their head', *betin* 'his face', *kursaγın* 'his stomach', *kızın* 'his daughter', *mujnın* 'his neck', *üjlerin* 'their house', etc.

-nIñ (Genitive case ending), e.g. *atnıñ* 'of the horse', *kartnıñ* 'the old man's', *padišanıñ* 'the padisah's', *terekniñ* 'of the poplar', *üjnıñ* 'of the house' etc.;

-Iñ[51] (< *-nIñ*) (shortened Genitive case ending), e.g. *kızıñ* 'the girl's', *čobanıñ* (< *čobannıñ*) 'the shepherd's', *išleriñ* 'of things';

-nI ~ -nU[52] (shortened Genitive case ending emerging by disappearing of the consonant *ñ* from the end of the affix *-nIñ ~-nUñ*), e.g. *kızımnı* 'my daughter's', *jüzügimni* 'of my ring', *kamčımnı* 'of my whip', *onu bašın* 'his head', *padišanı közüne* 'to the eye of the padisah', *teregimni ǯimišini* (Accusative) 'the fruit of my tree', *ulannı* 'son's', *üjnü eki katın* 'the two floors of the house (Accusative)', etc.

-I (shortened Genitive case ending (*-In*), it can be added to Sing1 possessive suffix or personal pronouns), e.g. *fesimi püsküliniи uǯı* 'the end of the tassel of my *fez*', *χanımı üji* 'the woman's house'; In the text of songs the similar forms are often used, e.g. *kamančemi (aγacı)* 'the (tree) of my fiddle', *hanımı (tügüni)* 'the (wedding) of the woman', *meni* 'my, ... of mine', *seni ~ sini* 'your, ... of yours', *onı* 'his, ...of him', *bizni* 'our, ...of ours' etc.[53]

-GAn ~ -KAn (verbal suffix of indefinite past tense), e.g. *tapkan* 'found', *bergen* 'gave', *ketken* 'went', *kürgen* 'saw', *ütürgen* 'killed' etc.

-j (forms present tense of vowel-final stems; when suffixing the final vowel of the stem *-A* contracts with the verbal suffix *-j* as follows: *-A+-j* > *-Ij*; other vowels do not behave like that), e.g. *azirlıj* 'he prepares, he is preparing', *bašlıj* 'he begins', *beklij* 'he is awaiting, *dij (dıj)* 'says', *jarašıj* 'fits, is fitting', *istej ~ istij* 'he wishes, he wants', *išlij* 'he is working', *kelij* 'he comes, he is coming', *kokıj* 'smells', *süjlij* 'he is speaking', *tamlijıj* (!) 'it is dropping', *tanıj* 'he recognises sb', *tilij* 'he wishes', etc.

Note: In the songs of Kúnos, as in the Crimean Tatar dialect of Polatlı (Turkey), this verb form is used in the simple present tense. Consequently, some of the translation errors that Zühal Yüksel (1994: 20) criticised in Zsuzsa Kakuk's edition of the text should not be considered errors. However, in the Turkish translation of the Crimean Tatar songs in the work in question, the Tatar verb forms *-ıj ~ -ij* of continuous present tense were, in most cases, translated into Turkish – incorrectly – in simple present.

---

51 Medzhitova (2017b: 55) lists it among the northern dialectal features. In a dialect of Turkmen, there is a similar possessive case ending, which is considered an Oghuz peculiarity there (Tenishev (ed.) 1988: 61–62).

52 It is also a phenomenon known from other sources, e.g. *arslannï* 'of the lion' (Hesche-Scheinhardt 1974: 238.)

53 Examples of this morphological phenomenon can also be found in other texts, e.g. *anamï* 'my mother's', *babamï* 'my father's' (Hesche-Scheinhardt 1974: 238), etc. In Karachay-Balkar and Kumuk, and even in the modern Uyghur Lobnor dialect, there is a *meni, seni,* in Kumuk *munu ~ bunu* pronoun form, which is a Kipchak peculiarity (Tenishev (ed.) 1988: 67).

-Am (Sing1 person verbal suffix, forms present tense of verbs ending in a consonant),[54] e.g. *alam* 'I am taking', *atam* 'I am throwing, I shoot', *baram* 'I am going', *berem* 'I am giving', *bilem* 'I know', *bolam* 'I am, I will be', *čıkam* 'I am going out', *edem* 'I am doing', *jatam* 'I am lying', *jazam* 'I am writing', *jürem* 'I am going, walking', *kalam* 'I am staying', *kelem* 'I am coming', *ketem* 'I am going', *kürem* 'I can see', *otıram* 'I am sitting, I dwell', *sarılam* 'I am embracing', *satam* 'I am selling', *tujam* 'I feel', *turam* 'I am standing', *tutam* 'I am taking, I am holding', *uzatam* 'I am reaching sg to sb', etc.

-mAn (Sing1 person verbal suffix used in present and simple tenses, and as the suffix of the copula of nominal predicates in nominal sentences), e.g. *ajtaman* 'I am saying', *alaman* 'I am taking', *bataman* 'I am sinking, I am diving', *juklajman* 'I am sleeping', *kajtarman* 'I am returning', *alırman* 'I am taking, I am buying', *berirmen* 'I give (usually), I give (then)', *kızıman* 'I am the daughter of', *Azrāilmen* 'I am Azrail', *musapirmen* 'I am a guest', etc.

-Ij (forms present tense of consonant-final stems; in fact, this suffix emerged by contracting of two suffixes of the present tense, namely *-A* and *-j*), e.g. *ačılıj* 'it opens, it is opening', *ačulanij* 'gets (angry)', *akıj* 'it is flowing', *alıj* ~ *alij* 'he is taking, he takes', *bakıj* ~ *bakij* 'he is looking', *berij* 'he is giving', *janašıj* 'he is approaching', *kalıj* 'he is staying', *kelij* 'he is coming', *sallanıj* 'it is swinging', *sündürij* 'he is putting the fire out', *urušıj* 'he is fighting', ütirij 'he is killing', etc.;

-mIj (forms present negative ten se of all kinds of stems since the negation particle *-mA* can be attached both to consonant-final and vowel-final stems, thus, the suffix *-mIj* is contracted of particle *-mA* and present tense verbal suffix -j); e.g. *ačılmıj* 'it does not open, it will not open', *bilmij* 'he does not know', *bolmıj* ~ *bolmij* 'it is not ..., it will not be', *bulunmij* 'it cannot be found, there is not', *etmij* 'he is not doing, he does not do', *kelmij* 'he is not coming', *kürmij* 'he can not see', *olmij* 'it does not exist',[55] *tajanmıj* 'he does not resist', *tojmij* 'he is not gorging', *ülmij* 'he does not die', *ütürmij* 'he does not kill', etc.[56]

-man (ending of Comitativus), e.g. *apakajımman* 'with my wife';

-daj (comparative suffix or case ending meaning 'like, similar to')[57], e.g. *bondaj* 'like this, similar to this', *ajγırdaj* 'like a stallion', *Čoradaj* 'like Chora, similar to Chora', *χijardaj* 'like a cucumber', *kündüzdaj* 'like daylight', *šondaj* 'like this', *ulakdaj* 'like a goat', etc.;

54 It is relatively common in songs.
55 It only occurs in the songs.
56 Most of such verb forms occur in songs.
57 Jankowski (1992: 274, 2010: 209) lists it among the case endings.

Dajın	(comparative postposition 'like, as'),[58] e.g. *bala dajın* 'like a child', *bolyanı(n) dajın* 'as it was, as it happened', *bondajın* 'like this', *bulut dajın* 'like a cloud', *degeni dajın* 'as he/she said', *euelki dajın* 'as before, earlier', *kümes tajın* 'like a hen house', *ondajın* 'like that, like him/her', *sen dajın* 'like you, similar to you', *zip tajın* 'like tar/pitch';
anau	'yonder, that one', *bo* ~ *bı* 'this', *mınau* 'this (one), of this place', *šo* 'this'; typical feature is the use of these demonstrative pronouns.
maya	(Dative case of Sing 1 personal pronoun *men*) 'for me, to me';
saya	(Dative case of Sing 2 personal pronoun *sen*) 'for you, to you';
oya	(Dative case of Sing 3 personal pronoun *o*) 'for him/her, to him/her';
buya	(Dative case of demonstrative pronoun *bu*) 'for/to this';
šoya	(Dative case of demonstrative pronoun *šo*) 'to/for this';
kaj	'what, which', *kajda* 'where', *kaj dajın* 'how', *kajdan* 'from', *nasıl etip* 'how', *ne daj* 'how', *ne dajın* 'how, how'; Common use of these interrogatives.
berli	(postposition 'since, for'); e.g. *beš ǯıldan berli* 'for five years', *ajdamaklık etkenden berli* 'since (we) have been robbing';
alaj	'so', *barabar* 'together', *šaj* 'like this'; typical use of these adverbs.

Typical lexical items

ajt- (370) 'to say'; *alaj* 'so', *aša-* (157) 'to eat'; *ata* (37) 'father', *čal* (24) 'grey', *čap-* (12) 'to run', *büri* (2) 'wolf', *ǯau* (2) 'enemy', ǯau- ~ *ǯav-* (6) 'to rain (snow, rain)', *ǯuvur-* (9) 'to run', *di-* (1024) 'to say, to tell'; *it* (5) 'dog', *kara-* (61) 'to look', *kart* (127) 'old'; *kaškır* (2) 'wolf'; *kiškene* (5) 'small'; *kop* (3) ~ *köp* (5) ~ *küp* (15) 'many'; *kurtka* (21) 'old woman'; *mañlaj* (2) 'forehead'; *süj-* (9) 'to love'; *šaj* 'like this', *tap-* (102) ~ *tab-* (3) 'to find', *tabıl-* (3) 'to be found'; *tanau* (1) 'nostril'; *tolkun* (1) 'wave'; *totaj* (4) ~ *tutaj* (2) 'sister, wife', *tuvar(čı)* 'cattle (shepherd)', üjlen- 'to marry', *apaj* (22) 'wife', *apakaj* (37) 'woman, wife',[59] *akaj* (29) 'husband', *kada(j)* 'sister; brother'[60] etc.

Southern (coastal and mountain, Oghuz) dialect

In some subdialects the Oghuz influence varies in degree but is generally more stressed than in the other two dialects. Initially, this dialect was also northern (Kipchak) in character. Still, the influence of a large number of Anatolian and Balkan immigrants has led to the predominance of a southern (Turkish) element. According to Radloff's (1837–1918) observation (1887: 267), old (i.e. northern, Kipchak) forms appear everywhere

58 Jankowski (1992: 274) considers it to be a dialectal variant of the comparative case ending *-DAy*.
59 We only see it 2 times in the lyrics.
60 It occurs only once in tales, and ten times in songs.

alongside the predominant southern elements, depending on the speaker. The grammar of the south coast is entirely Oghuz (Samojlovich 1916: 7).

The geographical area of the southern dialect is the districts of Foros (Forus), Nikita (Nikta), Balaklava, Orlinoe (Bajdar), Yalta, Privetnoe (Üsküt), Gurzuf, Tuvak, Morskoe (Kapsihor), Sudak and Feodosiya (Kefe).

Phonetic features

These features of the southern dialect are compared here to those of the northern dialect.

t- → d- (word-initial), e.g. *daš* (30) 'stone', *degil* (2) 'no (negation of the nominal predicate)', *dügün* (10) '(wedding) feast', *duj-* (6) 'to feel, hear' etc. For examples of *t ~ d* alternation, s. p. 21.

b- → v- (word-initial), e.g. *var* (1), 'there is', *var-* 'to go, arrive', *ver-* 'to give' (7);

m- → b- (word-initial), e.g. *boyun* (1) 'neck', *burun* (5) 'nose', *burungu* (2) 'earlier, old';

bol- → ol- (different forms of the conjugated substantive verb *ol-*); e.g. *oldoñ* 'you have become', *olγan* 'became', *olmaz* 'will not be', *olsın* 'let him/it be' etc.; The vast majority of our data has been taken from the lyrics.

f- (preserved in words of foreign origin), e.g. *fajton* (Fr.) (18) 'carriage', *felan* (13) 'a certain', *filǯan* (5) (Ar.<P.) '(coffee) cup', *furun* (6) (Gr.) 'furnace', *fukara* (2) (Ar.) 'poor man', etc.

k- → g- (typical word-initial voiced consonant of the dialect), e.g. *gül* 'flower, rose', *gajb* 'loss', *gün* 'sun', *gene* 'again' etc.;

-γ (uvular voiced plosive consonant), e.g. *baγ* (2) 'garden', *daγ* (8) 'mountain; forest', *saγ* (6) 'healthy', *jaγ* (4) 'fat, oil';

k → χ (uvular *k (q)* is replaced by *χ*),[61] e.g. *ajaχ* (2) 'leg', *ajtajıχ* (1) 'let us say', *čıχ-* (104) 'to go out', *χadar* (5) 'to, till, until (postposition)', *χajt-* (3) 'to return', *χara* (3) 'black', *χara-* (1) 'to look', *barmaχ* (3) 'finger', *χırmızı* (1) 'red', *χız* (15) 'girl', *joχ* (50) 'is not, there is not', *kırχ* (16) 'forty', *tuχsan-tuχız* (4) 'ninety-nine', etc.;

χ → k (breathed velar spirant *χ* of foreign (Arabic-Persian) origin is replaced by uvular *k* in some southern subdialects),[62] e.g. *χabar → kabar* 'news', *χalk → kalk* (31) 'people', *χatır → katır* 'memory', *χajır* (2) → *kajır* 'blessing, goodness', *χasta → kasta* 'sick', *χızmet → kızmet* 'service. work', *χoš → koš* 'pleasant, good', etc.

---

61 It also took place in the western subdialects of the Sudak district, while the eastern subdialects of the district retain the *k* (Kurtbilal 2007: 505).

62 In the subdialects of the villages of Suvuk-Su, Taraktaš, Koz, Tokluk and Otuz belonging to Sudak (Kurtbilal 2007: 505).

h (typically preserved);[63] In some subdialects, however, the disappearance of *h (h >Ø)*[64] or alternation of *h* with *Ø* can also be observed.[65] The examples of alternation can be seen in ch. *Disappearance of consonant h.*

-h- → *-j-* (change inside the word); e.g. *kajve* (46) ~ *kajva* (1) 'coffee', *kajveǯi* (8) 'café-owner';

j- (word-initial *j-* is typical of this dialect); e.g. *jaχšı* 'good (1)', *jajla* (6) 'summer pasture', *jala-* (7) 'lick', *jaman* (5) 'bad, evil', *jaš* (36) '(age) of life; tear', *jaša-* (28) 'live', *jiket* (12) ~ *jiget* (139) 'lad, youth', *jok* (118) ~ *joχ* (41) ~ *juk* (18) 'is not, there is not', *jüzük* (68) 'ring'; For examples of *j* ~ *ǯ* alternation see ch. *Pairs of words alterning of the word–initial j– ~ ǯ–.*

ñ → *n* usually *ñ* is replaced by *n*,[66] but in certain subdialects, the consonant *ñ* is retained. The examples of alternation *ñ* ~ *n* are listed in ch. *Alterning of ñ ~ n.*

-up ~ u̇p ~ -op ~ -ȯp (the linking vowel of the suffix of *-p* gerund is mainly labial (*u, u̇, o, ȯ*) when attaching to verb-stem with a labial vowel), e.g. *kojup* 'laying down', *turup* 'erecting' *ču̇ru̇p* 'rotting, ruining', *kȯru̇p* 'seeing', *su̇ru̇p* 'folding, spreading', *tu̇šu̇p* 'falling', *bolop* ~ *olop* 'being', *korkop* 'being afraid of', *totop* 'holding', *tottorop* 'having held', *oturtop* 'having planted', *čıkop* ~ *čıχop* 'going out', etc.;

Note: The consonant system of the Crimean Ottoman dialect of the southern (Oghuz) dialect group also shows a solid northern (Kipchak) influence, which is most evident in the alternation of *d* ~ *t*, *g* ~ *k*, *j* ~ *ǯ*, *b* ~ *m*, *b* ~ *v* and *g* ~ *j* consonants (Selimova 2006: 35).

Morphological features

-I ~ -jI (Accusative case ending), e.g. *almajı* 'apple', *balajı* 'child', *čalγıjı* 'music', *duajı* 'prayer', *gu̇li* 'flower, rose', *kapıjı* 'door, gate', etc.

*-nI*n (Genitive case ending), e.g. *padıšanın* 'the padisah's', *tereknin* 'of the tree' etc.

-In ~ -Un (Genitive case ending), e.g. *mülkün* 'of the wealth, property', *terekin* 'of the tree', *kamanderin* 'captain's, of the commander', etc.

-nI - (Accusative case ending with 3rd person possessive suffix); e.g. *ǯukusını* 'his sleep', *mujnını* 'his neck', *kabuyını* 'its shell', *kenǯesini* 'the young (one)', *ajaklarını* 'their feet, legs', *ǯılkılarını* 'their horses', *šijlerini* 'their things', etc.

gibik (comparative postposition); e.g. *ku̇pek gibik avla-* 'to bark as a dog'.

63 Izidinova 1997: 308.

64 Selimova 2006: 35.

65 It was observed in the eastern subdialects of the Sudak district by Kurtbilal (2007: 506).

66 Izidinova 1997: 308.

-(A)j ~ -mAj	(present (continuous) tense, both affirmative and negative of all kinds of stems, since the negation particle *-mA* can be attached both to consonant-final and vowel-final stems, thus suffix *-mAj* is contracted of particle *-mA* and the present tense verbal suffix -the *j*; the suffix *-Aj* emerged of two suffixes of present tense, namely *-A* and *-j*; these forms are prevalent in songs), e.g. *ača* 'opens', *ačmaj* 'does not open', *bakaj* 'looks', *čıdaj* 'tolerates', *čıkaj* 'goes out', *ǯılaj* 'is weeping / weeps', *etej* 'is doing, does', *javaj* 'rains / is raining (snow, rain)', *ketej* 'is going, goes', *saklaj* 'is protecting, protects, (preserves)', *beklej* 'waits / is waiting', *istej* 'wants', *tilej* 'wishes', *tujaj* 'feels', *tůšej* 'is falling, is landing', *tůšmej* 'is not falling, does not land', *ůpej* 'is kissing, kisses' etc.
-(I)j ~ -mIj	(forms present (continuous) tense, both affirmative and negative of all kinds of stems, since the negation particle *-mA* can be attached both to consonant-final and vowel-final stems, thus suffix *-mIj* is contracted of particle *-mA* and the present tense verbal suffix *-j*; suffix *-Ij* emerged by contracting of two suffixes of the present tense, namely *-A* and *-j*), e.g. *bašlanıj* 'is beginning, begins', *bolmıj* 'will not', *darılıj* 'is offended', *gelij* 'is coming, comes', *istıj* 'he wants, wishes', *nazlanij* 'is behaving coquettishly', etc.
-(I)jIr	(continuous present tense, occurs only in lyrics), e.g. *jatijir* 'he is lying', *kelijir* 'he is coming', *sůnijir* 'it is going out (fire)', *ůlijir* 'he is dying', etc.
-(A)jIr	(continuous present tense, occurs only occasionally in tales, and quite a few times (about 50 times) in songs), e.g. *bašlajir (~ bašlijor)* 'he is starting', *durtajir* 'he is pushing', *dejir* 'he is saying', *ujnajır* 'he is playing, dancing';
-(I)jor	(continuous present tense), e.g. *bašlijor (~ bašlajir)* 'he is beginning', *beslijor* 'he is nourishing', *gečijor* 'he is passing', *janıjor* 'it is burning', *parlijor* 'it is shining', etc.
-m ~ (j)Im -	(Sing 1 suffix of the copula of nominal predicates in nominal sentences), e.g. *ačım* 'I am hungry', *čıplakim* 'I am naked', *dertlijim* 'I am desperate', *jigetim* 'I am a lad/fellow', etc.
-mIš	(indefinite past tense), e.g. *ačmiš* (1) 'opened', *čıkmıš* (1) 'went out', *důvmiš* (1) 'beat', *kalmıš* (2) 'stayed', *kalkmıš* (1) 'rose', *kırslamıš* (1) 'stolen', *tapmıš* (1) 'found', *terlemiš* (2) 'sweated', etc. This past tense is used only a very insignificant proportion of the total material.
-dikče	(gerund meaning 'so long as, the more'), e.g. *dedikče* (4) 'so long as he said', *keldikče* (1) 'the more he came, the more …';

-nAn (Comitativus case ending),[67] e.g. *adamnen* 'with a man', *atnen* 'with a horse', *čeleknen* 'with a bucket', *inenen* 'with a needle', *julnan* 'on/by road', *kašıklarnan* 'with spoons', *kartnen* 'with an old man', *mennen* 'with me', *sauluknan* ~ *sauluknen* 'with health', *sokornan* 'with a blind (man)', *tajnen* 'with a colt', *topalnan* 'with a lame', *ularnan* 'with them', *ulunen* 'with his son', *zornen* 'with difficulty, hard', etc.

-AǯAk (definite future tense), e.g. *bereǯek* 'he will give', *keleǯek* 'he will come', *čıkaǯak* 'he will go out', *sarılaǯak* 'he will embrace', *seveǯek* 'he will love', *tujaǯaχ* 'he will hear, feel', *tüšeǯek* 'he will fall', etc.

Typical lexical items

baba (170) 'father', *bak-* (246) 'to look', *čok* (21) ~ *čoχ* (17) 'many', *čizme* (1) 'boots', *de-* (680) 'to say', *je-* (3) 'to eat', *gizlen-* (4) 'to hide oneself', *kapu* (30) ~ *kapı* (129) 'door, gate', *kemik* (1) 'bone', *küčük* (22) 'small', *it* (5) ~ *küpek* (15) 'dog', *parmak* (3) 'finger', *sev-* (4) 'to love', *sıyır* (8) 'cattle', *süjle-* (16) 'to say', etc. The word *tata*, meaning 'sister, woman', which is part of the Oghuz vocabulary, occurs once in the lyrics.

Central (mixed) dialect

The Central dialect[68] is spoken between the northern and southern dialects, i.e. between the steppe and the coastal strip. It is a typical transitional dialect. Some scholars, e.g. Doerfer (1959: 369), call it Middle Crimean Turkish or simply Crimean Turkish. It has a strong and roughly proportionate mix of Oghuz and Kipchak features.[69] In 1928, at the first Crimean Tatar linguistic conference in Simferopol, the central (middle) dialect was chosen as the basis for the literary language of Crimean Tatar (Gafarov 1973: 154).

The central dialect is spoken on the territory of northern districts of Bakhchisaray (Bagchasaray) and Simferopol' (Akmesdzhit), the area around Staryj Krym (Eski Kyrym), north of Sevastopol' (Akjar), Feodosija (Kefe) and Belogorsk (Karasubazar), and south of Oktjabr'skoe (Büyük Onlar) (Sevortyan 1966: 257; Mazinov 2012: 33).

Phonetic features

These features of the central dialect are compared here to those of northern and southern dialects.

67 Jankowski 1992: 273–74, 2010: 209; Izidinova 1997: 308.

68 Radloff (1887) calls this dialect *orta yolaq* 'middle band'. More recently, it is also called „Kipchak-kuman"; see Useinov–Ganieva–Seydametova 2011: 9.

69 According to Mazinov (2012: 33), the Kipchak elements are already beginning to dominate.

b- → *v-* (word-initial consonant, as in the southern dialect, see there);

b- (word-initial consonant, as in the southern dialect; see there for examples);

b- → *p-* (in word-initial position), e.g. *parmak* (3) 'finger', *pek* (8) 'very', but *bek* (8) and *barmak* (4) ~ *barmaχ* (3) also occur;

-p- (not voiced in intervocalic position); e.g. *čapık* 'quick', *japalar* 'they are making/closing sg', *kapa* 'he snatches', *tapup* (2) 'found', *tapar* 'he finds', *tapıl-* (5) 'to be found' (suffering verb form);

-γ (in word-final position, as in the southern dialect; for examples see there);

-γ- (in intervocalic position; as in the Northern dialect, for examples see there);

h (partially preserved as in the southern dialect, the examples see there):

h- → Ø (the disappearance of the consonant *h* can also be observed among the peculiarities of the northern dialect, the examples of changing *h-* > Ø see in ch. *Disappearance of consonant h*;

f (does not change; for examples see above at the southern dialect);

t- (word-initial) as in the northern dialect; for examples see there;

d- (word-initial) as in southern dialect; for examples see there;

j- (word-initial) as in the southern dialect; for examples of *j* ~ *ǯ* alternation see *Pairs of words alterning of the word–initial j– ~ǯ–*, p. 20;

k- (word-initial, palato-alveolar) as in the northern dialect; for examples see there;

k- (word-initial, uvular) as in the northern dialect; for examples see there;

χ- (word-initial), e.g. *χabar* (53) 'news', *χalk* (3) 'people, people', *χanım* (7) 'wife, mistress', *χas* (9) 'real, true', *χoǯa* (1) 'husband, teacher', *χoš* (8) 'pleasant';

k → *χ* (instead of uvular *k*), as in the southern dialect; for examples see there;

-u *(~-v)* (=*-u̯*, *-w*) (in some subdialects voiced palatovelar semivowel); e.g. *aldau* (1) 'cheating', *anau* ~ *anavı* 'yonder, that one', *ašau* (1) 'eating; food', *etu* (1) 'doing, making', *jazu* (1) 'writing', *kušu* (2) 'running', *kijeu* 'bridegroom', *kijeuvi* ~ *küjevi* 'his/her bridegroom', *sılau* (1) 'lime-painting, glazing', *tau* 'mountain' ~ *tavın* 'the mountain's (Genitive case)' *tüleu* (1) 'paying', *oklau* ~ *uklau* 'rolling spin', *uklavı* (rolling spin, Accusative);

inside the word before vowels and sonorant consonants *m, n, ñ, l, r;* the soundness (voiceness) or soundlessness (voicelessness) of the adjacent consonants remains unchanged (Izidinova 1997, 300, 308), e.g. *bolgačnı* (1) 'leafy pirogue (Accusative)', *χılıčnen* (1) ~ *kılıǯnen* (3) 'with, by the sword (Comit.)', *terekte* (4) ~ *terekde* (2) 'on a tree', *jigirem* (3) ~ *jikirem* 'twenty', *Fatma* (2) ~ *Fadmeǯik* 'little Fatma, Fadme', *aγačlar* (4) ~ *aγaǯlar* (2) 'trees', *küpri* (2) ~ *kübri* 'bridge'.

Morphological features

This dialect contains varying proportions of northern (Kipchak) and southern (Oghuz) linguistic features and is therefore called a mixed or transitional dialect.

-Am	(present tense Sing1 suffix attaching to consonant-final stems);[70] for examples see above, in the morphological paragraphs of the northern and southern dialects;
-nI	(Accusative case ending), as in the southern dialect, for examples see p. 39;
-nI	(Genitive case ending), as in the northern dialect, for examples see there;[71]
-nAn	(Comitative case ending), as in the northern dialect, for examples see there;
Dajın	(comparative postposition 'like, as'), as in the northern dialect, for examples see there;

Ttypical lexical items

In the central dialect, there are a significant number of Oghuz lexical items in addition to the northern (Kipchak) ones. Examples of lexical items used in parallel are given below:

– *aša-* (157) ~ *je-* (3), *ata* (37) ~ *baba* (170) 'dad', *kop* (4) ~ *kȯp* (6) ~ *čok* (21) 'many', *čizme* (1) ~ *etek* (1) 'boots', *de-* (680) ~ *di-* (1024) ~ *ajt-* (370) 'to say, tell', *kurt* (1) ~ *kaškır* (2) ~ *bůri* (2) 'wolf', *it* (5) ~ *kůpek* (18) 'dog', *bol-* (980) ~ *ol-* (155) 'to be (copula), to become', *sůjle-* (7) ~ *ajt-* (370) 'to say, speak'.

Summary

Of the dialectal features presented above, 47 have been examined according to the tales in which they occur one or more times. The results are illustrated in the bar chart below.

It can be seen that, in general, there are no sharp differences between the differentiating features of each dialect in the different tales. It is particularly instructive to compare the dialectal features of tale 39. We know for certain that this tale was recorded by Kúnos from a certain Mehemmed, a former inhabitant of Akmedjid (Akmesdjit, now Simferopol), which belongs to the central dialect area. Therefore, this story's dialect can be rightly classified as belonging to the central (mixed) dialect. The test results are consistent with this, since the northern and southern features are present in almost equal proportions (77–78%). At the same time, 50% of the features of the central dialect can also be detected in the text of the tale.

70 Relatively common in lyrics.

71 According to Izidinova (1997: 308), the homonymy of the Genitive case ending *-nI* and the Accusative case ending *-nI* is a characteristic feature of the central dialect.

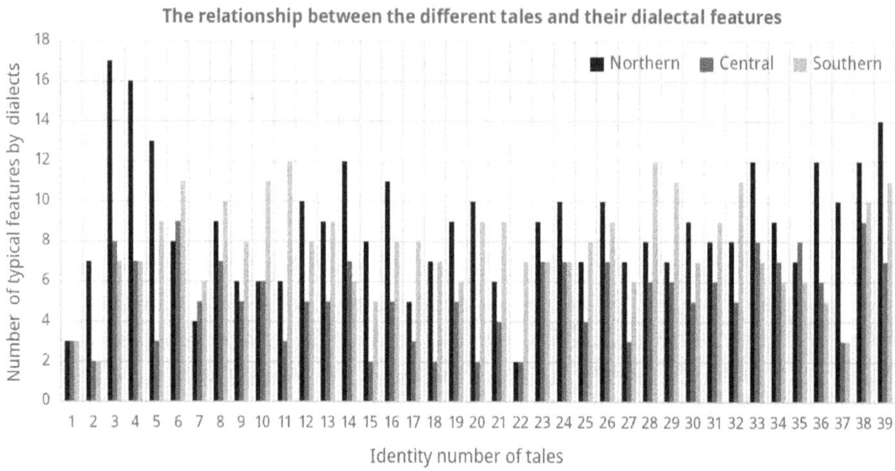

However, only cautious conclusions can be drawn from the comparative analysis. One of these is, for example, that 14 tales [nos. 2, 3, 4, 5, 14, 15, 16, 19, 23, 24, 33, 36, 37, 39] show a slightly prominent northern (Kipchak) character. With a similar reservation, nine tales [nos. 6, 10, 11, 17, 21, 22, 28, 29, 32] can be considered to have a southern (Oghuz) character. The remaining 15 tales [nos. 1, 7, 8, 9, 12, 13, 18, 20, 25, 26, 27, 30, 31, 35, 38] show an almost equal proportion of northern and southern dialect features, so they can be classified in the central dialect group.

Looking at the diagram as a whole, the professional cliché that the Crimean Tatar dialects are highly mixed in character, and that they are distinguished only by the different proportions of northern (Kipchak) and southern (Oghuz) elements, is also true.

Abbreviations

Abl.	Ablative
Acc.	Accusative
Adj.	Adjective
Adv.	Adverb
Ar. -	Arabic
cf.	Confer (Latin conferre)
Comit.	Comitative (Latin Comitativus)
Conj.	Conjunction
Cop.	Copula
Dative	Dative
demonst.	Demonstrative
dial.	Dialectal variant
dim.	Diminutive
Ecp.	Existential copula particula (e.g. *bar/var/jok*)
En.	Ethnical name
fact.	Factitive (verb)
fem.	Female
Fr.	French
Gen.	Genitive
Ger.	Gerund
Gr.	Greek
hypoc.	Hypocoristic, endearing form of a name
Imp.	Imperative
Inf.	Infinitive
Inf(ma).	Infinitive suffixed with *-mA*
Inf(mak).	Infinitive suffixed with *-mAk*
Inf(uw).	Infinitive suffixed with *-Uw*
Interj.	Interjection
Ir.	Iranian
It.	Italian
indef.	Indefinite
Int.	Interrogative
Loc.	Locative
Ms.	Manuscript (Latin)
N.	Noun
Neg.	Negative
Nom.	Nominative
Num.	Numeral
Nvoc.	Vocative form of noun
Onomatop.	Onomatopoetic word
P.	Persian
Part.	Particle
pass.	Passive form of verb
pers.	Person(al)
Plur.	Plural
Plur1–3	Plural first, second, third person
Pn.	Personal name

Poss.	Possessive
Postp.	Postposition
Pron.	Pronoun
Prop.	Proper name
R.	Russian
recipr.	Reciprocal (or co-operative) form of verb
s.	See
Sing.	Singular
Sing1–3	Singular first, second, third person
Sp.	Spanish
Tn.	Toponym, place name
V.	Verb
Vaux.	Auxiliary verb
Vfact.	Factitive form of the verb
Vpass.	Passive form of verb
Vulg.	Vulgar

Signs and symbols

~	a sign standing between alternating data (letters, suffixes, words, etc.) in the linguistic introduction;
/	is used to separate alternating words, phrases and abbreviations with the same or nearly the same meaning in the Glossary;
→	indicates the direction of change from the original data to the changed data;
<	points from the original sound, morpheme and word to the new version of it;
( )	the number in round brackets indicates the total number of times the word or linguistic phenomenon (phonetic, morphological feature) in question occurs in Ignác Kúnos' Crimean Tatar texts (tales, songs and riddles);
[ ]	1. in the introduction, the identity (serial) numbers of the published tales are given in square brackets; 2. a letter, a word or a sentence that was presumably omitted from the manuscript text and has been replaced by the editor of the text is shown in square brackets;
|	a vertical line separates English and Russian meanings in the Glossary;
*	in the introduction, a sign preceding a hypothetical, inferred sound or word form;
Ø	indicates the absence of the sound or suffix (form, sign, verb) in question;
→	is used to refer to the main entry whose headword is a linguistic or dialectal variant of the headword in question;
s.	used in reference entries for pointing to the *headword* of the *referenced entry* (namely another main entry) where the variant(s) of the referring headword is/are found.

Bibliography

Akbaba, Dilek Ergönenç. 2009. *Nogay Türkçesi Grameri. Ses ve Şekil Bilgisi* [Grammar of Nogai Turkic. Phonetics and morphology]. (Grafiker Yayınları 46). Ankara: Grafiker.

Aydın, Ayşe & Reshide Adzhumerova. 2014. Kırım Halk Yırlarında Belli Başlı Ses Özellikleri [Main phonetic features of Crimean folk songs]. *TÜRÜK Uluslararası Dil, Edebiyat ve Halkbilimi Araştırmaları Dergisi*. 2(4), 197–218.

Baskakov, Nikolay Andreevich. 1960. *Tyurkskie yazÿki* [Turkic languages]. Moskva: Izdatel'stvo vostochnoy literaturÿ.

Baski, Imre. 2004. I[gnác] Kúnos Kırım Tatarcası Folklor Metinlerinin Ağız Özellikleri Üzerine [On dialectal features of Ignác Kúnos' Crimean Tatar folklore texts]. In *V. Uluslararası Türk Dil Kurultayı, Ankara, Turkey, 2004(I)*, 391–403. Ankara: Türk Dil Kurumu.

Baski, Imre. 2019. Folklór szövegből szótár. Kérdések és megoldások Kúnos Ignác tatár gyűjtésének közzététele kapcsán [Folklore text to dictionary. Questions and solutions for the publication of the Tatar collection of Ignác Kúnos.]. *Keletkutatás* 2019 (Ősz). 97–121.

Chatskaya, O. I. & Nikolay Konstantinovitch Dmitriev. 1926. II. Chansons tatares de Crimée. Recueillies et traduites par Mme O. Chatskaya, avec introduction de N. K. Dmitriev. *Journal Asiatique* CCVIII. 341–369.

Çolak, Melek. 2019. Macar Arşiv Belgelerinin Işığında I. Dünya Savaşı Sırasında Avusturya-Macaristan Topraklarında Bulunan Tatar Esir Kamplarında Yaşam, Türkoloji Çalışmaları ve Ignác Kúnos (1915–1918) [In Light of Hungarian Archive Documents, Life in Tatar Prison Camps Located within the Borders Of Austrian - Hungarian Lands, Turcology Works and Ignác Kúnos during The World War I (1915–1918)]. *Türk Dili Araştırmaları Yıllığı. Belleten* 67 (Haziran). 231–256.

Dmitriev, Nikolay Konstantinovich. 1936. Krÿmskaya jazÿkovaya ěkspeditsiya [Crimean linguistic expedition]. *Revolyutsiya i pis'mennost'* 2. 158–164.

Doerfer, Gerhard. 1959. Das Krimtatarische. In Jean Deny et al. (eds), *Philologiae Turcicae Fundamenta*, Tomus Primus, 369–390. Wiesbaden: Aquis Mattiacis Franciscum Steiner.

Gafarov, Basÿr Gafarovich. 1973. Orfografiya krÿmsko-tatarskogo yazÿka [Orthography of Crimean Tatar language]. In Kenesbay Musaevich Musaev (ed.), *Orfografii tjurkskikh literaturnÿkh jazÿkov SSSR*, 153–160. Moskva: Izdatel'stvo Nauka.

Hesche, Wolfram & Hartwig Scheinhardt. 1974. Eine krimtatarische Wörterliste. *Central Asiatic Journal* 18. 227–250.

Işıktaş Sava, Işılay. 2008. *W. Radloff'un derlemelerinde Kırım Tatar Türkçesinin özellikleri* [The characteristics of Crimean Tatar Turkish in W. Radloff's folk literature collections]. Ankara: Master's Sc. thesis.

İzidinova, Sevilya Refatovna. 1997. Krÿmskotatarskiy yazÿk [The Crimean Tatar language]. Ědgem Rakhimovich Tenishev et al. (eds.). *Yazÿki mira. Tyurkskie yazÿki*, 298–309. Moskow: Indrik.

Jankowski, Henryk. 1992. *Gramatyka języka krymskotatarskiego* [Grammar of the Crimean Tatar language]. Poznań: Wydawnictwo Naukowe Uniwersytetu im. Adama Mickiewicza.

Jankowski, Henryk. 2010. *Język krymskotatarski* [The Crimean Tatar language]. Warszawa: Wydawnictwo Akademickie Dialog.

Kakuk, Zsuzsa. 1963. Poésie populaire tatare recueillie par I. Kúnos. *Acta Orientalia Scientiarum Hungaricae.* 16. 83–97.

Kakuk, Zsuzsa. 1978. Ignác Kúnos' Nachlass in der Orientalischen Sammlung der Bibliothek der Ungarischen Akademie der Wissenschaften. In Éva Apor (ed.) *Jubilee Volume of the Oriental Collection, 1951–1976*, 115–126. Budapest: Library of the Hungarian Academy of Sciences.

Kakuk, Zsuzsa. 1980. *Kasantatarische Volkslieder. Auf Grund der Sammlung von Ignác Kúnos.* (Oriental Studies 4.). Budapest: Library of the Hungarian Academy of Sciences.

Kakuk, Zsuzsa 1987. Tatarskiy material Ignatsa Kunosha [Ignác Kúnos' Tatar material]. *Sovetskaya Tyurkologiya* 3. 53–57.

Kakuk, Zsuzsa. 1991. Ein krimkaraimisches Wörterverzeichnis. *Acta Orientalia Scientiarum Hungaricae* 45. 347–401.

Kakuk, Zsuzsa. 1993a. *Kırım Tatar Şarkıları. I. Kúnos'un derlemesinden yayımlayan Zsuzsa Kakuk* [Crimean Tatar songs. From I. Kúnos's collection published by Zsuzsa Kakuk]. Ankara: Türk Dil Kurumu.

Kakuk, Zsuzsa. 1993b. Krími tatár találós kérdések. Esztergom 1915 [Crimean Tatar riddles. 1915 Esztergom (Hungary)]. *Keletkutatás* (Ősz). 116–130.

Kakuk, Zsuzsa. 1994. Krimtatarische Rätsel (Esztergom 1915). *Acta Orientalia Scientiarum Hungaricae* 47. 143–172.

Kakuk, Zsuzsa. 1996. *Mischärtatarische Texte mit Wörterverzeichnis.* Aufgrund der Sammlung von Ignác Kúnos herausgegeben von Zsuzsa Kakuk. (Studia Uralo-Altaica 38). Szeged: Universitas Szegediensis.

Kakuk, Zsuzsa. 2012. *Krimtatarisches Wörterverzeichnis.* Aufgrund der Sammlung von Ignác Kúnos zusammengestellt von Zsuzsa Kakuk. (Studia Uralo-Altaica 50). Szeged: Universitas Szegediensis.

Kakuk, Zsuzsa und Imre Baski. 1989. *Kasantatarische Volksmärchen. Auf Grund der Sammlung von Ignác Kúnos.* (Oriental Studies 8.). Budapest: Library of the Hungarian Academy of Sciences.

Kúnos, Ignácz. 1916. Tatár foglyok táborában. Jelentés a mohamedán fogolytáborokban végzett tanulmányokról. Különlenyomat [In a camp for Tatar prisoners. Report on studies in Muslim prison camps. Offprint.]. *Budapesti Szemle*. 165.

Kúnos, Ignácz 1918. Tatárfoglyaink táborában [In the camp of our Tatar prisoners.]. *Magyar Figyelő*. 8.

Kúnos, Ignácz 1978. *Türk Halk Edebiyatı.* Hazırlayan: Tuncer Gülensoy [Turkish folk literature. Edited by Tuncer Gülensoy]. İstanbul: Tercüman.

Kúnos Ignác. 1999. *A török népköltés: Kúnos Ignác török nyelven tartott egyetemi előadásai.* Tasnádi Edit fordítása [Turkish folklore: the university lectures of Ignác Kúnos in Turkish. Translation by Edit Tasnádi]. Budapest: Terebess Kiadó.

Kurtbilal, Niyar. 2007. Kırım Tatarcası: Sudak Ağızları üzerine Araştırmalar [Crimean Tatar languge: Studies on the dialects of Sudak]. *Turkish Studies / Türkoloji Araştırmaları.* 2/2 (Spring). 502–513.

Lach, Robert – Herbert Jansky. 1930. *Volksgesänge von Völkern Russlands. II. Turktatarische Völker.* 1. Abt[eilung] *Krimtatarische Gesänge.* Transkription und Übersetzung von Dr. Herbert Jansky. Wien und Leipzig: Hölder–Pichler–Tempsky A.– G.

Lach, Röbert – Herbert Jansky. 1952. *Volksgesänge von Völkern Russlands. II. Turktatarische Völker.* 2. Abt[eilung] *Kasantatarische, mischärische, westsibirische-tatarische, nogaitatarische, turkmenische, kirgisische und tscherkessisch-tatarische Gesänge.* Transkription und Übersetzung von Dr. Herbert Jansky. Wien: Rohrer.

Liebl, Christian, Gerda Lechleitner & Remmer Ulla (eds.). 2018. *Recordings from Prisoner-of-War Camps, World War I. Turk-Tatar Recordings.* Band: Series 17/4. CD 1–2. Wien: Verlag der Österreichischer Akademie der Wissenschaften.

Mazinov, Ahtem Seit-Ametovič. 2012. Dialektnaja sistema krÿmskotatarskogo jazyka. Struktura gradatsionnÿkh yedinits [Dialect system of the Crimean Tatar language. Structure of units of comparison]. *Uchennÿe zapiski Tavricheskogo natsional'nogo universiteta im. V. I. Vernadskogo. Seriya "Filologiya. Sotsial'nÿe kommunikatsii."* 25 (64). 32–34.

Medzhitova, Ĕdibe Nazimovna. 1997. O foneticheskikh osobennostyakh severnogo (stepnogo) dialekta krÿmskotatarskogo yazÿka. *Vostokovednÿy sbornik.* 1.

Medzhitova, Ĕdibe Nazimovna 1998. Nekotorÿe grammaticheskie osobennosti severnogo (stepnogo) dialekta krÿmskotatarskogo yazÿka. *Vostokovednÿy sbornik.* 2.

Medzhitova, Ĕdibe Nazimovna 2017a. O foneticheskikh osobennostyakh severnogo (stepnogo) dialekta krÿmskotatarskogo yazÿka [On phonetic features of the northern (steppe) dialect of the Crimean

Tatar language]. In L. S. Selendili, A. M. Memetov, E. N. Medzhitova & Ja. Gibadullin (eds.), *Materialÿ po dialektologii krÿmskotatarskogo yazÿka*, 49–54. Moskva: ANO IIC "Insan".

Medzhitova, Ĕdibe Nazimovna 2017b. Osnovnÿe chertÿ severnogo (stepnogo) dialekta krÿmskotatarskogo yazÿka v oblasti slovoobrazovaniya i slovoizmeneniya [Basic features of the northern (steppe) dialect of the Crimean Tatar language in the field of derivation and inflexion]. In L. S. Selendili, A. M. Memetov, E. N. Medzhitova & R. Ya. Gibadullin 2017. *Materialÿ po dialektologii krÿmskotatarskogo yazÿka*, 50–59. Moskva: ANO IIC "Insan".

Memetov, Ayder Memetovich. 1993. *Krÿmskie tatarÿ* [Crimean Tatars]. Simferopol': Anajurt.

Memetov, Ayder. 2012. *Krÿmskotatarskiy yazÿk. Fonetika. Uchebnoe posobie* [Crimean Tatar language. Phonetics. A study guide.]. Simferopol'. Krÿmuchpedgiz.

Németh, Gyula 1911. Kumük és balkár szójegyzék. (Kumükisches und balkarisches Wörterverzeichnis). *Keleti Szemle* XII. 91–153.

Németh, Gyula. 1970. *Die türkische Sprache in Ungarn im siebzehnten Jahrhundert.* (Bibliotheca Orientalis Hungarica XIII). Budapest–Amsterdam: Akadémiai Kiadó.

Radloff, Wilhelm. 1887. Vorläufiger Bericht über eine wissenschaftliche Reise in die Krim im Jahre 1886. *Mélanges Asiatique* 9. 262–275.

Samoylovich, Aleksandr Nikolaevich. 1916. *Opÿt kratkoy krÿmsko-tatarskoy grammatiki* [An essay on short Crimean Tatar grammar]. Petrograd: Tipografiya I. Boroganskogo.

Savran, Hülya. 2003. *Kırım Tatar Türkçesi. Dil özellikleri – Metinler – Sözlük* [Crimean Tatar Turkic. Language features – Texts – Glossary]. Balıkesir: Balıkesir Üniversitesi.

Selimova, Leniyara. 2006. *Kırım Tatar Türk Ağızları. (Akmescit, Bahçesaray, Güney Kıyı Bölgesi) Ses Bilgisi* [Dialects of Crimean Tatar Turkic. (Akmesdzhit, Bahchesaray, Southern Coastal Area). Phonetics]. Ankara: Ankara Üniversitesi doctoral dissertation.

Seres, István. 2003. Csora Batir: Egy nogáj hősepos Kúnos Ignác gyűjtéseiben [Chora Batir: A Nogai heroic epic in the collections of Ignác Kúnos.]. In Ferenc Csirkés, György Csorba, Balázs Sudár & Zoltán Takács (eds.), *Függőkert. Orientalisztikai tanulmányok*, 174–242. Budapest: Argumentum Kiadó.

Seres, István. 2006. Újabb adatok Kúnos Ignácnak az I. világháborús hadifogolytáborokban folytatott turkológiai kutatásaihoz [New data on Ignác Kúnos' turkological research in the POW camps of World War I.]. In Mihály Dobrovits (ed.), *Az előkelő idegen, III. Nemzetközi Vámbéry Konferencia*, 152–194. Dunaszerdahely: Lilium Aurum.

Seres, István. 2010. A Crimean Tatar Variant of the Čora Batir Epic. *Acta Orientalia Scientiarum Hungaricae*, 63(2). 167–178.

Sevortyan, Ĕ. V. 1966. Krÿmskotatarskiy yazÿk [Crimean Tatar Language]. In V. V. Vinogradov (ed.), *Jazÿki narodov SSSR. II. Tyurkskie jazÿki*, 234–259. Moskva: Nauka.

Szilágyi, Szilárd. 2007. *İgnác Kúnos Türk Folklor Araştırmalarında Bir Öncü* [Ignác Kúnos, a forerunner in research of the Turkic folklore]. Ankara. Ankara Üniversitesi doctoral dissertation]

Tenishev, Ĕdgem Rakhimovich (ed.). 1988. *Sravnitel'no-istoricheskaya grammatika tyurkskikh yazÿkov. Morfologiya* [Comparative–historical grammar of Turkic languages]. Moskva: Nauka.

Useinov, K., E. S. Ganieva & N. S. Seydametova. 2011. *Qïrïmtatar tili. Fonetika. Leksikologiya. Frazeologiya. Leksikografiya* [Crimean Tatar language. Phonetics. Morphology. Phraseology. Lexicography]. Simferopol': Qïrïmdevoquvpednešir.

Useinov, Seiran Memetovich. 1994. *Krÿmskotatarsko-russkij slovar'* [Crimean Tatar – Russian vocabulary]. Ternopol'. SMNPP+Dialog.

Yüksel, Zühal. 1989. *Polatlı Kırım Türkçesi Ağzı* [The Crimean Turkic dialect of Polatlı]. Ankara: Türk Kültürünü Araştırma Enstitüsü.

Yüksel, Zühal. 1994. Kitabiyat: I. Kúnos'un derlemesinden yayınlayan: Zsuzsa Kakuk; Kırım Tatar Şarkıları, Ankara: TDK, 1993. *Emel*, 200, 18–21.

Structure of the glossary

Alphabetical order of headwords

In our glossary, the headwords are arranged in the following alphabetical order:

a, ā, b, ǯ, č, d, e, ė, f, g, ġ, γ, h, χ, i, ı, j, k, l, m, n, ñ, o, ȯ, ö, p, r, s, š, t, u, u̇, ü, v, z, ž

Transcription system

The transcription system of our glossary fully follows the transcription system used by Ignác Kúnos (see the *Comparative table, p. 7*). It follows that the glossary can be used in the study of the entire Crimean Tatar manuscript of the heritage, including the tales that have also been published in the original transcription system.

Entries, headwords, references

Glossary *entries* are arranged in *alphabetical order of the headwords*. *Main entries* (4,880 items in total) contain the meaning of the given Crimean Tatar headwords in English and Russian, the part of speech of the headword, the variants in the source with their number of occurrences, as well as idiomatic phrases and collocations, with their English and Russian meanings. *Reference entries* (about 1,080) ensure the different headword variants are found.

Types of entries and their structure

The glossary contains data of two types of entries: 1. the *main entry*, 2. the *reference entry* (or *reference headword*).

Main entry consists of *a headword in bold* at the start of the entry (see the samples below), followed in brackets by the *total number of occurrences* of the headword and its variants, a list of *variants* with their occurrence numbers, abbreviation of part of speech (e.g. *N, V, Adj*, etc.), an indication of the origin of the headword (e.g. *Ar., P., R.*, etc.), the meaning of the headword in English and Russian, then idioms (collocations) and their meaning in English and Russian. English and Russian meanings are separated by a vertical line ("|"). One or more headwords may appear after the "→" arrow at the end of the entry, where appropriate, to refer to another main entry whose headword is a linguistic (dialectal) variant or semantic parallel of the headword in question.

Samples of main entries

doru (5) *N/Adv/Adj/Postp* true, right; toward, towards | верный, правильный; прямо; (послелог) по направлению к, мрямо к; к, под, около (о времени) → **dora, dorı, torı, toro, toru**
kavil (6), **kavıl** (3) *N Ar.* agreement, accord | слово; договор, пари; **kavil jap–** to agree (on sg), to make an arrangement with sb | договариваться, прийти к соглашению
šindi (57), **šinde, šindė** (8), **šindı** (2), **šındė, šındi** *Adv* now, at present | теперь, сейчас;
šindijeǯek until this time | до сих пор, до сего времени → **šimdi**
tiñle– (3), **tinne–, tiññe–** *V* to listen to; to pay attention to, to obey | слушать, внимать; слушаться
tıñ– (2) *V* to listen to; to pay attention to, to obey | слушать, внимать → **tiñ–, tiñe–, tiñle–, tıñla–**
tıñlat– (2), **tıñat–** *Vfact* to make sb listen (to) | дать/заставить (вы)слушать

Reference entry is really two headwords side by side: the first one is the *referring headword* and the second one the *headword* of the *referenced* entry (another main entry) in which the *referring headword* (as a variant of the referenced headword) is found. The abbreviation s. (see) is used for directing to the referenced headwords. See the samples below.

Samples of reference entries

kavıl s. **kavil**
šinde s. **šindi**
šindė s. **šindi**
šindı s. **šindi**
šındi s. **šindi**
tinne- s. **tiñle-**
tiññe- s. **tiñle-**
tıñat- s. **tıñlat-**

Samples of homonymous entries

The homonyms headwords are distinguished by Roman numerals (**I, II, III**), as shown in the sample entries below:

aǯı I (9) *Adj* bitter, sour | горький
aǯı II (4) *N Ar.* one who has performed the rites of the pilgrimage at Mecca; pilgrim, hadji | хаджи (человек, совершивший паломничество в Мекку) → **aǯi, haǯi**
it– I (3) *V* to push | толкать, отталкивать
it– II (11) *V* to do, to make | делать, совершать, производить; **aff it–** to forgive | извинять → **et–**
kan I (28) *N* khan, ruler, sovereign | хан, правитель
kan II (27) *N* blood | кровь; **kanɣa buja–** to paint with blood | красить кровью
kan III (1) *N* house; café, inn | дом, кофейня, гостиница → **hāne**
kůl I (26) *N* ashes | зола, пепель; **kůl–karıštıran** stirring ashes | смешивающий пепель

kůl II (13) *N* rose (Rosa), flower, blossom | роза, цветок → **gůl**

kůl III (1) *N* lake | озеро → **kȯl**

Demonstrating quantitative features of lexical data

The numbers in parentheses, seen above in the samples of main entries, indicate the number of times a given *headword* and *its variants* occur in the whole collection (songs, riddles, stories).

The system is as follows: the number in brackets after the headword indicates the *total number* of times the headword (in this case, for example, **šindi**) and its variants occur in the different texts. The numbers after each variant, on the other hand, naturally indicate the number of times that variant appears in the collection. If there is no number after a variant, it means that it *occurs only once*. The number of occurrences of the headword itself is obtained by subtracting the number of occurrences of the variants from the number after the *headword* (which is the total number of occurrences).

Let us see the entry **šindi** again. The five variants of the headword **šindi** (**šinde**, **šindė** (8), **šindı** (2), **šındė**, **šındi**) occur a total of 13 times (1+8+2+1+1) in the texts. If we subtract the number of occurrences of the variants (13) from the total number of occurrences (57), the result is 44, which shows that the headword **šindi** itself is a highly dominant variant in this case. The purpose of providing the occurrence numbers is to show the user of the glossary the frequency with which each variant appears in the texts. It is hoped that the frequency data provided will allow far-reaching linguistic and dialectal conclusions to be drawn.

Crimean Tatar folk tales (original)

1 Bir bar eken, bir jok eken

(Tekerleme)
(246) Bir bar eken, bir jok eken, zaman zamanda ekende, arpa sabanı ekkende, arık úgúz ʒekkende, karılyač kādi ekende, upúpuš amamʒı ekende, babam bojdak ekende, nenem kız ekende, babamnıñ tujında botka tašıp ʒúrgende, kırmıskanı koj etip, kesertkini at etip, kujuya ajdep ʒúrgende, kapıl-kupul baġalar kanatlanyan učmaya, teñızdeki balıklar kıra tutkan kúčmege, kubernatın tarašı tiregen balahların Kara-teñızden útmege, bat bataman, bataman dep kum ajtkan, karap turyan kiši ajtkan, kišini [karap] turyan taj ajtkan, taj ajtkan dep kum ajtkan, taj anası bije ajtkan, búјıp turyan čal at ajtkan, čal at ajtkan dep kum ajtkan? Šandır-šındır šırpıntısınen, kútınıñ tırpıntısinen br kart ʒadı kelip ajtkan. Šo ketišnen ne ketken, tere túpe túmtúz ketken, belden balčık tizinedek ketken, aj da ʒarık zipdaj da karanyı eken, kúpekler, kaškırlar uluša, batır ʒigetnen oronıñ kútú mort kesken, alyan kúrknıñ pamukları tıšta tura, jıgeremge alyan čizmelerıñ barmakları ičte tura, br zamanda bar iken, bir zamanda jok eken.[1]

2 Altın pazvan

(248)
- Altın pazvan[2] tilij bilejzigim alıp kulıña sarılsam, ne degenıñ bar?
- Altın pazvan tilij bilejzigiñ alıp kulıma sarılaʒak bolsañ, ben bir šāhin kušı bolop avaya učsam saña ne degenıñ bar?
- Sen bir šāhin kušı bolop avaya učaʒak bolsañ, ben bir kúk kúgerčın bolop pišıñe tüšsem ne degeniñ bar?
- Sen bir kúk kúgerčın bolop pišime túšeʒek bolsañ, ben bir ayır kasta bolop túšekke tüšsem saña ne degeniñ bar?
- Sen bir ayır kasta bolop túšekke túseʒek bolsañ, ben bir ekim bolop melhem sarsam ne digeniñ bar?
- Sen bir ekim bolop melhem saraʒak bolsañ, menim vaχtım tamam bolop túnjadan ketsem, ne digeniñ bar?

---

1 Words or expressions standing here below in the footnotes without any specific abbreviations are taken from the manuscript (Ms.) in their original forms.

Letter "R" (abbreviation of Latin recte 'correct(ly)') refers to words, wordforms or expressions that seem to be (more) correct in the given place of the text.

2 In the list of titles of the tales compiled by Kúnos, the title of the tale is "Ne digeni bar?" [What has he/she got to say?]

https://doi.org/10.1515/9783111442891-002

- Seniñ vaχtıñ tamam bolop túnjadan keteʒek bolsañ, ben bir aχ kefin bolop tenıñe sarılsam, saña ne digenıñ bar?
- Sen bir aχ kefin bolop tenime sarılaʒak bolsañ, ben bir kúmúš balık bolop teñizge ketsem, saña ne digenıñ bar?
- Sen bir kúmúš balık bolop teñizge keteʒek bolsañ, ben bir usta talʒı bolop taldırıp čıkarsam, saña ne digenıñ bar?
- Sen bir usta-talʒı bolop taldırıp čıkaraʒak bolsañ, ben bir kırmızı gúl bolop bakčende pitsem ne degeniñ bar? (249)
- Sen bir χırmızı gúl bolop bahčemde piteʒek bolsañ, ben bir búlbúl bolop talıña kunsam saña ne digenıñ bar?
- Sen bir búlbúl bolop talıma kunaʒak bolsañ, ben bir al kelin bolop karšında tursam saña ne degeniñ bar?
- Sen bir al kelin bolop karšımda turaʒak bolsañ, ben bir šāhin kújeu bolop beliñe sarılsam, saña ne degeniñ bar?

3 Kúlbastı masalı

(252) Zamanın evvelinde aχir zeman šehrinde kapıl-kupul bakalar kanatlanγan učmaγa, tiñizdeki balıklar kıra tutkan učmaγa, br zamanda bar eken br zamanda ʒok eken br ʒarlı apakajnıñ br ulu bolγan. Šo ulannıñ adına Kúlbastı kojγannar. Kúl obalarında újde otorγanda oʒak bašinda er vakėt[1] kúl karıštırėp[2] otorγan, šonın‿čún[3] adını[4] Kúlbastı kujγannar.

Kúlbastı undort-unbiš jašına kele ič br jakka čıkıp ta br hizmet etmij, dunjaγa čikıp da ič br šij úrenmij. Er nasıl bolsa da kúnnerde br kún anası bunı br aldatıp kapunıñ katına ketire: ulum, sen de baška kalknıñ balaları tarzı čıksan da br az kizmet etseñ, újde kapalıp turmasañ dij. Lakin Kúlbastı kúl karıštırmaktan baška br šijden dat-lezzet almaγan ide, ič br iš eteʒeji kelmij ide. Nasıl bolsa da anası bonı kapıdan aldatıp tıšarγa čıkara, kapını da kapaj, Kúlbastı tıšarda kala, nišlijʒegin bilmij. Ómrınde tıšarγa čıkmaγan Kúlbastı tavuk kórse tavukdan kurkıp kaltıraj, čipče kúrse čipčeden kurkıp kaltıraj, kúpek kúrse kúpekden korka, mıšık kúrse mıšıktan kurka, kateʒegin bilmij. Šašıp šašmalap kapınıñ ʒaklavına sújkenip otora ʒılaj-ʒılaj umızbašın kisep šınkıj-šınkıj, anası da kapını ačıp ičerge ʒibermij.

Fakir Kúlbastı šo jerde ike kún (253) ike keʒe otora. Br zam[an]dan son janına kiši kele: ulum, nege otorasın sen bu ʒerde dij, maña ırgat ʒüresin mi di. Kúlbasti bu kišine kórep korkkan ide. Kúlbastı: meni ašamazsın mı di, o kiši de: men sini nige ašajım di.

1 vaket

2 karıštirep; R karıštırıp

3 šonınčün; R šonın ičün

4 adını; In the texts of the tales we see several instances of the Acc. (and the shortened Gen.) case ending *-ni* in velar words (e. g. *boni, uni, buni, oni, padišani* etc.).

Ajsa ašamasañ, men de saña irgat bolıp išlerim dij. Kiši: ulım, men sinė[1] özüme evlad tarzı etermen dij. Anası da üjden: šaškın, bar ket, o kiši seni ašamaj, men seni üjge almam dij. Anasından üjge almajǯaγın ešitkenden son kizmetke ketmege ümütlenij. Nasl[2] da bolsa kete.

Bu kiši bonı üjine alıp bara, kizmetke bašlamazdan burun kursaγın ejiǯe tojdora, üstüne bašına urba, ajaγına etek kijdire. Aligi Külbastı jene de ašar mij‿eken[3], ašamaj mij‿eken[4] dep kurka-kurka bašlaj išlemege.

Akšam bolγanda išlettirgen kiši kelip: inde jeter išlegenin, čoram, boldı dij. Külbastı toktaj, akaj čıkarıp Külbastınıñ kulına jikirem biš kapek akča bere. Akšam üjlerine kele, anası ulım išten kele dip kuvana: nišlediñ ulum dep soraj. Külbastı čıkarıp anasın[ıñ] kulına kazanγan akčaji bere. Anası o akčanen uluna ekinči künine ašamaγa et ütmek alıp azirlij. Kuvanγanından: meni ulım[5] da kalkın balaları dajın kizmetke kete. O künı de išlij, kene jikirem biš kapek bere išletken kiši. Ana, akča kazanıp kele, kene de anasına čıkarıp bere.

Učunče[6] küni kene kete, išlep kelgen vakitta ǯol üstünde balalar ujnašıp otoralar. Külbastı boların katlarına ǯaγın bara, baksa balalar br sıčkannı tutkannar, kıjnap turalar. Balalarγa dij ki: bonı nege šaj etesiñiz dij. Balalar: (254) isteseñ saña satarmız al diler. O da šo sıčkannı alıgı kazanγan akčanı berip ala.

Ekinči küni kene šonıñ tarzı br sıčavul[7] ala, üčınči küni br ǯılan balasın kıjnap turγanların köre, šo ǯılan balasını alıp kele. Anası: balam, bolar ni bolaǯak, kateǯeksin boların di. Külbastı: bolar minem mallarımdir di.

Br kün bo malların alıp br obanıñ üstüne čıkıp ǯibere ǯajılmaγa, üzi šu jerde ǯatıp ǯuklaj. Bo ǯatkandan soñ sıčkan kete br jakka, sıčavul kete br jakka, ǯılan kete daha br jakka. Külbastı[8] haman atılıp turıp baka, algı mallar taralıp ketkenner. Hemen pırıldap turup ǯuγurup ǯılannı mojnından tota. Ǯılan šo‿ssāt[9] tilge kelip ajta: ej akam, ajsa sen mini ötürme, min siniñ br keregine jararmen, sen mine ötürme dij. Külbastı: ni kerek pitirirsin maña, kulından ne kelij ajt karajım di. Ǯılan: min br padiša kızım, sin mine balalar kıjnap turγanda kelip kazanγan akčañnı tölep[10] alγan idiñ, ajsa šindė pılan ǯerde koba bar, sen šindė meni pešimden kel, barajık di. Lākin anda barγanda nišlermiz dij. Ǯılan dij ki: anda barγanda atam seni konak eter, sen de ič br šij alma, jangız atamnıñ kolında br jüzük bar, šo jüzükni tile, berse al, jüzükin bermegenček br šij alma. Sen kapıdan čıkkanda men de seni artından ketemen. Atam mine sinnen ketkenimi karap: jüzükni al da (255) kızımni[11] tašla der di. Ana bolar šulaj etip, söz br etip šü‿jerden[12] bašlajler ǯılan padišasına ketmege.

---

1 sine

2 R nasıl

3 ašarmij eken; R ašar mi eken

4 ašamajmij eken; R ašamaj mi eken

5 ulim; All (about 25) *uli* data have been changed in this way.

6 R učunčė / üčünči

7 sıčaval

8 Kulbasti

9 R šo saat; All similar expressions (e. g. *šossāt, šos-sat, šussat, os-saat* etc.) have been rewritten in this form.

10 tölep

11 R kızımnı

12 šü jerden; R šu jerden

Br vakitten son ana baralar, ǯılan aldında, Kúlbastı artında br kobanıñ kapısından kireler, kobanıñ iči zip tajın karannık, ike balaban ǯılan kapını katında sakči tura ekenler. Ulan bolarnı kúrgeni tarzı: aǯeba bolar meni ašamaz mı eken dep teumetlene. Aligı ǯilan dij ki: sin korkma, men sinin katında bolγanda breu saña pıski čertėp bol‿almaz[1] di. Šo ike ǯılan bolarγa: kajda barasız di. Aligi ǯilan: men padišani kızım, kajda tilesem anda bararmen, bu jer minem úzėm[2] ǯurtımdir di. Sakčı ǯılannar bondan bo laplarnı tujıp ekisi de padišaların katına alıp ǯuvuralar: múžde, meni padišam, kızın kelgen dep. Padiša vezirleri bolγan ǯılanlarnen kızını karšılap čıka. Ulannen ǯilan br kapıdan da selam birep[3] kireler.

Padiša kızını kelgenin korep kuvana, kuvanγandan kateǯegini bilmij. N‿ise[4], br vakitten son: balam, koš geldiñiz, kajda kaldıñ, kajda ǯúrdúñ bo vakitke dij. Jer kósterip otorta ikesini de. Jā, bu ulan bizim uruvımızdan bolmaj, bo beni-adem balası kimnerden bola ve sennen nėge keldi dep soraj. Kız bašina kelgennerin, nasil balalar jol ičinde kıjnap torγannarın, ulan da kazanγan akčasinen satıp alıp balalardan kutarıp ǯılannı azad etkenni atasına br br añata. Atası bik sújúnep[5] (256) kuvanıp ulanγa: Allah rāzi bolsın minem balam ǯanını kutarγanına di. Nijse, padiša ulannı br kač kúnner musapir ete, bik jaχši sijlaj.

Br kač kún son padiša ulana: ulum, sen mine kızımnı úlúmden kutarėp katıma alıp keldiñ, seni etken ijgiligine karšı men de senin karuvuñnı kajtarmak kerekmen, ulım, šindi sin de minden tile ne tilijsin di. Ulan dij ki: padišam, min sinden ič br šij tilemijim, jangız šo kulıñdagi júzúkni berseñ, rāzi bolormen dij. Padša: ulım, men saña šo kadar kúmúš akča, altın akča, šo kadar ǯevairler berijim dij, tek šo júzúkni tileme, unı birep bol‿almam[6] dij. Ulan: min boten šijlerni katijim, maña kerekmij, júzúkni berseñ alırmen, baška šij almam dep keteǯek bola. Kız jirinnen torıp atasına: atajım, bo ǯigėt mini úlúmden kutardı, min seni kulındaki br júzúk kadar jok mijim, mini kutarıp saña ketirgeni ičin br júzúkni tilij, sen de oña bermege kıj‿almajsın[7]. Ajsa, min de onunnen brge ketermen dij. Kız ǯigetniñ[8] artından keteǯek bola. Kız šaj degende padiša baška túrlú ebini tap‿almaj[9], barmaγından júzúkni čıγarıp dij: bermesem bolmajǯak, ulana bere. Ulan júzúkni alıp sauklašıp sau esen boloñız dep kuvana-kuvana kete.

Kız bonı artından čıγıp ulana júzúknıñ únerlerin añlatmak tilij. Bu júzúkni únerlerin bilesin mi dij. Jok, bilmijim, júret dij. Kız dij ki: ajsa, sin boš-bošına kitme, bo júzúkni saña br jardım[ı] (257) bolor, sen abdas alıp ike rekjet namaz kılarsın, son júzúkni tilinnen br ǯalarsın, júzúkniñ ičinden kırık tene kap-kara arab čıkar, leppek, tórem dijler, ǯanıñ ni isterse, tavıp ketireler, ne istersen jasajler dij. Ulan kızdan bo laplarnı tujıp kete.

---

1 bolalmaz; R bola almaz

2 územ; R úzim

3 R birėp; The gerund on *-p* of the palatal verbs in the tales afixed with *-ep* (e.g. *birep, kúrep, ketirep, kitep, kirep, minep* etc.) has been retained without change.

4 nise; R ne ise

5 R sújúnėp

6 bolalmam; R bola almam

7 kıjalmajsın; R kıja almajsıñ

8 The expected shape is *ǯigėt*, but the shape *ǯiget* dominates the collection.

9 tapalmaj; R tapa almaj

Az kete uz kete, br kün br toγajlık ʒerde br aγaʒ keletkesinde br ekin korıʒısi otorıp[1] ütmek ašaj. Ulan korıʒınıñ katına bara, baksa korıʒını kulında br kamčı bar. Ulan salam bere, korıʒi salamın alıp ulandan br az ütmek tilej. Ulan: ütmek taparım ama sin de braz su tapıp bolsañ dij. Korıʒı su tapıp ketire. Ulan aptas alıp, eki rekjet namaz kıla, kopaja-kopaja jüzükni tilinen br ʒalaj, šo‿ssāt jüzükin ičinden kırık tene arab čiγıp: leppek, törem dijler. Ulan dij ki: tez bolıñız, kırık türlü aš isterim, jetištiriñiz dij. Arablar kajdan tavıp ketireler, ketirip ortaγa kujalar. Korıʒi šašıp kala. Ulan korıʒıni aška čaγırıp ašajlar. Ašap tojγandan soñ korıʒı ulanı jüzügine telmire, nasıl etip šo jüzükni alıp bolor ekemen[2] dij. Korıʒi ulanı aldataʒak bola, br mātu bellep, dij ki: jüzügiñni šo minem kamčımnen deñišmijsin mi dej. Ulan suraj: min jüzügimni üneri br ʒalaγanda kırık arab čıγıp ne tileseñ šonı (258) eteler, jā seni kamčınnı üneri nedir dej. Korıʒi: minem kamčımnı üneri breuni bar, urıp kıjna, deseñ kıjnar, alıp ketir deseñ alıp ketirir dij. Ulan: ajsa, šaj bolsa deñišijik dij. Korıʒi ulandan jüzükni alıp ʒalaγandan son üzinėñ kamčısını de arablarγa aldıraʒak bola, lakin ulan tuja. Korıʒı jüzükni alγanı tarzı aptas almaγa, legen kuman kıdırıp aptas alγanʒık. Ulan kamčını[3] alıp biš un adım artına tartılıp kamčıγa dij ki: bar, korıʒını kulından jüzükni al da kel dij. Šu jerden kamčı br čabdırılıp korıʒınıñ kulına oralėp parmaγından jüzükni čırmap alıp ketirep[4] ulana bere. Korıʒı ulanı aldataʒak bolıp, jüzükden mārim bolγandan son üziniñ tobulγusın da kaptırıp ʒibere. Ulan kamčını de alıp kete.

Kete-kete br kün jolda br odamanγa rastkele. Odaman bondan braz ütmek tilij. Ulan: min saña ütmek taparım, sin de maña braz su tapsañ dij aptas almaγa. Odaman šo‿ssat barıp kumannen su tapıp ketire. Ulan aptas alıp ike reket namaz kıla da jüzükni br ʒalaj, ičinden arablar čıkalar, leppek, törem dep. Ulan emr ete, araplar eveligi tarzı ketirep, ortaγa aš kujalar. Ulan odamanγa ajtıp, konaγa jakın kele ašap tojalar. Odaman ulannıñ jüzügine korıʒi dajın u da tilmire. Ulanı br matu jerine kujıp jüzükni aldap alaʒak bola. (259) Odaman dij ki: jüzügiñni satmajsın mı? Ulan: ne tölersıñ dij. Odaman: br čokmardan gajrı šijim juk, šonı birermen dij. Ulan: jā seni čokmarıñnı üneri nedir dij. Odaman dij ki: meni čokmarımnı üneri, nedajın balaban kiši bolsın, tilese ažda bolsın, bar urıp desem šo‿ssāt urıp jıkar. Ajsa, šaj bolsa deñišemen dij. Šondan ulan jüzükin čıkarıp odamanγa bere. Odaman legen kuman kıdırıp abdas alγančık namaz kılmak šulaj torsın, ulan biš-un adım artına tartılıp čokmarγa emr ete: bar, šu jüzükni, odamannı ur jık te alıp kel dij. Čokmar šü‿jirden[5] odamannı suγıp jıγıp jüzükni odamandan tartıp ulana bere. Jüzükni parmaγına takıp kamčınen čokmarnı[6] koltuγına kısıp, andan üzleriniñ ʒurtlarına kajtıp kele. Anası: balam, sen katıp ʒüresin, kajda kaldıñ šindijeʒek di. Ulan: ej anajım, men šindijeʒek kȯb‿išler[7] pitirdim, br kamčıγa, brde čokmarγa brde jüzükke sahib boldım.

---

1 ȯtorip

2 Dialect form of *ekenmen.*

3 kamčıni

4 R ketirėp; Each of 37 *ketirep* versions have been retained.

5 šü-jirden; R šu jirden

6 čokmarni

7 kȯbišler; R kȯp išler

Kel zaman, ket zaman, aradan br talaj vakėt[1] keče, ulan br kizmet išlemi. Anası br kùn jene de šo tembel bolomsız iken, esini bašına toplajʒaγı ʒok eken, dep: balam, sen br iš išlemijsin, bulaj itėp bolor mi, balalık čaγın kečti, ašajʒak ičeʒek joktır dij. Ulan: sen, anajım, ič te kajγırma, ašajʒak ičeʒek men saña tavarım, sen muγajma dij. Parmaγından jùzùgini (260) čıγarıp br kerre ʒalaj, šo‿ssāt kırık arab čıkıp emr et, tórem, dijler. Ulan emr ete, šù‿jerde kırık tùrlù aš ketireler, konaγa kujıp salalar. Anaj, dij, šte seniñ tilegenin aš tugel mi ide, šte, aš. Ašajlar, kursakları tojγandan son, anası ulannıñ etken išine kuʒurlanıp kala.

Eppi[2] br vakit ótkóze. Kùnlerde br kùn anasına dij ki: anaj, meni inde ùjleneʒek vaktım ʒetti, sen de bek kartsın, seni de bakmaγa br kelin kerek, baña padišani kızın ajt, padišaγa barıp bersin di. Anası: ulım, bz br ʒarlı-pakir kišilermiz, padiša kızın tilemege ne addımız dep baksa da baška[3] čık‿almaj[4]. Anasına zorlaj: kene de barmasañ bolmaz dep, anası čare tap‿almaj[5] bara. Ʒarlı katın sarajγa ʒuvuk barγanda padišani kizmetčileri bunı kùrep[6] ne kıdırasın munda dijler. Kurtka: padišaγa keldim, tilegim bar, dep baksa da bolmaj. Tap sonında padišaγa barıp ajtalar: mında br pakir-ʒarlı apakaj kelgen, padišam dijler. Padiša da: beriñiz kazneden br avıʒ altın ketsin dij. Kazneʒi br avıʒ altın ketirep bere, ʒarlı apakaj ùjine bara. Ulı: kattıñ anaj di. Šte, ulım, br avıʒ altın berdiler, padišanıñ katına kirsetmediler dep iši[n] nė[7] [dajın] bolγannı añata. Ulı: men akča tilemim, padišanı kızın tilejım, bar bir dā (261) ajt, kızını[8] bersin maña dij. Kurtka br dā bara, kene de br avıʒ altın berip ʒibereler. Ulan kene de anasın ʒibere. Ùčùnʒi sefer barγanda padiša: bu nasıl kurtka eken, ne pikirnen kegen[9] iken, ajtıñız kelsin katıma, br tıñajım[10] tilegini dij. Kurtka padišani katına barıp kire, salam bere, padıša: ajt, anaj, ne ojnen keldiñ dij. Kurtka: padišam, menim br ulım bar, šo ulıma allanın emrinen pajgambarniñ kavlinen kızınızın tilij keldim dij. Padša: men seniñ ulına katijim de kızımnı berejim di. Ajsa, šaj bolmaz, meni tilegenimni kım berip bolsa, šoña kızımnı bereʒekmen, jangız šo kavılnen berirmen, bašta ózùniñ jatkan urnından minem sarajlarıma kigenʒek[11] eki jaklap gùl bakčaları, tùrlù češmeler jasatır, eki bakčanı urtasına melmer tašından ʒol jasatır, ondan son menim sarajlarımdan jaχši sarajlar jasatır, eki tùje ʒùkù kızıma ʒez akčası[12] altın berir, bolaj bolsa bu šijlerini etip bolsa, kızımnı berirmen, oña kırık kùn rukset, kırık kùnge kadar u šijlerni etse jaraj, etip bolmasa, bašın čaptırıp aldırırmen dij. Kurtka: ajsa, padišam meni ulımnı br šijčigi juk, barıp br ajtıp karajim, tólemek ondan, br ojlansın dij.

Kurtka ùjine kelip ulına ajta: šte, ulım, padiša mıña bu šijlerin tólep bolsañ, kızını bereʒek, tóle bolmasañ, kırk birinʒi kùni mojnıñı[13] čaptıraʒak dij. Bek jaχši, dij, onın

1 vaket
2 R epij
3 bašγa
4 čıkalmaj; R čıka almaj
5 taplamaj[!]; R tapa almaj
6 R kùrėp; The *kùrep* variant has been retained as it dominates the whole collection.
7 išinė; R išin nė
8 kızıni
9 R kelgen
10 tiñajım
11 R kilgenʒek
12 akčasi
13 mojnıñi

tilegenleri (262) bik kulaj, sen ič te kajγırma dep üzi ukını umzına tayınıp avγa čıkıp kete. Ulan ič br šij ojlanmaj ǯüre, ne saraj jasatkanı bar, ne gül bakčası, ne de ǯez bermege br akčası bar.

Aradan kırık kün keče, anası: bu minem balam ni matu šij eken, üzin eǯelin üzi tilij dep keǯe-kündüz kajγırıp kasavetlene. Kırık kün öte, kėrėk birinǯi küni padiša bašın čavıp üzdüreǯek. Kırkınǯı küni ulan avdan kajtıp kele. Anası: balam, sen katıp ǯüresin, mına vaktėñ[1] ǯetti, sen mından baška br jakka kač, barım padiša bašıñnı čavıp aldırmasın dij. Ulı: ič te ojlanma.

Akšam bola, jatalar, jarı keǯe vaktında jokudan turıp, jüzükni br ǯalaj, kırık arab čıka, emr ejle, törem dijler. Ulan di ki: miña šondajın saraj, jollar altından bolsın, terekler sırmadan bolsın dep emr ete. Arablar ojle br sarajlar jasajlar, ojle bir bakča ve jollar azırlajlar, tilnen ajtıp bolmaz. Padišanıñ sarajından kırık kat jaχšı bola. Padiša juklaγanda jarı keǯe jukudan tura, penǯerege br karaj, dunjanı[2] šavle baskan kündüzdaj bolγan ortalık. Nege šaj kečkeǯek juklaγaman[3] dep abdıraj. Ejlenip karasa, ne karasın, özün sarajlarından güzel sarajlar jasalγan, šo sarajların šavlesi eken, penǯereden urıp padišanı[4] sarajların jarıklandırγan. Padiša erten tura, kavesin ičmege bakmaj, ulanı čakırtıp toj etip kızına bere.

Kel zaman, (263) ket zaman, ulan zauk-safasında ǯüre, br kün avlanmaγa kete. O avlanmaγa ketkende čıfutıñ bri ulanda bondajın br jüzük bolγanın tujıp, kalaj etip almasın bil‿almaj,[5] ulannıñ apakajına kelip br küp akča berip jüzükni aldap ala. Čuput jüzükni br legen kuman tapıp abdas alıp, betin kulın čajχap, eki reket namaz kılıp, jüzükni br ǯalaj, kırık arap čıkıp emr et, törem, dijler. Bo da: aman sarajlarnı apakajnen baraber Kaf tavını artına alıp ketıñız üzümnen dij. Araplar šo‿ssāt degenın eteler. Sarajın bolγan jeri bomboš kala. Ulan avdan kajtıp kele, ne sarajlar bar, ne de apakaj, ne de jüzük bar. Čokmarnen kamčını kujγan jerinde tura mi eken dep, anasınıñ eski üjlerine bara, karasa alıgı čokmarnen kamčı rafta kalaj kujγan bolsa, šolaj tora.

Padiša br vakitler kızını üzine kelmegeninden[6] teümetlene, ne bolγan eken bolarγa dep. Br kün br kiši ǯiberep[7]: bar, kızmnı katkennerni, kalaj bolγannarnı tıñlap kel dij. Kiši kele, baksa ne baksın, ne sarajlar bar, ne de padišanın kızı. Son barıp padišaγa išniñ nedajın bolγanın añata. Vaj, meni kızımnı bašına ǯetti bo ajlak erif dep padiša br talaj askerin toplap ulana ǯibere: barıñız šo eripni tutıp alıp keliñiz, üzi kulımnen[8] üzėm[9] üturijim, kelmese mujnın čavıp bašını ketiriñiz dij.

Asker bišjüz kadar, ulanėn bolγan ornına keleler. Kurtka boların kēgenin[10] kürep bašlaj ǯılamaγa: vaj, išimiz ne bolor eken dep. Ulan: anaj, sen ǯılama, (264) men uların išin šindi pitiririm dij. Asker jetišti degende, kamčınen čokmarγa emr ete: šo askerlerin

1 vakteñ

2 dunjani

3 Dialect form of *juklaγanman.*

4 padišani

5 bilalmaj; R bile almaj

6 kelmegenden

7 ǯiberėp

8 kulimnen

9 üzem; R üzim

10 R kelgenin

ur, breu kalmaγan dep. Šo‿ssāt kamči br jaktan, atlėp[1] tüše, čokmar br jaktan sıkırep[2] tüše, askerlerge čokmarnen kamčı ekisi eki jaktan bašlajler tokmaγa, biš júz askerin keregi dajın epsini ótórėp jerge ʒajıp tašlajler. İčlerinden breu kačıp kutulıp padšaγa bara: padšam, br kórüngen kiši juk, muna šolaj bir kamčinen br čokmar čıγıp az br vakitte epimizin jirge ʒajıp tašladı dij.

Padiša un biñ asker toplap ulana ʒibere: barıñız, katseñiz katıñız, jā kızımnı tavıp ketirsin, jā mojnını alıp keñız[3] di. Asker ulana jakınlaškanda, jene emr ete, evelki tarzı askerin telep etip pitireler. Askerlerden čıdaj almaj breu kačıp barıp padišaγa išin añata. Padiša ulana br elči ʒibere: bar ajt, jā kızımnı ketirsėn, jā men de uni[4] ʒurtını[5] ʒaγarman. Ulan: o menim bileʒegim iš, tilesem úzimni apakajımnı taparım, tilemesem onı iši túgel, ondan kurkum ʒok dij.

Kúnnerde br kún ulan anasınen sauluklašıp esenlešip anasına čokmarnen kamčını tašlaj, dij ki: šte, anaj, bo šijler bašıñızın er br sıkıntıdan kutarır dij. Saulıklašıp kete, úzinnen barabar katına mıšıkčıγı bar eken, šonı da ala.

Kete-kete br jerge bara, barγan memleketi sıčkan padišalıγı[6] eken. Sıčkanlar šo‿ssāt bolarnı kúrep, barıp padišalarına (265) ajtalar, dijler ki: padišam, munda bizin memleketge bzin Azrail kelgen, tukumızın[7] jok etmege. Sıčkan padišahı ne kadar kart sıčkanlar bolsa, epsin toplap mušavere ete, nišlijik bz? Br kart sıčkan di: ne kadar askerimiz bolsa, epsini ʒiberip olarnı útürtijik dij. Brsi br túrlú, brsi br túrlú ajta. Tap sunında en kart sıčkanı dij ki: bz šaj itep[8], br šij išlep bol‿almamız[9], ajsa bzge bo išni br ojlanıp etmek kerekmiz dij, mıšıknı čaγırtıñız, pikrini añlap biliñiz dij. Ana ulannen mıšıknı alėp[10] keteler padiša katına. Mıšık padiša katına kelgenče [unı] kapkač kete. Padiša: aman tórem, mini útúrme, ne tileseñ šunı išlijim, tek útúrme dij. Mıšık: men sinė[11] útúrmem, jangız menim senden bir tilegim bar, šonı išleseñ seni ʒiberėrim[12]. Sıčkan: ajt, tuγanım, ni tilegiñ bar dip. Mıšık dij ki: Kaf tavın artında br saraj bar, anda brisiniñ kulında šondajın br júzúk bar, šonı almaγa jardım etersin dij. Sıčkan rāzi bola.

Andan sıčkan mıšıknen br bolıp keteler, sarajγa baralar. İčerge kirgenden son baksalar ne baksınlar, čıput čalka-čalka túšep jatkan, katında apakaj de juklajlar, júzúkni avızına kapıp siʒimnen tavanγa bajlaγan br uʒın, sıčkan tavanγa čıka siʒimnı kese. Ondan dolafta braz bal kıdırıp tapa, kujruγın balγa batıra. U jerde braz karabiber tapıp (266) bulγalaj, alıp barıp čuputnıñ burnunı tanauna sukėp[13] ʒibere. Čuput br aksıra, júzúk pirıldap kete. Mıšık aman kapkač kete, júzúkni ulana alıp baralar.

1 R atılıp
2 R sikirėp ~ sekirėp
3 R keliñız
4 R unı
5 R ʒurtıni
6 padišaliγı
7 túkúmizin
8 R itėp, but the *itep* forms are dominant and retained.
9 bolalmamız; R bola almamız
10 alep; R alıp
11 sine
12 ʒibererim
13 sukep; R sukıp

Ulan jüzükni br ʒalaj, arablar čika, bujruñız törem dep, sarajlarını jerinde bolγanı istij. Arablar ketirep urnına kujalar. Čuput jukusından turmaγan, apakajnen kučaklašıp ʒuklajler. Ulan padišaγa barıp: kel, kızıñnı halın kür dij. Padiša kelip küre. Külbastı čokmarγa emr etmezden burun, kamčı barıp ekisini de suγıp ütüre. Padiša kiškene kızın ulana bere, toj etip, üzine kijeu ete. Küp zamanlar sefanen jašap ömer eteler, müradlarına asil bolıp kalalar.

4 Temir Bolat masali

(282) Burungu zamanda Šah-islam degen br padša bola. O padšanıñ üč tane ulu bola, en kičkenesin adı Temir Bolat bola. [Bolattan da kor bakılgan bolmij.][1] Šo balaların ičinde u zamanlarda ʒılkıγa[2] da bek aves ekenler. Padšanıñ tokoz toγaj ʒılkısı[3] bola.

Künnerde br kün bu tokoz toγaj ʒılkı ʒojula. Padšanıñ üjken ulunen ortanʒı ulu babalarından rühset tilejler, ʒılkıların tavup kelmege. Atajları rüχset bere, keteler. Br vakitler ketkenden son br šeherge kireler, ekisi eki kanγa tüšeler, oda tutalar, [ʒılkılarnıñ ʒokluγundan annatėp kalka][4] bekem tavarmėz dep ümütlenüp ʒüreler. Bolarγa br eki ʒanaj kelip: sultanım, ependim, bu ʒirlerde ne katesınız, sizler br kan balalarısınız, sizni br uršen ʒirge alıp barajım dijler. Boların br kabakka alıp baralar. Bolarıñ kiselerinde akčaları pitkenʒek, esirep ʒüreler, hatta atların de satıp ičeler. Padša balaları akčaları pitkenden son alγı dost ʒanajların kıdralar, bekım br kol jardımda tabılırlar dep. Šeherde sokakta ʒürerken ʒanajlar rast keleler. Padša balaları: aman arkadašlar nišlijik, akčamız pitti deseler de alıgi (283) ʒanajlar: akčanız pitkenden son nasap katijık sizleri dijler. Šondan bu eki padša balaları brsi barıp amamʒida amam süpürmege ʒüre öbürüsı br furunʒija un elemege ʒüre.

Bolar mında ʒüre torson, bz kelijik üjde kalγan Temir Bolatka. [Bolatnı eski sanγa breu katmaj, onı br sersem matu jerine kojup ʒüreler, anase da, atası da.] Br kun Bolat atasından izin istej, dij kı: atajım, maγa da izin berseñız, men de ʒılkılarımızın kıdırmaγa čıkar edim di. Atası bondan bu lapların ešitep bek ačulana di ki: menım bo kadar eki akıllı balalarım kettiler, br iš išlep bol‿almadılar,[5] kollarından br šij kelmedi, özleri de kajtıp kelmediler, sen ni išlejʒek bolasın, matu dep, ket jıkıl, meni ačulandırma dep kuva. Bolat moγajėp jerine kelıp jata, bašlaj ojlanıp ʒılamaγa: bo kadar ʒılkımız ʒojıldı, andan son eki aγajım ʒojėldė, endi men katijim, atajγa baram, o da tazirlep kuva, ej kudajėm, sen bilirsın dep moγajėp jata. Br zemandan son jokusında br tüš köre, br kart čal sakallė: balam, Temir Bolat, nečın ʒanıp ʒakılasın, sın[6] iš te kajγırma, muna šo

1 The sentence has been crossed out in the Ms.

2 ʒılkiγa; The *ʒılki* forms occur 8 times here, all changed to R *ʒılkı* variant.

3 ʒilkisi

4 Crossed out in the Ms.

5 bolalmadılar; R bola almadılar

6 R sin

jatkan odañnı[1] artėnda[2] br kapı bar, šo kapını ač ta ičinde br sandık bardır, šo sandıknı[3] ičindekin alsañ saña jeter, senıñ er br (284) muradıñ asıl bolor dij.

Bolat jukusından tura, kartın degeni dajın šo kaponı[4] ačip baka, baχsa ne baχsın, br sandık, ačıp baka, sandıknıñ ičinde br kılıǯ, br kat ǯenk urbaları, sandıkıñ katında karčγada br jigernen brde ǯügen, brde šėbırtkı[5] ilevli tora. Üjnıñ br kösesinde br kapı dā bar, onı de ačip karaj, karasa br aχır, aχırda br at, kulakları altın sırmalı, tırnakları kümüšten, ǯalını[6] sırmadan, özının de br közınden kan kete, brinden jaš kete. Bolat dij ki: ej ajvan, sen nege ǯılajsın, jatkan turγan ornın jaχši, aldında ǯemın bar, ǯılamaγa br šij jok. Mına meni bašımdan bo χadar sėratlıklar pelaketler ötti, men ǯılamajıp ta sen nege ǯılajsın dij. Altın kulak ajγır tilge kelıp ajta: men ǯılamajıp ta kim ǯılasın, men özėm ičın ič te ǯılamajım. Jā nečın degende: men senı bašına keleǯek seratliklarnı ojlap ǯılajmen dij. Ne de bolsa, šu jerden atnı ǯigerlep kılıǯnı[7] mojnına taγınıp šıbırtkını koluna alıp atın arkasına utėrėp[8] jolγa čıka, ǯılkılarnı, aγajlarnı kıdırmaγa.

Br kač künler kete. Künlerde br kün aγajlarnı barγan šehirge bo da sürıp bara, alıγı ǯanajlar bonı da köre: vaj törem, χoš keldıñız, jaχšı misıñız dijler, (285) boña türlü sijlar eteler. Törem, mında filan ǯerde jaχšı kabaklar bar, jaχši gastinčalar bar dij. Lākin Temir Bolat: maγa kirek ǯernı men özım tabarım, karnım ačıksa ašamaγa aš ta tabarım, siz menim‿čın[9] ič te kajγırmañız dij. Ǯanajlar: bu erıf obırlerı tarzı dugıl eken, bzge bondan avlak dolašmak kerek dep kajtalar. Temır Bolat br kač kün šeherde dolaša, br kün furunγa ütmek almaγa bara. Akası furunda ǯuvunduk metilerın tögıp turγanın köre. Andan ajlanıp indemi kajta, üsti de kirlengen edi. Br molčaγa barmak kırek bola. Molčaγa bara, ekınǯi akası molčanı silep süpürıp ǯürgenın köre, gene de indemij, ajlanıp ornına kire.

Son üzi bašına ojlana dij ki: munda jatunen bolmajǯak, ajsa, men ketıp kıdrajım ǯılkımnı dij. Šondan sırmalı-kulak ajγırγa minıp ǯolγa čıkıp kete. Kete-kete br kun br kopir bašına kele, kopirge kelgende ajγır šart toktaj, Bolat: ej ajvan, nege ǯürmıjsın dij. Ajγır tilge kelıp di ki: šindı bu kopırden beš bašlı ažda oteǯek, sen menı kulaklarımnı[10] jaγlıγınnen jakšı etıp nıktap bajlarsıñ, o senıñ kokoñnı[11] tujıp seni ašamaγa entelip kelgende men br kišnerım, men kišnegende aždanı[12] esı bašından ketıp ǯıγılij, (286) sen ǯuγurıp[13] barıp dürt bašın čavıp alırsın, kulakların kesıp ijbeñe kujarsıñ dij. Bolat tiz atnıñ kulakların nıktap jaγlıkinen bajlaj. Brde bašlaj ǯir kıbırdap, kupır sallanıp, ažda kelmege bašlaj, šu jerde sırma-kulak ajgır br kerre kišnij, kišnegen vakıtte aždanıñ esi bašından ketıp, essizlenıp audarla.[14]

---

1 odañni
2 artenda
3 sandıkni
4 kaponi, R kapunı
5 šėbirtkı, R šıbırtkı
6 ǯalıni
7 kılıǯni
8 uterep; R utırıp
9 menimčın; R menim ičın
10 kulaklarımni
11 kokoñni
12 aždani
13 ǯuγurip
14 R audarıla

Bolat šo sāt barıp aždanıñ dort bašın čavıp ala, brsi kala. Ažda dij ki: ej ǯigıt, er bolsañ br bašımnen menı kaldırma, bu br bašımı de kes dij. Bolat dij ki: babamnı br balası tugılsın, bar ǯan čekiš dij. Anda ažda tırıša-tırıša ülıp ǯıkıla. Bolat andan aždanıñ kulakların kesıp ijbesine tottorıp ǯolına kete.

Kete-kete br unbiš čakrım[1] kete, karšėsına[2] br balaban sarajlar kürıne. Šo saraj-larga sürip bara, kire, ičerde br dunja güzeli br kız naγıš[3] išlep tura. Bolat salam berıp kire. Kız salam ala dij ki: ej ǯiget, kuš kanatinen, ǯılan bavrunnen[4] kelıp bol‿almagan[5] ǯirge sen kalaj boldı da keldiñ dij. Ǯigıt ničın dep soraj. Menı br koǯam bar, biš bašlıdir, bu jaklarga ič br kiši ǯibermij. Bik uzak ǯirlerden ǯalmap ǯutar, sen kalaj bolgan da kutulγansın[6] onu kulından? Emdi tez ket mından, šindi kelse senı ȯtürür dij. Ǯigıt dij ki: ej güzel, men senı koǯannı ȯtürdüm, šte, koǯandan saña nišan, dep (287) kulakların ijbeden čıkarıp kıznıñ aldına kuja. Kız da: ajsa enten son sen menımsin, men senınım dep ǯigetnıñ mojnına sarıla. İšte ǯiget dij ki: men ǯojuk kıdırėp[7] čıktım, ǯojuγımnı tap-maγanǯek sen meni, men senı bolop bol‿almam[8]. Ajsa maña ruχsat, men ketejım, barıp ǯojuγımnı kıdrajım di. Kız ajta: ǯojuγıñ nedir, maña añlat, bekım bile torγan šejimdir di. Bolat ǯılkısını[9] kıdırıp čikkannın añlata. Kız da: ajse sen šindi mından ketkende jolda br kopırge rast kelirsin, u kopır menim aγamnı[10] mülkün sıñırındadir, bek batır, bek de jaramaz deudir, jedi bašlıdir, onıñ kolundan avede učkan kušlar bile kutulıp bol‿almaj,[11] bek sak olmalısın, ondan gajrė[12] senı mat eder di. Bolat: o maña br šej etıp bol‿almaz[13] dep šu ǯerden altın-kulak alašasına minep joluna kete.

Kete-kete šo kopırge[14] rast kele, evelki tarzı at toktaj, kene evelki tarzı atnıñ kulak-ların nıktap baγlaj, jedi bašlı deu kıčkırıp, karañık tuman koparıp kele. Altın-kulak alaša br kišnij, kupırın tašları pırıldap barıp devniñ bašına čačila, devın bašindan esi ketip tüše. Bolat šo sāt devın bašın barıp čavıp kese, kulakların alıp ǯebine kojıp kete. Br balaban kȯkke aralašken br sarajγa kele, sarajnı ičerisine kire, baksa br güzel kız, aldında kona, kolında oklau, kamır ǯajıp tura. Bolat salam bere, kız salamın alıp: ej ǯigıt, sen bu jaklarga kalaj (288) boldo da katıp tüštıñ? Menım br jaramas akam bar, bu jaklardan onıñ kurkusından avedeki kušlar učıp ketmege korka, sin nasıl boldo da kurkmaj keldiñ dij, kajdan kelip kajda ketesin dij. Ǯigıt Bolat bašına kelgenlerın brer brer ajta, kız šašıp kala. Ajsa ǯigıt, senın ǯılkıların šo menım jaramas akamdadir, seni ǯılkılarını oγorlap ketken šo akamdır. Bolat di ki: men bu ǯılkıların alıp ketmege keldim. Kız di ki: vaj ǯiget, saña o mallarnı alması[15] bek kıjın, bek balaban dejudir ki sen onı br jokosında tüšürıp de ȯtürsen bilmem, ondan gajri seni mat eter di. Bolat: kajdadir

---

1 čakrim

2 karšesına

3 nañıš

4 R bavrunen

5 bolalmaγan; R bola almaγan

6 kutulgansın

7 kidirep; R kıdırıp

8 bolalmam; R bola almam

9 ǯılkısıni

10 aγamni

11 bolalmaj; R bola almaj

12 gajre R gajrı ~ γajrı

13 bolalmaz; R bola almaz

14 kopırγa

15 almasi

šimdi o deju? [dep] soraj. Šte muna šo taunıñ artında br bajırnın üstünde ʒuklaj di. Dejunıñ ʒoklamaya ʒatkanı[1] da otoz tokoz gün bolyan eken, tamam kırk gün ʒoklaj iken, o jukudan toryančık kardaši oña keče-kündüz kobete, beleš piširıp azırlaj eken. Ana Bolat šo ʒerden br jıgırım čakrım kadar kete, taunıñ artına bajır kürüne, lākin bajırın[2] üstünde eki selbi tarzı ši uzanıp kökke sozolgan. Bu de dejunıñ eki közın šavlesi eken. Bolat dejuge jakın bara, bašından ajayına kadar br ʒüre, šo ortasıdır[3] dep br kılıʒ ura, dejunıñ jedi bašın da brden čavıp tüšüre. Ʒılkılar šo ʒerde br toyajlıkta ʒajılıp[4] ʒüre ekenner. Barıp ʒılkıların ajdap, kızın sarajına alıp kele.

Br kač künner sarajda rahatlanıp, kızıñ[5] (289) nikadar palı, kijmetli šijleri bolsa, arabalarga tijıp andan jolya čıkalar. Andan öbür kızga keleler, anda da br kač künner konak bolgandan son jolga čıkalar. Kızlarnı jannarına alıp ʒılkılarnı ajdap aligı agajların bolgan šeherge keteler. Bolat šeherge kelgennen son br jaχši oda tuta, kızlarga jaχši kat urba sumarlaj, urbalarnı ketireler, kızlarga kıjdıreler. Kızlar brer jaχši ʒiget bolalar. Urbalar erkek urbası eken.

Andan üjken akasınıñ bolyan purunga barıp akasına balaban br karavana bulgača sumarlaj. Piškenden son filan ʒerde minem odama alıp kelirsın dij. Lākin akaları Bolatnı br türlü tanıp bol‿almajler.[6] Ondan molčaga barėp ʒuvungandan son, ortanʒı akasına bazardan eki jaχši kat ruba sumarlap ala. Ortanʒa akasının algan urbalarnı odasına ketirte. Odaga kelijler, akaları šijlerın tašlap tıška čıkıp kajtaʒak bolalar. Bolat: ʒok, ʒok, otorınız, bu šijlerın men jangız özüme almadım, bu epmiz ičin dij. Ana šo ʒerde otoralar, bolgačnı jaχši etıp ašajlar. Son eki akasının ekisini de eki jaχši kat urba kėjdire[7], traš ettire, kene eveliki tarzı padša balalarına ošajlar. Kāve ketireler, kave ičkende bolardan Bolat bašlaj soramaya. Sizler kimlersıñız, kaj jerli bolasıñız dij. Bolar da bašlajler ajtmaya: šte bizler pilan jakta Šahislam degen padišanıñ üč ulı edik, ekimiz bolaj br kün ʒılkılarımız ʒojıldı, bzler (290) ʒılkılarımızın kıdırıp čıkkan edik tap soñında munda keldik, munda da br eki ʒanajlar bašımızga musibet boldı, akčalarımız pitkenden son katmege bilmedik. Brimiz molčaya ırgat boldık, brimiz purunda irgat boldık, ʒurtımızga kajtıp bolamadık, monda kaldık dijler. Bolat: jā, sizın Bolat degen br kičkene iniñiz da bolaʒak, o nišlij dij. Bolar dijler ki: o br matunın bri, onı bzlerge pajdasi tijmij, atajga da br pajdası tijmij, kudajnıñ br tentegi. Šo kadar ʒılkımız ʒojıldı, o tentekın šondaj bolomsız buluyı da br balaban janık dijler.

Bolat bolardan bo lapların tujıp, özi ičin šondajin tentek, bolomsız dep ajtkanlarına bu ne kuʒur laplar dij. Lākin Bolatnıñ kekelın tübınde bala ekende br tamya kalgan iken. Bolat dij ki: jā siz ʒurtınızdan[8] čikkanınız kob‿boldo mi[9] dij. Akaları dijler ki: jedi ʒıl[10] boldı. Jā siz kičkene iniñizın körseniz tanıj misınız dij. Elbette, dijler, tanır-

1 ʒatkani
2 bajırin, R bajırnıñ
3 ortasidır
4 ʒajılip
5 kizıñ
6 bolalmajler; R bola almajler
7 kejdire; R kijdire
8 ʒurtinızden
9 kob boldo-mi; R köp boldo mı
10 ʒil

mız. Nesinden tanırsıñız dij. Balalıyında, dejler, ʒıkılıp kekelin tübınde br tamɣa kalgan, ondan tanırmız dejler. Ajse, bondaj tamga mı dij, bašın ačıp körsete. Šo sāt šašıp kalalar: Aman tuɣanım, bz señ‿čin[1] šondajın jaman laplar ajttık, bzim kusurlarımızın baɣıšla dijler, Bolatnıñ[2] ajaklarına ʒıɣılalar. Bolat bolarɣa: zarar ʒok, dünjade er br šej bolop kete torɣan, šte men sizin kardašıñız Bolatım, men de (291) sizden son munda ʒılkılarımızın[3] izlep čıktım, bašıma bondajın išler keldi dep bašından kečken gurbetliklerın ajta, nasıl dejulernen suɣušıp[4] ʒılkılarnı alɣanın akajlarına br br añlata, šte bu ʒigetler er urbasınen kördögınız šu ötördıgım dejulerın kızkardašlarıdır, bo kızların ekisi sizge bolsın, bolarnı sizin‿čün[5] ketirdim dej. Kızlarɣa da soraj: muna bo ʒigetler meni tuɣan akajlarım dij. Kızlar: bz padiša balalarından abadanɣa baraʒaɣımız jok, tek toj etip alsın bzin dijler. Ajsa šaj bolɣanda atamızın katına barɣanda nika‿kıjdırėp[6] toj etermiz dijler, šaj bolsın dep söz bereler.

Nejse br talaj vakėt[7] Bolotnen šeherde ʒürüp rātlanalar. Künnerde br kün šeherden ʒolga čıkalar, atajların katına kelmege. Bolat akajlarına eki jaχšı ʒorgalar alıp minsete, ʒılkıları[8] aldlarında, üzleri artta üč ulan üč kız Šehislam padišaga kajtıp keleler. Šehirden čıkıp jikirem biš otoz čakrım kadar ʒüreler, alıgı Bolatnı akaları artta kalıp lap eteler brbrlerinen dijler ki: bz šindė[9] atajmızga tukum uruɣımızga[10] kaj jüznen barırmız, bzin br matu inimiz bu kadar išler etip bzin de bo serātlıktan kutarıp atamızga alıp bara jatır, boña nasıl čıdarmız, bizge atamızɣa, ʒurtımızga šaj etip barudan ülmek dā lajık dijler. Son ortanʒıları ajta: ajsa bz šaj (292) etmijik, Bolatnı aldatėp[11] artta alıp kalajık ta unı[12] öturijik, o vakıt bz ondan kutulėp[13] ʒılkımıznen kızlarnen atamızɣa sau bararmız dij. Üjkenı bu iške kajl bola. Andan Bolatnı katına barıp: enekejim, bz azčık artta kalajık ta saña br ajtaʒak lapımız bar dijler. Bolat aɣajların digenin totıp artta kala.

Üč kızlar ʒılkılarnen ketip bara tursınlar, padıšanın üč ulı artta kalıp, büjük akaları belinden kılıʒın čıkarıp Bolatka br ura, eki ajaɣın da tizinden šort-šort üze[p] ala. Bolat ʒirge ʒıkılıp ʒürüp bol‿almaj: ej tuɣanlarım, aɣajlerim, meni šaj etip ničin tašladınız dese de bolar Bolatnı br iš pitirep bol‿almajʒaɣın bilep[14] üturmij tašlap keteler, ʒılkını ardından ʒeteler.

Bolar künnerde br kün öz ʒurtlarına baralar, padišaɣa müždeʒi bereler. Padiša balaların nekadar vezirleri batırları bolsa, epsin toplap ulanların aldına koš keldige čıka. Ona br zamandan barėp[15] ʒeteler. Padiša ulanların karšilap ala, akšam sarajɣa vezirler, aʒılar, oʒalar ʒıjılalar, ʒıjın balaban bola. Aš ašaɣandan, kave ičkenden son bolar bašlajler ajtmaɣa: šte filan ʒirge bardık, filan dejunı ötördık, filan ʒirge šundajin[16]

1 señčin; R seniñ ičin
2 Bolotnıñ
3 ʒılkilarımızın
4 suɣušip
5 sizinčün; R sizin ičün
6 nikak kijdirep, R nikah kıjdırıp
7 vaket
8 ʒılkiları
9 šinde
10 uruɣimızga
11 aldatep; R aldatıp
12 uni
13 kutulep; R kutulıp
14 R bilėp, but *bilep* forms are dominant.
15 R barıp
16 šundaji

aždanı mat ettik, šo aždanın ʒerinden ʒılkılarımızın taptık, šte bo üč kızın da ekisın üzümızge, brin de kiškene enimiz Bolatka ketirdik dijler. Padiša: matu sizin artınızdan men de ketem akajlarımnı em de (293) ʒılkımıznı kıdırmaya dep ketken edi, körmediniz mi šo matunė, kajdalarya ʒüre‿ken[1] dij. Bolar da ič te χabarımız ʒok dijler. Bolar Bolatnı jamanlap onı etken batırlıkların bz ettik dep maktanıp kalka toñatmakta bolson, bz kelijik bo jakta Bolatka.

Kačan Bolat akajları šaj itep tašlap ketken ideler, o küni Bolat ajakları avurta, amma katsın kalaj etse bolmajʒak, anau jaktan mınau jaktan ješil ülenler toplap ajaklarını ʒaralarına kuja. Br kač künler šaj itep ajakları jaχši bola, u jakka bo jakka ʒüre. Br kün br ayaʒ tübinde köletkede karnı ač otoryanda, boña br sokornen brde čolak katına kelip, sokornen čolak topalya selam bereler: i topal arkadaš, sen mında ne ojlanıp turasın dijler. Bolat bašından kečken čilleni brer brer ajta. Čolaknen sokor dij ki: ajsa bzler üčümız br bolajık, bzniñ zenatımız tilenmekdir, bz tilenep kečinirmiz, sen de bznen barabar kečinirsėn[2] dijler. Bolat kajlı bolıp keteler.

Kete-kete br tau ičine kireler, tauda čalaš jasajlar, ana jakka muna jakka ʒüreler, čolaknen sokor tilenmege keteler. Topal Bolat bolar ketkende tırpın-tırpın auya čıka, anı jakka munı jakka ajlana br karaʒa tapa, tarz atıp soya, čalaška alıp (294) kele, u jaktan, bo jaktan br ike koraj ayačlar toplap ketire. Bolar kegenʒek ankıta-ankıta karaʒani piširep[3] azirlıj. Br vakitten son ular da kelirler amma mubarek šaj jakšı pišken, pütin taular ankıjtap. Nise, bolar keleler, otėrıp[4] ašajler. Sokor dij ki: ej, arkadašlar, bo šaj bolmaz, bzge br apakaj kerek almaya dij. Čolaknen topal ajtalar: bz üč ermiz, üč akajya br apakaj ni kate, bolaj ši bolor mı dijler. Sokor dij ki: bz üzımız tarzė[5] br ʒarlı kız tapsak, bzge tünja akret kardaš etsek, o da bzge aš piširir, kır ʒuvar, χizmetimizi išler edij di. Obirleri bu išni lajik köreler. Šonda čolaknen sokor br köjden br ʒarlı kız ketireler. O ʒarlı kız bolarya aš piširgen, kır ʒuyan bola. Ana, šaj etip ʒüreler.

Künnerde br kün topal Bolat oñuzına[6] tüfegın alıp tauya kete, obirleri tilenmege keteler. Čalašta tek kız kala. O vakitlar ot bek az tapıla eken, šėrpė jok eken, otnı söndürmez ekenner. Šu künnü nasil da bolsa kız aš piširgende kazan tašıp artılıp otın söndüre. Bolat ta bek ačulı[7] bolyanından kız Bolattan korka eken: Endė Bolat-akam kelse, maña kim bilir katir, ende kalaj etijim dep čalaštan čıkıp kete ot kıdırmaya.

Az kete (295) küp kete, br tauda ʒılpın-ʒılpın ot šaulesi körüne, kız‿se[8] šaulege toru kete. Šaule bolyan üjge barıp kire, baksa br kurtka otora. Kız: salam kart anaj di. Kurtka: vaj, kizım koš keldiñ dij, mında kajdan keldiñ dij. Kız añata, dij ki: kart anaj, men pılan ʒerde br čalaš bar, šo čalašta otıram dı. Kurtka ʒadı: anda jangız üzıñsıñ mi dij. Jok, üš tene akajlarım bar dij. Ular ne iš išlijler dij. Kız ajta ne išlegenlerin. Jā mında nasıl tüštiñ

1 R ʒüre eken
2 kečinirsen
3 R piširėp, but only *piširep* forms can be found in the texts.
4 R otırıp
5 tarze; R tarzı
6 oñuzina
7 ačuli
8 kız-se; R kız ise

degende, di ki: meni Timer Bolat degen akam bek ačulė[1] ʒiget, šte olarγa aš piširgende, su taštı[2] otnė[3] kalaj da bolsa sündürdi. Šindi akajlarım aššız kaldı, kajtėp kelseler ne ajtırım? Čolak aγajimnen sokır aγajėm[4] bekim pr[5] lap ajtmaz, topal Bolat akamdan bek teumetlenim, urar dep. Miña sizın ʒurtınızdagi šauleni körüp sizden ot tilemege keldim di. Ʒadı[6] kurtka: ajsa šaj bolsa, meni bašım bek avurta, azgınačık bašımı[7] türt[ėp] kara da, son[8] ot alıp ketirsėn[9] dij. Kız bek jahši tep[10] otora. Kurtka ʒadı kızın tizine bašın kujıp jata.

Br ajlı vakitten son: ajde, kızım, ʒeter dep tura, kızγa ot birmege. Lakin šu kız üzınde br takatsızlık tuja, aligė ʒadı onėn takatın emgen eken. Kızın kaberi bolmaj, ne de bolsa kızγa ot bere, lakin kız ot kujmaγa abdıraγanından bi šij almaγa unuta. Kız: kartıjem, ot kujmaγa šijim ʒok dij. Kurtka kızın peštmelinde[11] br tešik (296) köre: kızım, peštmelini tut, tübine kül kujajım, andan üstüne otın korların kujak[12]. Ana šaj itep kızın peštmeline kül totra[13], üstüne de kordan kuja. Kız kete, ketkende jol buji kül čubura-čubura kete. Kız bara čalaška, ot ʒaγıb evelki tarzı ašın pišire, son alıgı akajlar keleler, ašap tojalar, ertesı künı kene de išlerine keteler. Bz kelijik kurtkaγa.

Kız ot alıp ketkenden son, bu da köp kečikmij kizın artından bašlaj izlep ketmege. Čubırγan külden izlij-izlij, kızın bolγan čalaška ʒete, kire, selam alejkum, kızım dij: Balam, bašım šaj avura bükün, obırısi kün sen bašımnı[14] türtkenden son br jaχši bolγan edim, bükün gene bašım avurta, saña kene de br parča bašėmnė türttürmege keldim dej. Kız: bek jaχšı kartanaj dij. Bonı bašın bakıp br vakittan son obur: jeter balam, dep tura. Alla rāzi bolsın dep tırpın-tırpın kete. Br kač kün šaj etip, er kün kelip kızγa bašın türttüre[15], lakin kızın künden künge takatı[16] pitep[17] tavusla.

Br kün Bolat kızγa soraj: a kız, saña ni bola, šaj azdıñ dij, kız ajtmaj. Br kün gene soraj, kene de ajtmaj. Bolat di ki: endi de ajtmasañ mojnını üzerim, bo sendeki χastalık tegıl, sende br kızlı ši bar dij, zorlaj. Kız ajta: mına mında šolaj br kurtka apakaj kele, maña bašin baktırta, turıp ketkenden son tujam minde br takatsızlık bola. Bolat: ajsa šaj bolmaz, bo kortka oburdur, bo obur seni eʒeliñe ʒeter, ojlanajık üturejik dij. Kız: katp‿ütürırmız[18] di. (297) Men erten čıγıp ketken kiši bolorım, sen de mında čalašta kalırsın, obur kene kelij, sen onı[19] tizine ʒatırıp bašın kararsın, bašın karaγanda čašların jaχši itep, orap tutkandan son br ʒütkürürsin, bz de tujarmız ki sen oburnı[20] begitkeniñi. Bz ičerge kirep,[1]

---

1 ačule; R ačulı
2 tašti
3 otne
4 aγajem; R aγajım
5 R br
6 ʒadi, occurs 3 times here.
7 bašımi
8 R sonra
9 ketirsen
10 R dep
11 R peštemel
12 Shortened, maybe dialectal variant of *kujajık* 'let us put (into)'.
13 R totıra
14 bašımni
15 tüttüre
16 takati
17 R pitėp
18 R katėp ütürırmız
19 oni
20 oburni

onı begitirmiz, tek sen ʒiberme. O senden kutulıp kačmaya kòp tırıšır,[2] lakin sakt bol dij. Kız jaχši di.Ondan bolar čolak topal sokır účeu čiyıp[3] ketken kiši bolıp čalašın katında saklanalar.

Br vaχitten son obur-ʒadı kene kele tırpın-tırpın etip, belin búkrejtep ičerge sukılıp salam bere. Vaj kızım, dej, kačan kelsem, seni ʒangız kúremen di. Kız: kartijem, akajlarım bek ʒarlılar, išlemege keteler, mında úzim ʒangız kalam dij. Kurtka dij ki: gene de br parča bašımı[4] túrttúrmege keldim. Kız: bek jaχši kartijim, ʒatınız túrtijim dep, tizine oburnı ʒatkıza. Bašın túrtken kiši bolıp čašlarinden orap čırmap alyandan son br ʒútkúre. Saklanėp oturyannar[5] súrúp kireler. Obur kòp tırıšıp baka kačmaya, ama br túrlú kačıp bol‿almaj. Bolatnıñ epdešleri bunı jaχši kòteklijer, dijler ki: bo kızın sen bolaj azdırıp inʒe-maraz itep tašladın, aru et dijler. Obur: bek jaχši, aru etijim dep bašlaj kızın bašından ajayadek tilinnen ʒalamaya, kız kene evelki tarzı bolıp túzele. Sokor di ki: menim kòzlerim kòpten (298) sokor boldı, kòrep bol‿almajim, menim kúzlerimni de savult dij. Ʒadı etmijʒek bolsa da čolak solakaj kolınen[6] br šamar ora, šo‿ssat bek jaχši etijim dep anı da eki kòzin ʒalap aru etti. Anadan tuyma kúzleri ačıla.

Epsi jaχši bolalar, tek br Bolat kala. Arkadašları oburya dij: šte bzin kardašımız kaldı topal, onı de aru etseñ son seni ʒiberirmiz, tilegen[7] jakına ket dijler. Obur dij ki: men biñlernen topal ketirseñiz jaχši eterim, tek bo topalnı ajakların jaχši etmem dij. Nečin dep sorajlar. Nečin bolsın, menim úč ulumnı bašların čavıp, kanların ayıdıp újlerin ʒiyıp[8], úč te kızlarımnı alıp kettė, men bonı kalaj jaχši etejim. Siz meni òtúreʒeginizi bilsem, kene itep bol‿almam dij. Bo kurtka Bolatnıñ òtúrdigi dejulerniñ anası iken. Šu ʒerde Bolatnıñ ipdešleri oburı òtúreler, kursayın ʒarıp, ʒúregin čıkarıp Bolatnıñ ajaklarına kujıp bajlajler. Bolatın ajakları anadan tuyma jakši bolıp kete. Bolar epsi sap-sau bolalar.

Br kún dòrteu de ʒolya čıyıp Bolatnıñ ʒurtına kelijler, padišanıñ sarajlarında davul zurna ur patlasın čala. Toj bola, bolar da toj újine baralar. Bolarya br ike pakir tojya kelgen dep, túrli-túrli ašlar ašatalar, kave ičeler, sorajlar: bo kimniñ tojı bola jatkan dijler. Ʒıjındaki bolyan kalk dijler ki: bizin padišanı[9] úč ulu bar idė, (299) ikesi bulaj br kún ʒılkıları ʒojılıp br kún čiyıp[10] kettiler, ʒedi ʒıl degende úč ažda-diju útúrep, úč te diju kızı alıp ʒılkılarinen barabar keldiler, šte šindi šo ketirgen kızların tojların etip toralar. Bo keče ferdege kireʒekler dijler. Jā kiškenesi ne bolyan dep sorajler. Ʒıjın kalkı: kim bile, o zāten br matu edi, kim bile kajda kaldı, šte mınavu tojları bola jatkan akaları oña da br deju kızı alıp kelgenler, lākin unı kajda ʒúrgeninden breúniñ kaberi ʒok dijler. Deju kızları ant ičkenner: kırık kún bzin jasımız bar, kırık kúnge kadar beklegen br kišimiz bar, kırık kúnden kelmese son bz sizge apakaj bolırmız degen idiler, Bolatnı kúte ediler.

1 R kirėp, but the *kirep* variants are dominant in the whole collection.
2 tirıšır
3 R čıyıp
4 bašimi
5 òturyannar
6 R kolınen
7 tilegin
8 R ʒıyıp
9 padišani
10 R čıyıp

İšte uların da jasların son künleri iken, bolar ʒıjın kalkından bo lapların ešittiler. Nišlesinler, Bolatın büjük akajları jannarına kelip kete ama ič br türlü tanıp bol‿almaj. Bolatın ipdeši sokır di ki: bz šaj etip tojγa keldik, üzi de br padiša balaların tojı br parča jaraštırmak kerekmiz dij. Ajakka turıp dij ki: ej ʒıjın kalkı, bz padiša balaların tojı bolγanın tujıp toj ʒaraštırmaγa keldik, bar ajtınız toj kalkına kelsinner ʒıjılsınnar, padišaγa da ajtıñız, tojnı[1] ʒaraštıraʒaγımızın dij. Ana pütün toj kalkı toplana, padiša da kele. Sokır bašlaj šindė[2] annatmaγa, išin evvelden nasıl bolγanın. Padišadan izin tilej: padišam, ište dünjade ešitilmegen br ikaje ajtaʒaγım, izin berseñiz dij. Padiša izin bergenden son bašlaj ajtmaγa: br zamanda br padiša (300) bolγan, onı da üč ulu bolγan. Künnerde br kün tokoz toγaj ʒılkıları ʒojılıp, nasil čıγıp ketkennerin, bolardan son en kičkenlerin[3] ketkenin, andan ujkenlerin ʒanajlarnen nasil esirkep akčaları pitkenden son biri molčada, biri purunda ırgat ʒürgenlerın, Bolat nasil dejulerin ötürgenin, kajtıp akalarına urba alıp kijdirep ırgatlıktan kuturγanın, andan bolar ʒolda kalaj etip aldatıp Bolatın ajakların čavıp keskenlerın, andan nasıl bolar br kıznen üč erkek arkadaš bolγanların, nasıl oburnen išler bolγanın, nasıl ajakların savulγanın, nasil kilep[4] padišanıñ katında masal itep ajtıp tıñlatkanların br br ajta kele.

Padiša šo ʒerden čıdaj almaj, daraγači jasatıp ekı üjken ulanların astırıp, eki üjken kızın en ujkenin sokorγa bereler, ortanʒısını[5] čolakka bereler, en kiškenesin Bolatka bereler. Aligi üzlerinen beraber kelgen kızın da padišanı baš vezirniñ ulına bereler. Šaj itep kırık kün, kırık keʒe toj itep dörteu de br keʒede ferdege kirep murad alıp murad bereler. Kudaj bzlerin de muradlarimızın asl etsin, Timer Bolat dajın.

5 Baj-terek masali

(318) Br zamanda br padiša bola, onın da dürt ulı bola. Padišanıñ azbarında br balaban terek, onın adı Baj-terek bola. Terekniñ ʒemiši pišep vaktına kelgende br kere üzlerı ʒimišinden ašap bol‿almajler. Endi pišti vaktına keldı dep torγanda er ʒıl br geče ʒėmiši gajb bolıp kete. Padišanıñ balaları br ʒerge tuplanıp mušavere eteler, dijler ki: babaj, bz terekimizge br keče karavıl čıkıp karajık, ne etkenin añajık, šo bizin Baj-terekniñ ʒimišı ne bola eken, šonı br bilijik dijler. Söz bir bolıp šo ʒıl gine de terekde ʒimiš pišep, inde vaktı ʒetti degende, en üken[6] ulı ukını janına alıp terekniñ katına barıp ʒatıp ʒuklaj. Terekke ne bolγanın bilmege dep br talaj vakit ʒata. Br ajlı vakitten son jarı keče vaktı bola jazdı[7] degende brde karasa ne karasın, kökten šatırdap[8] güdürdep kijamet koparıp br ši kele, ama kurkusından az kala ödü patlap öle jaza. Esi bašından ketep[9] ʒıkılıp ʒata.

1 tojni
2 šinde
3 R kičkenelerin
4 R kilėp
5 ortanʒısıni
6 R üjken
7 bola-jazdi
8 šatirdap
9 R ketėp

Br vakitten son turıp karasa terekin ǯimišiniñ br pajı bitep[1] ketken. Ana šo ǯerden turıp urnına kajtėp kele, išniñ kalaj bolyanın atasına ajtıp añata.

Ekinǯi kúni ortanǯa ulı oyını[2] katına alıp karavıl turmaya varıp ǯatıp ǯuklaj. Jarı keče bolyanda aligi karañık[3] bolot šatırdap kele. Ǯiget jokodan šašıp tora, kateǯegin bil‿almaj,[4] esi bašından kitep ǯıkıla. Br talaj vakitten son torop[5] u jakka bo jakka karaj, br šij (319) kór‿almaj.[6] Terekke ǯakın barıp karasa, ǯimišniñ jarısı ǯoy‿olyan[7]. Ȯzi ǯimišin ni alıp ketkenin kórúp bol‿almaj, újrúlep u da urnına kajta. Kelıp išniñ kaj dajın bolyanın atasına añata.

Ůčúnǯi keče kúčuk ulı dij ki: bu kečede min barıp karajım, bekim ne ekenin bilep bularım dij. Keč[e] bola, ulan umzına ukını tayınıp tereknıñ katına bara, ǯokısi kele ama ǯoklamaj. Br vakitten son šatırdap gudurdep kókten br bolot kelip terekke sarıla. Ulan šo bolotka toru okını[8] br toratıp[9] ǯibere. Bolotnıñ ičinden br ses čıka: er bolsañ br dā at dij. Ulan br uk dā ata. Br ses dā kele: ǯiget ujnı účkeǯekdir, br dā uk ček dij, ulan br uk tā ata. Bolotnıñ ičinden ǯerge br šij újrúlep túše, katına jaklašıp barıp karasa, eki karıš bujı, dúrt karıš sakalli kiši desen kiši túgel, ǯin desen ǯin túgel, ne ekenin bil‿almaj[10], suraj: sen ins misin, ǯins misin dij. İns ekenin, ǯins ekenin senı išin túgel dij. Ulan: ajsa menim išim bolmasa, seni kardašlarıma, atama alıp barajım, uların iši bolsın dij. Andan dórt karıš sakallını alıp ketip atasına bere: šte ataj, bznin Baj-terekniñ ǯimišini pitirgen muna šo eken dij. Atası bo kišini br siǯimnen sakalından čırmap újnıñ katında br bayanaja bajlattıra[11]: šte menim šondaj Baj-terek adlı teregimni ǯimišini bolaj br kuǯurlı kiši ašaj iken, ište totıp bajladık.

Kalk ǯıjılmakta bolsın, (320) bo jakta kičkene ulan dórdúnǯisi br kún azbarda uk atıp ujnayanda oyı čaptırıp kete, bo sakallınıñ[12] sakalına barıp kire kala. Ulan barėp istep amma bermej, dij: men saña okını berirmen, lakin sen minem sakalımnı pıčaknen šo bajlanyan ǯerinden kesip tašlarsın. Kičkene bala barıp anasından: anaj, maña pıčak ber dij. Anası pıčak bere, kulına alyandan son šo‿ssāt ǯuvurıp, sakallınıñ sakalın bajlanyan jerın túbinden kesip ala. Anau sakallı kesilgeni tarzı, šo ǯerden havaya pirıldap újrúlıp evelki tarzı karañık tuman bolıp, ǯok bolıp kete. Bala da katkenin bil‿almaj.

Aradan br eki zaman ȯte, padiša úzin kalkına kiši čayırtıp, bu sakallını kȯstereǯek ide, kalk her bir jaktan ǯıyılıp kele. Padišadan sorajler nege čayırtkanın. Dij ki: muna mında bizin bakčada br Baj-terek bar ide, er ǯıl šaj etep[13] br šij bıla‿jdı[14], ne ekenin bz bil‿almaj idek. Šindi tuttık, sizlerge šonı kústermege čayırttım, bu br kuǯur šij. Padišam, kajdadir šo šij, kúrijik diler. Mına aranda bayanada bajlı tura di. Ajdınız, barıp karajık

1 R bitėp
2 oyıni
3 karañik
4 bil-almaj; R bile almaj
5 tȯrȯp
6 R kȯr-almaj; kȯre almaj
7 čoyalyan [sic!]; R ǯok olyan 'disappeared'
8 okıni
9 toratip
10 bilalmaj; R bile almaj
11 bajlattira
12 sakallıniñ
13 R etėp ~ etip
14 bılajdı; R bola edi

diler. Karamaүa baralar, padiša karaj, ůzniñ bajlaүan ӡirinde br šij ӡok. Šondan dij ki: ajse šaj bolmaz, siz taralıñız, men bonı kim čečip kačırүanın añlap tıñlap bilijem[1] di.

Brin čakırıp soraj sen etkensin dep, basturukka ӡibere. Ekinӡisin ӡibere, kim etkenin bil‿almaj br türlü. Son eñ kičkene ulu babasına di ki: babaj, u kišini men čečip ӡiberdım, dij, boten halkın kıjnamañız dij. Padiša: nečin (321) ӡiberdiñ dep suraj. Ulan di ki: ataj, min uk atıp ujnap ӡürgende minem ukım onı sakalın tübinde tañmajında kaldı bermedi: kačan pıčak ketirep[2] sakalımnı kesersin, šo vakit berirmen dedi. Men de barıp anıñ sakalın kestim, pırıldap ůjrılıp bulut dajın avaga siñėp ketti di. Atası bo laplarnı ešitkeč, vezirlerine emr ete, daraүaӡ jasattıra, ulını astıraӡak bola. Vezirleri dijler ki: padišam, sin bu ulıñnė ütürme, diler, sen onı kulına padiša ulı bolүanına kulından br kaүıd[3] berirsin, kajda barsa ač bolmaz, hürmet eteler, üzini timezler, mında uturmaz kuvarsıñ, oña bondan boten ӡeza kerek mi dijler. Padiša ulunı kulına padiša balası bolүanına br kaүıd jazıp bere di ki: ajde seni ӡanıñı saña baүıšladım, munda utırma, kajda ketseñ ket dep kuvalaj. Ulannıñ anası da br jaχšı majlap kalakaj piširep bere, ketkende ašamaүa. Bo kete.

Bašın alıp ӡolda boña br taz ulan rast kele: ej arkadaš, kajdan kilep kajda ketesin deje. Taz ulanүa bala bašına kelgenlerni br br añata. Taz ulan: seni atañ küүan, meni de anam küdi, inde olar bzni br iš etip bolamaz belliler[4], bz ekimiz br bolıp ketejik, inšallah bz ularүa mohtač bolmamız boүaj di. Padiša balası: ajsa maña arkadaš bar dep kete.

Kete-kete br ӡirde aүač kuletkesin tübinde utėrıp ӡal alalar. Ulan: men braz [barıp / ketip] karajım dij. U čıkıp ketkende taz üzin čalmandan br tuzlı kalakajin čıkarıp arkadašın majlı kümečin alıp, üzini tuzlı[5] (322) kalakajin onı čalmanına kuja. Ulan kelip utıra di ki: endi br parča ašajık dep čalmanın ačıp baksa, üzinıñ kümečiniñ urnına br kalakaj kujılүan, ašamaүa bašlaj ama tuzdan zeher ašalmaj, lākin indemi. Taz majlı kümečnen kursaүın tutura.

Šo ӡirden torıp keteler, br ӡirge baralar, br kujudan[6] su ičmege tilijler, lākin kujunıñ čelegi juk. Taz di ki: sen mınau menim kušaүından‿dut[7], men sarkıp ičep čıkajım, son sen de sarkıp ičep čıkarsin dij. Taz sarka, ičep čıka, son padiša balası sarka. Su ičkenden son: inde ček čıkajım di. Taz čekmem di. Ej arkadaš, nege šaj ettıñ degende: sen maña laki bolırsın, üzinni urbalarıñıznı maña berirsen, men padiša balası bolormen, muna šoña and eterseñ, breuge de tujdurmajıӡaүına[8], men de seni čekip čıkarırmen di. Bala and‿ete[9], taz ulan čıkara. Taz ulannıñ urbaların kijep, onı atasından alүan kaүadın kulına alıp, bala da bašına unı eski takjesın kijep keteler.

Br kün br padišanıñ sarajlarına baralar, taz ulannı aldına čıkıp karšılap alalar. Padša bolarүa si etip sijlaj, taz ulanүa si balaban bola. Tazdan padiša soraj: ulım, bu

1 R bilejim
2 R ketirėp, The more common version is *ketirep*.
3 kaүid; 2 times occurs here.
4 R bellijler
5 tuzli
6 kujuden
7 R kušaүımdan tut
8 tüjdurmajıӡaүına
9 R ant ete

jaklarya nedaj keldiñ di. O da halinden kozyap di ki: meni atam šulaj etip kūdı, men de lakijımni alıp čıktım, sizėn[1] katıñıza kelip, vezir jā br paša bolormen. Padša: ajsa šolaj bolsa, men de seni ůzime vezir etijim dep baš vezir (323) ete. Padša balası bolyan ǯigetni katijik dep suraj. Unı ǯılkıǯi et di. Bek jaχši ǯılkıǯı bolır dij. Andan padiša aligi ǯigetke br tabın ǯılkı bere, di ki: šte balam, mınau br tabın ǯılkını bakarsın, lakin kůn tuvıška alıp ketep otlatırsın, kůn batıya alıp ketersin, sırtka alıp ketersin, jangız Kibla jakka alıp ketmezsin dı. Žiget bek jaχšı dep borkini maññajına basıp, čalmanın tůrůp, katına kalakaj alıp, ufak br jasman su totırıp ǯılkısını ajdap kete.

Br kůn, ike kůn šaj. Kůnnerde br kůn kene de ǯılkısın alıp ketıp ǯaja ǯatėp ǯuklaj, lākin Kiblaya ketken iken, tabın unı ujlanmaj ka‿jakka[2] ketse ketsin dij. Br zamandan son avadan br ši kele, bunı ǯuklayan ǯerinden kůterep alıp kite, ůzi ǯukusını bozmaj. Bonı alıp ketken, alıgı důrt karıš sakalı, ike karıš bujı kiši eken. Bonı alıp kete, tora ůzın mayarasına. Ůzınıñ jata toryan br salınǯayı bar eken, šoya jatkıza. Bo kišiniñ de kırık kızı bar eken. Kızlarya di kı: bo ǯigetni terbetiñiz, salınǯakta jaχšı jukusına tojsın di. Kızlar bonı terbeteler. Br talaj ǯuklap tora. Jukudan ajındıyı tarzı baksa katında kırık tene kızlar otoralar. Aligı sakallı: nasıl ulım, tůštůn kapkanya di, min seni ayalarına kůmenlim, šimdi senı ȯtȯrir edim, lākin seni men ȯtůrmiǯegim, sen (324) meni kutaryan idiñ di. Šondan ǯigetnı kızlarnen sefa ettirėp, br kůn ǯigetke der ki: endi seni braz dolaštırajım[3] dep, mayarasındaki ǯirlerin ǯursetip kȯrsete.

Br odanı ača, ičinde čokrak bar, br jayından kůmůš aka, br jayından altın. Žigetke der ki: bašıñnı šo čokrakta br batır di, ǯiget bašın batıra, bašın jarısı kůmůš, jarısı altın bola. Ondan kırık kızların katına kele: endi bo ǯiget kete, sizlerin brer nišan bermeñizin munasib kůremen, sizler de brer baχšiš beriñız. Andan kıznıñ birisi čıkarıp br altın tarak bere, brsi de br gȯzgů bere, obirsi baška br šej bere, er bri brer baχšiš bereler. Tarak bergen kız di ki: bu taraknıñ ůneri, artından důšmen kuvıp kelgende, artına taš-lasañ, br kalau bolıp, tůšmen ȯtıp bol‿almaz di. Kůzgůni bergen: bo kůzgůni ǯebinden čıkarıp aldına kujsañ, ičinde kırık tůrlů čalyı čayana čalıp, kırık tene de ujınǯı čıkıp ujnarlar di. Sakallı: kızlar er brsi brer šij berdiler, men de saña šonı ajtajım, menim bayšım šo bolsın, keče karañıkta bašıñnı[4] ačarsıñ, bašıñnı jarıyından[5] čırak kerek bolmas di. Endi saña ruχset, kajda ketseñ ket dij.

Šondan ǯiget sauklašıp kete, ǯılkını alıp sarajya bara, aranga kapaj ǯılkını, ůznıñ ǯata toryan tauk kůmesi kibik urnına kirėp otıra, aldına ketirep braz umač čorbası[6] kujalar. Ašaj, ašayandan son (325) aldına čıyarıp u kůzgůni sala, bašın ačıp utura. Kůmesnıñ iči ǯap-ǯarık bolıp, bašlaj kůzgů de kırık tůrlů čalyı[7] čayana čalıp, kırık ujunčı čıkıp ujnamaya. Padıšanıñ kičkene kızı tšarya[8] čıka, čalgı sesi ešite ama kaj jaktan kelgenın bilep bol‿almaj. Nejse de ǯůre-ǯůre sesin kelgen jayına toro ketep kůmeske

1 sizen
2 ka (hangi) jakka; Recte: kaj jakka
3 dolaštirajım
4 bašıñni
5 jarıyınden
6 čorbasi
7 čalgi
8 R tıšarıya

bara. Kapını kakıp ičerge kireǯek bola. Ǯiget: sen kumsin dep suraj, padšanıñ kızıman dij. Ǯiget: padıša kızı tugel, kum bolsañ bol dij, men šimdė[1] juklajman, erte turıp išime keteǯekmen di. Kız ajlanıp kajta.

Br kün šaj, ike kün šaj, alıgı taz ulan padišanın veziri bolγan čočka, bo ǯigetni bulaj bolγanın tujıp, bonı ütürtmesem, maña rātlık bolmaz, mında šonı ütürtijim di. Br kün padišaga di ki: bzim ǯılkıǯı šondaj br ǯigettir, ne tileseñ šonı eter dij. Padiša dı ki: ne kelir unıñ kulından, šte oña aldında ajdap ǯürgen tabunnen ǯılkını altın etip ketir deseñiz šonı eter dij. Belli ki onı etep[2] bol‿almaz, jā etip bol‿almasa katermiz di padiša. Etip bol‿almasa mujnını alırsıñız di. Br kün padiša ǯigitke: ej ǯiget, šo aldındaki tabunnen ǯılkını altın etıp ketiresin ketire almasan mujnını alırman di. Ǯiget alıgı ǯerge sakallıga bara, išniñ kaj dajın bolγanın ajta. Sakallı: ajde bu jakka, čokraknıñ katına ǯılkını di. Ǯiget (326) čokrakın katına ǯılkısını ajdap bara. Sakallı ǯigetke: ulım, mına saγa br čelek, šo čokraktan su alıp čeleknen ǯılkıların üstüne serp di. Ǯiget čokraktan su alėp[3] čeleknen ǯılkıların üstüne serpe, ǯılkılar epsi altın bola. Andan ajdap sarajγa kele, padiša baksa ne karasın, ǯılkılar epsi altın bolγan. Üziniñ kümesine kirep ǯata.

Kız kene de kelip ǯigetke di ki: sen padiša ulı kerčekten ekensen, laki[j] sen tugıl ekensen, u vezir eken di. Bo nekadar da padiša ulı ekenin saklajaǯak bolsa da, kız ǯigetke ašik bolıp, urılıp čidaj[4] almaj. Šondan tora atasına barıp išin bilgeni kibik añata. Padiša ǯigetni čakırta, kızγa di ki: kızım, unı padiša balası bolγanın katıp bilirmiz, neden bilesin unı? Kızı dij ki: šte ataj, sen unı br bašını ačıp kararsañ bilirsiñ. Padiša ǯigetni čakırta, biš-un kiši čakıra, čakırınız taz ulanı dij. Barıp čakıralar, u da kele. Br konušma bolıp torγanda, padiša di: taz ulannıñ bürkini küteriniz, kujaš tijsin dij. [...][5]. Kütereler, ne baksınlar, ulannın bašının jarısı kümüš, jarısı altın. Šondan padiša išniñ nedajı bolγanın suraj, ǯiget te añata bašına kelgennerin brer-brer.

Padiša taz ulannı ütürtep, bonı evladlik ete. Br vakitten son padiša öle bo onıñ urnına padiša bola. Br kač ǯıl son atasına silaa[6] bara, atasinen kürėšėp[7], gene de urnına kelip üzin tahtında utėrėp[8] sefa ile ömür ete.

6 Mejzin masali

(336) Br zamanda Hasim isminde br adam nejet ede haǯa gitmege. Julγa čıka. Köjnıñ mejzini ozγarγanda hajėrli[9] jul bolsın ama kızıñ kimge emanet ettiñ dij. Hasim ajta: kızımnı Allaγa emanet ettim. Mejzin ajta: Allah kökten tüšüp seniñ kızıñnı bakmaz,

1 šimde; R šimdi
2 R etėp
3 alep; R alıp
4 R čıdaj
5 Missing part of the text saying something like: “but they see he’s bald”. Then the padisha has the fellow’s cap taken off.
6 silaha; R silāja
7 küresep
8 uterep
9 hajerli; R hajırlı

kerek br kimge emanet etmege dij. Hasım: ejse kerek šileri bolsa, sen alıp ketirirsin, penǯereden verirsin, ičeri kirmezsin. Mejzin bir vakit et alıp kete, penǯereden tašlaj, kız kapıjı ačmaj. Mejzin bir ǯadı[1] katınya jalvara: bu kız ile minem aramı jasa, minden ne alsañ al dij. Katın kıza kelip, nasılsa aldatıp kapını ačtıra[2]: kel kızım, Ahmadıñ kızınıñ dügüni bar, barajık ta braz eglenesin dij. Katın mejzinge čukur ičindeki hamamı üč küne iǯare ile tutmasını emr ede. Köj ičinde kırχ baj bar iken. Kırχınıñ kırχ tene kızı bar iken, kırχınıñ enterisı bar, hamamın kapısına[3] asajler.

Ertesi kün katın kelip kızı alėp dügüne dep čıχa. Kız üjden čıkkanda eki ǯebini arpa toldurdu. Hamama gelinǯe töke-töke keldi. Hamamγa keldıler, baχsalar kapuda kırχ enteri bar. Ej kızım, kızlar bunda bar, kel bz de junalım[4] da, son kitermiz dügüne dij. Katın hamamıñ kapısın ačıp kızı hamama itepjüvere, üzi χajtėp ketti. Mejzin hamamda. Kız mejzini kürinǯe bašından aklı kitti, aldau boldıγını bildi. Kız kurtulmanın čarası[5] (337) bolmadıγından mejzinge: men seni br güzel juvundėrajım[6] da digen. Mejzin köbek-taška mındi. Kız güzel spınlap[7] jumaγa bašlaj, bašın küzin sapunlaγannan sun, kız ajaγından nalın čıχarıp mejzinniñ bašina ura-ura aklını[8] alγan. Kız undan čıχıp arpalar tükülgen juldan üjlerine keldi.

Hamamnıñ küni toldı. Mejzin hālā hamamda ǯata, hamam sahibi kelip mejzini hamamdan kötürep attılar. Mejzin katına keldi: seniñ maña eteǯegin bu iken mi, ben jarın br kün haǯi Hasim keldikte ne jüznen uña kürünijim dij. Katın: buχadar ālim bolıp, munıñ čarasını tapmazsıñ mı digen. Hasımγa br mektüb jaz ki: kızıñ munda baštan čıχtı. Köjde kırχ kiši bar, munnarnıñ kırχ balası bar, kırχı da kızına barıp keleler. Mejzin mektübi jazdı. Hasim bu jazıjı haǯlikten kajtıp kelgende julda alγan. Mektübi oχurken bašindan esi kitep jıχıldı, brazdan turıp oγlunı köjine jolladı[9]: bar, kız kardašın suj da külmegin kanına boja da maña ketir dij. Uγlı ketti, köje keldi, kardašina kapuji[10] ač dedi. Kız: men kapıjı ačmam dij. Jigėt: men senin tuγan aγañım, mennen kurχma dej. Kız: eger menim aγam bolsañ, jüzügıñ üstüne adıñ jazılı bolaǯaχ, penǯereden tašla, baχajım soñ ačijmen dij. Jiget[11] jüzügi tašlaγan, ukıp baktı, aγanı boldoγını bildikten sun kapıjı ačtı[12], hal χatir surıštılar. Sonra: kardašım, (338) babamız hačtan kele, ajde karšısına čıχajık dej. Aldap üjden čıχarıp tauγa kittiler. Br ters jerge barγan sun: kardašım, men bu išlerın saña boχtan boldıγın añladım, lākin babamız emr etken, seni sujıp külmegın kanγa bujap ketirmegi, šimdi nasil etmek[13]? Sen külmegın čıχar, küčük barmaγın kanatajıχ, külmegin bujap alıp kitejim, sen bu jirde kal dij. Jigėt külmegi alıp ketken, babasına kürsetti. Hasim üje kajtıp keldi, mejzin buña χoš keldige kelgen, hal χatir surıštılar.

1 ǯadi
2 ačtira
3 kapısina
4 R jūnalım ~ juvunalım
5 čarasi
6 javundarajım
7 R sapınlap
8 aklıni
9 jolladi
10 R kapujı
11 R jigėt; Since the *jiget* version is used 162 times, it has been left unchanged.
12 ačti
13 ėtmek

Bėz kelijik tauda kalγan kızγa. Aulanmaγa čıχχan br bejuγlı[1], kızı kürep ašχa tüšep kızı alėp[2] üjlerine kitti. Br kač vakitten sun bejuγlına: sen üjlen dediler. Jigėt: eger men üjlensem, šu taudan taptıγım kızı alırmen dij. Anası, babası degenner: taudan tapılγan insanγa jar bolmaj, sen jaχši bir jirden kız begen de unı al dejler. Jiget: hajr, men üjlensem munnan bašχasın almijmen dij. Nasılsa da nikah χıjıp bu kızı alγan. Kač vakitler jašadılar[3], üč balaları bolor. Bu vakta kadar jiget katından heč surmaγan, seniñ babañ bar mi dep. Xatın buña ačuvandı. Bir kač vaχitlar til jašırdı. Jigetke ajttılar: biz saña baštan ajtkan‿dık[4], taudan tapılγan šejden χajr bolmaz dep. Šindi jiget buña čoχ χasavetlendi, juχusı kačkan. Bir geče χatın jangız balalarinen jatχanda (339) aγlaγan balasına türkü jırladı[5]. Jiget bunı četten išitken. Jiget: sen čoχ vakitten beri tilsiz bolγan‿dın[6], bunı sebebi nedir, maña ajt dej. Xatın: bunǯe jıldan beri bir jirde omer etemiz, sen maña surmadıñ[7] anañ babañ bar mi dep, buña ačulanıp men laf ejtmedim, ... [Jiget:] seniñ anañ babañ barsa jarın babaña jullajım dij.

Sabah turıp üč balasınen χatın birde vezir fajtonγa kujıp julladı[8]. Vezir χatınγa küz kujdı, bir tau ičinden kečkende toχtadı, menimle iške rāzi bol dij, eger bolmasañ, balaların birin sujarım dij. Karı[9] rāzi bolmadı. Vezir balaların birin sujdı. Biraz daha barγanda, jene surdı[10], χatın jene rāzi bolmadı, balaların birin daha sujdı[11]. Biraz ketkenden sun jene rāzi bolmadı, birin daha sujdı[12]. Sonra: indi men seniñ üzini de sujarım, eger rāzi bolmasañ di. Kadın čare bolmaǯaγın tujγannan sun: sen meni jüver, tıšarı[13] čıχajım da soñ iš bolor degen. Vezir jüvermege istemedi, χatın: eger kurksañ, belime bir jip bejle de jüver dedi. Vezir beline jip bejlep jüverdi. Xatın čal‿artına[14] ajlandı, jipi terekke bejlep, χatın kačıp kitti. Xatın čoχ ejlendi, jipıle kitep baktı[15] jip terekke bajlı, χatın juk. Pajtunγa minep artχa kajttı, dedi: bir taudan kečkende uč balasinen učıp kittiler. Herkes buña ınandılar, taudan tapılγan‿di,[16] jene tauγa gajb boldı degenner.

Xatın ise bı‿jerden[17] (340) tau artına čıχtı,[18] bir čobanγa rast kele, čobanγa bir kuj sujdırıp kojnıñ kursaγın juvıp bašına kečirep küjge kite. Küjde mejzin bunı kürdi: sen kaz bakmaγa jürmezsıñ mi dej. Jürürmen degen. Mejzin de Hasim ičin tuttı[19]. Bir kač vakitler kaz baktı, son annan čıkıp šo küjde bir elvaǯı tükjenine jürdi.

Künnerde bir kün bejoγlı[20] aulanıp bu küjge keldi, Hasimγa kunak boldı. Asim aχšam mejzini ve ǯadı[21] χatını čakırta[22], kunak janına otormaγa. Son helvaǯını de čakır-

1 bejuγli
2 alep; R alıp
3 jašadilar
4 ajtkan-dık; R ajtkan edik
5 jırladi
6 bolγandın; R bolγan ediñ
7 surmadiñ
8 julladi
9 Kari
10 surdi
11 sujdi
12 sujdi
13 tišarı
14 čal artına; R čalı artına
15 bakti
16 R tapılγan edi
17 R bu jerden
18 čıχti
19 tutti
20 bejoγli
21 ǯadi
22 čakirta

dılar. Aštan sudan son herkes bašına kelgenni ajttılar[1], bu oγlan kaldı. Hasim: ulum, sen de ajt bašiña kelgenni dip. Ulan: men ajtırım ama üjden čıkkan kirgen bolmaz dij.

Bašladı: bir zamanda Hasım degen bir adam haǯge kite, χızını üjine tašlaj, mejzin buña āšıχ bola, bunı aldamaχ istej degende mejzin: maña ruχset beriñ, men bir čıχıp sijip kelejim dij. Joχ, olmaz, jeriñe otor dijler. Hasim: ajt ulum ajt dij. Ulan: mejzin üč kün hamamın kıra tottı, kızın amamγa alıp kittiler. Kız mejzinniñ bašın küzin sabunlap, nalını čıχarıp bašına urıp tašlap kačtı dij. Son mejziniñ ǯadınıñ vezirnıñ japχan išleriñ hepsın ajtkan son, bularıñ üčün de[2] üč atnıñ kojroγına bajlap jüvereler. Bejoγlı janadan bonı toj-dügün etip jašajler.

7 Faχirnen molla masali

(347) Bir zamanda bir faχir adam bar iken. Her kün sabā turıp, elin betin juvup kismet istegen de bu adam ne jirde čöplük barsa, onda dolanıp jürür‿ken[3]. Bunı br molla küre: bu püsür jirlerge ne kıdırasın[4] dij. Faχir kismet χıdıram dij, ǯanım, bu čöplükler arasına[5] kismet olor mu, kel maña irgat jür dij. Min saña ırgat jürsem, ne iš japarmen dij. Menim dürt ögüzım bar, buların teñizge suvarıp kiterirsın dij. Faχir ajta: eger ögüzlerin teñizge kite p buγulsalar, minden surmazsın mi? Molla: joχ, ular alıšχan iken, teñizge kitmijler dij. Bular bazarlıχ eteler.

Bir kün, biš kün ögüzlerin suvarıp kele. Bir kün ögüzlerin kojroχların kesip teñizge sančıp kuja, ögüzlerin bašχa bir jirde ajdaj. Mollaγa kelip: ögüzler buγıldı, ajda barıp čıχarajıχ dij. Teñiz katına kelip baχalar, kerčekten ögüzlerin kojroχları kürünip tora. Molla: bar, arandan kazma kürek ketir de kazıp čıχarajıχ dij. Faχir: men saña ögüz baχmaγa jürdim, kazma kürek tašımam dij. Molla üzi kite.

Molla kitkenden soñ faχir ügüzleri alıp bir jajla bašına čıχa, baχsa bir čoban kıčırėp[6] jata. Čobanγa selamdan soñ halin suraj. Čoban bašım avura dij. Üjle ise bar bir kısır koj sujıp ketir, men seniñ bašın avurusına čāre tapajım dij. Čoban: menim turmaγa halım joχ, bar sen üziñ istegen kojın suj dij. Faχir bir koj sujıp kursaγın jarıp (348) čobanıñ bašına kečire, sujılγan kojın da čobanıñ eline tottorıp, bu ečki ne vakit kıčırsa, u vakit bilirsin ki kojlarıñ kačkandir, turėp[7], kıdırırsen[8] dij. Üzi ügüzlernen kojları birge ajdep kite. Karšısına bir jılkıǯi čıχtı: dürt atım kačtı[9], kürmediñ mi dep surdı. Kürdim, bu tarafga ketti dep aldatıp jüverdi. Üzi atlarına tapıp ajdap kitti. Bir küjge bardı, bir üj aldı, bir mıχtar tofraχ aldı.

---

1 ajttilar
2 üčun-de
3 jürürken; R jürür eken
4 kıdirasin
5 R arasında
6 kıčırep; R kıčırıp
7 R turıp
8 kidırırsen; R kıdırırsėn
9 kačti

Bir kač vaχitlerden soñ molla ügüzlerın χıdırėp[1] jajlaja čıχtı, čobanı[2] kürdi, bašından kursaγı čıχardı, bunı kim jasadı dep surdı[3]. Čoban: dürt tene čontuχ ügüz ajdap bir adam kildi, menim bašım avura‿jdi[4], maña ilaǯ japıp ketti dij. Üjle ise ügüzler menim, seniñ kojların[5] da u alıp kitkendir. Ajde ketip χıdrajıχ[6] dij. Jılkıǯıγa rast keleler, surıštılar[7], u da jılχıların u adam alıp kitkenın bilgen soñ uči de bu adamın χıdırmaγa kettiler, küjge bardılar[8]. Molla ügüzlerin tanıdı, čoban kojların[9] tanıdı, jılχıǯı atların tanıdı, lākin bular ajttılar:[10] biz böjle üz bašımıza alıp kitmejık, u adamın üjine barıp konuk bolajıχ, soñ rāziliχnen mallarımızın alıp kitermiz dejler.

Surıp munın üjine bardılar, buların konuk etip odajı kirsetti. Mollanıñ ügüzlerin birin sujıp aš azırladı[11]. Ašap ičken[den] soñ bular mollanıñ [ügüzlerin] suradılar[12], mollaγa: ügüzlerin semirmege bašladı[13], čobanγa da: kojların kozlama[14] bašladı, jılχıǯini de bijeleri kolunlamaγa bašladılar dep ǯavab berdi. (349) Jataǯaχ vaχit keldi. Bu adam ič tutulmaγan, tüšek, jurγan ketirep[15] jajdı, baχınız, eger bu tüšeklerge siseñiz, mallarınızni[16] [ne] alėr[17] [ne de] kitersıñiz dej. Bular jatkanlar. [Faχir] bir kač sāttan soñ čobannıñ astına biraz su ketirip χojdu. Čoban ujanıp baχtı, tüšek ıslanγan. Jılkıǯını ujatıp: men sıjgenım dij, jılkıǯı da karmalap baχtı[18], u da sıčχan. Mollajı ujattılar[19], baχsa unın da küteni čıχχan. Šindi bular mušavere ettiler, malγa bola [da] asılıp ülmeden ajdiñız, kačajıχ dejler de molla küteni koltuγına alıp kačıp kitteler. Mallar hepsi faχirge kalγan.

8 Muradina İrmegen Dilber

(353) Zamanın birinde bir faχir jiget bar idi, bu jiget bir nalband katına irgat idi. Bir kač vaχitlerden soñ bu nalband jigetni üjlendirdi. Jul janında bir eski üjde jašadılar. Altı aj, bir jıldan soñ bunıñ χatını bala tuγuraǯaχ zamanı keldi. Xatın hoǯasına: bar nalbanddan bir uda iste, bu jul kenarında utırmaγa utanam dij. Jiget barıp istedi, nalband kajtarmajıp bir uda birdi. Šindi bala tunjaga kelgen. Üč kün boldı, balanıñ adını kujmaχ kerek boldı. Bu jigetge bir para joχ ki br mıχtar et maj alsın da adamlar čakırsın. Bir kasabγa barıp bir rublelik et alγan borčka. Aš pıširdiler, mollajı čakırdılar[20], mollanıñ[21]

1 χıdırep; R χıdırıp
2 čobani
3 surdi
4 avurajdi, R avura edi
5 kojlarin
6 χidrajıχ
7 surıštilar
8 bardilar; All (about 15) *bardi* ~ *bardilar* data have been changed in this way.
9 kojlarin
10 ajttilar
11 azırladi
12 suradilar
13 bašladi
14 R kozulamaγa
15 R ketirėp ~ ketirip
16 mallarınizni
17 aler; R alır
18 baχti
19 ujattilar
20 čakırdilar
21 mollaniñ

artından jedi sufu kelgen. Utėrdılar[1], ašadılar[2]. Sufularıñ biri: bunda ne ǯemijet bar dep suradı[3]. Br jañı dúnjamız bar, unı adını kujaǯaχmız diler. Ůjle bolsa, balanıñ adın men χojajım dij. Bek güzel, χoj dediler. Sufu: ulan mi, kız mı dep surdı. Kız degenner. Munıñ adı Muradına irmegen dilber bolsın dij. Ekinǯi sufu: bu Ůlmez-χan bolsın dij. Ůčünǯı sufu: bu kız aγlaγanda küzlerinden inǯiler aχsın dij. Dúrdúnǯi sufu: bunı bašı juvulγanda[4] lejen tulı[5] altın aχsın dij. Bišinǯi sufu: külgende eki janaγında eki gül ačıla dij. Altınǯı: bu kıznıñ sau kulına menden bir dua dij. Jedinǯi sufu: bu kız jürgende ajakları tübine ut bitsin dij. Dua ettiler, χajttılar[6].

Kız üsti. (354) Bunı br padiša ulı išitken, anasına: bar, falan küjde bır kız ajtalar, eger sāhi bolsa, maña nišan ber, men barıp alırmen dij. Anası bardı baχtı, kerček, nišan berdi χajttı, kırχ künden soñ toj bašlandı, paraχodlar kelin almaγa kelgen, kırk keče toj boldı, kelını paraχodγa kujdılar, kelinnen barabar kızın tezesın, tezesinıñ kızları paraχodγa minep kettiler. Biraz ketken soñ kelin su istedi, tezesi: bir küzin čıkarıp berirseñ, saña su berirmen dij. Kız küzin brin čıχarıp berdi. Brazdan gene susadı, gene br küzın čıχarıp verip bir su daha ičken. Bir daγga jakın keldiler, tezesi: hajdınız br tüšip daγda ejlenejik dij, tüštiler. Tezesi keliniñ urbaların üz kızına kidirep, kelini tauγa tašlap, paraχodγa minep kitkenler, padišanıñ sarajına bardılar.

Merdimenden čıkkanda padišanı[7] χatını[8] baška kız boldoγın bilse de seslemedi. Kijeu ketirdiler, kijeu kirdi, baχtı kıznıñ janaγında gül joχ: hana, seniñ janaγında gül bolaǯak, jok ta dij. Bular kač vakitler omur ettiler.

Kelelim muratina irmegen, tauda jatkan dilberge. Br kün br kart faχir kiši tauγa otunγa bardı, baχsa br kız, küzlerinden akkan jaš inǯi bolγan jata. Bu kızı kart alıp üjine keldi, kartnıñ χatını: biz üzimizi baχmaγa alımız joχ, bu kür kızın da nasil baχarmız[9] dij. Kız: ana, sen korkma, tez menim bašım juvıñız dij. Judular, ortalıχ altın tuldı. Bir kün bu kız kartγa: babam, sen menim tapχan jeriñe br türbe japtır, eki kapısı[10] bolsın, hem de (355) müratina irmegen dilber dep süjlensin dij. Kart japtıra. Kart baba, men külejim, menim janaγımda eki gül bolor, bu güllerı falan küjge alıp kit, kim de ike küz berse, bu gülleri oña berirsin dij. Kart gülleri alıp küjge bardı. Kızıñ tezesi eki küzi berip gülleri alγan, üzi kızına julladı. Padišaniñ χatını eki güli eki janaγına χujdı. Jiget kelip gülleri kürep sevindi, hem χoχladı[11]. U sātta muradine irmegen dilber balajı kaldı.

Kelelim kızıñ tezesine. Tezesi kızınıñ[12] kulından duasin almaγan unutχanın bilep bir ǯadı[13] χatın jüverıp kızın[14] elinden duasın aldırdı. Kız šu‿ssāt ülgen. Kızı šu türbege alıp kitep kümdiler. Kabr ičinde kızdan br er bala toγa. Kızın barmaγın emıp jata. Bir

1 utėrdilar
2 ašadilar
3 suradi
4 juvalγanda
5 tuli
6 χajttilar
7 padišani
8 χatinı
9 baχarmiz
10 kapısi
11 χoχladi
12 R kıznıñ
13 ǯadi
14 kizın

kůn padiša ulı šu tauɣa auɣa čıχa, birden bir ses ešite: bu men išitkenım dilber bolmasın dep, seske dorı barıp baχsalar bir kabir, ačıp bakalar bir ȯlů, janına bir bala, balajı alıp kete.

Bala biraz bůjůgen son naslsa[1] anasının duası eline keče, χatın bermege istemij. Bala aɣlaj, jiket[2] kelip suraj. Katın čāre tapamadıɣından duājı ketirep[3] hoǯasına bere. Hoǯası duajı alıp kabra bara, ačıp ůlůniñ kulına baɣlaj, u‿ssāt kız tirile, alėp ůjine kele, soñ išin temelin bilgen son kızın tezesin čakıra. Ůz kolınen[4] ekisin de eki at kojroɣına baɣlap parčalatıp, jañıdan toj-důgůn etip jašajler.

9 Biš altın masali

(360) Zaman zamanın euelınde Jenǯ Osman namında alɣannı bermez br adam bar idi. Br kůnden br kůn bu adama biš altın kerek boldı, kuñšusından barıp istedi. Kuñšısı[5] munın χararını bildi, alɣannı bermegenni. Juktır[6] dej. Ů‿jerden[7] ačulanıp čıktı. Julda buña haǯi Ameša aɣa rast keldi, halin aña ajttı[8]: maña biš altın lazim dep. Haǯi: bik gůzel, ben saña biš altın berirmen, lākin degen kůnın alırmen. Ma, saña biš altın, jarın ůjlede alırmen[9] dij.

Kıskası, ekinǯi kůni ůle keldi, haǯi Ameša baktı[10], altınlar kelmegen. Tajaɣın kulına alıp buña keldı, bu da unı keleǯegini tujup[11], tůšek hastası bolėp[12] jatkan. Bunı apaji ůleǯek dep kurkıp akrabalarına χabar birdi, hepsi bunun bašına ǯıjıldılar. Šu arada Ameša salam birep ičerige kirdı, baktı aɣır kasta. Ne japsın, ůler de maña burǯı kalɣan dep javaštan janına barıp kulaɣına: benim biš altınım nas[13] bolaǯak dep surdıkta, χasta: sen bu halde para istejsin dep bašladılar[14] haǯıni vurmaɣa. Nejse, kullarından aǯi kutulėp[15] kačtı, kasta da haǯiden kurktıɣından ne vakit bolsa da ůlmegenǯi bu haǯiden kutulmam dep enkazdan ůldi. U kůjde adet ůlůni kůmmej ekenler, mezerleri joktır eken. Bir maɣara japtırɣannar, o maɣara ičerisine ůlůlerni [ilaǯlap][16] kuja ekenler. Bunı[17] de ilaǯlap u maɣaraɣa ketirep kujɣannar. Haǯi bunı ůlgenni (361) ešittiginde, u da enkazdan ůldi, barım biš altın jerine kefinni alırmen dep. Bunı de ilaǯlap maɣaraɣa ketirep bunın janına kujdılar. Bir birinden χabarları jok ide enkas ůldůklerinden.

---

1 R nasılsa
2 R jikėt; Since the *jiket* version is used 18 times, it has been left unchanged.
3 R ketirėp, but the more common version is ketirep.
4 kolinnen
5 kuñšisı
6 juk-tir
7 ů jerden; R u jerden
8 ajtti
9 alirmen
10 bakti
11 tůjůp
12 bolep; R bolıp
13 R nasıl
14 bašladilar
15 kutulep; R kutulıp
16 Crossed out in Ms.
17 bunın-de

Keǯe boldı, ekisi de indemejip jata‿jdı[1], kıbırdanmaya kurkıp. Bir daha baktılar, šatır-gudur mayaranıñ kapısı ačılyan, baktılar küz kıjıyınnen, beš tane ajdamak, kullarında bir turba altın, birde kılıǯ, utırdılar. U küjde ǯın padišahiniñ üjini baskannar. Padišanıñ janında da diuler ǯinsinden Genč Amed nāmında bir pejliven bar eken. U pejlivenni ütürep kılıǯnı alyannar. Šunıñ uturdıyı udasından br turba altın tapıp alyannar. Kıskası,[2] altınnan tenha jerde paj etijik mayaraya kelgenner. Aralarından biri: en bašta kılıǯnı[3] kim alaǯak degende, biri: kim karaman, u alır dej. Jā neden bileǯeksiniz karaman oldıyın, bu jerde köresmek olmaz. Buña küreš lāzim tügül, muna bu kenardaki ülüni turyuzıp ajak üstüne tajandırıp kujarmız, kim bir kerre kılıǯnan[4] uryanda [. . .], kılıǯnı[5] u alır dij. Muna karamannık böjle bolor. Anǯak kenardaki ülü de enkastan ülgen haǯinıñ ülisı iken. Haǯı da kurkusından[6] [...] alıp bermege bašladı. Bunlar[ın] eki danesı haǯının ülüsinı tutıp tilediler, endi arajık dep sülegenge haǯi kurkusından bir ses etti: ja, ülüler, daha bakıp ne uturasız[7] turmajıp degende (362) bašta engasten ülgen ülü de muna: ha, tutarımız, dep kıčkırıp köt üstüne tordı[8]. Bunlar da bellediler, sankım hep ülüler tırılup[9] turdı, aman hič bir šije bakmajıp br birini taptap kačtılar[10]. Altın da, kılıǯ da mayarada kaldı.

Artlarından bunnar ekisi de külüše-külüše torıp, bašta mayaranı kapısını japtılar. Sonra bašladılar[11] altınları paj etmege. Br koš avıǯ[12] birisine, [br] koš[13] avıǯ birisine. Üjle paj etip bitirdiler. Haǯi: kana, seni biš altın borǯım nas bolaǯak, ularnı ber dedı. Bu da ačulanıp: ajvan erif, hiš te namusıñ jok iken, bu kadar altın sābi bolıp otorasın, jine de közıñ tojmaj. Kıskası bunlar ekisi isporlašmada olsın, biz kelijik hajdamaklarya.

Bunlar da birbirini tazerlemege bašladılar: ülülernen šaka olur mu dep. Bereket bersin artımızdan kumadılar, kuyan olsalar da išimiz bar ide dediler. Biri daha: ben hič bir šije janmajım, tek kılıǯ urada kaldı, ben uña zijade ükünem, nas edejik de u kılıǯnı alajık, degende biri jüreklenıp: ben ketijrim dep kajtıp javaštan mayaraya jakın keldi. Baksa ičerde šamata balaban. Bunlar da hep kauya etip utura‿dilar[14]. Bu javaštan kapıdan kirmege kurkıp, kabırयada[15] penǯeresi bar iken. Barajım javaš, penǯereden bakajım dep bašını uzatıp bakkanda, bunı fesnıñ püskülni uǯında br balaban altın bar iken. İčerden alyannı bermegen kürep aman čapkılıp (363) üzüp aldı da hemen aǯıya: muna saña biš altın dep attı. Bu jürekli ajdamak bek kurkıp kačkan, nefesı totolıp arkadašlarına keldikte: u sizlernı kürgeniñiz bir iš mi, šimdi hep ülüler tırılgendir[16], bašlayannar birer altın paj etmege. Bir denesine bir altın jetmegen onı töpelep ütirdiler,

1 jatajdı; R jata edi
2 kiskası
3 kiliǯni
4 kiliǯnan
5 kilıǯnı
6 kurkusindan
7 ütürasız
8 tordi
9 tirılup
10 kačtilar
11 bašladilar
12 koč-aviǯ
13 koč
14 uturadilar; R utura ediler
15 kabiryada
16 tırılyan dir

barajım penǯereden uzanajım de karajım dedim. U vakit kılıǯnen bašımı[1] keseǯek olıp, br kılıǯ čektiler, u da maña timejip fesimi púskúlini uǯındaki altınγa tijdi, unı kesıp attı da ben de kačtım. İnanmazsañız bakıñız, púskúlni uǯında altın jok didi. Bular bójle gajb olıp kittiler.

10 İbrāhim padiša

(368) İbrāhim padišanıñ Abdurrešıd degen br pašası bar eken. Pašanıñ da Abdussadiχ digen br ulu bar eken. Padiša kanda kitse Abdussadiχka br baχšiš ketire iken. Br taha kitkende muña br altın sandıχ ketirgen. İčinde br padiša χızınıñ sureti bardı. Paša bunı ulına kústermedi, haznege kujdı. Jiget túšúndi, padiša her vaχit ketirgen bakšišların maña berirdi, bu sefer ničin bermedi dij.

Kúnlerden br kún jiget anaχtarnı čalıp haznedeki sandıγın ačıp, sureti alγan, šu‿ssāt aška túšep jıχıldı. Kač sātten aχlın toplap turıp újlerine kelgen, ol ola kimsege tujırmadı. Čoχ vaχitler túšúnep júrgen. Pašanıñ hizmetčisi jigetten: nečin sen bujle saraldıñ, ne kajγuñ bar? Aǯab bu suretniñ sahibi saγ mi‿ken[2] dij. Eh, u kolaj, sen babaña ajt, gezmege dep kitermiz. Ulan babasına br paraχod japmasını reǯa etti, paraχod japıldı. Bular keregi kadar azık hazirlaγannan soñ kitmege izin istej. Babası: kimnen kiteǯegini surdı[3]. Jiget: hizmetčnı ajtkan soñ rāzi boldı. Hizmetči jigetke: bar padišadan suretniñ sāhibi kanda bolγanın sura dij. Jiget padšanıñ utasına kirep, kúrdi padša balık ašaγan, balıknıñ kemigi buγazına kalγan, úlúm halına jata. Jiget uña kirmejıp kajttı. Bular paraχodγa utırıp kiteler.

Kitejler, kitejler, az mı čoχ mu altı[4] aj, br kúz mi kiteler. Jigetniñ babası ulı bu suretni (369) sāhibin χıdırmaγa kitkenın tujdı, ama kajtarmaγa čāre juχ, allaγa amanet bolsonlar dij. Bular kite-kite, suları bite, jalıda br kúčúk újden su istemege jiget bara. Baksa br χart adam utıra. Kart jigetten ne išnen bu jirlerge kelgenin surdı. Jiget sureti čıχarıp bunı χıdırmaγa kitem dij. Kart: bu Memmet padišanın kızı[5] bolγan, ama bilmijmen, sen barıp taparsın mı dij. Jiget su alıp jullarına kitkenner. Kite-kite jene br jalıγa[6] čıχχannar, paraχodnı bejlediler, baχtılar br jiget balıχ tutıp tura. Bu jiget balıχčıγa: sen bújle gúzel jiget, bu balıχčılıχnen ne kazanasın, kimge bularnı satasın dij. Balıkčı: ej dustım, men bu balıkların Memmet padšanıñ kızına[7] satam dij. Jiget sureti čıχarıp kústerdi. Balıkčı bu kız boldoγını ajttı. Jiget balıχčıden br uda istegen, balıχčı bularnı úz újlerine alıp kitte.

Bu χız aftada eki kún gezmege čıχar iken. Bu χız jene de bahčaγa čıχtı. Bu jiget bahčada utırγanda bunı ǯebine úč alma kelip kirdi. Jiget gezindikten son kajtıp keldı, balıkčınıñ anasına bunı sújledi. Kart χatun dij ki: ā balam, bu kız kimi sújse, uña úč

1 bašimi
2 miken; R mi iken
3 surdi
4 alti
5 kızi
6 jaliγa
7 kızina

alma bereǯek dep añlašıldı[1] seni sújdi, lākın anası babası unı saña bermezler. Men saña br tılsım japajmen, ular ašamazdan evel gezmege čıχajlar, sen barıp munı kazanγa tašlajsın. Ular ašaγan son bajılıp (370) jatırlar. Sen u arada kirep kızın alıp kačarsın dij.

Jiget bújle japıp, alıp kačkan, paraχodka minep[2] kittiler. Kıznıñ anası babası ujanıp baχtılar kız joχ, ne bolγanın bilmediler. Jiget kıznen újlerine kelip tuj-dúgún etip aldı, muradina erdiler.

11 Karı koǯa masali

(374) [Zaman zaman ikišti, deve dellal ikišti, kaplı-kuplu bakalar kanatlanγan učmaγa, tengizdeki balıklar kıra tutγan kóčmege, Ak-meǯidniñ minaresi ejgildi Salγurdan su ičmege, kamanderin taraši [tiregen] balakların[3] Kara-tengiz kečmege...][4] U jalan, bu jalan, filin juttı br jılan, taušanıñ[5] sırtına mindim, devenı kučaγıma aldım, o da jalan mı dedi, ikajedir bunı adı, tiñlemeniñ kelij dadı, bunı tiñlemegen újle bolsın, bújle bolsın.

Zamanıñ evel vaktında br bajnıñ br ulı bar ide, lākin akıldan br parča kire eken. Bu ulan kırk jašına kadar omur kečirdı. Babası ajttı: bu ulannı bāri újlendirejim. U jak, bu jak bir kz bulγanlar, úzi kibik. Šu‿ssāt dúgún tedariki kúrep ulanın újlendirdiler. Lākin ne kız br šej piširmege bile, ne ulan. Br kún utka kazan astılar, suvnı kajnattılar, ičine un attılar. Ne kadar tuz kujılaǯak bilmijler. Kelin ajttı: bar anama sur da kel, ne kadar tuz kujajık kazanγa. Ulan kitte, anasına barıp: ana, bz aš piširemiz ama nekadar [tuz] kujılaǯak bilmejmiz. Vaj ulum, bújúk kazanγa koš avıǯ, kúčúk kazanγa sıñar avıǯ.

Ulan unutırım dep julda sıñar avıǯ koš avıǯ dep ǯele-ǯele[6] kelgen. Julın kenarinde erifin birisi arman basar iken, keldi unıñ janına sıñar avıǯ koš avıǯ dep. (375) Armanǯı: aj erif, armanγa sıñar avıǯ koš avıǯ derler mi dep buña br kač jumruk urdı. Jā ne ajtajım dij. Bereketli bolsın dep ajt dij. Ulan gine bašladı bereketli bolsın dep gitmege. Julın kenarında br mezerlik bar iken, mezerlikte br úlúni ǯenazesin kılalar. Bu da kutlı bolsın dedi. Ulanı dúvdiler. Allah rāmet ilesın dediler. Gene ulan bašladı Allah rāmet ejlesin dep ǯele. Julda kiterken julın kenarında br tujγa rast kelgen. Allah rāmet ejlesın dij. Aj erif, tujγa Allah rāmet ejlesin dejler mi dep bir kač jumruk taha urdılar. Jā ne ajtajım dej. Úziñ čal, úzıñ ujna de dijler. Gene ulan bašladı úziñ čal, úzıñ ujna dep kite. Br bakčanıñ kenarında[7] auǯınıñ brisı tüfegını alγan taušanı atajım dep tura. Bu da úzıñ čal, úziñ ujna dep jul bojı kelıj. Auǯının auėni[8] kačırdı. Auǯı bunı tutıp br kač jumruk urdı. Jā men ne ajtajım dij. Pusıp-pusıp kit dej. Gene ulan bašladı pusıp-pusıp kite. Julnıñ kenarında br bustan bar iken. Bustanǯi da bunı kúrdi, aman aldına čıkıp bunı tuttı, br kútek

1 anlašıldi
2 R minėp
3 Balaklanıñ
4 A part of *tekerleme* being crossed out in the text of the tale.
5 taušaniñ
6 R aǯele-aǯele
7 kenarinda
8 aueni; R auını

urdı. O belledi bustan kırslamıš, bakti bunda br šij jok. Andan de kutuldı, üjine keldi, tap kečeniñ jarısı boldı, bašladı kapını čalmaɣa, karısı da kapını ačmaj. Kitte ǯamige jattı. Čok suvuk da (376) iken.

Sabaha kadar ǯāmi ičinde kalɣan. Sabah mejzin kelip, ǯamınıñ kapısın ačtı[1], baksa br kiši jatır. Mejzin ajttı: tur ǯıjıl kit, üčkündür kibik ne jatasın dej. Meger ulannıñ karısınıñ adı Üčkündür iken. Keǯe kelip kapını čalɣan zemanda adın tapmaɣan, unınčın karısı kapını ačmaɣan. Mejzin buña üčkündür degennin bašladı taptım taptım dep üjine čapa. Mejzin de bunın bir šij tapmıš sanıp u da unın artından čapa, ne taptıñ dep.

Ulan üjge kelip jetti, Üčkündür, kapın ač dej. Karı turıp kapını ačkan. Ulan baktı kazan ele de utta tura. Ulan uklavın alıp karıštırajım dep bašladı kazanın karıštırmaɣa. U arada artından mejzin de jetip kelgen. Ulanın kulında uklavı kürep ajlanıp artına kačtı, ulan da kazannıñ bulamıɣın alıp kačkan bellep artına tüšti. Mejzin kača, ulan kuva, tokta mejzin, br batırajım dij. Mejzin de meni uraǯak dep kača. Mejzin üjine barıp kirde[2], ulan da unın pešinden mejzinın üjine kirde. Šimdi mejziniñ karısı bu ulannı üjine mejzin br fenālik ičin barɣan. Üjle bolɣanda mejzin de bek jaχšı adam tugıl iken dep mejzin bašladı karısınen kavga etmege. U jak, bu jak derken kauga balaban boldı, mejzinın karısi ajrıldi, (377) mejzın tul kaldı, ulan kajtıp üjine keldı.

Eki kün, bir keǯede br kazan aš piširep ašap bolamadılar. Akšam boldı, bularnı kelin küjeu musafirlikke čakırdılar. Kıznıñ anası dij: kızım, barɣan jirimizde maskara etme, aš ašaɣanda toktaǯak vakitte, men saña išmar etermen, sen toktarsın, musafirlikte az aš ašarlar.

Bunlar bardılar utırdılar, hošbešten son kajve tütün ičkenden son kefleri tamam buldı, sufra ketirdiler, tuplašıp sofra[ɣa] utırdılar[3], keliniñ janından bir mıšık kelip türtti. Šimdi artına čekildi [...][4].

12 Balıχčı jiget masali

(382) Br zamanda br balıkčı bar iken. Balık tutıp kečinij iken. Br kün bunıñ aumna br balık tüše, jarısı kız jarısı balıχ. Bu jiget anasına ketirep berdi. Anası bunı üjıñ dulabına kujdı. Munnan son her kün vaχtı vaχtıle bularɣa aš hazır bolıp kele, kim japχanın bilmejler. Bu jiget bunı deñip baka küremej, soñ dij: ǯin bolsañ da čıχ, insan bolsañ da čıχ. Bir kız čika[5]. Kızı kürgen kibi jiget jıχıla. Kač sāttan turıp kızɣa sarılɣan, kız dij: maña nikāsız sarılma, bar nikā kıjdır, soñ sen meni, men senı dij.

Bular nikā kıjdırdılar, uč biš kün omur ettiler, aš su bol, lākin üjleri eski iken. Jiget dij ki: šukür, dunjada her šejim bol, bir üj kerek. Hatın: bar deñizge Fatma ana dep kıčkır, u

1 ačti

2 R kirdė; *kirde* shapes alternate with *kirdi* ~ *kirdė* shapes.

3 utırdilar

4 Missing ending.

5 R čıka

čıχar, sen undan br fındık kabuyı iste, u saña kel berejım dej, sakın barma, barsañ gajb bolorsıñ, at, alajım derseñ, u atar, alıp kelirsin dij. Jiget barıp, findık kabuyını ketirgen, üjlerı üstüne attılar. Erte turıp baksalar, padšanıñ sarajından jaχšı br saraj.

Kač künden soñ jiget χatınına: sen bu balık kabıyını kijme dij. Katın: joχ, men bu urbajı tašlasam, erkes meni kürer, saña düšman bolor, rātlık kürmejsin dij. Jiget ačulanıp urbajı ateše jakaj. Br kün bu jiget padišanı kunakka čakırdı. Padiša vezirlerinen kelgenner, ajttılar. Vezir padišaya: bu katın (383) anǯak padišaya lājık. Bu jigetke bir ayır iš küsterijik, japamassa[1] bašını kesıp hatınını alırmız dij. Padša bunı čakırıp üč künde dunjade ič bolmayan br saraj japarsıñ, japmasañ bašıñ kesilij dij. Jiget tüšüne χatınya bildirij. Katın: men saña ajttım, menım urbamı jaχma, diñlemediñ. Nejse, kasavet etme, kel sarılıp jatajıχ, soñ baχarmız dij.

Erte turdılar, katın koǯasına: bar Fatma anadan br čeviz kabuyını ketir dij. Jiget barıp ketirdi, keče bunı padišanıñ azbarına attı, sabā baχsalar, dünjada bolmayan br saraj. Jiget kelip begendin mi dep surdı. Padiša: begendim dij, lākin vezir turıp: jaχšı ama bu sarajya ǯın zili kerek dij, kırk kün saña mulet, ketirmeseñ bašıñ kesermiz dij. Jiget kurka-kurka katınına kelip ajttı. Katın: kurkma, kel sarılıp jatajıχ dij. Erte turdılar, katın julluχ azirladı: kitersıñ, tauya barmazdan tikenlikke barırsın, jalan-ajak kurkmajıp čapıp kečersıñ, br česmege barırsıñ, andan bismilla dep br su ičersıñ. Česme artına kapı bardır, ačarsıñ kojun bardır, ügünden eti alırsıñ, br kapı taha ačıp küpege barırsıñ, üginden pičeni alıp eti berirsiñ, pičeni kojya ketirirsėñ[2]. Br kapı taha ačarsıñ, onda ǯin bardır, kulajnen jastıyı tübinden zili alıp kačarsıñ dij.

Jiget kitte, bardı, zili alyan. Ǯin küpege tut dedi, küpek tutmayan: (384) kırk jıldır munda jatkanım, koj üginden eti alıp bermediñ dij. Koj, tut dep bayırdı[3], u da kırχ jıldır munda jatkanım, küpek üginden pičeni alıp bermediñ dij. Česme, sen tut dep bayırdı, kırk jıldır japtıryanıñ kelip br kerre su ičmedin dij. Bularnıñ biri de tutmayan, kečip kitken, ketirep padišaya bergen.

Padiša begendi, lākin vezir indi ǯin padišasınıñ kızı kerek dij. Üč kün mület berip julladılar. Jiget katınga kelip ajttı. Hatın: kolaj, kel sarılıp jatajıχ dij. Sabā bunı julladı, nasıl barıp ketireǯegini ügretti. Jiget kitti, tauya bardı, br üjge kirde, baχsa bir katun, eki memesin uñuzdan artka atkan, un elip utıra. Hemen čapıp emmeje bašlayan, son salam birep utėrdı. Hatun: sen menim süt uylım boldıñ ama šindi ayañlar kelip, brden kürseler, seni ašarlar dij de ulanı br pilǯan japıp rafya kujdı. Dejuler kelgen utıryanlar. Bri: mında br insan kokıj dej. Anaları: mında insan bolır mi dej.

Brazdan soñ bri taha büjle surdı, bri taha, zurladılar. Anası bu jiget emčegın emip süt ulı bolyanın ajttı. Bular: küster, bz uña br zarar etmemiz dijler. Filǯanı alıp jire kujdı, jıget salam berip utırdı. Bular ne hizmetnen kelgenin surdılar, kanda kitkenın ajttı. Bular epsi bašlarından brer tel berdiler: bašın zurya kalsa, kızdırırsın,[4] (385) biz saña jardımya barırmız dijler.

1 R japamazsa
2 ketirirseñ
3 bayırdi
4 kızdırirsin

Jiget kitte, br čobanγa ras kelgen, čoban kaval čala iken. Salam, Ahmad čoban, digen. Ekisi arkadaš bolıp kitteler. [...uba tüzlegen Ahmedga rast kelip, unı da aldılar.][1] Andan kitkenner, suγa tujmaγan Ahmedge ras kilep unı de alγannar. Biraz taha kitep br uzun ajaklı[2] Ahmadγa ras kilep dürdü de bunı arkasına minep eki atlamada ǯin padišasınıñ üjine bardılar. Kıza χudalıχ ičin kelgenlerin ajttılar. Ǯin padišası: Jaχši, size üč iš küsterirmen, japsañız berirmen degen. Kızım čapıp uzarsınız, bu kujunıñ suvın ičep bitirirsiniz, kırk čuval baklanen arpanı br keǯede ajırırsınız, son berirmen dij.

Uzun ajaklı Ahmad kızı čapıp uzdı, ekinǯi Ahmad kujunı suvın ičep bitirdi, uba tüzlegen Ahmad da karınǯa duāsı ukıp karınǯaları čakırdı, sabaha kadar karınǯalar arpaları baklalardan ajırılγannar. Padiša aš piširtep ičine zehir kujγan, buların ütüreǯek iken. Kaval čalγan Āhmad bunı dujdı, sofra kelgen soñ čoban: aštan evel kaval čalmak adetimizdir dep br kıčkırttı, padiša bajıldı. Čoban padišanıñ ügınden aši alıp zeherli ašı kujdı. Padša turıp ašadı üldi.

Kızı alıp uzun-ajak Ahmadnıñ sırtına minep kačtılar[3], küje jakın kelgende dürdi de kajttı. Br jiget kıznı padšaγa ketirep berdi, padša begendı (386) ama vezir dij: inde kırk tajların anasın ketirsin dij. Üč kün mulet berip jigeti julladılar. Üjine kelgen hatıninen[4] jattılar, sabā hatın dij: muna bu munǯaklarnı al, jajlaγa čıkkanda bir čokrakka barırsın, bu munǯaχlarnı su bašına kujarsın, kırk tajlar anası kelir, su ičkende munǯaχlarga šašıp kalır, sen čoχraχ artından brden čıkıp sırtına minersın, u seni ataǯak bolır, lākin sen šačlarından bek tut jiberme. Eger üč kerre siliktikte atamazsa, saña teslim bolor, miner kelijsin dij.

Jiget kitep ketirdi, padšaγa berdi. Vezir erte kelip: bunı saγarsın. Jiget hatınına kelip ajttı, hatınnen jattılar, sabā turdılar, koǯasına: bar saγ, son saña bu süti kajnat, ičinden keč dijler. Al bu jüzügni, süt kajnaγan soñ kurkmajıp kečejsin dij. Bardı saγdı, kajnatkan; ičinden keč digenler, kečken. Padišaγa: men senin istegen šejlerin japtım, sen de meni kajtarmajıp sütten keč dij. Padiša aman kečeǯek bolıp sütte pišken. Vezirge sen de keč degende, vezir jüzügnı istedi. Bu ačulanıp veziri zurnen sütke atıp, kendi kajtıp hatınnen rahat jašadılar[5].

13 Ahmed-batır masali

(393) Bir zamanda bir padiša[6] bar iken. Padišanıñ ne ulu, ne de kızı bar. Biš namazın ič tašlamajıp kılajken, namazdan soñ Allahdan üzine bala istej eken.

Künlerden bir kün buña bir derviš kele, bir alma bere: bu almanıñ jarısın[7] üzıñ aša, jarısın χatınıña ber dij. Bular almanı ašaγan soñ χatın balaγa kala. Toχoz aj un kün

1 Crossed out in the Ms.

2 ajakli

3 kačtilar

4 hatininen;R hatıninen ~ hatınınen

5 jašadilar

6 Crossed out of the title: Bir padiša masali [Tale of a padisha].

7 jarisın; This variant occurs five times in the taxts of the tales.

degende kırk küne kadar her kün günde bir bala tuyura. Xatın ajta: hoǯama kırk bala berdiñ, üzime ajrı br er bala ber, ama hajırlısından bolson dij.

Künlerden bir kün jene šu derviš kele, padišanıñ χatınına bir alma bere, bu almanın jarısın üze ašaj, jarısın hoǯasına bere. Derviš: bala tuγγan son, men üzėm[1] kelip adını kojarmen dij. Bala tuγa. [...][2] Bunı tutėp[3] bir kafes ičine kujdılar, adı kujılaǯakta derviš keldi, adını Ahmed-batır kujdı kitti. Bašta tuγan kırk ulan büjüp jiget čaγına keldiler, analarına digenner: ajt babamızγa bizı üjlendirsin. Anaları: men birden babañızγa ajtamam, jarın üjlede ašamaγa kelgende siz ašamazsız, kašıklarıñıznı tašlap turıp kitersiz. Babañız ničin ašamajler dep suryanda, men ajtırmen dij.

Anası padišaγa ajta. Padiša: men ularnıñ kırkına kırk padiša kızın kanda tapıp alajım, jiget bolsalar, barsın üzleri (394) tapsınlar dij. Ulanlar babasından izin alıp kız kıdırmaγa kitkenner. Biz kelijik Ahmed-batırγa.

Šu künė[4] Ahmedge üle ašını keč berdiler, buña ačulanıp kafesi parladı, aš ketirgen hizmetčige surdı: ničin keč ketirdiñ? Hizmetči: seniñ kırχ aγan bar, ular üzlerine kız kıdırmaγa kitteler, uların uzγardık, unın‿čin[5] kečke kaldım dij. Hizmetčige: sen meni anam babam janına alıp ket dide. Hizmetči alıp bardı. Üjlerine barmazdan bularγa derviš jetišti. Bu jiget üjlerine kirdi, babasın[6] anasın[7] kulın üpti, aγaları artından ketmege ruχset istedi. Derviš: joχ ulum, sen üč künden un sekiz jašına keleǯeksin, u vaχit kitersin dej. Ulan: joχ, men šindi kiteǯegim dedi. Jaχšı, üjle bolsa, bu künden munda kunarmız, jarın küjnıñ kiblesine čıχarsın, onda bir oba bar, obaγa salam berirsiñ, selamın alıp kapı ačılıj, kirersiñ ičeri, anda senıñ egeriñnen silālarıın bar, ularnı maña alıp ketir[ir]sin. Erte sabā küjıñ sırtına čıχarsın, andagi obaγa salam berirsin, kapı ačılıj, senın düldüliñ bar, alıp ketir[ir]sin, son üčünǯi kün kitersin dij.

Jiget bularnı ketirgen, üčünǯi kün babasından izin alıp, minep kitti. Aγalarınıñ kırk künde (395) kitken julı bir künde kitep artlarına jetti, ularnıñ kardašları bolγanın bildirdi, aγalarinen brge kittiler, aχšamγa bir čimenlik jerge bardılar. Ahmad-batėr[8] aγalarına br keče mında kunajıχ dij. Šu jerde atların jüverıp jattılar. Kečeniñ bir vaχtında Ahmed abdıradı, bašın küterep baχtı, uzakta bir jarık kürindi. Üz üzıne: bu jarıγı barıp bak, istej. Atına minep kitti, jarıkka bardı, baksa bir saraj, kirmege ne kapısı bar, ne penǯeresı. Düldülge bir kamčı urėp[9] bir teptirdi. İčerden brı: Ahmed-batır sen misiñ dij. Evet, menim, ač kapıñı dij. Kapı ačılγan, baksa br kart karı, u karı Ahmed-batırγa kırk tene anahtar berdi, kırk uda küsterdi. Ahmed kırkın da ačıp baktı, br oda kaldı. Keldi karttan u udanıñ anahtarın istedi, unı de ačıp baktı, unda jangız inǯiler tulduryanlar. Jene kartka kelip: anahtarları alıñız, čok beklediniz, az taha bekleñiz, kajtkanda kelip alırmen dep, aγalarına kajtıp keldi. Aγalarına ajtmaγan.

1 üzem; R üzim

2 Crossed out in the Ms.: Üzini elge birmejip kačıp kitken.

3 tutep; R tutıp

4 küne

5 unınčin; R unın ičin

6 R babasının

7 R anasının

8 Ahmad-bater; R Ahmad-batır

9 urep; R urıp

Erte turıp jullarına kitteler. Akšam jene br čimenlikke kirep jattılar. Keče jene batėr seskendi, bašın kúterep baktı, uzakda jene br jarıχ kúrindi, uña da bardı, u da újle br saraj, ičine kirdi jene kart kadın buña: (396) ma, ulum, emanetin dep br anahtar berdi. Bir oda kústerdi, batır kirde, baksa kırk kız ukėp[1] otoralar, kırk birinǯi o[r]tada sejr etip tura. Kırk kız aγalarına, biri de buña, ulara da br šij ajtmajıp, anahtarı χatınγa[2] berip kajtıp aγalarına keldı, jene kettiler.

Akšam jene čimenlikke bardılar, anda kundılar. Keče jene aptıradı, jene uzakta br jarık kúrindi. Ata minep kitte. Kete jatχanda dúldúl: ústúmden túš dij, men ketkende br tirelirim, sakın meni urma, ursāñ bašın sikletge kalır dij. Ahmed urmaǯak bolıp dúldúlge minde. Dúldúl br tirelgende, alt‿kúregine[3] br kamčı urγan. Šo vakitta kırk bašlı br deu kilep bunı juttı. Jine túkúrıp: ah kahpe ulan, menim tupraγımda ni išiñ bar dij. Ahmed: aman sultanım, men jañıldım, kabātım baγıšla dij. Deu: eger sen barıp ǯin padišasından túnja gúzelini alıp kelsen, seni aff ittim dij. Batır br kač kún mūlet alıp ketti, aγalarına barıp kızlarınen mallarnı berip újlerine julladı, úzi kajtıp keldikte dúldúl jene batırı arkasından túšúrep dij: deu beni saña birmez, lākin úzi silañdan ajrılma, ma, menim de bir túkúmi, zurγa kalγanda χızdırırsın[4], men saña jardımγa barırmen dij.

Deu muña br χara at berip (397) dúldúli alıp kalγan. Batır atnen ketti, kač kúnnen son br tauγa bardı, šunda atın terekke baγlap juχuγa jattı. Juχuda eken kulaγına br ses keldi, turıp surdı: ins misin, ǯins misin? Ǯin tugelim, senin kibik adam ulıjım, kel bu tarafγa dij. Bardı. Senı deu júverdi mi dij. Batır: evet, túnja gúzelini ketirmege júverdi. Ajse sen de menim kibik bolorsın, men de seniñ kibik iken, bundan atım meni tašlap kačtı, ǯinler kúzlerim čıχardı[5] dedi. Batir: joχ, bu at kačsa, menim úz atım kelir, dij de šu keče anda jattılar. Sabā turıp dúldúlnıñ čačın kėzdėrdė[6]. Dúldúl keldi, kiteǯek te batır: sen dúnja gúzelini alıp kelgende mundan keč, men úlúrmen, meni kúmersin de ketersin dij. Bular sauluklašıp kitte.

Kite-kite dúnja gúzeli bolγan šehere bardı, br kāvege konuk boldı, undaki jašlardan dúnja gúzeliniñ dúgúni bolduγını añladı. Erte turıp dúnja gúzeli baraǯaχ hamamγa kirep saχlanıp turdı. Dunja gúzeli kırk kıznen keldiler. Kırk kırk tarafγa kittiler. Bu gúzel batırnıñ katına kirde. Bular br brine āšik boldılar. Batır kıza: sen jarın babañdan bahčaγa izin iste, men seni annan alıp kačarmız dij. Kız babasından izin aldı, kızlarnen bahčaγa bardılar. Batır bunı alıp kačtı, [batırγa][7] kelgen, baktı u úlgen. Bunı kúmep, kızın (398) deuγa ketirip birdi. Deu bunı bošadı[8]. Šimdi kız deuγa: menım babam úlgeni br kúndir, men kırk kún jas tutaǯakmen, soñ kelin kijeu bolırmız dij, em de kırk put demirden br kafes japtır, sen ketseñ meni kelip almasınlar dij.

1 ukep; R ukıp

2 χatinγa;

3 alt kúregine; R ald kúregine

4 χızdırirsın

5 čıχardi

6 kėzdėrde; R kızdırdı

7 Word "batırγa" crossed out in Ms, and mistakenly "Dúldúlge" was written in instead.

8 bošadi

Deu böjle br χafas japtırdı, üzi kitte čölge. Akšam keldikte kız deuγa: sen tauda jürijseñ men χorχam, seni üldürirler dij. Deu: joχ, sen χorχma, menim ʒanım seniñ babañ jajlasında br sandıχ bar, sandıχ ičinde χas, tübine üč jumırtka bar, bularnıñ bri belim takati, bri küzim jarıγı, bri de ʒanımdir, unı kimse alamaj dij. Kız bunı batırγa ajtkan, bu da düldülge minep kitte, jajlaγa bardı, sandıγın čevresi küke kadar ut. Šeherden üč kat urba alγan, bri balaγuz, bri kijiz, bri temir. Bularnı kijep düldülge minep br kamči urıp ut ičinden kečip sandıγı parladı, jımırtkalarnı aldı, brin šu jerde urıp parladı, deunıñ belinden takat kitte, ekisin kızγa alıp keldi. Deu batırı kürep ekisini ber, avızıma sıtajım, malım seniñ bolsın dij. Batır bireʒek boldı, ama kız elinen urıp jımırtkaların brin taha kırdı, deuniñ eki küzı kür buldı. Batır: malıñ tilsim etip, br alma jasap maña ber dij. Deu japıp birdi. Batır almajı ʒebine kujdı, kıznıñ üjlerine kajtıp keldiler. Anası babasınen kürüštı, kırk kün dügün etip safada kaldı.

14 Jüzükniñ marafeti

(405) Br zamanda bar eken, br zamanda jok eken,[1] br ʒarlı apakaj bar eken, unıñ da br ulu bar eken. Ašamaγa ütmeklerı juk eken. Nenesi künde br čorap ürüp ulunen bazarγa ʒiberip sattıra eken, u akčaγa ütmek aldırıj‿ken.[2]

Br kün čorapnı alıp bara sata un biš kapekke. Balanıñ biri br mıšık balası sata eken, un biš kapekke mıšıχ balasını ala. Ekinʒi küni bir taraš ala, üčünʒi küni br ʒılan balası alıp kele. Anası dij: eki ʒan eken biš ʒan boldok, ašamaγa ütmegimiz juk, balam, bolarnı kateʒeksin? Balası: anaj, bolarnıñ kismetin kudaj bere di.

Ʒılan balası tilge kelip: ʒiget, meni balalar küteklep turγanda kutarγan ediñ, men de seniñ br ejiligine jarajım di, tur menimnen bizge kelijik br obaγa, anda menim atam anam bar, meni alıp barγanıña atam anam bek kuvanıj, kuvanγandan saγa čülge sımaγan mal, üjge simajʒak akča berijik dijler, lākin sen alma sakın. Atamnıñ kulında br jüzük bar eken, šunı iste dij. Soñ šu obaγa baralar, ʒılannıñ kobasına kireler. Bašinda altın taʒlı br ʒılan patšası, ʒilandan vezirlerınen bu ʒigetni karšılap alalar. Üziniñ balasını ʒigetnen kelgenini kürep, bu kumdır dep suraj. Ʒılan balası: meni balalar kütekliginden kutarıp alγan bu ʒiget, meni küp zamannar üzini ašajʒak ütmeginen bakkan. Ʒılannıñ atası: menim balamı ülümden kutarγansıñ (406) tile menden ne tilejsıñ, saña čülge sımaγan mal berejim, üjge sımaγan akča berejim dij. Ʒiget: čülge sımaγan malıñ kerekmij, üjge simaγan akčañ da kerekmij, berseñ kulındaki jüzükni ber, bermeseñ sau bol dij.

---

1 The original title of the story ("Jüzükči masalı") has been changed to better reflect the content of the story. Only in the Tatar title of the tale is the word *jüzükči* 'maker or seller of rings' used, not elsewhere. In the tale there is a Jewish watchmaker who also sells rings, but he is only a minor character.

2 aldırıjken; R aldırıj eken

Arkadašı bolγan ʒilanγa: tur, ketejik dij. Ketmege turalar. Ʒilannıñ anası bašlaj ʒılamaγa: menım balam br jüzük kadar kürünmij, jedı ʒıl balamnı kürmedim, sen balañnı ič aʒımajsın mi dij. Šaj degende babası čaγırıp jüzükın ʒigetniñ kulına bere. Jüzüknıñ marafeti bir ʒalaγanda kırk ajdamak čıka, ne tileseñ, šulaj eteler. Jiget br kač kün konak bola, son kajtıp arkadašı ʒilannen üzlerin ʒurtına kele.

Anası koš keldiñ balam, kajda kaldıñ dij. Anasına ajta: bar kānnı kızın iste. Anası: balam, biz pukaramız, kān bizge kız bermij di. Bersin, bermesın, sen bar di. Anası papıčın šıpırındısınen[1] kānγa bara: Kān ependim, saña keldim ʒüz sürip, br ʒiget balam bar, kızıñızın uluma tilejım di. Kān čıkar: beriñiz ʒarlı totajγa br avıʒ altın dep kaznadarγa emr ete. Ʒarlı totaj: men akčaγa kelmedim, kudajnıñ emrinen, pajγamberniñ kavlınan uluma kızıñı istep keldim dij. Šaj degende Kān ačulanıp batırlarγa emr etip [kulaγı dajın][2] kırk parča etip šo ʒarlını čuvalγa kujıp al da ket[ip] üjine tašlañız dij. Batırlar kulaγı dajın kırk parča etip, čuvalγa kujıp, üjine alıp ketip tašlajler.

Ʒiget anasın[3] (407) halını kürep ʒüzükın ʒalaj, ʒüzükten kırk ajdamak čıkıp jasajık mi, jıkajık mi dijler. Ʒiget: anamı burungu tarzı sau etiñiz dij. Ular da šo‿ssāt degeni dajın eteler. Br taha kānγa ʒibere, kān kürep bu kurtka tegin tügül eken dep šona karar bere: men seniñ uluña kızımnı berijim, ama menim sarajımnı bosaγasından altın tüšelir, menim sarajlarım tarzı saraj jasatır, kırk küngeʒek bu ajtkanlarım bolsa, beririm, etip bolmazsa mojnını[4] χijardaj kesejım dij.

Ʒarlı totaj kajtıp kele, uluna kānnıñ ajtkanların ajta. Bolarda kümesdaj ʒartı üjden baška šıj jok eken. Aradan otoz tokoz kün keče, urtada br šıj jok. Kırkınʒı küni keʒe ʒiget jüzügın ʒalaj, ajdamaklar čıkıp jasajik mi jıkajık mi degende: menim bosaγamdan kānnı[5] bosaγasınaʒek altın töšeñiz, kānnı sarajlarından jaχšı sarajlar jasañız di. Bir küz ačıp ʒumγanʒek bu degennerin hepsi hazir bola. Kān sabā turıp baχsa ne baχsın, kān sarajları kümes tajın kala, bu ʒigetnıñ sarajnın katında. Kān baka küre bolmajaʒaχ, ʒigetni čaγırıp kırk kün, kırk keʒe toj etip kızın bere.

Kıznen ʒiget bajakı vakitlar ʒašajlar, lākin br ajnaʒı ʒıput bu ʒigette šondajın ünerli jüzük bolγanın bile. [Er kün br taran balıγı ašaj turγan bolsa, balıknıñ jarısını ašap bol‿almaj][6] ujlanmaktan: šu jüzükni katıp alır ekemen[7] dep, (408) kelinčekni jüzüknıñ ünerlerinden χabarı jok eken. Ʒuput bir jüzük jasata, šunıñ üzi dajın. Br takım eski jüzükler taha ʒıjıp, jüzük satıp ʒigetnıñ sarajına kele. Ʒiget jüzügın üjde tašlap br ʒıjınγa kete, čıput lokata tapup kelinčikke uʒuz ʒüzük satam dep kele. Kelinček bonoñ kulında ʒüzükni kürep: bu bizın ʒüzük mi dep kursi üstünden jüzükni barıp baka, ič brbrinden ajrılmaj. Čuputnıñ da tilegeni šo ʒaramazlık eken. Üzin jüzügın urnına munıñ bergi jüzügın ala.

1 šıpırındisınen
2 Crossed out in the Ms.
3 R anasının
4 mojnıni
5 kānni; R kānnıñ
6 Crossed out in the Ms.
7 Dialectal form of *ekenmen*.

Kulya aldıyı tarzı br ǯalaj, jüzükniñ ičinden kırk ajdamak atlıp[1] čıka: jasajık mi jıkajık mi dep. Čuput ise: ačık teñıznıñ artında er barmayan jerde kırk kat saraj tilejim, bu ǯigetni de kijizge urap bosaya tübine tašlañız, bu sarajlarnı ǯıkıñız[2] dij. Br küz ačıp ǯumyanǯek kırk ajdamak čıpıtnıñ ajtkanların epsin jasajler, ǯigetnı kijizge urap tašladılar, čıpıtın de alıp ketkenler. Br küz ačıp ǯumyanǯek ačık teñizniñ artındaki sarajlarya bardılar. Čıpıtnıñ ǯanı ne tilese, tabıldıra.

Bz kelejik bizın ǯarlı totajnıñ uluna. Br vakitten jukusın ačtı[3], brde karasa ne karasın, ne sarajlar bar, ne de ǯüzük bar, kateǯejin bilmegen iken. U jakka karadı, bu jakka karadı [ašlıktan da üzegi talyan].[4] Tarašnen mıšık br jerge ǯıyılıp brbrinnen ajtıšalar: biz bu šorbaǯımızıñ ütmeginen üstük, bz de šimdi uña jüzügin (409) kajda bolsa da alıp kelmek kerekmiz dijler. U jerden mıšıknen taraš ǯulya čıkalar, bašlajler ketmege. Kete-kete teñız kenarına barıp toktajler. Mıšık tarašnıñ üstüne minep teñızden üteler, biraz ǯul ǯüreler, br sıčan memleketne[5] baralar, mıšık bašlaj kürgen sıčanın boyıp tašlamaya, taraš da bašlaj ülülerın uba etip tašlamaya. İčlerinden br eki sıčan sıptırılıp kačıp padšalarına χabar bereler. Dij ki: padšamız, bizim azrailimiz kelgen, Ali-pašanı de ütirdi, Veli-pašani de ütirdi, katijık? Šulaj ketse bo iš, br kišimizın kaldırmajaǯak. Padša dij ki: br jokları barsa, sorañız[6] dij.

Barıp surajlar. Mıšık ajta: Kenarsız deñızın artında bir saraj bar, anda br čıpıtta br jüzük bar, šonı alıp berseñiz, sizlerge tijmem, [undan γajri br teneñizni kaldırmam dij][7]. Sıčanın biri: men barıjım dij, bir sokor sıčan, eki bojdak sıčan: bz de baramız dijler, tap soñında keteler, sarajγa jakın baralar, bašlajlar laγım kazmaγa. Br vaχitten soñ laγımnı tešip duvardan čıkalar, brde bakajler ki čıput jüzükni sıžımγa bajlap, tavanγa asıp juklaγanda, avızını[8] tišlep jata. Sokor sıčan nasap üzine dujdurmaj[9] alsın, baka rafta bir kutu kızıl[10] biber tapa. Šü‿jerde[11] braz sije, kujruγın ıslata da ketirep büberge bolγalaj, aman čiputnıñ janına barıp bu[r]nıñ[12] tešiginden sukıp (410) ǯibere kojroγın. Ǯıput aksıra, aksırdıγı dajın jüzük avızından pırıldap atlıp[13] kete, sıǯımnı[14] da tüpege čikup tišinnen[15] kırkkan iken. U jerde sokor sıčan jüzükni kapıp laγımdan bašlaj kačmaγa, čıput ta aksıra-tüksıre kala.

Bz kelijik tarašnen mıšıkka. Ular beklep utırγanlar. Bajram χabarın beklegen dajın brde baktılar-ki ǯanaj sokor sıčannıñ avızında jüzükni alıp kelgen. Šimdi tarašnen mıšık jüzügın alalar sıčanların da azat eteler. Andan tarašnen mıšık teñız ǯakasına baralar. Bri men alıp barajım dij, übiri de men alıp barajım dij. Mıšık: seniñ tıšıñ sijrek, tüšürirsiñ, taraš ta dij: sen ǯaldamaγa bilmejsin, ajsa menim ünerimnen boldo bo iš,

1 R atılıp
2 ǯıkıñiz
3 ačti
4 Crossed out in the Ms.
5 R memleketıne
6 Künos hesitated between the *sorañız* and *soranız* variants.
7 Crossed out in the Ms.
8 R avızına?
9 düdürmaj
10 kizıl
11 R šu jerde
12 R burnınıñ
13 R atılıp
14 šıǯımni da
15 R tišinen

men alıp barajım dij. U jerde ur tut kavγa eteler, jüzügın teñızge tüšüreler. Zavallılar ekisi de br kač kün teñız bujlarında ʒüreler. Bir kün br auʒı kelip teñizge čilme tašlaj, auγa br balık tüše, ketire, ač ʒürgen tarašnen mıšıknı küre, balıγın bolarγa bere. Ašap otorγanda mıšıknıñ ašaγan jaγından jüzük čıka. Taraš kürep kapkač kete, mıšıkka da min arkama dep, mine, teñizden kečıp šorbaʒılarına ketirep jüzükni bereler.

U da aman br ʒalaj, kırk ajdamak čıka, ʒavajık[1] mi, ʒıkajık mı dep ʒıkkanıñız kerekmij, japkanıñız kerek iken. (411) Ӡıputın ütırep, taran balıγı tarzı kurutıñız dij. Euelkinden un kat jaχši sarajlar istejim dij. Ajdamaklar da baš üstüne dijler. Br küz ʒumıp aškanʒek epsin jasajler. Ӡiget kene de euelki dajın jašaj, padša da kārindan br tostoγan aγu ičep üle, jerine de bo ʒigėtın padšalıkka lājik küreler. Bu ʒiget padša bolop küp vakitler sepalarnen umurler ete, muratına āsıl bola. Allah bizlerin de muratlarimızın asıl bolson. Āmin.

15 Šangeri kān ikejesi

(419) Kānnıñ br kün ʒanı sıkılγan, janında br akıldānesi bar iken. Čaγırıp ajtkan: Nogaj-Ahmad[nı][2] čakır dij, biraz nakırdı etejım dep, ama ne atnen kelsin, ne jajan kelsin, ne ičeri kirsin, ne tıšarda utırsın[3]. Kabar bergenner. Ahmad-aka šimdi men nasıl barajım dep tüšünmege[4] bašlaγan. Akanıñ ulı: baba ne tüšünesin dij. Tüšünmeji nasajım, zulumγa rast kelgenim. Šangeri kānγa barmak kerek, meni čaγırγan, ne atnen kelsin, ne jajan degen. Bala ajtkan: bo bir kulaj iš, kasvat etmeñiz, maña br tüje arabası, un tene bala, brde ečki, jüz de āšiχ, men barajım dij.

Tüje arabasın ʒekken, balaların arabaγa tudırγan[5], Šangeri kānγa barγanlar. Jarı keʒe bar iken. Tüjelerın tuvarıp, ıšıklık ičine ʒiberdiler. Sabā turγannar, kānnıñ kišileri ıšıklık ičinde iki tüje utlap ʒüre, on da bala āšik ujnijler. Kānniñ kišileri [ıšıklık ičinde][6] kelip: balalar nišlejsiz mında, sızın br balabanıñız juk‿em[7] dij. Ana balabanımız dep tüjeni küstergenner. Juk, juk, br sakallınız juk‿em dij. Munı dep ečkini küstergenner. Kānnıñ adamı ardına kajtıp kanγa barıp süzlerin ajtkan. Čaγır, kelsin dij. Ahmadın ulı[8] ečkiniñ arkasına minep tora kānnıñ kapısına barγan, ald ajakları kapınıñ bosaγasından ičeri kirgen, ard ajakları tıšarda kalγan. Kān baka, kapıda br ečki mingen bala utıra. Sen nige kelgensıñ dij. Bala ajttı: babamnı čaγırγansız[9], unıñ jerine men kelgenim dij. Sen nasıl istediñ üjle (420) keldim. Tüš ečkinden kel otor, braz sennen nakirtı etejim degen.

1 A variant of *ʒabajık*.

2 Noγaj Ahmadın ulın [false!]; As the story goes on, it is clear that the khan invited to talk not the boy, but his father.

3 utırsin

4 tüšümmege

5 tudirγan

6 Crossed out in the Ms.

7 R juk mė

8 uli

9 čaγırγansiz

Ulum, jā sizın kůjde bir kart jok mu iken, seni ǯibergenner munda? Üzim babam‿čin[1] kelgenim. Kān ajttı: sen br balasıñ, ne kůrdiñ, ne bilesin dij. Bala ajttı: julda kelgende un tene ulak kůrgenim, jıyırım tene ajyır kůrgenim dij, otoz at kůrgenim dij, kırk tene ůgůz kůrgenim dij, elli tene kuj kůrgenim dij, altmıš toryaj kůrgenim dij. Šangeri kān bašladı tůšůnmege, bu ne demek, bu šıjlerı čızamayan, tap terlegen. Bala ajttı: kān efendi, kara tauya kar javup, tamašasından su akkanı kač kůn bolyan dij. Kān taha bek utanyan. Kānnıñ br kızkardaši bar iken. Bek akıllı iken, bundajın nakirdiların čezer‿ken[2]. Kabar ǯibergenner: Šangeri kān bir balanıñ aldında maskara bolyan, tez kele.

Kız kelgen. Bejim, dij, saña tārif etken br adamın ömrůn. Bir kiši un jašında bolyanda ulakdaj bolor, jıgırım jašına kelgende ajyırdaj bolor, otoz jašına kelgende br atdaj bolor, kırk jašına kelgende br tuvardaj bolor, elli jašına kelgende br kojdaj bolor, tek tilinde kalır kuveti. Kara tauya kar demek šačları čallanyan, tamašası da su akmak demek, beti terlep suları čuburyan. Kān dij: balam, kač kišige karšı salasın? Üzim dajın br kišige karšı salajmen dij. Kān bularya br talaj alma, armud, fundık[3], akča bayıšlayan, br arba šıj tuldıryannar[4]. Saulašıp kitkenler.

Bunlar br otoz čakırım jir kitken son, bularnıñ artına kān kırk tene pelivan salyan: barıp jetišiñiz, (421) balaların ůtůr, arabaların al da kel dij. Pelivannar jetiškenner. Balalar ajttılar: nasajıχ? Ahmat ulı ajttı: Sizler utırıñ araba ičinde, men tůšejim onlarya, erkekligi kůsterejim dep arabadan atkan. Otoz beš tanesin supanen suyıp bašların parlayan, biš tenesı ajlanıp kačkanlar. Kānya barıp kabarın ajtkanlar. Kān: aru aru, ajtıñız, kajtsın kelsın dep, baryanlar artına, kān ister seni degen. Kajtıp kelgen, kānnen kůriškenner. Kān ajttı: jā sen ůzim dajın br kišige karšı bolormen digen iken, men saña kırk kiši ǯiberdim, ular seni baš etemediler, āferim ulum dep jerine kajtaryan.

Kān bu baladan ůzi kurkıp bu balanı ůttirmege istej. Pelivanya[5] χabar ǯiberdi: jůz čakırımda[6] uzak ketsinler, ajgırların davuši munda išitile, meni rahatsız ete dij. Halk tuplašıp ajtkanlar: ǯanım pehlivan, balanı gine de ǯiber, barıp kānnen nakırtı etsin. Bizler bu kadar balaban koj[7] kajda ketijik dep jalvaryanlar. Balaya ajtkanlar: Šangeri kānga br taha bar. Bala: men barijim, ama tokoz bala, tokoz kenǯal berijsiz dij. Kenǯalların balalarya berip uzdıryanlar[8].

Bunlar kānya kelgenler, boların aldına itler ůrůp čıkkanda balalar itlerin ůtůrgenner. Balalarnı tutıp nejčın itlerın ůtůrgensiñiz dep surayan. Onnar ajttı: sizin itler ůrgende ůjlerimizi juklap bulamajmız. Kān ajttı: otoz čakırım jirge itnıñ davuši barij mı? Bala ajttı: jā otoz čakırım jerge ajyırıñ davuši kelij mi dij. Aferim, ulum, sen adam bolaǯaksın, ket, jiriñde rahat bol digen, dep kajtkanlar.

---

1 babam‿čin; R babam ičin

2 čezerken; R čezer iken

3 fundik

4 tuldiryannar

5 Based on the context, the word pelivan 'wrestler-champion' means here the boy's father, Nogaj-Ahmad.

6 čakirımda

7 Dialectal variant of *kòj* 'village'.

8 uzdiryanlar

16 Uf-arab ekjeti

(426) Eski vakıtta bir adamnı br ulu bar iken. U ulan da babasın[1] su̇zin‿dutmaj‿ken[2]. Atası čara tapmaɣannan u ulannı br erifge ırgat bereʒek bola. U̇jden br u̇tmek alıp čalmanɣa kujıp ulannı janna[3] alıp julɣa čıkıp keʒe-ku̇ndu̇z kiteler, [Kısχası bu adamnı boldormaktan čarası[4] kalmaj....][5] br uva janna keteler. Atası dij ki ulanɣa: bı‿jerde[6] toχtap braz boldordım, u̇tmek ašajıχ dep, uf dep tajana. Ama u ubanı ičersinde Uf degen fokusčı br arab bar eken. Bu adamnı jurılıp uf dep ajtχaninen oba ʒarlıp[7] ičinden br arab čıka. Ne kıčrasız maña dep. Bu adam dij ki: men seni kıčırmadım. U da dij ki: uf edip kıčkırdıñ. Men uf dedim ama kerček boldorɣanımdan ajttım. Kısχası birbirinden χabar surašalar, kısχadan bajlajık, ben balamı irgat birmege ketem. Arab u su̇zni išitken son bek kuvandı. Men tutajım dep kačɣa bolaʒak jıllıɣı [dep] suradı. U adam tuχsan ku̇mu̇š dep ajttı. Arab kuvanɣannan parasını ȯgine berdi. Akaj paranı alıp ketti.

Arab ırgatnı alıp ičeri kirdi. Ulan ičerde učsız, pajsız čoχtan čoχ, uzundan uzun br sarajlar gu̇rdi. Arab munı tašlap kitte. Ulan br kapını ačıp ičeri kirde, ku̇rdi (427) br kız naɣıš tike. Kız de munı ku̇rgennen sıčrap turıp bujnın[8] kučakladı. Vaj, tuɣanım, sen munda nasl[9] keldiñ dep surdı. U da bašına kelgenneri ajttı. Kız buña tembeledi, soñ penʒereden ku̇sterdi, tuχsan tuχız adam bašı sajıp, bašları da kazıkka kečip sıranan tura. Kız bu bašlarnı dep tembeledi. Uf-arab tuχsan tuχız baš kesken, brde bunıñ bašın kesse ju̇z bolaʒak, soñ bu kıznı u̇jleneʒek, u̇jle kavil japkannar. Kız bu balaɣa ajta: u arab saña fokišli marafetler u̇gretir de senden surar bildiñ mi dep. Eger taptım deseñ bašını kese, joχ deseñ kesmij. Kısχa laf, ne aχran u̇gretse, sen de bilseñ, kene de bilmem dep ajt, soñ bašın kesmej, tılsım buzılıj.

Jıl tola, kene ulan bildım br šej dep ajtmaj. Ulannıñ babası jıl tolɣan son kene u ubaɣa kelip uf dep toχtaj. Arab emen čıka, ne χajırsız balañ bar eken dep balanı bojunnan u̇tep ʒibere. U̇zi kajta jirine. Bala arabdan kapkan hissesin br br babasına ajta. Babası ınanmaj. U̇‿jerden[10] kočıp[11] keteler.

Ketkende ku̇rijler julda br auʒı. Bu bala babasına ajta: baba, men br tazı bolajım, bašlajım kujan ajdamaɣa, siz de br auʒı boloñız. Anʒa[12] meni bujnımdaki ku̇mu̇š kanʒamı satmazsız, bunı satsañız, men jere (428) kajtamam. Atası jaχšı dep kuvana. Bala barmaχlarnı brbrıne ura, babası auʒı bola, balası tazı bola, bašlar kujan ajdamaɣa ama unın kibik tazı tu̇njada jok, u̇jle gu̇zel ajdaj. U̇bir auʒı bunı ku̇re, aman buların janına kelip satmajsın‿ım[13] tazını dep suraj. U da ničin satmajmen[14], akča čok berseñ dep kırk ku̇mu̇š istej. U da br šej ajtmajıp, bazarlık etmejıp parasın saja.

---

1 R R babasının
2 dutmajken; R tutmaj eken
3 R janına
4 čarasi
5 Crossed out in the Ms.
6 bı jerde; R bu jerde
7 R ʒarılıp
8 bujnının
9 R nasıl
10 u̇-jerden; R u jerden
11 R kȯčip or kačıp?
12 R anʒak
13 R satmajsıñ mı
14 satajmen

Bu tazı bašlaj kujan ajdamaya. Unıñ babası kite küjine. Akšam bola, auǯı tazını alıp üjine kele, čapık aštır dep katlama piširmege bašlaj. Tazı uradan kačıp kete, babasına kele: kürdiñiz mi baba, para nasıl kazanıla. Ende men br šılab‿ajgır[1] bolajom, Tüpesi-ačık küjde br aχot mejzin bar, beni uña satarsız, ne deseñiz u paraya alır, anǯe bašimdaki kümüš ǯügenni satmañız, satsañız tılsım buzula. Bir ǯorya ajgır boldo. Babası mindı arχasına ama emr et dep utıra. Son Tüpesi-ačıχ küjge kitte.

Küjge ike vers kalyanda atnıñ takıldısı išitilmege bašladı. Mejzin u sesnı išitkennen čıyıp bakkan kürdi br ǯorya kele, amma barım kele. Jandaki adamlarya maktanmaya bašlayan, eger kelgen akaj satsa alırmen dep karar berdi. Adam ǯorya ile u jerge kelgen (429) salam berdı. Mejzin salam almaya bakmadı, akaj satasın‿ım[2] atıñ dep surdı. U da akča güzel berseñ ničin satmajım dep ajttı. Mejzin ne istejsin, dep u da avzını tuldırıp jüz elli altın dedi, anǯa ǯügenni satmam dej. Mejzin kajl boldı, ǯoryanı aldı. Birde minep bakajım dej. Mında atın ǯürišin bek begendi, kajtıp keldi, parasın birdi. U adam paranı alıp jere kajttı, mejzin de čülge kezinmege kitken.

Ekindi vaχtına hāklar[3] ǯamige tuplašχanda mejzin kajtıp keldi, ǯami üginde atnı kezdirdı. Ü‿jerde[4] bir češme bar iken, suvarajım atın da dep. Ketti češmege suvarmaya. At bašladı su ičmege, iče-iče češmeni uluyına kirdi, mejzin šaštı kaldı, bašladı kıčırmaya akajler akajler, keliñ munda atım češmenı uluyına kettı, čıkarajıχ dep jalvardı. Šašχın mejzin, ǯorya češme uluyına sıyar mı dejler, mınau bizem mejzın at delisi degil mi dejler. Mejzen kene inanmajsız dep ajttı: bakın nasil kıčıra. Adamlar čapuχ keldiler, baχtılar kerčekten kıčra[5]. Bašladılar češmeni buzmaya.

Ular češmeni buzmada bolson, ulan kajtıp babasına keldı: babam, ben šindi bir hamam bolajım, anǯa tünjada unı kibik büjük hamam hič te bolmaz dep ajttı. Torχa berip satıñ, anǯa šu hamamnı anaχtarnı satmazsız (430) dedi. Tılsım anaχtar üzerinde. Babası bašladı torχ etmege. At ajayı jetken jerlerden adamlar kelgen. Bu ulannı čorbaǯısı bolyan Uf degen arab de keldı, bašladılar torχ etmege, torχ da čıktı biš bin altınya. Son ulannı babası ajttı: lākin anaχtarı satmazmen dij. Šaj dep ajtχan ile arab tılsımını añladı. Sen biš biñ istejsin, men saña berejım un biñ, lākijn anaχtarnı da ber dij. Paranı da küzi siǯak degil mi? Babası šejtanya ujıp bar χajrın kür dep χajırlaštı.

Anaχtarnı arabka uzata jatχan vakitte anau hamam br turyaj bolıp pırıldap uštı. Šu arab da br kıryıj bolıp artından kuvalap[6] uštı. Turyaj šašmaladıyından urada br padša bar iken, unın kızları uturyan küške barıp penǯereden br bülbül bolıp kirde. İčerdeki kızlar čapıštılar bülbülni tutmaya. Kıryıj da jetip keldi, keldi ama penǯereni kapatkan ediler. Kıryıjya kirmege jir kalmayan. Bašladı kıčırmaya. Padišada br kozo alaǯayım bar dep. [Padiša ešitken bu sesnı, ü‿jerge[7] keldi su[r]dı.][8] Sen nasil kozo istejsin degende dij: benim br avım bar ide, kačıp küške kirde, unı istejim dij. Padiša küške čıktı, surdı

1 R šılap ajyır
2 R satasıñ mı
3 R halklar
4 ü-jerde; R u jerde
5 R kıčıra
6 kuvalıp
7 ü jerge; R u jerge
8 Crossed out in the Ms.

kızlardan. Kızlar da unaǯı kadar bülbülnı tutχan ediler. Seveǯek bolıp kergef üzerine kujγanda br jazılı kahat boldı, üstünde jazılγan: ben čın[1] padišahin balasıjım, bir (431) šeddetnen aramız bozık tüšti, u beni kuvalap keldi, lākın kulına berseñız, beni ašajǯak dij, kulına bermejıp küterip atsañız, ben undan kutlaǯaγım[2] dij, soñ ben de sizin br dāneñizni alaǯaγım degen.

Šu süzni ular ešitkennen künnemege bašladılar. Padiša kelip: beriñ kitsin, unı bülbülini dij. Bular kıjmaj alıp kahatnı penǯereden ataǯak boldılar. Jel kajtarıp ičerde tüšürdı, jere tüškende tüj bolop sepılıp ketti. Padiša šaštı, kızlar da. Kırγıj amam‿br[3] kuluk čipče bolıp bašladı kıt-kıt ašamaγa. Tüjın bir tānesi kergef arasına kısılėp[4] kalγan iken. Kuluk tüjlerni aša[p] bitirdi, baktı u jaχta bu jaχta kalγanı bar mi‿ken[5] dep. Kürmedı kergef arasında kalγan tüjni. U tüj br tilki bolop čipčeni külükni ašamaγa bašladı. Vajvajladı padiša, külük silkendi br arab boldı, sim-sija čırajlı boldı. Tülkü silkindi, ajın un dürti kibi br ǯivan boldı, bašlarına kelgenneri padišaγa ajttılar. Padiša: her kabahatnı arabγa kaldırıp, arabnı zıldanγa kujdı, ǯivanγa üken[6] kıznı kırk kün, kırk keǯe toj-tügün etip aña bergen. Kükten tüštı üč alma, birisi masal ajtkanγa, biri diñlegenge, biri de kene maña.

17 Kurku masali

(438) [Zaman zamanda ikešti, deve tellal ikešti, kaplı-kuplu bakalar kanatlanγan učmaγa, teñizdeki balıklar kira tutkan köčmege, Ak-mešǯidni minaresi egilgen Salγırdan su ičmege. Gubernatırnı tarašı balaklarnı kütergen jedi terjanı kečmege.][7] Br zamanda bir kartnıñ br ušaγı bar iken. U ušak ta biš altı jašında bir ušak bar iken. Ušaknı anası berir uχutmaγa. Uχutmadan kajtkanda balalar āšik alijler. U balalar[γa] dej ki: beriñiz beni ašiklarımı. Balalar bermejız dijler. Son tutmıš, u balalarnı düvmiš, kimisin elini kırmıš, kimisin küzini čıkarmıš. U čoǯukların babası gelmiš dij ki: eger čoǯuγını[8] begitmesen, sonra gider üldürirmiz dej. Sonra [anası] aña dij ki: balam, sen büjle šejler japma, seni kurkuzu[r]lar. Bala dij ki: aman, kurku dedigin nedir. Anası dij ki: kurku dedigın uralar senı, düverler seni. Bala dij ki: pišir baña jol ašlık, kitermen kurku kudurmaγa. Anası biširmiš bermiš jol ašlık, u da ketmiš kurku kudurmaγa.

Akšam bolγan, baγar ki br čadırda kırk bir ajdamak utırγan, konušijler. Sajlam[9] berij, kirej ičerige. Ular dij ki: ište bizge br šiš-kabab kelgen. Bala uların bašınγa[10] dej

---

1 R čin
2 R kutulaǯaγım
3 amam br; R aman br
4 kısılep; R kısılıp
5 miken; R mi iken
6 R üjken
7 Crossed out in the Ms.
8 čoǯuγıni
9 R salam
10 R bašıγa / bašına?

ki: bakalım ušayından ne čıkaǯak. [Ušak dij ki][1] Beriñız (439) aš jejim, kusaǯayım[2] ačtı[3]. Dijler ki: bır kazan alva bišir, sen de jersıñ, biz de jermiz. Nerede biširejim [dep] suraj. Bu jerde pišireǯeksin, urada bir mejzerlik iken. Kazan asa, bašlajir biširmege. Bir kul kele kazanıñ ičine, dij ki: benimnen šaka japma, čık[ar] kulıñı, eger bir taha uzatırsın kulıñı, sonra kırarmen. Bir taha uzatir kulını. U da dutaj kırıj, kazannı alij, gütirij ajdamaklarγa. Dıj ki: geñız[4] ašajıχ.

Bizim de br reǯamiz bar saña, unı da japsañ ne de müradıñ bolorsa, saña berirmiz. Süjlejiñ dij. İšte saña br čokıš[5], brde čüj, šu üjge kak da čık. Ušak bašlijor čüji kakmaγa, br [kijinmiš][6], kušanγan, altın dakmıš br kız gelij: kardašım, senı bujıñ güččük, ben jere jatajım, sen üstüme minep šo čüjni kak. Ušak jüzine bakmaγan bir šamar dürtejir. Kız jedi kat jer dibine gidejir, anasından emgen süti burnından fitil gibik kele. Kız dij ki: ešχ bolsın, šo güččük ušaγı[7] šamar dürttıgıne, kulımdan bilezigim tüšti. [...][8]

18 Bir baj kiši masali

(442) Bir baj adam üleǯekti, üč ulını čaγırıp vasijet ajta. Büjük ulına dij: men ülgen soñ malların üčüniz de bir alıñız, delidir dep kardašıñızı inǯitmeñiz dij. Baj ülgen son büjük aγaları mal büleǯekti, br jañı aran japajıχ, malları ajdep ketirirmiz, jañı aranγa kirgenleri seniñ bolor, eski aranγa kirgenleri bizim bolor dij. Deliniñ aχlı azlıγından rāzi boldo, hajvanları ajdap kelgenner. Hajvanlar epsi eski aranγa kirdiler, eñ soñında br arık tana kalγan. Unı zurlap jañı aranγa kirsettiler. Deli ulan: bu χadar maldan maña bunı tašlaγan dep, babasın sügıp üjge kirep jattı.

Sabā turıp tanajı čölge čıχarıp sujdı, etin tašlap terisin satmaγa alıp kitte. Julγa kitkende boldorıp teriji baši astına kujıp jukladı. Bu juklaγan soñ, χarγalar kelip bu terini čoχumaγa bašlaγan. Ujanıp brde χarγanıñ brin avuǯladı, teriji tašlap χarγanı alıp kitte. Kiterken bir küjge uγradı, br üjge bardı, baksa br jaš kelinǯek. Kelinge: meni konuk al dep jalvarγan. Hatın: hoǯam üjde jok dep baksa da čara tapmadıγından kunuk alγan.

Aχšamγa buña umač čorbası piširde, ašattı, ičirde. Jatar vakit boldokta bir čette br jastıχ birep konuγı jatırdı. Jiget χarγanı baši tübine χojıp jattı. Juχuγa kiter kitmez, bri kelip penǯereni (443) urdı. Hatın kapını ačkan, baχsa bir jiget, ičeri alγan, zijāfet azırlamaγa bašladı. Br saγan čebürek piširep ašajıχ degende, hatınnıñ[9] üzi hoǯası kelıp kapunı kaktı. Ujnaši[10]: men šindi kanda barajım dep. Hatın: tolapγa kir, u jatχan soñ čıχıp kitersin dej. Ujnaš tolabγa kirgen, hoǯası üjge kirgen ve köšede jatχan konuγı

1 Crossed out in the Ms.
2 R kursaǯıγım
3 ačti
4 R geliñiz
5 R čökıš
6 Crossed out in the Ms.
7 R ušaγın
8 Missing ending.
9 hatinniñ
10 ujnaši

surdı. Hatın: br julʒı adam, čara tapamadım aldım dij. Hoʒası utırdı[1], hatın hoʒasına konuүa piširgen umač čorbasından ketirdi. Kunakbaj musafiri de ašχa ujattı.

Konuk turdı[2] ve karүanıñ buүazınden sıχıp baүırttı. Baj: ulum u nedir dij. Ulan: karүa rafta bır saүan čebürek bar dep ajta. Kunak-baj hatınүa[3]: ničin ketirmijsin degende: ā men unutχanım dep turıp ketirdi. Ašaүannar, ašaүan soñ bir taha karүanıñ boүazından sıχtı, karүa kıčırdı. Ne ajta dep surүanda: dulapta br jiget bar dep ajta dij. Konak-baj eline baltasın alıp dulab kapısın ačıp baktı ne baχsın, bir jiget. Baltanen bašına urıp ütürdi. Ulan bajүa: men sabā barıp seni ukumette bildirirmen, šu ademi ničin ütürdın dep. Kunak-baj: aman ulum, men bir jañıšlık[4] ettım, sen ajtma, saña br tulup altın berejim, bunı alıp ket, br četke küm dij.

Ulan altunlarnen ülüji alıp čıχtı, küj kenarine br numaj suүa atkan. (444) Üjlerine kajtıp keldi. Keldigi kibi altınları üj ičine daүıttı. Büjük aүaları ekinʒisine: bar baχ, deli tıšta br šij parladı dij. Ekinči aүası čıχıp: deli, unda ne japajsıñ dij. Eh, men sizin kibik deli tegilim, menim br tanam bar iken, sujdım, terisin bazarүa alıp kittim, tüjine bir [bu kadar] altın aldım. Sizin kibik ajvanım čoχ bolsa, men taha čoχ altın ketirirdem dij. Bular buña šaškannar, ve kajsı šeherde satχanın surap hajvanların hepsin sujүannar, terilerin br arabaүa jüklep bazarүa bardılar, buların terisine kimse ajlanıp baχmadı. Bular brbrine: bu deli bizni aldap maldan ajırdı, barajıχ üzini čuvalүa kujıp teñizge atajıχ dijler.

Kajttılar, bunı čuvalүa kujip teñız kenarına alıp kitteler. Teñizge barүan soñ bunı büjle atsaχ uzaχүa kitmez, barıp br uzun aүač ketirejik te uña baүlap atarmız dijler. Bular aүač ketirmege kitteler. Deli bašladı čuvalda: men χan χızını almajım dep χıčırүan. Bir čoban išitep kelgen, čuvalnı[5] čezdi. Čoban ničin seni ataʒaχlar dep surdı, maña χan χızın bereler, men istemejim, unıñ_čin[6] dij. Čoban üjlese men kirejim dij. Čuvalүa kirde. Deli küjine kitken.

Čoban bašladı: men χan χızını alırmen dep χıčırmaүa. Bu jandan aүaları keldiler, čobannı teñizge attılar. Br kiʒe bu deli kujları ajdap üjlerine keldi. Aүaları kürep čıχıp baχtılar ne (445) baχsınlar, br sürü kuj, deli de saү. Vaj, deli, sen bu kujları χannan alүan[sın] dijler. Siz meni teñizge attıñız, anda aχča, kuj, ʒanıñ ne istese bar, menim ʒanım kuj istegen, kuj aldım dij. Aүaları: üjlise jarın sen bizni de at, biz de kuj ketirijik dijler. Birer čuval alıp teñiz četine barүannar, büjük aүaların attı, bu bašladı kırkıldamaүa. Kırk bolsa da alıp čıχ dep ekinʒi aүasın da attı, üzi kajtıp üjlerine kelgen, küj ičinde brinʒi adamın kızın alıp mal-mülk sahıbı bolıp kalүan.

---

1 utırdi
2 turdi
3 hatinүa
4 jañıšlik
5 čuvalni
6 unıñčin; R unıñ ičin

19 Bir noyaj masalı

(450) Br zamanda Kırımda Jalı-bujında[1] br Abdurrahman kādi bola. Ve lākin bek azir-ʒavab kiši bola. U vakitler Kırımda, Tarkan[2] ičinde noyajlar da bek čınčılik etıp ʒürgenler, em de bek azir-ʒavab bolyanlar, üzleri okomaχ jazmaχ bek az bilgenler. Jalı-bujın[3] kalkı iš noyajlarnen laf etip bolamayanlar. Kādi Abdurraman noyajlarnen[4] erišlešmege bek aves bolyan.

Künnerde br kün Čongar tarafından Tarkan ičlerinden br noyaj kādınıñ jaklarına tüše, kādi da br jaktan kele, noyaj da kādije rast kele. Noyaj bir küjni suraʒak bola. Kad‿akam[5] salum alejkum dij. Kadi selam ala. Noyaj: kad‿akam, kajdan kelesiz dij. Kadı: ʒennetten kelem, noyaj, dij. Šaj dij de aksıra. Noyaj: šufalar bolson kad‿akam dij. Kadi ajta: noyaj, ja su ičkende ne dersiñiz dij. Elamdulilla dermiz dij. Ja ʒellengende ne dersıñız dij. Noyaj: Jalı-bujında Abdurraman kadiniñ murnına kokusu dermiz dij. Au, noyaj, ja kajdan kelesin šundi dij. Noyaj dij ki: ʒennetden kajttım kad‿akam. Kadi ajta: ej sersem noyaj, ʒennetke kirgen kiši br taha ʒennetten čıkar mi undajın jaχšı jirden dij. Noyaj ajta: kad‿akam, ʒennetke ʒıvuk bardım, (451) kirep bolamadım. Nečin kirep bolamadıñ degende: efendim, ʒennetin kapısın janına Abdurraman kadi adlı br balaban küpekin bajlayannar, kelgenni de tilij ketkenni de tilij, baktım kürdim kirmege ebini tap‿almadım[6], ajlandım kajttım dij.

Abdurraman kadı br türlü ebin tapıp da noyajın mat etip bolamadı. Ajse, noyaj, bo keče mende konak bol da jarın ketersin, seni bir sıjlajım dij. Abdurraman kadi noyajnı üjine alıp kete, msapir[7] ete, [koj sujdırıp[8] sıjlaj][9] Lākin kojnıñ bašinen kojroyın br tavaya alıp kuja, sabaha kaldıra. Noyaj dij ki: kad‿akam, menim bilgenim kunakka kojrokın ašatırlar, sen sabaya üziñe alıp kaldıñ dij. Kadi dij ki: noyaj, bo geʒe nijet etip jatarmız tüške, sabā kajsımız jaχši tüš kürsek, kojroknı o ašar, bašını jaramaj kürgen ašaj dij. Noyaj: bek jaχši, kad‿akam, dij. Br vakitten soñ tüšek kujula, jatalar juklamaya. Noyaj keʒe joχodan tura, karnı da kereki dajın ačıkkan; nišlesin, bašnen kojroknı sıjırıp ašaj, kene de jata juklaj.

Son sabā bola turalar, bet kul ʒuvıp kajvege otoralar, lākin kürgen tüšni bašta kadi ajtaʒak, kavilleri büjle iken. Noyaj kadiden suraj: kad‿akam, nasıl tiš[10] kürdiñ dij. Kadi dij ki: bo keʒe tišimde üzimni ʒennette kürdim, ʒennet (452) zeuklerinen zeuklendım ve Allahnen biñ bir kelam ettim. Kadi noyajdan suraj: jā sen nasıl tiš kürdiñ dij. Noyaj da: kad‿akam, men de seni ʒennette aršalada Allahnen bin bir kelam etkeniñni kürdim. Kadi dij ki: ajsa menim tišim senkinden jaχši, kojroknı men ašajʒayım di.

---

1 "Jalı bujı", the name of the coastal region of the Crimea, is crossed out from the title.

2 A village in the Perekop region (oblast').

3 Jali-bujın

4 noyajlernen

5 kad-akam; R kadi akam

6 tapalmadım; R tapa almadım

7 R musapir

8 sujdirıp

9 Crossed out in the Ms.

10 R tüš

Noγaj dij ki: kad‿akam, men kujrukın da başın da aşadım dij. Nečin aşadıñ degende, noγaj: kad‿akam, men seni tüšimde ʒennette aršalada Allahnen biñ bir kelam etkeniñni kürdim, ajsa endi aršaladan jer jüzine tüšüp de ʒennet ašlarından vaz gečip te jalan tünjanıñ boklu küt kojların kojroγın bašın ašamaz dedim, ašadım[1] dij.

Noγaj kadini er tarafdan mat etip musapir bolop bašın kojroγın ašap Jalı-bujın[2] en ütkürlerin mat ede. Ālā bükün Jalı-bujın[3] kalkı noγajlarıñ janına kelgende braz sakča bolop laf eteler.

20 Üč kardaš masali

(456) Bir zamanda üč χardaš bar iken. Buların eñ büjük aγaları üjli iken. Küjniñ mullası bu χatunγa kele iken. En küčükleri taz iken. Bu taz bunı tujγan. Bir kün aγalarinen iške baralar. Kač sāt išlegennen soñ taz: men χastalandım dep kajta, üjleriniñ tavanına mine. Mulla kele. Bular ujnaj küleler, soñ kiteʒekte mulla: men jarın felan tarlamda sabanγa barırmen kelemem, sen bir tabaχ čebürek pišir da kelijsin, menim ögüzlerim biri čubar, biri de kara iken dij.

Ertesi kün kardašlar iše kittiler. Bu χatun čebürek pširep mullaγa kitken. Xatun keleʒek vaχtını taz bilep, ügüzlerin biriniñ sırtına külmegin, briniñ sırtına da čekmenin japtı. Hatun doru bularγa kelgen, χatunıñ hoʒası: nasl[4] šej, sen bir vakit aš ketirmezdiñ, ne boldo saña dij. Hatun: ej tüšindim, sizni aʒıdım[5] da ketirdim dij. Taz iši bile ama tiñmadı[6]. Ašadılar, ičtiler, χatun kitken ama χatun da bunı tujdı. Ertesi kün bular jene išlerine kittiler, mulla kelgen. Dün ničin kelmegenin surdı. Hatun nasl[7] bolγanın ajttı, lākin u vaχıt taz jene tavanda iken. Hatun mollaγa: jarın ular iške kitmejler, sen falan tükandan braz zeher al da men ularγa aš piširep berejim, hepsi de ülsin, soñ rahat bolormız dij.

Taz bunı išitep, [aman][8] mulla baraʒaχ tükanγa barıp: tükanʒı, br sāt tükanını ve urbalarını maña ber, saña jüz ruble dij. Rubaları kijinėp[9] utırdı[10]. Molla (457) kelip tükančıden zeher istej. Taz buña zeher jerine braz kara biber berip julladı, üzi braz zeher alıp kajttı. Mulla zeheri χatınγa ketirep verdi. Hatın aš piširep biberleri sepken soñ bular ašχa kelgen. Euelden taz bu išni añattı, braz kašık kapıp juvarlanırsıñız dij. Bular üčü da brer kašiktan juvarlanγan. Hatın mullajı čakırdı, buları sakladılar, akšam kümermiz dep mulla kitte. Keʒe jene kelgen, χatun mullaγa kajγanaχ pšireʒek bulıp br kazan[γa] jaγ kujdı, jaγ kajnamaγa bašlaγan. Mulla ateš karšısına[11] tajanıp juk-

1 ašadim
2 Jali-bujın
3 Jali-bujın
4 R nasıl
5 aʒıdim
6 R tıñmadı
7 R nasıl
8 Crossed out in the Ms.
9 kijinep, R kijinip
10 utırdi
11 karšısina

laүan. Hatun ekinӡi üjden jımırtka ketirmege kitkende, taz turıp kazanın jaүın mullanıñ avzına tükti. Mulla üldi. Taznıñ aүaları da turdılar. Taz mullanı arkasına alıp üjine bardı, mullanıñ sesi kibik kapunı ač dep kıčıryan. Mullanıñ hatunı: sen er vaχit kitep bújle keč kelesıñ, men saña kapı ačmam, jat azbarүa dij. Bu jene: ač, ačmasañ baӡadan tüšermen, seni düverim dij. Hatun tüsseñ tüš, ačmam dij. Taz aman tamүa čıχıp mullanı baӡadan atkan, mulla ičeri tüšti, χatun šaštı. Taz: mulla‿kaj[1] dep kirde. Hatun: aman saña bir kise altın berijmen, sen bunı ajtma, em de al kit, bir jerge küm dij.

Taz altınları, mullanı jüklenep küj četinde br bustan bardı, bustanıñ urtasına tiklep kujdı, ügine br kač kavın tuplap kitte saχlandı. Bustanӡi baχsa bustan urtasında br adam, sen kim dep tüfenk alıp (458) urdı. Mulla bir taha ülgen. Bu taraftan taz bustanӡiden karpuz almaүa kelip kürdi. Bu bizim mulla, sen ničin ütirdiñ, men seni šindi bildirejim dij. Bustanӡı: aman saña šo kadar para, ajtma, bunı al kit küm dej. Taz paraları alıp, üliji jüklenėp kitte, bardı br degirmen suvını kesti, üliji su kesilen jerge ajak üstüne utırtıp kitte, tegirmen tuktadı. Tekirmenӡi čıχıp baksa bir adam suvı kesken otura: sen nečin suvı kestiñ dep br čapa vurdı, üli jıχıldı. Brde baksa taz braz boүdaj alıp kelgen. Bunı kürdi: vaj sen bizim mullanı nečin ütirdiñ, šindi seni ӡujdurajım. Tegirmenči: aman bo tegirmeni saña berejim, bunı al kit, br jerge küm dij. Taz jene mullajı uñzuna urıp, br numaj suүa attı, tegirmeni de aldı. Soñ kelip aүasınıñ χatının ajırtkan. Bular üčü de bojdaχ kalүan.

Künlerden br kün br padiša čoban χıdırup[2] kelgen. Taz buña čoban jürdi, kojları baχarken keӡe taudan br kürülti čıχtı, taz saχlandı. Kürültiden bir kara at kelip kojları taүıtkanda taz brden čıχıp atnıñ čačlarından‿dutkan[3], jüvermedi. At tilge kelip: meni jüver, ma, saña br kıl, sen zurүa kalүanda kızdırırsen[4], men saña kelirmen dij. Ertesi keӡe jene šu küriltiden br čal at čıχıp kele, bunı de tuta. Munnan da br kıl ala. Üčunӡı keӡe jene bir sarı at kele, (459) bunı de tutıp br kıl alıp jüvere. Padišanıñ üč kızı bar iken. Čoχ jiketler istediler, padiša üč šej küsterdı: kušuda atlarım kečkenge küčük kızım, atnen bu sarajnıñ üstüne čıχıp tüškenge, ortanӡı kızım, bu atešten kečkenge büjük kızım dij.

Erkes ӡıjıldı, kušu bašlanүan. Taz kač kün üjlerine sılaja kelgen, aүaları da atların minep kitkenler. Tazүa: ajde, sen čoban jürgen padišanıñ kušusı bar, kitejik, sejr bakarsın dijler. Taz: joχ, men üjle jerde jarašmam, siz de barmañız dej. Bular tiñmediler[5], ata minep kitteler. Bular kitken soñ taz kara at kılın kızdırup, buların artına jetip brer kamčı urıp kitte, bardı kušujı kazanүan, aүalarından euel üjlerine kajttı. Akšam aүaları kelgen, taz halların suradı. Bu kün julda bizni br jiket urıp kitte dijler. Taz: men sizge barmañız dedim tiñemediñız[6], bu kün br urdılar, jarın ekı urırlar dij. Ertesi kün bular jene kitkenner, taz čal atnı čakırup minep artlarına jetip ekišer kamčı urıp uzėp[7] kitte.

1 mullakaj; R mulla akaj
2 χıdirup
3 čačlarından dutkan; R čačlarından tutkan
4 kızdirırsen
5 R tiñmadılar ~ tıñmadılar
6 R tiñamadıñız / tıñamadıñız
7 üzep; R uzıp

Barıp úč kat sarajγa minep túšti, jene kazanıp kajttı. Akšam jene aγaları kelgen. Taz bularγa ekišer kamčı urγanın ajttılar. Taz: jarın úč kamčı urır dij. Erte jene kitkenner. Taz sarı atnı čakırıp aγalarına účer kamčı urıp kečip kitte. Barıp atešden kečtı, kavıller bitkenden soñ padišanıñ kúčúk kızın úzi, ekinǯisin de eki aγasına alıp padiša bolıp jašadılar.

21 Bir fukara adam

(464) Bir adamnıñ úč ulı bar iken. Bulardan briniñ adeti kúl karıštırmak iken. Bu adam kúnlerden br kún hastalana, balaların janına čakıra: men úlgen soñ účúnúz de brer keǯe kelip janıma jatarsız dij. Eki ulı br šej ajtmajler, kúčúgi kúl karıštıran bek jaχši babam, jatarmız dij. Babaları kač kúnnen úle. Bújúgini adı Hasan iken. Sen bújúksin, bu keǯe nebet seniñ dijler. Hasan dij ki: bu vakta kadar neǯe baj adamlar úlgen, kimniñ ulını kúrdiñiz barıp kabirde karaul boldoγını. U ǯan sıχındısından ajtkan dij. Ajsa siz barmasañız men barırmen dep kúl karıšdıran[1] babasından kalγan čokmar tajaγı alıp kabirge bardı.

Kabir ústúne tajaknen urdı. Babası kabirden: Hasan, sensiñ mi dij. Žavab: jok. Veli, sensiñ mi dij. Žavab: jok. Kúl karıštıran, sensiñ mi degende menim, babam dij. Eh, sen bolsañ χoš keldiñ balam, otur dij. Sabā erte br taha tajaγı kabirge urdı de ajttı: babam tur, sabā namazın kıl de jat. Babası turdı. Soñ saulıklašıp kajttı, keldi uǯaχ bašına, jene bašladı kúl karıštırmaγa. Aγaları: bardıñ mi, neler kúrdiñ, ne ešittiñ dep surdılar. Mennen surmañız, bar úziñiz bekle de kúrıñız dij.

Ekinǯi kún nebet Veliniñ iken, uña: sen bar dij. Deli kit, bašımızdan ǯujul, baraǯak bolsañ bar úzıñ dij. Kúl karıštıran jene čoχmar tajaγı alıp kitte, jene eulki kibik tajaγı kabre urdı. Baba: Asan, sensiñ mi? (465) Veli, sensiñ mi? Žavab: jok. Kúl karıštıran, sensiñ mi? degende evet menim, babam, dij. Sabā namazında[2] jene tajaγı urıp babasın namazγa ujattı. Babası turıp namaz kılγan soñ saulıklašıp jene kajttı.

Účúnǯi keǯe nebet úzinin iken. Kimseje br šej ajtmajıp tajaγı uñzına[3] urıp bardı, kabirge tajaγı urdı. Asan, sensiñ mi, Veli, sensıñ mi? Žavab: joχ. Kúl karıštıran degende menim, babam dij. Jattı, sabā jene babasın ujattı. Babası: sabr et, balam, tašlap kitme, namaz kılajım, saña br ajtkanım bar dij. Namaz kılγan soñ: balam, felan jerde br uba bar, šo uba ústúne eki reket namaz kıl, ubadan saña kırχ arab čıχar bojor, akam, dijler. Sen: šonda menim úč atım bar, čıχarıñız kúrejim dep, atlar čıkkan soñ účiniñ kojroγından brer tel alırsıñ, ne vakit seniñ bašına br [belā][4] zur kelse, ǯebinden čıχarırsıñ, telleri jaχarsıñ, ular saña kelijler dij. Býjle japıp jene kitep oǯaχ bašına kelıp kúl karıštırép[5] jatkan.

1 R karıštıran
2 namazinda
3 uñzina
4 Crossed out in the Ms.
5 karıštirep; R karıštırıp

Aγaları turγan soñ munnan surdılar. Mennen ne surajsız, babamıznıñ sizge čok χatırı kalγan dep bašχa šej ajtmadı. Bular turdılar, kajve piširdiler ičtiler, buña birmediler. Karnı bek ač iken, aldı bašın čıχtı ubaja kitte, eki reket namaz kıldı, arablar čıχtılar, bojor, akam dijler. Maña kırχ türlü aš dij. Šu‿ssāt arablar češit ašlardan ketirdiler, bu ašadı, ičti de jene üjge kelgen. Aγaları: sen kajda jüresiñ dep surdılar. Heč br jerde, anda mında[1] (466) jürem, aš jok su jok, nišlejim dij. Biz seni bu vakta kadar baχtıχ, ašattıχ, ičerdik, ende bar üz üziñi baχ dep üjden de kudılar[2]. Babasından kalγan br üjleri bardı, u da siziñ bolsın dep čıχıp kitte, jene ubaγa bardı, namazın kılγan son arablardan aš suradı. Ašap ičgenden[3] soñ čal atnı egerlep minde.

Kite-kite br padišanı kapısına tüštı. Padšanıñ kızı bunı kürep ašıχ boldo, jaš kajtıp kite, čölge čıχχan soñ atı jüvere, kele br adamγa, ırgat jüre. U akaj: men tujγa kiteʒejim, sen men kelgene kadar bu kapustalarnı čapalap bitir dij. Jaš jatıp juχladı, turdı baχsa aγasınıñ keleʒek vaχtı kelgen. Čal atnıñ kojroχ jünin jaka, at kele bojor akam dep. Bu kapustalar ičine čapkalap astın üstke čevir, urtasına aunap kit dej. At üjle japtı kitte. Aγası keleʒek vakit bu jaš bašladı brer brer kapustaları jañadan jere otortmaγa. Aγası kelip bu nasıl šej dij. Jaš aγlajarak br čal at keldi, talaštım čıkaramadım, büjle japıp kitte dij. Adam bunı urmaγa bašlaγan. Jiget br tarafγa, üzi br tarafγa jıkılėp kaldılar. Padišanıñ kızı[4] bunı kürep aʒıdı, vezirlerin jüverıp jašı čakırttı, kač jıllar padiša kapısına jürdı.

Br kün padišanıñ kızların üjleneʒek zamanı kelgen. Ülkede ne kadar jigėt barsa, ʒümlesin čakırdı. Kızlar kimni begenseler almanen uraʒaχlar, urulγan jigetke bereʒekler. Padišanıñ kızlarınıñ ekisi bu kül (467) karıštıranıñ[5] aγasına urdılar, ülkede bolγan jašlar epsi kečti, br tene ırγat kalγan. Kelip bunu čakırdılar, barmadı. Meni ütüreʒekler mi digen. Padiša emr etti, zurnen bunı alıp keldiler. Kapıdan kečkende kız almanı buña urdı. Ʒemāt jañlıš dep baγrıštılar. Büjle üč kerre kečti, üčinde de alma buña urıldı. Sevmij-sevmij padiša kızın buña berdi, ajırı br eski uda berdiler, kıznen unda jašadılar.

Künlerden br kün ekinʒi padiša bu padišaγa ʒenk ačmaχ istegen. Bunın χatını kelip haber berdi: babamızıñ ʒenki bar, aγalarıñ kiteler, se‿ne[6] japajsın dej. Üjle ise bar, babaña ajt: br at ta maña bersin, men de kitejim dij. Hatın barıp ajttı, padša: šunda bir ešek bar, barsın alsın da kitsin dij. Bu jiget ešegi minep kite, bulardan euel barıp ešegnı[7] br čamırγa batėrup[8] čıχardı[9], bular kelip jettiler. Bunı kürep külüštiler: munda nišlijsıñ degende uzėp[10] kittiler. Bular kitken soñ bu čal atnıñ kojroγın kızdırdı, čal at keldi. Atγa minep ʒenk mejdenine janaštı. Padiša bunı kürdi ama tanımadı. Jiget: men sennen taraf bolıp, ularnı jeñsem ne berirsiñ dej. Padiša: sen[i] bašvezir etermen dij. Bu bašladı tüšmenı kesmege. Birazdan kulınıñ küčük parmaγı jaralanγan. Padiša ʒebinden

1 minda
2 kudilar; R küdılar
3 R ičkenden
4 kızi
5 Ms karıštiranıñ
6 se ne; R sen ne
7 R ešekni
8 R batırup ~ batırıp
9 čıχardi
10 R uzıp

jaulıyın čıχarıp bunıñ parmayın bejledi. Ʒenk bitte. Bular χazanıp kajttılar, (468) kimse bunı tanımadı. Bulardan euel kajtıp üz utasına kirep jatkan.

Xatını barmayındaki jaulıyı juvıp kurutmaya jajdı. Bu χatınıña: bar babaña ajt: maña vezirlik bersin dij. Hatun barıp babasına ajtkanda, ayaları: bu deli šaštı, seni masχaralıkka almaya bašlayan dıjler. Padiša: bar čakırıñ, bašın kesejim dij. Bardılar baksalar, bu juχlaj. Hatın degmeñiz uña, šimdi boldorıp kelgen, ujansın[1] son barır dij. Kajtıp padišaya ajtkannar. Padiša: tez bar, sürep ketiriñ dij. Keldiler, bu ujanyan utıra. Bularya dij ki: bar, ajtıñız, uña ʒenkte kim jardım etken, br tüšünsin degen üzi čal atı čakırup[2] minep padišanıñ utası ügine barıp turyan. Padiša bu kım bolyan dep surdı. Bu küčük barmayın küsterdikten soñ iši añlaj, soñra jeñiden dügün etip bašvezirlik berdi.

22 Br hırsız ekjeti

(475) Br zamanda čok usta br hırsız bar iken, bunıñ un eki jašında br ulu bar iken. Br kün ulına [nigedir ačulanıp][3]: sen br iš japamazsıñ, senden adam bolmaz dij. Ulan babasınıñ zanātın suryan, babası hırsız boldoyın bildirdı. Ulan: üjlese sen maña üz ünerni küster dij. Menim ünerim šu terekte jatkan karyanıñ astından jımırtkasın alıjım, karyaya tujdurmam dij. Ulan: hade baχajım, čıχ al dij. Bu adam terekke čıχıp ujanıñ astın tešip jımırtkalarını alyan, karya da[4] tujmadı. Babası jımırtkalarnı almada bolson, ulan babasınıñ kütinden[5] ištanını čıχarıp alyanın da tujmadı. Terekten tüšep: muna karyanın jımırtkası dedikte, ulı: ja babam, kütinden[6] ištanıñnı kim alyan dij. Baba baktı ištanı jok, joksa terekte kaldı[7] mi‿ken[8] dep terekke bakkanda kulından jımırtkaları[9] da aldı. Baksa ištan[ı] ulınıñ[10] belinde, ej ulum, sen mennen de usta hırsız[11] ekensın dij.

Babası br kün: hajde balam, barıp padišanı haznesın čalajıχ dij. Kittiler, haznege bardılar, hazneniñ üstüne čıχıp br tešik ačtılar, ipten basamak jasap ašaya tüštiler. Altın alırken altınların sesi čıχıp šatırdadı, bekči tujdı, ulı baskıčdan uyarı[12] čıχtı, bekči de kelip jetken. Babası tüpeden čıχarken artından (476) bekči tutkan. Ulan babasın alamadı, kim boldoyın bilmesinler dep bašın kesip kitte. Üjlerine baryan, anası babasından χabar surdı, ulı: babanen hırsızlıkka baryan‿dık[13], babam tutılyan, tanımasınlar dep bašın kesıp ketirdim, üldü dep kümijik dep kümdiler.

---

1 ujansin
2 čakirup
3 Crossed out in the Ms.
4 de
5 kutından
6 R kütiñden
7 kaldi
8 mi-ken; R mi iken
9 jimirtkaları
10 uliniñ
11 hırsiz
12 R juyarı
13 baryandık; R baryan edik

Padiša χabar bergen: bu adamın bašın kimi de kesip alγan bolsa kelsin kevdejı de azbarın küšesinden alıp kačsın, jigetlijin bilejim dij. Jiget: anam, men bunı barıp ketirirmen, maña ruhset ber dij. Anası: jok balam, babañ bùjle ʒujıldı, seni de totop kesseler, menı kim baka, biz ʒansız kevdeji ne jasajıχ dij. Joχ, men inšallah saγ-salam kajtıp kelirmen dij de jiget šeherden barıp eki at terisi alγan, bri kara bri čal. Čal teriji üz atınıñ br jakına jajdı, kara terini br jakγa. Sarajnıñ ügінde kevde asılγan, eki jakına eki de bekči. Jiget bularnı kürgen, br ajlanıp kelgen, kılıʒnen ipi kesip, kevdeni alıp kačtı. Bekčiler artından[1] kuvıp jetemediler. Padiša bekčilerden nasil boldoγın surγan. Saγda utırγan čal atlı br adam, soldaki kara atlı degen. Bular ikeu bašladılar dāvājı. Padiša ıč biriñızınki tora tügel, dep ekisin de bašların kesti. Ulan babasın kümdi.

Padiša jene χabar etti ki saraj ügine br čuval altın, (477) kevdeji alγan bunı de ala dij. Jiget jene anasından izin surap kitte. Bazardan br balıχ aγı alγan, ičine pičen toldırdı. Br boran küni ičine kirde, juvarlana sarajγa barγan. Kečerken bekčiler kürseler de pičen hisap etip bakmadılar. Altın janına bardı, altunları aldı, jene juvarlana četke čiχıp atına minep kajttı, anasına ketirep[2] altınları berdi. Akšam bekčiler baχsalar, altın joχ, kelip padišaγa χabar bergenner. Padiša u jerden kimse kečken mi dep suradı. Bekčiler br top pičenden baška kelmegen dijler.

Padiša jene χabar berdı: šu adam kelip kızımnen jatsın, kızım da berejim, padišalık da. Jiget jene anasından izin istep, [bunı de jasamaγa karar berdi][3] Anası: seni anda ütürirler dij. Jiget: eger meni ütürseler, men saña ketirgen altınlar jetej dep kitte. Kiterken kabakka uγradı, at üstünden br šiše raki istep alγan. Kabakčı akčajı alaʒakta belindeki kılıʒnen[4] kabakčınıñ kulın kesıp alγan. Jiget sarajγa jetištı, akšam boldı, padišanıñ br sevgili ečkisi bar iken, ičeri tıšarı čıχıp kire‿jdi[5]. Jiget ečkini tutıp br az lampa jaginen[6] jaγlap br serniknen ečkiji jakıp jüverdi. Ečki ičeri tıšarı čapmaγa bašlaγan. Kürenler padišanıñ sarajı jana dep čapıštılar. Šu arada (478) bu suvlık[7] turubadan kız bolγan udaγa kirdi, ol keʒe kıznen[8] jatkan. Sabāγa br sāt kalγanda br simege barajım dij. Kız: joχ, munda odada sıj degen. Üjlese kulımdan tut, men odadan čıχıp sijejım dej. Jiget kıza kabakčınıñ kulın tutturıp, üzi čıχıp ketken. Üjlerine bardı.

Padiša bunı da išitken, baktı tutmaγa čara bolmajʒak. Habar berdi, bu išlerni japχan adam kelsin, br ʒeza jok, kızım da berejim, padišalık da. Bu jiget išlerni japkan menim, japaʒaγın jap degen. Padiša: aferın balam, kjāmil ikensin dep kırk kün toj etip kızın berdi.

1 artindan
2 R ketirėp, but the more common version is ketirep.
3 Crossed out.
4 kiliʒnen
5 kirejdi; R kire edi
6 jaγinen
7 suvlik
8 kiznen

23 Akıllı kız masali

(483) Br zamanda br jaš padiša bar iken. Künlerden br kün čülge čıkkan ve bir adam bašı tapkan. Bašta: χatıdan χatı ne bar, aʒıdan aʒı ne bar, tatlıdan tatlı ne bar dep jazılγan. Bu bašı alıp kelgen, vezirlerine: kim buña ʒavab tapsa, br tulup altın dij. Vezirler ʒıjıldı, baχtılar, kimi taš, čelik, kimi kemik dijler. Aʒıdan aʒı: limon, tuz, büber, suγan dijler. Tatlıdan tatlı: bal, šeker, jüzüm degenner. Padiša buña rāzi bolmadı. Ülkeniñ er jakına χabar bergen, bilen kelsin dep. Br küjde tura iken br kart, unı brden br kızı[1] bar iken. U kız akıllı iken. Babasına: padiša br baš tapχan, undaki jazıγa ʒavab bergen bolmaγan, sen bar, χatıdan χatı at tujaγı, aʒıdan aʒı ülüm, tatlıdan tatlı jigetnėñ[2] üjlengen keʒesėdir[3] degen.

Kart sakalın sallaj-sallaj kapı ügine kelgen, bekči ičeri jüvermegen. Kart dij: bar padišaγa χabar ber, men keldim unı suvalına ʒavab birmege. Padišaγa χabar berdiler. Kirsin baχajım dej. Kart kirde, kartnıñ ič de ukudan jazıdan χabarı jok. Bašı alγan, evirep-čevirep baχtı da kız degenı kibik ʒavab bergen. Padiša: sen bunı kimden ügrendiñ dej. Kart üzim bilem dese de padiša: jok, senı bilmeʒegın men añadım, (484) saña bunı kim de bolsa ajtχan, dorısın ajt, jāki bašın kesejım degen. Kart šaštı, korktı, kızım ajttı dej. Padiša kartχa: sen kit, men kızıña br tel ipek jüverirmen, maña bu telden br ʒübbe išleje, japmazsa senin bašıñ da, unın bašı da kesilij dij. Kart aγlaj-aγlaj kajtıp kızına: sen menim bašım belaγa kujγan, ıč japılmaʒaχ šej emr etken dij. Padiša eki vezirinen jipegı julladı, ketirdiler. Kız alγan son, bularγa br kibrit kibik čüp verip: bizler faχir kišiler, bundan padiša ʒübbeni toχumaγa br tezja japtırıp jüversin degen.

Vezirler čüpi padišaγa ketirgen. Padiša bu kız delidir, bu čüpten tezja jasılıj mi, bar ajtıñız japsın, japmasa bašin kesermen dij. Kıza kelip ajttılar. Kız: eger bu čüpten tezja olmasa, men de br telden ʒübbe japamam degen. Jene kajtıp padišaγa ajtkannar. Padiša kartnı čaγırttı: bu üküzi al, kızıñ bu ügüzi altı ajda buzaulatsın dij. Kart jene aγlaj-aγlaj, kelgen. Kız: sen jene nege aγlajsın dij. Kart: padiša altı ajda ügüzi buzaulatmajı emr etken dij. Kız: aγlama, u ügüz bizge altı ajγa ašamaγa jeter, son ʒevabi menim bujnıma dij. Ügüzi ketirdiler. Kız: bek güzel, altı ajdan son kelip alırsız dij. Vezirler kajttı. Kız babasına: bar, ügüzni suj dij. Kart kurχa-kurχa sujdı, ašamaγa bašlaγannar.

Künler kečken, altı aj kelip jetken. Kız babasını hamamγa kirsetti. (485) Vezirler kapıdan kirep utırγannar, surγannar. Kız: babam hamamda bala tuγıra dep vezirlerni kajtarγan. Vezirler kajttılar, padišaγa χabar berdiler. Padiša: er adam bala tuγurıj mi, u jalan, men unın bašın kesejım, bar kartı ketiriñ dij. Vezirler kızγa kelip, er bala tuγurmaʒaγın ajtkan. Kız: er adam bala tuγurmaz da ügız buzaular mi? Bz ügizi sujıp ašadıχ degen. Padiša jene kartı čaγırγan: seniñ kızıñ bek akıllı iken, men unı χatınlıkka alajım dij. Kart: men bilmem, üzinden surajım dep kajtıp ajttı. Kız: biz faχir adamlar, padišaγa lājıχ tegil, padiša bızni masχaralıkka almasın dep, bir kahad jazıp babasinen padišaγa julladı.

1 kizi
2 jigetnen
3 keʒese dir

Kahatγa jazγan: ej padiša, sen büjük adamsıñ, bizi külme, biz seniñ suvaline ǯavab berdik, vāde kılγan malın ber degen. Padiša ukıp baχtı. Kartka: juk, külmek tegil, kerčekten alaǯaγım dij. Kart kajtėp[1] kızγa ajtkan, kız rāzi boldı, χudalar kelip nišan bergenner. Biš un künden soñ dügün bašladılar, padiša jaχšı rubalar ve pajton jüverip kızı aldırγan. Kız kijinėp[2] pajtonnen padiša sarajına kelgen, čalginen[3] karšılap čıktılar, sarajγa kızlar janına kirsettiler. Kızlar ujnap küleler. Bu kız čalγıjı sejr etip šašıp tura. Čünki üz ömründe üjle čalγı kürmegen. Kač sāttan kızlarγa (486) čıkmaγa emir berildi. Padiša kelip kirdi, kürüšdiler. Aš su kelgen ašap ičtiler, son padiša kızγa: sen šindi menim χatınım boldoñ, menim emrinden baška šej japmajsıñ, menden surmaj hüküm etmejsıñ, ejer etseñ kapımdan kuvarmen dij. Bular kelin kijeu boldılar.

Kač zamandan soñ padiša üjde joχ iken. Hanım penǯereden bakıp utırırken karšıda kajve-hāne bar iken. Bu kajvege kač konuk kelip tuktadılar. Brsınıñ atı bajtal, ekinǯisiniñki ajgir[4]. Bular kajve ičep turγanda bajtal χolumladi, taj kelip ajγır astına jattı. Bu adamlar čıχıp baχsalar, ajγır astında br taj jata. Ajgirnıñ[5] sābi: menim ajγırım kolumlaγan dij, bajtal sābi: joχ, menim bajtalım kolumlaγan dij. Bular dāvalaštılar, padišadan surmaγa kelgenner. Padiša üjde jok. Hanım: dāvaların ne [dep] suraγan. Bular ajttılar. Hanım kürep tura‿jdı[6], ajgir[7] kolumlamaz, taj bajtalnıñdir degen. Bular kajta jatχanda karšılarına padiša rast kelgen, surdı. Bular hanım dāvaların kestigin ajttılar.

Padiša sarajγa kirdi, hanımγa[8]: sen nečin mensiz dāva kürdiñ? Men saña uruχset bermegen‿dim[9], sen inde bu sarajdan istegenıñ al da kit dij. Hanım: men jañıldım, kel braz kejf sürejik, ašap ičejik, soñ kitermen dij. Ašap ičep turγanda (487) türlü ičkiler bar iken. Padiša ičep sarhošlanıp jattı. Hanım br fajton jigilmesine emr etken, fajton keldi. Hanım kijindi, padišanı ketirep[10] pajtonγa jatırγan. Azbardan cıχıp šeher julların kač kerre tulašıp, üzi küjine jullandı. Julda barγanda padiša ujandı: meni kanda alıp kitesin dep suradı. Hanım: sen maña istegen šejıñ alırsın dediñ. Menim sennen γajrı istegenım joχ, seni alıp kitermen degende padiša bunın dora boldoγın añlap kabātin baγıšlap[11], kajtıp kelgenner, brge jašap bala-čala sābi boldolar.

24 Baj kiši ekjeti

(493) Vaχtinen br zengin adam bar iken. Her zaman sufasın üstüne utırıp kajve ičer iken. Jene br kün kajve ičep turγanda üj ögündeki terekke br kuš kelip kundı. Bu adam bunı kürer kürmez, bašından[12] esi kitep jıχılγan. Kač sāttan turdı, baχtı, ügine br kanat

1 kajtep
2 kijinep, R kijinip
3 R čalγınen
4 ajγir
5 Ajγirnıñ
6 turajdı; R tura edi
7 ajγir
8 hanimγa
9 bermegendim; R bermegen edim
10 R ketirėp, but the more common version is *ketirep*.
11 bayišlap
12 bašınden

tüšep kalγan. Kuš joχ, alıp kanatı saklaγan. Birazdan üj ičinden bunıñ χatını čıχχan. Bu adam kajvesin ičmegen. Saña ne hal boldı da bu vakta kadar kajveñı ičmediñ dij. Adam: sen surma, men ajtmajim maña ne bolγanın dij. Hatın zurlaγan, jok ille ajtaӡaksıñ dij. Men akaj bolıp tajanamadım, esim kitte, sen kadınsıñ, eger kürseñ ütıñ patlaj ülijsin degen. Jok, ister istemez küstereӡeksin degende, akaj ӡebinden čıχarıp kanadı küstergen. Hatın kürdi de jıχıldı üldı. Bahčada üč ulı bar iken, χabar bergenner, kelip anaların kümdiler. Babaları neden üldigin ajtmadı.

Künlerden br kün bu adam: muna, anañız bu kanadı kürgende üdi patlap ülgen, siz bu kušı tapıp ketiriñiz dij. Üč ulına ekišer jüz ruble para bergen, üčü de kitkenner. Bular br jerge bardılar, üč jul [avzında] tahtada jazılγan[1]: bu julnen kitken falan šehirge čıχar, bunınnen kitken br küjge čıχar, bu julnen kitken kajtmaz. Büjügı šehirge kitken, (494) urtanӡısı küjge kitken, küčügi kajtılmaӡaχ julnen kitken. Kite-kite br tauγa bardı, baksa ügine tuman kibik [...][2] eždeler kele, artık bunı jutaӡaχlar. [Uların][3] artından br ežder taha čıχtı, ügindekilerge kečip: beniadamγa dejmeñiz, men teñızde boldoχta, bu jiget br bala‿j‿iken[4], menı tolkun četke atkan, baška balalar ütireӡek bolγanda bu alıp teñızge tašlap ӡanım baγıšlaγan dij. Jiget kutulıp jula kitte.

Kite-kite br sarajγa barγan, čevresin jürüp baktı ič kirileӡek jer jok. Nasılsa br čare etip sarajγa kirgen, baχtı, sarajda br güzel kız, baharda ačılγan kül kibik, sufra kujılγan, aš hazır. Jiget utırıp ašamaγa bašlaγan. Kız: ej jiget, šimdi munda bir deju kelij, seni de halak etej, meni de, men seni saχlajım dij. Jiget: jok, maña saklamak kerekmij, kelse bakajıχ dij. Laf biter bitmez, deju kelip jetti: bu sufra seniñ‿čın[5] kurulγan mı, sen bašta menimnen küreš[6] te soñ ašasın dij. Jiget: tuχta, men ačım, ašajım da soñ küreširmiz[7] dij. Jiget ašaγan, tujγan, dejunen kapıštılar, dejuji kılıӡnen tuγraγan. Kız: artıχ bunı elak ettiñ, meni alıp kit dej. Jiget: joχ, men büjle br kuš χıdırmaγa kitem, kajtχanda seni alıp kajtırmen dij. Kız barmaγından jüzügin čıχarıp jıketge bergen: men senıñ, sen menı dep, saulıklašıp čıχıp kitken.

(495) Kite-kite ekinӡi sarajγa bardı, baχsa eki kız. Bu kızlar ežder keleӡegin ajtır‿ken[8], ežder kelgen, tutıštılar, bunı de elak etken, kızları kajtkanda almaγa süz berip kitken. Br azdan bir terek tübine barıp jatkan. Jukuγa kitej kitmej, kuš balaları čivildemege bašlaγan. Jiget baχsa, bir jılan terekke čıχa, aman turıp jılanı duγrap jene juklamaγa jatkan. Kuš kelgende: men čoχ jıldan beri mında bala asrajamadım, meger bu jiget ala eken, bunıñ küzlerın čıχarajım degende, balaları: joχ anam, bizi ašaӡak šu jılan iken, bu jiket biznı kutardı degenner. Ajse juklasın dep betine kün urmasın‿čın[9], bunıñ betine kanatların jajkan.

1 jazilγan

2 Crossed out in the Ms.: br kara bulat-kaj [-taj!] br ežder čıχχan.

3 Crossed out in the Ms.

4 balaj iken; R bala iken

5 seniñčın; R seniñ ičın

6 küriš

7 küriširmiz

8 ajtırken; R ajtır eken

9 urmasın-čın; R urmasın ičın

Jiket brden abdırap ujandı, baχsa ústúne br kuš utıra. Kuš tilge kelip: jiget, isteǯegin iste dij. Jiget senıñ saulıγıñ dij. Jok, menim saulıγımdan saña fajda jok, úziñ‿čın[1] iste degende: újlese, menim isteǯegim muna, bu kanatnıñ sābi dij. Kuš: vaj, kartlıγımda rast keldiñ ama nejse bakajıχ, bar kırk metej šarab úč júz put da et, men saγga bakkanda et, sulγa bakkanda šarab berijsin dij. Jiget ketirgen. Kuš hepsin júklenip učıp kitken.

Bara-bara br teñızden kečkenner, kuš kunγan, jiget kuštan túšken. Kuš: ajda ket, šındė[2] sen br suγa barırsıñ, bismilla dep br avıǯ[3] su (496) ič, su ekı janγa ajrılıj, kečıp kitersin, br sarajγa barijsın, andaki ačık kapılarnı kapa japıχların ač. Asılı kılıǯlar bar iken, ularnı alıp jaγlap jene asarsıñ. Šundan br kapı taha ačajsın, kuš asılıdır, hemen alıp kačajsın dij. Jiget bújle etip kušı alγan. Arttan deju tujıp čıχχan, kılıǯlarγa kes dij. Kılıǯlar: bunǯa jıldan berli asılı turamız, bı‿kerre[4] jaγlamadıñ, barsın kitsin degenner. Jiget kitej, kušγa kelgen. Jene sırtına minep julγa čıkkanlar.

Sudan čıkmaγa az kalγanda et bitken. Jiget baǯaγından braz et kesip kaptırγan. Kuš bildi jutmadı, čitke čıkkanlar. Kuš: ajde imdi túš kit dej. Jiget topallıγın bildirmeǯek bolıp: baštan sen kit dej. Bular sen kit, men kit, čara bolmaγan. Jiget kitken ama topal. Kuš jigetke ajaγına ne bolγanı suradı. Jiget et bitken‿di[5], kesip saña kaptırdım dij. Kuš gaχ etip avızından etni čıχarıp, ajaγına japıštırıp, saulıklašıp kitkenner.

Kitkende kúrgen úč kızı alγan, aγalarından ajrılγan jerge kelgenner. Bu túšingen: bız mınnan kitkende úč kardaštık, šindi jalγız kajtmaχ jaχšı tegıl, dep bújúk aγası kitken julnan kitti. Barıp br šeherge kirgen, baχsa aγası parasın bitirgen, em de júz ruble burǯlanγan, kajve kúšesinde utıra. Bunın burǯların bergen son kittiler ekinǯi aγasın χıdırup taptılar, unı de alıp kelgen. (497) Úč kız, úč ulan kajttılar. Aγaları bunı nasl‿útirep[6] kušı alıp čaresin χıdırdılar. Kújge jakın barγanda aγaları: siz javaš javaš ketiñ, biz šu kújden etmek alıp čıχajıχ digen. Bular kitkenner, kújden ašaǯak ičeǯek, birde kijız alıp čıχtılar. Br kujunıñ avzına jajdılar. [Arttan bular da jetip kelgen ašap ičtiler.][7] Aγaları kijıznıñ eki četine utırγannar. Kúrmek ičin kušı istejler, ve jigetke urta jerde utırmaγa izin bergenner. Jiget utıra. Kušı alij almaj, ekisi brden turdılar. Jiget hemen kujuγa kitken, kızları ve kušı alγannar. Üčúnǯi kza[8] da zurladılar, ama kız: men úlúrem, munnan kitmem dij, bu da kalγan.

Bular babalarına kelip: muna kušı ketirdik dijler. Ja kana kardašlarıñız dep suraj. Biz kitkende úč jul avzına bardıχ, tahtada bu julγa ketken kelmij deje jazılı iken. U bizni diñemejip kitep kelmegen dijler.

Jiget kujıda iken esine kızın júzúgin jalaǯaγı túštı, jaladı, úc deju kelip: jakajıχ mi, jıkajıχ mi dep surajler. Jakmak jıkmak iste[n]mez, menı túnja júzine čıχar dij. Šu‿ssāt[9] dejuler čıχarγan, brde baksa kız hālā unda, kızı alıp, úz kújleriniñ br četine kelip br

1 úziñ-čın; R úziñ ičın
2 šınde
3 aviǯ
4 bı-kerre; R bir kere
5 bitkendi; R bitken edi
6 R nasıl útirėp
7 Crossed out in the Ms.
8 R kıza
9 šussāt; R šu sāt

altınǯı tükjanı ačkan. Aɣaları tügün etmege istegenner. Kızlar tügünde jarašık‿čin[1] br altın horas japtırmak istegenner. Küjde bu jiketten baška altınǯi jok, buña barɣannar. Jiget: güzel japajım, ketirsiz bir put fındıχ, findıɣı ašap bitėrgene[2] kadar horaz bitij dej. Jiget hem de japkan. Bu jiget bašına kelgen šejlerni bir (498) bir kahatɣa jazıp horaz ičine kujɣan.

Aɣaları kelip horazı alɣannar. Horaznıñ güzelligine šašıp, ıčin ačıp baχtılar. Ne baχsınlar, kahat horazdan čikkan[3] ukudılar. Babaları bunı išitij-išitmej, aman jigetni čaɣırɣan. Jiget kelgen. Babaları aɣalarınıñ bašların kesmege izin bergen. Jiget: joχ, ular etse de men etmejim dej sučların baɣıšlaɣan son üčeu üč kızı alıp jašadılar.

25 Tuvarčı masali

(505) Br zamanda padišaniñ biriniñ br kızı bar iken. Birde tuvarčısı ve bir jılkıǯısı bar iken. Kız jılkıǯını sevgen, bunı tuvarči tujɣan. Tuvarčı bularnı hep deñep jüre. Künlerden br kün kız jılkıǯıɣa: eki at ketır, birine šejlerimizni, birine de üzimiz minep χačarmız dij. Tuvarčı bunı išitgen. Keǯe br šiše raχi ketirep[4] jılkıǯını esirttı, atları alıp kızɣa χabar bergen. Kız mallarnı atɣa jüklep, brine minep kačıp kitteler. Kız jigetni tanımadı. Kite-kite sabā bolɣan. Kız baktı, jılkıǯı tegil, tuvarčı. Ej meded, men buña χısmet ekenım dep sesın čıχarmaɣan. Kač künner ketken, br padišanıñ šehrine barɣannar, u šehirde üzlerine jer alıp üj jasatkannar. Tuj etep brbrin alɣannar.

Kač vaχitlerden soñ padišahi üzi üjlerine čaɣırɣanlar, χatun br tepsi turšu patılǯan hazırlaɣan, hoǯasına: halk kelip sufraɣa utırɣan soñ bu turšunı de kujarsın dij. Bu adam turšu kujmaɣa unutkan. Ašaɣanda padiša: ah bir turšu bolsa da ašasaχ dij. Hatun ekinǯu[5] üjden ešitken, kapını ačıp koǯasına turšunı küstergen. Padiša hatınnıñ kulın kürep āšik bolɣan. Ašı tašlap turıp üjine kajttı, vezirlerın čaɣırıp: aman bu adamnı ǯujmaɣa br čara tapıñ, son χatının men alajım dij. Vezirlerinden biri: Munar tauında[6] br čokraχ bar, čokraχ bašında br terek, u terekniñ jafraχlarından biši čalar, biši ujnar, bunı ketirmege jüveriñiz (506) anda br deju bar, unı parlaj dij.

Tuvarčını čaɣırıp ajttılar. Bu garib aɣlaj-aɣlaj üjıne kelgen. Hatın kürep: nečın aɣlajsıñ, tünjada munnan kulaj šej juχ. Jarınkı kün deju juklamaɣa jataǯaχ, u juklaɣanda kırk kün bolmajınǯa turmaz, sen tauɣa barıp čokraχtan abdas al, eki reket namaz kılajsın, aman teregni umzına urıp dora üzi üjimize ketirirsıñ. Sakın munda ketirmej padišaɣa alıp barma dij.

---

1 jarašık-čin; R jarašık ičin
2 bitergene
3 R čıkkan
4 R ketirėp, but the more common version is *ketirep*.
5 R ekinǯi
6 tauinda

Sabā erte turup kitte, čokraχγa bargan, abdasdan son güzel teregni umzına alıp kajtgan. Kúj kenarina kelgende padiša tereknıñ šuvultısın ešitken, br adam jüverep dora munda ketirsin dij. Bu adam karšı čıχıp ajtχanda tuvarčı: joχ, men terekni üzim‿čın[1] ketirdim, padiša‿čin[2] tegıl dep újine alıp kelgen, penčeresı úgine utırtkan, χatınınnen[3] čıχıp terek tübine kajva ičkenner.

Padiša bularnı kürep ačulandı, tez bu adamnı ǯujmalarını emr etken. Vezirlerden biri: biz bunı jajlaya jüverijik, anda br kora bardır, kora ičinde üč kočχar bolaǯaχ, ularnı almaya koraja kirij. Kočχarlar tılsımlı, unı ašajlar dij. Tuvarčını čayırkannar: bar, felan jerden kočkarlarnı ketir dijler. Bu jene aylap újine kelgen, χatunı kürep külgen: kurkma kolaj, jarın barıjsıñ, ular koranıñ br četine utlasalar, sen bašχa br četine kirep brden kičırıjsıñ, ular abdırajler. Seni kürep saña čapajlar (507) sen ularya karšı barma, kač koradan atıp tüš, ular da koradan atajler, tılsımları buzılıj[4], jene újge ajdap kelijsın dij. Bújle japıp alıp kelgen.

Padiša kürep jene say kelgen dep ačulanyan, vezirlerin čayırıp: tez buña br čara, jā sizin de útirirmen dej. Bular tüšündiler. Vezirlerniñ biri dij: bu adam tünjada bolyan küč išlerin epsın japaǯaχ, biz unı obir tünjaya senıñ babañdan χabar ketirmege jüverijik digen. Čayırdılar, padša buna obır tünjaya barmajı emr etken, kırk kün de mūlet, kırk birinǯi kün bašıñ ajrılıj dej. Jiget: bolor dep aylap kajtkan. Hatunı kürep: bújle kolaj šejden kurχma, sen bar padišaya, br jerge kırk araba toban jıysın, üstüne br direk utırtsın, seni zınǯėrnen eki ajayından tarttırıp baylasın, tobanları jaχsınlar, sen aχrette barıp unıñ babasından χabar ketirijsin dij.

Bu adam taha zijade kurksa da čara juχ, padišaya barıp ajttı. Japyannar, tobanları jakkanlar. Hatın br küveršın bolıp učkan. Koǯasın tarı etep kačkan. Sabā bir mektüp jazıp[5], koǯasını padišaya jüverdı, tuvarǯı mektubı bergen. Padiša šašıp ne bar dep surayan. Mektübde: ej menim balam, sen bu fakirleri išinden alıp jullarya kujaǯayına[6], bújle etip üziñ kel, hem kürüširmiz, hem de halim bilep kitersin dep jazılyan. Padiša tüšingen, barmaχ mümkin olyanda barıp kürüšmek gerek dep kırχ araba toban tüktirdi, üzini direkke baylatıp tobanları jaktıryan, tobanlarnen üzi de janyan. Tuvarčı unıñ jerine padiša bolıp kalyan.

26 Derviš almasi

(512) Bir zamanda bir baj adam ulyan, unıñ da ne ulı ne kızı bar iken. Sabā akšam bu adam evlad istep Allaya jalvara eken. Br kün br derviš[7] kelip buña br alma berip: jarısın üzıñ, jarısın χatınıñ ašasın, sizden br kız bala bolor. Bu kızya br temir kafas japtırıjsıñ,

---

1 R üzim-čın; üzim ičın

2 padiša-čün; R padiša ičün

3 χatininnen

4 buzilıj

5 jazap

6 kujaǯayina

7 In Kúnos' list of content, the title of the tale reads "Bir derviš almasi" [The apple of a dervish].

kafastan čıχarmajsın, ülgenʒe anda jašaj. Eger čıχarsañ, šu‿ssāt unı kırk bašlı deju alıp kačaj dij. Bular almajı ašaγannar, toχoz aj un kün degende br kız boldo, adın Gülzar kujγannar. Kız čaγına kelgen, kafasta tora.

Bunı padišalar ešitken ulanlarına almak‿čın[1] χudalar jüverdiler. Baj: men kızım berijmen ama kafastan čıχarmajsız dij. Hudalar rāzi bolmajıp kajtıp kiteler. Künlerden br kün Čin-mačin padšası bunı ešıtken de čıχar[ma]maγa rāzi bolop, χuda jüverdı, nišan berdi, kajttılar. Kač ajdan son dügün etip kızı padša alγan. Kač jıllar jašaγanlar, ularnıñ üč ulı boldo, büjügını adı Ali, urtanʒasını adı Hasan, en küčügınıñki Ahmed.

Künlerden br kün bu ulanlar jetišep auγa ketkenner. Hatın koʒasına: men bunʒa jıldan berli kafasta utıram, tünja jarıγı kürmegen, kart boldom, endi meni deju begenmej, ajde braz bahčaγa čıχıp ejlenejıχ dep čıktılar. Bahčaγa barij-barmaj deju kelip, karlı buzlar jaγdırıj, karanγı tuman kestirij, hatunı alıp kačkan. Akšam boldı. Bunıñ eki büjük ulı kajtıp kelgenner, baksalar babaları aγlaj, anaları (513) jok. Suradılar. Babaları: anañızın kırk bašlı deju alıp kačkan dij. Ulanlar izin istep analarını kıdırmaγa ketkenner.

Birazdan küčük ulı kajtıp kelgen, bu da anasın surap, u da izin istej. Babası: jok balam, sen taha jaššıñ, aγañlar tapıp ketirijler dij. Ahmad: jok, bolmaz, men de kitejmen dep zurlaγan.

Babası baχtı čara jok, üjlese bar, arandan br at al da kit dij. Ulan aranγa kirgen, atların bašlarından sıpap, terilerın sıjırėp[2], kojroγından čıχarγan. Kajtėp babasına: sen at beslemegensin, domoz beslegensin, men čülden barıp kıdırıp[3] ketirejım degen bardı. Sürüde br kotor taj bar iken, unı begenıp alıp kelgen. Babası[4] külgen: aman at tapkansin, men seni kitemeʒegın bildim, hajde bar, ne japsañ jap, Allah išnıñ kulaj ketirsin dij, saulıklašıp[5] čıχıp ketken.

Kiterken jul kenarında br ʒadı čıχıp: Čin-mačın padšasınıñ ulı Ahmad, sen bilep kitesin mi juχsa bilmejip kitesın mi? Jiget ajlanıp barγan, unın karšısına: ne bilem, ne bilmejim, kararnen kitem dij. Ʒadı[6]: ajsa sen jañıš kitesın, kajt babaña, kırk kün seni beslesin, son krk[7] batman gurz alıjsın, eki ajaγına da seksen batman čarıχ[8] japtėrėp julγa čıχajsın dij. Ulan kajtıp, büjle japtırıp baškadan julγa čıkkan.

Bara-bara br čojun šehrine bar‿ken[9], šeher četinde bir češme janında atın jüverep konuk boldı. Keʒenıñ br vaktinde br čojun küpri (514) tübine barıp gizlendi. Br vakitten son kırk bašlı br deju ajkırıp kelgen. [Kele-kele bunı atı süründı.][10] Deju atına: sen br vaχit bu jerde sürünmejip [...] bükün saña ne boldo, joksa Čin-mačın padšasınıñ ulı Ahmedni kürgen misiñ dij. Ahmat bolsa, sen oña ne japajsın dep küpür tübinden atılıp čıkkan. Deju Ahmatγa: atıš kerek mi tutıš kerek mi dij. Ahmat: atıš babañ jurtına, kel

1 almak-čın; R almak ičın
2 sıjirep
3 kidirıp
4 babasi
5 sauliklašıp
6 ʒadi
7 R kėrk ~ kırk
8 čariχ
9 barken; R bara eken
10 Crossed out.

tutušajıχ dij. Deju küpürden ašaγa tüšti, tutıštılar. Deju Ahmatnı[1] tizine kadar jerge urdı. Ulan suvırlıp[2] dejunı br tutkan, beline kadar jerge kirsetken, otoz toχoz bašın kesip birin kaldırγan. Deju: jiget bolsañ, bunı de kes. Jiget: babamnıñ čift ügüzi tegilsıñ, bar talaš dep dejunıñ atına minep kitken.

Bir kač künnen soñ kümüš šeherine aγaları artına jetken. Šehernıñ kenarında[3] konuk boldolar. Keʒenıñ br vaχtında aγalarına: siz jatıñ, menim br baraʒak jerim bar dep br kümüš küpür tübine barıp gizlengen. Br vakitten seksen bašlı deju kelip kečeʒekti. Unıñ da atı süründi, o da euelki deju kibik ajttı, bunuñnen de tutıšıp jetmiš toχoz bašın kesıp atın aγalarına ketirgen. Büjük aγasın atın tašlatıp bu devniñ atın minep kitken. Kite-kite altın šeherine uγradılar, jene šeher četine jatkanlar. Keʒe altın küpür tübine gizlendi. Br vaχitten soñ jüz bašlı deju kelgen, unıñnen[4] de tutıštılar, (515) lakin bu dejuniñ atı da Ahmadnı tepelemege bašlaγan. Bunıñnen čok talaštılar. Jiget jurılγan, ajaγından br čarıγın čıχarıp aγalarına attı, čarıχ aγaları janına tüšken. Bular: vaj bu jer ʒinli dep baška tarafγa kačkanlar.

Ahmat bakij-bakij, aγaları kelmij. Br čaruγın taha atkan, u da česme janına tüšken ama aγaları onda joχ iken. Aγalarından fajda bolmaʒaγın bilgende gurzın jüverdi[5] üz atına, gurz at janına tüšken. At šart etip ipi jüzken, jetıšep keldi de jüz bašlı dejunıñ atın tepelemege bašlaγan, dejuniñ atın jengen. Ahmat dejunıñ toχsan [toχoz] bašın kesip, atın alıp česme janına kelgende baksa aγaları joχ, jattı. Sabā aγaları kelgen, Ahmat: siz keʒe kanda ikensiz dep suryanda: bu jer ʒinli, bz kurkıp kačtık dijler.

Uzatmajıχ, jüz bašlı dejuniñ atın ekinʒi aγasına birep kitergen jüz bašlı dejuniñ atı tilge kelip Ahmatka: men seni endi bilgenim, Čin-mačin patšasınıñ ulı Ahmat-batır bolγanın, senıñ anañı alıp kačkan dejunıñ atı menim balamdir, men seni anda alıp kitejım dij. Kitkenner. Künlerden bir kün üksek br kaja tübine barγannar. Ahmat-batır aγalarına: siz munda kalıñ, men kaja üstüne čıkajım, anamızı alıp kitken dijunen tutıšırmen. Eger u meni jeñse, men uña menim bašımnı kajadan ašaγa atajsın derim, eger menim bašım munda tüšse, siz kajtıp kitersiz, eger men unı jeñsem, unıñ (516) br bašın atajım. Siz braz beklersiz, men tüšermen degende jüz bašlı dejuniñ atın minep kırk kün kırk keʒe degende kaja üstüne čıkkan.

Attan tüšep, atnı kajtarıp jüverdı, üzi anası bolγan üj ügıne barγan, gurzin bir urdı, üjın üg dıvarı ardına, ard duvarı ügıne avuškan. Unıñ[6] kaltıramasından anası: bu nasıl kahp ulıdir[7], menim üjım dep br kade zeher čıχarγan, tanımajıp Ahmadka bergen. Ahmad alıp artına tükep anasından nasil munda kelgenin suradı. Anası baštan ajaγa kadar bašına kelgenleri br br ajttı. Ulı: sen küčük ulıñ kürseñ tanıjmısın degende evet tanıjım dij. Nesinden degende, jaš bolγanda jerge jıkılup mañlajnı tešilgen‿di[8], andan

---

1 Ahmatni
2 R suvırılıp
3 kenarinda
4 uniñnen
5 juverdi
6 uniñ
7 ulidir
8 tešilgen-di; R tešilgen edi

tanıjım dej. Ulan šu‿ssāt bašın ačıp küstergen. Anası: aman ulım dep üjge kirsetti. Ahmad ičeri kirej-kirmej baktı br jaš kız utıra.

Braz utırdıktan[1] son anası: sen inde mundan kit saklan, šindi deju kelij, seni de bizni de alak etej dij. Ulan: jok, sen unıñ šindi ne jerde bolγanın ajt, men barıp unı ütürep sizni alıp kajtajık degen. Anası: bu julnen kitejsin, felan jerde u šindi kırk bašın kumγa kümep jatadır dij. Ulan barγan. Deju bašların kümgen jata. Tur, kjafir, men seniñnen küresejim[2] dij. Deju turdı, tutıštılar. Deju bunı küterep beline kadar jerge saplaγan. Jigėt ačulanıp brden küterep dejuni boγazına kadar jerge kirsetti, otoz toχoz bašın kesken, biri kalγan. Bunı de kes degende: babamnıñ čift ügızi tegilsıñ bar, talaš dep, br bašın alıp kajadan aγalarına atkan. Baš aγaları janına tüšken. (517) Aγaları bekleǯeklernı unutχanlar, jene kurkıp kačkanlar. Birazdan büjük aγasınıñ esine kelgen, barıp bakijler dejunıñ bašı. Bular da bašlıj dejunıñ bašın ezmege. Bular bunı ezej iken, biz kelijik Ahmatka.

Ahmat anasına kelgen, azırlan, kitejik dij. Bular kajadan tüšmege ip alıp kelgenler. Ahmat baštan analarin baγlap sarkıttı, aγaları Ahmatka beklejler. Kız Ahmatka ajta: sen tüš, eger men tüssem, ipi čekejler, seni munda kaldırıjler dij. Jiget: joχ, sen munnan kitmege istemejsın, sen tüš dij. Sen tüš, men tüš, kız čara tapamaγan, ǯebinden jüzügini Ahmatka bergen: men bilem, senı munda kaldırijler, bu jüzügni jalajsın, kırk arab čıkaj, istegenını japtırıjsın dij. Amat kızı bajlaγan, kız ašaγa tüšken. Aγaları kıznı Amattan kıskanıp[3] brden ipi Ahmadın kulından tartėp[4] alγannar. Ahmat kalγan[5] tübede.

Bular anasınen kızı alėp[6] üjlerine kelgenler. Kızı büjük ulana nišanlaγanlar. Kız: men kırk künge kadar nišanlı[7] utırıjim, kırk künnen soñ tügün bolor dij.

Bu tarafdan Ahmat ǯebinden jüzügni čıkarıp jalaγan, kırk arab čıkkan: ne isterseñ emr et dijler. Ahmat: istejim, meni babamıñ šeherine alıp kitejıχ dij. Bular: jum küziñ, ač küziñ šeherge kelgenner. Ahmat kasaptan br kursak alıp bašına kečirgen, br terzige (518) barıp šaγird jürgen. Kırk kün boldo, tügün bašlanaǯaχ da kız: sinden br fereǯe istejım, makasnen kesilmegen, inenen tikilmegen bolsın dij. Baj šeherde bolγan terzilerni hepsın čakırγan, br kimse bujnına almaγan. Ahmat ešitep ustasına: bar al, men japajmen dij. Usta barıp japaǯaχ bolop kelgen.

Ahmat kumašnı alγan, ičeri kirep makasnen parča parča toγraγan. Ustası kürep: vaj, bujnıma urdım dep tarılmaγa bašlaγan. Ahmat jüzükni čıχarıp jalaγan, kırk arab čıχχan, ne keregın surγannar. Ahmat: maña inenen tikilmegen, makasnen kesilmegen br fereǯe kerek dij. Arablar aman ketirep[8] bergenner. Amat ustasına berıp: ajda, al kit dej. Usta alıp kitep bergen. Kiz: ende tügün bašlansın degen. Tuj bašlanγan, tujnıñ

1 utirdiktan
2 kürišejim
3 kiskanıp
4 tartep; R tartıp
5 kelgen
6 alep; R alıp
7 nišanli
8 R ketirėp, but the more common version is *ketirep*.

üčünǯi küni küreš vaχtında Ahmat jüzügnı jalaγan, arablardan br kat urba, brde at istegen. Ketirgenner, Ahmat kijinėp[1] kušanėp[2] atka minep küreš[3] mejdenine barγan.

Attan tüštı, bašladı pelivanlarnen küréšmege. Br dene kaldırmajıp jıkkan, pelivanlar jıkılγan son kijeu bolaǯak jiget ačulanıp üzi čıχχan, tutıškan. Ahmat kijeunıñ ukın [...] ajaγından tutıp eki tarafγa ajėrėp atkan da üzi ǯujılup[4] kitken. Bunı tanımaγanlar. Kijeu ülgen, kümgenner. Kač ajlardan soñ kızı ekinčı ulan Hasanγa nišanladılar. Kız jene kırk kün utırmak istegen. Kırk kün tamamında (519) kız: br sini japtırıñız[5], üstüne br tavuk, tavuk artında br horas, bularnıñ[6] artından da bir jımırtka, brbrin kuvalasınnar dij.

Patša ustalarγa χabar bergen, jene kimse bujnına almaγan. Kız: fereǯeni japkan usta munı da japsın dij. Ahmadın ustasın čaγırγannar. Usta kitkende Ahmat: ne türlü iš bolsa da, alıp kel, japması menim bujnıma dij. Usta bazarlıχ etip alıp kajtkan. Ahmat jene jüzükni jalaγan, arablar kelip jasaγannar. Usta alıp kitken, dügün bašlanγan. Üčünǯı kün kušu vaχtında Ahmat jüzügın jalap arablardan češit urbalarnen at alıp kušuγa barγan, atını bir balaγa mindirep jüvergen, atların hepsin uzγan. Kušudan son ǯemāt ǯāmige duā‿čin[7] baraǯaχta ǯemātten euel [Ahmat] ǯāmige kirep kapını beklegen, ǯemāt kelgen talaštılar, talaštılar kapı ačılmaj. İčinden Ahmat: bar, ajtıñ patša kelsin, son ačajmen dij.

Barıp čakırgannar[8], padša kelgen [i]ti‿bergen[9], jene kapı ačılmaj. Amat ičten: kač ulanıñ bar ajt dij. Padiša: üč ulım barγan, üčünǯısi felan vakit anasın kıdırmaγa kitep üzi ǯujıldı dej. Ahmat: eger ǯujılγan ulıñ šindi kürseñ tanıj misin dej. Padša anası ajtχanı kibik ajtχandan son kapını ačıp bašın kürsetken. Padiša: vaj balam dıp sarılγan, üjine alıp kelgen. Ahmat bašına kelgen šejlerın, aγasın de ütirgenın ajtkan. Ahmat aγasına: kel sennen uk (520) atıšajıχ dij čıχtılar. Ahmat aγasına: jul büjüknıñdir, bašta sen at dij. Bunıñ ukı Ahmatnıñ eger kašından kečip kitken. Nebet Ahmatka kelgen, atkan, bunıñ ukı hem eger-kašından, hem de jürek bašından kečıp ütirtken. Kajtıp üjlerine kelgen, tuj-dügün etip kıznı alγan.

[Ahmed batır masali][10] Künlerden br kün Ahmat-batır auγa čıkkan, kıdırγan-kıdırγan, br kijik tapamaγan. Kükke baktı, br kuš učıp kite. Munı atajım degende kuš kelip bunıñ kujnına kirgen. Ahmad kušnı[11] kujnından čıkarıp baksa, her jeri jaralı. Šenkten kajtkan kuš tilge kelip: aman Ahmat, sen bilijsıñ, meni kıjnama, azad et, men senıñ br künine kerek bolormen dij. Ahmat kušı üjine ketirep jaraların ilaǯlap bakkan, üzinıñ malı nekadar barsa, satıp bunı baktırγan. Artık satmaγa br šej kalmaγan son, kuš buña: sen meni baχtıñ, malıñ sarf ettiñ, šindi min menim sırtıma, bizniñ üjge

1 kijinep, R kijinip
2 kušanep; R kušanıp
3 kireš
4 R ǯıjılup
5 japtırıjñız
6 bularniñ
7 duā-čin; R duā ičin
8 čakirgannar
9 R iti bergen
10 Crossed out in Ms.
11 kušni

babama barajıχ, men de seniñ χatrıñ χošlajım dij. Ahmat bunı sırtına mingen. Kite-kite br teñizden kečeʒekde kuš Ahmadı teñizge tašlaγan. Ahmat buγulγan degende jene kuš kapıp alγan. Uzatmajıχ teñız kenarına čıχtılar.

Kuš Ahmadnı sırtından[1] tüšırgen, em de nasıl Amad kurktıñ mi dep surγan. Evet, kurktım dij. Kuš: sen meni ataʒaχta men de üjle kurkkan‿dım[2] dij. Kuš: šindi men seni üjimize alıp barırmen, bašıma (521) kelgen hallarnı ajtırmen, bu adam bolmaγan bolsa, men üleʒektim, bu meni baχtırγan, kurtarγan dejim. Babam senden: mennen ne‿stejsıñ[3] ulım dep, u vakit sen babamdan altın kujγan sandıknı istejsin, u sandıχka bir altın kujsañ jüz, un altın kujsañ biñ bolor dij. Babam unı saña berir, lākin alıp kajtχanda julda juklama, juklasañ topal šejten χırsızlap kačıj dıj.

Bular kušlar üjine barγannar, ajtkannar, sandıχnı istegende: joχ, eger sen balam kibik beš bala taha ketirseñ[4], sandıχnı bermem, bašχa ne istejseñ berejim dij. Ahmad: bermezseñiz ümrıñiz čok bolsın, baña baška šej kerekmij dep kajtkan. Kuš babasına: eger sen bu jigetniñ istegenin bermezseñ, men de unınnen kitem dep üjden čıχtı. Kušnıñ χatunı: vaj sen menim balamdan zijāde kürgensin, sandıγın[5] bermezseñ men de kitem dep u da čıkkan. Kušnıñ babası kürdı čara joχ: bar, kajtarıñ berejim dij. Ahmat kajtarıp sandıχnı alıp kačkan.

Ahmat kite-kite boldorıp br jerge jatıp juklaγan. Topal šejten kelip bunıñ baši tübinden sandėknė alıp kačkan. Brazdan Ahmat jukudan turıp baksa[6] sandık jok. Topal šejten kelip: ne abdıradıñ dij. Ahmat altın sandıγınıñ[7] čalγanın ajtkan. Šejten: eger men senıñ sandıγın[8] (522) tapıp ketijsem, maña ne berijsıñ dij. Ahmat: šindi menim uña karšılık br šejim joχ degende, šejten: eger sen üjde untup kaldırγan amanetın berseñ, sandıγın berirmen dij. Ahmad tüšüngen, tüšüngen, üjinde χatınından baška kaldırγan šeji jok. Eger χatınım bolmasa berirmen dij. Šejten buña jemin ettirep sandıknı ketırep berdi, bergen soñ topal: sen čıkkanda χatınıñ[9] aγır-ajaklı iken, šindi Bāri degen br ulıñ bolγan, menim senden alaʒaγım udur dej.

Bu Ahmadnıñ ulı šejten alaʒaγın tüšınde kürgen. Bular Ahmatnıñ azbarına kelgenner. Ulan aman kapudan čıkıp saulıknen kal babam dep šejten artından kitken. Ahmat šašıp kalγan. Ahmat sandık bereketinen euelki malların kajtarıp alγan. Biz kelijik uγlına.

Topal šejten Barini üjine ketirgen, šejtannıñ kızı bar iken, bu da ulana āšik boldı. Šejten kızına: al kit, bu ulannı br jerge kapa dij. Kız: kapamak kanda, dora üziniñ utasına alıp kitken, em de: babam seni čaγırıp ne iš bujursa, maña kelirsıñ dij. Künlerden br kün šejten bunı čaγırıp: muna, saña br čal at, bunı minmege alıštırasın, üleden soñ kelip al dij. Bāri kızγa kelip ajttı. Kız: u at tegildir, babam üzi at bolor, miñerseñ bašın güzel etıp čekip bajlarsın, kamčinen bašlajsın bašına küzine urmaγa, son u seni alıp kitep

1 sirtından
2 kurkkan-dım; R kurkkan edim
3 nestejsıñ; R ne istejsıñ
4 kelijseñ
5 sandiγın
6 In the version crossed out the awated dialectal variant *baχsa* can be seen.
7 sandiγıniñ
8 sandiγın
9 χatinıñ

(523) kenefke tašla ʒak bolor. Kenef janına baryanda men artından br gurz ururmen, sen üstünden tüšep üjge kirejsıñ, u sen baryanda küšede utıryan [bolor], kirep sizniñ iši bitirdim, taha japılaʒak iš bolsa ajtıñız dej.

Bāri büjle japkan. Topal šejten: kit jat, jarın ülede kelijsıñ dij. Ulan kızya baryan. Sabā ülede Bāri iš surap baryan, šejten buña br avıʒ tarı bergen, bu tarını sačajsın, orajsın, dermenge un etip ketirijsın jarın üjlegeʒe. Bāri tarıları kızya küstergen. Kız bularnı hazirlayan, Bāriniñ habarı bile bolmayan. Ertesi kün Bāri: alıñ unı dıp šejtenge bergen. Šejten: eji kajt jarın jene kelijsiñ dij. Sabā jene baryan, šejten: kırk kėz učurajım, kajsısını begenseñ unı alıjsıñ dij. Bāri kızya kelip ajttı. Kız: ularnıñ bri men bolormen, men ajayım sallap učarmen, sen meni begenıjsıñ, saχın jañılma dij.

Ertesi kün kırk tene kız učkan, šejten suryan. Bāri ajayın sallayan kıznı küstergen. Šejten kızlarya: karıšın dij, karıškannar. Jene surdı, kız ep ajayın salladıyından jiget jañılmajıp jene bunı küstergen. Kızlar tüškenner, begenilen kızya: bunı utaña al kit pišir aša, jarısın[1] maña ketir dij. Kız ulannen utaya barıp kerek šejlernı alyannar. Kapıdan čıkkanda kız jerge tükürgen de tükürükke: eki sāt ē-ē dep turıjsın dip üzleri kitkenler. Topal šejten aš keleʒek dep beklej, (524) baka-baka aš jok. Ačulanıp kapuya baryan, kızına kana ničin ketirmijsin dep kıčkıryan. Tükürük ē ē dep kajtıp kitken. Braz taha beklegen, jene joχ, jene kapıya kelip ačulanıp kapını parlayan, baksa kimse jok. Soñ eki kijeuvın čayıryan, bunlarnıñ bri braz akıldan nuχsan iken. Brer at miniñ kitiñ, julda sizge ne rast kelse alıp keliñ dij. Bular jullanganlar.

Kız tujdı, kulayın jerge kujıp diñegende yaliba eništemler keleler, ular šindi bizge jetijler. Men br sıyır bolajım, sen tuvarčı bol, meni bak dij. Kız sıyır bolyan, jiget tuvarčı. Jul kenarında[2] [kulına tajayı][3] sıyır baka, eništeleri kelgenner. Mundan br kıznen br jaš kečmedi mi dep suradılar[4]. Tuvarčı: hajr, men čoχtan bu jerlerge jürem ama kimseni kürmegenim dij. Bular bu juldan kajtkanlar, bašχa jerlerde kıdırėp[5] tapamayanda kajtıp ajttılar ki: tuvarčiden baška kimse jok digenner. Šejten: akılsızlar, tuvarčı ulan, sıyır da kız iken, tuvarčını alıp kelgen bolsañız sıyır unı artından kelirdi dij. Jene kitkenner, julda ne kürseñiz ketiriñ dij. Kız jene tujyan: eništemler kele, men terek bolajım, sen eždeгha, sen maña sarıl, fısıldap tur dij. Büjle boldolar. Eništeleri kelip eždeгhanı kürep kurkıp kajtkanlar. Jene topalya kelip: julda eždeгhadan baška šej jok, kürdik, janından kečmege kurkıp kajttıχ dijler.

Bu jene ačulanyan: eždeгha jigėt, (525) terek te kız iken, jılanı ketirgejdiñız, terek anda čürüp kalyajdı dep bularnı[6] kuvyan: barıñ ne kürseñiz alıp keliñ dij. Jene kitkenler. Kız tujıp üzi česme boldo, jiketni tas japkan. Bular česmege kelgenler, brer su ičkenler akılsız kijeu su ičken soñ tasnı alıp kujnına sukkan [alıp kajtmak istegen. Akıllısı bunı kürgen: tez bir adamın hajrını jerine kuj dep tasnı tašlatkan kajttılar.][7] Kelip šejtenge

1 jarisın
2 kėnarinda
3 Crossed out in the Ms..
4 suradilar
5 kidirep; R kıdırıp
6 bularni
7 Crossed out in the Ms..

češmeni ve tasnı ajtkannar. Šejten: tasnı ketirgejdiñiz češme kurıp kalγajdı dej. Soñ sız de adam bolmaǯaksız dep χatının jüvergen. Hatın julγa čıkkan. Kız kulaγın jerge kujıp diñedi de: vaj, anam kele, seni ašaj, meni de dep kız üzı köl boldo, ulanı papı japkan. Anası kelgen, küli kürgende ne bolγanın bildi, bašladı papını čaγırmaγa, čaγırγan-čaγırγan kelmegen. Bašladı külnıñ suvın ičmege. İčti, kuruttu, papını ašajım degende kız anasınıñ ičinde br eżderha bolop anasın[1] kursaγın jarkan, ütırip čıkkan.

Bular kite-kite, künlerden br kün Bārinıñ babası Ahmat üjine kelgenner, kırk kün kırk keǯe[2] toj-tügün etip br brin alγannar. Eki jaš br brinen ve baba ulaninen[3] muradlarına irgenler.

27 Hizir (pejgamber) masali

(541) Bir zamanda bir padša bolγan.[4] Ülkesinde bolγan χalkka χabar bergen: kimi da maña Hiziri tapıp kelse, kırχ biñ altın, taparım dep kelgenge jıkerme biñ altın pešin, kırχ kün de mület dij. Fakirniñ biri men taparım padšam dep jikerme biñ altınnı alıp kajtkan. Bol-bol ašamaγa bašlaγan, Hizirni χıdırmak esınde joχ, tapmaγa ümüdi de jok. Xatını: sen bu iše bujnıña alγan, bar ič bolmasa br kač kün dulaš, [Hizir][5] joksa bašıñnı kesejler dese de herif: adam sen de, biz otoz toχoz kün ašap ičejik, soñı ne bolsa bolor dij. Kırk kün kečkende padšaγa ketejmen dep julγa čıkkan. Kitkende bu fukaranıñ artından br kavuklu bala kušılıp kitkenner, salam birep utırγanlar.

Padiša ügıne kelgende padša: hana, Hiziri tapkan misıñ dep suradı. Bu fakir bujnın bordı[6], baška ǯavab jok. Padiša vezirlerin čaγırγan, eñ büjük vezirge: eger sen padša bolsañ, saña biri jalan ajtsa, senı aldasa uña ne ǯezā berirdiñ dij. Vezir: eger men bolsam, furunı jakardım, munı furunγa kujıp piširep čıχarıp eki jul arasına asardım, kürgenler ibret alırdılar degende, fakir jene bujnın bordo. U vakıt kavuklu bala kullu šejın jerǯiju ila-aslihı[7] dij. Padiša ekinǯi vezirden surγan, u da: men bolsam, koj kibik (542) sujardım, kırk čengelge asardım degende, jene kavuklu bala: kullu šejin jerǯiju ila-aslihi ajtkan. Üčınǯi vezir: men bolsam, Hurasandan[8] duvar čekerdim, ekı kulın, eki ajaγın tarttırıp baγlardım degende kavuklu bala jene: kullu šejin jerǯiju ila-aslihi digen. Dürtinǯi vezir: eger men bolsam, jikerme biñ altın beri‿jdim[9], jikerme biñ taha birėp āzād eterdim dij. Kavuklu bala jene: kullu šejın jerǯıju ila-aslihi dep ajtkan.

1 R anasının

2 kirk [!] kün kırk keǯe

3 R ulanınen

4 In Kúnos' list the title is "Hızır–nebi masali" [Prophet Hizir's tale].

5 Crossed out in the Ms.

6 bordi

7 "kullu šejın jerǯıju ila-aslihı" (Ar.) 'Everything returns to its origin' (Courtesy of G. Karaağaç)

8 R: Hurasan harǯından '(made) of Horasan mortar', cf. Trk. *Horasan harcı* 'Horasan mortar is a kind of very stiff mortar made of Horasan lime', cf. https://tr.wikipedia.org/wiki/Harç_ (mimarlık), (accessed 2 December, 2023).

9 berijdim; R bere edim

Šimdi padša kavuklu balaүa: men bulardan suradım, bular epsi brer ʒeza ajtkanda sen: kullu šejin jerʒiju ila-aslihi degensiñ, bunıñ aslı ne bolor dej. Bala: bu birinʒi veziriñ babası furınʒı iken, aslına čeke. Ekinʒi veziriñ babası kasabčı eken, u da aslina čeke, üčünʒisi de duvarʒı ulı iken, dürdinʒisi vezir ulı vezir iken, muna, men de Hizir-a-lejet-salamım dep aman үajb bolop kitken. Padša fakirge: jā sen Hizirnı ketirgende ničin maña ajtmadıñ dij. Fakir: eger men ajtkan bolsam, u vakit kačardı, u kullu šejin degende men bašımnı burardım, munı tutıñ demekti, siz bilmegensiz dep jikerme [biñ] altın taha alıp rahat bolүan, vezirlerin de üčün čıχarıp dürdinʒisin bašvezir etkenler.

28 Bir kart adam ekjeti

(546) Bir zamanda br kart adam bar iken, unıñ da bir ulı. Bularnıñ üč ta ečkileri bar iken. Kart karısinen müsafirlikke kitkenler, ečkileri ulına kala: ulım, katık čok ašama, juklarsın da büri kelıp ečkilerni ašaj dij. Ulan diñemedı, tujүanči katık ašap juklap kalүan. Büri kelip birer-birer ečkilerni ašap kitken. Jukudan ujanıp baktı ečkiler jok, tüšüngen: hep bir, babam kelse, meni ütirij, munnan kitejım degende kačıp kitken.

Kiterken julda buña eki jiket rast kelgen: ajde arkadaš olajık dijler. Bu rāzi bolүan, üčlešip kitkenner. Akšam boldo, keʒe karanlıχ, jaүmur jaүa. Bular aʒıχtılar, br sürü jılүıүa rast kelgenner. Bularnıñ br sāhibi bar, χıdırajıχ, braz ütmek alajıχ dijler, hem de kıdırıp[1] tapkanlar. Baksalar br küzlı adam. Bular halların, ačlıχların ajtkannar. Jılkıʒı bularүa br bije bergen: barıñ felan jerde menim üjime, bunı sujıñ, χatın piširsin, ašañız degen. Bular alıp kitep sujүanlar, χatın pišırgen, ašap jatkanlar. Keʒe br küzli kelip χatınına: bu keʒe bularn išlerın bitirmek kerek dij. Xatın: bilmem, ne japajsın jap, men kirišmem dij. Jaš ulan bunı ešitken, bularn ekisın bašın kesken, üčünʒige keldikte ulan χılıčnen br küzlinıñ, hem de χatınınıñ bašların čapıp alүan kujuүa atkan, mallarүa ve üjge sāhib bolop kalүan.

Br kač vakit (547) kečken soñ, bunıñ ʒanı sıkılүan, tüšündi, babasın[2] anasın ketirmek istegen. Azirlanıp br atүa minip küzın jumup ačχanʒe üjlerine kelgen. Baksa babası ülgen, anası da ülüm halına. Jaү ütmek ašatıp anasın tırıltgen, alıp üziniñ üjine ketirgen, anasına altmıš tene anahtar bergen: bularnı al, šunda altmıš kuju bar iken, ellı toχozın bak, birin ačma degende üzi ajvanlarnı baχmaүa kitken. Br kün, biš kün büjle kečken.

Künlerden br kün bu kart ananıñ ʒanı sıkılүan: ajde šu bir kujunı[3] de ačıp baχajım, aʒeb ne bar dep ačkan. Ne baksın, br deju jata. Xatını kürdigi kibik: uүulsızүa uүul, χızsızүa χız, ersizge er degen. Xatun: uүul, χız kerek tegil, maña koʒa kerek ama men seni kujudan nasıl alajım degende, deju: felan aranda čal ajүır bar iken unıñ kojrok kılından uč kıl ketir, men uña sarılıp čıχajım dij. Kart χatın ketirgen, deju de čıkkan. Bular safa

1 kidırıp
2 babasin
3 kujuni

su̇rdiler. Deju birazdan: senıñ ulıñ kelij, u bizni br jerde ku̇rse ačulanıp menı u̇tirij dij. Xatun: u̇jle‿se[1] sen bu sandıkka kir, seni ku̇rmij dep sandıχγa kirsetken. Kač vakitlar bu̇jle gizli jašaγanlar, bulardan br bala da tuγγan. Xatın dejuge: šindi balanı ne japajıχ, bunı sandıkγa saklasak ta aγlaj, ulım tujar dij. Deju: jok, sen bunı sar, ulıñ keleǯek jul kenarına kuj, u taptım dep alıp kelij. Sen bašta kerekmij derseñ, u jok sen buna bak, mañа (548) arkadaš bolor dej bakarsıñ degen. Bu̇jle japkan.

Ulı balanı alıp kajtkan. Anası: jok balam, men endı kartajdım, buña bakamam dese de: jok emzip bak, belkim su̇t čıkaj dep zurlap emzirtken. Čok vakitler kečken. Bala u̇sken. Deju χatınγa dij: men ne vakitγa kadar bu sandıχta saklanıp jašajım, bašta br ulıñ bar‿ken[2], jene br ulıñ bar, ajde senın ulıñnı u̇tirejik dep torγan. Xatın: jaχši ama nasıl u̇tirejık dep sordo. Deju: barmaχımdan braz et kesip: munnan aš jasa, menim etimde zeher bar‿ken, u akšam ašaj u̇lij degen. Ku̇ču̇k bala da bunı diñlep ešitken. Bala aγasın bek sevgen. Kart χatın ašnı piširgen, azırlap kujγan, ulını kelmesine bakıp tura.

Bala hemen azbarγa čıkkan, aγasın karšılaγan em: aγam, anamız aštan ašamajınǯe sakın ašama dij. Jiket ičeri kirgen, sufraγa utırγannar. Anası ašamaj. Ulı nečin ašamaγanın surdı. Anası: jok balam, men ačtım, saña beklemedim, čok ašadım, ič ašamaγa alım jok dese de: ille br kapım kapaǯaksın degen. Xatın čaresizden kapkan da u̇lgen. Bala aγasına: sandıχta deju babam bar, dij. Sandıknı ačıp unı da u̇tirgenner, kujuγa atkanlar.

Kač vakitler ekisi jašaγanlar. Aγası kardašına: eki er adam ne vakitγa kadar utıraǯakmız, biz ende u̇jlenejik dej. Hazirlanıp kız kıdırmaγa[3] kitken, kardašına un ku̇n bekle, ben kajtmasam, meni χıdırmaγa čıkajsın dij. Kite-kite br šeherge barıp jetken. Šeherde br karaman adam bar eken, u̇zini jıkkan (549) adamγa kızım berem, jıkmasa bašın γajb dep χabar taγıtkan. Jiket bunı jıkıp tuj-du̇gu̇n etip kızın alγan. Baška br šeherde br karaman da jene bu̇jle χabar bergen, unı[4] de ešitken, jıkıp kızın kardašına almak istegen. U̇č ku̇n talašγan jıkamaγan. Du̇rtinǯi ku̇n karaman bunı apıska atkan. Un ku̇n boldo, aγası jok. Kardaši čıkkan bunı χıdırmaγa.

Kiterken julda aγasın apısta bolγanın ešitken, aman karamanγa kitep jetken. Karaman, aγamnı apuska atkansıñ, kel menimnen tutıš baχajık dij. Karaman: aγañ sıγγan jerde sen de sıγajsıñ dep tutıškannar. Bala karamannı br kakmadan jerden u̇č aršın tu̇pege ku̇terdi. Karaman: aman atma, aγañı da čıχarajım, kızım da berejım, tu̇njada mennen ku̇čli jok bellerdim, jiket ekensıñ degen. Jiket tašladı, aγasın čıχarγannar, tuj-tu̇gu̇n etip kızı alγan, aγası u̇z u̇jine kitken. Bular čok vakit jašaγanlar. Br ku̇n ku̇ču̇k jiket kajın-atasına: sen bizni uzγar, biz ende u̇zimiz tuγγan u̇jimizge kitejik dij. Bunı uzγarıp aγası u̇jine kelgenler, unıñ kajn-atası da košılıp jiketlerın χatınlarinen uzγarγannar, jiketler u̇z u̇jlerine kelip muratlarına irep jašap alγannar.

1 u̇jle-se; R u̇jle ise

2 barken; R bar eken

3 kıdırmaγa

4 uni

29 Baj kızı hekjeti

(555) Bir zamanda bir baj olγan. Bunıñ br χatını, birde kızı bar iken. Kız kararına kelgen. Bu baj üz kızına aveslengen. Kač vakitlar utanıp jürse de tajanamajıp br kün kızına: sen menimnen jatmaγa rāzi bol dij. Kız: joχ babam, sen šašma, šejtanγa ujma, dese de čare bolmaγan. İlle jā bolaǯak ja seni ǯajarım dij. Kız babasına: üjlese, men kitabγa baχajım, eger gunası saña bolsa, bolor dij. Kız kitabγa bakkan, elbette babası kart adam boldoγından gunası uña tüšken. Kız istemej-istemej, rāzi bolγan.

Kač vakitler kečken, kıznıñ kursaγı büjümege bašlaγan. Kart baχtı, ıš majdanγa čıχaǯaχ, χatunı tujaǯaχ, ırgatına: bu kızı daγγa alıp kit, sujıp tašla dij. Irgat kızı daγγa alıp kitken, sujaǯaγın ajtkan. Kız buna jalvardı: sen menı ütirme, felan jerde br eski aran bar, anda koj, kirgende čıkkanda aš u su ketirijsin bakajsın[1] dij. Irgat tüšüngen, baktı elbette br kabātı bolmaγan adamın ütirmek jazıχ dep alıp aranγa kujγan. Kač vakitlerden soñ bu kız br er bala tuγırγan. Irgat hep kelip, bularnı bakıj. Br kün ırγat jene aranγa kirgen, kapı japılγan, üzi ičeri kalγan. Kız buña: eger sen bu balanı alıp čıχsañ, kapı ačılıj, eger almasañ sen de munda kalasın dij. İrgat: men (556) munı kanda alıp kitejim? Kız: sen bunı babam azbarına kuj, üzıñ kirep munda bir bala jata derseñ, uların balası joχ, alıp bakajler, kimse bilmij dej.

İrgat alıp kitken. Bu kartlar bunı sütnen bakkanlar, bala büjügen[2], χızmetge jaraγan. Kart: ende bizge ırγat kerekmij dep čıχarγan. Bala br kün: bu eski aranda ne bar iken dep kirgen, baχsa br χatın: ıns misiñ, ǯıns misiñ dep surγan. Katın: jok, men senıñ anañım, kel kurkma dij. Bala tüšüngen: menim anam babam bar, sen nasıl anam bolaǯaksın, br adamnıñ eki anası bolor mı dij. Kız buña: balam, senin babañ menim de babamdir, büjle büjle boldo dij. Br mıχtar uturıp laf etken soñ, bala arandan čıχaǯaχta anası buña br peškir bergen: ma bunı, ne vaχıt ašamaχ isteseñ jajarsın, her türlü aš kelij, ašar ičejsin dij.

Bala alıp čıkkan, lākin kart babasınıñ etken išine rāzi bolmajıp müšküllengen. Kartlar buña: sen nečin büjle müškül jüresın, juksa üjlenmek istejsin mi dijler. Bala: jok, maña üjlenmek kerekmij, men sızin janıñıza utırmaχ istejmim, čıχıp kitejim dij. Bular jüvermege istemegenner: seni jašlıχta baktık asradık, tamam iške jaraγanda bizni[3] tašlap[4] kitejsin deseler de diñemegen kitken.

Julda kitkende bunıñ ügine br top sıčkan čıkkan, bu ularnı kuvγan: ja ajaklarım tübinden kiteñ[5], ja sizni ütirirmen dij. Sıčkanlar: sen bizni ütirme, bzden birer tük al, biz saña (557) kerek bulıjmız dijler. Bu sıčkanlardan brer tük alıp kečken.

Birazdan ügine bir sürü kanarja čıχtı, bašlaγannar tišlem[e]lge. Bu kulinen[6] bašlaj ularnı ütirmege. Bular da jalvarıp: bizni ütirme, saña kerek bolırmız dijler. Braz taha

1 bakajsin
2 büjüken
3 bizim
4 tašlıp
5 kiteñ
6 R kulınen

kitken baχsa tamam bunıñ kečeʒek jul ičinde br oba buz, kečmege jul jok. Bu buzlarγa ačulanγan, urıp parčalaʒak bolγan. Buz da tilge kelip: meni parlama, men saña kerek bolijmen [men]nen br parča al, ʒebiñe kuj degen. Alıp kitken.

Andan barıp, bir kart anačıχ üjıne musafir bolγan. Kart ana: balam sen kanda kitesın dij. Ulan: perinıñ kızın almaγa degende kart ana: vaj balam, nafile talašma, sen unı alamajsın, anda kitken adam kajtmij dij. Ulan: ne bolsa bolor, ülsem ülejmen, ille barıp bakajım dij. Kart ana: balam, čara jok, sen baraʒaksın, men saña ügretejim, u saña üč bajtal berij bakmaγa, bajtal tegil, kızlarıdır. Sabā alıp janlarından ajrılmajıp[1] akšam kajtara bilseñ ʒujmasañ alıjsın dij.

Jiget peri üjine barıp jetken, salam berip kirgen. Periler karšılap: bizge kismet kelgen dep alγanlar. Biraz utırγan soñ peri: ne išnen kelgensıñ dep sordı. Ulan ič čekinmejip: men senıñ küčük kızın almaγa keldim dij. Peri: jaχšı, bolor ama men saña jarın üč bajtal berijmen, eger sen ʒujmajıp akšam kajtarıp ketirseñ bolor dij. Sabā boldı, bajtalları alıp daγga kitken. Daγga barij barmaj, bunı kulından kurtılıp kačkannar. (558) Bu šaštı, neteʒegın bilmegen iken. Ep bir, meni ütireʒekler, ič bolmasa br ašap tujajım dep peškirni jajγan. Ašap turγanda julda rast kelgen kanarjalar esine kelgende čaγırγan[2]. Kanarjalar sürišep kelgenler. Ne tilegıñ bar deje surγannar. Jiket: tauda menim üč bajtalım bar, bularnı kuvalap čıχarıñız maña dij. Kanarjalar aman tauγa kirep bajtallarnı tišlemege bašlaγanlar, bajtallar topunen taudan čıχıp janına kelgenler, tutıp alıp kajtkan.

Peri sejirlengen: üč išnıñ brini japkansıñ[3], eki taha kalγan dij. Jarın šu kırk kebenden danelerini ajırıjsıñ, kebeni taγıtmajsıñ, degirmenge čektirep un etejsın, undan ütmek japajsıñ dij. Jigėt tüšüngen, esine sıčkanlar tüšken, tüklerın kızdırγan, ʒıjılup kelgenler. Bašladılar[4] buγdajlarnı ajırmaγa. Sabaha degin ajirıp[5] de un etip ketirgenner. Sabā jiket ütmek piširep perinıñ ügine ketirgen. Peri kürep esi kitken. Br šartım taha bar iken, unı de japsañ kızımnı alijsıñ dij. Perı hamamı tajanmaʒak kibik jaktırγan, jiketni jakaʒak bolop amamγa kirsetken. Jiget kapuγa barγan, ič ičeri kirmege čara jok. Ʒebinden buz parčasın čıχarıp atkan, aman süngen. Jiget kirgen juvunγan.

Birazdan peri: ende (559) janγan dep amamγa kirep baχtı jiget sau. [Peri sejirlengen.][6] Üč šartnıñ üčun de japkansıñ, čara jok, kızım alıjsıñ, son tuj japarmız digen kitken. Aγasına barıp bu bolγan šejlerin bir bir ajtkan, aγası: üjlese sen uña bir zarar ketirmeʒeksin, kıznı ber de üjınden ku, senin üzin de ʒujar dij. Peri kajtkan, jigetke: balam, kızımnı alıjsıñ ama munda turmazsıñ, bizim tüšmenimiz čoχ, seni br kazaγa kujajler dep aldatıp üz memleketine kajtarkan[7]. Bular karı koʒa kajtkanlar, üz babası üz üjine kelgen, baχsa kart baba ülgen, kart anası jangız kalγan. Bu jiget üz anasın da arandan čıχarıp ketirgen, peškirden ašap ičep jašaγannar.

---

1 ajrilmajıp
2 čaγirγan
3 japkansiñ
4 bašladilar
5 R ajırıp
6 Crossed out in the Ms.
7 R kajtarγan

30 Bir taz ulan ekjeti

(565) Bir zamanda bir kart ana bolγan, unıñ de bir taz ulanı bar iken. Bularnıñ mal degende br kazları bar iken. Künlerden br kün taz ulan anasına: men bu kaznı bazarγa al‿kitep[1] satajım, u da kalmasın degen de alıp bazarγa barγan. Kaz satam dep ügine bir jaš jiget čıχıp: bu kaz tegil, čıpče taha dij. Taz ajlanėp[2] bašχa čaršıγa barγan. Jene ügine bu jaš čıχχan, ne tarafγa ajlansa bu adam čıχıp kaz tegil čepče dij. Taz kutulmajıp čipče jerine satχan, üzi de kaz alγan adamnı artından kitep üjini belledi, br jerge saχlandı. Bu jaš anasına barıp: bak ana, men bükün br taznı aldatıp čıpče dep kaz alγanım dij. Taz bunı išitken. Bu aldatγan jaš ta br patša kijeuvı iken. Anasına: men jarın χajn-atam üjınde bolormen, sen bu kaznı jaχšı jaγγa kızartıjsın, ičıni altın tudurıjsın, men bir bala jüverijım, alıp kelij dij. Bunı de taz ešitken.

Ekinǯi kün üjlege kadar saklanıp turγan. Jašnıñ anası kaznı kızartγan, ičini altın tudurtkan. Birazdan taz kapıdan: kart ana, kaz hazır boldo mo dep kıčırγan. Kart ana kaznı čıχarıp bergen. Taz kitkende kapıγa kazım, kazım, kaz sābi, küčük ujnım bu iken, büjügė[3] son kelıj dep jazıp kitken. Bu taraftan patša üjinden de br adam kelip kaznı istegen, ičten kart ana: ja men kaznı jüverdim, alıp kitkenner (566) dij. Bu bala barıp jigetke ajtkan, baχtılar kaz kelmegen. Kajtıp anasından: kaznı kimnen jüvergensiñ dep sordo, brde baktı kapuda jazını kürep kim bolγanın bilgen.

Taz kaznı ašap, altınlarnı üjlerine ketirgen. Padiša askar tükken, kıdırmaγa[4] baš-laγan. Bu taz aman br tükančige barıp: saña jüz ruble, br sāt tükenine utırajım[5] dep. Tükenči rāzi bolγan. Taz kijinep kušanıp tüken üginde utıraj‿γan[6]. Bu tarafdan baχsa askar kele bašlaγan. Taz buların ügine čıχıp, zabitlerden kanda kitkenlerın surγan. Zabitler: br taz ulan kıdıramız, bu jerden kečmegen mi surajler. Taz bularγa: e siz büjle sılanen kitesiz, sıznı kürgen χırsız saklanıj, siz sılalarıñızın bu māzege kujıñ, silāsız kitėñ[7], kimse umut etmesın dij. Bular bu süznı mākul kürep sılaların māzege kujγan soñ üzleri kitkenler. Taz māze kapısın kirtlep kapıγa kazım, kazım, kaz sābı, küčük ujnım bu iken, büjügi son kelij dep jazıp kačkan.

Bular čok vakit kıdırγannar[8], tapamajıp kajtkannar, baksalar māze kirtli, silālar jok. Utana-utana patšaγa barıp ajtkannar. Bizniñ kulımızγa tüšken ama tutamadık, u kjamil br adam bolmalı dijler. Künlerden br kün baška br patša bunı ešitken: bu nasıl patšalık, üziniñ mülkinde bolγan br χırsıznı tutamaj, dep külgen. Eger bu (567) išlerni japkan χırsız šu patšanıñ br tavuγın čalıp ketirse, uña χızım berejim dij. Taz ulan ešidip kelgen: padšam, men unı tavuγın tegil, üzini isteseñ ketirijmen degen, kač kün mület alıp kitken.

1 al-kitep; R alıp kitėp
2 ajlanep; R ajlanıp
3 büjüge; R büjügi
4 kıdirmaγa
5 utirajım
6 utırajγan; R utıra eken
7 kiteñ; R kitiñ
8 kıdirγannar

Julda br čobanγa rast kelgen. Čobannıñ br ečkisın sujdırıp[1] tulup čıχartıp, jalıngız terini un beš rublege alγan. Šeherde bir čañ ustasına barıp teriniñ tùki bašına brer čañ taktırγan. Kitep padišanıñ ùlkesine barγan, br aj utırγan. Padišaγa at bakmaγa jùrùp, azbarında ni jerde nesi bar, epsin ùgrengen. Bu keǯe čañlı tulubnı kijep padša kapısına barγan. Bekči: toχta, kanda kitesin degende: kıčırma, men Azrāilmen, šındi ǯanıñ alırmen dep, tulubnı šanγırdatkan. Bekčiniñ kulından tùfegı tùšken, ekinǯi kapıγa barγan, unda da bekči ùgine čıkkanda, uña da ùjle japkan, son padša jatχan jerine kirep kùjsi ùstùne čıχıp utırγan. Padša brden ujanıp: kim u degende: kıčırma, men Azrāilmen dep tulubnı šanγırdatkan. Padiša kurka-kurka: aman Azrāil, maña ùč kùn mūlet ber dij. Taz: berejim ama men saña kelgenım kimsege ajtma, ajtsañ ǯanınnı katı alırmen dij de kajtkan.

Padiša balasın čaγasın ǯijep[2] elallaškan. Ertesi keǯe jatχanda taz jene sùrùp kirgen, brde (568) sandıχ ketirgen. Padšanı ujatıp: kir šu sandıχka, men senın ǯanın munda almijmen, jedi kat kùkke čıχarıp alajım, em da kitkende men meleklerge salam berijim, sakın sen χıbırdanma, sesiñni čıχarma, eger χıčırsañ, seni jùverirmen, patlap ùlijsıñ dij. Sandıχka kujıp gùzel kirtlep, bir atγa jùklep kitken. Kite-kite baštaki padiša ùjine jakın kelgen ajtkan: men sandık ùstùne jumruknen ursam, sen kùpek kibik avlarsıñ.

Padiša ùjine kelgen, [sandıknı] azbarγa tašlap ùzi kirgen. Padša: hane ne ketirgensıñ degende: sen istegenın ketirdim dep sandıγnı jukarı čıχarıp ùstùnden urmaγa bašlaγan, bu da ičten kùpek gibik avlaγan. Sandıγnı ačkanda padiša čıkıj, baksa ùzinıñ sarajı kibik br sarajda. Ùj sābi padiša buña: [bak bu adam senin kijeviñe bir χırsızlık japkan, ben seni kùlgenim][3] ama šindi seniñ ùzini čalγan degende bunı ùti patlap ùlgen, kùmgenler. Taznı aldatγan jigetke berileǯek kıznı taz ulanγa bergenler. Taz muradına erip sefada kalγan.

31 Bir bajnıñ eki balasi

(573) Bir zamanda bir baj adam olγan, unın da br kızinen br ulı bar iken. Baj ùleǯekte ulına[4] ve kızına: men ùlgen son briñiz χatın almaz, briñiz de erge barmaz, menim kazanγan malım sizni ùlgenǯe bakaj, ašar jatırsız dij. Kùnlerden bir kùn baj ùlgen. Kač vakıtlar kečken, bular ùjlenmij. Šehirnıñ χalkı bu ulanγa: sen tamam čaγında jigėt oldoñ, ničın ùjle[n]mijsiñ, ùziñ χatın alıp kızkardašıñ[nı] da erge ber dijler. Bu kajtıp kızkardašinen mùšavere etken. Kız: men bilmim, ùziñ bilijsın, ùjlenseñ ùjlen, men babamnı[5] vasijetin tutam dij. Bu ulan tùšùngen, tùšùngen de tap sonı ùjlengen.

1 sujdirıp

2 R ǯijėp ~ ǯıjıp

3 The roles of the two different padishahs in the story were confused by the narrator at the end of the story. The confused fragment is as follows: "bak bu adam menim kijeuvime bir χırsızlık japkan, sen meni kùlgensin ..." The logically correct words are given in square brackets.

4 ulina

5 babami ~ babamni?; R ~ babamnıñ

Kač ajlar χatinnen[1] jašaγanlar. Baška bir šeherde bunı br māzesi bar iken, una joχlamaγa kitkendi. Bunıñ χatıninnen kızkardašı eki ara kavgalaškan. Hatın ùj ičinde olγan kijmetli šejlerni kırıp koǯasına kahat jazγan: senın kızkardašıñ[2] delirgen, ùjde br šej kaldırmajıp parlaγan. Jiget kajtkan, šejler parlaγan kerček ama zarar jok, mal bizim ekimizın, babamızdan kalγan, parlaγan parlaγan dij. Kardašına br sùz ajtmaγan.

Jene kač kùnner kečken, χatından br bala bolγan. Jiget jene baška šeherdeki tùkenge kitken‿di[3]. Xatın ùz balasın sujγan, jene kardašıñ sujγan dep kahat jazγan. Jiget kahatnı ukuγan, ınanmadi‿se[4] de kajtıp kelgen, baktı ki bala gerčekten sujulγan. (574) Kardašından: ničin sujdıñ dep surγan. Kız: jok, men kùrmedim, sujmadım, χatınıñ sujγan dej.

Jiget tùšùngen: adam, χatın ùz balasın sujmaz, dep χatınγa inanγan. İrgatγa araba jektirdi, kardašına: sen kijin, šeherge kitep gezejik dij. Kız kijingen, araba kelgen, kardašinen minep kitkenner, šeherni dulandılar, son jiget kardašına: šu taunı da barıp kùrejik dij. Kittiler tauγa kirgenler. Kıznı tùšùrgen, ùzi arabada azırlap kujγan baltasın alıp br terek jarıp kıznıñ eki kulın terek arasına kıstırγan: sen menim balamnı sujdıñ, bar talaš dep kajtkan. Kız aγlaj-aγlaj kalγan.

Kač vakitlar turγan son kıznıñ eki kulı čùrùp tùšken. Kız u jerden turıp kitken, br baj adam bahčasına barγan. Kùndùz saχlanıp keǯe bahčada alma ašap kečinir‿ken[5]. Br kùn baj bahčaγa kelip bakkan, terekte almalarnıñ brer tarfı ašalγan. Bekčige kùsterip: unı kim bùjle ašaj dep kitken. Bekči juklaγanda kız jene kelip ašaγan. Baj sabā kelip, baksa jene čoχ almalarnıñ brer tarafı jok. Karaulγa tarılγan. Ertesi keǯe ulını de jùvergen: br tarafdan sen baχ, br tarafdan bekči baχsın, ille bunı tutuñ digen. Keǯeniñ br vaχtında ulan br šatırdı ešitken, kıčırėp[6] kim u dej. U tarafdan: men de seniñ kibik adam (575) balasım ola men ačım, em de čıplakım[7], kimsege kùrinmege vaktım jok, utanam, sen maña br kat urba ketir kijinejim, son saña kùrınirmen, halım ajtırmen dij.

Jiget urba ketirgen, kız kijingen, kız da pek jaχšı iken. Baj ulı buña urılγan. Jiget: men seni ùzime χatınlıkka alajım, kız da: jok, menim babam ùlgende maña koǯaγa barma dep vasjat etken, em da zāifim, saña jarašmam dij. Jiget: sen bilijsin, ajda, men seni br kart katınγa berejim, anda jat, sabā babam kùrij, soñ kabātıñ baγıšlasa kitejsin dep bir kart χatın ùjine alıp kitken. Kartčık munı ašatkan ičerdi jatırdı. Ulan barıp babasına almalarnı kim ašaγanın ajtkan, em de šu kız maña kelse, men unı χatınlıkka alıjdım dej. Babası: sen šaštıñ mı, asli-nesli bilinmegen, keǯe anda munda jùrgen kıznı almak saña jakıšmij dese de, jok alaǯaγım dij. Tap suñı babası jıp buγazına: ne jasasañ jasa dij. Ulan kart katınγa barıp: čok šejler berijim saña, ille bu kıznı kandır, maña kelsin dij.

---

1 R χatınnen
2 kizkardašıñ
3 kitkendi; R kitken edi
4 ınanmadise; R ınanmadı ise
5 kečinirken; R kečinir eken
6 kıčırep; R kıčırıp
7 čıplakim

Xatın bu kıza: üjle bolmasa büjle bolor, kulıñ jok, seni kim bakaǯaχ, bu bir baj ulı, sen buña bar, babañ ülgen kitken, unıñ vasjatin tutmak kerekmij dep kandırγan. Kız mejl bergen, tuj-tügün[1] etip alγan, ömr[2] etkenler. Bu jiget [alıš-verišnen][3] br tarafγa čıkıp kitken. Xatunı (576) br güzel er bala tuγırkan. Babaları ulına: falan kün kajtıp kel, br güzel ulıñ bolγan, adın kujaǯakmız dep kahat jazγan. Bu jiget kıznıñ aγasın üjınde utıra iken. Bu kahadnı ketirip bereler. Kahat bašta euelki bajnıñ ulınıñ χatınına dejgen. Ačıp ukıp bakkanda üzinıñ baldızınıñ sau olγanın bilep jazγan: senın χatınıñ br bala tapkan, insan deseñ insan tegil, ajvan desen ajvan tegil, ne japajıχ, sen kajtkanča tursın mi, joksa kajb etijik mi dep jazγan.

Ekinǯi bajnıñ ulı kelgen, χatın kahatnı bergen, jıget ukuγan. Xatın ič br šej bilmegen kibik ne jazγannar dep suraγan. Jiget: br šej jok, br tükanımız janγan, baška šej jok dij. Jiget üzi babasına: men barγanče hatınımnı de balanı de ǯujıñ dep jazγan. Kahatnı babası alγan, lakin bunu ütirmege aǯıdij: al balañ da mundan kit, hoǯañ büjle jazγan dep, balasın beline bejlep üjden čıχarıp jüvergenler. Xatın kitken.

Künler bek ısıǯaχ boldoγonnan χatın susaγan, br kujıγa tüšken, bala buγılmaγa bašlaγan. Šu arada bunı kulaγına: kulıñ uzat ta al balanı degen br ses kelgen. Xatın kulın uzatkan, kulı euelki kibik bolγan, balasın da küterip alγan.

Javaš-javaš üzınıñ tuγγan šeherge aγası üjine barγan, musafir alıñ[4] dep jalvarγan. Bular tanımajlar[5]. Kıznıñ[6] aγası ve koǯası: eger sen bizge masal ajtsañ alırmız dejler. Bu ajta: masal bilmem, bašima kelgen šejlerni ajtıjım dij. İngesi bunı (577) sezgen: adam sen de, šunı[7] kibik tilenǯi ne ajtaǯak, üjle püsür, üjge alma dij. Aγası[8]: jok, kajtarmak bolmij, jazıkdır, balası da bar, kalsın bir četke jatsın dij. Musafirnıñ balası[9] anda munda čapkalanmaγa bašlaγan. Kıznıñ aγası bu balajı seugen: bunı bizge sat, sen zaifsiñ, sana bunı jüklenep jürmek zur bolij, men bakıp büjütijim dij. Xatın: jok, menim malim budır, ǯanım sıkılγanda men bunınnen eglenem dij. Bular balanen eglenıp χatından suraǯak šejlerni unutkanlar. Kajve čaj ičep kiteǯekte χatirlerine tüšken: vaj, jā biz bunı bašına neler kelgenın suraǯaktık degenner de kajtıp utırγannar.

İngesi jene ačulanγan: siz išiñizin baχın, unı bašına kelgen šejler sizge kerekmij dij. Bizim išimiz de seniñ išiñ tegıl, ajt degenner. Xatın: ajtırmen ama arada kimse laf etmij dej. Bašladı babasınıñ vasjatinden ajtmaγa. Epsin ajtkan soñ bular išni añlaγannar. Šu‿ssāt[10] aγası χatının bašın kesip alγan, ekinǯi baj ulı da χatıninen balasın alıp üjine kitken, muratlarına erip jašaγanlar.

---

1 tüj-tügün
2 emr[!]
3 Crossed out in the Ms.
4 aliñ
5 tanimajlar
6 kizniñ
7 šuni
8 aγasi
9 balasi
10 šussāt; R šu sāt

32 Jedi kız masali

(583) Bir zamanda br kart adamnen kart χatın olγan, ularnıñ jedė kızları, br kùpekleri, brde mıšıχları bar iken. Karı koʒa müsapirlikke kiteʒek olγanlar. Kiteʒekte kızlarına: biz kitken soñ sakın bu mıšıχnı urmañız, ursañız epiñiz alak olorsuz dep kitkenler. Bular kitken soñ br kùn mıšıχ penʒere ùstùne sijgen, kızlarnı biri bunı urγan. Mıšıχ kitep ateš ùstùne sijep atešın sùndùrgen. Eki kùn atešsiz utırγannar. Bu jerde ataš joχ. Kızların biri azbarγa čıχıp, u jana bu jana bakıp turγanda, uzak br jerde tùtùn kùringen. Kıdej ataš ketirmege. Barγanda baksa br kart ana utıra, salam berip kirgen. Kart ana: balam, gùzel kelgensıñ, tiziñe jatajım, braz bašımγa bak dij. Kız bakij. Bular laf etkenner. Kız: kart ana, bizge utırmaγa kel, bizım anamız babamız ùjde jok, seniñ de kimseñ jok, eglenirmiz dij. Brazdan kart kızγa ataš bergen, br kulına da braz kùl, bu kùllerni tùke-tùke kitijsıñ, men kùllernen kitep sizge barırmen dij.

Kız kitken, ùjlerine barγan, kardašlarına: muna, men em ataš taptım, em de br kart ana dij. Bularnıñ birde kùpeklerı bar iken. Kart kelgen, lākin kùpek ič jakın almaγan. Kart kajtıp kitti, kızlar da beklejıp kelmegen, jatıp juklaγanlar. Sabā olγan, eki kız kartγa barıp: ničin akšam kelmediñ dijler. (584) Kart ana: men bardım ama siznıñ kùpek azbarγa jùvermegen, siz kùpegın eki ajaγın kırıñ, jùrmege jaramasın, son barırmen dij. Kızlar kajtıp kùpeknıñ ajakların kırıjler. Akšam kart ana jene kelgen. Kùpek ajaksız bolsa da jene azbarγa jùvermegen. Sabā kızlar jene barıp surdılar, kart ana kùpegiñizı ùtirip kùmmeseñiz, men sizge barmijmen dij. Kızlar kelip kùpegın ùtirep kùmgenner. Akšam kart ana sùrùp kelgen, ašadılar ičtiler jattılar.

En bùjùk kız kart ana janında jatχandı. Sabā turdılar, baχsalar br kėz jok, kart ana ašaγan. Kızlar apteleŕın surγanlar, ana: u sizın kibi juklap jatmıj, siz juklaγan soñ hačke kitken dij. Jene akšam olγan, jatkanlar. U keʒe jene ekisın taha ašaγan. Sabā jene surdılar, kart ana euelki kibi ʒavab bergen. Akšam jene jatkanlar, kart ana bašladı kızların birin taha ašamaγa. Kùčùk kız ujanγan, bakij kart br šej ašaj. Kart ana, nedir ašaγanın degende: koʒamdan kalγan katı-kutular. Kız baška šej ajtmaγan ama tujγan. Kart ana juklaγan. Bu kız aptelerin ujattı, br kajrak, brde tarak, brde kùzgù alıp kačkanlar. Kart ana sabā turıp baksa kızlar jok, br keleni at etip mingen, tokmaγın da kamčı etip kulına alγan, kızların artından kumaγa bašlaγan.

Kızlar kitken, kart kitken. Kart kızlarγa jakınlašγan. Kızlar kùrep taraknı artlarına atkanlar, (585) tarak bir kara tau olγan, kart ana otorıp kajraχ jasaγan, tùkùrep balta, bašladı taunı kesmege. Az kalγanda br tilki kelip: kart ana, sen boldorγansın, bar, ende men kesejim dij. Kart ana baltasın tilkige bergen de ùzi de jukuγa jatkan. Tùlki braz kesej, son baltanı kırγa tašladı da ùzi kačıp kitken. Kart ana turıp baktı, tùlki jok. Jene baltasın alıp kesmege bašlaγan. Brazdan br tùlki taha kelgen, u da balta istegen. Kart: jok, briñiz taha kelip, keseʒek olup, aldatıp kitken dij. Tùlki: jok, u kızıl tùlki, men kara, men aldatmam dese de kart ana: siz epiñiz de br dep tùlkiniñ bašın kesip atkan da ùzi kese-kese taunı bitirgen. Kızlarγa jakın barγanda kajraknı tašladılar, bùjùk bir kaja bolγan. Kart čıχtı da tùšken. Čok vakitlar kečken.

Nejse kač künlerden soñ talaša-talaša čıχtı, kızlarγa jeteʒekti. Kızlar küzgüni jerge tašlaγannar, br büjük teñız olγan. Kızlar šunuñ u tarafında, kart ana bu tarafında. Kart ana kızlardan surıj: nasıl u tarafγa čıkkansız dep. Kızlar: biz ajaklarımıza taš baγlap kaburčaγγa minep čıχtıχ dijler. Kart ana da eki ajaγına balaban tašlar baγlap kaburčakka utırγan da aman teñızge kitep buγılγan. Kızlar kutılγanlar. Bular kite-kite br tau ičine kirgenner, kijik almalar ašaγannar da čıχıp terek üstüne utırdılar[1].

(586) U künnü patša vezirlerinen tauγa čikkan‿dılar[2]. Tamam bu kızlar utırγan terek tübine utırγannar. Kızların birin sıjmesı kelgen, sijıp jüvergen, jaγmur kibik patšanen vezirler bašına jaγdı. Bular kerčekten jaγmur bellediler. İčlerinden biri: joχ, büjle ačık avada jaγmur bolmij, terekte insan bar dij. Askerlerden birin terekke mindirep baχsa üč kız. Asker χabar bergen, kızlarnı tüšürgenler. Padiša bularγa urılıp üčini de üzine χatınlıkka alγan, brde üzinıñ χatını dürt olγan.

Br kün bu patša baška bir šeherge gezmege kitkende χatınların janına čaγırıp: men felan jerge kitejmen, kajtkanda siz ne azirlajsız dij. Büjük χatını türlü ašlar, ekinʒi χatın tülki terisinden don, üčünʒisi bit terisinden kolčaχ, dürdinʒisi: men saña altın bašlı br bala tuγurırmen dij. Patša kitken. Keleʒek vaktı jakınlaγanda erkes üziniñ süz bergen šejlerin azırlamaγa bašlaγanlar. Dürtinʒi χatın br altın bašlı bala tuγırdu ama üč χatın paχıllap mušavere etkenler. Balanı čalıp, tauγa alıp kitkenler. Jerine br küčelek kujγannar. Taudaki balanı br obur tapkan, üjine ketirep bakkan‿di[3].

Patša kajtkanda katınlar karšılap čıχtılar, azırlaγan šejlerni čıχarγannar. Dürtinʒi χatının surγanda, χatınlar altun bašlı (587) bala jasaʒak χatının küčelek tuγırγan degenler. Patša: tez balasinen tauγa al‿kitep[4] tašlañız, küzime kürünmesin dij. Alıp kitkenner.

Čoχ vaχitlar kečken, küčelek br kün tauda kezgende altın bašlı bala kürep tilge kelip: ajda, seni anaña al‿kitejim[5] dij. Bala: jok, menim anam munda dij. Küčelek: jok, seniñ anañ u tegildir, baštan nasil olγanın ajtıp anasına ketirgen. Lakin obur da bunı tujγan, kuvıp artlarından kelip: benim balam ber dij. Bular kaugalaškan. Tap sonu obır bilgen bala bu χatınnıñ olγanın, balanı anasına baγıšlaγan, em da felan jerde menim üjimden istegeniñiz kadar altın alıñız[6] dij. Barıp čok altın alγanlar, büjük saraj jasaγanlar.

Bir zaman patšanıñ br veziri u jerden kečkende bu sarajnı kürep χatını tanıγan‿dı[7]. Vezir musapir olγan, χatın bašına kelgenin ajtkan, vezir patšaγa bildirgen. Patša kimsege ajtmajıp barıp baktı, baksa kerčekten üziniñ χatını. Alıp kelgen. Üč χatınγa türlü türlü ʒezalar etip ütirgen, dürtinʒisinen jašap kalγannar.

1 utırdilar
2 čikkandilar; R čikkan ediler
3 bakkan-di; R bakkan edi
4 al-kitep; R alıp kitėp
5 al-kitejim; R alıp kitejim
6 alıñiz
7 tanıγandi; R tanıγan edi

33 Beš-salkım-bejniñ masali

(594) Zamanıñ euelınde, aχir zaman šehrinde [kapıl-kupul bakalar kanatlanyan učmaya, teñizdeki balıklar kıra tutkan köčmege, bij-akamnıñ tarašı tiregen balakların Kara teñizden kečmege][1] anam bešikte iken‿di[2], babam kučakta iken‿di, men tünjada jok ikende, babamnıñ tojunda ajakčı bolop ǯürgende, br Beš-salkım-bıj[3] bolyan, unıñ de beš salkım jüzümi bola iken, her jıl baška türlü malı jok iken, šu jüzümni satıp kečine‿ken[4]. Beš-salkım-bijniñ jüzümleri piše kelip[5] čalašna[6] kuja. Üzi ašamazdan euel jüzümlerge br tülkı tata. Br kün jüzümlerin barıp ašaj, Beš-salkım-bijni ač kaldıra. Bij bir-kač künler tülkını aulaj tutmaya, ama br türlü tutalmaj, tap soñında tülkini aula-aula tuta.

Tülki ajta: bij-akam, sen meni ǯiber, men senıñ br kerregine[7] jararmen dij. Bij: sen br tülkisıñ, ne keregıme jarajsıñ dij. Tülki dij ki: ne kerek istejseñ, unı pitirejim dij. Ajsa felan jerde br patša bar, šu patšanıñ kızına urulyanım, almaya küčım jetmij, u br baj kišige bereǯek, mende ič te mal jok, šunı kızın alıp berseñ boten šej iste[me]jmen dij. Tülki de: Beš-salkım-bij akam, menden bil, men bu išni saña bitėrejim dep süz bere. Beš-salkım-bij de: šu išni maña bitireǯek bolsañ, men de seni āzād edem dep ǯibere.

Šu ketišnen (595) altı aj, br küz kete, tülki dere tüpe tüs‿kete[8], balčık belden kete, avası jarık zıpt dajın karañık, kaškırlar küpekler uluša, üzi de šu χadar batır ǯiget mort kese.[9]

Tap soñında patšanıñ sarajına bara, kırk basamakdan čıka, basamaknıñ janında patšanıñ er kün čıkıp utıra turyan br kursisi bar eken. Bara da šu kursıga[10] utıra, unı da padšanıñ vezirleri kelip küre. Bara patšaya ajtalar: patšam, altın kursiniñ üstünde br tülki utıra dijler. Patša da: bar siliñiz bir kamčı, ketsin dij. Vezirler kuvmaya keleler. Tülki dij ki: siz meni kuvmañız, patšaya br ajtaǯak lapım bar dij. Patšaya barıp ajtalar. Patša dij ki: kelsin ajtsın dij, kürejim nedir pikiri dij. Tülki bara: salam, padšam dij. Patša salamın ala, tülki ınnım[11], ne tilijsıñ, nege keldiñ dij. Tülki ajta: patšam, bizim jakta br Beš-salkım-bij bar, kaznesindeki kümüšlerin ülčeǯek br jerde kümüš ülčegen merka tapıp bol‿almadı[12], meni sizge ǯiberdi, sizin kümüš ülčegen merkañizin istetti.

Patša šaša: bu nasıl kiši iken, bu χadar akčası bar dep ajta: ej tülki ınnım, šu ǯigetni maña kijeu etip bolamazsın mi dij. Tülki dij ki: patšam, unı sizge kijeu etmek bek kulaj, lākin janıñızya ič-kijeu alıjsıñız[13], undan gajri[14] almaj kızıñı. Nečin dep suraj. Padšam, unı utıryan jerın avası bek jaramaj, unın‿čin[15] br patšanıñ janına ič-kijeu barmaya pikrindedir (596) dij. Ajsa ınnım, sen bu išni maña jasa dij. Tülki: menden bil, patšam

1 Crossed out in the Ms.
2 ikendi; R iken edi
3 In Kúnos' list the title of the tale reads "Biš-Salkım-bijniñ masali" [Bish-Salkım-bij's tale].
4 kečineken; R kečine eken
5 piše kelirip
6 R čalašına
7 R keregiñe
8 tüs kete; R tüz kete
9 This sentence is a part of a *tekerleme*.
10 kursıya; R kursige ~ kürsige
11 R inim
12 bolalmadı; R bola almadı
13 alıjsıñiz
14 yajri
15 unınčin; R unın ičin

dij. Patša kúmúš úlčmege merkanı berip ʒibere. Túlki de merkanı alıp čúlge čıka, patša-dan merkanı alγanda merkanıñ kenarinda br kúmúš akča kıstırılup kalγan eken. Túlki túšúne, kateʒejin bilmij, kalaj etsin de patšanı br aldatsın. Merkanı u jaγın ajlandıra, bu jaγın ajlandıra, kúmúšnı kúre. Kúmúšın merkanıñ arasından čıkarıp, ičine kujıp patšaγa alıp bara.

Patša baka, merkanıñ ičinde bir kúmúš akča kalγan. Túlki akam, merkaniñ ičinde br kúmúš akčañız kalγan, tašlamañız, alıñız dese de, túlki: patšam, unıñ icinde bizim Beš-salkım-bijın bondajın neʒe-neʒe kúmúšleri jullar bujında túkúm-sačım bolop kite, nasajʒaγım alıp dep tašlap kete. Br kač kúnler ejlenip, kene de patšaγa bara, vezirlerge dij ki: men patšanı kúreʒegim dij. Patšaγa χabar bereler, patša čıka, túlki patšadan bu kere altın úlčegen merkasin istij, patša bere. Túlki alıp kete, gene de euelki tarzı u jaγın, bu jaγın ajlandırup baka, merkanı arasında br altın kıstırılup[1] kalγan. Altınnı túlki merkanıñ ičine kujıp patšaγa alıp kele: patšam bojroñoz, išimi pitirgenim dij. Patša baka merkaniñ ičinde br altın kalγan. Túlki akam, merkada br altın kalγan dese de túlki: patšam, nasajʒaγım unı alıp, undajın niče altınlar túkúm- (597) -sačım bolop, jullar bujında čubırıp túkúle dij.

Son čıkıp Beš‿salkım-bijge[2] bara, akšam kapını kaka, bij čıkıp kapıγa kele, kim u dep suraj. Túlki menim dij. Bij kapını ača, túlki salam bere, bij salam ala: koš keldiñ, akam, sapale keldiñ bojor dij. Túlki de torga[3] čikıp[4] otora, al-katır suraškandan soñ bij túlkige ne kizmetnen túštúñ bu jaklarγa dij. Túlki de Allanıñ emrinen saña patšanıñ kızın alıp bereʒek bolam, rāzi bolsañ, br vakitlar men seni beš salkım júzúmıñni ašaγanda sen meni kabātımnı azad etken‿diñ[5], men de saña karuγıñnı kajtarıp br jaχšılık jasarmen degen‿dim[6]. Patša da men unıñ azbarına tavuk ašamaγa barganda meni artımdan útir-mege kuvalaγan eken, sende šindi beš salkım júzúmden baška mal jok, kel indi men saña šu patšanıñ kızın alıp berejım dij. Bij ajta: Túlki akam, men dajın pukaraγa patša kızın bermij dep baksa da túlki: sen tek almaγa rāzi bol, patša berij, men alijım undan dij. Bij rāzi bola.

Túlki zavallı ʒajau-ʒalpı julγa túše, patšanıñ sarajına kele, kırk ajak basamaktan jukarγa čıka, altın kursınıñ ústúnde utıra. Vezirler kelip kúreler, kursıniñ ústúnde br túlki utıra, barıp patšaγa ajtalar. Patša: čaγırıñız, janıma kelsin dij. Túlki padšanı janına bara, padša kuš keldiñ (598) dij. Túlki: Alla rāzi bolson patšam dij. Patša ajta: ne kizmet-nen kelgensiñ, akam dij. Túlki dij ki: patšam, men sizge bızın bijniñ kudalıγına kelge-nim, Allanıñ emrinen kızıñızın istejim dij. Patša rāzi bola, tuj kuralar. Túlki ondan čıkıp[7] bijge kele, patša seni kijeu kabul etken, seni kúreʒek. Endi bizge anda ketmek kerekti[r]. Bij: bu kijafetnen nasıl barajım oña dij. Túlki dij ki: ʒúršen[8] ketijik, unın ebini tapar-

1 kistırılup
2 R Beš-salkım-bijge
3 torγa; R tórge
4 čikip
5 etken-diñ; R etken ediñ
6 degen-dim; R degen edim
7 čıkip
8 ʒúrsen

mız dij. Julya čıkalar, patšanıñ sarajlarına jakın kelgende tülki dij ki: sen mınau jerdeki sazya jaχšı br avuna, men patšaya barajım dij. Bij češme janında[1] sazya jaχšı bulana.

Tülki patšaya bara: patšam, kijeuñizin Bes‿salkim-bijnı[2] fajtonnen ketirgende atlar br šijden ürktiler, kijeuñizin arabasın br češme janında br sazlıkka alıp kelip šıpırtkannar[3], tez boloñoz, üsti bašı saz balčık boldo, atlar kačtı, ka‿jakka[4] ketkeni belli tegil. Beklep utıra. Br kat urbanen üč atlı br fajton beriñiz, alıp kelejim dij. Patša šo‿ssāt üč atlı br fajtonnen jaχšı br kat urba bere, hizmetčilernen urba berip pajtonya mindirep ǯibere. Bij zavallı brde baka ki üč atlı br pajton kele, češme janına keleler. Tülki rubalarnı bijge šü‿jerde[5] berip, češmeden br aru ǯuvındırup[6] kijdire. Brde: sakin, sen bu urbalarnen baryanda urbalarına baka berme, kürmemiš kiši dajın dij.

Nejse fajtonya mineler, (599) patšanıñ sarajına ǯayın keleler. Patša kijeu balanı karšılap čıka. Nejse vezirler koš kelgen ajtalar, bijni patšanıñ kızına kijeu kapajler. Bij patšanıñ kızına da bir salkım jüzüm ašata, tujnı atkaralar, son patša ǯez altı jasaj, kijeu balanı kürmege keleler, lakın [kürmemišniñ kününe lānet degen dajın].[7] Bes‿salkım-bij urbaların tuzın parmaginen[8] kaka bere. Patša bunı eslij: askem, bu bir šij kürmegen iken bu dij. Zavalli tülki bunı šaj degenın ešitken, šo‿ssāt kele, bijge: urbalarına baka berme dij. Andan gajri patša seni ǯellat ettireǯek dij. [Patša da bondaj kürmemiš, kalıklı[9] kišiden süjmegen kišisi jok eken.][10] Tülki šo‿ssāt patšaya bara: patšam, bij kijeuñiz jasattıyınız urbaların tikilgenin begenmij, baška br kat ruba istij, bolmaz mi‿ken[11] baškasın tiktirmege dij. Padša: bek jaχšı, kijeume br kat ruba tiktire. Bij u urbalarnı tašlap taha güzel rubaların kije, patšanıñ üzi tarzı jašaj.

Br kün tülki bijniñ utasında breu bolmayanda suraj dij ki: ej benim bijim, senin beš salkım jüzümden baška br šejıñ bolmaj iken, šunnen omur ete‿jdiñ[12], sen dajın bu ülkede jašayan breu jok iken. Men seni patšaya kijeu ettim, šindi erkesten abadansın, erkesten rahat jašajsın, šindi menim etken ijgiliklerime karu kajtarmaya ne etejsiñ maña dij. Bij: ej tülki akam, men seni nasıl karuyuñnı kajtarajım, sen ülgenden son seni men jerge kümmem, altından kafas jasatıp jatkan jerimniñ üstüne tavanya asıp (600) seni er zaman añarmen dij. Tülki bir kün enkasten tüše üle kala.

Bijniñ hizmetčileri küreler tülkiniñ ülgenin, kelip bijge ajtalar. Bij: bar ajtıñız ıryatlarya, bir čuvalya kujıp šeherden tıšarya bir čet jerge alıp atsınlar, sasıp turmasın dij. Kizmetčiler ıryatlarya barıp ajtalar. İrgatlar šo‿ssāt bir čuvalya kujıp šeherden tıšarya[13] čıkarıp br čet jerde ǯar ičine tašlajler. Zavalli tülki, irgatlar andan ketkenden son javaščıktan tura, ularnı artından u da kete, torı bijniñ sarajıne bara. Bij kurulyan utıra,

1 janinda
2 R Beš-salkım-bijnı
3 šıpirtkannar
4 kajakka; R kaj jakka
5 šüjerde; R šu jerde
6 ǯuvindirup
7 Crossed out in the Ms.
8 parmayinen
9 kalikli
10 Crossed out in the Ms.
11 miken; R mi eken
12 etejdiñ; R ete ediñ
13 tišarya

salam alajkùm, bij akam, sùziññiñ sābi ekenıññi bildim, bu mi iken maña eteǯegin dip. Bij n‿ajtaǯaγın[1] bilmij. Br vakitler tùlki gene de bijniñ janında ǯùre. Kùnlerde br kùn tùlki učundan ùle, lākin bij tùlkisiniñ ùlgenine inanmaj. Gene de meni br sıñemekdir dep, bir altından kafas jasata da ùzniñ[2] jatkan jerine tavanγa asıp kuja. Bu Salkım-bij patšanıñ kızın ala, tùlki de altın kafasta jata. Muradlarına hasil olalar.

34 Kırsız masali

(608) Br zaman bar iken, br zaman jok iken, br kırsız bar iken, unın da br sevgilisi bar iken. Kırsız kùndùz ùjinde jata, akšam karañıklanγan, kete kırsızlık etmege. Bùtùn keǯe ùjge kelmij, sabā kelij. Sevgilisi br baška adam da tapa ùzine, u da jankesiǯi, keǯe ùjde otora. Sabā ačılγan, čıχa kete. Bularnıñ ekisi de br brinden χabarı jok. Br talaj vaχit bùjle jašajler.

Bir kùn kırsız ajta apajna[3]: men baška šeherge ketejim, munda kırslamaγa šij tapılmaj, maña aš pišir dij. Apaj da piširij čùberek. Čùbereklerin ala čıkaj, julγa kete. Jul da br kač kùnlùk bar. Eki kùn jul ǯùre, br češme rast kele, janında balaban terekler bar, tereklerın tùbinde jata.

Jankesiǯi de kele, apajna ajta: min baška bir šehirge ketejim dij, baña aš pišir. Katın da kırsızγa nasil aš piširgen bolsa, jankesiǯige de unnan pišire, ašlarnı ala, jula čıka, u da tora kırsıznıñ jatkan jerine kele. Salam bere, utıralar. Sen nerelisiñ dip suraj, u da ni jerden bolγanını[4] ajta. Kırsız: ekimiz de bir kùjden ekenmiz dij. Braz otoralar, čıkaralar aš ašamaγa. Krsız[5] kùre jankesiǯiniñ ašı da bonoñ ašından. Bu aš meni apajnıñ piširgen ašı dij. Jankesiǯi de meni apajın piširgen ašı dij. Surajlar birbirinden: sen kajsı mällede otorasın, aññadılar[6] ki ekisi br jirde otoralar. Bašlajler čekišmege.

Artlarına kajtajler, keteler ùjge. Kırsız dij: meni apajım, jankesiǯi de dij: meni apajım. Son apaj olarγa ajta: men ekiñizin de (609) apajım ama sizniñ brbrinizden χabar[ı]ñız joχ eken. Annar: ni japajmız šindi dijler. Apaj ajta: sizler imtam‿boloñoz[7], kajsıñız [...][8] ùstùn čıksañız, men unı[9] bolorom dij. Kırsıznıñ bir kùmùši bar eken, bu kùmùšni alıp satmaγa čıkkan, jankesiǯi de bunı kùrep janına bara, bu kùmùšin ǯebinden čala, apajna kùtùre, ajta: sakın, kırsız kelip menim ketirgenimni ajtma, munda kaldırγansın dep ajt.

Kırsız baka, ǯebinde kùmùši jok, ùjge kaldırγanım dij. Kele ùjge apajdan suraj, apaj da dij: mına, mında kalγan. Kırsız kùmùšni ǯebine kuja, gene de kete satmaγa. Jankeseǯi kene de čalaj. Braz jer kete, baka ki kùmùš jok. Vaj, jankeseǯi čalγan dij, u

1 najtaǯaγın; R ne ajtaǯaγın
2 R ùzınıñ ~ ùziniñ
3 R apajına
4 bolγanıni
5 R kırsız
6 aññadilar
7 R imtan boloñoz
8 Crossed out here in the Ms.: bir kùmùšı bar iken, bu kùmùšni alıp.
9 uni

da apajγa ketirep bere. Ajtma sakın benim bergenimni dij. Kırsız de kelij, apaj čıχara kümüšni bere. Kırsız ala, tizin tibine baγlaj, bunı de jankeseǯi küre. Kırsız julda keli‿jatkanda[1] jankesıǯi bara, kırsızın kalpaγın ala, üjın tepesine ata. Kırsız ničın atasın meni kalpaγımnı dij. U da šaχ‿etkenim[2] dij, al da ber kalpaγımnı dij. U da: meni ajaklarım bulašıχ, sen čıχ menı umzıma da al kalpaγın. Kırsız čıχa jankeseǯini umzına, kalpaγını ala jatkanda jankeseǯi kümüšnın[3] čala.

Kırsız kalpaγını ala, bara kümüšin sataǯaχ adamın janına, baka kene de kümüš jok. Kajta üjine, jankeseǯige ajta: bu senin marafatıñ, akšamda menimkini küresin. Otoralar akšamγaǯıχ. Keǯeniñ bir vaχtinda čıkıp keteler. Kırsız tora patšanıñ azbarına tüše, jankeseǯi korχa. Kırsız: korχma kel degen. (610) Kırsız dora kaz kümesinden br kaz ala, ketire, suja jankeseǯini kulına bere, br ataš jaka, ajta: men patšanıñ sarajına kirejim, padšanen braz nakırdı etejim, u vaχit sana da ajtıjım: ajlandır kazın küjmesin, u vaχit sen kaznı atašta kızarttırsıñ[4]. Kırsız kire padšanıñ sarajına. Padša juklaj, janında lakiji de juklaj, lakijın avızın bajlaj, kulların bajlaj, ala, baška br utaγa tašlaj, üzi kele lakıjın jerinde jata.

Brazdan padša ujana. Padišam, dij, ǯanıñız sıχıla mi juχsa sizge br ikeje ajtajım dij. Padša ajt dij. Vaktinen dıj br kırsıznen br jankeseǯı bar iken dep üz ikejesın ajtaj. Bitirdikten soñ kırsız: ajlandır kazın küjmesin, kirgen čıkkan bolmasın dij. Ajlandır kazın küjmesin degenin nas[5] šıj dep padiša suraj. Nakirdinı šebetiji, dij padšaγa, ikeje ajta iken jankeseǯi de azbar ičinde kaz kızarta[6] iken. Büjle olγan vakitte apaj kimni bola? Padiša da: kırsızın mārefeti zijade, apaj kırsıznı bolor. İkeje büjle pite, padiša juklaj, kırsız da sarajdan čıka. Jankeseǯige: ajde, ketejik dij. Keleler üjge. İšitkensin mi padišanıñ ajtkanını dep, jankeseǯi išitkenim dij.

Sabā padiša tura, baka lakiji juk, bašlaj kıčkırmaγa, lakij kelmij. Üzi tura kıdıra, baka br utanıñ ičinde avızı kulı bajlı jata. Bunı čeze: ne boldo saña dij. Bilmem patšam, akšam siznen breu jatkanda, ujanıp bu alda kürdim üzimni. Padša añnaj ki keǯe kelgen kırsız eken. Tallal kıškırta (611) šeher ičinde: bu keǯe padišanı sarajına barγan kim bolsa da kelsin, padiša čaγıra, br zijan etmijǯek. Bunlar padišaγa baralar, ikejesini nasıl olγanı br br ajtkanlar. Padiša apaj kırsızıñ dij, čıχara bešer jüz kümüš bere. Jankeseǯige dij: sen bu šeherden ket, br taha kelme. Kırsızγa da: sen de br taha kırsızlık japma dij, apajın nıkāsına[7] kıjalar, kırsızγa berilir, kırsız da ömrünü sefānen kečire.

1 keli-jatkanda; R kele jatkanda
2 R šaχa etkenim
3 R kümüšnı
4 kizarttırsiñ
5 nasıl
6 kizarta
7 R nıkāsını ~ nıkāsın

35 Eki ulan ekjeti

(616) Bir vakitta br adamnıñ eki ulı bar iken. Bunlar ekı künnü brisinde[1] pelevanlık[2] japa‿kenner[3]. Babaları büjügine br kün tarılıp üjinden kuva. Bu da bašın alıp Jaltaja kite, bašlaj aχbarʒılık etmege. Keʒe sajın br adamnıñ kajvesinde jata. Künlerden br kün čuval kütürmekten bezip kajveʒige alın süjlep baña akıl ügret dep, kajveʒi de br zamparaʒı iken: ben saña akıl ügretejim, lakın kulamıtañnen para kazanıjsıñ, ma, saγa biš para, biš paralıχ langa-χıjarı al, sonra ket sarı amamnıñ janına. Urada aγaʒlar üzerine otor, u χijarnı ištanıñıñ ičerisine kuj, χanımlara karšı otor, amamdan čıkan χanımlar kürsünler. Ü‿jaγıni[4] kendiñ bil.

Ulan kitej amam janına. Beš un χanımlar amama kelip kirejler, amamdan čıχtıkta χanımnı bri bunı kulamıtasın kürep, al bu ʒemidenni kitejik saña kač para berem dej. Bular kitejler, χanımı üjine barγanlar. Bu ʒemidenni tašlajıp kajtaʒak bolγan. Xanım: ne ašχasın, seniñ kursaγın ačtır[5] dij. Kir ičerige, br az aš aša dij. Kıskası, χanım bunı kulamıtasına sıklandıγından, bunlar išlerin ujdırdılar. İšten turγan soñ χanım uña un kümüš berip: sen bundan kit, čünki akajım keleʒek zamanı olγan. Jarın kündüz sāt unda kelijsin dij. Bu χanımnıñ akajın da zanāti kajveʒilik eken. Er dajım sabā kajvege kitep akšam (617) kelij‿ken[6].

Ulan u χanımı üjinden čıkıp kajveje kele, [kajveʒi] hana ne kazanγansıñ dep suraj. U da un kümüš dep čıkarıp üstül üzerine kuja. Nas-nas[7] boldıγını surdıkta bu kajveʒi tüšünip baka kendi apajı iken. Tudaγını tišlej, lākin buña br šej dimej. Jarın da barıjsın‿[ım][8] dep surγanda elbet barijmen sāt unda dij.

Ekinʒi küni bara, unı artından kajveʒi de kite, baka kapu bekli. Kapunı kaka. Xanım ičerden aman akajını kelgenini tujıp, ulanı balaban zembil ičersine saklaj, soñra kapunı ača: vaj, vakıtsız kelgeniñiz neje? U da br masal ujdırıp ajta. İčerge kirep aman bašlaj etrafin aktarmaγa, br šij tapmaj, üzi de pek susaγan iken sačmaladıγından penʒere ičerisinde tuχsan altı gradosluχ keskinliginde alma suvından šıra japılγan sirke tura iken. Ak boldoγondan su bellejip iče. Aman oχšumaγa bašlaj da u jerde legen ičerisinde ulanı arkasını ʒuγan suvını iče de čıkıp kajvege kite.

Xanım artından kapunı beklep zembilden ulanı čıkarıp tüšek kefi japalar, sonra un kümüšni alıp kajvege kele, nasl[9] boldoγını ajta. Jarın da oraγa baraʒaksın‿ım[10] dep suraj. Elbet barijim dej. [...][11] Üčünʒi küni bara χanımıñ üjine, bunı artı sıradan kajveʒi de [...][12] kele. (618) Kapu bekli, kakmaγa bašlaj. Xanım aman akajım kele dep. Bunın

1 R birisinde

2 pelevanlik

3 japa kenner; R japa ekenner

4 ü jaγıni; R u jaγını

5 ačtir

6 kelijken; R kelij eken

7 R nasıl-nasıl

8 Missing particle: *ım* ~ *mı*. R barıjsın mı

9 R nasıl

10 baraʒaksınım; R baraʒaksın mı

11 Crossed out in the Ms.: Jene kendini χaberi boldoγını ajtmajıp anʒa barmali dij.

12 Crossed out in the Ms.: jine tutaʒak bolop

jerde tıyırčıklanyan[1] hasirler bar eken. Unın ičerisine sarıp duvarya karšı tiklep[2] kuja. Čabık barıp kapunı ača. Kajveǯi ačulanıp bašlaj kamasını čıkarıp kıdırmaya. Bašta barıp dünki saklanyan zembilge kamanı utırta, br šij jok. Kıskası üjini epsını kıdıra, ulanı tapmaj. Xanım bašlaj tarılmaya akajına.

Bu kendi bašına: aǯajıb, juksa baška hanımı degen zanya barıp, kendi hanımını jel tarafından alıp uplamaya bašlaj, br kajve piširtep ičtikten sonra jine čıkıp tüšüne-tüšüne kajvege barıp išine baka. Hanım da ulanı čıkarıp jine iš küreler, sonra jikerem biš ruble taha bere. Ulan alıp tolaša-tolaša kajvege kele. Kajveǯi turıp bunıñ janına kelip kana, bakajım ne japkansıñ dep suraj. U da hanımnıñ akajı jine de kelip naslı[3] kıdırdıyını ajta. Bu šaša, čıχarıp parasını küstere. Br az vakitten soñ taha baraǯaksın‿ım [dep] suraj. Ulan hebet dep ajta. Sabā jine šu vakitta bara, artından bu dinlep kele. Xanım bunı azbarında br avuz bar iken, u avuznıñ ičerisine tüšürıp bašına br ojuk kabak kečirip su ičine saklaj, barıp kapunı ača. Akajı zijade ačulanıp ičerge kirgenden soñ zembil ile χasırlar ičerisinden kıdırıp[4] baka, hič br jerden tapmaj. Kendi kendine (619) ačulanıp juχsa bu ıš bašχa br adamnı üjinde mi degen zanya barıp karısını hoplap jine üjden čıkıp kajvege tora kite. Bz kelijik ulanya.

Xanım unı havuzdan čıkarıp jine muradlarına kavuškandan soñ un ruble alıp jine kajvege bara. Kajveǯi bunın kürdüginden nı japkansıñ dep suraj. Ulan da naslı havuzda saklanyannı br br ajttıkta, bu χanımya ačulanıp suraj: üjimde br ǯemjet japsam, māle halkını de čayırsam, bu bašına kelgen šıjlerın torasını ajtırsın‿ım[5], naslı[6] sen br kapeksiz bolop para kolajnen kazanyanın ajtırsın‿ım degende ulan: hebet ajtijım dij. [Baryan kajvesi bunın kajvesı boldoyından χanımın da χaberi jok.][7]

Bu adamnıñ kajvesine ǯemjet japıp adamlar čayıra, mālenı adamları de čayıra, apajnı[8] ajıraǯak bolor, buna šatlataǯak bolop jemekler baktıra. Ulanı akšamdan sonra ters kapıdan alıp üjine kire, br utada utırıjlar. Aš ašadıktan[9] soñ bundan surmaya bašlajler. Bu naslı boldoyını ajtmaya bašlaj.

Adamlar bašlaj külmege, χanım da kapını aralıχ etip baksa, iš jañıš[10]. Ni japsın. Üjinde br mıšık bar iken. Mıšiknı ičerge ǯibere. Ulan mıšiknı kürgenden aklı bašina kele, keldiji üji bolyannı. Brden χalklarya tujdurmij son sizge ajtaǯayım küzümi ačıp baksam, sabā ačılyan amam ayačlarnı üstünde juklayan kalyanmın. Büjle ben br tüš kürgenim dij. Xajr ejlikke bolson dij.

1 tiyirčıklanyan
2 tikilep
3 R nasıl
4 kidirıp
5 ajtırsınım; R ajtırsın mı
6 R nasıl
7 Crossed out in the Ms.
8 apajni.
9 ašadiktan
10 jañiš

36 Kȯse masali

(625) Bir vakitte bir padšanıñ br ulu bola. Padša apakajına kes ete: men ůlgenden soñ ulumnı jigirem biš jašınaʒek okot, okıp čıkkandan soñ sokornen topalnen kȯsenen salam alıp salam bermesin dij. Bir vakitten soñ padša ůle, ulannıñ jašı jigiremge kele, ukıp pitire. Botem jerde br padiša emʒesi bar eken, kete emʒesin kůrmege.

Julda ketkende br kȯse rast kele, kȯse salam bere. Ulan bašın ajlandıra boten jakka. Kete jatkan julından kajta, ůjge kele, anasına ajta: anaj, men bůkůn br kȯse kůrgenmen, emʒeme keti‿jatkanda[1] julda maña salam bergen, men de salamın almadım, boten jakka bašımnı ajlandırdım emʒeme ketemedim, kȯse maña rast kelgen‿čůn[2]. Anajı de bek jaχšı etkensin balam dij.

Ekinʒi kůni gene de emʒesine ketmege teumetlenip julγa čıka, gene de keti‿jatkanda[3] šo kȯse rast kele. Ulanγa salam bere. Salamın almaj kene de ůjge kajta, anajına ajta: anaj, bůkůn šo kȯse rast kelgen, men salamın kene de almadım, ůjge kajttım dij. Anajı: bek jaχšı etkensin balam dij.

Ůčůnʒi kerre kene de šo kȯse rast kele, salam bere. Ulan kȯsenıñ sakalın sůge. Kȯse ajta: balam, men seni teumetlengen kȯseñ tůgulim, nečin šulaj menı sůgesın dij. Ulan dij: jā menim babajımdan maña kes etu (626) bar, sokornan topalnan kȯsenen laf etme dep dij. Kȯse ajta: seniñ ajtkanıñ bek učun, lākin seniñ degen kȯseñ botem kȯsedir, men tůgulim. Senı babajıñ mennen bek dust eken, maña er vakit akča berir‿ken[4], menı breulerge mojın ektirmez‿gen[5], meni bek sůjer‿gen[6].

Kȯse ʒigetten suraj: ulum, šindi kajda barasın dij. Ulan ajta: felan jirde padša menim emʒemdir, šimdi anda ketem, emʒemni kůrmege dij. Kȯse ajta: ajsa siz br padša ulu ikensiz, men de br pukara kišimen, meni de katıñızγa alıñız, sizin alašañızın kašaγalarım, durbasın berijmen, χizmetiñizde tabılıjım dij. Ulan boña br at tapıp mindire, ekeu barabar keteler.

Az mi keteler, kůp mi keteler, br kůn br jerde češme bašında otoralar. Jiget ůstůnden kalatın kılıčın čıkarıp rahatlana, köse de unıñ urbasın kije, ůzinkin tašlaj ʒıgetnı atına mine, ʒıgetniñ aldına[7] čıka: men šimdi padša ulu bolam, sen kȯse bol, undan gajri[8] bašıñ alam dij. Ulan namazın kılıp pitire, bašlaj kȯsege ʒalvarmaγa: etme, ejleme kȯse dustım dij, lakin kȯse br tůrlů ʒigetniñ jalvarγanna[9] br šej etmij. Ulan ebin tapıp bolamaj, katijım, padša ulu sen, kȯse men bolsam bolorom dij. Kȯse jigetke and‿ičtire[10]: ůlip tirilmegenʒek breuge kȯse kum, padša ulu kum ajtmaʒaγına. Ulan da ant ete ajtmaʒaγına.

1 keli-jatkanda; R kete jatkanda
2 kelgenčůn; R kelgen ičůn
3 keti-jatkanda; R kete jatkanda
4 berirken; R berir eken
5 ektirmezgen; R ektirmez eken
6 sůjergen; R sůjer eken
7 aldina
8 γajri
9 R jalvarγanına
10 and ictire; R ant ičtire

(627) Ů‿jerden[1] atlarına minep ǯulɣa kitejler. Br kün jigetni emǯesi bolɣan ülkege ǯetkenler. Patišanıñ da ennesiniñ ulu üzine musapir bolop keleǯeginden kabarı bar iken. Akšam salkınnıginen[2] üjinde perǯeresinde[3] julɣa bakıp otora‿jdı[4], birde akasınıñ ulu kele. Aman šo sātta čalɣı čaɣananen ǯigetni karšlap[5] almaɣa čıkalar, brde baksa ne baksın br čirkin kȯse erif kelgen aɣasın ulu artındaki kizmetčisi undan biñ kat jaχšı. Padiša: aǯajb[6] menim aɣamnıñ bondaj ulu bar mi‿ken[7], bonda kelip te meni maskara etmege, šte šo kizmetčisi padša ulu bolmaɣa lājik. Šu ǯiget meni aɣamnı ulu bolɣan bolsa ne bola‿di[8], šolaj br čirkin kȯse bolmajıp ta dij. Vezirleri ajtalar ki: padišam, biš barmaknıñ epsi bir bolmaj, katmalı, er nekadar čirkin bolsa da sizın akañızın uludır. Padša katsın[9] kabul ete.

Kȯsenen jiget biš un künler vakit kečireler, lākin padiša er vakit dij ki: šte padiša ulu bolmaɣa kim lājik pilis-pidandaj ǯiget šu kȯsege kizmet ete dij. Kȯse ičinden kurka, menı tujdırıj dep. Bu ǯigetten nasıl bir kutulsın, baška türlü čara tapamaj, jigetni ütirmege karar bere ama nasıl ütirtsin, br türlü ebin tapamaj. Bir kün padšaɣa dij: emǯem, men χas padšanıñ kızın istejim dij. Padiša dij ki: be ej aram kȯse, unı nasıl alajsın, χas padišanı kızın almaɣa barɣan ǯigetler er birisi sen dajın (628) jüz kȯse eteǯek, šondaj ǯigetlerniñ bašlarından kırk kat saraj jasattırɣan, šimdijeǯek ne kadar jüz biñlernen ǯigetler u kıznıñ aškına[10] kurban ketkenler, breu alamaɣan, sen mi alaǯaksın, nasıl alaǯaksın degen. Kȯse dij ki: menim kizmetčim šolaj br ǯigetdir, u alıp kelij, tek sen emr et, alıp kelmege dij.

Padša ajta ki: aɣamnı ulu tügül, kim bolsañ bol, jazık[11] tügül mi, šondajın ǯigetni anda ǯiberip sendaj bir kȯse‿čin[12] ütirtmege dij. Ů‿jerde kȯse bašlaj ǯılamaɣa: sen meni br emǯem bolop ta men kȯse bolɣan‿čin[13] sen menden meni kizmetčimni aǯınɣanın ne dep, vezirler de bašlajler jalvarmaɣa: padišam, ǯiberiñız šo jigetni ketsin, bekim akañızın uluna šo kızın alıp kelij dep. Jiget atka minep kete. Kulına br kamčı bereler.

Zavalli ulan ǯılaj-ǯılaj julda ketkende karšısına br kart kiši rast kele, dij ki: nege ǯılajsın ulum, ne dertiñ bar eken dij. Ulan dij ki: katijim, kalaj etijim, men ǯılamaj da kım ǯılasın, men šimdi ketemen χas padišanıñ kızın almaɣa, men unıñ kızın alıp bolor mijım dij.

U kart Kıdir-nebi iken: ulum, dij, χas padišanıñ kızın alması zurdir, unı almaɣa barɣan ǯigetlerniñ bašından kırk kat saraj jasatıp, üstünde uturıp emr ete, lākin men saña br avıǯ tufrak berejım, bu tufraknı χas padišanıñ sarajlarına ǯuvuk barɣanda, jarısın sepejsın, jarısın de sarajnıñ kapısının ičinden kirgenden soñ sarajlarɣa sepej-

1 ü-jerden; R u jerden
2 salkınnıɣinen; R salkınnıɣınen
3 R penǯeresinde
4 otorajdı; R otora edi
5 R karšılap
6 R aǯajıb
7 mi-ken; R mi iken
8 boladi; R bola edi
9 katsin
10 aškina
11 jazik
12 kȯse-čin; R kȯse ičin
13 bolanɣan-čın; R bolɣan ičin

sıñ dij. Padišanıñ kızı (629) gülbakčada sijirge čıkkan bolor, ekindi salkıninnen[1] kızın alijsın, lākin padšanıñ sarajından elli čakırım ajrılyanʒek ajlanıp artına bakmajsın, andan gajri[2] telef bolorsıñ dij.

Jiget ni bolsa da andan kete, saraja jakın kele. Sarajnıñ er tarafı u kadar üksek kala ki asıl ičerge kirmesıni ebi jok iken. Tufraknıñ jarım avuʒın bir sepe, ü‿jerde dıvarlarya dora duvarlar tübinden ʒıkıla, üzi de šašıp kala. Bir vakitten soñ ʒiget ičerge kire, padišanı sarajınıñ er janında kırk tene eżderhalar sakčı turalar, padišanın kızın saklap. Šü‿jerde[3] tufraknıñ kalyanın da sepe, sarajnıñ ičindeki tašındaki epsi ülü dajın jıkıla kalalar. Šü‿jerden ʒiget gülbahčadan padišanı kızın atnıñ üstüne ala, alašaya br tepkende elli čakırım jerin birden ala. Kız julda ʒigetten suraj: ej ʒiget, meni šaj etip kimge alıp ketesin dij. Ʒiget ajta: men seni padišanıñ ulu kösege alıp baramen dij, lākin kız: ej ʒiget. men senden yajrısına barmam dese de ʒiget: jok, men seni üzim ičin almadım, seni padiša uluna kösege [alıp baram dij],[4] ketirmege süz berdim, šimdi senı alıp baram dij.

Br vakitten soñ brde bakalar ki ʒiget χas[5] padišanıñ kızın alıp kele. Köse zavallı nišlesin bilmij: bu ʒigetten maña kutulmak jok eken, br kün bu meni masχaralıklarımı mejdanya čıkara dep nišlemesini bilmij. Ʒiget kıznı ketire patšaya: ište padšam, emriñizin iʒra ettim dep bere. Köseni u keʒe kijeu kapajler kelin janına, lākin köseni: ket, χınzir (630) köse, meni aramlama dep kuva, köseni katın jakın bile almaj. Köse nišlesin, aman bara padšaya, ajta: bu ʒigetın bašın almasañ bolmajʒak dij. U nečin degende: men bu ʒiget sau bolyanda jašap bolmijmen, ište kız meni janına jakın almaj, oña küz etken dij. Padša: ulan köse, jezid, bolor mo bondajın iš, šte bonʒa išler jasayan, insan bašindan kırk kat saraj jasatkan χas padšanıñ kızın alıp kelgen, taha ne istejsıñ dese de köse toñoz bašladı ʒılamaya: sen menim emʒem bolop ta meni bir ırgatımdan ašaya mı küresin, br ıryattan[6] eserkijsin dep ʒılaj.

Padša čara tapıp bolmaj: χınzır köse ne ete ete, ʒigetni ütürte, lakin ülmij. Kidir-nebi oña ülmesin dep duva etken‿di[7], ularya ülgendaj kürüne. Nejse ʒigetni alıp ketirip ʒıjalar. Köse kutıldım dep kuvana, kutluyından[8] kateʒegin bilmij. Ʒiget kümülgenden soñ kız u keʒe tišinde br derviš kelip: kızım, tez bar ʒigetni mezerden čıkar dij. Kız bara ʒıgetın mezerden čıkara. Mezerden čıkkana dajın br ıryat rubası kijep padšanıñ kapusına ıryat ʒürmege kiši kerekmij mi dep suraj. Padša bu ʒigeti kürgenine dajın baka ki köseniñ julına kurban bolop ketken mator ʒigetke pek ošaj. Padša ülgen ʒigetke bek ošayanı ičin süje, ʒigetni ıryat tuta. Köse jašamada ičmede ašamada davam eter‿di[9], ʒigetın ütürdim dep (631) kuvana. Aradan bir vakitler kečken. Künlerde bir kün padšaya musapirler kele, ularnı ejlendirmek ičın burunki zamanlarda masal ajtmak pek makbul iken.

1 salkininnen; R salkıninen

2 yajri

3 R šu jerde

4 Crossed out in the Ms.

5 has

6 iryattan

7 etken-di; R etken edi

8 R kutluhyından

9 eterdi; R eter edi

Padša šeher ičinde ne kadar masalǯılar bar ise, epsin tuplattıra. Br kün kunaklarγa masalǯılar tarafından masal ajttıra[1], ǯiget te anda kelgen ketken kunaklarγa baka eken. Erkes bilgenine küre masal ajtalar, lakin padša bu ǯigetni bek süjer iken, ülgen ǯigetke ošaγan‿ičin[2]. Padiša ajttı: ǯiget, sen tünja kürgen bir ǯigetke ošajsın, bilgen jaχši masalın jok mı dij. Ǯiget: padšam, men masal ajtarım, lākin kapıların kapatijsiñ, breusi tıšarγa[3] ǯibermejsiñ dij. Padša: bek jaχši dep emr ete, ǯiget bašladı masal ajtmaγa, bašindan kečken seratlıklarnı ikejesin, baštan nasıl akajı kösenen, topalnen, sokornen laf etme dep salam alıp salam berme degenin, ve nasıl köse: babañ menim dustım iken degenin, lākin ǯigetnın br köse bolγan dep añlata masalni. Sonra nasıl köseni üzerine ırγat kabul etkenin, undan soñ nasıl köse ǯigetniñ urbasın kijip namaz kılarken bašın uǯuna kelip, nasıl ǯigetke üzine ırγat bolmasın teklif etken, lakin köse išni tüšünip kele: ǯanım, bu erif nasıl šij, bizim janımızda beraber mi ǯürdi, juksa bu erif üzi mi iken.

Köseniñ vuǯudına[4] ot tutıša ama tıšarı[5] čıkmaγa bolmaj. Ondan soñ padša nasıl bolarnı karšılap čıka, padiša nasıl süjıne. Bondaj ǯiget ırγat, bondajın köse (632) menim emǯemıñ ulu bolγan, šte padša ulu bolmaγa kim jaraša dep süküngenın süjlej. Ondan nasıl köse ǯigetni ütürtmesi rahat bolmajǯaγın[6], andan nasil padšanı aldatıp köse χas padšanıñ kızına ǯigetin ǯibere. Köse išın añamaγa bašlaγan. Karnım bura, kursaγım avurta dep baka ama kapınıñ janında ǯallat bir kimseni ǯibermij. Ǯiget vire añata, kösege kulak bile bermij. Nasıl julda Kidir-nebi rast kelgen, nasıl da kızın alıp kelgen, kösege kız ajlanıp bakmaγan, köse kene de ǯigetten kutulmadım dep nasıl padišadan ütürtmesin tilegen, padša nasıl köseni süggen, köse de nasıl padšanıñ tamurlarına[7] kirgen kandırγan, br br ikejeni añatkan.

Padšanıñ küzleri kan čanaγı tarzı ajlanγan, kıp-kırmızı[8] boldo, išni tojdı. Ǯiget gene de añlata, nasıl ǯigetni ütürgen dep mazarγa kümdiler, nasıl kız barıp čikarγan, nasil ǯiget padšaγa irγat ǯürgen, padišanı ǯamātnı nasıl boldo da masal ajtmaγa kelgenler. Ǯiget: šte köse dustım, sen maña ülıp tirilgenden soñ ajtırsın dep and ettirgen‿din[9], šte kösejım, men šindi ülüp tirildim, sırasi kelip ajttım dij. Padšaγa da: köse menim irgatım iken, men seniñ akanı ulujum, sen menim emǯemsin dep turıp kulından üpe. Padiša tuγanǯıγım dep, küzlerinden üpıp ǯılaj, köseni de bir terekke bajlap br kün kulaγın kesip, übir kün murnını kesip ütüriler.

Ǯiget ülip tirilep emǯesinen küriše, köse de etkenin karšısın tapa. Allah bizlerin de muratımızγa[10] erištirsin. Bu masal bondajın bolop pitken.

1 ajtıra
2 ošaγan ičin; R ošaγanı ičin
3 tıšarγa
4 vuǯudina
5 tišarı
6 bolmajǯaγin
7 tamurlarina
8 kip-kırmizi
9 ettirgen-din; R ettirgen ediñ
10 müratimizγa

37 Eki ǯanaj ekjeti

(642) Bir zamanda Kırımda bolγan bir ǯanaj, Mısırda bolγan bir ǯanaj. Mısır ǯanajı Mısır jaklarında bek maktavlı ǯanaj bolγan. U jaklarıñ ǯanajları hepsi undan destır alaj‿kenner[1] kırsız bolmak ičın. Kırımdaki kırsızların atamanı bolγan ǯanajnen Mısırdaki kırsız bir birlerın etkennerini ešitip kürišmege istegenner, kahat jazγannar rast kelišmek ičin. Mısır ǯanajı Kırım ǯanajın Mısır jaγına čaγırta. Kırım ǯanajı: ajde, barajım, br kürišejim, di.

Künlerde bir kün julγa tüše, bašlaj ketmege. Kete-kete bir kün Mısır čülinde bir češme bašinda bir terek salkınında bir ǯıget otora. Kırım ǯanaji salam alejküm, dep bara. Mısır kırsızı otor ǯiget di. Kırım ǯanajı otora. Mısır ǯanajı tütün jak dep tütün bere. Suraj: ǯiget, kajdan uγur dij. Kırım ǯanajı Kırımdan kelem di. Ka‿jakka[2] uγur dij. Mısırγa baram dij. Mısır ǯanajı teumetlene: bu mu eken Kırım ǯanajı dep, anda kimge barasın dep suraj. Mında bir ǯanajnıñ nāmın ešittim, šonoñ ünerlerin karamaγa keldim dij. Ajsa u ǯanaj menim dij. Bir-birlerin kürgenlerine bek kuvanalar. Soñ Kırım ǯanajı ajta: bir ünerıñizin küsteriñız kürejim. Mısır ǯanajı balaban bir üner küsterejim dij.

Kületkesinde otorγannarı terekniñ (643) üstünde bir kušbala jataj‿kan[3], bak ǯigėt dij, šu terekdeki bala jatkan kušnıñ juvasından ǯımırtkaların alijmen, üzın kabarı bolmaz dij. Kırım ǯanajı kana et kürejim di. Mısır ǯanajı terekke tırmašıp čıka, jımırtkaların alıp ištanıñ ǯebine kujmaγa bašlaγanda Kırım ǯanajı da unı artından tırmašıp Mısır[4] kırsızın ištanın čıkarıp ala, üzıne dıjdırmaj. Mısır ǯanajı bir üner küsterdim dep kelip jerine otora. Nasıl, begendiñ mi ünerimi dep Kırım ǯanajından suraj. Kırım ǯanajı da dij ki: kušnıñ ǯuvasındaki ǯımırtkanı almak üner tügül, unı[5] almaγa čıkkan kišiniñ ištanın kırslamaktır, üzine dıjdırmaj dı. Men senden bir üner, bir kırsızlık jürenıjım dep Kırımdan[6] munda keldim, sen ǯımırtka kırsızlıγı küsterdiñ. Men bondaj ofak kırsızlıkların biš jašında bala eken euel ǯımırtka kırslap bašlaγan edim dij, sen jañı bašladıñ kırsızlıknı jüremmege[7] dij. Mısır kırsızı Krım kırsızın kulın üpe, zanātıñı sābi bolγansın dep. Aman ataman, maña de ügret šo ilmini dij. Kırım ǯanajı da munı janına alıp Kırımγa ketire.

Bir kač vakitler kırsızlık jürete, u da Kırım kırsızından ašaγı bolmaj. Bir vaχitten soñ Kırım ǯanajı kırsızlarnıñ jataγı (644) bolγan Or šehrinde bir kırsızlıkka barγanda atıp üldüreler, šergetleri munı Or šehrine ketirip kümeler. Kırımda ne kadar ǯanajlar bolsa, epsi atamannarıñ ülgen‿ičin[8] jas tutalar. Son epsi Mısır ǯanajın munasip kürep ataman eteler. Mısır ǯanajı Kırım kırsızın jerin[de] baš bolop kala. Ele bükün Kırımda Or šehrinde bolγan metinni ǯanajlar kibik hič bir jerde jok.

1 alajkenner; R ala ekenner
2 ka-jakka; R Kaj-jakka
3 jatajkan; R jata eken
4 Misır
5 uni
6 Kırımden
7 R jürenmege
8 ülgen ičin; R ülgeni ičin

38 Čora batırnın ekejesı

(648) Evel zamanda Narık bar iken. Gājet bir akıllı fikırlı, em de auǯı bir adam iken. Narık kajda kartların kürse, katına barıp unnarın nakirdisın tinnerdı. Kartların biri: ne zaman bir kannıñ janına jürse, adam bolor, lākin bir baj kapısına jürse, bir šıbuk urıp ǯiberir. Narik bu süzlerni ešidep tora ketip bir kannıñ janına jürdü, kesen-kes jıllık salmadılar. Anda poštaǯı kujdılar. Narik bir kün pošta alıp keli_jdi[1], julda čoban košında musafir boldı. Akšam čobannar jıylıštilar, ašau ašadılar, kāve ištiler[2]. Kefler tamam boldıktan son avıǯılık nakırdısını bašladılar. Čobannıñ birisi ajttı: egerkiñ[3] janından er kün sabā br bejaz kojan turıp kača. Narikka küsterijik, bekim tazıya tottorıj. Sabā turdılar, egerekın janına bardılar. Kojan kene torıp kačtı. Tazını[4] kojannıñ artına saldılar. Narık atına minep kojannıñ ardına tüštı. Kova-kova bir urmanya čıktı, urmanda da bir büjük koba bar edi. Kojan kobaya kirdı, tazı[5] da onıñ artında.

Narik karadı, kobanıñ iči kara duman, üz üzine: karañık koba, bejaz kojan, bekim tutarım dedi. Narik kobaya kirdi, lākin koba čok karañık edı, kobanıñ iči bütün čakıl ešlij[6], ǯüremedi. Javaš-javaš artına (649) kajttı, bir avıǯ čakıl alıp ǯebine tašladı. Kobadan tıšarı[7] čıktı, lākin avuzı[8] burnı kıjıštı[9], üzi oraya tüšti, essız jatır. Čobannar karadılar Narik kelmedi, aǯeb kajda kaldı. Bir čoban ǯiberdiler kıdırmaya, u barıp Narikın taptı, kobanıñ aldında essiz jata. Arkasına alıp košya ketirdi. Kaber berdiler kanya. Kan ekim ketirdi, mollalar keldi, ukıdılar, laǯ[10] ettiler. Narik ajındı, esi bašına keldi. Surdılar: kajda kaldıñ, ne boldo? Narik da bašladı bašına kelgeni ajtmaya. Kojan bir karañık[11] kobaya kirdi, men de artınnan kirdem, koba iči bütün čakıl, ešildim-ešildim jıkıldım. Br avıǯ čakıl ǯebime saldım, mınau čakıldan dedi, ǯebinden čıkardı, karadılar ǯebindeki tamamnen altın.

Šimdi toplaštılar, kobaya bardılar, baktılar kara duman. Bašladılar ukumaya, tılsımın taptılar, kara duman čezildi, bašladılar altın bošaltmaya. Kırk katır altın jüklediler. Kannıñ kaznesine ketirdiler. Kanya: biz bu Narikden balaban fajde taptık, biz de buña bir ijlik[12] etijik. Kan ajttı: bašta eulendirijik, ama kimin kızın tilese, onı[13] alėp[14] verijik. Üzlerinde de birer kızları bardı: seniñ kızın tilese sen de ber, menim kızım tilese, men de berejim. Üč pare küj bayıšladılar. Narik üjleneǯek, kaber berdiler, ne kadar kız barsa, (650) gülbahčaya tolsunlar. Kızlar toldılar, Narikke kaber berdiler: mında ičinde kimnıñ kızın begenseñ unı alaǯaksın ajttılar. Kızlar ičinde vezirın kıznė[15] begendı, kanya kaber berdiler. U da mubarek bosson[16] dedi. Narikni üjlendirdiler.

1 kelijdi; R kele edi
2 R ičtiler
3 R egerekiñ
4 Tazinen
5 tazi
6 R ešilij
7 tišarı
8 avuzi
9 kıjıšti
10 R ilaǯ
11 karañik
12 R ijilik
13 oni
14 alep; R alıp
15 R kızını
16 R bolson

Lakin u kannıñ ulu da bu kıznı süjer idi. Altı aj kadar karadı, kıznıñ asretligi jüregini urdı. Bir kart katınγa ajttı: bar Nariknıñ apakajına ajt: bu keǯe barırmen, kaberi olson. Kart katın Nariknıñ apakajına barıp ajttı. Narik küj ičinden keldi, bu keǯe pazarγa ketirmen dij. Apakajı: sen bu keǯe bir jere ketme dedi, üjini sakla. Akšam boldı, kanıñ ulu doru Nariknı üjine ketti, kapını kaktı[1]. Apakaj kapını ačtı[2], kan ulu ičeri kette uturdı. Bir parča uzγanda nakırdıγa bašladı. Karı otγa kajve saldı, lākin ottan uzakča kujdı ǯezveni. Karaj-karaj, kajve pišmij. Men saña kajve ičmege kelmedim, tur, tüšege jatajık. Bu nedajın süz, nedaj maskaralık bu senin ajttıγın šejler, sen bir kan ulu bolıp bir kara tatar karısına zinalıkka keldiñ, bunı kan babañ tojsa ne ajtır, hem de kazi bej babam dujarsa, o ne ajtar. Jak feneri ketejım dij. Karı feneri jakıp eline berdi. Kapıdan čıkarkan Narik de anda utıra‿jdı[3], hemen bunı bašını kesti. Čukur kazdırij[4] orda kümdiler.

Lakin bu kanniñ ulu aftada üjlerine bir kelir iken. (650/a) Vezirıñ kızı üjlengen son eki atnada üjlerine bir kelgen, son ič kelmegen. Kan tujdı ki balası gajb oldı, vezirnı čakırdı: menim ulum ǯojıldı, sen unı tap dij. Vezir: bir adam bir adamnı üttürse, küzine bakkanda tanır. [Milletı ǯıjdılar][5], bašladı erkezin küzine bakmaγa. [...][6] (654) küjevin jandan gečtikte baktı küzine, küjevi ütürmiš. Keldi üz üjine, küjevine χabar ǯibergen, tez mından ketsinner.

İkesi ike at azırlaγan, arkasına mingenner. Ekindi vaktinda üzenden atlar ǯaldatıp Etli kāna üttüler. Etli kanγa barγan son dāva darab bi‿šej[7] jok. Orada da etin ukkası bir altın, küp pukara boldolar. Biz kelijıχ kānγa.

Akildanenen kān ekisi küp muhabat edijler. Kānıñ ulu ülgen son kān vezirden suvudı, vezir de ajttı: ničin menden küstiñ? Kān nasil küsmejim, benim dar-dünjada bir balam bar iken, seniñ küjeviñ ütirdi. Menim küjevim ütirdi ama unıñ apakajına zinalıkka barγan, unın‿čın[8] ütirken. Hā, dij, ničin ajtmadıñ maña. U bir jerden keskendir, men biš jerden keser idim. Tez χabar ber, kajtsın kelsin dij. Üč pare küj de bahš ettim ularγa. Šimdi mektüp jazγannar, kabātları bašıš etkenner, üč pare küj de baγıšlaγannar. Mektubi julladılar.

Narik audan kelgen. Mahale ičinden kelirgen, aldına postaǯı čıkıp mektub kulına bergen. Narik mektubi ukuγan, balaban küzajdın bulγan. Üjine keldi, kapıdan külep ičeri kirgen. Turıp apakajı ne külesiz, küjden (655) χabar bar mi dij. Kabātıñız bahšıš bolγan, küjge čaγıralar dij. Erte sabā turdılar, bir arabačık takıšturdılar. Bularnıñ birde atları bar iken, araba ǯektiler, ketkenner. Julda ketkende bir takım koč[9] māǯire rast kelgen, ularnan arχadaš bolγan kitejler. Šimdi bunnar bu kadar ǯarlı bolγannar, ašaǯak ičeǯek jok. Narik kitep bir kujan suγıp kelse ašardılar ama tapmasa ač kalır ikenler. Narik turdı, kujan kıdırmaγa ketken. Mında da karı bašladı zurlanmaγa, baktık balačık

1 kakti
2 ačti
3 utırajdı; R utıra idi
4 kazdirij
5 Crossed out in the Ms..
6 Missing some part of the text.
7 bišej; R bir šej
8 uninčın; R unın ičin
9 Northern dialect variant of *köč*.

tünjaya kelgen. U koč kalklarından bir kart apakaj ebe boldı. Nariknıñ br jesiri bar iken, čap Narikın čayır, ŝte mužde ajtkannar. U da: Narik-akam, bayıryan, ulıñ bar dij. Narıknıñ üstünde br akčası jok, nasasın, kalaj etsin. Adını (656) ber de azad bol dej.

Balanıñ adın Čora kujdılar. Jesir boldı azad. Jek koš arabañ, jekıp arabanı üstüne mindiler, üzlerin küjlerine kelgenner. Baktılar kabātları bayıšlanyan. Üč para koj[1] bayıšlanyan bunlarya. Bunlara karšılap čıktılar, brbrinin betinden küzinden üptiler, χos‿safa[2] keldiñiz ajttılar.

Br jıl, eki jıl biš altı jıl omur etkenler. Čoranı mektebge berdiler, üč jıl okoyan boldo, br küp ünerli balaban bolyan, ep balalardan üstün bolyan. Balalarnıñ ičbirinden košmadı, balalar bondon korktular. Br kün Čora balalarya ajttı: men jerimden küčkende, sizler maña bajrak küteriñız, men sizge tijmem. Balalar: jā biz bajrayın kajdan alajık? Men sizge bajrak tabarım dij. U jerde bir buzauǯu bala bar iken, tuttı onıñ br buzaunın sujdı, terisın sėjırdı[3], balalarya berdi. Sız bajrak etıñız de küteriñız. Balalar bajraknı ketirgen. Čora balalarya tijmedı, balalar teslim boldolar.

Buzau bakkan (657) bala ǯılaj-ǯılaj üjge kelgen, bizım bızaunı[4] Čora sujdı dep. Bızaunıñ[5] sābı: bar ketirıñ Čoranı munda. Čoranı čayırkannar, kelgen. Bızaunıñ sābı: Čora, ničın sujdıñ menim bızaunı dej. Čora ajttı: kütüñe kirsin baǯakları degen, amma bızauyın bar iken dij. Bızaunıñ sābınıñ košuna keldi küldi. Čora, men saña birde at bayıšlajım dij. Xudaj rāzi bolson bayıšlasañ dij. Šimdi ǯılkıǯınen čülge kitkenner. Jılkıǯı at totop bereǯek. Ǯılkıǯıya bardılar, bašladılar ǯılkı karamaya. Jılkıǯı bašladı eñ jahšı alašaların totop Čoraya ketirgen. Čora birisin begenmedı, bašladı üzi karamaya. Jılkını artından izlerın kıdırdı[6]. Kıdıra[7]-kıdıra br doru jız taptı, jiznen br kotor taj taptı: men mınaunı süjdim (658) dij.

Jılkıǯı ajttı: bizim küjnıñ kalkı külej dij, büjle atın küjge men ap[8] ketmem dij. Čora ajttı: berseñ ber bu atın, bermeseñ kütüñe kirsin baǯakları. Arandan[9] da br tasma taptı, bajladı tajıñ bujnına, munıñ adı Tasmalı-bujı[n] bolson dej. Ketirde üjge. Erkes bunı zauklandılar. Biš jüz ǯılkı ičinde br kotor taj tapkan dijler. Čora bašladı karamaya, biš jüz attan üken bolyan. Janaša küjde br metinni alaša tā bar iken, adı da Ǯel-ǯetmez iken. Čora bu atın tujdı: ajtıñız oña, u atın ǯibersin maña. Baryan ajtyanlar, bu at uña jarıj dep. Berip ǯiberdi. Šimdi bir taha janaša küjde bir at taha bar iken, bek ǯüjrük eken. Čora aña da χabar ǯibergen, atın berıp ketirsin dij. Barıp ajttılar. Men unıñ‿čin[10] asramajım atımnı dij. Čoraya ajtkannar: atın bermej.

1 Northern dialect variant of köj.
2 χos safa; R χoš safa
3 sejırdı; R sıjırdı
4 bizaunın
5 bizaunıñ
6 kıdırda
7 kidıra
8 R alıp
9 Aradan
10 uniñ-čin; R unıñ ičin

Čora bek ačulandı, mindi Tasmalı-bujın[1] sırtına, jerinden küčkende başın tüpesinden br ut čıχtı. Šımdi atın sahibisi kürdi, karadı Čora keli‿jata[2]. Aman Kuj-bakarnı alıp julγa čıktı, Čora da kelip jetti. Atın istegensın, alıp keti‿jata‿jdim[3]. Čora: sizge zāmet bolmasın, men üzım ap[4] ketirmen dij. Alıp atın ketken. Šimdi u jerden kač kunaklık aulaγında bir karaman taha bar iken, (659) adına Ahtaǯı Ali-bej degenner, üzinden başka kırk tene taha pelivanı bar iken. Bu Ali-bej čok ačulanij Čoraγa. Pezevenk, atnen batırlık ete, men barajım šonıñ atların alajım, atı elden ketken son, br šej kelmij kulından dep.

Br kün mindi atına, Ali-bej. Pelivannar tujdılar, kete: sen Čora-batırγa[5] barma, barsañ barabar barajık degenner. Ali-bej ajttı: men anda barmajım dej, bunlardan jizledi. Janına eki arab batırı alγan, tora Čora-batırnıñ üjine kelgen. U küni de Čora tojγa ketken, üjde jok iken. Šimdi Ali-bej ekindinen akšam arasında (660) keldi baχtı Čora joχ. Šo‿ssāt[6] konan dünen sujdırdı, susumγa balnen rakı čıγarttı. Tasmalı-bujını tilej turdı. Čoranıñ babası Tasmalı-bujını bermedı. Küjın ǯemātları Čoradan büjük bar eken, Čora anası fallandı dediler. Čora anası bullarγa[7] br bejt ajttı:

Kıl kojroγı kajtandaj
ǯuvurγanı šejtandaj[8]
[. . .]
kajsıñızın ulu bar
Narik ulı Čoradaj.

Šimdi Ali-bej aχırγa kirdi, Tasmalı-bujınγa japıškan. Čoranıñ kızkardašı Kün-kürmez Meñaru ǯıvırıp kelip, atın başından tuttı, atın bermegen, lakin Ahtaǯınıñ Ali-bejı čirikkin[9] jaramaj süzler ajttı. Kız ujalγanından atın tašladı, Čorañ babasın[10] bašına tarttı. [Arab digen apšıkka jaman aǯıtıp tarttı, anañ degen apšıkka šoγon süzler ajttı.][11] Kün-kürmez Mejnaru kardašinnen[12] čok atıštılar, čok tartıštılar, šoγon süzler ajtıštılar. Šimdi Ali-bej atın alγan ketken. Čoranıñ babası atına minep ketken Čoraγa χabar bermege. Julda ketkende surajlar: kajda ketesin dejler. Narik bularγa bejt ajttı:

Aj kajdadır[13] alim[14]
Kelsin ajtajım alım
(674) Siz de išitirsiz – dep ketken.

---

1 Tasmalı-bujin; everywhere changed to *Tasmalı-bujın*.

2 keli-jata; R kele jata

3 keti-jatajdim; R kete jata idim

4 R alıp

5 The name *Čora-batir* has been everywhere changed to *Čora-batır*.

6 šossāt; R Šo sāt

7 R bunlarγa

8 These two lines are about the qualities of Chora's horse.

9 R čirkin

10 R babasının

11 Crossed out in the Ms.

12 R kardašıñnen

13 kajdadir

14 R ālim

Tujγa[1] barıp Čoranı[2] suradı. Čora-batırγa χabar berdiler. Tıšarda seni br kart suraj ajttılar. Čora tıšarı[3] čıktı, karadı babası kelgen. Babası Čoranı kürgenʒe bu bejtın ajttı:

Aj Čoram, Čoram
Čonkajıp atγa mingende
Kart atañıñ bašına
Türlü korlıklar ketirdiñ.
Čora babasına bir bejt süledi:
Aj atam, aj atam,
Ajtkanıña küjermen
Tüšmenımı ajtkanın
Ajlakta bek süjermen.
Šindi Ahtaʒınıñ Ali-beji kelıp katıp Tasmalı-bujın aldıγının süledi:
Ekindijnen[4] akšam[5] arasında keldi
Tübetauγa kunan dünen sujdırdı
Susumγa balnen rakı šıγarttı
Tasmali-bujnı tilej turdı.
Babañ degen bajgušın
Bašına küzine tarttı
Anañ degen apšikka
Šoγon süzler ajttı
Arab degen apšıkka
(661) [jaman aʒıtıp tartti][6].

Kün-kürmez Mejnaru kardašinen čok ajtıštılar. Tasmalı-bujnın[7] alıp ketti. Čora ajttı: ka‿jakka[8] ketti? Babası ajttı: Kük-kügem‿bujlarına[9], Kük-čokrak bujlarına ketti. Čora ačulanıp atına mindigi vakit atnıñ ajakları bakkajnaʒek[10] jerge battı kirdi. Jerinden küčkende bašın tübesinden br ot čıχtı. Ali-bej Čoranıñ jerinden küčkenın[11] tujdı. Janına kırk tene arab batırı bar iken, epsi kurktılar. Batırlarnıñ birisi: biz Čoranen bašγa čıkamamız, bar sen etkenıne tüleu bersen bekim baγıšlar. Šimdi Ali-bej Tasmalı-bujnına minep Čoraγa karšı kelgen, Čora da kelip jetišken, bu bejtin ajttı:

1 R tujdan
2 Čoranın
3 tišarı
4 ekindijnıñ
5 akšami
6 Crossed out in the Ms.
7 Tasmali; everywhere changed to *Tasmalı*.
8 ka jakka; R kaj jakka
9 R Kük-kügen bujlarına
10 R bakkajınaʒek
11 küčkennın

Aj, Ahtaǯınıñ Ali-beji, ǯau beji
ǯau kürgende kurkak beji.
Ali-bej:
Akšamnen ekindi arasına bardım,
Tübetauya kunan dünen sujdırdım
Susumya balnen rakı čıyarttım
Unı de tülermen barıšırım[1].

Čora dij: Babam degen bajyušın bašini küzıni tarttıñ, anam degen apšikka šoyon süzler ajttıñ, arab digen apšıkka jaman aǯitip tarttıñ, Kün-kürmez Mejnaru kardašımnı čoχ ajtıštıñız, čoχ (662) tartıštıñız, unı kalaj tülejsin dij. Ali-bej ajttı: añadım bayıšlamajsın[2], bāri at sen dij. Čora ajttı: seniñ bašıñ da balaban, üzin de karaman, nübet senı[3], at dij. Šimdi Čora küttı. Ali-bej bir uk čektı, Čora šamarın tutyan, uk Čoranıñ[4] šamarına tijdi, jerge tüšti. Čora ajttı: er ujnı üčkeče[k], eki sefer taha at degen. Ali-bej eki sefer taha atkan, br šej etmegen. Šimdi nübet menim, azır obor[5] dij. Čora br uk čekken, Ali-bejniñ eki ajayı atın arkasında kalyan, belden jukarısın uk ap[6] ketti. Čora Tasmalı-bujının ǯetip küjge ketirgen, Ali-bejnıñ ajakları Tasmalı-bujınıñ arkasında.

Küjıñ halkın br jerye tuplayan, suradı Čoranıñ anası fallandı mı[7], juχsa Ali-bejnıñ anası fallandı mı[8]? Kulına aldı ukın: tez ajtıñız, bir uktan ǯanıñızın[9] alam. Halk bašladı ǯılašmaya, bardılar babasına, kulına ajayına tüškenner. Kel bizge bayıšlat ajttılar. Men bayıšlatamam[10] dij, üziñ barıp ajtıñ. (670) Šimdi anasına bašlayannar jalvarmaya. Kel sen bu kalkın kırdırma, senıñ süzin tutaj. Čoranıñ anajı: Kel balam, bayıšla menim katırım‿čin[11] u kalknıñ kabātnı dejse de Čora diñemij. Kulına br külük balta alıp küksıne ike balta uryan: eger sen bu kalknı kırsañ emgen südüñi aram etermen dij. Čora ajttı: bir kalknıñ br uktan ǯanın alyanı br šej tegıl, lākın br ananıñ katırın tapmak büjük br šej. Čora kulından ukını jerge tašladı: al ukı[12] da bar jerine salıñız dij.

Šimdi jašlar tuplašıp bašladılar ukın kusku ile kapıya juvarlatıp ketirirge. Küterip jerine salamadılar. Šimdi Čoranıñ Meñaru kardašı küle-küle kapıdan ejilip ukın alyan: jazık sizlernıñ jašlıyıñızya dep ejilep bir kulinen[13] tutıp jerine salyan. Ondaj bir talaj gečti, büjle ümür karamanlıknen üttirdiler.

Br talajdan son Kazan memleketinde suyıš bašlanaǯaχ dep χabar kelgen. Čoranıñ br arkadašı bar iken, adına Kulumčak-batır degenner. Arkadaš bolyannar. Šimdi suyıška keteǯekler ama Čoranıñ babası: balam, men seni nišanlajım da son ket dij. Čora ajttı: men senın ajtkan süzinden kajtmam. Čoranı nišanladılar.

1 barišırim
2 bayıšlamajsin
3 R senıñ
4 Čoraniñ
5 cf. Russian *oborona* 'defence'?
6 R alıp
7 fallandi-mi
8 fallandi-mi
9 ǯanıñizın
10 bayišlatamam
11 katırım-čin; R katırım ičin
12 ukiñ
13 kulinnen; R kulınen

Kač künden son Kulumčaknen (671) brge kettiler, un kunaklık jer jürdiler, bir küjge baryannar. U küjde kırk batır bar iken. Kulumčak batır ajttı: ajda Čora-akam, bu küjge br ferište küsterijik. Bunlar başladılar, br ferište küstergen[1], lākin kırk batırdan br tanesi kelip, buların sizler kimsiz dep suramaya kurktılar. Šimdi batırlarnıñ bašı: buların kim barıp surasa kim bolyanın, br avıʒ altın berirmen. U jerde br sajil bar iken. Men barıp surajım dij. Čora-batırya barıp suradı: kaj jerlisiz, aj ulu (673) [ajtıñız maña kajtırmam dij][2] Šimdi Čora: men Kazanya ketemen, Kazanya baryanʒe Kazannı kazak almasa, utka tahtasın ʒakmasa, men Kazannı kazakka ber‿almam[3], aj.

Sajıl kelip bu χabarnı batırların bašına süjledi. Batırların bašı čakırėp[4] onda konuk ettiler, küp rābet, küp sij[5] sıjladılar. Batır bašı ajttı: Čora, senıñ batırlıknı küp bolyanını tujdık, [...][6] šükür seni de kürdik, lākin bizge br uk at, ukıñı davıšın ešitejik. Čora ajttı: atajım ama küj ičine χabar ʒiberiñiz, erkezın χabarı bolsın ok atılaʒayından, uk davıšından kurkajler.

Küj ičine χabar berdiler, küjnıñ kalkı ajttı: zar biz uk davıšı išitmedik, munda batırlar er kün atajlar. Čora kušlık vaktı majdanya čıχup, br uk atkan, ukın gürüldisinden jekirmi tokoz kišiniñ üdi patladı, üjler tüpesinden kirametler tüküldi, penčereniñ ʒamları parčalandı. Eki kün taha ütirdiler, batırlar da suyıška karap turalar. Čora ajttı: mında br eki kün turaʒakmız, sen bir kojnı[7] ʒokla da kel.

(672) Kulumčak batır Tasmalı-bujını arkasına minep u keʒesi küjge kelgen. Bašta Čora-batırnı nišanlısına[8] barıp ʒukladı, penʒereden karadı, üjın ičinde Čoranıñ nišanlısı br jašnen jumurlašıp ujnajır. Kulumčak bo išge šašıp kalyan, emen artına kajtıp tura Čoraya kelgen, kabarın ajttı, nišanlinen bir jaš bondaj bondaj jumurlašıp ujnaj dep. Čora-batır bu χabarın išitkenʒe atına minep artına kajttı, küjlerine keldi, üjlerine kirde. Čora-batırnıñ Miñero[9], ayasınıñ bujnına sarılıp[10] betinden küzinden üpti. Son ajttı: akam, men saña br jaramazlık ettim, seni küp sayınyan iken. Senin urubalarıñı[11] kijep[12], seniñ nišanlıña bardım, akšamdan jar‿keʒeje[13] kadar jumarlašıp ujnadık. Čok aru, dij, sen šindi ajttıñ, min unıñ bašın keserdim.

Sabā turıp gene de suyıška ketken, Kulumčaknen küristiler. Suyıštan χaber alyannar, felan küni suyıš bašlanaʒak. Kulumčakka kelip jetišken, kurkunıšlı[14] tüš hajırlı bolor dejler, u menimki de hajırlı (675) čıkkan. Šimdi br kahat kelgen, tez-tez buların suyıška kelıp ʒetiñiz dep jazyannar. Kulumčak batır ile julya čıkkanlar. Čora ajttı: br uk atajıχ, bizim julda kelgenmizin χabarları bolson. Čora br uk atkan, ukı Kazan memleketın ʒāmisin minaresi tübinde, bir melmer taš bar iken, uk u melmer tašya urıp tašya

1 R küsterilgen
2 Crossed out in the Ms.
3 R bere almam
4 čakirep; R čakırıp
5 R sıj
6 Crossed out in the Ms.: üzıñ kürmege
7 kojnıñ; *koj* is a Kipchak (northern) dialectal form of *köj* 'village'.
8 nišanlisina
9 A variant of the name of Chora's sister.
10 sarilıp
11 urubalarıñi
12 R kijėp
13 jar keʒeje; R jarı keʒeje
14 kurkunišlı

kirep kalyan. Anda seksen ike batır taha bar iken, ǯılıšıp ukın tutıp tarttılar[1] taštan čıkaramadılar. Batırlarnıñ baši Sarı-batır[2] bar iken, anası kilep ukın kùrdi: ulum, sizler bu uknı bu taštan čıkaramazsıñız, bu Čoranıñ ukı. Karaul julda salıñız, bekim bu kùnnerde Čora mında kelij, suyıš ta ùč kùnden bašlanaǯaχ, azırlık kùreler. Čora kelip suyıš majdanına br čadır kordo, atnı čadır janında bajlayan.

Br kimse Čoranı tanımadı. Čora ajttı: sen atlarnıñ janında otor, men barıp karajım ne bar ne joχ. Tora ǯemijetke kelgen. Br ùjniñ ičine tolyanlar, epsi batırlar bal ičeler. Čora ıčerige kirep salam bergen, br jerčikke otoryan. Čoranı breu tanımadı, em de salamın de almadılar. Bal sùzlùp[3] kelgende Čoranı atlatıp kečtiler. Šimdi Čora bunlarya br bejt ajttı:

Sırasinen kùrem sırasınen (676)
Jarın bašıñızya kıjyınnık tùškende
atıñızıñ bašın tartıp kalyanda,
tanırsız akañı.

dıj. Sarı-batır ajttı: bu nasıl tilenǯi, kùter atıñız tıšarı[4] degen. Čora tez jerinden turıp tšarı[5] čıkkan, dorı čadırya keldi, jatkan. [Suyıš] bašlandı, bu jaktan kazaklar aldına ǯùrdiler. Bular da suyıška[6] davrandılar. Sarı-batırnın anası kùre kaldı, Čora kegen[7]. Tez, dij, barıñız Čora kelgen, kùziñiz ajdın bolson dep, ketiriñız munda degen. Dùrt-beš batırlar barıp Čoranıñ eline ajayına tùškenner. Bz seni tanımadık, bzge ajib etme, dep jalvardılar. [Čora da: unlar kabātsiz, ùz-ùzini tanıtmadı,][8] jerinden turıp dorı[9] ǯemijetge kelgen. Epsi batırlar Čoranın betinden kùzinden ùptiler. Bz senı[10] tanımadık, ničin bizge ùzın tanıtmadıñ dijler. Bašladılar bal ičmege. Br bardak bal Kulımčak-batırya ǯibergenner, u da iče. Bašladı Sarı-batır erkezge brer edije bermege, kimsene[11] kılıǯ, kimsene uk, kimsene kalkan bergen. Čoraya br tabaka bergen. Čora kotonı ǯebine salıp čadırya kelgen kùterdi attı at ajakların tùbine. Epsi batırlarına brer suyıš alatı berdi, maña br tùtùn kotosı bergen, men tùtùn ičmejim, tùtùn kotosın nasajım dij.

Šimdi kùn tuyudan bašladı (677) suyıš, bu jaktan bular da bašlayannar. Čora bašın kùterıp karamaj, tùtùn kotosını darılyan. Šimdi batırlar kelıp bunıñ kulına ajayına tùškenner, nejčin darıldıñ Čora dej jalvardılar. Sen epsi batırlarına brer suyıš alatı berdiñ, maña br tùtùn kotosı berdiñ dij, men tùtùn ičmejim, seni kotoñı nasajım dij. Sarı-batır ajttı: kotonı nasadıñ dij. Jerde tura, at ajaklarınıñ[12] tùbinde. Sarı-batır jerınden sıčrap

---

1 tarttilar

2 "Sari-batir" has been replaced with *Sarı-batır* everywhere.

3 R sùzılùp

4 tišarı

5 R tıšarı

6 suyiška

7 R kelgen

8 Crossed out in the Ms.

9 dori

10 senın

11 R kimisėne ~ kimisine

12 ajaklarnıñ

turıp kotonı alyan, kapayın ačkan, ičinde bir Zilfikar bar, ketirip Čoranıñ kulına bergen. Čora Zulfıkarın kürgennen kuvanıp Zilfikarın kulına alıp Tasmalı-bujnının arkasına mindı. Jerinden küčkende atın ajakları jerge bakajnaǯı[1] kirdi. Üč kün üč keǯe askerın bujına ketti, kamalap Kazan kalasını üginde aldında tuplap u‿yadar[2] suyıš etti, at kišejinden kan kelgen.

Jedi kün ičinde takımı[3] kırıp, kalyanı[4] jesir alyan. Suyuš bitken. Br kač kün rātlandılar. Sonra memleketine kajttılar. Küjnıñ kalkı χoš keldikke keldiler, bašladı suyuš χabarın ajtmaya. O kadar suyıš bolyan, toχsan jašına kelgen kartlar bondaj suyıš ne išitken ne kürgen.

(678) Soñra gene χabar kelgen, suyıš tazeden bašlamaya azėrlanalar[5]. Čora-batır Kolumčakka χabar ǯibergen, jene de suyıška ketmek kerek boldı dij. Ketijik dep azėrlanıp[6] ukını gürsini alıp Tasmalı-bujını minep sauklašıp jula čıkkanlar. Julda ketkende, rast kelgen kišiler surajlar: Čora-batır, bondaj aǯele kajda ketesıñ. U da ajttı: küp nakırtı sırası tegil, br bejt ajtajım, añasan añarsın[7]:

Ben Kazanya ketem
Ben Kazanya baryanǯı
Kazannı[8] kazak almasa
Utka taktasın ǯakmasa
Kazanı[9] kazakka ber‿almam[10] ej.

Julda ketkende aldına br atlı čıka kaldı, karadı eki kši[11] bek ajbetli keteler. Kajda ketesın dep suradı. Čora da üzini bejtnen Čora bolyanını añnattı. U atlı da Čoraya kelgen, suyıš χabarın ap[12] ketirgen. Čoraya ajttı: br kün taha jetišmeseñ Kazanın kazak alıp jatır, tez jetiš. Čora bu χabarın išitkenǯi jerinden tirelep kečkende, Tasmalı bujını ajakları bakajnaǯi jerge kelgen. Uradan suyıšnıñ urtasına tüšejim dep koštu. Kalanın[13] aldıramajėp kaleden atlap teñizge tüšken, bakajlar Čora teñizge ketti, suyušın[14] kazaklar (679) kazandılar, Kazanın[15] aldılar.

Čora jene de kelij dep ülgennı ınanmajıp teñizge ketken jerine karaul kujdılar. Hala bukün anda karaul utıra, Čora teñızge kitken jerde. Suyıš pitken son χabar ǯiberdiler Čoranıñ anasına, Čora teñizge ketken dep. Čoranıñ anası ajttı: u teñizge ketse de [br šej] bolmaz, Čora tuyanda menim avızımdan br ut čıkkan, suya tüštı, sünmedi. Son men efendige suradım, br ut avızımdan čıχχan, suya tüšti, sü[n]medi, bu ne demek.

1 R bakajınaǯı
2 uyadar; R u kadar
3 takımıñ; R takımı or takımın
4 kalyanıñ; R kalyanı or kalyanın
5 azerlanalar; R azırlanalar
6 azerlanıp; R azırlanıp
7 añarsin
8 Kazannın
9 Kazanın
10 ber-almam; R bere almam
11 R kiši
12 R alıp
13 R Kalanı
14 R suyušı
15 R Kazanı ~ Kazannı

Efendi ajttı: senden tuγan erkek balanıñ eʒeli sudan bolaʒak, suγa tüšeʒek, ülmeješek degen. Čoranıñ anası unın‿čin[1] küp janmadı, kulların küterep bir duva etip, Allah erkezın de kürüštirsin, menim balam da kürejim, inšallah br vakit zamanı kelij, Kazannı gene de alij.

39 Kič'kene ajdamak masalı

(700) Br zemanda br ʒarlı[2] br kart bar eken.[3] Bo kartnıñ üč kızınen br ulu bar ėken. Kartnıñ ulu jedi sekiz jašına jetiškende medresege berip sabak okotmaγa aves ete. Kart bazarγa barıp br talaj kitablar alıp kele. Kojlerinden elli čakrım kadar avlakta br medrese bar edė, balasın šo medresege alıp kete. Lakin šehirge ketkende jol bojında balaban taular bar eken, bolar jolda ketkende, tauların ičinden[4] aldılarına kırk br ajdamak čıka. Bo ajdamaklar boların öttörmege istejler, atların arabaların alup ketijik dep ajtıšalar. Šo vakıt geʒe bek karangı bolγanından ajdamaklar arabada kum bolγanın eslep bolamaγanlarından katlarına kelėp, šırpı čertip karajler, arabada br ʒarlı kartnen br bala bar. Bolar kartnen balanı körüp br talajı öttörejik dij, br talajı öttörmejik bu pakirlerni dijler. İčlerinden[5] breu čıkıp balasın bizge berse öttirmejik, eger bermezse öttörörmüz dep ajta. Karttan balanı tilep karajık dijler. Son karttan sorajler, balanı bizge beresiñ mı dijler, balanı bzge berseñ bzde em okor em de akča berermiz dijler. Kart bolardan korkop balasın bere. Özö ajlanıp ʒurtına kajta, üjge (701) kelgende apakajine bašına kelgennerin ajta. Apakajı di ki: alla soñun χajır etsin, eger kismet bolsa br kün bola körermiz balamızın dij. Bz kelejik bo jakta balaγa.

Aradan jedė sekėz ʒıl keše[6], bala bu ajdamaklarnen tauda kaja kobalarında oko jazu ne bolγanın bilmij jüre. Kırk ajdamakların kırk brinʒilerė eñ bašları eken, bala da bašları bolγan atamanlarına čırak bolop ʒüre. Br kün balanıñ aklına šo kele: men bolarnıñ katlarında jedė sekėz ʒıl turdum, menė üjge atajım anajımnen körüšmege br kerre kajtarmadılar, endė men özöm br kačıp ketmesem bolmajʒak dep šono ojlana, šaj etip ojlanėp turγanda ajdamaklar brer-brer epsė išlerinden kajtalar. Akšam ʒıjılıp aš ašaγanda bribrlerinen ajtıšalar: bzler mında bu taularda on beš ʒıldan berli ajdamaklık etemėz, basmaγan ʒerlerėmėz, sojmaγan kšilerimiz[7] kalmadı, lakin šo bzim jašaγan[8] taumızγa ʒuvuk šeherde br baj kši[9] bar, šo[10] kšinėn[11] malınıñ akčasınıñ esabı ʒoktur, bizler šoγa[12] br kerre akılmızγa[13] kelmedi barmaγa. Bz šoγa barmasak bolmaz dep lap

1 unınčin; R unıñ ičin

2 Every letter <i> with ring above in this tale has been changed to dotless <ı>.

3 In the manuscript, above the title line there is a note saying: A Crimean tale by Mehemmed from Akmeʒid (Akmedjid).

4 ič'inden

5 ič'lerinden

6 keš'e; R keče

7 ks'ilerimiz

8 jasaγan

9 ks'i

10 s'o

11 ks'inėn

12 s'oγa

13 R akılmızγa

eteler, bala da bolarnıñ lapların kapınıñ artında tıñnap[1] tora edė. Šaj dep lap etėp ašap ičėp kursakların totorėp alašalarėna mınıp[2] sılaların takınıp kalaj da bolsa, bzler bu keče šunuñ (702) sarajın basıp aχšasın malın alırmız dep keteler, lakin orınlarında atamanları ůjde kalyan edė.

Šindė bu kırk ajdamaklar ketmekte bolson, ůjde bala atamannıñ katına kelip dij kė[3]: ej ataman akam, men szin[4] katıñızda on ʒıldan berli ʒůremen, bulaj etėp sizn[5] katıñızda ʒůrsem ič br šejge sahab bolmajʒayım, ajsa sen maya da br atnen br kılıʒ berseñ, ben de ȯzůmů br tiñep bakar edėm, bekėm men de šo ajdamaklar tarzı br ajdamak bolop br iške ʒarar‿dım[6] dij. Ataman: bek ijgi dep, balaya br kotır tajnen br toñka kılıʒ ketirėp bere. Bala kırk ajdamakların pešinden kalmaj kete. Bala kırk ajdamakların artından ketmekte bolson bz kelijik [ajdamaklarya].

Ajdamaklar šeherde šu kšini kelip ůjnů basalar, ȯzȯn ȯttȯrȯp karavullarnı ȯttȯrȯp tamam kumekteki akčalarnı bolar torbalarya totormaya bašlayan edėler, bala da boların artından jetišėp keldi, kazneniñ janėna atėn bir ʒerge bayanaya bayladė, ȯzů bašladı karavul dep bakırmaya. Ajdamaklar da: bzlerin askernen bastılar dep teumetlenip akčadan alup bolyan[ı] ala, alıp bolamayanları br akčasız o ʒerden kačtılar[7], bala da bolarnıñ kačkanların eslep ičerge kirdė, bašladė kapčıklarnen akča totorop[8] almaya. Šo ʒerden br arabaya akčalarnı tijep (703) atı tartıp bolyanʒek, andan jolya čıkmakta bolsın, kırk ajdamaklar šo kačıšnen tora ʒataklarına barıp atamanlarına halların anlatalar, bzėn talımız jaman edė, zornen kutulıp kaštık, šůkůr ʒanımızın sau kalyanına dijler. Bala da bolarnıñ artından kop kečikmij, arabanen akčanı ketirėp ʒatakları bolyan ůjlerin katına kelėp: ustam tur kel, akčaların bošatmaya, kol ʒardım et dep bakıra. Ataman da šondan čıka, karasa ne karasın, ȯzlerin onbeš ʒıldan berli toplayan akčaların malların bo br kerre baryanda alıp kelgen, apperėm, asabam, tůyanım dij, sen br kičkene[9] bala bolop bokadar mal akča ketirdiñ, kırk ajdamak ʒigitler kettėler, br šij ketirėp bulamadılar dij. Andan akčaların ůjge ketirėp altınların br oba, kůmůšlerėn br oba etėp ajrı ajrı ʒıjıp obalap kojdular.

Akšam da tamam bolop ʒitken edė. Kırk ajdamak aχšam er bri brer jaktan kelėp ůjge kirėp otordılar, andan aš ašamaya keldėler, atamanları da o ʒerde olarnen brle aš ašayanda atamanları 40 ajdamaklarya balanıñ ketirgen akčasından anlatıp: sizler ajdamaklık etkenden berli ketirgen akčañızdan šo bala br sefer čıkkan da sizden kop ketirdi, szler[10] br kičkene bala kadar joktursız dep ajdamakların tazerlij. (704) Ajdamakların ičinden breu dij kė: bzler šeherde bajın kaznesin basıp akčaların kapčıklarımızya tottorop turyanda, bzni breu kelip korkuttu, keče de karangı edė, ne bolyanın ojlap

1 tiñnap
2 R minip
3 R ki
4 R sızin
5 R sizın
6 ʒarardım; R ʒarar edim
7 kač'tılar
8 totorȯp
9 kič'kene
10 R sızler

bolamadık, tašlap kaštık[1], bznė korkotkan bu bala eken, galiba bzin korkotėp akčaların ȯzȯ arabaya tijep alıp kelgen. Eger ǯigėt bolsa bukun de ketėp šo kadar mal ketirsėn ȯzȯ bašına, o vakıt ǯigit bolyanın bilirmiz dij, eger ketirėp bolmasa bz de onu bašın kesėp čavup alırmız, br brlerimizėn kuturtup ačılaštırıp ǯürmesėn dij.

Bala bolarėn šaj degenlerėn tujup: ajsa men bugun ȯzȯm čıkop[2] ketejim, br šej ketėrejim, ondan gajri bular menė ottoreǯek[3] dep algı tokal kılıǯın beline takıp kene ȯzėn kotor tajına minėp jataklarından čıkıp br kȯjge ǯuvuk keldė. Lakin koluna br šij kečiremedė, tap soñunda brisin aranından br atın χırslap keti‿jatkanda[4] artından kırsız atımızın kırslap alıp keti‿jatır[5] dep ǯuvuralar. Bu olan kateǯeyin bilmejėp atnı tašlap, ȯzȯ kutulup kača. Andan baška br kȯjge ǯuvuk bara, sahatın čıkarıp baka tañ ataǯak vakıtlar ǯuvuklayan, šondan katsın, kolondan br šij kelmijǯeyin duja, artına kajtsa ajdamaklar ȯttȯreǯekler, ǯavaš-ǯavaš ȯzlerin ǯatkan ornına (705) kajtėp kele, akırtından atnı ketirėp ȯjlerin aldındaki bayanaya bajlaj, ȯzȯ ajatka kirėp ȯjnȯñ ičine baka, kırk br ajdamaklar ise epsi de bolyanı[6] dajın ǯuklajler. Bo katsın, ajdamakların brin adın ajtıp bakıra: batır akam, sandıknen mal ketirdėm, bek avur, kȯterėp ičerge kirsetip bol‿almajım[7], kel, ǯardım etsen[8] dij. Joklap jatkan ajdamak jukusından turup kȯzlerin uva-uva, tıška čıka, tıšarsı da zıp tajın karangė edė. Bo ajdamak tšarya[9] čıkkanı tarzı, bala tokal kılıǯın čıkarıp ajdamakka br siltij, bašın čavup br četke alıp ketėp tašlaj. Bondaj etėp 40 ajdamakların brer-brer ptire, atamanya sıra kele, atamanya da bolarya etkeni dajın etėp davušlaj. Ataman jukudan turup: apperėm, kadajım, sen ajdamak bolaǯak ekensin[10], dep tšarya čıka jatkanda onın da bašın čavup ala, ȯzȯ de bek boldoryan edė. Ornona kirėp ǯatėp ǯuklaj, tañ da keregė tarzı ačılyan edė. Šo jatıšnen kȯndȯz ȯlege karšı ekė saat kadar kalganda turup, betėn kolun ǯuvup, tšarya čıkıp karadı. Ajdamaklar epsė kanlarya bulanıp ǯatalar. Bala br ojlandı, katijim dep, o‿ssaat[11] (706) kolona br ayaǯ kalak tapup ȯjȯn bosayasın tȯbine br čukur uru tarzı kazıp kırk br ajdamakın ketirėp alıyı uruya totordo, kanların da kırıp ketirėp oroya kujup, ȯstȯnden topraknen jaχšı etėp bastırdı, lakin ȯzȯ bašına ojlanmaya bašladı: boların bekėm br ǯaktan br tȯyanları ja br ǯuvukları kele kalır da menė ȯttȯrȯrler, maya ondan kačmak kerek dep, šono ojlandı.

Ertesi kȯnȯ erten turup ellė arabaya ellė čipt ȯgȯz hazırlap arabaların kȯmȯsine kȯmȯš tijedė kȯmȯsine altın tijedė, kȯmȯsine de baška tȯrlȯ kijmetlė šijler tijedė, jolya čıkup kudajya tevekkel bolop kettė, ȯzȯmȯ tȯyan ȯsken ǯurtuma barajım dep.

Br kač kȯnler šaj kettė, kȯnlerde br kȯn ȯzȯnȯñ tuyup ösken ǯurtuna keldė. Son ȯzȯnȯñ atajnıñ anajnıñ kaj jerde bolyanın soradı, buya kosterdiler. Bunu br ȯjnȯñ

1 kaš'tık
2 Its varinant *čıkup* occurs 3 times in the tale.
3 R ȯttȯreǯek
4 keti-jatkanda; R kete jatkanda
5 keti-jatır; R kete jatır
6 bolyani
7 bolalmajım; R bola almajım
8 etse
9 R tıšarıya
10 ekėnsin; R ėkensin ~ ekensin
11 os-saat; R o saat

katına ʒuvuk alup keldiler: mına sizin ůjůñůz budur dep kosterdiler, ůjlerine br baktı, ůjlerin penʒeresi kalmaγan, kapusu ʒıγılγan, duvarları ʒıkılajım dep tura, azbarda ʒaz kůnů kogetler olenler ósken, olenligin ičinde br kartnen br kurtka (707) ʒarlılar koget čöplep ašap turalar.

Ʒigit selam alejküm dep kartların katına bara. Garib ʒarlılar ulannıñ selamın alalar, lakin ȯzlerin balası bolγanın dujmajlar. Ulan: kart anam, bu otların ašamasañız baška šijñiz ašamaγa jok mı dij, ʒarlılar dijler ki: katijik balačıγım, br kimsečigimiz ʒoktır bzėn karajʒak. Üč kızımız bar, lakin kız baladan ne fajda bolaʒak, er bala dajın kızmet etėp bol‿almajler[1], ȯzlerin bašlarına pajdaları ʒok, bzge nasl[2] pajda etsinner, br ulumız bar edė, ajdamaklar alıp kettėler, kajda ketkenėn kajda bolγanın bilmijmiz dijler. Ulan dij ki: ajsa siz uluñıznı korseñiz men dajın etėp sizėn katıñızda lap etėp tursa tanıp park etėp bulur mısınız dij. Kartlar: ebet tanırmız ulum dijler. Ulan: ajsa siz menė tanıp bol‿almadıñız[3], men sizn[4] uluñız bolam dij, ʒarlı kartlar šu ʒerde: vaj balam, sen ekensin[5], χoš keldėñ balačıγım dep kučaklap balasın balalı babalı ʒilašalar.

Šu ʒerden alıgı akčaların bosaγanı katına ketirėp ʒešil ȯlenlerėn üstüne bošaltıp, arabalarnen čevre jaγın koralap tašlaj, breu tijmesėn dep, ȯgüzlerin de ȯlenge ʒibere otlasınlar dep. Šindė ȯzů alıgı akčalarnı saklamaγa br ʒer kerek, dep šeherden en brėnʒė br usta tavup (708) ketėre, bosaγasın tůbinden ölčettirėp azbarnıñ ičine br jaχšı ůj jasatıp aχčaların jerleštire. Andan ustaγa ajta: men br saraj saldıraʒaγım, lakin šulaj br saraj bolaʒak, ičinde pečlerė mermer tašından bolaʒak, ȯzů jedi kat bolaʒak, br katı altından, br katı kȯmušden, br katı mermerden, br katı kerbardan jedi katı da kijmetlė tašlardan jaχšı brer turlu kijmetli taštan bolaʒak, jasap bolor mısın dij. Usta jasarım dep bojnına ala, šondaj br saraj bolson ku padša korgende suklansın, padšanın sarajından on kat ijgi bolson dij, jasap ptirmesin[6] usta baši mojnına ala, üč ajda jasap ptirmege sȯz bere.

Üjnü eki katın jasajler. Bu ulan jine de arabaların ʒekėp kalgan akčaların alup kelmege kete, algı ajdamakların akčaların bolγan jerine kele, gene de jigirim beš čipt ȯkuzlerge kalgan malların tijep ptire[7], šonda ȯzlerin otorγan oronların[8] katında br balaban saraj bar edė, šo sarajga iš te[9] kirgeni jok edė, šo sarajga kirdė, sarajnıñ ičinde br balaban sandık tura edė, sandıknı üstünde br anaktar tura edė, anaktarın alıp sandıkın ačıp karasa, sandıkın ičinde br kškene[10] sandık daa bar, kiškene sandıknı ačıp karadı, ičinde guzel br jüzük bar, jüzükni koluna alıp bakkanda nasl[11] da bolsa (709) avuzuna kaptı, kaptıγı tarzı ʒigit bir mıšık boldo, šu ʒerde šulaj br mušuk suretinde kȯrüp ȯzün bek jaman korktu, avuzından jüzüknů tašladı, jüzüknů tašlaγan tarzı kene de evelki tarzı ʒigit boldo. Bu alların kȯrüp br ojlandı, bu jüzükte br iš bar eken dedė,

1 R bolalmajler; bola almajler

2 R nasıl

3 bolalmadıñız; R bola almadıñız

4 R sizın

5 ekėn-sin; R ėkensin ~ ekensin

6 R pitirmesin

7 R pitire

8 orȯnların

9 iš' te; R ič te

10 R kıškene ~ kiškene

11 R nasıl

jüzüknü kesesine kojup malların keregi tarzı tijep šu ǯerden özün ǯurtuna kajtıp sau esen kajtıp kele.

Bo malların alıp kajtmakta bolsın, bz kelijik bú‿jakta[1] üjin beš katın išlep ptirip eki katı kalgan edė. Šo memlekette br balaban padša bar edė, šo memleketten eki ǯüz čakrım kadar avlakta br baška padšaγa msapirlikke[2] kete‿jdė[3], bu padša obir padšaγa keti‿jat-kanda[4] bu ulannı jasata jatkan sarajların aldından ötüp kete jatkanda padišanı közüne bu čardaklar körüne. Padša uč atlı pajtonın toktatıp šu köjnün kalkın toplap kalktan soraj: bu čardaklarnı kim jasata dij. Šu jerdeki kalk: mında ǯarlı bir kart akaj bar, šunuñ br ulu bar, šu ulan jasattıra dijler. Padša šo saat ustaların toplap čakırtıp bu sarajların kimin izninnen jasajsıñız dij. Ustalar: padišam, bzler mında ırγatmız, akčanen išlijmiz, bunu jasattırγan ǯarlı kartnıñ uludir dijler, oγa kim izin bergen, menden izinsiz bonda-jın sarajler (710) jasatmaγa dij, išlemeñiz tašlañız, bo saraj menėm sarajlarımdan jaχšı, men br padiša bolop menkinden jaχšı sarajler jasattırmaγa menim iznėm joktor, kim bolaǯak menden balaban sarajler jasatıp dij. Kajda šu ulan, özü bar čakrıñız[5] kelsėn mında dij. Alıgı kalk dijler ki: padšam, o ulan br jakka ketken edė, daa kelmedė, kelse sizge ǯiberėrmiz[6] dijler. Padša da kelgenė tarzı ajtıñız maγa kelsin dep ačulanıp šu jerden kete. Padišaγa msapirlikke[7] kete jatkan ǯolundan artına kajta.

Padša ačuγundan kajtmakta bolson, bz kelijik bergi ǯatta[8] padša kajtkandan son ulan ėki kün son kele, jigirim beš ǯük malların tijep[9] kele. Alıgı malların kene evelki tarzı bošalta br uruγa. Alıgı kart babası padišanıñ kelgenėn, nasl[10] ačulanıp kajtka-nın uluna kerekė tarzı anlatkandan son: ulum, balajım, bu mallarnıñ bu sarajlarnıñ sebebinden bzėm bašımızγa kop kijametler kelir buγaj, bo padiša mında bizge tınıšlık bermez, em de senė čakırttı, maγa kelesin dep kettė dij. Ulu kartka: atajım, sen bulaj šijlerden ište korkma, men mında kalıp irgatlarγa bakajım, sen padišaγa bar añnaš dij. Kart: ulum, men anda barıp padšaγa senėñ ulun bu mallarnı akčalarnı kajdan aldı dese men n‿ajtarım[11] dij. Ulan: sen onın‿čün[12] iš te[13] kajγırma, (711) sen išnėñ torasın bolγan dajın ajt, menėm br ajdamak kırsız ǯanaj ulum bar, onıñ kırslap kötürgen mallarıdır, dep ajt dij.

Kart šu ǯerden čıkıp kete, padšanıñ sarajine barıp kire: selam alejkum padšam dij. Padša selam alıp: otur kartım dij. Kart otora. Ej, kartım, sen br ǯarlı tilenüp ǯürgen br kart edėñ, šindė sen menėm sarajlerimden jaχšı sarajlar jasatıp jatasın, bu kadar aχčanı kajdan taptıñ, ajt maγa torasın dij, eger torasın ajtmasañ, bašıñı čavup alırım dij. Kart dij kė: padšam, menė[14] br ulum bar, br kırsız, bek ajdamak, šunu kırslap ketirgen akča-larıdir, akčanı kajda bermege bilmegenden padišanıñ sarajlarından jaχšı bolson saraj-

1 R bújakta; bu jakta
2 R musapirlikke
3 ketejdė; R kete edė
4 keti-jatkanda; R kete jatkanda
5 R čakırıñız
6 ǯiberermiz; R ǯiberirmiz
7 R mısapirlikke
8 R ǯakta
9 tijėp
10 R nasıl
11 najtarım; R ne ajtarım
12 onıñčün; R onıñ ičün
13 R ič te
14 R meni ~ meniñ, Gen. of *men* 'I'.

larım dep jasata dij. Padša dij ki: sen oγloño bek kırsız dep ajtasın, bu geǯe men apakajımman[1] br tȯšekte jatkanda bzlerni joχodan tursatmaj, menėm tȯšegimni kırslasa, jaχšı kırsız ekenėn bilirim, kırslap bolmasa ertenge ekėñiziñ de bašıñızın čavup alırım dij.

Kart: barıp uluma ajtajım dep čıkup ulunıñ katna[2] kele, uluna dij kė[3]: ulum, bu ủjlerni botenden jasatasın, padiša saγa mına bondajın br iš etsin dep ajttı dij, eger etėp bolmasa bašını čavup alırım dedi di. Ulan: atajım, sen onun‿čün[4] kajγırma, ondajın br kırsızlık etmesė iš tugul dij. Padiša da: eger kırslap (712) bolsa, dar-i dünjade br kızım bar, šo kızımnı bererėm uluña degen ėdė.

Ulan kündüz u jakka bủ‿jakka[5] ǯürüp vaχıt kečire. Keǯe karangı bolgandan son algı jüzüknė ǯebine kojup padišanėn ủjın katından br kečüp kete, br daa baksa ne baksın, padša sarajların er tarafın karavullar sakčılar totorgan. Bu ulan sarajnėñ br jaγına kete, jüžükni br čıkarıp avuzuna ala, šu ǯerde br mıšık kijafetine kirėp bekčilerin ajakların tübinden padišanıñ sarajın ičine kirėp kravatın tübinde ǯata. Akšam bola, padša apakajinnen[6] tȯšek östönde lap eteler, padša dij kė: apakaj, bu keǯe ǯarlı kartnıñ ulu bzn[7] tübimizdeki ǯataγımızın kırslajǯak, bu keǯe bizler katıp katıjık ǯuklamajık, eger ǯatakın kırslasa, em kızımızın alaǯak, em menė maskara eter dij. Bolar šaj dep lap etkende tıñlap tura ėdė. Padšanıñ apakajı: adam sen de, bondajın iš bolmaz, er jaγımız sakčılarnen tolo, br kišiniñ tübindeki jataγını kelėp bo kadar sakčılardan kutulıp kırslar degen šij bolasu di. Endė saat eki boldo, juklajık dep tatlı-tatlı jukuγa dalalar.

Bolar juklamakta bolson, ulan šu jerde avuzından jüzükün čıkara, jüzükėn čıkardıγı tarzı jene br ǯigit bolop ȯz kijapetne[8] kire, katsın (713) jataknın almasın ebin tap‿almaj[9], šü‿jerden[10] ủj ičindeki br sandıknı ačıp baka, padšanıñ apakajın saklıγa kijgen urbaları bar eken, šolarnı kijip padšanıñ apakajı dajın bola, ajatka čıka, br čuval un bar ėken, šonda br sahannı ičinde braz un čajup kamır tarzı etėp, padišanı tȯšekin ủstüne tȯke, apakajın de javaščıktan koterėp[11] ispande br minder ủstüne kuja. Son ȯzü padšanın katına kelėp tursata, padša turajım degende kolu kamırga kire, padiša bu nasl[12] šijdir, kajsımız šaj astımızga bulaštırganmız dij. Aligi ulan padišaγa dij kė: sen ištanıñı kolmegini čıkarıp ber, ǯuvup ketirejim, breu kȯrmesin dij. Padša čıkarıp bere, ulan kapıdan alup čıkkanda sakčılar ǯibermijǯek bolalar, kelėp padšaγa dij ki: bak, bekčilerėn ǯibermijler dij, padša čıkup bakıra ǯiberiñiz, sizn[13] padša anañızdr[14] nege ǯibermijsiz dij. Nebetčiler ǯibereler.

Ulan česme bašında ǯūγan kši bolop, šo jerden br sıptırlıp[15] kača, padša br eki saat kadar beklij, ne ištan bar ne kolmek bar, ne de jatak, šu jerden bu nasil šij, bu išnė bu

---

1 Northern dialect variant of *apakajımnan* 'with my wife'.
2 kat'na; R katına
3 ke
4 onunčün; R onun ičün
5 bủjakka; R bu jakka
6 R apakajinen
7 R bėzėn ~ bizin
8 R kijapetėne ~ kijapetine
9 tapalmaj; R tapa almaj
10 šü jerden; R šu jerden
11 kotėrėp; Kipchak dialectal variant of *kȯtėrėp.*
12 R nasıl
13 R sizın
14 R anañızdır
15 R sıptırılıp

ulan maγa etti mi dep u jakka bo jakka kıdrınsa[1] da ispan ičinde padišanıñ apakajı juklaj, padša apakajın tursatup soraj (714) bu nasl šejdr[2] dij, apakaj men de bilėp bol‿almajım, nasl šij bolγanın dij, padšaγa sandıktan temėz ištan kolmek čıkarıp bere, padiša kijėp otora[3], saba bola vezirlerėn bašna[4] toplap išniñ bolγanı dajın añnata, vezirlerin ičinde br vezir bar edė, šo vezir de padišanı kızın alırım dep teumetlene edė, bo iš boγa avur kelėp: padšam, ondajın br kırsızγa kızıñnı bermek jaramaγan br iš dij. Padša nasl etijim dep soraj, vezir dij kė: sizn[5] aranıñızda br čal atımız bar, šo atnı kapuñızın bosaγasın janına taχtınızın kujıp, tizgininden tutėp oturursuñuz, her jakka da sakčılar karakollar kujarsıñız, unu da kelėp kırslasa son br čaresine daa bakarmız dij.

Bolar šaj demekte bolson, ulan padšanı jataγın ištanın kırslap jerine kelėp otorγan edė, atasın padišaγa ǯiberdė, atası padišaγa kelėp: endė ne emr etesiñiz dij. Padša: kartım, uluñ tamam kırsız eken, lakin šondajın menė bir čal atım bar, her jaγın karakollarneŋ saraǯaγım, br kümsege tujdurmaj tahtımnı u̇stu̇nde kolumnen tutup turγanda ȯzu̇me de bildirmij alıp ketėp bolsa, son kızımnı berirėm, ondan baška ekiñizėn de bašıñızın čavup akčañızın ȯzu̇me alırım dij. Kart uluna kelėp išėn añlata, ulan: atajım, sen bo iške kajγırma, men čal atka minėp kuvanıp[6] ku̇lu̇p kajtarım dij.

(715) Šondan ulan kene de algı jüzükni alıp padšanı sarajine ǯakın bara, lakin sarajnė her jaγėn čırmap alγanlar. Ulan katsın, avuzına jüzüknė kapa, br mıšık bola, padšanı kıravatın tu̇bine kene de kirėp jata. Padša tšarda[7] beklij; ulan kelėp atın alaǯak dep, sabaa jakın padšanıñ jokusı kele, vezirlerden brėn čakırıp dij kė: mına šo atnı menim tahtımnı u̇stu̇ne oturop tutup tur, men br tınıš joklap čıkajım, menden baška breuge berme dep ičerge kire de kereki tarzı češinėp joklaj. Ulan aman padšanı urbaların taǯın kijėp padišaγa ošap tšarγa[8] čıka, lakin padšanı jerine kalγan vezirniñ adını bellep kalgan edė. Vezirėn adėn ajtıp kıčkıra, vezir najtasın padišam dij, ulan padišanı ȯzu̇ tarzė etėp menėm čal atımnı jigerle de ketėr, men barajım ǯarlı kartnen ulun bašėn čavup alıp kelijim dij. Vezir atın jigerlep ulanγa alıp bara, ulan atka mine, bekčilerge kapuların ačıñız dij, šo‿ssaat[9] ačalar, ulan padšanın ȯzu̇ tarzı at čaptırıp kete.

Erten bola. Padša ǯukudan turup azbarγa čıka, baksa ne baksın, ne at bar ne askerler bar, ne de vezirler: aman šo‿ssaat vezirlerėn čakırtıp: sen nišlediñ, men saγa menden baška breuge berme degen edėm, at dij. Vezir: padšam, sen ȯzu̇ñ atnı jigerlettiñ, kartnen balasın bašėn čavup alaman dep kėtken (716) edėñ, menė ne kabātım bar dij, padša u̇jge kirėp baksa ne taǯı bar ne de urbası bar, šo saat vezirlerėn ǯıjıp nešlijim katijim, bu baladan nasl[10] kutulajım dep mešavre ete.

1 R kıdırınsa
2 R šejdır ~ šejdir
3 otȯra
4 R bašına
5 R sizın
6 kuvnap
7 R tıšarıda
8 R tıšarıγa
9 šos-saat; R šo saat
10 R nasıl

Bz kelijik bu jaχta. Ulan atėn padišanıñ kolunda nasl rahat bolsa, bo da šondajın baktırıp rahatlaj, padšanıñ urbaların da jaχšı br sandıkka biklep kuja, kartın kene de padšaγa ʒibere, kart padšaγa bara selam alejküm padšam dep kire, padša selamın ala: kartım, uluñnı kırsızlıγına ište sözüm jok, lakin br ojun da köstersin, son kızımnı br lapsız berėrėm[1] dij. Padšam, nasl ojun di. Padša dij ki: men sizėn šeherge eki jüz čakrım avlaγındaki padišaγa sizėn sarajın katından kečėp jatkanda menėm apakajımnı özümni arabaʒımnı dujdurmaj üč atlı arabamnı da kırslasa šo vaχıt br sözüm kalmaj, kızımnı berėrėm dij.

Kart ulanγa kelėp ajta, ulan sarajnın penʒeresinden bakıp otora‿jdı[2], padša üč atlı pajtonnen bu ulannı kapusun aldından ʒel dajın učup kėte, ulan ėndė kalaj etsėn, ojlana br türlü ebėn tap‿almaj[3], šo saat u jerde bilegėn balakların tirep padišanıñ arabasın artından ʒuvura.

Bo šaj ʒuvurup ketmekte bolsın, padiša ulannı sarajın aldından jıgırım čakırım ajrılγandan son br ʒal almaγa (717) toktajler. Braz rahatlanėp oturalar, ulan šu ʒıvurunen buların artından ozop[4], bolarγa körünmij, aldına čıχop jolda balaban br taunıñ ičinde ʒol katında ėki balaban terek bar edė, ulan taudan br talaj ješil ülenler ʒıjıp üstüne orap, insan bolγanı belli bolmajıp özün jolγa köndelenden aγaʒlarγa astı, padša da fajtonnen ajdatıp keli‿jatkanda[5] bu asılıp turγan šijni körüp fajtonnı toktattıra bundan soraj: ins misin, ʒins misin dij. Ulan: men kijik insanim dij, padša: bondajın br ulannı maγa tutup ketirėp bulur mısın dij, o da ketirirmen dij, sizler šü‿jerde[6] azčık kütüp turuñuz, men šindė ketirirmen dij, ʒol katında ʒijlekler ösken tamam pšken[7] edė. Padša fajtonnen ʒijlek bolγan ʒerge ajdatıp, šü‿jerde[8] br ʒal alajık dep otoralar. Ulan da oların janında: men sizge šu ulannı ketirejim dep ajlanıp, boların fajtonın tübine kirėp otora. Padšanı apakajı tšarga čıka, padša ʒijlek ašajım dep tüše, fajtonʒi da br su tökmege dep tüše. Ulan arabaγa minüp ajdap kača, özün sarajına kelėp padišadan kijip kačkan urbaların kijėp atasın anasın fajtonγa minsetėp (718) özi de padšanıñ čal atına minėp dülber br arabaʒıγa ajdatıp ʒajau ʒalpı kalγan padišaga karšı baralar.

Šü‿jerden[9] padšanı kartın janına padišanıñ apakajın ʒarlı kurtkanıñ janına oturtop[10] sarajγa keleler. Andan kırk kün 40 keʒe toj-tügün etėp padša kızėn ulanγa bere. Ulan kız brbrlerine murad alıp murad bereler, šeftalilernėñ pškenin[11] tüšürėp ašajler. Eki murad br bolalar. Kudaj bzlerėn de muradlarımızγa kavuštursun.

1 berėrėm

2 otorajdı; R otora edi

3 tapalmaj; R tapa almaj

4 özöp

5 keli-jatkanda; R kele jatkanda

6 šüjerde; R šu jerde

7 R pišken

8 šü-jerde; R šu jerde

9 šüjerden; R šu jerden

10 oturtöp

11 R piškenin

Crimean Tatar folk tales (translation)

1 Once There Was, Once There Was Not

(Tekerleme*)

(246) Once there was, once there was not; the time was in the time, [and] there was a time when the barley sowed plough with a lean ox when the swallow was a judge, the hoopoe was a bath attendant, my father was unmarried, my mother was a maiden, I served porridge at my father's wedding, I turned an ant into a sheep, [and] I made a horse out of a lizard and drove it into the well. The frogs suddenly grew wings to fly, the sea fish hired [a carriage] to migrate, the governor's foxhound rolled up his trouser leg to cross the Black Sea; dive! "I'm diving; who said I'm diving?" said the man who watched it, said the foal who watched the man; "Who said that?" the foal said, the foal's mother, the mare said, the boastful grey horse said; "Who said that?" the grey horse said. A witch came angrily, grumbling, shaking her ass; it was she who said. With this going, what was gone, she kept going incessantly uphill and down dale; she was in waist-high mud, knee-up in mud; it was pitch dark even by moonlight, dogs and wolves were howling together, the ass of valiant lad and thief turned weak [soft], the lining of the purchased fur coat was facing outwards, the toes of the boots bought for twenty were facing inwards: once there was, once there wasn't.

2 Gold Bracelet

(248) – If you wanted a gold bracelet, [and] I turn [myself] into a bracelet and wrap myself around your arm, what would you say?

– If I wanted a gold bracelet, and you were to turn into a bracelet and wrap yourself around my arm, and I became a hawk and flew up into the air, what would you say?

– If you turned into a hawk and wanted to fly up into the air, and I turned into a grey pigeon and followed you, what would you say?

– If you were to become a grey pigeon and follow me, and I [became] seriously ill and [fell] into bed, what would you say?

– If you became seriously ill and fell into bed, and I became a doctor and put ointment on you, what would you say?

– If you were to become a doctor and anoint me with ointment, and my time was up, and I were to leave the world, what would you say?

– If your time was up, and you were to leave the world, and I were to turn into a white shroud and fall upon your body, what would you say?

– If you were to become a white shroud and fall on my body, and I were to become a silverfish and go into the sea, what would you say?

https://doi.org/10.1515/9783111442891-003

- If you turned into a silverfish and went into the sea, and I became a master diver, dived after you, and pulled you out, what would you say?
- If you were to become a master diver, dive after me, and pull me out, and I became a red rose and grew in your garden, what would you say? **(249)**
- If you were a red rose and grew in my garden, [and] I would be a nightingale, and I would fly to your branch, what would you say?
- What would you say if you were a nightingale and came to my branch, and I stood before you as a bride in red?
- If you were to stand before me as a bride in a red dress, and I were to turn into a hawk-bride and embrace your waist, what would you say?

3 A Tale of Kulbasti

(252) In olden times, in the month of the end of the world, frogs grew wings to fly, fish in the sea hired [a carriage] to fly; once upon a time, once not,[1] a poor woman had a son. This son was named Kulbasti. At home, he sat on the ash heap by the stove and was constantly stirring in the ashes, so he was named Kulbasti.[2]

By the time Kulbasti was fourteen or fifteen, he had never gone anywhere [away from home], done [anything], gone out into the world, and learned [anything]. But somehow or other, one day, his mother deceives him and takes him to the gate.

"Son," she says, "you could go out and work like other people's children instead of being shut up at home all the time."

But Kulbasti found no pleasure in [any kind of work except] digging in the ashes. So his mother somehow deceives him and takes him outside the gate, locks the gate, and Kulbasti is stuck outside, not knowing what to do. [When] Kulbasti, who has never been out of the house in his life, sees a hen, he shivers with fear of hens; when he sees a chick, he shivers with fear of chicks; when he sees a dog, shivers with fear of dogs; when he sees a cat, shivers with fear of cats: he doesn't know what to do. So, wondering and feeling lost, he sits against the gatepost and weeps and cries bitterly, his neck between his shoulders, and his mother won't open the door or let him in.

Poor Kulbasti sits there for two days **(253)** and two nights. After a while, a man comes to him. "My son, why do you sit in this place?" [he says.] "Will you serve me?"

Kulbasti saw that man and was afraid of him. So Kulbasti saith unto him, "Wilt thou not eat me?"

And the man said, "Why should I eat you?"

---

1 Fragment of a *tekerleme**.

2 The original form of his name is *Kulbastı*, meaning, "boneless meat roasted (over coals or ashes)". Kúnos translated it as 'Hamuházi'.

"If so, and you will not eat me, then I will work for you as a servant," [he says.]

"My son," says the man, "I will treat you as if you were my child."

And his mother says from the house, "Go, you fool, that man will not eat you; I will not let you into the house." So, hearing that his mother won't let him into the house, he takes it upon himself to be a servant. Anyway, he'll go.

This man takes him to his house, and before he begins to work, he makes him well-fed and puts clothes on his body and shoes on his feet. And then Kulbasti, saying, "Will he eat, or will he not eat me?" with great fear begins to work.

In the evening, the man he is working for comes and says, "You have worked enough, my child; that's enough." Kulbasti stops [working], and the uncle takes out and gives Kulbasti twenty-five kapeks*[3] of money. He comes home to their house in the evening, and his mother, [seeing her] son coming home from work, asks him, rejoicingly, "What have you worked for, my son?" Kulbasti takes out the money he has earned and puts it in his mother's hand. His mother uses the money to buy and prepare meat and bread for the second day's meal. In her joy, [she says,] "My son, like the children of the people, goes to serve." He works that day too; again, the man who makes him work gives him twenty-five kapeks. Here he comes, having earned money; he takes it out again and gives it to his mother.

On the third day, he leaves again, and as soon as he comes [home] from work, there are children playing on the road. As Kulbasti approaches them, he sees that the children have caught a mouse and are torturing it. So he says to them, "Why are you doing this?"

The children [say], **(254)** "If you want, we will sell it to you, buy it." So he buys the mouse, giving them the money he had earned in the same way as before.

On the second day, he buys a rat, again in the same way; on the third day, he sees a baby snake being tortured; he buys this baby snake and brings it [home]. "My child," says his mother, "what about them? What are you going to do with them?"

"They are my property," says Kulbasti.

One day, he takes his fortune, goes up a hill with [it], lets [it] go, and he lies down on the spot and goes to sleep. As soon as he falls asleep, the mouse goes in one direction, the rat in another, and the snake in yet another. Kulbasti immediately jumps up and sees that his former fortune has been scattered. He quickly springs up, runs after them, and grabs the snake by the neck.

The serpent immediately starts speaking and says, "Oh, brother, don't kill me, you may need me, don't kill me."

Kulbasti says, "In what need of mine can you help me? What can you do for me?"

The serpent says, "I am the daughter of a Padishah; when the children were tormenting me, you came and bought me, paying with the money you earned; well, there is a cave in a particular place, now you follow me, let us go there."

3 * – Words marked with an asterisk in the text are explained in the "Explanatory glossary" at the end of the volume.

"But when we get there, what shall we do there?" he asks.

The serpent replies, "When we get there, my father will host you, and you don't accept anything from him, but my father has a ring on his hand: ask for that ring; if he gives it to you, accept it; until he gives you the ring, accept nothing. As soon as you step out of the door, I will follow you. If my father sees me leaving with you, he will say, 'Take the ring **(255)** and leave my daughter here.'" Behold, they are thus reconciled and depart from this place to the snake's Padishah.

After a while, they arrive, the snake in the front, Kulbasti behind; they enter through the cave gate; the cave's inside is as dark as pitch, and two giant snakes stand guard next to its gate. The boy, seeing them, says anxiously, "Will they not eat me?"

The former[ly] [baby] snake says, "Don't be afraid, for if I am with you, you won't even be flicked by any of them."

These two snakes ask them, "Where are you going?"

"I'm the daughter of the Padishah," says the former [baby] snake, "this place is my home; wherever I want to go, I go."

Hearing these words from her, the sentinel snakes go running to the Padishah. "Good news, Padishah, your daughter has come," they say to him. With the snakes who were his Viziers, the Padishah goes out to meet his daughter. The boy and the snake girl say a greeting and go through the door.

Seeing that his daughter has come home, the Padishah is delighted, not knowing what to do in his joy. After a while, he ends up saying, "Welcome home, my child. Where have you stayed; where have you been until now?" He shows them a place and sits them both down. "This boy is not of our kind; of whom is this son of man, and why has he come with thee?"

One by one, the girl tells her father what happened to her, how the children tortured her on the road, and how the boy bought her from the children with the money he earned, saving and freeing the snake girl.

Her father is very pleased **(256)** and says to the boy, "May Allah be pleased [with you] for saving my child's life." In the end, the Padishah hosts the boy for a few days and feeds him very well [with food and drink].

After a few days, the Padishah says to the boy, "Son, you saved my daughter from death and brought her to me; I must return your favour to you. Ask me now for whatever you wish." The boy says, "My Padishah, I desire nothing from you but this, the ring on your hand; if you give it to me, I will be content."

The Padishah says, "My son, my son, I will provide you with so much silver money, so much gold money, so much treasure, but don't covet this ring; I cannot give you that."

The boy says, "What should I do with the unnecessary things? I don't need them; if you give me that ring, I will accept it; I won't take anything else," he says and is about to leave.

The girl also rises from her seat and says, "Father, this boy has delivered me from death; am I not worthy of the ring in your hand, for my deliverance, and for bringing me to you, he desires a ring, and you are sorry to give it to him? If so, I will leave with him." She is about to go after him.

As soon as the girl has said this, the Padishah, finding no other way, pulls the ring from her finger and says, "I cannot help but give it to you" and gives it to the boy.

The boy takes the ring, says, "Goodbye, good health to you all," and leaves with great joy.

The girl goes out after him because she wants to explain the ring's skill. "Do you know," she asks, "the skill of this ring?"

"I don't know," he says, "teach me."

The girl says, "Then, do not go away in vain; this ring will be of help to thee **(257)** ; after performing ablution, thou shalt say two prayers, then thou shalt lick the ring once with thy tongue; from the ring will come forth forty pieces of jet-black Arabs; 'command, sir', they will say, and whatever thy soul desires, they will find and bring; whatever thou wilt, they will do."

The boy listens to the girl's words and leaves.

Little goes, much goes; one day in a grassy place, in the shade of a tree, a ranger sits eating bread. The boy goes up to the guard and sees the guard holding a kamchi* in his hand. The boy greets him, and the guard accepts his greeting and asks him for some bread.

"I can find bread," says the boy, "if you can find some water."

The guard finds water and brings it. The boy performs his abdas*, says two prayers, and boastfully licks the ring with his tongue, and at that moment, forty pieces of Arabs come forth.

"What is your command, sir?" they say.

The boy says, "Be quick, I want forty kinds of food; bring them here."

The Arabs fetch it from somewhere, bring it, and lay it before them. The guard is amazed. The boy calls the guard to eat, and they eat. After they have eaten and were] satiated, the guard sets his eye on the boy's ring. "How could I possibly take that ring?" he says to himself. Thinking the boy foolish, the guard wants to trick him and asks, "Will you exchange your ring for my kamchi*?"

The boy says, "The skill of my ring is that when I lick it with my tongue, forty Arabs come out, and what you wish, they **(258)** do," and then he asks, "What is the skill of your kamchi?"

The guard says, "The skill of my kamchi is that if you tell him to go and beat somebody, he beats and thrashes him; if you tell him to bring him, he brings him here."

The boy says, "Well, if that's so, let's change."

The guard takes the ring, thinking that if he licked it, he would make the Arabs take his kamchi back, but the boy guessed [he would do so]. As soon as the guard takes the ring, he looks for a washbasin and a pitcher to perform the aptas*.

The boy takes the kamchi, follows the guard five or ten paces behind, and says to the kamchi, "Go, take the ring from the guard's hand and bring it here."

The kamchi immediately dashes against the guard's hand, twists the ring off his finger, [and] brings it and gives it to the boy. The guard tried to trick the boy, and after the ring was lost, even his kamchi was gone. The boy took the kamchi too.

He walks, he saunters, and one day on the road, he meets a head shepherd. The head shepherd asks him for a piece of bread. “I will find some bread for you,” says the boy, “if you find me some water, I can perform the aptas*.”

Immediately the shepherd goes with the pitcher, finds water, and brings it.

The boy performs the aptas, says two prayers, and licks the ring once; the Arabs come out of it, and, “What do you wish, sir?” they say. The boy commands, and the Arabs, as before, bring the food and place it before him. The boy calls to the head shepherd, who comes nearer to the table, and they eat and are filled. As did the guard, the shepherd also sets his eye on the boy’s ring. Thinking the boy foolish, the head shepherd wants to get the ring by cheat[ing]. **(259)** The head shepherd says, “Will you not sell your ring?”

“What will you pay for it?” asks the boy.

The head shepherd says, “I have nothing but a cudgel; I will give it to you,” he says.

“What is the skill of your cudgel?” asks the boy.

The head shepherd says, “The skill of my cudgel is that no matter how big a man is, be it even a dragon, if you tell it to go and hit him, it will go hit and knock him down.”

“Well, if so, I’ll change,” says the boy. Then he takes out the ring and gives it to the head shepherd. In the meantime, the head shepherd find[s] a washbasin and a pitcher for performing the aptas* and say[s] prayers; the boy, sneaking up five to ten paces behind him, orders the cudgel, “Go, hit and knock the head shepherd down, and take the ring.” The cudgel immediately hits and knocks the head shepherd down, beats him, pulls the ring off, and gives it to the boy. He puts the ring on his finger, tucks the kamchi and the cudgel under his arm, and from there, returns to their home.

His mother says, “Son, what have you been doing, where have you been until now?”

“Oh, mammy,” says the boy, “I have done many things up to now; I became the owner of a kamchi*, a cudgel, and a ring.”

Times come, and times go, and a lot of time passes; the boy does not do any work. One day, his mother, seeing again that he is lazy and useless [and] has not yet come to his senses, says to him, “Son, my child, you do nothing. Can it go on like this? Your childhood is gone; there is nothing to eat and drink.”

“Nothing to worry about, mammy,” says the boy, “I’ll find you something to eat and drink; don’t grieve about it.” He pulls the ring **(260)** from his finger, licks it once, and immediately the forty Arabs appear and say, “Command, sir.” The boy gives the order, and forty kinds of food are brought to this place and placed on the kitchen board. “Mammy, behold,” says he, “Did you not wish for food? Behold, here is the food.” They eat, and when their bellies are full, his mother is astonished at what the boy has done.

He spends a lot of time like this. One day, he says to his mother, “Mother, now it is time for me to get married, you too are very old; you need a daughter-in-law to take care of you; ask me for the Padishah’s daughter: go to the Padishah and ask him to give her to me.”

His mother says, “My son, we are poor and destitute people; on what basis would we propose to the daughter of the Padishah?”

He sees that there is no other way out. He urges his mother, “There’s no way you won’t go,” and his mother has no choice but to go.

When the poor woman comes close to the palace, the servants of the Padishah see her and ask, "What are you doing here?"

"I have come to the Padishah," says the old woman. "I have a request to make to him," but she sees it will come to nothing.

At last, they go to the Padishah and say to him, "Here is a poor-destitute woman come to see you, my Padishah."

The Padishah says, "Give her a handful of gold from the treasury and let her go."

The treasurer brings a handful of gold, gives it to her, and the woman goes home.

"What have you done, mother?" says her son.

"Here, my son," she says, "they gave me a handful of gold; they wouldn't let me go to the Padishah," and she tells how it happened.

The son says, "I do not ask for money; I ask for the daughter of the Padishah; go again **(261)** and tell him to give me his daughter."

The woman goes once more. Again, they give her a handful of gold and send her away. The son sends his mother again.

When the third time she comes, the Padishah says, "What kind of old woman is this? What kind of idea has she come up with? Tell her to come to me; let me hear her wish."

The old woman goes before the Padishah, and she greets him.

"Tell me, gammer, with what intention you came," asks the Padishah.

"My Padishah," says the old woman, "I have a son; by the command of Allah and the word of the Prophet, I have come to ask for your daughter."

"Why should I marry my daughter to your son?" asks the Padishah. "Well, that is not possible like this. Still, whoever will fulfil for me what I desire, I will give him my daughter, but only on condition that he first, coming from his dwelling to my palace, makes rose gardens on both sides, various fountains, makes a road of marble stone in the middle of the two gardens, then makes palaces more beautiful than my palaces, gives my daughter as a trousseau as much gold money as two camels can carry. When these things are done, if he does these things, I will give him my daughter. Forty days respite: if he does these things in forty days, fine; if he doesn't, I will have his head cut off."

"Well, my Padishah," says the old woman, "my son has nothing, but I'll go and tell him; he'll pay for it; he should think about it."

The old woman goes home and says to her son, "Behold, my son, the Padishah will marry his daughter to you if you pay these things; if you don't pay them, on the forty-first day, he will have your neck cut off."

"Very well," he says, "what he wishes, **(262)** that is very easy; you have nothing to worry about," he says; putting his bow on his shoulder, he goes hunting. The boy walks about without a thought. He makes neither the palace nor the rose garden nor has a single akcha* for the trousseau.

Meanwhile, forty days pass, and his mother says, "What a foolish thing my son is; he wishes for his own fate," and day and night, she grieves for him. "And the fortieth day is past, and on the forty-first, the Padishah will have his head struck." On the fortieth day, the boy returns from the hunt. “My son,” says his mother, “what are you doing? Behold, your time has come; run away from here to somewhere else, just so your Padishah doesn’t have your head cut off.”

Her son [says], “Don’t think about it.”

It will be evening, and they go to bed. At midnight, he rises from sleep and licks the ring; the forty Arabs come out, “Command, sir,” they say.

The boy says, “Look, let there be such and such a palace, let the roads be of gold, and the trees of silver thread,” he commands them. The Arabs have built such a palace and made such gardens and roads that it is beyond words: forty times more beautiful than the Padishah’s palace.

The Padishah sleeps, but at midnight he awakes from his sleep, looks at the window, and it is as if the world were filled with daylight. “I wonder why I slept so late,” he wonders. He looks around, and what do his eyes see: There are palaces built more beautiful than his own ones, and these had that great brightness, which shone through the window and lit up the palaces of the Padishah. The Padishah got up early, thought nothing of drinking coffee, summoned the boy, made a wedding feast, and married him to his daughter.

Times come, **(263)** and times go; the boy lives in joy and happiness. One day, he goes hunting. While he is hunting, a Jew hears what ring the boy has, but he doesn’t know how to take it. He goes to the boy’s wife and buys the ring by deception, paying a lot of money for it. The Jew, finding a basin and a pitcher, performs the abdas*, rinsing his face and arms, says two prayers, and licks the ring once; the forty Arabs come out: “Command, sir,” they say.

This [Jew] saith, “The palaces, with the woman and me: take them straight away behind the mountain of Kaf*.”

The Arabs do as he says. The place of the palace remains completely empty. The boy returns from the hunt: no palaces, woman, or even ring. Wondering if the cudgel and the kamchi are in the old place where he put them, he goes to his mother’s old house and sees that the old cudgel and the old kamchi are on the shelf, just as he left them.

The Padishah is suspicious because his daughter hasn’t visited him for a while. So he is wondering what happened to them. One day he sends a man there: “Go and find out what my daughters have done and how they are,” he tells him.

The man comes, and what does he see: no palaces nor the Padishah’s daughter.

Then he goes and tells the Padishah how things are. “Oh my, this useless fellow has brought trouble to my daughter,” says the Padishah and gathers a bunch of soldiers and sends them to the boy. “Go, seize him and bring him here; with my own hands, I want to kill him myself; if he won’t come, strike off his neck and take his head.”

About five hundred soldiers go to where the boy is. As soon as the old woman sees them coming, she begins to cry, “Alas, what will happen to us now?”

“Don’t cry, mother,” **(264)** says the boy. “I’m going to fix their problem right now.” As soon as the soldiers arrive, he commands the cudgel and the kamchi: “Beat these soldiers so that none will be left!” Immediately the kamchi falls on them from one side, the cudgel jumps on them from the other side, and the cudgel and the kamchi, the two from both sides, begin to knock the soldiers [out]. The five hundred soldiers, as they should, are all killed and spread on the ground.

Only one of them escapes by running away and goes to the Padishah. "My Padishah, there is not a single visible man; suddenly, in this way, a kamchi and a cudgel came out, and in a short time, they laid us all on the ground," he says.

The Padishah gathers ten thousand warriors and sends them to the boy, "Go, do what you will, or find my daughter and bring her, or bring his head]!" he says.

As soon as soldiers approach the boy, he again gives orders and has the soldiers killed.

One of them, no longer able to resist, runs to the Padishah and tells the story. The Padishah sends an envoy to the boy: "Go and tell him: either he brings my daughter, or I will burn his house."

The boy says, "I will know my own business; if I want to, I will find my own wife; if I don't want to, it is none of his business; I am not afraid of him."

One day, the boy says goodbye and farewell to his mother, leaving her the cudgel and the kamchi, and says, "Behold, mother, these things will relieve you of all your troubles and needs." He took his leave and went with his little cat with him.

They go; they go; they come to a place: the place they come to is the land of mice. The mice immediately see them; they go to their Padishah **(265)** and say to him, "My Padishah, here, in our land, our Azrail* has come to destroy our kind."

The mouse-Padishah, as many old mice as there are, [gathered] all together, takes counsel: "What should we do?"

An old mouse says, "Let us gather as many warriors as we have and have them killed."

One says one thing, the other another.

At last, the oldest mouse says, "So we can't do anything; we must think this through. Call the cat here to understand his intentions." They bring the boy's cat and take him before the Padishah. When the cat reaches the Padishah, he grabs him.

"Oh, my lord," says the Padishah, "don't kill me; I'll do what you wish but don't kill me."

The cat says, "I will not kill you, but I have a request to you, and if you do it, I will let you go."

The mouse says, "Tell me, brother, what is your wish?"

The cat says, "There is a palace behind the mountain of Kaf*: there is a man with such and such a ring on his hand: you will help me to bring it."

The mouse-Padishah agrees.

From there, the mouse goes with the cat, and they arrive at the palace. When they go in, what do they see: the Jew is lying on his back, the wife is lying next to him, and they are asleep. He holds the ring in his mouth [tied on a thread], and the other end of the thread is fastened to the ceiling. The mouse climbs up to the ceiling and [chews through] the thread. Then he finds some honey in the cupboard and sticks his tail in the honey. He also finds a little pepper **(266)** in the same place and rolls his tail in it, then goes and sticks it in the hole in the Jew's nose. The Jew sneezes once, the ring flies out [from his mouth], the cat catches it immediately, and the ring is taken to the boy.

The boy licks the ring once, and the Arabs come out: "Command, sir," they say. He wants the palaces to be in their [former] place. The Arabs bring them and put them in their place. The Jew has not yet awakened from sleep; he and his wife are asleep, embracing each other. The boy goes to the Padishah [and says], "Come and see your daughter's state!" The Padishah goes and sees. Before Kulbasti can give the order for the cudgel, the kamchi goes and strikes them both to death. The Padishah marries the younger daughter to the son. They have a wedding feast and make him his son-in-law. For a long time, they live happily ever after.

4 Temir Bolat's Tale

(282) In the old days, there was a Padishah called Shah Islam. That Padishah has three sons, the youngest of whom is named Temir Bolat. Among these children in those days was a great love of horses. The Padishah had so many horses that they filled nine pastures.

One day the horses are gone] from these] nine pastures. The eldest and middle sons of the Padishah ask their father for permission to find their horses and bring them back. Their father permits them, and they leave. They go for a while, then come to a town, stay in two separate inns, rent a room [to talk to people about the loss of their studs],[4] and hope they may find it.

A couple of rogues come up to them. "My sultan, my lord, what are you doing here?" They say, "You are the children of a Khan; we will take you to a place of amusement."

So they take them to an inn. Until their purses run out of money, they get drunk and even sell their horses to buy more. When the Padishah's sons run out of money, they look to their former rogue friends to help them. Wandering the streets of the city, they find the rogues.

The Padishah's children say to them in vain, "Alas, my friends, what shall we do? We have run out of money."

The former **(283)** rogues say, "What shall we do with you if you have run out of money?"

Then one of these two Padishah children goes into service as a bath-keeper to sweep the bath, and the other goes into service at a baker to sift flour.

Let them just serve here. Let us go[5] [back] to Temir Bolat, who stayed at home. [Nobody, not even his mother and father, appreciates Bolat for his rank; they take him

4 Crossed out in the manuscript.

5 In this narrative turn, Tatar storytellers usually use the Tatar verb *kel-* 'to come', which is quite foreign to the formal logic, so we use the word 'go' instead of 'come' in the translation.

for a stupid booby.] One day, Bolat asks his father for permission. He says, "Father, if you would allow me, I would also go in search of our horses."

On hearing these words from this, his father gets furious and says, "So far, I have had two clever children go away; they couldn't do anything, they haven't been up to anything, and they themselves have not returned. What do you want to do, you fool? Go away, do not annoy me," he tells him and chases him away.

Bolat goes sadly to his place, lies down, thinks, and begins to weep. "So many of our horses have been lost, and then two of my brothers have perished. What am I to do now? I go to my father, who scolds me and chases me away; oh my God, you know," he says and saddens. After a while, he sleeps, and he has a dream; an old man with a white beard says to him, "My child Temir Bolat, why are you so worried? Do not distress yourself a bit; behold, behind your room where you are lying, there is a door; open that door: inside there is a chest, if you take what is in that chest, it will be enough for you; all your **(284)** goals will be fulfilled."

Bolat wakes from his sleep, and as the older man told him, he opens the door and looks in; what do his eyes see: there is a chest; he opens it, and in the chest is a sword, a set of battle clothes, and on the wall next to the chest is a saddle and a rein and also a whip. In one corner of the house, there is another door; he opens it and looks inside; he sees a stable; in the stable, a horse, his ears decorated with gold thread, his nails of silver, his mane of gold [thread]; he himself has blood coming out of one eye and tears dripping from the other.

Bolat says to him, "Hey, animal, why are you crying? Your resting place is nice; you have fodder in front of you and nothing to cry about. Behold, I have suffered so many exiles and calamities. If I do not weep, why weepest thou?"

The golden-eared stallion starts speaking [in human language]: "If I do not weep, who shall weep? I weep not for myself," he says.

"Why, then?" he asks him.

"I weep, thinking of the exile that awaits you." Anyway, at that point, he saddles the horse, ties the sword around his waist, takes the whip in his hand, and, mounting the horse, sets out to find the horses and his brothers.

A few days go by. One day, this one also comes to the town where his brothers have gone, and the former rogues see this one also. "Oh, sir, welcome, they tell him. Are you well?" **(285)** They are going to host this one in various ways. "Sir, there are good taverns and good hostelries in a certain place," they tell him.

Temir Bolat replies, "I will find the place I need myself; if I am hungry, I will find something to eat; you need not worry about me at all."

"This fellow seems to be unlike the others," say the rogues, "we must keep away from him," and with that, they go away.

Temir Bolat walks around the city for a few days. One day, he goes to the bakery for bread. He sees his brother in the bakery, pouring out the slop barrels. He turns out of there and goes back without a word, but he himself becomes dirty. He needs to go to the bathhouse. He goes to the bath, sees his other brother there cleaning and sweeping the bathhouse; again without a word, [he] turns around, and returns to his place.

Then he thinks to himself like this: "Lying here is not going to do any good, so I'm going to go and find my stud."

And then he mounts on the golden-eared stallion and goes on his way. He walks and walks, and one day he comes to a bridge, and as soon as he reaches it, the stallion stops short, and Bolat says, "Hey, you beast, why don't you go on?"

The stallion says, "Soon a five-headed dragon will cross this bridge: you will cover my ears with your shawl tightly, he will smell you and rush to eat you, I will whinny once, and the dragon will collapse in a faint, **(286)** and you will run up and cut off his four heads, and cut off his ears and put them all in your saddle-bag." Bolat quickly ties the horse's ears tightly with his shawl. All at once, the earth rumbles, the bridge sways, and the dragon approaches. At this moment, the golden-eared stallion neighs, and as he neighs, the dragon collapses unconscious.

Bolat immediately goes and cuts off the dragon's four heads, one of which remains.

The dragon says, "Hey, lad, if you're a man, don't leave me with one head; cut this one off too."

Bolat says, "You are not my father's only child. Go and be tormented!"

The dragon fell down dead after much agony. Bolat then cuts off the dragon's ears, puts them in his saddlebag, and sets out on his journey.

He walks, he walks, he walks for about fifteen chakirim*, with large palaces in front of him. When he reaches these palaces, he goes in, and there is a worldly beautiful maiden embroidering. Bolat enters and greets her.

The girl receives his greeting and says, "Hey, lad, how could you come to a place where a bird with its wings and a snake crawling on its belly cannot come?

"Why?" says the lad.

"I have a five-headed husband who does not let any man come into this land. He swallows you from afar; how have you escaped his hands? Now get thee hence quickly, for if he comes now, he'll kill thee."

The lad says, "Oh, beauty, I have slain thy lord; behold, here is a sign from thy husband," he says **(287)** and from his saddle-bag takes out his ears and puts them before the girl.

"If so," says the girl, "thou art now mine; I am thine." Saying [this,] she falls into the lad's neck.

"Well," says the lad, "I went to seek the lost one; till I find the lost one, you cannot be mine; I cannot be yours. So let me go and seek the lost one."

The girl says, "What is it that thou hast lost? Tell me, perhaps it is something that I know of."

Bolat tells her that he has gone to look for his stud.

"If so," says the girl, "if you go from here, you will find a bridge on the road; that bridge is on the border of my brother's estate. He is a very brave and wicked devil, seven-headed; even birds flying in the air cannot escape from his arms; you must be very careful; otherwise, he will defeat you."

Bolat [says], "He can do nothing to] me," and then he mounts on his golden-eared steed and goes on his way.

Walking and pacing, he comes to the bridge mentioned, and there, as before, his horse also stops again; as before, he fastens his horse's ears tightly; the seven-headed devil comes roaring, shrouded in a dark mist, but the golden-eared steed gives a whinny, the stones of the bridge fly up and fall on the devil's head, and the devil collapses unconscious. Bolat goes and instantly strikes the devil's head off, puts his ears in his pocket, and goes away. He comes to a large palace reaching to the sky, goes into the palace, and sees a beautiful girl with a kitchen board in front of her, a rolling pin in her hand, and she is stretching the dough.

Bolat greets her, and the girl receives his greeting. "Hey, lad, how **(288)** and why did you come to this land? I have a wicked brother, towards this way; in fear of him, birds in the air dare not fly; how is it that you have come here not afraid?" She asks, "where do you come from, and where are you going?"

He tells her what happened to him, and she is amazed. "If it is so, lad," she says, "your horses are with my wicked brother; this brother of mine stole your horses."

Bolat says, "I have come to take the horses."

The girl says, "Alas, lad, it will be very hard for you to take them: he is a great devil; if you should kill him in his sleep by surprise, knocking him down, maybe, I don't know, otherwise he will defeat you."

"Where is that devil now?" asks Bolat.

"Look, behind a mountain is a hill, on the top of which he sleeps."

It is thirty-nine days since the devil went to sleep, and he sleeps for exactly forty days; until her brother wakes from sleep, his sister makes kobete* and belesh* day and night for him. Behold, Bolat covers a distance of about twenty chagirim*; the hill is visible behind the mountain, but on the top of the hill, two things like cypress trees stretched up towards the sky. And this was the beam of light from the devil's two eyes.

Bolat walks close to the devil, moving from his head towards his feet; he says, "This is its middle," strikes at it with his sword, and strikes off the seven heads of the devil at once. The horses were scattered in that place, in a pasture. He goes and drives the horses to the girl's palace.

He rests in the palace for a few days, and in the meantime, he puts [all of] the girl's **(289)** precious and valuable belongings [on carts], and they set off. From there, they go to the other girl, and after spending a few days there as guests, they set out on their journey. They take the girls with them and, driving their horses, go to the town where their brothers are. Bolat, on arriving in the town, hires a fine room, orders the girls suits of fine clothes, and the clothes are brought, and the girls are dressed in them. The girls became two handsome lads. Their clothes, in fact, were men's clothes.

Then he goes to the bakery where his elder brother used to be and orders a big bowl of bulgacha*. "When it's cooked, you'll bring it to my room in this and this place," he says. But his brothers do not recognise Bolat at all. Then he goes to the bath and gets his middle brother to order two nice sets of clothes from the bazaar after he has washed. He makes his middle brother bring the clothes he has bought to his room. They come into the room; his brothers put the things down and want to leave.

"No, no, sit down," says Bolat. "I did not buy these things for myself alone but for all of us."

They sit down in that place and eat the bolgach* with a good appetite, and then he puts the two fine clothes on his two brothers and makes them shave so they look like padishah youths again, as they did in the old days.

They bring coffee, and while drinking it, Bolat starts asking them questions. "Who are you; where are you from?"

And they begin to say, "Well, we were three sons of the Padishah Shahislam of such-and-such place; we two [. . .] on a certain day our horses were lost, we **(290)** went in search of them until, at last, we came here; here, one or two rogues brought misfortune on our heads, and when our money ran out, we did not know what to do. One of us became a servant in the bath, and the other worked as a servant in the bakery; we could not return home, so we stayed here."

Bolat says, "As if you have a little brother called Bolat: what does he do?"

They say, "He's a half-wit: he's no use to us, he's no use to our father too, he's a fool of God. We have lost so many horses; the uselessness of that silly is a great pain to us."

Bolat, hearing them say these words, says [to himself] that it was strange that he himself was told that he was lazy and good-for-nothing. But on the back of Bolat's head, at the roots of his hair, he had a cicatrix from his childhood. Bolat says, "And has it been a long time since you left home?"

His brothers say, "It is seven years now."

"When you see your little brothers, will you know them?" he asks.

"Of course, we shall," they say.

"What would you know him by?"

"When he was a child," they say, "he fell down and left a mark on the back of his head."

"If so, is it such a cicatrix?" he asks, pointing to his head.

At once, they are amazed. "Woe, brother," they say, "we have said so much evil against you; forgive us our trespasses," they say, and they fall down at Bolat's feet.

"Alas," says Bolat to them, "many things happen in the world; behold, I am your brother Bolat; I too **(291)** have gone after you in search of our horses; I have passed through many such things," he says, and he tells them all that has happened to him in his exile: how he fought with the devils and how he took the horses; he tells his brothers in detail. "Behold, these lads whom you see in men's clothing are the sisters of the devils who were slain; these two girls shall be yours; I have brought them to you," he says. He also asks the girls, "These are my brothers."

The girls [say], "We cannot go to anybody bigger than padishah's sons," they say, "but let them marry us having a wedding feast."

"Well, if that is so, when we arrive at our father's, we will have an engagement and a wedding," they say and give their word that they will.

Then they rest for a while with Bolat in the city, walking around. One day they leave the city to go to their father. Bolat buys two beautiful ambler horses for his two brothers, puts them on, their horses in front, and they themselves, the three boys and the three girls behind, are coming back to Padishah Shah Islam. Coming out of the city, they go about twenty-five or thirty chakirims*, their brothers leaving the said Bolat behind, talking to each other and saying, "With what face shall we now stand before our father and our kinsmen; our half-mad brother has done so much, and having relieved us of our troubles, is bringing us to our father, how can we bear it: to go home to our father in this way is worse than death," they say.

Then the middle one says, "If it is so **(292)** let us not do it, let us deceive Bolat; let us stay behind with him and kill him, thus, having got rid of him, we shall come to our father with our herd and the girls in good health."

The eldest one agrees to this.

Then they go to Bolat and say, "Brother, let us stay a little behind; we have something to tell you."

Bolat, obeying his brothers, stays back.

Let the three girls go on with the horses; the three sons of the Padishah stay behind; their eldest brother draws his sword from his belt, strikes one blow at Bolat, and cuts off his two legs with a great crack below the knee.

Bolat falls to the ground, unable to walk. "Alas, my blood brothers, tell me, why have you treated me so?" They, knowing that Bolat will not be able to do anything, do not kill him, leave him there, and catch up with their herd.

One day, they come to their homeland and send a carrier of joyful news to the Padishah. The Padishah gathers all the children, as many Viziers*, batirs* as he has] and goes out to welcome them. The Padishah receives his sons, and in the evening in the palace, the Viziers, the hadjis*, and the hodjas* are gathered; the crowd is great.

After eating food and drinking coffee, they begin to tell their stories: "Behold, we went here and there, we killed such and such a devil, we defeated such a dragon in a certain place, we found our horses in the place where this dragon lived, here, we brought two of these three girls for ourselves and one for our little brother Bolat," they say.

The Padishah says, "That fool has gone after you, saying that 'I am going to look for my brothers and our stud.' Haven't you seen that fool? I wonder where he wanders about?"

They answer that they have not heard of him.

Let them discredit Bolat and brag] to the people that they did his brave deeds; we should go to Bolat at this point.

When Bolat's brothers left him like this and went away that day, Bolat's feet hurt, but what to do? Whatever he does, he will not succeed. He picks green grass from here and there and puts it on his feet. After a few days of doing this, his legs mend, and he walks up and down. One day he is sitting in the shade of a tree, hungry, when a blind man and a cholak* come up to him.

The blind man and the cholak* greet him: "Hey, lame friend, tell us what you are thinking about."

Bolat tells them one by one the sufferings he has been through. "If so," says the cholak* and the blind man, "let us three be one; our trade is begging; we live by begging; with us, you too can live." Bolat agrees, and they set off.

They go, they walk, and once they reach a forest, they make a hut in the forest, and they walk this way and that. The cholak* and the blind [man] go begging. The lame Bolat, as soon as they go away, goes hunting with a hissing shuffle, turning this way and that, finds a deer*, knocks her down, and brings her to the hut **(294)**; from here and there, he collects one or two dry branches and brings them back. Till these come back, he roasts the deer*, giving out a very nice smell. Soon they also came back, but the roast became so wonderfully good that its smell filled the whole forest. So they come, sit down and eat.

The blind man says, "Hey, friends, it will not do; we have to take a woman."

The cholak* and the lame man says to him, "We are three men; what good is one woman to three men? Is such a thing possible?"

The blind man says, "If we could find a poor girl of our own kind and take her as our sister forever, she would cook us food, wash our laundry, and serve us."

The others see fit to do this. Then the cholak* and the blind man bring a poor girl from one of the villages. That poor girl cooked food for them and did their laundry. This is how they live.

One day, the lame Bolat takes his rifle on his shoulder and goes into the forest; the others go begging. Only the girl remains in the hut. At that time, there were few places where there was fire, no matches, and the fire was not extinguished. That day, as the girl was cooking the food, somehow, the cauldron boiled over] and put out the fire.

Bolat [was] of an angry nature; the girl was afraid of him. "If my brother Bolat came now, who knows what he would do to me; what should I do now," she says, and she goes out of the hut to look for fire.

She walks little **(295)** or a lot, and there is a glimmering firelight in a forest. As for the girl, she walks straight towards the light. She goes into the house where the light is and sees an old woman sitting there.

The girl [says], "Salam*, gammer."

"My daughter," says the old woman, "welcome; where did you come from?"

The girl explains, "Gammer, there is a hut in a certain place; I live in that hut."

"Are you there alone, by yourself?" asks the witch old woman.

"No, I have three brothers," she says.

"What do they do?" she asks.

The girl tells her what they do.

And when she asked, "Well, how you got here?" she replied, "My brother called Timer Bolat is very angry; well, while cooking food, the water boiled and someway or other the fire was put out. My brothers are now without food; when they return, what will I tell them? My brother cholak* and my blind brother may not say a word; my lame brother Bolat, I hang in doubt, will beat me. So seeing the light in your house, I have come to ask you for fire, she says."

The witch old woman says, "Well, if it's so, my head hurts so much: do rub my head a little, and then you will take some fire."

"Very well," says the girl, and [she] sits down. The witch gammer lays her head on the girl's knee.

After a long time, she says, "Well, that's enough, daughter," and gets up to give the girl some fire. But the girl feels powerless because the former witch has sucked her power. Unbeknownst to the girl, she gives the girl fire, be that as it may, but the girl, in her confusion, has forgotten to bring something to put the fire in.

"Grandmother," says the girl, "I have nothing to put the fire in."

The old woman sees a hole **(296)** in the girl's apron. "My daughter," she says, "hold your apron, let me put ashes at the bottom, and then let us put the live coals on it." Mother, doing so, pours ashes on the girl's apron and puts the glowing embers on it. The girl walks away, and as she goes, the ashes spill out on the roadside. The girl reaches the hut, lights a fire, and as before, cooks the food; then her brothers come, eat, get well fed, and the next day they go about their business again. Let us go [back] to the old woman.

After the girl with the fire has gone, without much delay, [the old woman] begins to follow the girl's trail. Following the scattered ashes, she reaches the hut where the girl is, enters, and says, "Selam aleykum, my daughter. My child, my head hurts so much today; the other day, after you rubbed my head, I was so well; today, my head hurts again; I came to have my head rubbed a bit by you again."

"Very well, gammer," says the girl.

She takes care of her head, and after a while, the witch says, "That's enough, my child," and stands up. Saying, "May Allah be pleased [with you]," she goes away with a shuffle.

For a few days, she acts] in this way. Every day she comes, she has her head rubbed by the girl, but day by day, the girl's strength wanes, and she weakens.

One day, Bolat asks the girl, "What's the matter, hey, girl, you have wasted so much," he says, but she won't tell him. Another day he asks her again and again, but she won't tell him. Bolat says, "If you don't tell me this time again, I'll wring your neck; what's wrong with you is not a disease, but something is hidden," he says, forcing her.

The girl says, "Look, it so happens that an old woman comes here; she has her head cured by me, and after she gets up and leaves, I feel such a lack of strength in me."

Bolat says, "If so, it can't be like this; this old woman is a witch, this witch is going to put you to death, let's think of killing her."

"In what way will we kill her?" asks the girl.

"I'll be the man who leaves early [in the morning]; you stay here in the hut; the witch comes again, you lay her on your knees and nurse her head, and while you are nursing her head, you hold her hair in a braid and cough; we hear [and will know] that you have held her down. We go in, tie her up, but you don't let go. She will try very hard to get away from you, but be careful."

"Alright," says the girl.

Then these three, the cholak*, the lame, and the blind, all three go out and hide near the hut.

After a while, the vampire witch comes again, shambling and shuffling, double hunched, pushing in, and greets [the girl]. "Oh, my daughter," she says, "whenever I come, I see you alone," she says.

"My gammer," says the girl, "my brothers are very poor; they are going to work, so I stay here alone."

The old woman says, "I've come again to have my head a bit rubbed."

"Alright, my gammer," says the girl, "lie down and let me rub." She lays the vampire on her knees. While she is rubbing her head, she braids his hair and rolls it up, then coughs. One by one, the hiders come in. The vampire tries hard to run, but there's no way she can escape. Bolat's companions beat her badly and say, "You have made this girl so weak; you have made her sick; cure her."

"I'll cure her," says the vampire, and she licks her from head to foot with her tongue, and she is cured and becomes as she was before.

The blind man says, "My eyes have long **(298)** been blinded: I cannot see; heal my eyes also."

The witch at first would not do it, but as soon as the cholak* strikes her in the face with his left hand, she says,] "I will make amends at once," and licks] his two eyes; she has cured them. His eyes became as when he was a newborn.

All are well; only Bolat remained [disabled]. His friends say to the vampire, "Look, our brother remained lame. If you cure him too, we will let you go, go where you like," they say.

And the vampire says, "If you bring thousands of lame men, I will heal them all, but I will not heal this lame man's leg," she says.

"Why?" they ask her.

"Why, he has cut off the heads of three of my sons, spilt their blood, destroyed their houses, and taken three of my daughters; how shall I heal this one?" she says, "I could not do it even if I knew you would kill me." This old woman is the mother of the devil Bolat killed. On the spot, Bolat's companions kill the vampire, slit open her belly, take out her heart, and tie her to Bolat's legs. Bolat's legs became as good as his mother had given birth to him. All are safe and sound.

One day, all four set off and arrive at Bolat's home. In the palaces, the drum and zurna are playing with all their might. It is a wedding feast, and they also enter the house of the feast. Thinking that one or two poor people have come to the wedding, they feed them all kinds of food. They drink coffee and ask, "Whose wedding is this?"

The people of the crowd say, "Our Padishah had three sons, **(299)** two of them, since one day their horses were lost, went away one day, killed three dragon-devils in seven years, and took even three devil-girls, and came [back] together with their horses;

behold, now they have their wedding feast with the girls they brought. Tonight they go behind the curtain."[6]

They also ask what happened to the youngest.

The assembled people say, "Who knows: he was half-witted anyway; who knows where he struggles; behold, his brothers, who are having their wedding, have brought him a devil-girl too, but none of them knows where he wanders. The devil-girls swore, 'We have forty days of mourning; there is one man for whom we wait forty days; if he does not come for forty days, then we will be your wives,' they said. They were waiting for Bolat."

So it was also the last day of their mourning; these words were heard by the people gathered. What should they do? Bolat's elder brothers came to them[a few times], but they didn't recognise him at all. Bolat's companion, the blind one, says, "So we have come to a wedding feast, which itself is the wedding feast of the sons of a padishah, [thus] we must do something to honour it." He rises and says, "Hail, assembled people, we have heard that the wedding of the Padishah's sons is taking place. We have come to honour the wedding; go and tell the wedding people to come and gather, and also tell the Padishah that we are going to honour the wedding." So all the wedding people gather, and the Padishah comes too. Then the blind man begins to tell the whole thing from the beginning. He asks the Padishah's permission: "My Padishah, I will tell a story the world has never heard before if you allow me."

Once the Padishah has given his permission, he begins to speak: Once upon a time, there was a Padishah; (300) he also had three sons. One day his nine studs are lost from his nine pastures area, he tells how they [the elder brothers] went after him, later how the youngest went off, then how the older ones, drunk with the rogues, ran out of money, and how one worked in a bath, the other in a bakery, how Bolat killed devils, then how he bought his brothers clothes, dressed them, and freed them from slavery, then how they had deceived and cut off Bolat's legs on the way, then how these three men and a girl became friends, what things happened to the vampire, how his leg was healed, and how he came to the Padishah to listen to stories: one by one, [the blind man] tells them all.

From this point, the Padishah, no longer patient, having a gallows made, has his two eldest sons hanged; the eldest of the two elder daughters is given to the blind man, the middle one to the cholak*, and the youngest to Bolat. The former [servant] girl they bring with them is given to the son of the Grand Vizier of the Padishah. Thus doing forty days [and] forty nights of wedding feasting, all four go behind the curtain, reach their goal, and get [their mate] to their destination. May God bring us to our destination, as he did Temir Bolat.

6 "To go behind the curtain" means "To spend the wedding night."

5 The Tale of Bay-Terek

(318) There is a Padishah, and there are also four sons of him. In the Padishah's yard is a big tree called Bay-terek. The fruit of the tree is ripening, but they themselves cannot eat any of its fruit. It is ripe now, they say, but every year one night, the fruit disappears. The children of the Padishah gather in one place and say, "Father, we will go out to the tree at night to stand guard, to understand what has happened, what is happening to the fruit of our Bay-terek, let us know." This year again, the fruit ripens; now the time has come; since they agreed, the eldest boy, taking his bow and arrow, goes to the tree, lies down, and falls asleep. To find out what is the matter with the tree, he lies for a while. After a long time, it was almost midnight; what did his eyes see? Something coming from the sky, rattling and thundering, causing doomsday, but in his fear, he is almost scared to death, almost dies. Unconscious, he collapses. After a while, he gets up and sees that part of the fruit of the tree has gone. Then he returns home and explains to his father how it happened.

On the second day, his middle son takes his arrow, goes to stand guard, lies down, and falls asleep. At midnight, the former dark cloud comes snapping. The boy is startled out of his sleep, not knowing what to do, and collapses unconscious. After a long time, he wakes up, looks that way, looks this way, and he can see nothing **(319)**. Going near the tree, he sees that half the fruit disappeared. He himself, having not seen what had taken the fruit, turns around and also goes back to his place. Having come [home], he explains to his father in what way it happened.

On the third night, his third son says, "Tonight, I'll go and try and see if I can find out what it is." The night falls; he shoulders his arrow, goes to the tree, and gets sleepy but does not sleep. After a while, a cloud comes down from the sky, rattling and rumbling, and covers the tree. The boy shoots his arrow, pointing it at the cloud. From the cloud comes a voice: "If you are a man, shoot again." The boy shoots another arrow. Once more, a voice comes, "The [true] lad's game is until the third [turn]; shoot another arrow," he says, and the boy shoots another arrow.

From the cloud, something falls to the ground, spinning and rolling, [the boy] approaches him, and what his eyes see: If you say he is a man, he is not a man; if you say he is a genie, he is not a genie; you cannot tell what kind: he is a two-span-statured man with four span-long beards; the boy asks him, "Are you a man or a genie?"

"It's none of your business whether I'm a man or a genie," he says.

"Well, if it's not my business," says the boy, "I'll take you to my brothers and to my father; let it be their business." Then he takes the four-span-long bearded [man] and gives [him] to his father, "Behold, father, this is the destroyer of the fruit of our Bay-terek*. His father has this man tied with a rope by the beard to a post by the house, saying, "Behold, the fruit of my tree called Bay-terek is eaten by such a strange man, and now we have taken him and tied him up."

Let the people be gathered; **(320)** the youngest boy, the fourth, was playing one day with his arrow in this place in the courtyard when his arrow missed and pierced the beard of this bearded one and lodged in it. The youth goes and asks for the arrow, but he will not give it, saying, "I will give you the arrow, but you will cut my beard with a knife where it is tied."

The little child goes to his mother. "Mother, give me a knife," he says. His mother gives him a knife, and as soon as he takes it in his hand, he runs and cuts his beard at the knot. And the bearded man, as soon as his beard is cut off, flies up into the air, spinning, and disappears as before, turning into a dark mist. And the child cannot know what he has done.

In the meantime, some time passed, and the Padishah called for his own people; he wanted to show this bearded man to the people. People are gathering from all sides. They ask the Padishah why he has summoned them. He says, "Well, here in our garden, there is a Bay-terek, every year something has happened, and we didn't know what it was. Now we have taken it; I have called you to show it to you [because] it is a very strange thing."

"My Padishah, where is that thing? Let us see it," they say.

"It's tied to a post here by the stable."

"Come on, let's go and see it," they say.

They go to look, the Padishah looks, but there's nothing where he tied it himself. Then he says, "It can't be like that; [now] you disperse; I'll find out who untied it and let it slip away."

He calls someone, asks, "Did you do it?" and throws him in the dungeon. He sends the second one [there too, because] he can't find out who did it at all. Finally, the youngest son says to his father, "Papa, I untied that man; don't torture other people."

His father asks him, "Why **(321)** did you let him go?"

The son replies, "Dad, as I was playing arrow shot, my arrow remained at the base of his beard, in his palate, and he would not give it back, 'If you bring a knife and cut off my beard, I will give it to you,' he said. And I went and cut off his beard, whereupon he flew up like a cloud into the air and spun and disappeared."

His father, hearing these words, gives orders to his Viziers, has a gallows made, and is about to have his son hanged.

His Viziers say, "My Padishah, do not kill this son of yours," they say, "give him writing in your own hand that he is the son of a Padishah, wherever he goes he shall not starve, he shall be honoured, he shall not be harmed, he shall not live here; you shall banish him; is there any greater punishment for him than this?" they say.

The Padishah writes a paper that he is the son of a Padishah, puts it in his son's hand, and says, "Go on, I have donated you your life, do not stay here, go wherever you like," he says, and chases him away. The boy's mother bakes kalakay*, a good fat pirogue, and gives it to him to eat on the way. This goes away.

On the road, he meets a bald boy, "Hey, my friend, tell me, where are you coming from? Where are you going?" says he. The child tells the bald lad in detail what happened to him.

The bald lad says, "You were driven away by your father; I was driven away by my mother. Now they believe that we are not capable of anything, let us two go together, inshalla*; we will not come upon them, probably."

The Padishah's son says, "So, I have a friend," he says and goes on.

They go, walk, sit down in the shade of a tree, and rest. The boy [says], "I'll look around a bit," he says. As soon as he has gone, the bald lad takes out a salty kalakay* from his coat and takes his friend's greasy kumech* from his coat, then puts his own salty kalakay **(322)** into his coat. The boy comes back, sits down, and says, "Now let's have some food," so as soon as he opens his coat, he sees that in place of his kumech, a kalakay has been put, he starts to eat, but the salt makes it like poison, inedible, but he doesn't say anything. The bald [lad] fills his belly with the greasy kumech.

Then they get up and leave this place; they come to a place where they want to drink water from a well, but the well has no bucket. The bald [lad] says, "Now you take my belt, I will hang down and drink and come out, then you hang down and drink and come out." The bald [lad] hangs down, drinks, and comes up, and then the Padishah's son hangs down. After drinking, he says, "Now pull me up; I'm coming out."

"I won't pull you out," says the baldhead.

"Hey, my friend, why are you doing this?" he asks him.

He [the baldhead] says, "You will be my servant and give me your clothes, I will be the Padishah's son, and you will swear not to tell anyone, then I will pull you up." The boy swears, and the baldhead pulls him up. He puts on the boy's clothes, takes the writing his father gave him, and after the boy puts [the baldhead's] old hat on his head, they go on.

One day they arrive at a Padishah's palace, [the people] come out to meet the bald lad. The Padishah honours them with a hospitable reception; it was a great reception for the bald lad. He asks the baldhead, "Son, how did you come to this land?"

And he, getting right inside his situation, says, "My father drove me away, and I took my servant and went away, and came to you, and will be a Vizier or a Pasha*."

The Padishah says, "If it is so, I will make thee my Vizier," and so he makes him the chief Vizier. **(323)** What shall we do with that fellow [the true son of the Padishah]," [the Padishah] asks.

"Make him a horseherd. He'll make a very good horseherd."

Then the Padishah gives the former fellow a stud of horses and says, "Well, my son, you will look after this stud of horses, only you will take it eastwards, graze it, take it westwards, take it northwards, but you will not take it in the direction of Kibla*."

The boy says, "Very well," then he puts his fur cap on his forehead, rolls up his coat, takes some kalakay* with him, fills a small wooden flask with water, and drives his horses away.

One day, two days [pass] in this way. One day he takes his stud again and goes away, stretches out and falls asleep, but [the stud] has gone towards Kibla; he doesn't care, let the stud go where it wants. After a while, something comes out of the air, lifts him from his sleeping place and takes him away. He doesn't even wake up from his sleep. The one who took him away was the two-span-statured man with four span-long beards. He takes him straight to his cave. He had a porch swing where he himself used to lie, and there he lays him down. This man had forty daughters. He tells the girls to rock this fellow so that he may sleep well enough in the porch swing. The girls rock him. He sleeps for a long time. As soon as he wakes up from his sleep, he sees forty girls sitting next to him. [The bearded one says], "What is it, my son? You are trapped," he says. "I am angry with your brothers, I would kill you, but I will not kill you; you **(324)** have freed me." Then he has the fellow entertained by the girls, and one day he says to the fellow, "Now let me show you around [here] a little," and, entering the places in the cave with him, he shows him.

He opens a room with a fountain, silver flowing from one side and gold from the other. He says to the fellow, "Dip your head in this spring!" The lad dips his head in; half of his head is silver, and the other half is gold. Then he comes to the forty daughters and says, "Now this fellow is going away, I would consider it worthy if you would give him a gift [a souvenir]." Then one of the girls takes out and gives him a golden comb, one gives him a mirror, another gives him something else, and each gives him a gift.

The girl who gave the comb says, "The skill of this comb is that if you are chased from behind by an enemy, and you throw it behind you, it will become a fence; the enemy cannot pass through it."

The mirror-giver says, "This mirror, if you take it out of your pocket and put it in front of you, will have forty instruments playing in it and also forty dancers dancing in front of you."

The bearded one says, "Every girl has given you something, and I tell you this, my gift to you shall be this: 'In the dark of the night if you uncover your head, the brightness of your head will make a torch unnecessary.' And now you are free; you may go where you will."

The boy says goodbye and leaves. He takes his stud and arrives at the palace, locks the stud in the stable, and he himself goes to a place like a henhouse and sits down. They bring him and put some grated noodle soup in front of him. He eats, and after eating **(325)** he puts the mirror in front of him, uncovers his head and sits down. The inside of the henhouse becomes completely light, forty instruments begin to play from the mirror, and forty dancers come out and dance. The Padishah's youngest daughter comes out and hears the music but doesn't know where it's coming from. However, pacing back and forth, she goes straight to where the sound is coming from and arrives at the chicken coop. She knocks on the door and wants to go in. "Who are you?" the boy asks her. "I'm the Padishah's daughter," she says. He says, "Not only the Padishah's daughter, whoever you are, but I'm also sleeping now, and I'll get up early and go to my business." She turns around and goes back.

One day this way, two days that way, the first bald lad, the pig who had become the Padishah's Vizier, hearing what the case with the fellow is, said, "If I don't have this one killed, I'll have no peace; I'll have this one killed right away."

One day he says to the Padishah, "Our horseherd is such a fellow who will do anything you wish."

"What will he be apt to do?" asks the Padishah.

"Well, if you tell him to turn into gold the stud of horses driven on in front of him he will do it."

"Sure, he can't do that, and if he can't do that, what do we do?" asks the Padishah.

"If he cannot do that, you will have his neck cut off," he replies.

One day the Padishah says to the fellow, "Hey, fellow, turn the tabun of the horses in front of you into gold and bring it to me; if you can't bring it, I'll cut your neck off."

The boy goes to the former place to the bearded man and tells him how things stand. "Drive the stud this way, next to the spring," says the bearded man.

The fellow **(326)** drives the stud next to the spring.

The bearded man says to the fellow, "Son, here is a wooden bucket for you; draw water from this spring with the bucket and water your horses," he says. The boy draws water from the spring and pours it from the bucket on the horses, and each horse becomes golden. From there, he drives them to the palace; the Padishah looks on, and what does he see: each horse is gold.

The girl comes again and says to the fellow, "You are really the son of Padishah; you are not the servant; it is the Vizier." And as much as he tries to hide the Padishah-son status, she falls in love with him, can't stand it, and falls for him. From here, she goes straight to her father and tells him what she knows.

The Padishah summons the fellow and says to the daughter, "How do we know he is a Padishah's child? How do you know?"

"Well, father," says the daughter, "when you uncover his head, you will see and know."

The Padishah calls for the fellow; [another] five or ten men calls. "Call for the bald lad," he says. They go and call him, and he also comes.

While talking, the Padishah says, "Take off the bald lad's hat; let the sun shine on him." [They take it off, and what do they see: bald.][7] They take off the boy's hat, and what do they see: that half of his head is silver and half gold. The Padishah asks this fellow how it happened, and the fellow tells him how it happened to him one by one.

The Padishah has the bald boy killed and adopts this one as his child. After a while, the Padishah dies, and this one becomes the Padishah instead. After a few years, he visits his father, meets his father, and then returns to his place again and sits on his throne, living his life happily.

7 Missing part of the text saying something like, "But they see he's bald. Then the Padishah has the fellow's cap taken off."

6 The Tale of the Muezzin

(336) At one time, a man named Hasim intends to go on a pilgrimage. He sets out on his journey. The village muezzin, on seeing him off, says to him, "Have a good journey. To whom have you entrusted your daughter?"

Hasim says, "I have entrusted my daughter to Allah."

The muezzin says, "Allah will not descend from heaven to look after your daughter; she must be entrusted to someone."

Hasim says, "If she needs things, you bring them, you put them through the window, you don't go in."

One day the muezzin brings meat and puts it through the window; the girl doesn't open the door.

The muezzin begs an old witch woman, "Set me up with this girl; I'll give you whatever you want," he says.

The woman comes to the girl, and with some cunning, she makes her open the door, "Come, my daughter," she says, "Ahmad's daughter is having her wedding; let's go, at least you can have some fun." In the meantime, the woman orders the muezzin to rent the village bathhouse in a pit for three days. There were forty rich people in the village. These forty had forty daughters; all forty had outer garments; they hung them on the gate of the bath.

The next day, the woman comes and takes the girl to the wedding. On leaving the house, the girl stuffs some barley in both her pockets. On the way to the bathroom, she threw them on the road. When they came to the bath, they saw forty dressing gowns [laid out, left there] at the gate. "Oh, my daughter," says [the woman], "the girls are here [inside] already, come, let us wash, and then we will go to the wedding breakfast." The woman opens the door of the bath, pushes the girl into the bath, and she goes back. The muezzin was [already] in the bath. As soon as she saw the muezzin, the girl went out of her mind; she realised she had been deceived. Since there was no way **(337)** to escape, she told the muezzin, "I will give you a nice bath." The muezzin lay down on the sweat stone. The girl soaped him thoroughly and began to wash him, and after soaping his head and eyes, the girl pulled the wooden slipper off her feet and beat the muezzin's head until he fainted. Then the girl went out, and along the barley scattered, she went home.

The bath [rental] period has expired. The muezzin is still lying there, the owner of the bath comes, and the muezzin is thrown out of the bath. The muezzin goes up to the woman, "Is that all you have done for me?" he asks. "When Hasim hadji* comes [back] tomorrow or one day, with what face shall I show myself to him?"

The woman answered, "Could you not find a way, wise man that you are?" she said. "Write a letter to Hasim saying, 'Your daughter here has been debauched. There are forty people in the village, forty male-bairns of these, and all forty are going to your daughter.'" The muezzin has written the letter. Hasim received this writing on the way back from his pilgrimage. After reading the letter, he collapsed in a faint, got up a little later, and sent his son to the village, "Go, slaughter [stab] your sister, dip her shirt in her blood and bring it to me!"

His son went and came to the village and said to his sister, "Open the door."

"I will not open it," she says.

"I am your brother," says the lad, "don't be afraid of me."

"If you're my brother," she says, "your name will be written on your ring. Throw it through the window, let me see it, and I'll open it."

The lad threw the ring in; [she] read it, and, finding out it was her brother, [she] opened the door, and they inquired after each other's health and circumstances. Then the lad says, "Sister, **(338)** our father is coming from a pilgrimage, come on, let's go and meet him." Tricking her in this way, he took her out of the house, and they went into the woods. Having come to a desolate place, the lad says, "Sister, I understand that these are shitty [false] things for you, but our father has ordered me to slaughter you and bring him your shirt dipped in blood; now, how shall we do? Take off your shirt, we'll bleed your little finger, put it on your shirt, I'll take it away, and you stay here," he says. The lad took the shirt, went away with it and showed it to his father. Hasim came home, the muezzin came to welcome him, and they inquired about each other's health and circumstances.

Let us go to the girl left in the forest. A younker went hunting, and as soon as he saw the girl, he fell in love with her, took her and brought her home. After a while, they said to the younker, "Get married!"

"If I marry, I'll marry the girl I found in the forest," says the lad.

His mother and father said, "A woman found in the woods cannot be a lover; catch the fancy of a girl from a good place and marry her," they say.

The lad says, "No, if I marry, I will marry no one but this one [her]."

Anyhow, they got engaged, and he married this girl. They lived together for a long time and had three children. During this time, the fellow never asked, "Have you got a father?"

She was angry about that. She held her tongue for a long time.

[His mother and father] said to the lad, "We told you at the beginning that something found in the woods does not bring good luck."

Then the lad became very sorrowful; his sleep was gone. One night, when the woman was alone with her children **(339)** she sang a song to her weeping child. The fellow listened to it from a distance. The fellow says, "You have been silent for a long time; what is the reason for this? Tell me."

The woman replied, "We have lived in the same place for so long, and you did not ask me if I have a mother or father, so I got angry and did not speak."

[The fellow said,] "If you have a mother or father, I will send you to your father tomorrow."

In the morning, he got up, put the woman and the Vizier in the carriage with his three children, and sent them away. The Vizier set his eyes on the woman, and as they passed through a forest, he stopped and said, "Agree to the deal with me, or I will kill one of your children." The woman did not agree. The Vizier killed one of the children. After they had gone a little further, he stopped again and asked again; the woman again did not agree, and he killed another of her children. As they went a little further, again, she did not agree, and he killed another one. Then he says, "Now I will kill you yourself too if you do not agree."

When the woman felt there was no other way out, she said, "Let me go aside [to pee], and then it will be done."

The Vizier did not want to let her go.

"If you are afraid," said the woman, "tie string around my waist and let me go."

The Vizier tied a cord around her waist and let her go [aside]. The woman turned behind a bush, tied the cord to a tree, and the woman fled. The woman has already had much fun, [thought the Vizier], and he goes along the string and sees that the string is tied to a tree, and the woman is not there. He got into the carriage, returned, and said, "As we had been passing through a forest, she had flown away with her three children." Everybody believed it, and they said that the one they found in the forest was the one who had disappeared again in the forest.

As for the woman, she goes out of the woods from this place **(340)** and meets a shepherd; she has a sheep slaughtered by the shepherd, washes the sheep's rumen, and puts it on her head; she goes to her village. In the village, the muezzin sees her and says to her, "Would you take charge of geese?"

"I will," she said. The muezzin hired her for Hasim. He guarded geese for a while, then stopped guarding them and went to work in the same village in the shop of the helva-seller*.

One day, the younker came to this village while hunting. He stayed at Hasim's. In the evening, Hasim summons the muezzin and the witch woman to sit with the guest. Then they summoned the helva-seller too. After eating and drinking, everyone told them what had happened to him, and only this boy [the helva-seller] was left. Hasim says, "Son, tell us what happened to you too."

"I will tell you," says the boy, "but no one is to go out of the house or come in."

He began: at one time, a man named Hasim goes on a pilgrimage and leaves his daughter at home; the muezzin falls in love with her and wants to get her out of the house; as soon as he says this, the muezzin says, "Let me out for a while, I'll pee and come back."

"No," they tell him, "sit down."

Hasim says, "Say, my son, say."

The boy [says], "The muezzin rented the bath for three days, and the girl was taken to the bath. The girl soaped up the muezzin's head and eyes, pulled off his wooden hood, beat him to a pulp, and fled." Then, having told the deeds of the muezzin, the witch-woman, and the Vizier, all three were tied to the tails of three horses and let loose. The younker arranges for her another wedding feast, and they live [together].

7 The Tale of the Poor Man and the Mullah

(347) Once, there was a poor man. Every morning he woke up, washed his hands and face, and asked for good luck, and this man, wherever there was a rubbish heap, was crawling around. A mullah sees this and says, "What are you doing in these rubbish places?"

"I'm looking for my good fortune," says the poor man.

"My dear, is fortune found in the trash? Come and serve me," says the mullah.

"If I serve with you, what will I do?" asks the poor man.

"I have four oxen; you take them to the sea to drink.:"

The poor man says, "If your oxen go into the sea and drown, will you not ask me for it?"

The mullah replies, "No, they're used to it; they don't go into the sea." They agree with each other.

One day, five days in a row, he goes to water the oxen. One day he cuts off the oxen's tails and thrusts them into the [bottom of the] sea and drives the oxen to another place. He comes to the mullah* and says to him, "The oxen have drowned! Come on, let's go and pull them out!" When they reach the shore of the sea, they see that the oxen's tails are seen [protruding] indeed.

The mullah* says, "Go get a shovel and a shovel from the stable so that we can dig them out!"

The poor man [says], "I came to you to look after oxen. I don't carry a shovel or hoe." The mullah goes away himself.

Once the mullah has gone, the poor man takes the oxen and goes with them to the top of a [mountain] summer pasture. He sees that there a shepherd is wailing. He asks the shepherd, after greeting him, what is the matter with him. "My head hurts," says the shepherd.

"Go then," he says, "slaughter a barren sheep and bring it here, and I will find a solution for your headache."

"I don't have the strength to stand up," says the shepherd, "so you go yourself and slaughter the sheep you want."

The poor man slaughters a sheep, cuts open its rumen **(348)** and puts it on the shepherd's head, gives the slaughtered sheep into the shepherd's hand, and [the shepherd] says to him, "If this goat begins to bleat, you will know that your sheep have run away, then you get up and go look for them." And he himself drives the oxen and sheep away together.

A horseherd came opposite him and asked, "Four of my horses have run away. Haven't you seen them?"

"I have," he says, "they were going this way," he said, tricking him. Then he himself found his horses and drove them away. He went to a village, bought a house and bought a certain amount of land.

After a while, the mullah, searching for his oxen, came up to the summer pasture. Seeing the shepherd, he pulled the rumen off his head and asked, “Who did this?”

The shepherd says, “A man came driving four short-tailed oxen. My head was aching; he made me some medicine and left.”

“Then those oxen are mine; he must have taken your sheep too. Come on, let’s go and find him!”

They met the horseman and questioned him too. When he also found out that the man had taken his horses, the three of them went in search of that man and came to [that] village. The mullah recognised his oxen, the shepherd recognised his sheep, and the horseman recognised his horses, yet they spoke thus: “We should not just take them away, but go into that man’s house and be his guests, and then with his consent, take our animals and go.”

Asking him, they entered his house, and he welcomed them, took them to their room, slaughtered one of the oxen of the mullah, and prepared a meal for them. After eating and drinking, they asked him about the mullah’s [oxen, the shepherd’s sheep and the horseman’s horses], and he said to the mullah: “Your oxen are getting fat.” To the shepherd: “Your sheep started to yean lambs.” And to the horseman: “And your mares began to foal,” he answered. **(349)**

It’s time for bed. The man, unconfused [as if he suspected nothing], brought mattresses, and blankets, laid them down and said, “Listen, if you piss on these mattresses, you will neither take your cattle nor leave here alive.” They went to bed. After a few hours, [the poor man] poured some water under the shepherd. When the shepherd woke up, he saw that the bed was soaked. So he wakes the horseherd and says, “I seem to have wet myself.” The horseherd also touches himself and feels that he has also shitted. The mullah is awakened and sees that his ass has come out too. Then they consulted thus, “Even if it costs us our fortune, come on, let us escape before we are hanged and die!” they say.

The mullah [and his mates] got their arses under their armpits, and they all escaped. All their cattle remained with the poor.

8 Beauty with an Unfulfilled Wish

(353) At one time, there was a poor lad; this lad was a servant to a blacksmith. After some time, this blacksmith married the lad. They lived in an old house by the road. After six months, a year, the time came for this woman to give birth. The woman says to her husband, “Go and ask the blacksmith for a room [i.e. a nicer, new house]; I am ashamed to live here by the roadside.” The lad went and asked; the blacksmith didn’t refuse; he gave him a room. The child was born. Three days passed, and the child had to be named. This lad has not even a para* to buy some meat and fat and to invite people [guests]. He went to a butcher and bought a rouble of meat on credit. Food was cooked, the mullah was invited, and after the mullah* came seven sufus*. They sat down and ate.

One of the sufus spoke up and asked, "What kind of gathering is this?"

They answered, "We have a new world [newborn]; we will name it."

He says, "If so, let me name the child."

"Very well, name it," they said.

The sufu asked, "Is it a boy or a girl?"

They say, "It's a girl."

"Her name shall be Beauty with an Unfulfilled Wish."

The second sufu [says], "Immortal-han* shall be."

The third sufu [says], "If this girl weeps, let pearls flow from her eyes," he says.

The fourth sufu [says], "When she washes her head, let the basin be filled with gold."

The fifth sufu [says], "When she smiles, let two roses blossom on each of her cheeks."

The sixth [sufu says], "On the right arm of this girl shall be a talisman from me."

The seventh Sufu [says], "When this girl walks, under her feet [i.e. in the footsteps of her feet] shall grass grow," he says.

They prayed and went away.

The girl grew up. **(354)** The son of a Padishah hears this and says to his mother, "Go, they are talking about a girl in such and such a village, and if it is true, engage her to me, I will go and marry her." His mother went and saw [her]; it was true, betrothed her and came back, forty days later the wedding began, steamboats came to take the bride, forty nights the wedding lasted, the girl was put on a boat, her aunt and her aunt's daughters were put on the boat with her.

After they had gone a little way, the bride asked for water.

Her aunt said, "If you take out your half-eye and give it to me, I will give you water."

She took out one of her eyes and gave it to her. Soon she was thirsty again; she took out her eye again and gave it to her. She drank some more water.

They came near a forest, and her aunt said, "Well, let's get off [the boat]; let's have some fun in the forest," they got off. Her aunt gave the bride's clothes to her own daughter and left the bride in the forest. They got on the boat, left, and arrived at the Padishah's palace.

As she [the aunt's daughter] was going up the stairs, the Padishah's wife, even if she knew that another girl [was coming], said nothing. They brought the bridegroom, and the bridegroom came in and saw that there was no rose on her cheek, "Well, there ought to be a rose on your cheek, but there is not," he says. They lived a long time [together].

Let's come to Beauty with an unfulfilled wish who lived in the forest. One day, a poor old man went into the forest to fetch wood, and he saw a girl whose tears flowing from her eyes become pearls. The old man takes this girl home, and the old man's wife says, "We can't even take care of ourselves; how are we going to take care of this blind girl?"

The girl says, "Mother, don't be afraid, wash my head quickly." They washed it, and the place was filled with gold.

One day this girl says to the old man, "Father, in the place where you found me, build a mausoleum, let it have two doors, and **(355)** let it be called [be written on it] 'Beauty with an Unfulfilled Wish,'" she says.

The old man has it made.

"Old father," saith [the girl], "[now] I'll laugh, and on my cheek shall be two roses; these roses thou shalt take to such-and-such a village, and to whom that gives two eyes shall thou give the roses."

The old man took the roses and went to the village. The aunt of the girl gave the two eyes, received the roses and sent them to her daughter. The Padishah's wife put the two roses on her two cheeks. The lad came, saw the roses, and was glad and smelled them. At that moment, Beauty with an Unfulfilled Wish became pregnant.

Let's come to the girl's aunt. Her aunt found out that she had forgotten to take the prayer [talisman] from the girl's arm, [so] she sent a witch woman and had the prayer [talisman] taken from the girl's arm. The girl died at that moment. The girl was taken away and buried in that mausoleum. In the grave, a son is born from the girl. He is sucking the girl's finger. One day the Padishah's son went hunting in this forest, and suddenly he heard a voice: "What I hear, don't be Beauty," he said, and they went straight to the voice and saw a mausoleum and opened it and saw a dead body with a child beside it, and took the child and carried it away.

After the child has grown a little bigger, he accidentally gets his mother's talisman, but the woman does not want to give it to him. The child is crying, and the lad comes and asks [why]. The woman knows no other way out; she brings the talisman and gives it to her husband. He takes the talisman to the grave, opens it, and ties it to the dead woman's arm; the girl comes to life right away; he brings her home, and after he has heard the gist of the thing, he calls the girl's aunt. With his own hands, he ties them both [the aunt and her daughter] to the tails of two horses, has them torn apart, and they have a new wedding feast and live [happily] ever after.

9 The Tale of the Five Gold Coins

(360) In the time before time, there lived a man named Yendj Osman,* who never gave [back] what he took [borrowed].

On one of the days, this man needed five gold coins, so he went and asked his neighbour for them. His neighbour knew the habit of not giving back what this [man] takes. He tells him, "There aren't any." On the spot, he got angry and left.

On the way, he met Hadji* Amesha-aga and told him his problem: "I need five gold coins."

The Hadji [says], "Very well, I will give you five gold coins," he says, "but I will get them on the appointed day. See, here are the five gold coins for you; I will get them back tomorrow at noon," he says.

In short, the noon of the second day came, and Hadji Amesha saw that the gold did not come. She took her staff in her hand and came to this one, and he, sensing her coming, went to bed sick. His wife, fearing he would die, sent word to his relatives, who all gathered around him. In the meantime, the aga came, greeted him and entered, seeing that he was seriously ill.

What should he do? He would die and be left with his debts, he said], slowly approaching him, whispering in his ear, "So, what about my five gold coins?"

The sick man replies, "In such a state, you ask for money," and [the others] begin to beat the Hadji. So with great difficulty, he escaped from their hands and ran away, and the diseased, in his fear of the Hadji, said [to himself], "Whenever it may be, but until I die, I will not get rid of this Hadji," he has died pretending [he has become seemingly dead]. In that village, it was the custom that the dead were not buried; they had no cemetery. A cave was built, and the dead were put in that cave [embalmed]. This [man] was also embalmed and put in that cave.

When the Hadji heard that this one was dead, (361) he also became seemingly dead, thinking, "In place of the five gold pieces, I will at least take the shroud from there." They embalmed this one, too, brought him into the cave, and put him next to that one. They didn't know about each other or their faked deaths.

Night fell, and they both lay soundless, afraid to move. Once they looked, the door of the cave creaked open, and out of the corner of their eye, they saw five robbers [come in], holding a bag of gold and a sword, and they sat down. They had robbed the house of the Padishah of the jinn in that village. And next to the Padishah was a champion of the race of devils called Gench Amed*. They killed that hero and took his sword. In his room where he lived, they found a bag of gold and took it. In short, "We should share the gold in a shady place," [they said and] came to the cave.

One of them asks, "First of all, who gets the sword?"

The other says, "He who is the hero will have it."

"And how will you know who is the hero when you cannot wrestle here?"

"Wrestling is not necessary for that; here, the dead man on the edge is put on his feet, supported, and whoever with a sword stroke [will cut its head] gets the sword. Now, heroism is like that."

Nonetheless, [. . .][8] the one on edge was the corpse of the seemingly dead Hadji. And the Hadji in his fear began [. . .][9] get and give. Two of them wanted to take the corpse of the Hadji, and as soon as they said, "Let's now look for one," the Hadji, in his fright, cried out, "Hey, dead men, what are you looking at and sitting still, not getting up?" (362)

[Hereat] the dead, who had firstly pretended to die], also spoke up: "Well, what about our debt," he cried out, and sat up on his bottom. These perceived as if all the

8 Crossed out in the Ms., "envious devil".

9 Because of the missing part of the text here, the meaning of the sentence is unclear.

dead had risen, and, looking at nothing, they ran away, trampling each other. Both the gold and the sword remained in the cave.

The two of them laughed together for a long time, and then they closed the gate. Then they began to share the gold. A handful for one and a handful for one. In this way, they have shared.

Hadji [says], "Now, what will happen to your five gold debts? Give me that too."

Hereat this one got angry, "You beast," he says, "have you no honour, you are the owner of so much gold and your eyes still desire it?"

In short, let these two argue among themselves; let us come to the robbers.

They also began to scold each other, "You can't joke with the dead. Thank God they didn't chase us; if they had chased us, we would have been in trouble."

And one of them said, "I don't mind anything else," he said, "only the sword is left there, I mind that most, what shall we do to take that sword," he said, and one of them took courage and said, "I will take it," and returning he came near the cave cautiously.

He sees that there is a great noise [quarrelling] inside. They are still quarrelling. This [robber] was afraid to go carefully through the gate. [The cave] had a window on the side. He says, "I will go carefully and look through the window." He stretches out his neck and looks in; only a big gold button is at the end of this fez's* tassel. From inside, he, who had not given back what he had taken, saw this, immediately reached out for it (363) and tore it off and threw it at the Hadji, saying, "Look, here are your five [coins of] gold."

This brave robber was very much frightened and ran away, scarcely breathing, when he came to his companions.

"Is that something that you have seen?"

"All the dead now came to life and began to share the gold one by one. One of them was missing a gold coin; [that's why] he was beaten to death. I thought I'd go and climb up and look through the window. Then they drew the sword and tried to cut my head off, but it missed me; only the gold at the end of the tassel of my fez* was hit, and it was cut off, so I ran away. If you don't believe me, look, the gold is not here at the end of my tassel," he said. These disappeared in this way.

10 Ibrahim Padishah

(368) Ibrahim Padishah had a Pasha called Abdurreshid. And the Pasha had a son named Abdusadik. Wherever the Padishah went, he brought Abdusadik a gift. Once he went out again, he brought him a gold box. In it, there was an image of a daughter of a Padishah. The Pasha did not show it to his son but kept it in the treasury. The boy wondered, "Whenever the Padishah had brought a gift, he had always given it to me, so why didn't he give it to me this time?"

One day, the lad stole the key, opened the chest in the treasury, grabbed the picture, immediately fell in love, and fainted. After a good few hours, he pulled himself together, got up and went home, but no matter what, he didn't let anyone know. She walked around for a long time thinking. The Pasha's servant asked the young man, "Why are you so yellow [pale]; what is wrong with you?

"Is the owner of this picture alive?"

"Oh, that's easy, tell your father we're going on a trip."

The boy asks his father to have a [steam] boat made; the boat is ready. They have prepared the necessary amount of provisions for the trip, and he asks permission to go.

His father [asks], "Who will you go with?" asks him.

After the lad said, "With the servant," he agreed.

"Go," says the servant to the lad, "ask the Padishah where the owner of the picture is."

The lad enters the room of the Padishah, and sees that the Padishah is eating a fish; the bone of the fish is stuck in his throat and is about to die. The lad did not go in [to him] but went back. These get on the boat and leave.

Do they go, do they go, a little, a lot, six months, an autumn? The lad's father has learned **(369)** that his son has gone to seek the owner [i.e. the sitter] of the picture, but cannot bring him back, "Let him be entrusted to Allah!" he says.

They go, they go, their water runs out, and the lad goes to a small house on the shore to ask for water. He sees an old man living there. The old man asks the lad what business he has come there for. The lad pulls out the picture. He says, "This is what I'm going to look for."

"This is the daughter of Memmet Padishah," says the old man, "but I don't know if you'll find her."

The lad gets water, and they continue their way.

They go, they go, then went to the shore again, they moored the boat, and they saw a lad fishing. The lad says to the fisherman, "A handsome lad like you, how much do you earn from this fishing? Who are you selling them to?"

The fisherman says, "Hey, my friend, I am selling these fish to the daughter of Memmet Padishah." The lad took out the picture and showed it to him. The fisherman said that was the girl. The lad asked the fisherman for a room, and the fisherman took them to his house.

This girl goes for a walk two days a week. This girl went out in the garden again. As this lad was sitting in the garden, [somehow] three apples found their way into his pocket. The lad came back after his walk and told to the fisherman's mother. The old woman said, "Oh, my child, this girl whom she loves gives him three apples: it seems she has loved you, but his mother and father will not give him to you. I will prepare a talisman for you; before they eat, they will go for a walk, and you [then] go and put it in the cauldron [pot, kettle]. After they have eaten, they will lie faint **(370)**. Meanwhile, you go in, take the girl and run away with her."

So the lad did, took her, ran away with her, got on a boat and left. The girl's mother and father woke up and saw that there was no girl; they didn't know what had happened. The lad returned home with the girl, had a wedding feast, married her, and their wish was achieved.

11 A Couple's Tale

(374) [It was the time in time: the camel was a crier, the frogs suddenly grew wings to fly, the sea fish hired [a cart] to migrate, the minaret of Akmedjit bent down to drink water from Salghur*, the commander's foxhound [rolled up] his trouser leg to cross the Black Sea].[10] That is a lie, and this is a lie; a snake swallowed an elephant; I sat on the rabbit's back; I took the camel in my lap; "Is that a lie," he said; it's called a fairy tale; hearing it will taste good; whoever doesn't listen, so be him like this, so be him like that.

In the time before time, a rich man had a son, but a little bit was backward in his mind. This lad was well into his forties. His father said, "I should at least marry this boy." [They were searching] this way and that way, [finally] they found a girl like himself. Immediately they started to prepare for the wedding, and the boy was married. But she could cook nothing, and neither could he. One day they hung a cauldron over the fire, boiled the water and threw flour into it. They don't know how much salt to put in it. The daughter-in-law [young woman] said, "Go and ask my mother how much salt to put in the kettle and come back."

The boy went away and came to the [girl's] mother. "Mother," he says, "we are cooking a meal, but we don't know how much salt to put in it."

"Oh, my son," she says, "in a big kettle, two handfuls; in a small kettle, one handful."

The boy [says], "I forget," he says, and on the way, he repeated: one handful, two handfuls, and hurried back. By the roadside, a man was threshing grain, and as he came near, he said, "One handful, two handfuls." **(375)**

"Hey, man," says the thresher, "Is it customary to say, 'One handful, two handfuls', when threshing?" and he hit him a few times with his fist.

"Well, what can I say," he asks.

"Say: 'let it be abundant [blessed],'" he says.

The lad goes on and repeats, "Let it be abundant [blessed]."

There was a cemetery on the side of the road, and in the cemetery, a dead man was being buried. This one says to them, "Let it be happy [blessed]. The boy was [well] beaten [for it].

"May Allah have mercy on him!" they say [to him].

The boy again starts saying hurriedly, "May Allah have mercy."

As he continues his way, he comes across a [wedding] feast on the side of the road. "May Allah have mercy," he says.

"Hey, you guy, do they say may 'Allah have mercy' at the wedding," he is told, and got several fist punches again.

"Well, what shall I say?" he asks.

"That you should play [make music] yourself; dance yourself."

He starts to say again, "Play [make music] yourself, dance yourself," and he goes on.

---

10 A part of *tekerleme* being crossed out in the text of the tale.

At the edge of a garden, a hunter is holding his rifle, about to shoot a rabbit. "You [make music] yourself; dance yourself," he says as he comes down the road.

The hunter's prey is [thus] scared away.

The hunter caught the lad and struck him several times with his fist.

"What can I say?" he asks him.

"Walk crouching [hiding]."

The boy again [repeats], "Crouching," and goes on crouching. There was a vegetable garden by the road.

The gardener saw this [going crouching], went right out in front of it, caught this one, and beat him up good. He thought he had stolen from the garden, but he saw that he had nothing. He got rid of that too and came home; it was just about midnight, he started knocking on the door, but his wife wouldn't open it. He went to the mosque and laid down there. It was also very cold **(376)**.

He stayed in the mosque until the morning. In the morning, the muezzin comes, opens the gate of the mosque, and sees a man lying there. "Get up," says the muezzin, "get out of here; why are you lying there like a red beet?" By the way, the boy's wife's name was Red-beet. When he came in the evening and knocked on the door, he didn't remember her name; that's why his wife didn't open the door. When the muezzin called him a red beet, he began to repeat, "I have found, I have found and ran home."

The muezzin thought he had found something; he also hurried after him and asked him, "What had you found?"

The boy came home and shouted, "Red-beet, open the door." The woman got up and opened the door. The lad sees that the kettle is still over the fire. The boy takes the rolling pin to stir it and begins to stir the cauldron. Meanwhile, the muezzin catches up with him. As soon as he sees the rolling pin in the lad's hand, he turns and runs away. The lad also jumps up and runs after him, thinking he has stolen the bulamik* from the cauldron.

The muezzin runs, and the lad chases him and says, "Stop, muezzin, let me dip [the rolling pin into the cauldron]." The muezzin thinks he wants to beat him and runs on. The muezzin goes into his house, and after the muezzin, the lad also goes into the muezzin's house. Now, the muezzin's wife thought the muezzin went to the lad's house because of some bad deed. As it was, the muezzin was not a very good man either, so he started to quarrel with his wife. This was said, and that was said, the quarrel became great, the muezzin's wife divorced him, **(377)** the muezzin was left a widower, and the lad turned back and went home.

Two days and one night, they were cooking a cauldron of food, but they couldn't eat of it. When evening came, they were invited as a young bride and bridegroom to a feast. The mother of the girl says, "Daughter, in the place where we are going, do not be ashamed to eat; when it is time for you to stop, I will signal to you, and you stop, [for] one eats little being a guest."

Then they went and sat down, and after the usual greetings, questions, coffee and smoking, they were in a good mood; they brought in the sofra* and they all sat down to the sofra. A cat walked past the daughter-in-law and nudged her [leg]. Now she retreated [. . .][11].

12 The Fisherman's Tale

(382) Once, there was a fisherman. He caught fish and lived by it. One day he caught a fish in his net: half girl, half fish. This lad takes it and gives it to his mother. His mother put her in the cupboard of the house. Then every day, in due time, they get a cooked meal, and they don't know who made it. The lad tries to look, but he cannot see, and then he says, "If you are a genie, also come out; if you are a man, also come out." A girl comes out. As soon as he sees the girl, he falls down fainted. After a few hours, he stands up and embraces her, but she says, "Don't embrace me without a wedding; go and arrange the marriage, then you are mine, and I am yours."

They married, lived their lives for three or five days, and had plenty of food and drink, but their house was old. The lad says, "Thank God, I have plenty of everything in the world; I need a house."

The woman says, "Go to the sea, and shout, 'Mother Fatma!' She will come out, and you ask her for a peanut shell, but she will say to you, 'Come, I will give you,' but be careful: don't go; if you go, you will perish; if you say, 'Throw it here, I will take it', she will throw it there, and you will take it."

The lad goes away; he has brought the peanut shell they threw on their house's roof. The next day, as soon as they wake up, what do they see: a palace even better than the Padishah's palace.

After a few days, the lad says to his wife, "Don't take these fish scales on yourself."

"No," says the woman, "if I leave this dress, everyone will see me, they will become enemies to you, you will have no peace."

The lad gets angry and burns the dress in the fire.

One day, the lad invited the Padishah as a guest. He was told that he would come with his Viziers. The Vizier says to the Padishah, "This woman **(383)** is only worthy of a Padishah. We will give this lad a difficult task, if he does not do it, we will cut off his head, and we will take his wife."

The Padishah summons this one and says, "In three days, build a palace like no other in the world; if you don't do it, your head will be cut off."

The lad thinks and tells his wife.

"Did I not tell you," says the wife, "not to burn my clothes; you did not listen to me. But don't worry, come on; let's hug each other and go to bed, and then we'll see."

The next day they got up, and the woman said to her husband, "Go and ask Mother Fatma for a nutshell and bring it here." The lad went and fetched it and threw it into the

11 Missing end of the tale.

courtyard of the Padishah at night, and in the morning, they saw a palace like no other in the world.

The lad came and said, "Is it to your liking?"

"It is to my liking," said the Padishah, but the Vizier stood up.

"In this palace, you must have a bell of the genie;" he said, "you have forty days; if you don't fetch it, we will cut off your head."

The lad went in great fear and told the woman.

"Don't be afraid," says the woman, "come, let's lie down and cuddle." The next day they got up, the woman made provisions for the journey and said, "You will go, and before you reach the forest, you will go to a thorny place, you will run through it barefoot without fear; you will come to a spring, there you will say bismilla* and drink water. Behind the spring, there is a gate; you open it, you see a sheep, you take the meat from it, you open another gate, you come to a dog, you take the hay from it, you give the meat to it, and you take the hay to the sheep. You open another gate, and there is the genie; you easily take the bell from under his cushion and run away."

The lad went and got there and took the bell.

"Take it!" the genie says to the dog. The dog didn't take it. **(384)**

"Forty years I've been here," he says, "you didn't take the meat from the sheep to give it to me."

"Sheep, take it," he cried, but it also said, "forty years I've been lying here, you didn't take the hay from the dog to give it to me."

"You take it, spring," he cried, and she said, "forty years since you have had me made and never once came to drink water." So none of these took him, got through them, brought and gave it [the bell] to the Padishah.

The Padishah liked it, but the Vizier says, "Now he wants the daughter of the genie Padishah." They gave him three days' delay and sent him on his way. The lad went to the woman and told her. The woman, "That's easy," she says, "come, let's lie down cuddled together," she says.

In the morning, she gave him a lecture on how to get there and how to bring it and then sent him away. The lad went, got to the woods, went into a house, and there was a woman, throwing her two breasts over her shoulders, sifting flour. He immediately rushed up to her and began to suck her breast, then greeted her and sat down. The woman [says], "You became my milk-son, she says, but your brothers will be here soon, and if they see you here, they will eat you," she says, and she conjured the boy into a findja* and put him on the shelf. The devils came and sat down.

One of them [says], "There is a human smell here, he says. Is it usual to find a man here?" says their mother.

A little while later, another asked the same question, and another pushed. Their mother tells them this boy sucked her breast and became a milk son.

"Show us!" they say. "We will not harm him." She took off the findja and laid it on the ground. The lad saluted them and sat down. They asked him what business he was on, and he told them where he was going. Each of them gave him a hair from their heads and said, "If you get into trouble, light it, **(385)** and we will come to your aid."

The lad went on and met a shepherd, who was playing the flute. "Salam*, Ahmad shepherd," he says. The two became friends and went right on. [They met Ahmed, who was able to smooth hills, and took him along with them.] From there they went on and met Ahmed, who cannot satiate himself with water and took him along with them. They went a little further, and met long-legged Ahmad, and the four of them mounted him, and in two leaps came to the house of the jinn-Padishah. They said that they had come to propose to his daughter.

The genie-Padishah says, "Very well, I will show [give] you three tasks; if you complete them, I will give her to you. If you catch up with my daughter running, drink the water of this well, sort out forty sacks of beans and barley in one night, then I will give her to you." The long-legged Ahmed caught up with the girl at a run, the second Ahmad drank the water out of the well, and Ahmad, who was able to smooth hills, called the ants together saying the prayer of the ants, and the ants separated the barley from the beans by morning. The Padishah had had food cooked for them and put poison in it; he wanted to kill them. Ahmad, the flute-playing, smelt it, and after the sofra* was brought in, the shepherd saying that it is customary to play the flute before eating, made the flute cry once, and the Padishah fainted. The shepherd takes the food from the Padishah and [instead] puts the poisoned food in front of him. The Padishah got up, ate, and died.

They took the girl, put her on the back of the long-legged Ahmad and ran away, and when they got close to the village, the four of them went back. The lad alone brought the girl to the Padishah, and the Padishah liked her, **(386)** but the Vizier says, now bring the mother of the forty foals. They gave three days delay and sent the lad on his way. He went home and slept with his wife.

In the morning the wife says, "Here, take this necklace; when you get out to the meadow you will come to a spring; you put this necklace by the spring; the mother of the forty foals comes out, and while she is drinking the water she wonders about the necklace, you then quickly come out from behind the spring, sit on her back, she wants to throw you down, but you grab her mane firmly and don't let go. If she shakes three times and can't throw you off, she surrenders; you sit on her and come."

The lad went and fetched the horse and gave it to the Padishah. The next day, the Vizier comes and says, "Milk this horse!" The lad came to his wife and told her; he and the wife went to bed; they got up in the morning; [the wife] said to her husband, "go and milk her, then they will tell you to boil the milk and go over it. Take this ring, and as soon as the milk is boiled, do not be afraid; you will go through it."

He went, milked her, boiled it; they told him to go through it, and he went through it. Then he said to the Padishah, "I have done as you wished; you too must not refuse my request and go through the milk."

The Padishah immediately wanted to go through and was boiled in the milk. When he told the Vizier that you should now go through, the Vizier asked him for the ring. He became angry and threw the Vizier into the milk by force, he himself went back (to his house), and he and the wife lived in peace.

13 Ahmed-batir's Tale

(393) There was once a Padishah. The Padishah had neither son nor a daughter. Not missing one of the five namaz*, after [each] namaz he wished to have a child by Allah.

One day, a dervish comes to him and gives him an apple and says, "Eat half of this apple yourself and give the [other] half to your wife." After they ate the apple, the woman became pregnant. After nine months and ten days, she gave birth to a child every day for forty days. The woman said, "You have given my husband forty children, give me a son separately, but let him be of the goodly [lucky] ones."

One day the dervish comes again and gives the Padishah's wife an apple, half of which she eats herself, and half of which she gives to her husband. The dervish says, "When the child is born, I will come and name it myself." The child is born. They took him, put him in a cage, and at the time of naming the dervish came and named him Ahmed-batir, and went away.

The forty sons who had been born before grew up and came to their young manhood and said to their mother, "Tell our father to marry us."

Their mother says to them, "I cannot tell your father just like that. Tomorrow at noon when you come to eat, you don't eat; you leave your spoons, you get up and go. If your father asks why you have not eaten, I will tell him."

Their mother tells the Padishah.

The Padishah says to her, "Where can I find forty Padishah daughters for all forty of them; if they are [so big] lads, let them go **(394)** and find them themselves." The boys got permission from their father and went to look for a girl. Let us come to Ahmed-batir.

That day Ahmed's midday meal was served late, so he became angry, broke his cage, and asked the servant who brought the food: "Why did you bring it late?"

The servant replied, "You have forty brothers, and they went to find a girl for themselves; we saw them off, that's why I came late."

He said to the servant, "Take me to my mother and father!" The servant took him there. Before they got to their house, the dervish caught up with them. This lad entered their house, kissed his mother's and father's hands, and asked permission to follow his brothers.

"No, my son," says the dervish, "you will be eighteen in three days, and then you will go."

"No," said the boy, "I will go now."

"Very well, if that's the case," [says the dervish], "from today we'll sleep here [stay overnight here]; tomorrow you'll go out to the south side of the village, there's a hill, you'll greet the hill [you'll say salam* to it]; when the greeting is received the gate will open, you'll go in, there are your saddle and your arms, you'll bring them to me. The next morning, you go to the north side of the village, greet [say salam to] the hill there, the gate opens, your magic steed is there, you bring it, and on the third day, you set off."

The lad brought these, and on the third day, having asked his father's consent, mounted his horse and set out. He made the journey his brothers had made in forty days **(395)** in a day, caught up with them, let them know that he was their brother, and went on with his brothers, arriving in the evening at a grassy place. Ahmad-batir says to his brothers, "Let us stay here for one night." In this place, they let the horses go free and lay down [to sleep]. At some point in the night, Ahmed roused [from sleep], lifted his head, and saw a light in the distance. "Go and see this light," he tells himself.

He got on his horse and went to the light; he saw that it was a palace, with no gate or window to go in. He slapped his kamchi at the steed and made him kick once. From within, someone calls out, "Is that you, Ahmed-batir?"

"Yes, it's me; open the door," he says. The door opened, and he saw an old woman; that woman had given Ahmed-batir forty keys and showed him forty rooms. Ahmed opened all forty, [leaving] only one room. He came and asked the old woman for the key to that room, and he opened it and saw that it was full of beads. He came to the old woman again. "Take the keys, you have waited for a long time, wait a little longer, I will come back [for you] and take [you] away," he said, and came back to his brothers. He did not tell anything to his brothers.

The next day, they got up and continued their journey. In the evening, they arrived at another grassy place and laid down [to sleep]. In the night, the batir heard a sound again, and lifting up his head, he looked around, and saw that there was a light in the distance, and he went there, and there was a palace, and he went in, and an old woman said to this [him]: **(396)** "See, my son, here is your deposit," and handed him a key. She showed him a room, the batir entered, and what did he see: forty girls reading [studying] in it, the forty-first standing in the middle, watching [keeping an eye on] them. The forty girls for her brothers, the one for this one [himself]; he says nothing to them too; he gave the key to the woman, returned to his brothers, and they go on again.

In the evening, they came to a grassy place again, and they stayed overnight. In the night he roused [from sleep] again, and again, a light was seen in the distance. He mounted his horse and set off. On the way, the steed says, "Get off me,; while going I will be a little stubborn [I'll stop], but be careful not to hit me, because if you hit me, you will be exposed to danger."

Ahmed sat on the horse, not intending to hit him. The steed suddenly resisted, and he hit the horse's front shoulder blade with the kamchi [whip]. In a flash, a forty-headed devil comes and swallows him. But he immediately spits it out, saying, "Oh, you son of a bitch, what are you doing here on my land?"

Ahmed says, "Oh, my sultan, I have erred; forgive me my sin."

The devil says, "If you go and fetch from the genie's Padishah the World-beauty, I will forgive you."

The batir was given a few days' respite, and went away, went to his brothers, gave them the girls and the wealth, and sent them home, and he turned back, and the steed threw the batir off his back again, and said, "The devil will not give me to you, but you must not part with your weapon; see, here is a hair; if you get into trouble, burn it, and I will come to your aid."

The devil gave him a black horse **(397)** and took away his steed. The batir set out with the horse, and after a few days, he reached a forest, where he tied his horse to a tree and went to sleep. In his sleep a voice struck his ears; he stood up and asked, "Are you a son of man, or a genie?"

"I am not a genie, I am a son of man, like you, come to this side," he says.

He went there.

"Did the devil send you?"

"Yes," says the batir, "he sent me to bring the World-beauty."

"So then you will be like me; I was like you; my horse ran away leaving me, and the genies put out my eyes."

"No," says the batir, "if this horse runs away, my own horse will come."

They slept there that night. In the morning he got up and burnt the hair of his steed. The steed came, and on setting out he [the blinded batir] says to the lad, "If you have got the World-beauty and are going back, pass here, I will die, you will bury me, and then go on." They bade each other farewell, and he [the batir] went away.

He went and went and got to the city of the World-beauty, stayed in a café, and learned from the lads there that the World-beauty was having her wedding. The next day he got up and went to the baths where the World-beauty went and hid. The World-beauty came with forty maidens. Forty went forty ways. This Beauty went next to the batir. These [they] fell in love.

[The lad said] to the girl, "Tomorrow, you ask your father's permission to go to the garden; from there, I will take you, and we will escape."

The girl asked her father for permission and went with the girls to the garden. The batir took this and ran away; they came [to the blinded batir] and saw that he was dead. He [the batir] buried him and took the girl **(398)** to the devil. The devil forgave the lad. Now the girl says to the devil, "One day since my father died, I will mourn for forty days, and then we will be bride and bridegroom, and you shall make a cage of forty pounds* of iron, so that when you go away, lest someone comes and takes me away."

The devil had such a cage done, and he himself went into the meadow [steppe]. In the evening, as soon as he came back, she said to the devil, "If you go into the woods, I am afraid you will be killed."

The devil saith, "No, do not fear, my soul is mine, in thy father's meadow is a chest, in the chest is a goose, under it are three eggs, one of which is the strength of my waist, one of which is the apple of my eyes, and one of which is my soul, no man can take it away."

The girl told this to the batir, and the batir mounted the steed and went away, he came to the field, and around the chest, a fire [flames] was burning to reach the sky. In the city, he took three orders of garments, one of wax, one of felt, one of iron. He took these, mounted the steed, struck one with his kamchi, and passing through the fire, broke the chest, took the three eggs, broke one on the spot, the devil's strength went out of his waist, and the other two he brought to the girl.

The devil saw the batir and said, "Give me those two, and I'll break them in my mouth, and you'll have my fortune." The batir was about to give it to him, but the girl slapped it with her hand on it, so she broke another egg, and the devil's two eyes went blind.

The batir says, "Make a talisman of your wealth, make an apple and give it to me." The devil did and gave it. The lad put the apple in his pocket, and they returned to the girl's home. He met his mother and father; they had a forty-day wedding and lived happily ever after.

14 A Tale of the Magic Ring[12]

(405) Once upon a time, there was; once upon a time, there wasn't a poor woman; she had a son too. They had no bread to eat. His mother knitted one stocking a day, sent it with her son to the bazaar and sold it, and with the money, she had some bread bought.

One day he takes the stocking and sells it for fifteen kapeks*. One child was selling a kitten; he bought the kitten for fifteen kapeks. On the second day, he bought a hound, and on the third day, a small snake. His mother says, "We were two souls, now we are five souls, we have no bread to eat, my child, what are you doing with these?"

Her child says, "Mother, God will determine the kismet* of these."

The little snake started speaking, "Lad, when the children beat me, you delivered me; I will repay your kindness," it says.

"Get up, come with me to a hill, my father and mother are there if you take me and bring me there, my father and mother will be very happy, they will say with joy that we will give you livestock that cannot fit on a pasture, money that does not fit in a house, but be careful not to take it. My father has a ring on his finger; ask him for it."

Then they go to the hill and enter the snake's den. The serpent's Padishah, with a golden crown on his head, and his serpent's Viziers welcome this lad. When he sees his own child coming with a lad, he asks, "Who is this?"

The little serpent replies, "I was saved from the beating of children by this lad, and for a long time, he fed me with his own bread to eat."

The father of the snake says, "You have delivered my child from death; **(406)** desire what you will of me: I will give you livestock that cannot fit on a pasture, I will give you money that does not fit in a house."

12 The title of the story had to be changed. Only in the Tartar title of the tale is the word "Ring-maker", "Ring-seller" used, not elsewhere. In the fable there is a Jewish watchmaker who also sells rings, but he is only a minor character!

The lad says, "I don't want your livestock that cannot fit on a pasture, I don't want your money that does not fit in a house: if you give something, give me the ring on your finger; if you don't give it, good-bye," he says.

He saith unto his serpent friend, "Come, let us go!" They are getting up to go.

The snake's mother starts crying, "My child doesn't seem [isn't worth] as much as a ring, I haven't seen my child for seven years; don't you feel sorry for your child at all?" As soon as she says this, the father calls the lad and puts the ring in the lad's hand. The skill of the ring is that once licked, forty bandits come forth, and what you wish will be done. The lad stays there for a few days and then returns with his snake friend to his own house.

His mother greets him and asks him, "Where have you been?"

He says to his mother, "Go and ask me for the Khan's daughter."

"My child," says his mother, "we are poor; the Khan will not give us his daughter."

"Whether he gives her or not, you go," he says.

The mother goes to the Khan with a shuffle of her slippers. "My Khan, sire, I have come to you, falling on my knees before you; I have a lad child, I want to ask your daughter for my son."

Going out, the Khan orders the treasurer, "Give this poor aunt a handful of gold."

The poor woman [says], "I have not come for money; by order of Allah, by the word of the Prophet, I have come to ask your daughter for my son," she says. Hearing this, the Khan becomes angry and orders this old woman to be cut into forty pieces the size of her ear, put into a sack, and taken to her house. The batirs cut her into forty [ear-sized] pieces, put her in a sack and take her home.

The lad, seeing the state of his mother, **(407)** licks his ring; from the ring, forty bandits come out and ask, "Shall we build, [or] shall we destroy?"

The lad says, "Make mother as she was before, just healthy."

They immediately do as he says.

Once more, he sends her to the Khan; the Khan sees that this old woman is whole [intact] again, so he decides, "I will marry my daughter to your son, but from the threshold of my palace [to the other palace, everything] will be covered [spread] with gold, and he will have a palace as mine built if in forty days there will be what I said, I will give her to you, if not, I will cut her neck like a cucumber," he says.

The poor woman comes home and tells her son what the Khan said. They had nothing but a poor house, which was like a henhouse. In the meantime, thirty-nine days pass, and nothing is seen in the middle. On the fortieth day, at night, the lad licks his ring, and the bandits come out and ask, "Shall we build or shall we destroy?"

Then the lad says, "From my doorstep to the doorstep of the Khan, cover with gold, and build a more beautiful palace than that of the Khan's."

In the blink of an eye, all that he said was ready.

The Khan wakes up in the morning; what do his eyes see: the Khan's palaces are as big as a hen house next to the lad's palace. The Khan looks, but he cannot stand seeing it [the bigger palace]. He summons the lad and holding a wedding for forty days and forty nights marries his daughter.

The girl and the lad lived together for a long time until a cunning Jew* found out what a ring of great skill this fellow had. [Otherwise, he eats a whole roach every day, but now he can't even eat half of the fish] because [he is] thinking, "How could I take that ring?" (408) The young wife had no knowledge of the ring's skill. The Jew has a ring made just like that one. She gathers up another bunch of old rings and comes to the lad's palace with the ring to sell. The lad has left his ring at home and goes to a meeting. The Jew finds an excuse and comes to the young wife, "I sell cheap rings," he says.

The young wife sees the ring on his hand [finger] and says [to herself], "Is this our ring?" She goes and looks at [fetches] the ring from the bedside; you can't tell the difference. And that's exactly what the Jew expected. Instead of his own ring, he takes the ring belonging here.

As soon as he takes it in his hand, he licks it once, and forty bandits spring out of the ring, "Should we build or destroy?" they ask.

"Beyond the open sea," he says, "where no man goes, I desire a palace forty storeys high, and this lad wrapped in a felt [felt sheet]; put him in front of the threshold, and these palaces you shall destroy" he says. In the blink of an eye, the forty bandits do everything the Jew says, the lad is wrapped in felt, and the Jew is taken away. In the blink of an eye, they arrived at the palace behind the open sea. Everything the Jew's heart desired, he found.

Let us come[13] [back] to the poor auntie's son. After a while, he woke up from his sleep, and what he did see was there were no palaces, no ring, and he didn't know what to do. He looked there, he looked here, [his strength was gone from hunger].

The hound and the cat gather in one place, talking to each other: "We have grown up on our master's bread, and we must take back his ring **(409)** wherever it is." The hound and the cat immediately set out and start going on their way. They go, they go, and then they stop when arriving at the shore of the sea. The cat sits on the hound, and they cross the sea. They travel a short distance [on foot] and arrive in the realm of the mice. The cat begins to drown the mice he sees, one by one, and the hound piles up the dead ones. One or two of the mice slip out [from his clutches] and run away and give the news to their Padishah. They say, "Our Padishah! Our Azrail* has come, has killed Ali-pasha, has killed Veli-pasha, what shall we do? If this goes on, not a single living one will be left."

The Padishah says, "Ask them if they are lacking [anything]."

They go and ask.

The cat says, "There is a palace beyond the shoreless sea, there is a Jew with a ring, if you bring it to me I will not touch you, [otherwise I will not leave a piece of you]."

One mouse says, "I will go," then a blind mouse and two more single mice say, "We will go too," and they leave immediately. When they get close to the palace, they start digging a tunnel. After a while, having bored the tunnel, they come out of the wall and see the Jew lying with the ring tied to a thread, tied to the ceiling [hanging from the

---

13 In this narrative turn, Tatar storytellers usually use the Tatar verb kel- 'to come', which is quite foreign to the Hungarian fairy-tale texts, so in the translation we use the word 'go' instead of 'come'.

ceiling], and sleeping with the ring in his mouth. The blind mouse, thinking about what to do to take it away so as not to notice, sees that there is red pepper on the shelf. He urinates a little in that place, wets his tail, and dips it in the pepper, immediately goes to the Jew, and sticks **(410)** his tail in the hole of his nose. The Jew* sneezes, and in his sneeze, the ring flies out of his mouth; [the mouse] climbs [to the ceiling] and cuts the thread with his teeth. At that point, the blind mouse catches the ring and runs away through the tunnel, and the Jew keeps sneezing.

Let us come to the cat and the dog. They are waiting. As if they were waiting for the news of the feast, and suddenly they see the rogue, blind mouse bringing the ring in his mouth. Now the hound and cat take the ring, and the mice are set free. Then the hound and the cat get to the shore of the sea.

One says, "I will carry it across;" the other says, "I will carry it across."

"Thy tooth is rare," says the cat, "and thou dropest it;" and the hound says, "thou canst not swim, and that is my trade, and I will carry it," he says.

After bargaining [wrangling] hard there, they drop the ring into the sea. Poor things, they both walk up and down the sea for a few days. One day, a fisherman comes and throws his net into the sea, a fish gets caught in the net, sees the hungry hound and the cat, and gives them his fish. As they are eating, the ring comes out of the part the cat ate. The hound sees it, quickly takes it, and says to the cat, "Sit on my back!" It sits on him; they cross the sea, take the ring to their master, and give it to him.

He immediately licks it; the forty bandits come up [and ask], "Should we build or destroy?"

"Now you must not destroy; you must build," he says. **(411)** "Kill the Jew and dry him up like a roach. I want palaces ten times more beautiful than the former."

The bandits [say], "With pleasure," they say. They do everything in the twinkling of an eye. The lad lives again in the former manner, and the Padishah, in shame, drinks a cup of poison and dies, this lad being deemed worthy of being the Padishah in his place. The lad becomes the Padishah, and as the Padishah, he lives happily for a long time, having achieved his goal. May Allah make our wishes come true.

15 The Tale of Shangeri Khan

(419) One day, the Khan was bored, and he had an adviser with him. He called him and said, "Call Nogai-Ahmad[14], I will talk to him a little, but he should not come on horseback nor foot; he should not come in nor stay out." They have sent him the word.

Ahmad-aka begins to think, "How am I to go now?"

The son of Ahmad-aka asks, "What are you thinking about?"

---

14 According to the original text it should be interpreted as 'Nogai-Ahmad's son', but as the story goes on, it is clear that the Khan invited to talk not the boy, but his father.

“I can’t get anywhere with thinking; I have been in a difficult situation. I must go to Shangeri-Khan; he called me and told me not to come on horseback or foot.”

His son says, “It’s an easy thing, don’t worry; [give] me a camel cart [camel-drawn cart], ten children and a goat, a hundred [sheep] knucklebones, and I’ll go.”

He harnessed [the camels] into a camel cart, loaded the cart with the children, and they went to Shangeri-Khan. It was midnight. They unhitched the camels and let them into the courtyard. They woke up in the morning, and the Khan’s men saw two camels grazing in the courtyard and ten children playing knucklebones in the same place. The Khan’s men come up to them [in the courtyard], “Children, what are you doing here? Is there someone grown with you?”

“Yes, there is our grown,” they say, and point to the camel.

“No, no,” they say, “don’t you have a bearded one?”

When asked, they point to the goat.

The Khan’s man turned back, went to the Khan and told him their words. “Call him, come here,” says [the Khan].

Ahmad’s son mounted the goat’s back and went straight to the Khan’s gate, [the goat’s] front legs crossing the threshold of the gate, his hind legs left outside.

The Khan sees a boy sitting on the back of a goat at the gate. “Why have you come?” he asks him.

The boy replied, “You sent for my father, and I came instead. As you wished, so **(420)** I came.”

“Get off your goat, come and sit down, and let me talk to you a little,” he said.

“Son, isn’t there an old man in the village that you were sent here?”

“I came in my father’s place,” he says.

The Khan says, “You are a child. What have you seen, what do you know?”

The boy replied, “As I came along the road, I saw ten kids of goats, I saw twenty stallions, I saw thirty horses, I saw forty oxen, he said, I saw fifty sheep, I saw sixty sparrows,” he said.

Shangeri-Khan began to wonder what it all meant; these things he could not decipher, and he kept sweating.

The boy said, “My Khan, my lord, if it snows on the black mountain, how many days will water flow from that white sight?”

The Khan became even more ashamed.

The Khan had a sister. She was very smart, and she used to decipher such cryptic speeches. They brought her news: “Shangeri-Khan had been ridiculed by a child,” come quickly.

The girl came. “Sire,” she said, “he has told you the life of a man; a man is like a goatling at ten, a stallion at twenty, a horse at thirty, an ox at forty, a sheep at fifty, only his tongue has the strength left. On a black mountain, snow means his hair has turned grey, a spectacle means water, and sweat, and a stream of water means a sweating face.”

The Khan asks, “My son, how many men can you bear?”

“I can handle one man like me.”

The Khan gave them a lot of apples, pears, peanuts, and money, and filled a cart with them. They said goodbye and left.

After they had travelled some thirty chakirims*, the Khan sent forty pelivans* after them, saying, "Go, catch up with them, **(421)** kill the boys, and bring back their cart."

The pelivans caught up with them. The boys said, "What shall we do now?"

Ahmad's son said, "You sit in the cart, I will fall on them, I will show my courage," he said and threw himself off the cart. He smashed the head of thirty-five ones with a cudgel; five of them turned and ran back.

They went to the Khan and told him the news. The Khan says, "Well, well, tell him to come back."

They went after him and told him, "The Khan wants to see you."

He came back and met the Khan.

The Khan said, "You said, 'I will face a man like me,' and I sent forty men after you, and they could not have beaten you," said, "well done, my son," and sent him home.

The Khan himself was afraid of this boy and wanted to have him killed. He sent word to pelivan [i. e. Nogaj-Ahmad]: "Let them go a hundred chakirims* away, [for] the voice of the stallions is heard here and disturbs me."

The people are gathered and say, "Pehlivan, my dear, send the boy again to speak to the Khan. We are such a big village; where shall we go?" [they said,] begging.

The boy was told, "Go once more to Shangeri-Khan."

"I will go," says the boy, "but give me nine children and nine daggers."

They give the daggers to the children and let them go. They come to the Khan, and in front of them come dogs barking; the children kill the dogs. They catch the children and ask why you killed the dogs.

They said, "When your dogs bark, we cannot sleep in our houses."

The Khan said, "For thirty chakirim*, will the dogs' voices be silenced?"

The boy answers, "Will the sound of horses be silenced for thirty chakirim*?"

"Bravo, my son, you will become a man, go and be quiet in your dwelling place," he said, and they returned.

16 The Tale of Uf-arab

(426) In ancient times, a man had a son. This son would disobey his father. Since his father could find no [other] way out, he was about to give him to a guy as a servant. He took a loaf of bread from home, put it in his bag, took his son with him, and they set out on a journey and walked day and night until they came to a hill. [In short, this man was tired and had no choice.][15] The father said to his son, "Let us stop here, I am a little tired; let us eat some bread," and with a sigh 'uff!'* he leaned on the hill. But inside the

15 Crossed out in the Ms.

hill was a wizard Arab called Uf.* As soon as this man tiredly sighs 'uff!', the hill splits in two, and an Arab comes out of it.

"Why are you shouting at me," he asks.

This man says, "I did not shout for you."

And the other one says, "You shouted Uf."

"I said 'uff!' because I was really tired, that's why I said it."

And then they question each other, to make a long story short.

"I'm going to give my child as a servant."

The Arab, hearing these words, got very pleased.

"I'll take him," he said.

"How much will his yearly wages be?" he asks.

The man said, "Ninety pieces of silver."

The Arab, delighted, gave the money in advance. The man took the money and left.

The Arab took the servant and went inside. Inside, the boy saw endless, more than the many, longer than the long palaces. The Arab left this [him] and went away. The boy opened a door, went in, and saw **(427)** that a girl was embroidering. As soon as the girl saw this, she jumped up and embraced him. "Oh, my brother, how did you get here?" she asked. He also told her what had happened to him. She had forewarned him and then showed him through the window. He counted ninety-nine human heads, each on a stake [impaled] in a row. She had called his attention to these heads. Uf-arab had cut off ninety-nine heads, and if he cut off his head, he would have a hundred, and then he would marry this girl; they agreed on this. The girl says to this child, "That Arab will teach you magic tricks and will ask you if you know [have learned] them yet. If you answer that I have figured it out, he will cut off your head; if you say that you haven't [yet]; he won't cut it off. In short, however long he teaches you, and if you know it already, say you don't know it, it doesn't cut your head off, and the spell is broken."

The year is out, and the boy does not say he knows [has learned] something.

After the year is up, the boy's father comes to that hill again, says 'uff!', and stops.

The Arab immediately comes out and says, "What a good-for-nothing son you have?" takes the boy, and pushes him out by the neck. And he returns to his place.

The son tells his father, one by one, the teachings he has picked up from the Arab. His father does not believe him. From there, they run away.

On the way, they see a hunter on the road. The boy says to his father, "Father, I'm going to be a greyhound and start chasing a hare, and you're going to be a hunter. Only don't sell the silver collar around my neck, because if you do, I won't be able to come home **(428)** anymore."

"Good," says his father, gladly.

The child crosses his fingers, the father becomes a hunter, the son a greyhound, and begins to hunt the hare, but there is no greyhound like it in the world; he hunts so well.

The other hunter sees this and comes up to them and asks, "Will you sell the greyhound?"

And he says, "Why should I not sell it if you give me a lot of money for it," and he asks for forty pieces of silver. He counts out the money without a word, without bargaining.

The greyhound starts to chase the hare. And the father goes to his village. It will be evening; the hunter takes the greyhound, comes home, saying it's a quick meal; he starts baking cheesecake.

And the greyhound runs away, and comes to his father. "Have you seen, dad," he says, "how they earn money. Now I'll be a galloping stallion. There is a hunter muezzin in the Open-top village; you will sell me to him, as much as you say he will buy me for such money, only don't sell the silver bridle on my head, for if you sell it, the spell will be broken."

He has become an ambler stallion. His father sat on his back, but he was sitting there as if it was ordered [soldierly]. Then he went to Open-top village.

It was only two versts to the village when the horse's beating [hooves] could be heard. The muezzin, hearing the sound, came out and saw an ambler coming, but it was coming too. He starts bragging to the people next to him, saying, "If the uncle coming here sells it, I will buy it," he decided.

The man came to that place with the ambler **(429)** and, saying salam, greeted him. The muezzin did not even bother to return the greeting but immediately asked, "Brother, will you sell your horse? If you give me a good price for it, why should I not sell it," he said.

"How much do you want," he asked, and he greedily said, "A hundred and fifty gold pieces, but I will not sell the bridle."

The muezzin agreed and bought the ambler. "However, I want to mount on it; let me try it," he said; then he liked the horse's gait very much, turned back and gave the money.

The man went home with the money, and the muezzin went for a walk in the field [on horseback].

At the time of the afternoon prayer, when people were gathering at the mosque, the muezzin also came back and walked his horse in front of the mosque. There was a fountain with a spout [spring] at that place, and he thought, "I will water the horse too." He went to the fountain to water it. The horse starts to drink the water, and while drinking, it goes into the trough of the well. The muezzin was amazed and started shouting, "People, people, come here, my horse went into the trough of the fountain [in the well itself?]; let's pull him out," he says, begging.

"Foolish muezzin," they say; how can an ambler fit into the trough?

They say, "Our muezzin really is a horse fool, isn't he?"

"You still don't believe it," he said, "just listen to how it whines."

The people came quickly, and they heard it whinnying. They started to smash the fountain.

Let them destroy the well; the boy [meanwhile] came home to his father and said, "Father, I am going to be a bath, but there is no such a large bath like this one in the world. Put it up for auction and sell it, but don't sell the key to the bath," **(430)** he said. "The spell is in the key."

His father began to advertise. Wherever a horse's [hooves] could reach, people came from all over. The boy's former Arab master, called Uf, also came, they started auctioning it, and at the auction, the price went up to five thousand gold pieces.

Then the boy's father said, "I will not sell the key."

The Arab recognised his spell from his words. "You want five thousand," says the Arab, "I'll give you ten thousand, but give me the key too."

Money also has warm [tempting] eyes, doesn't it? His father, succumbing to the devil's temptation, agreed, "Go, enjoy the advantage and profits of it," he said.

When he held the key out to the Arab, that bath turned into a sparrow and flew up quickly. This Arab turned into a hawk and flew off after him, chasing him. The sparrow, confused, fled headlong in the form of a nightingale and flew through the window of the pavilion in which the daughters of the Padishah lived. The girls inside were running all over the place to catch the nightingale. The falcon got there too, but the window was slammed shut in front of him. The falcon had no gap left to get in. It started to scream.

He says, "A lamb is owed to me by the Padishah."

[The Padishah heard this sound, came to the place, and asked,] "What lamb do you want?" he asks.

"I had a prey," he says, "it escaped and went into the summer pavilion; that's what I want."

The Padishah went into the pavilion and asked the girls.

The girls, meanwhile, touched the nightingale. Thinking he would love [it there], they put it on the embroidery frame, but in the meantime, it became a written paper, on which it was written, "I am the Chinese Padishah's son, we had a quarrel about some **(431)** difficulty, he caught up with me in pursuit, if you give him a hand he will eat me," he says, "if you do not give him a hand but give me up and throw me away, I will get rid of him, and then I will marry one of you."

As soon as they heard these words, they began to be jealous.

The Padishah came. "Give him [his nightingale]," says the Padishah, "let him go." They took the paper and, without regret, wanted to throw it out of the window. But the wind turned and dropped it inside, and as it fell, it turned into millet and scattered. The Padishah and the girls were amazed. The falcon immediately turns into a brood hen with chickens and begins, knock-knock, to peck. A piece of the nightingale's millet is stuck in the embroidery frame. The brood hen finished eating the millet and then looked to see if there was any left. She didn't see the piece of millet stuck in the frame. That millet grain turned into a fox and started to eat the brood hen and the chickens. The Padishah wailed, and the brood hen shattered and turned into an Arab, all black-faced. The fox also shattered and became a handsome youth with a face as beautiful as the moon on the fourteenth day[16]. They told the Padishah everything that had happened to them. The Padishah laid all the blame on the Arab and threw him in a dungeon. He married his eldest daughter to the handsome fellow having forty days, forty nights of a wedding feast. Three apples fell from heaven, one for the teller of the tale, one for the listener, and one for me again.

17 A Tale of Fear

(438) [It was the time, the camel was a crier, the frogs suddenly grew wings to fly away, the sea fish hired [a cart] to migrate, the minaret of Akmesdjit bent down to drink water from the Salghir. The governor's foxhound rolled up his trouser leg to cross seven rivers.][17] Once, an old man had a son. And that son was a boy of about five or ten years old. The mother sent his son to learn. On his return from school, the children take his knucklebones from him.

He says to the children, "Give me back my knucklebones!"

"No," the children say.

Then he catches them and beats those children, whose hands he breaks, whose eyes he pokes out.

The fathers of the children came and said, "If you don't lock your child up, we will come and kill him."

And the boy's mother says to him afterwards, "Son, don't do such things; you will be intimidated."

The child says, "Oh, what kind of thing is that intimidation you said?"

His mother says, "Intimidation is that they will beat you and beat you up."

The child says to her, "Bake me provisions for the journey; I'll go and find intimidation."

The mother baked and gave him food for the journey, and he went to find intimidation.

It was evening, and he saw forty-one bandits sitting in a tent, talking. He greets them [saying salam] and goes in. They say, "Behold, roast meat has come to us."

The child says to their leader, "Let me see what your servant can do [cook]. Give me
(439) something to eat; my stomach is hungry."

They say, "Boil a kettle full of helva; you will eat; we will eat."

He asks, "Where shall I cook it?"

"You will cook it in this place;" there was a graveyard.

16 The full moon is the symbol of the beautiful face.

17 The scroll or text fragment is crossed out in the Tatar manuscript, but is included in the edition.

He hangs up the cauldron and starts cooking. [Suddenly,] a hand reaches into the cauldron; [he] says, "Don't joke with me; pull your hand out; if you reach your hand out again, I'll break it."

And he reaches it again. He catches and breaks it, takes the cauldron and brings it to the bandits. He says, "Come, let us eat."

"We also have a request to make of you," they say, "if you do whatever you wish, we will give it all to you."

"Say," he says. "Behold, here's a hammer and a nail for you; drive it into that house and come back."

The boy begins to hammer the nail, and a girl comes out, dressed up, girded, and adorned with gold, "My brother," she says, "your stature is short, I will lie down on the ground; you stand on me and hammer the nail."

The boy, without looking at her face, slaps her across the face. The girl goes down seven stories to the ground, and milk sucked from her mother comes out of her nose in a stream. The girl says, "Praise for this little boy's slap; my bracelet fell off my arm." [. . .]

18 A Rich Man's Tale

(442) A rich man is dying; he calls his three sons and makes a will. He says to his eldest son, "When I am dead, you three shall have my goods equally, and don't call your brother a fool, and do him no harm."

After the rich man has died, the elder brother, in order to distribute the cattle, says, "Let us build a new stable; we will drive the cattle there, and the one that goes into the new stable will be yours, and the one that goes into the old stable will be ours."

Being of little sense, the fool agreed, and the cattle were put there. The animals all went into the old stall; the last one left was a skinny calf. He was forcibly moved to the new stable. The foolish boy says, "Out of all the cattle, he left me only this one," and he scolded his father, went into the house, and went to bed.

In the morning, he got up, drove the calf out into the field, and slaughtered it, leaving the meat and taking the hide to sell. As he walked along the road tired, he put the hide under his head and fell asleep. After he fell asleep, the crows came and pecked at his skin. Then he woke up and caught a crow, left the skin, and took the crow away. He went on to a village, entered a house, and what did he see? There was a young wife. So he begged the young woman, "Take me in as a guest."

"My husband is not at home," said the woman, but seeing that she could find no way out, she took him in as a guest.

In the evening, she cooked him some grated noodle soup, fed him, and gave him a drink. When it was time to go to bed, he gave her a pillow and laid her down in a secluded place. The lad put the crow under his head and went to bed. No sooner had he fallen asleep than someone came and banged on the window. **(443)** The woman opened the door, and seeing that it was a lad, she let him in and began to prepare a feast. He was frying a dish of chebureks*, and as soon as they said, "Let us eat now," the woman's husband came and knocked at the door.

"Where am I to go now?" asked her lover.

"Go into the closet," said the woman. "You will go when he has gone to bed."

The lover entered the closet, and the master came into the house and asked who the guest was lying in the corner.

"A traveller man," said the woman, "I had no choice but to take him in."

The gentleman sat down, and the woman brought him some of the grated flour soup she had cooked. The landlord woke the guest up to eat.

The guest stood up and, squeezing the crow's neck, it squawked. "What is it, my son?" The landlord asked.

The boy replied, "The crow said a dish of chebureks was on the shelf."

When the farmer asked the woman why she hasn't brought it, she said, "Ah, I seem to have forgotten," and she got up and took it.

They ate; after eating, he squeezed the crow's neck again, and the crow squawked.

"What does he say?" asked [the owner], and he answered, "He says a lad is in the closet."

The landlord takes his axe in his hand opens the closet door, and what do his eyes see? There is a lad. Hitting him over the head with the axe killed him.

The boy [says] to the rich man, "I will go in the morning and report you to the authorities for killing this man."

"Alas, my son," says the landlord, "I have sinned; you must not tell. I will give you a bag of gold; take this away and bury it in a distant place."

The boy took the dead body with the gold coins and went and threw it into the deep water at the edge of the village. **(444)** Then he returned to their house. As soon as he returned, he spread the gold around the house.

Their elder brother tells the second, "Go and see; the fool outside has smashed something."

The second brother goes out and asks him, "Fool, what are you doing there?"

"No, I'm not a fool like you; I had a calf, slaughtered it, took its skin to the fair, and got some [so much?] gold for its hair. If I had [had] as many cattle as you, I would have brought even more gold."

They were amazed at this and asked in what town he sold it. They slaughtered [and skinned] all their animals, put their skins on a cart, and went to the market, but no one looked at the skins.

So they said to one another: "This fool has deceived us; he has dispossessed [robbed] us of our cattle [wealth]. Let us go and put him in a sack and throw him into the sea."

They returned home, put it in a sack, and took it to the sea. When they got to the sea, they said, "If we throw it like this, it won't go far; let's go and get a long tree, tie it to the tree, and throw it in." So they went to get a tree.

The fool in the sack started shouting, "I will not marry the Khan's daughter!"

A sheepherder heard that and went and untied the sack. The sheepherder asked why they wanted to throw him.

He said, "They will marry the Khan's daughter to me, and I don't want her: that's why."

"If it is so," says the shepherd, "I'll go in." So he went into the sack, and the fool returned to his village.

The sheepherder started shouting, "I will marry the Khan's daughter!" Then the brothers came and threw the shepherd into the sea. One night, this fool came to their house, driving [sheepherder's] sheep. When his brothers saw this, they went out, and what they saw **(445)** was a flock of sheep, and the fool was safe.

"Oh, fool, where didst thou get these sheep?" They ask him.

He saith, "Ye have cast me into the sea; there is money, sheep, all that thy soul desireth, all that thou hast. I wanted sheep, so I brought sheep."

His brothers say, "If that's so, throw us in tomorrow and let us get sheep."

So they took sacks and went to the sea. He threw his eldest brother in; [when drowning, he shouted "Buble! buble!"][18]

"If it is 'Forty,' then take them and come out," said the fool, who also threw his second brother in. He himself returned to their house, married the daughter of the first man in the village, and became a wealthy man.

19 The Tale of a Nogai

(450) At one time in the Crimea, in a place called Yali-Boju, there was a Kadi* named Abdurrahman. And he was a man of very good eloquence. In those days, in the Crimea, in the city of Tarkan, the Nogai were also very good at the art of chinchis*, and they were also very quick at repartee, although they themselves could not read or write very well. The coastal peoples did not communicate with the Nogai at all. Kadi Abdurrahman was very fond of bantering with the Nogais.

One day, a Nogai from the village of Tarkan in the region of Chongar came near the Kadi, and the Kadi was passing by, so the Nogai met the Kadi. The Nogai was asking about a village.

"Salum-aleykum*, my Kadi aga*," he says.

The Kadi receives his greeting.

The Nogai asks, "Kadi agam, where are you from?"

The Kadi replies, "I come from heaven, Nogai." So he says and sneezes.

"Bless you, my Kadi aka*."

"Nogai," says the Kadi, "if you drink water, what do you say?"

"We say praise be to Allah."

18 Words that sounded like number forty.

"What do you say when you poop?"

"We say," saith the Nogai, "in the nostrils of Kadi Abdurrahman in Yali-boyu* with its smell."

"Hey, you Nogai, and so where do you come from now?" he asks.

"I have returned from heaven," says the Nogai, "my Kadi aka."

"Oh, you foolish Nogai," says the Kadi, "will a man who has once gone to heaven come out of heaven again from such a good place?"

The Nogai says, "My Kadi aka, I went near to heaven, **(451)** but I could not enter."

"Why couldst thou not go in?" asks [the Kadi].

"Sir," he saith, "by the gate of heaven they have tied a great dog named Kadi Abdurrahman, and it bites the one that goes out, and it bites the one that goes in, I looked and saw that I could not find a way to go in, I turned and came back."

Kadi Abdurrahman couldn't find a way to overpower the Nogai. "If so," he says, "Nogai, be my guest tonight, and then tomorrow you will go and let me entertain you."

Kadi Abdurrahman takes the Nogai home, sees him as a guest, [slaughters a sheep, so entertains him]. But he puts the head and tail of the sheep on a pan and leaves it for the morning.

The Nogai says, "My Kadi aka, I understand that the tail is fed with the guest, you have put it aside for yourself in the morning," and the Kadi replies, "Nogai, we fall asleep this night with a [definite] purpose: the one who has had a better dream eats the tail in the morning, and the head is eaten by the one who has not had a [so] good dream."

"Very well, my Kadi," says the Nogai.

After a while, they make their beds and go to sleep. The Nogai wakes up from his sleep in the night, hungry as he should be; what to do; he whittles down the meat off the sheep's head and tail and eats it, then lies down again and goes to sleep.

In the morning, they get up, wash their faces and hands, and sit down to coffee [i.e. breakfast]; but the Kadi tells his dream first; that was their agreement. The Nogai asks the Kadi, "Kadi, what dream did you have?"

The Kadi says, "In the night I dreamed I saw myself in heaven, enjoying the pleasures of heaven **(452)** and exchanging a thousand and one words with Allah." The Kadi also asks the Nogai, "What dream did you dream?"

"My Kadi aka," says the Nogai, "I too saw you in heaven exchanging a thousand and one words with Allah."

The Kadi says, "Then my dream is better than yours, I will eat the tail."

The Nogai says, "My Kadi aka, I have eaten the tail and the head."

"Why did you eat it?" asks the Kadi, and the Nogai says, "My Kadi aga, I saw you in my dream in heaven exchanging a thousand and one words with Allah, I thought, now he will not come down from heaven to earth, and forsaking heavenly food, he would not eat the tails and heads of the shitty-assed sheep of the lying world, so I had eaten them."

The Nogai defeated the Kadi anyway, and as a guest, he ate the head and tail of the sheep, defeating the cunningest of Yali-boyu. So even to this day, when the folk of Yali-boyu comes among the Nogai, they speak a little cautiously.

20 The Tale of the Three Brothers

(456) At one time, there were three brothers. Their eldest brother was married. The village mullah came to see the woman. The youngest was bald. The baldhead guessed it. One day he goes to work with his brothers. After a few hours of work, the baldhead says, “I have fallen ill, I am going home,” and goes up to the attic of their house. The mullah comes. They play, laugh, and then, as he is about to leave, the mullah says, “Tomorrow, I am going to plough my land here, and here, I cannot come. You fry a dish of chebu-rek*and come to me; one of my oxen is mottled, the other one is black.”

The next day, the brothers went to work. This woman fries the chebureks and goes to the mullah. The baldhead knew when the woman would come, so he put his shirt on the back of one ox and his cloak on the back of the other.

The woman comes straight to them; the woman’s husband [says], “What a thing is this, you have never brought food before; what has come over you?”

“I have thought,” says the woman, “I took pity on you and have brought you.”

The baldhead knows what is the matter but says nothing. So they ate, they drank, and the woman went away, but she also suspected the thing.

They went about their business the next day, and the mullah came again. “Why didn’t you come yesterday?” he asked the woman.

She told him how it happened, but the baldhead was again in the attic. The woman says to the mullah, “Tomorrow they won’t go to work, but you buy a little poison in this and this shop, and I’ll cook food, give it to them, let them all die, and then we’ll have peace.”

The baldhead heard this and went straight to the shop where the mullah was going. “You, shopkeeper, give me your shop and your clothes for an hour; here are a hundred roubles for you.” He put on his clothes and sat down. The mullah **(457)** comes and asks the shopkeeper for poison. The baldhead gives him a little pepper instead of poison and sends him away, and he takes a little poison himself and goes back. The mullah takes the poison and gives it to the woman. The woman cooks the food, puts the pepper in it, and they come to eat. The baldhead explains to his brothers, “You take a spoonful or two, then you fall.” All three of them fell with a spoonful each.

The woman called the mullah; they hid them.

“We will bury them in the evening,” said the mullah and left. In the night, he came again; the woman was going to make scrambled eggs for the mullah, she put oil in a cauldron, and the oil began to boil. The mullah leaned against the fireplace and fell asleep. While the woman went out to fetch eggs from the cook-house[19]; the baldhead got up and poured the oil from the cauldron into the mullah's mouth. The mullah died.

The baldhead's brothers also stood up. The baldhead took the mullah on his back, went to his house, and called in the mullah's voice, "Open the door!"

The mullah's wife [says], "You always go out and come so late; I will not open the door for you; lie down in the yard."

This again: "Open the door, if you don't open it, I'll come down the chimney and beat you."

"If you come down, come down; I won't open it," says the woman.

The baldhead went up to the attic and threw the mullah down the chimney, the mullah fell inside, and the woman was amazed.

The baldhead goes in and asks, "[Is] uncle mullah [home]?"

The woman says, "Right now, I'll give you a bag of gold; don't tell anyone, but take it and bury it somewhere."

The baldhead took the gold and the mullah. There was a melon field at the edge of the village; he stuck [the mullah in the ground] in the middle of the melon field, gathered some melons in front of him, and then went away and hid.

The watermelon farmer saw a man in the middle of the garden; "Who are you?" he asked, then he took his rifle **(458)** and shot him. The mullah died once more.

Then the baldhead came out to buy watermelons from the melon gardener and saw him. "This is our mullah," he says, "why did you kill him? I'll report you right now."

The melon gardener says, "I'll give you so much money, but don't tell [anyone]; take this and bury it somewhere. The baldhead took the money, picked up the dead man and went away, came to the ditch of a mill, blocked it, set up the dead where the blockage was, and went away, and the mill stopped.

The miller goes out, and what does he see: a man has blocked the water, "Why have you plugged it?" he asked and hit him so hard with a hoe that he collapsed.

And he also sees that the baldhead is bringing some wheat. He [the baldhead] saw this. "Alas, why have you killed our mullah, now I will have you destroyed."

The miller [says], "I'll give you this mill immediately; take it and bury it somewhere."

The baldhead threw the mullah on his back again, threw it into a large pool of water, and the mill was given to him. Then he came and separated the woman from her brother. All three of them remained bachelors.

One day, a Padishah came by looking for a shepherd. The baldhead was serving as a shepherd, looking after the sheep, when one night, there was thunder from the forest, and the baldhead hid. Out of the thunder came a black horse, and when he had scattered the sheep, the baldhead suddenly came out, grabbing the horse's mane, and did not let go. The horse starts speaking: "Let me go, see, here's a hair for you; if you get into

19 Or "Cookshack", in the Ms., "E*kinǯi uj*" 'second house'.

trouble, burn it, and I'll come to you." The next night, out of this roar, a grey horse comes out and takes this one too. He gets a hair from this one too. And on the third night, a yellow horse comes out, **(459)** takes it too, gets a hair, and lets it go. The Padishah had three daughters. So many lads asked [wanted them], the Padishah set three things[20]: "The one who beats my horses in a horse race will have my youngest daughter; the one who goes up to the top of the palace with a horse and from there comes down will have my middle daughter; the one who goes through this fire, will have my eldest daughter," he says.

Everyone gathered, and the race began. Baldhead came home to visit for a few days, and his brothers got on their horses and left. They say to the baldhead, "Come on, it's the race of the Padishah for whom you are shepherding; let's go and see."

The baldhead says, "No, I'm not fit for such a place, and don't you go either." They didn't listen to him, got on their horses, and left. After they had gone, the bald boy burnt the hair of the black horse, caught up with them, and with his kamchis, he hit them one by one and then went on, winning the horse race, and reached home before his brothers. In the evening, his brothers came, and the baldhead asked them what had happened.

"Today, a lad hit us on the road and left," they say.

"I told you not to go," said the bald boy; "you didn't listen to me; you were hit once today, tomorrow you'll be hit twice."

The next day, they went away again; the bald boy called the grey horse, got on it, caught up with them, beat them two or three times with his kamchi and rode off.

He got there, went up to the three-storey palace, came down from there, and came home victorious again. In the evening, his brothers returned and told him that the lad had struck them twice or thrice with his whip.

The bald one [says], "Tomorrow, he will strike three with his kamchi." The next day they left again. The bald boy called for the yellow horse, hit his brothers thrice with his kamchi, and left them there. On reaching there, he crossed the fire, and when the conditions were met, he took the younger daughter of the Padishah for himself and the other two for his brothers; they continued to live as Padishahs.

21 A Poor Man's Tale

(464) One man had three sons. One of the [sons] had the habit of stirring the ashes.

This man fell ill one day, calling his children to him and saying to them, "When I am dead, all three of you come out one night and lie beside me."

The two boys don't say a word; only the youngest, the one who is stirring the ashes, says, "All right, father, we'll lie there."

20 I. e. conditions.

Their father died a few days later. The older boy's name was Hassan. "You are the biggest; tonight, it's your turn," they tell him.

Hassan replies, "How many rich men have died during this time? Whose son have you seen guarding the grave? He obviously said it out of boredom."

"Well, if you don't go," says the Ash-stirrer, "I will," and taking the cudgel inherited from his father, he went to the grave.

He hit the grave with the cudgel. His father speaks from the grave: "Is it you, Hassan?"

Answer: "No."

"Veli, are you?"

Answer, "No."

"Ash-stirrer, are you?" he answered when he asked, "I am, father."

"Ah, welcome; if it is you, my child, sit down," he says.

Then, early in the morning, he struck the grave again with his stick and said, "Father, get up, say your morning namaz* and lie down."

His father got up. Then they said goodbye [to each other]; and he went home, came to the stove and started to stir the ashes again.

His brothers asked him, "Did you go there? What did you see? What did you hear?"

"Don't ask me," he said, "go there yourselves, wait and see."

It was Veli's turn the second day, and he told him to go. "Go, you fool, get lost from among us, if you want to go, go yourself." The Ash-stirrer retook the cudgel and went away, and, as before, this time, he struck the grave with his cudgel.

The father asks, "Asan, is it you? **(465)** Veli, is it you?"

Answer: "No."

As soon as he said, "Ash-stirrer, is it you?" he said, "Yes, I am, father."

At the time of the morning namaz*, he struck again with his staff and woke his father for namaz. His father got up, and after performing the namaz, they said goodbye, and he returned home again.

On the third night, it was his turn. He said nothing to anybody; he put his cudgel on his shoulder and went and struck the grave with his stick.

"Assan, is that you? Veli, is that you?"

The answer is: "No."

And as soon as he had asked, "Ash-stirrer, is it you?" "I am, my father," he said.

So he lay down and woke his father again in the morning.

His father [says], "Be patient, my child, don't leave me. I perform the namaz; then I have something to say to you." After praying, he says, "My child, there is a hill in this-and-this place; on the top of that hill, perform the namaz twice, and forty Arabs will come out of the hill and say to you, 'Command, brother!' Thou shalt say, 'Here I have three horses, lead them forth, let me see them'; when they come forth thou shalt take a hair from the tails of each of the three, and when thou art in trouble thou shalt take it out of thy pocket, and shalt burn the hairs, and they shall come to thee." So he did, and then he returned to the stove and stirred the ash.

When they woke up, his brothers asked him again. "Why are you asking me," he said; "you have offended our father very much," that's all he said. So they got up, made coffee, drank it, and didn't give him any. He was very hungry, so he went to that hill. He performed the namaz twice, and the Arabs came out and said, "Command, brother!"

"Forty kinds of food for me!" he says. Immediately the Arabs brought different types of food; he ate, drank and came back home again.

"Where have you been?" his brothers ask.

"Nowhere," he says, "I wander here and there, **(466)** no food, no drink, what shall I do?"

"We have looked after you, they say, fed and watered you; now go and look after yourself," and they drove him out of the house.

They had a house left by their father. "That shall be yours," he said and went away, and went again to the hill, and after praying, asked the Arabs for food. After eating and drinking, he saddled his grey horse and mounted it.

He went and went on horseback and got down from his horse at the gate of a Padishah. The daughter of the Padishah saw him and fell in love with him. The young man turns and goes back; when he comes out into the meadow, he lets his horse free, comes to a man and serves him. The man says, "I'm going to a wedding, by the time I will have come back, you are to finish hoeing all these cabbages." The young man lay down and fell asleep, then he woke up and saw that the time had come for his master to return. He burns the hair of the tail of his grey horse; the horse comes and says, "Command, my brother!" He says to run among these cabbages, turn them upside down, and roll in the middle. The horse did so and went away. When his master returned, this lad began to plant each cabbage in its place again. His master comes and asks what this is all about. Crying, the boy says, "A grey horse came, I tried, but I couldn't oust him, so he did and left." The man began to beat him. The lad collapsed here, and he [the farmer] collapsed there. The Padishah's daughter saw this, took pity on him, sent her Viziers, and had the boy called. For some years, he served around the gate of the Padishah.

One day, the time came for the Padishah's daughters to marry. All the lads in the country were invited. Whomever the girls liked, they would throw apples at him, and whichever lad they threw, they would marry him. The two daughters of the Padishah threw their apples at the brothers **(467)** of this Ash-stirrer, the young men of the country all passed before them, and only one piece of servant remained. They came and called him, but he would not go. He said, "Will they kill me?" So they took him by force, by order of the Padishah. As he passed the gate, the girl threw the apple at him. "Wrong," cried the crowd. He passed like this thrice, and the apple hit him all three times. The Padishah, being unhappy at all, gave his daughter to him, and they were given a separate room [i.e. old house]; he lived there with the girl.

One day, another Padishah wanted to start a war with this Padishah. This one's wife came and told him, "Our father is at war, your brothers are going, what are you doing?"

"If that's so, go and tell your father to give me a horse, and I'll go too."

The woman went and told him.

"There's a donkey here," says the Padishah; "let him go, take it, and go."

The lad mounts the donkey and sets off. He goes before the others, dips the donkey into the mud, and from there, he tries to pull it out; the others catch up with him. Seeing this, they laughed at him, asked him what he was doing, and went on their way. After they had gone, the grey horse's tail hair was burnt, and the grey horse came. He mounted it and approached the battlefield. The Padishah saw this but did not recognise him.

"Thus," saith the lad, "If I stand by thee and beat them, what wilt thou give me?"

"I will make you a Grand Vizier," said the Padishah. So he began to slaughter the enemy. A little later, his little finger was wounded. The Padishah took his handkerchief from his pocket and bandaged his finger. The fight was over. They had won and returned; **(468)** no one realised him. He returned before these, went to his room, and went to bed.

His wife washed the scarf on his sleeve and laid it out to dry. He tells his wife, "Go and tell your father to give me the viziership."

The wife goes, and as soon as she tells her father, her brothers say, "This fool has lost his way; he has begun to mock you."

The Padishah says, "Call him, and I'll cut his head off."

They went to him, and they saw that he was asleep. "

Don't touch him," the wife tells them, "he came tired, let him wake up, and then he'll come." They return and tell the Padishah.

The Padishah says, "Go quickly and drag him here." They came, and he was already awake and sitting.

He says to them, "Go and tell him he should think who helped him in the battle," he said, and then he himself called the grey horse, mounted it, went to the Padishah's house [suite] and stopped there. The Padishah asked who that was. When he pointed his little finger, he understood, then he made another wedding feast, and gave him the grand viziership.

22 A Thief's Tale

(475) There was once a very skilful thief whose son was twelve years old. One day he says to his son [he was angry for some reason], "You can't do anything; you won't become a man."

The son asks his father what his father's trade is, and his father tells him he is a thief. "Then," says the son, "show me your skill."

His father says to him, "My skill is to take the egg from under the crow sitting in the tree and the crow doesn't even notice."

"So let me see," says his son, "go up and take it."

The man went up, pierced the bottom of the nest, and took the eggs; the crow didn't notice. While his father was taking the eggs, the son pulled his trousers off his father's arse and he didn't notice.

He came down from the tree. "Behold, the crow's eggs," he said; and the son said, "Who took your trousers off your arse?"

His father saw that his trousers were missing, "Were they left on the tree?" he said, and while he was looking up the tree, [his son] took the eggs from his hands. He sees that the trousers are around his son's waist. "My son," says he, "you seem to be a more excellent master thief than I am."

One day his father says, "Come on, my son, let's go and steal the Padishah's treasure." So they went, they got to the treasury, they went up to the top of the treasury, they opened a gap, they made a stairway out of rope, and they went down. As they were picking up the gold, the sound of the gold rang, and the guard heard it; the son went up the stairway, and in the meantime, the guard arrived. His father, as he was climbing out on the roof, was caught by the guard from below. **(476)** The son could not pull his father out and cut off his head and prevent his identity from being known. He went home; his mother asked him what was wrong with his father. The son says, "My father and I went out to steal; my father was caught, and so that they wouldn't recognise him, I cut off his head and took it away; let's bury it, and they buried it."

The Padishah made it public, "Whoever cut off this man's head and took it, come and take his body from the corner of the courtyard and run away with him, let me know what kind of lad he is."

"Mother," says the lad, "I will go and fetch him; give me leave."

"No, my son," says his mother, "your father is gone; if you are taken, and if you are cut down, who will take care of me, what shall we do with a lifeless body?"

"No," says the lad, "I shall return safe and sound, with Allah's help."

So the lad went to town and bought two horsehide, one black and one grey. He put the grey skin on one side of his horse and the black skin on the other. The body was hung in front of the palace, with a guard on each side. As soon as he saw them, the lad went round, then came back, cut the rope with his sword, and ran away with the corpse. The guards chased him but could not catch him. The Padishah asked the guards [who] it had been.

The man who had been sitting on the right [says], "It was a man riding a grey horse," and the man on the left [says], "It was a man riding a black horse." These two started arguing.

The Padishah says, "Neither of you is right," and he cut off both heads.

The lad buried his father.

The Padishah again announced that there would be a bag of gold in front of the palace, **(477)** and whoever took the body would take it. The lad went again, asking his mother's permission. He bought a fishing net in the bazaar and filled it with hay. On a stormy day, he slipped into the net and rolled to the palace. As he passed by, the guards saw him but thought it was [just] hay and ignored him. He got to the gold, took the gold, and again rolling, he went aside, got on his horse, rode home, and gave the gold to his mother. In the evening, the guards see that the gold is gone. They came to the Padishah and told him. The Padishah asked him if anyone had passed by that place. The guards say that no one but a bunch of hay passed by.

The Padishah proclaimed again, "Let that man come and lie with my daughter, and I will give him my daughter and my padishahship."

The lad asks his mother's permission again [as he is determined to do this too].

"They will kill you there," says his mother.

"If they kill me," says the lad, "the gold I brought you will be enough," and he is gone. On his way there, he stopped at a tavern and asked for a bottle of raki sitting on horseback. He cut off the tavern keeper's hand with the sword at his waist when he was about to take the money and took it [i.e. the tavern keeper's hand] with him. The lad arrived at the palace; it became evening. The Padishah had a favourite goat, and it [freely] wandered in and out. The lad took the goat, oiled it with lamp oil, lit the goat with a match and let it go. The goat started running around. All those who saw it ran together, saying the Padishah's palace was on fire. During this, **(478)** he went [climbed?] through the water pipe to the girl's room and spent the night lying with her. When an hour is left until morning, he says he must go out to pee.

"No," says the girl, "pee here in the room."

"If not, hold my hand, while I am out to pee," he says. The lad made the girl hold the barman's hand, and he left. He went home.

The Padishah heard this, too; he saw that he would not be able to catch him. So he made it public that the man who did these things should come forward with no punishment. I'll give my daughter to him, the padishahship too. This fellow says, "It is I who have done these things, do what you must do."

"Well done, my child," says the Padishah, "you are perfect;" forty days and forty nights, he has held a wedding and gave [him] his daughter.

23 The Tale of the Clever Girl

(483) There was once a young Padishah. One day he goes out into the meadow and finds a human skull. On the skull was written, "What is harder than hard, what is more bitter than bitter, what is sweeter than sweet?" He brings this head to his Viziers and says, "Whoever finds the answer to this for me will be rewarded with a bag of gold."

So the Viziers assembled and watched to see who would say stone, who would say steel, and who would say bone. "More bitter than bitter is lemon, salt, pepper, and onion," they say.

"Sweeter than sweet is honey, sugar, and grapes," they say.

The Padishah did not receive these solutions. So he sent word to all parts of the country, "Whoever knows [the answer,] come."

In one village, there lived an older man with his only daughter. That daughter was [very] clever. She says to her father, "The Padishah has found a skull, but there is no answer to the writing on it; go and tell him: 'The horse's hoof is harder than hard, death is more bitter than bitter, the man's wedding night is sweeter than sweet.'"

The old man, waving his beard, came to the gate, but the guard wouldn't let him in. The old man says, "Go and tell the Padishah I've come to answer his question." They took the news to the Padishah. "Let him come in," he says, "let me see." The old man went in; the old man knew nothing about reading and writing. He took the skull, looked at it, turned it round and round, and then answered what his daughter said.

"Who taught you that?" asks the Padishah.

The old man says, "I know it for myself."

"No," says the Padishah, "I have seen that you will not know it; **(484)** someone has told you so; tell me the truth, or I will cut off your head."

The old man was surprised, frightened, and said, "My daughter told me."

The Padishah says to the old man, "You go, I will send your daughter a silk thread, and from this thread, she will make me a cloak; if she does not do it, your head will be cut off, hers too."

The old man goes home crying and says to his daughter, "You have made trouble for my head; he has ordered me to do an unachievable thing," he says.

The Padishah sent the silk with his two Viziers; they brought it. The daughter took it and gave them a wood chip the size of a matchstick and said, "We are poor men; the Padishah should have a loom made of this and send it to me so that I may weave the cloak on it."

The viziers take the wood chip to the Padishah. "The girl is a fool," says the Padishah, "can a wood chip be made into a loom? Go and tell her to make it, for if she does not, I will cut off her head."

They came to the girl and told her.

The girl said, "If this wood chip cannot be made into a loom, I cannot make a cloak out of a piece of thread."

Returning again, they told the Padishah. The Padishah called for the old man, "Take this ox, and your daughter will have this ox calved in six months." The old man comes home crying again.

His daughter asks him, "Why are you crying again?"

The old man says, "The Padishah has ordered you to have this ox calved in six months."

"Don't cry," says the daughter; "that ox will feed us for six months, and the answer is my business."

They brought the ox. "It is very well," says the daughter; "after six months, you will come and take him away."

The Viziers went back. The girl says to her father, "Go and slaughter the ox." The old man, very frightened, slaughtered it, and they started to eat it.

Days have passed, and six months have passed. The daughter sent her father to the baths. **(485)** The Viziers went through the door, sat, and asked questions. The girl says, "My father is in the bath giving birth to a child," so she sends the Viziers back.

The Viziers returned and told the Padishah.

"Man," says the Padishah, "can he give birth to a child? That's a lie; I'll cut his head off; go and bring the old man here."

The Viziers come to the girl and say that no man will bear a child. The girl says, "Can an ox have a calf if a man cannot bring a child into the world? So we slaughtered the ox and ate it."

The Padishah calls the old man again, "Your daughter is very clever; I will marry her," he says.

"I don't know," says the old man, "I'll ask her myself," and he went home and told her.

The daughter says, "We are poor people; we are not worthy of the Padishah; the Padishah should not make fun of us," she said, and then she wrote a letter and sent it to the Padishah with her father.

He wrote in the letter, "Hey, Padishah, you're a big man, don't laugh at us, we've answered your question, give us the money you promised."

The Padishah has read it. He tells the old man, "No, I'm not laughing at her, I really want to take her."

The old man goes back and tells his daughter.

She agreed, and the matchmakers came and got her engaged. After five or ten days, the wedding began; the Padishah sent for the girl with beautiful clothes and a phaeton. The girl dressed and came in a phaeton to the Padishah's palace, where she was received with music and taken out to meet the girls there. The girls are dancing and laughing. The girl looks at the music and marvels, because this girl has never seen such music in her whole life. After a few hours, the girls **(486)** were ordered to go outside. The Padishah came in, and they met. Food and drink came, they ate and drank, and then the Padishah said to the girl, "You became now my wife; you will do nothing except my orders; you cannot judge without my permission; if you do, I will banish you from my door." These became now husband and wife.

After a long time, the Padishah was not at home. She looked out of the window; there was a café opposite. Some customers [on horseback] came to the café. One of the horses was a mare, the other a stallion. While they were having coffee, the mare foaled. The foal lay down under the stallion. These men come out and see a foal lying under the stallion. The stallion's owner says, "My stallion has foaled."

"No," says the owner of the mare, "my mare has foaled."

They argued, and they came to ask the Padishah. The Padishah is not at home; the woman asked what they were quarrelling about. They told her. But the woman had seen and said that a stallion cannot have a foal; the foal belongs to the mare. As they were leaving, the Padishah came to them and asked them. They told him his wife had done them justice [judged their case].

The Padishah went into the palace and asked the woman, "Why did you judge without me? I did not give you permission; now take what you want from this palace and go."

"I made a mistake," says the woman, "come, let's have some fun; let's eat and drink, and then I'll go." While eating and drinking, **(487)** several drinks were consumed. The Padishah drank and got drunk. The woman ordered a phaeton to be harnessed, and the phaeton arrived. The woman dressed took the Padishah out and laid him in the phaeton. They went out of the courtyard and made their way around the town's streets a few times, then started to her own village. On the way, the Padishah woke up and asked, "Where are you taking me?"

"You said," she replied, "that I could take away anything I wanted. I have nothing else to take but you, I am taking you," she said. The Padishah understood that [his wife] was right; he forgave her, and they returned, lived, and had family [children] together.

24 The Rich Man's Tale

(493) At one time, there lived a rich man. He used to sit on his sofa and drink coffee. Again, one day, as he was drinking coffee, a bird flew into the tree in front of the house. This man, as soon as he saw it, fell unconscious. A few hours later, he woke up and saw a quill fall before him. No bird, he takes the quill and hides it. Soon his wife came out of the house; her husband had not drunk his coffee.

"What's got into you," she says, "that you haven't had your coffee yet?"

"Don't ask," says the man, "I won't tell you what's the matter with me."

The woman insisted, "No, you will say to me."

"As a man, I could not stand it," he said, "I fainted; you are a woman, if you see it, you will be frightened to death, and you will die."

"No, you will show me, whether you like it or not," she says; and the man takes the quill out of his pocket and shows it to her. The woman saw it, collapsed, and died. Her three sons were in the garden, they were told; they came and buried their mother. Their father did not say how she died.

One day, this man says, "Your mother, behold, she died at the sight of this wing feather; find this bird and bring it to me." He gave his three sons two hundred roubles each, and the three of them set off. They came to a place, at a junction of three roads, on the board was written, "Whoever goes this way will come to such and such a town, whoever goes this way will come to a village, whoever goes this way will not return."

The bigger [elder] one went to the town, **(494)** the middle one went to the village, the little one went by the one from which there is no return. He goes on, he goes on until he comes to a forest. What he sees before him is like a mist. Dragons come, ready to swallow this one. From behind them, another dragon came out, and went before them, and said, "Do not touch the human being, for when I was in the sea, this lad then was a child, I was cast ashore by the wave, and when the other children were about to kill me, this one caught me, threw me into the sea, and donated my life." The lad got away and went on his way.

He walked, he walked, and he came to a palace; he looked around and saw that there was no entrance. Somehow, he found a way and got into the palace; he saw that in the palace, there was a beautiful girl, like a rose blooming in spring, the table was set, and the food was ready. The lad sat down and began to eat. The girl said, "Hey, lad, a devil is coming, he will destroy you and me, I will hide you."

"No," says the lad, "I don't need hiding; if he comes, we shall see."

No sooner had he said this than the devil came and said, "Has this table been laid for you? First, you wrestle with me, and then you can eat," he says.

The lad replied, "Stop, I am hungry; let me eat, and then we will wrestle," he said.

The lad ate, was satisfied, wrestled with the devil, and cut the devil in pieces with his sword.

She said, "You have destroyed this one; take me away from here."

"No," says the lad, "I am going in search of such and such a bird. I will take you on my way back."

The girl pulls a ring from her finger, gives it to the lad, and says, "I am yours; you are mine;" they say goodbye, and he goes away.

(495) He went, walked, came to the second palace, and saw two girls there. These girls, while saying that the dragon would come, the dragon came, and they clashed. He destroyed that too and gave his word that he would take the girls when he came back, and he went away. Soon he came to the base of a tree and lay down. No sooner does he fall asleep than the hatchlings begin to chirp. The lad saw a snake climbing up the tree, immediately got up, cut the snake to pieces and went back to sleep. The mother bird comes and says, "I have not been able to raise my chicks here for many years, and this lad has taken them away," she says, "I'll gouge out his eyes," but then her chicks say, "No, mother, this snake would have eaten us, this lad saved us."

"Then let him sleep," she says, and, to keep the sun from shining, she covered his face with her wing feathers.

The lad suddenly woke from his sleep, shook, and he sees a bird sitting on him. The bird says, "Lad, wish all you want."

"Health to you," says the lad.

"No," says the bird, "my health is useless to you; ask for something for yourself."

"If so," says the lad, "behold, this is what I wish, the owner of this wing feather."

"Alas," says the bird, "you have found me in my old age, but never mind, it will be done; go for forty barrels of wine and three hundred pounds* of meat; if I look to the right, you will give me meat, if I look to the left, you will provide me with wine."

The lad brought it. The bird took them all on his back, and they flew away. They went, they went, they crossed a sea, the bird landed, and the lad got off the bird. "Now then, go," says the bird to him, "first you come to the water, say bismilla*, drink a handful of water, **(496)** and the water will split in two, you will cross it and come to a palace, close the open doors there, open the closed ones. There are hanged swords there; you take them off, oil them and hang them again. Then you open another door; there is the bird hanging; you immediately take it and run away with it."

So the lad did and took the bird. However, the devil sensed it, came out from behind, and told the swords to cut him. "For so many years," said the swords, "we have been hanging; you have not once oiled us;" [and they] let him go. The lad goes on and gets to the bird. Again, they mount her back and are on their way.

The meat ran out when they had little distance to cross the water. The lad cut some meat from his leg and had it engorged [by the bird]. The bird knew she didn't swallow it; they got to shore. "Come on, get off me and go," says the bird. The lad, not wanting to show his lameness, says, "You go first."

"You go first."

"I'll go first," so they went on, unable to agree. The lad goes, but he is lame.

The bird asks the lad, "What has happened to your leg?"

"I cut it out and made you eat it."

The bird gags and pulls the meat out of her mouth, sticks it on his leg, and they say goodbye and leave.

Along the way, he picks up the three girls he met, and they go to the place where he parted from his brothers. This got him thinking, "When we left here, we were three brothers; now to return alone is no good," he says and sets off along the road his eldest brother had taken. He came to a town and saw that his brother had spent his money, was a hundred roubles in debt, and was sitting in the corner of a café. He paid off his debt, they left, found his second brother, and took him with them. **(497)** Three girls and three lads went home. His brothers were wondering how to kill him and how to take the bird from him. When they came near a village, the two older brothers said, "You go very slowly; we will bring bread from this village." So they went away, bringing food from the village and also a felt [rug]. They laid it on the opening of a well. [Meanwhile, from behind, the others also caught up with them, and then they ate and drank (together).] Two of his brothers sat on either side of the felt [rug]. They want to see the bird, so they let their brother sit in the middle. The lad sits down. He takes out the bird no sooner than they both stand up simultaneously. The lad immediately falls into the well, and they take the girls and the bird. They force the third girl, but the girl says, "I'd rather die, but I'm not leaving here." She also stayed there.

They came to their father and said, "Here, we have brought the bird."

"Where is your brother?" he asks.

"When we left," they said, "we came to a three-way fork in the road, and the sign on the board said that whoever goes this way would not return. He didn't listen to us; he left and never came back."

The lad in the well remembered to lick the girl's ring, licked it, and three devils came, "Should we burn or destroy?" they ask.

"Neither to burn nor destroy wanted," he says, "bring me up to the surface of the earth."

The devils immediately take him up, and he sees that the girl is still there. He takes the girl, comes to the edge of their own village and opens a goldsmith's shop. Meanwhile, her brothers want to have a wedding. The girls want a golden rooster made to be the jewel of the wedding. There was no goldsmith in the village except this lad, so they went to this one. "All right, I'll do it," says the lad, "bring me a pound* of peanuts; by the time I have eaten the peanuts, the rooster will be ready." The lad did it. This lad wrote down on a piece of paper in detail **(498)** all that happened to him and put it in the cock.

His brothers come and take the cock away. They are amazed at the beauty of the rooster; they open it and look inside. Well, what can they see: a paper comes out of the cock, and they read it. Their father, as soon as he hears this, calls the lad. The lad comes. Their father allowed him to cut off the heads of his two brothers. "No," says the lad, "even if they did it, I will not do it." So he forgave their sins, and then the three of them took the three girls and lived their lives that way.

25 The Shepherd's Tale

(506) Once, one of the Padishahs had a daughter. He also had a shepherd and a horseherd. The girl loved the horseherd, and the shepherd heard about it. The shepherd followed them everywhere. One day, the girl said to the horseherd, "Bring two horses, we will put our things on one of them, and on the other, we will get on ourselves and run away." The shepherd heard this. In the night, he brought a bottle of raki and got the horseman drunk, brought the horses and told the girl. The girl put her belongings on the horse, and they got on one and escaped. The girl did not recognise the lad. They went, they went. It became morning. The girl saw that it was not the horseman but the shepherd. "Oh, help, it seems to be my fate," [she thinks but] says nothing. They walked for a few days, and they came to a town of a Padishah. In that town, they bought land and had a house built for themselves. They had a wedding and got married.

After a while, they invited the Padishah to their own house. The woman prepared a pan of pickled aubergines and said to her husband, "When the guests come and sit at the table, you will put this pickled dish in front of them."

The man forgot to put the pickle on.

While eating, the Padishah says, "Oh, if only there were pickles, we would eat them too."

The woman overhears from the cook-house, opens the door and shows her husband the pickle. The Padishah saw her arm and fell in love.

He left the food, got up from his seat and went home, summoned his Vizier: "Find a way to destroy this man, and then I will marry his wife."

One of the Viziers says, "In the forest of Munar, there is a spring, and by the spring, there is a tree; of the leaves of that tree, five are making music, and five are dancing; send him to bring it. **(506)** There is a devil there; it will break him into pieces."

They called the shepherd and told him. This wretch came home crying very much.

The woman sees him and says, "Why are you crying? There is nothing easier in the world. Tomorrow the devil will go to sleep; he will not rise from his sleep for forty days. You go to the forest, perform abdast* at the spring, perform the namaz twice, throw the tree over your shoulder immediately, and come straight home. Be careful not to take it to the Padishah until you have brought it here."

In the morning, he gets up early, goes to the spring, takes the beautiful tree on his shoulders, and returns with it. When he reaches the village border, the Padishah hears the tree rustling and sends a man to take him straight there. The man goes there, and as soon as he says this, the shepherd says, "No, I brought the tree for myself, not for the Padishah." He takes it and goes home, plants it in front of his window, and he and his wife go outside and drink coffee together under the tree.

The Padishah, seeing all this, got furious and ordered that man to perish quickly.

One of the Viziers says, "Let us send this one out to the summer pasture; there is a paddock, and inside the paddock, there are three rams; go in there to fetch them. The rams are bewitched, and they will eat him."

They called the shepherd and said to him, "Go and bring the rams from such and such a place."

He came home crying again. His wife saw him, laughed at him, and said, "Don't be afraid; it's easy; tomorrow you'll go; if they're grazing at one end of the paddock, you go in at the other end and shout, they'll wince. As soon as they see you, they will run towards you; **(507)** you do not go against them, but run and jump out of the paddock, they will also jump out of the paddock, and their magic will break. Then you drive them back again." So he did and brought them home.

Seeing that he had returned safe and sound again, the Padishah was angry and called [his] Viziers. "[Find] a way quickly to this," he told them, "or I'll kill you too."

They were thinking. One of the Viziers said, "This man will solve all the difficult tasks in this world; we will send him to the other world to bring news of your father."

He has been summoned, and the Padishah commanded this [man] to "Go to the afterlife, forty days respite; on the forty-first day, your head will be cut off."

"It will be done," said the lad and went home crying.

His wife sees him and says, "Do not be afraid of so easy a thing, go to the Padishah, and say that he may pile up forty wagons of straw in one place, set a pole at the top, hang you by your two feet with a chain, put the straw on fire, you will go to the other world and bring news of his father."

This man, even more afraid but with no other solution, went to the Padishah and told him. They did it; the straw was lit. The woman turns into a pigeon and flies up, and her husband turns into a millet and disappears. In the morning, she wrote a letter and sent it to the Padishah's house with her husband, and the shepherd handed it to him. The Padishah was astonished and asked him what the matter was. The letter says, "Oh, my child, instead of taking these poor people out of their work and sending them on their way, come yourself, and we will meet, and you will know my situation." The Padishah thought if it is possible to go there, then he must go and meet. He had forty cartloads of straw piled up, had himself tied to the post, had the straw set on fire, and with the straw, he himself was burnt. The shepherd became the Padishah instead.

26 An Apple of the Dervish

(512) Once upon a time, there was a rich man who had neither son nor daughter. This man prayed to Allah from morning to night, asking for a child. One day a dervish comes and gives him an apple and says, "Half for yourself and a half for your wife, and you will have a daughter. You will make an iron cage for this girl; you will not let her out of the cage; she will live in it until she dies. A forty-headed devil will immediately take her away if you let her out." They ate the apple, and after nine months and ten days, they had a daughter, to whom they gave the name Gülzar. The girl lived in the cage until she reached adulthood.

The Padishahs heard this and sent suitors to take her for their sons. He says, "I'll give you my daughter, but you can't let her out of the cage." The suitors do not agree to this and go back. One day, the Padishah of Chin-machin heard this and, agreeing not to let the daughter out, sent a suitor and betrothed her. After a few months, they had a wedding, and the Padishah married the girl. They lived together for some years and had three sons, the eldest named Ali, the middle named Hassan and the youngest named Ahmad.

One day these boys grew up and went hunting. The woman says to her husband, "I've been living in a cage for so many years, I've grown old without seeing the light of day; now the devil doesn't like me, let's go out into the garden and have some fun," she said, and they went out. No sooner did they get out into the garden than the devil came, throwing snowy ices and casting a dark mist, and ran away with the woman. It was evening. Her two elder sons returned and saw their father weeping and their mother
(513) gone. They asked. "With your mother," says their father, "a forty-headed devil has run away." So the boys, asking permission, went to look for their mother.

A little later, the younger boy came, he too asked for his mother, and he too asked permission.

"No, my son," says his father, "you are still young; your brothers will find her and bring her back."

"No, that's impossible," says Ahmad, "I'll go too," he insisted.

His father saw that there was no way out, and he said, "If that's the way it is, then go, pick a horse from the stable and go."

So the boy entered the stable, stroked the horses' heads, rubbed their skins, and pulled their tails. Then, returning, he says to his father: "You have not fed a horse; you have fed a pig, I will go to the pasture, look for one there, and fetch one from there," he said and went away. There was a scabby colt in the stud; he liked it and took it.

His father laughed at him. "Well, you've found your horse too, I knew you wouldn't be able to leave; go and do what you like. May Allah make it easy for you," he told him, and they said goodbye and (the boy) left.

As they go, a witch comes out from the roadside and says, "Ahmad, son of the Padishah of Chin-machin, do you know where you are going, or do you not know where you are going?"

So the boy turns to him, goes up to him, and says, "I know, and I don't know, but I'm going with a firm intention."

"Otherwise, you are going the wrong way," says the witch, "return to your father, let him feed you for forty days, then take a mace weighing forty batman*, and having rawhide sandals weighing eighty batman made on your two feet, you will set out." So the boy returned, had everything done like that and set off again.

He went on, he went on, and once he came to the town of Cast Iron, there was a spring at the edge of the city, and he let his horse go and stayed [there]. At night, he hid under the Cast Iron bridge. (514) After a while, a forty-headed devil came crying. [As he was coming, his horse recoiled.]

"You have never baulked here before," says the devil to his horse. "What is the matter with you today, or have you caught sight of Ahmed, the son of the Padishah of Chin-machin?"

"And if it were Ahmat, what would you do with him," says the boy, dashing out from under the bridge.

[The devil says to him:] "Shall it be shooting, or shall it be grappling?"

Ahmat says, "Shot into your father's country; come, let's cling together."

So the devil came down from the bridge, and they clung together. The devil stroke Ahmat to his knees into the ground. The boy braced himself, grabbed the devil and stroked him up to the waist into the ground, cut off his thirty-nine heads, and left one.

"If you're a lad," says the devil, "cut that off too."

"Thou art not my father's yoked ox; go, torment thyself," said the lad, [mounting] the devil's horse and [riding] away.

He caught up with his brothers at the Silver City a few days later. They stayed on the outskirts of the city. At some point during the night, he tells his brothers: "You go to bed; I must go somewhere." So he went under the Silver Bridge and hid. Soon an eighty-headed devil would come and go through there. His horse also recoiled, and he said the same thing as the first devil; wrestled with him; cut off seventy-nine heads, and took his horse to his brothers. Leaving his own horse to his elder brother, he himself mounted the devil's horse and rode on. They went on and on, came to the Golden City, and lay down again on the city's outskirts. At night he hid under the Golden Bridge. After a little while, a devil with a hundred heads came, and he wrestled with him **(515)** but that [former] devil's horse also began to kick Ahmad. He struggled with it a lot. The lad got tired, pulled down one of his rawhide sandals and threw it at his brothers; the rawhide sandal fell beside his brothers. "Alas," they said, "this is an enchanted place" and ran away in the other direction.

Ahmad looks and looks; his brothers just don't come. He threw his other rawhide sandal, which had also fallen by the spring, but his brothers were no longer there. When he saw that his brothers were useless to him, he threw his mace towards his horse, and the mace fell by the horse. The horse suddenly broke his rope and, arriving in time, began to kick the horse of the hundred-headed devil, and the devil's horse was defeated. Ahmat cut off the devil's ninety[-nine] heads, took his horse, and when he came to the spring and saw that his brothers were not there, he lay down. In the morning, his brothers came, and when Ahmat asked them where they had been during the night, they said, "This is an enchanted place; we were frightened and ran away."

Let's not lengthen it any further; when he went to give the horse of the hundred-headed devil to his middle brother, the horse of the hundred-headed devil spoke and said to Ahmat, "Now I know who you are, son of the Padishah of Chin-machin, you are Ahmat-batir, the horse of the devil who took your mother, is my colt, I will take you there." They set out. One day, they came to the foot of a high cliff. Ahmat-batir says to his brothers, "You stay here, I will go to the top of the rock, I will fight the devil who took our mother. If he defeats me, I will tell him to throw my head off the cliff. If my head falls here, you will go back; if I beat him, I will throw one of his **(516)** heads off. Then you wait for a little, and I will come down," he said and mounted on the horse of the hundred-headed devil; he came up the rock in forty days and forty nights.

Dismounting from the horse, he let the horse back, and he went to the front of the house where his mother was, and with his mace struck one blow, whereupon the front wall of the house was put back, and the back wall in front. At the trembling of it, his mother [was frightened and] said, "What a son of a bitch; this is my house," she said and took out a cup of poison, and not recognising him, gave it to Ahmad. Ahmad took it, poured it behind him, and asked his mother how she came there. His mother told him what had happened to her from beginning to end, one by one. When he asks her, "If you saw your youngest son, would you know him?" she says, "Yes, I would."

He asks her, "About what?"

She says, "When he was young, he fell on the ground, and his forehead was pierced, I would know him."

The boy immediately uncovered his head and showed it.

"Oh, my son," said his mother and led him into the house. As Ahmad entered, he saw a young girl sitting there.

After they had sat for a while, his mother said, "Now you go away and hide; the devil will come and destroy you and us."

"No," says the boy, "tell me where he is now, I will go and kill him, and we will go home with you."

His mother says, "You go this way, and in this-and-this place, he lies with forty heads buried in the sand."

And the boy came there. The devil sleeps with his head buried. "Arise, thou infidel," he says, "let me wrestle with thee."

The devil got up, and they clung to each other. The devil lifted him and stabbed him into the ground up to his waist. The lad got angry, suddenly lifted him and pushed the devil into the ground up to his throat; thirty-nine of his heads were cut off, and one remained. "Cut this one off too," he said; [the lad] said, "You are not my father's yoked ox, torment yourself," took one of his heads, and threw it off the cliff to his brothers. **(517)** His brothers had forgotten what they were waiting for; they were frightened and ran away. A little later, the elder brother remembered, and they went and saw that it was the devil's head. So they also start to crush the devil's head. While they are crushing it, let's come [back] to Ahmat.

Ahmat returned to his mother and told her, "Get ready; let's go." They brought a rope with them to descend from the cliff. Ahmat first tied his mother to it and let her down; his brothers were waiting for Ahmat.

The girl says to Ahmat, "You go down, if I go down first, the rope will be pulled down, and you will be left behind."

"No," says the lad, "you don't want to leave; you get down."

You get off, I get off; the girl saw no other way; she gave Ahmat her ring from her pocket and said, "I know they will leave you here. If you lick this ring, forty Arabs will come out, and you can make them do what you want." So Ahmat tied the girl to it [the rope], and the girl went down. His brothers envied Ahmat's girl, and the rope was suddenly pulled out of Ahmat's hand. Ahmat remained on the clifftop.

They took their mother and the girl and came home. The daughter was betrothed to the older son. "I," says the girl, "will be engaged for forty days, and the wedding can take place after forty days."

At this place, Ahmat took the ring out of his pocket and licked it; forty Arabs came out and said, "Command whatever you want."

Ahmat says, "I want you take me to my father's town."

These say, "Close thine eyes, open thine eyes," and they came to the city. Ahmed took a rumen from the butcher, put it on his head, went to a tailor **(518)** and became his apprentice. Forty days had passed, and the wedding was about to begin when the girl said, "I want a pelerine from you, [but] one that has not been cut with scissors or sewn with needles." [She went to all the tailors in town], but none of them would do it. Ahmed hears this and says to his master, "Go and take it, I will do it." The master went and came back, saying he would do it.

Ahmed took the fabric, went in (to the workshop), and cut it into odds and ends with the scissors. The master saw this and said, "Oh, I'm finished," said the master and became furious. Ahmat took out his ring and licked it; the forty Arabs came out and asked what he wanted. "I want a pelerine not sewn with needles, not cut with scissors," said Ahmat. The Arabs immediately brought it and handed it over. Amat gives it to his master and says to him, "Go on, take it and carry it." The master took it and gave it to him [to the bridegroom]. The girl said, "Now let the wedding begin." The wedding starts; on the third day of the wedding, the time of the wrestling, Ahmat licks the ring and asks the Arabs for a dress and a horse. They brought it to him, Ahmat got dressed, got on a horse and went to the wrestling square.

He got off his horse and started wrestling with the pelivans*. He didn't miss a single one, (all of them) were knocked down. After the pelivans were knocked down, the groom-to-be got angry, and he himself came out to fight. Ahmat took the bridegroom's arrow [. . .][21] and then he grabbed him by the legs and tore him in two, and he himself disappeared. No one recognised him. The bridegroom died and was buried. A few months later, she was engaged to the second son, Hasan. Again she wanted to wait forty days. At the end of the forty days, **(519)** the girl says, "Have a bowl made, a hen on it, a rooster behind the hen, and an egg after them, and let them chase each other."

The Padishah told the craftsmen, but again no one would do it. Then the girl says, "The master who made the pelerine, let him make this one too." They sent for him. Ahmad's master was called. Before the master leaves, Ahmat says to him, "Whatever the work is, take it, I will do it." The master agreed with them, took it and came home. Ahmat licked the ring again, and the Arabs came and made it. The master took it, and the wedding began. On the third day, at the time of the [horse] race, Ahmat licked the ring and asked the Arabs for various clothes and a horse, and went to the race, put a child on his horse and overtook all the horses. After the run, the congregation was about to go to the mosque for prayer, but [Ahmat] entered the mosque before the community and locked the door; the assembly came, they strove, and strove, but the gate did not open. From inside, Amat says, "Tell the Padishah to come here, and I will open it."

They went and called him, and the Padishah came and pushed [door], but the door didn't open. Amat says from inside, "Tell me, how many sons do you have?"

"I had three sons," says the Padishah, "and the third went to look for his mother, and he himself was lost."

Ahmat asks him, "If you saw your lost son now, would you recognise him?"

After the Padishah said the same thing his mother had said, he opened the door and showed his head.

"Oh, my child," said the Padishah; [he] embraced and took him home. Ahmat told him, in turn, what had happened to him and that he had killed his brother. "Come, let us shoot arrows **(520)** at each other," said Ahmat to his brother, and they went out. "It is the bigger [i.e. elder] man's turn [first]," Ahmat says to his brother, "you shoot first." His arrow flew through Ahmat's saddle shaft. After that, it was Ahmat's turn to shoot, and his arrow went through his saddle shaft and heart and killed him. He returned to their house, they had a wedding feast, and he married her.

[Ahmat-batir's tale][22] One day Ahmat-batir went hunting, searching and searching but found no game. He looked up into the sky, where a bird was flying. He was about to shoot it when the bird flew into his arms [lap]. Ahmad took the bird from his lap and saw it was all wounded. The bird, returning from the fight, said, "Ahmad, you know, don't torture me; let me go, I will be helpful to you one day." Ahmat took the bird home, healed its wounds, nursed it, and sold what wealth he had to care for it. When he had nothing left to sell, the bird told him, "You have cared for me; you have spent your wealth; now get on my back, let us go home to my father, and let me try to please you."

21 Missing parts in the manuscript.
22 Crossed out in the manuscript.

Ahmat sat on his back. They went, they went, and when crossing a sea, the bird threw Ahmad into it. Ahmat was drowning when the bird picked him up again. Not to pull it out long, they reached the shore of the sea.

The bird takes Ahmad off its back and asks, "Ahmad, what's the matter? You were afraid, weren't you?"

"Yes, I was scared," he says.

"I was afraid too, when you tried to shoot me," says the bird. "Now, I'll take you to our home," says the bird, "I'll tell [them] what I've been through **(521)** and if it hadn't been for this man, I would have died, that he took care of me and saved me. My father will ask thee 'what thou wilt ask of me, my son;' then thou shalt ask my father for the golden propagating chest; if thou put one gold in that chest, it shall be a hundred; if thou put ten gold, it shall be a thousand. My father will give it to you, but on the way home, don't fall asleep when you take it home, for if you fall asleep, the lame devil will steal it."

Then they went to the bird's home and told him everything, but when he asked for the box, he [the father of the bird] said, "No, even if you brought five children like my child, I wouldn't give you the box, I'll provide you with everything else you want."

"If not," says Ahmad, "live long, I want nothing else."

So the bird says to his father, "If you do not give what this lad asks, I will go with him," and with that, it goes out of the house.

The bird's wife says, "Alas, you value something else more than my child; if you don't give him the chest, I'll go away too," and she goes out.

The father of the bird sees that there is no other way out; "Come back," he says, "I will give it to you." Ahmat went back and ran away with the chest.

He walked, he walked, he got tired, lay down, and fell asleep at one point. The lame devil came, took the chest from under the lad's head, and ran away. Soon Ahmad awoke from his sleep, and what did he see: the chest was gone.

The lame devil came and said, "Why were you surprised?"

And Ahmat says that his chest has been stolen.

"If I find your chest," **(522)** says the devil, "and bring it, what will you give me?"

"I have nothing to give you now," says Ahmat, and the devil says, "If you give me the deposit you left at home, I will give you the chest."

Ahmad pondered and pondered, leaving nothing at home but his wife. "If it is not my wife," he says, "I will give it to you."

The devil had him sworn to this and brought and gave him the box; after he gave it, the devil said, "when you left home, your wife was pregnant; now you have a son named Bari; he is the one I will take from you."

Ahmad's son saw in a dream that the devil would take him away. Finally, they arrived at Ahmat's court. The son immediately went out the door and said, "Good health [goodbye], father," and went after the devil. Ahmat got very surprised. With the blessing of the chest, Ahmat recovered his former wealth. Let us come to his son.

The lame devil brought Bari to his house, the devil had a daughter, and she fell in love with him. The devil said to the girl, "Go and lock this boy up somewhere." Lock him up, no way; the girl took the boy straight to her room and said to him, "If my father calls and orders you to do any task, you will come to me [and tell me]."

So one day, the devil summons this and says to him, "Behold, here is a grey horse for you; you will get it used to be mounted, and in the afternoon, you will take it away."

Bari came and told the girl.

"That's not a horse," says the girl, "my father himself will be that horse; if you sit on him, tie his head neatly and begin to beat his head and eyes with the whip, he will then take you and **(523)** want to throw you into the cesspit. When you get near the cesspit, I throw a mace after you; you get off his back and go into the house. When you go in, he's already sitting in the corner; you go in and say, 'I've finished the job you gave me; if there's anything else you need to be done, just let me know.'"

This is how Bari did. The lame devil says, "Go to bed; you'll come again tomorrow at noon." The boy went to her. The next day at noon, Bari went to ask for work.

The devil gave him a handful of millet. "You shall sow this millet, reap it, and make flour in the mill by noon tomorrow."

Bari showed the millet to the girl.

She made everything, but Bari didn't even know it.

The next day, Bari says, "Take the flour," he tells the devil and gives it to him.

"All right," says the devil, "go [back]; you'll come again tomorrow."

In the morning, he went to him again; then the devil told him, "I'd have forty girls flown. The one you like of them, you'll marry."

Bari came to the girl and told her.

The girl says, "I will be one of the girls; I will fly with my feet waving; you will like me; be careful, make no mistake."

Forty girls flew past him the next day, and the devil asked [which one he liked?]. Bari pointed to the girl waving her legs. The devil says to the girls, "Mix yourselves!" They are mixed up. He asked again [the boy]. As the girl was waving her legs again, the lad was not mistaken; he pointed to her again. The girls dropped [to earth]; [the devil] says to the chosen girl, "Take this to your room and cook food, bring half of it to me." The girl went into the room with the lad; they took out the things they needed. As they went out the door, the girl spat on the floor and said to the spit, "You will be e-e-ing (moaning) for two hours [here]," and with that, they themselves left. The lame devil waits for the food to come **(524)** waits and waits, but there is no food. Finally, he was angry, he went to the door and shouted, "What is it, why don't you bring it? The spit says, 'E-e-e,' [then] he's returning this way." He waited a little longer, again [with] no food, came to the door again and broke the door in anger, and saw that nobody was there. He then called his two sons-in-law, one of whom was a little simple-minded. He told them to ride one horse each and bring anything they met on the road. They went on their way.

The girl sensed it, put her ear to the ground, and heard, "My brothers-in-law might be coming; they'll catch up with us. I'll be a cow, and you'll be a shepherd; guard me." She became a cow, and he became a shepherd.

Along the road, her stick in hand, looking after the cow, and her brothers-in-law came.

They asked him, "Hadn't a girl and a young man passed here?"

"No," said the shepherd, "I've been walking in these parts for a long time, but I've never seen anyone."

They returned from this way, and having found nothing when they looked elsewhere, they returned and told him that there was no one but a shepherd. The devil said to them, "You foolish people, the shepherd was the lad, and the cow was the girl; if you had brought the shepherd, the cow would have come after him." They went again. He says to them, "Get whatever you see on the road." Again the girl sensed, "My brothers-in-law are coming; I'll be a tree; you'll be a dragon; you'll roll over me and keep hissing." So they did. Her brothers-in-law came, but as soon as they saw the dragon, they got scared and turned back. They came again to the lame [devil], "There was nothing on the road but a dragon; we saw it and dared not pass it, so we turned back."

He got angry again, "The dragon was the lad **(525)** and the tree the girl; if you had brought the snake, the tree would have rotted there," he said, and he chased them away, "Go, what you have seen, get it." They went away again. The girl smelt it, became a spring, and turned the lad into a drinker. They came to the spring and drank of the water, and the foolish son-in-law, when he had drunk, took the drinking cup and hid it in his bosom, for he wanted to take it. The wiser man, seeing this, quickly put it back as a human good deed, said to him, so made him leave the drinking vessel there, and they returned. Returning, they told the devil about the fountain and the drinking vessel. The devil says, "the spring would have dried up if you had taken the drinking vessel. You will never be a decent human being," he says and sends his wife after them. The woman went on her way. As she set out, the girl put her ears to the ground and listened. "Oh, alas," says she, "my mother is coming; she will eat you and me." She herself has turned into a lake, and he has turned into a duck. Her mother came, and she knew what it was as soon as she saw the lake. She started calling the duck, calling and calling, but he didn't come. She started drinking the water from the pond. She drank it, dried it up, and when she was about to eat the duck, the girl turned into a dragon in her mother's body, killed it by ripping her mother's stomach and came out of her.

They went on their way, and one day they came to the house of Ahmat, Bari's father. They had a wedding feast for forty days and forty nights and married each other. So the two young people reached their goal with each other, and the father with his son.

27 The Tale of Prophet Hizir

(541) Once, there was a Padishah. He told the people of his country, "Whoever finds the Hizir for me and brings him shall have forty thousand gold pieces. Whoever comes and says, 'I will find him' shall have twenty thousand gold pieces in advance, and forty days respite," he says.

A poor man said, "I will find it, my Padishah", picking up the twenty thousand gold pieces; he went home. He began to eat heartily and had no thought of looking for Hizir or hope of finding him.

His wife tells him, "You've taken it upon yourself; go and walk around for at least a few days; if you don't have Hizir, they will cut your head off."

But in vain, her husband replies, "Come on, we will eat and drink for thirty-nine days, and then what will be, will be."

After forty days, he says, "I'm going to the Padishah," and off he went.

While walking, the poor man was joined [from the back] by a child in a turban, so they walked [to the Padishah], said salam*, and sat down. When he arrived in front of the Padishah, the Padishah asked, "Well, have you found Hizir?"

The poor man [only] bowed his head; he had no other answer.

And the Padishah called his Viziers and asked the highest rank, "If you were a Padishah and someone lied to you or cheated you, what punishment would you give them?"

The Vizier replies, "If I were, I'd light the furnace, put them in the furnace, roast it, take it out, hang it between two roads [at a fork in the road], so that those who see it may learn from it."

At this, the poor man bowed his head again. Then the child in the turban spoke, "All things return to their origin."

The Padishah also asked the second Vizier, who said, "If it were me, I would skin them like a sheep, **(542)** hang him on forty [butcher's] hooks."

Then the turbaned child speaks again, "All things return to their origin."

The third Vizier says, "If it were me, I'd pull up a wall of Khorasan lime*, tie two hands and two feet to it, and spread them apart."

Then the turbaned child speaks again, "All things return to their origin."

The fourth Vizier says, "If it were me, I would give him twenty thousand more gold pieces and let him go."

The turbaned child spoke again, "All things return to their origin."

The Padishah says to the turbaned child, "I asked these, and they all said one punishment each, and you said, 'All things return to their origin.' What is the meaning of this?"

The child said, "Your first Vizier's father was a baker, and she returns to him. Your second Vizier's father was a butcher; he returns to him. The father of your third Vizier was a mason; the fourth Vizier is the son of a Vizier. Well, as for me, I am Hizir, peace be upon him," and immediately disappeared.

The Padishah says to the poor man, "Why didn't you tell me you brought Hizir?" The poor man says, "If I had told you, he would have run away. When he said [that] 'All things . . .', I always kept my head bowed, and it meant, 'Take this [i. e. him),' but you didn't seem to know it," he said and got twenty [thousand] more gold pieces and got satisfied. So the three Viziers were driven away, and the fourth one was made a Chief Vizier.

28 An Old Man's Tale

(546) Once upon a time, there was an old man, and he had a son. They had three goats. The old man went to a visit with his wife, and the goats stayed with the son, "Son, he says, don't eat too much curd, you'll fall asleep, and the wolf will come and eat the goats." The boy didn't listen, ate the curd until full, and fell asleep. The wolf came and ate the goats one by one and left. When he woke from sleep, he saw that the goats were gone. So he got to thinking, "Never mind, if my father comes, he will kill me, I will go away from here," he says, and he runs away.

As he walks along the road, he meets two lads. "Come on," they say, "let's be friends." So the three of them walk on. It's evening, the night is dark, and it's raining. They were hungry; they met a stud. "These have a master," they say, "let's look for him, let's buy some bread," and they look for him and find him. They see he is a one-eyed man. They told him about their situation and their hunger. The horseman gave them a mare, "Go to my house in this-and-this place, slaughter it, let the woman cook it and eat." So they go, slaughter; the woman cooked, they ate and went to bed.

In the night, the one-eyed man comes and says to his wife, "Tonight, the things of these [lads] must be finished [done]."

"I don't know," says the wife, "do what you want, I won't interfere."

The young son heard this, [the one-eyed man] cut off the heads of two of these, and when he came to the third [the son], the lad cut off the heads of the one-eyed man and his wife with his sword and threw them into the well, and the property [the animals] and the house became his.

After some time, **(547)** he began to get bored and thought he wanted to bring his father and mother. So he got ready, mounted his horse, and was home in a twinkling of an eye. He sees that his father is dead and his mother is about to die. Feeding his mother with fat and bread, he brings her back to life and takes her to his own house. He gave his mother sixty keys. "Take these," he says, "there are sixty wells here, look at fifty-nine, but don't open one." He himself goes to look after his animals. One day, five days passed like that.

One day, the old woman was bored, "Come on," she said, "I'll have a look at that one well too, and see what's in it." She opened it up, and what did she see: there lies a devil.

As soon as he sees the woman, he starts saying, "A child for the childless, a daughter for the daughterless, a husband for the husbandless."

The woman says, "Neither boy nor girl is needed; I need a husband, but how can I take you out of the well?"

The devil says, "In this-and-this stable, there is a grey stallion, bring me three hairs of his tail, and I will roll myself into it and get out of here."

The old woman brought it, and the devil emerged from the well. They lived happily ever after.

After a little while, the devil says, "Your son, is coming home; if he sees us together, he will be angry and kill me".

"If so," said the woman, "hide in this box; he will not see you," and she put him in the box.

They lived in secret for a long time, and they had a child. The woman says to the devil, "What shall we do with the child now; if we hide it in the box, it will cry there too, and my son will hear it."

"No," says the devil, "wrap it up in something and put it by the side of the road your son will come on, 'I have found it,' saying, he will bring it home. You say at first that you don't need it, and he will answer that 'you bring him up,' that 'he may be a friend to me,' **(548)** and you will bring him up." So she did.

Her son brought the child home.

"No, my son," says his mother, "I'm old, I can't take care of him," but she said no in vain.

Her son says, "Breastfeed him, maybe your milk will come." He forced her to breastfeed.

A long time passed, and the child grew up. So the devil says to the woman, "How long shall I live like this, hiding in the box? First, you had a son; now you have a son again, come on, let's kill your [elder] son, he kept saying."

"All right, but how shall we kill him?" asked the woman.

The devil cut a small piece of flesh from his finger and said, "Make a meal of it; my flesh has poison in it; if he eats it tonight, he will die."

The little boy heard this. The child loved his brother very much. The old woman cooked the food, served it and waited for her son to come.

The child immediately went out into the yard to his brother and said to him, "Brother, while our mother is not eating the food, be careful not to eat."

The lad went in, and they sat down at the table. His mother does not eat.

Her son asked why she did not eat.

"No, my child," says his mother, "I was hungry; I didn't wait for you; I ate too much; I can't eat any more."

"But at least you can have a bite," says her son.

The woman had no choice but to take a bite and die.

The child tells his brother: "My father, the devil, is in the box."

They open the box, kill him, too and throw him into the well.

For a long time, the two of them lived together. The elder brother says to the younger brother, "Two men, how long will we live like this, now let's get married." He gets ready and goes to look for a girl. He tells his younger brother, "Wait for ten days if I don't return, you'll set off looking for me." He goes, he walks, and he comes to a town. In the city, there was a mighty warrior, who made it known that whoever defeated him would **(549)** get his daughter, and if he did not defeat him, his head would be lost. The lad beats him, they have a wedding feast, and he takes his daughter. In another town, a mighty warrior also publishes the same news again. He heard it, too; he wanted to defeat him and take his daughter for his younger brother. They fought for three days; he could not defeat him. On the fourth day, the hero threw this lad into prison. Ten days have passed, and still no brother. His younger brother went to look for him.

On the way, he heard that his brother was in prison. He immediately went to the hero. "Hero," [he] tells him, "you have thrown my brother in jail, so come and fight me."

The hero says, "Where your brother can fit, you can fit too," and so they fight.

The boy lifted the hero three yards high with one blow.

The hero [says], "Oh, alas, don't throw me down, I'll let your brother out, and I'll give you my daughter, I knew there was no one stronger than me; you're an honest lad."

The lad let him down, his brother was brought out, a feast was made, the daughter was taken, and the elder brother went to his house. They had been living together for a long time. One day, the little [younger] lad says to his father-in-law: "See us off; we are going to our own birthplace." This [lad] was seen off; they came to his brother's house. His father-in-law joined in, the lads were seen off with their wives, and the lads came to their own house and could live having attained their wishes.

29 The Rich Man's Daughter

(555) Once, there was a rich man. He had a wife and a daughter. The daughter had reached adulthood. This rich man was attracted to his own daughter. Even though he was ashamed for a long time, one day, he could no longer resist and said to his daughter, "Agree to sleep with me."

"No, father," she said, "don't be aberrant, don't listen to the devil," but it was in vain.

He says, "It will be, or I will lay you out."

The girl [says] to her father, "If it is so, I'll look it up in the book; if thine be the sin, so be it."

The girl looked in the book; since her father was old, of course, the sin fell on him. The girl very, very unwillingly agreed.

A long time passed, and the girl's belly began to grow. The old man saw that the thing would come out, that his wife would find out, he said to his servant, "Take this girl out into the woods; kill and leave her there." So the servant took the girl into the woods, telling her he would stab her. The girl begged him, "Don't kill me, there is an old stable here and here, leave me there and come and get me something to eat and take care of me," she said. The servant thought it over, of course, he saw that it was a pity to kill a man without a crime and took her to the stable. Not long after, this girl gave birth to a son. The servant goes and takes care of them. One day the servant went into the stable again, the door was locked, and he himself remained inside.

The girl says to him, "If you take this child out, the door will open; if you don't, you will stay here."

So the servant says, **(556)** "Where shall I take him?"

"Put him down in my father's yard," says the girl, "go in, and if you say a child is lying here, they have no child, so they will take him and take care of him, and no one will know it."

The servant took him. The elders fed this with milk, and the child grew up and was fit to serve. The old man says, "We don't need a servant now, and they let him go out."

One day, the child, wondering what was in this old stable, said he went in, saw a woman, and asked her, "Are you a man or a genie?"

"No," says the woman, "I am your mother; come, do not be afraid."

The child thinks and says, "I have a mother and a father; how can you be my mother; can a man have two mothers?"

The girl replies, "My child, your father is my father too, and so it is so," she says. After they had sat and talked for a while, the child was about to leave the stable; his mother gave him a shawl and said, "Take this; when you want to eat, you put it down, and all kinds of food come out, you eat, you drink."

The child goes outside but is not at all pleased with his old father's actions and is sad. The old men ask him why you walk so sadly or do you want to get married. "No," says the boy, "I don't want to marry, I don't want to live with you, I'll go away."

They don't want to let him go; they say, "We have cared for you in your youth age, we have nurtured you; when you are fit for work, will you leave us?" they tell him, but he doesn't listen; he goes.

On the way, a bunch of mice come up to him, and he chases them away, saying, "Either you get out from under my feet, or I'll kill you."

The mice [say], "Don't kill us, take a hair from us, you may need us," they say.

It takes a hair from the mice and moves on.

A little later, a swarm of mosquitoes came out and started biting. He starts killing them with his hands. They beg him, "Don't kill us; you may need us." He walks a little further but sees an ice pile on the path he wants to take, but he has no way through. He gets angry at the ice and starts to punch and break the ice. The ice also responds, "Don't break me; you'll need me," he says, "take a piece of me and put it in your pocket." He took some.

He went on from there and became a guest in an old mommy's house. "My child," said the old mother, "where are you going?"

The boy [says], "To marry the fairy's daughter."

"Oh, my child," says the old mother, "fight not; you cannot take her; the man who goes there will not return."

"What will be, will be," says the lad, "if I die, I die, I will still go and see."

"My child," says the old mother, "[I see] all is in vain; you will go anyway, I will teach you. He [the fairy] gives you three mares to look after, but they are not mares; they are her daughters. In the morning, thou shalt take them, thou shalt not depart from them, and if thou do not lose them, in the evening, if thou canst bring them back, thou mayest take her [the fairy girl]."

The lad reached the fairy's house, greeted them, and entered. The fairies greeted him, "Luck has come to us," they said and welcomed him. After a little sitting, the fairy asked, "What business have you come for?"

Not at all shyly, the lad replies, "I have come to marry your younger daughter."

The fairy [says], "Well, maybe, but tomorrow I will give you three mares; if you don't lose them and bring them back by the evening, then maybe."

It was morning, and he took the mares and went into the forest with them. They had no sooner reached the woods than the mares got out of his hands and ran away. **(558)** He was amazed; he did not know what to do. All the same, he says, "They will kill me; at least I will eat and be well-fed," he says, and he lays down his shawl. As he eats, he thinks of the mosquitoes in his way and calls them. The mosquitoes came out in droves. "What is your wish?" they ask him. The lad, "In the forest, I have three mares; chase them and bring them to me," he says. The mosquitoes immediately went into the woods and began to peck at the mares. The mares came out of the forest together, came to him, took them and drove them home.

The fairy looked at him. "You've done one of the three tasks, two more to go," he said. "Tomorrow, you will select the grains from these forty stacks of wheat, but without scattering the stacks, you will make flour having it ground in the mill and then bake bread of it," he says.

The lad thinks, and then he remembers the mice; he has lit their hairs on fire, and they have come gathering. They began to sort out the wheat grains. Till the morning, they sorted them out, made flour and brought it back. In the morning, the lad baked bread and got it to the fairy. The fairy was astonished when he saw it. He says, "I have one more condition, and if you meet it, you can marry my daughter." The fairy heated the bath to an unbearable temperature and sent the lad in to be burnt. The lad went to the door; there was no way in. He took the piece of ice from his pocket and threw it in. It [the fire] went immediately out. The lad went in and washed up.

A little later, now he is **(559)** obviously burnt, the fairy thinks, and he goes into the bath and sees that the lad is healthy. [The fairy is very surprised.] "You have fulfilled three of the three conditions; there is no other way out," he said, "you marry my daughter, then we have a [wedding] feast, and he left. This [fairy] went to his brother and told him what happened in detail.

His brother says, "Then you can't hurt him anymore, marry your daughter and drive him out of your house; he'll destroy you too". The fairy returns and says to the lad, "My child, you will marry my daughter, but you cannot stay here. Our enemies are many; they will make trouble for you too." So deceived that he sent him back to his own country. They came back as husband and wife, he came to his own father's house and saw that his old father was dead, and his old mother was left alone. The lad brought his own mother out of the stable, and they lived eating-drinking from the tablecloth.

30 A Bald Boy's Tale

(565) There was once an old mother, and she had a bald son. Their whole fortune was a goose. One day the bald boy said to his mother, "I'll take this goose to the bazaar and sell it so that this one won't be left either."

"I sell goose," he says, and a young man comes up to him and says to him, "It's not a goose, it's only a chicken." The baldhead turns around the corner and goes to another market. He [i.e. the former young man] steps in front of him again, and wherever he turns, the man appears and says, "It's not a goose, it's a chicken." There was no escape, the bald boy sold it for chicken, and he himself followed the man who bought his goose, remembered his house and hid in a place. The young man went to his mother and said to her, "Look, mother, I have deceived a baldhead today and bought the goose from him, saying it was a chicken." The baldhead heard this. The young man who cheated [happened to be] the son-in-law of the Padishah, says to his mother, "Tomorrow, I will be at my father-in-law's house, you will roast this goose nicely in fat, stuff the inside with gold, I will send a child for it, he will bring it." The baldhead heard that too.

He hid there until noon the next day. The young man's mother roasted the goose and stuffed it inside with gold. Soon the baldhead shouted from the gate, "Old mother, is the goose ready?" The old mother brought out the goose and gave it to him. On his way out, the baldhead wrote on the gate, "My goose, my goose, owner of the goose, this was my little joke [trick], the bigger one will come after this," and with that, he left. In the meantime, a man from the Padishah's house comes and asks for the goose. "But I have already sent the goose; they have taken it away," **(566)** says the old mother from inside. This child went and told the lad; they looked, but the goose didn't come. When he got home, he asked his mother, "Who did you send the goose with?" At the same time, he looked around and saw the writing on the gate, so he knew who it was.

The baldhead ate the goose and took the gold back home. The Padishah sent soldiers everywhere [flooding the city with soldiers], and he started looking for him. This baldhead immediately went to a shopkeeper and said to him, "I'll give you a hundred roubles; let me sit in your shop for an hour. The shopkeeper agreed. The baldhead was sitting in front of the shop, nicely dressed. He saw soldiers approaching from one side. The baldhead approached them and asked the officers where they were going.

The officers [say], "We're looking for a bald boy. Hasn't he passed in this place?" they ask him.

The baldhead says to them, "Alas, [if] you go with weapons, the thief will see you and hide. So put your weapons into this shop, go without weapons, so no one will have any hope [of escape, of hiding]."

They saw sense in this word [speech], and after putting their weapons into the shop, they themselves went away. The baldhead locked the shop door and wrote on it, "My goose, my goose, owner of the goose, this was my little joke [trick], the bigger one will come after this," and he ran away.

They [the soldiers] searched for a long time, [but] when they could not find him, they came back and saw that the shop was closed and the weapons were not there. They were humiliated; they went to the Padishah and told him [what had happened]. He was in our hands, they said, but we couldn't hold him; he must be a clever, skilful man.

One day another Padishah overheard this. "What kind of realm is it," he says, laughing, "that cannot catch a thief on its own property [territory]. If the thief who has done these **(567)** things steals a hen of that Padishah's and brings it [to me], I will marry my daughter to him."

The bald boy heard this and went to the Padishah and said to him, "My Padishah, I will not only take a hen from him, but I will also take himself if you want." So he got a few days' respite and left.

On the way, he met a shepherd. He had the shepherd cut off one of his goats, had the skin of it pelted, and gave him fifteen roubles for the skin alone. He went to a master bell-maker in the town, had bells hung on the ends of the goat's hairs, and went back to the country of his Padishah, and lived there for a month. In the meantime, he went to his Padishah to look after the horses and memorised everything in his yard where it was. One night, he put on the goatskin bells and went to the gate of the Padishah. The guard said, "Stop, where are you going?"

Then he said, "Don't cry out, I am Azrail, I will take your soul at once, and he rattled the skin."

The rifle fell from the guard's hand, and he went to the second gate; when the guard stood before him there, he did likewise to him, then went to where the Padishah slept and sat on his breast.

The Padishah woke up right away, and when he asked who it was, he said, "Don't shout, I am Azrail," and shook the goatskin.

The Padishah said in great fear, "Oh, Azrail, give me three days' respite." I will," says he, "But tell no one that I have come to see you, for if you do, I will take your soul by force," and he went home.

The Padishah had his household called together and said goodbye to them. The next night, after he had gone to bed, the baldhead returned, bringing a **(568)** chest. He woke up the Padishah and said to him, "Go into this chest, I will not take your soul here, but I will take it up to the seventh heaven. But as we go, when I greet the angels, be careful not to move, make no sound if you cry out, I will let you fall, and you will die torn apart," he says. So he put him in the box and closed it up tight. Then, putting it on a horse, he went away. So they went and went, and when they came near the former Padishah's house, he said, "If I bang my fist on the crate, you will bark like a dog."

He came to the Padishah's house, left [the box] in the yard, and he went inside. When the Padishah asked, "Well, what have you brought?" he said, "I have got what you wanted," and he brought the box up and began to beat the top of it. That one inside barked like a dog. When he opened the box, the Padishah got out and saw that he was in a palace like his own. The Padishah, the house owner, says to him, "Look, this man stole from my*[23] [your!] son-in-law, you* [I!] laughed at me* [you!], but now he has stolen you yourself," he says. This [the 1st Padishah] [on hearing this], was frightened to death, died and was

23 The roles of the two different Padishahs in the story were confused by the narrator at the end of the story. The wrong word is marked with asterisk, the logically correct words are shown in square brackets.

buried. The girl who would have been given to the lad who deceived the baldhead was given to the baldhead. The baldhead achieved his goal and lived happily ever after.

31 Two Children of a Rich Man

(573) There was once a rich man, and he had a daughter and a son. When the rich man was about to die, he told his son and daughter, "If I die, one of you will not marry a wife, and one of you will not marry a man. My wealth will take care of you until you die; you will have something to eat and somewhere to live." One day the rich man died. A long time passed, and they would not marry. The people of the town say to the boy: "You have become a lad of the most beautiful age, why don't you marry? You yourself shall take a wife, and your sister shall marry." When he got home, he discussed the matter with his sister. She says, "I don't know, you know if you want to marry, marry, I'll keep my father's orders." The lad thought and thought, but in the end, he married.

He had been living with his wife for a few months. He had a shop in another town; he went to check it out. His wife and his sister had a fight. The wife broke all the valuable things in the house and wrote her husband a paper saying, "Your sister got crazy, left nothing at home, ruined everything." The lad came home. "Indeed, she broke things," he says, "but it's all right; the property is ours, left to us by our father; if she broke it, she broke it." He didn't say a word to his sister.

Another few days passed, and his wife had a child. The lad went again to his shop in the other town. The wife stabbed her own child, likewise wrote a paper saying, "Your sister stabbed him [i.e. the child]."

The lad read the letter, and even if he didn't believe it, he returned. **(574)** He asks his sister, "Why did you stab him?"

"No," says the girl, "I didn't see him; I didn't stab him; your wife stabbed him."

The lad thinks, "A man, a woman does not stab their own child," and give credit to his wife. So he made his servant harness to the carriage and said to his sister, "Get dressed; let's go for a walk into town." She got dressed, the cart arrived, and he and his sister got in and set off. They went round the city, and then the lad said to his sister, "Let's go and see this forest." So they went and entered the forest. He made the girl get off; he took the axe he had prepared for the cart, split a tree in two, put her two hands in the crack of the tree and said, "You stabbed my child; here is where you suffer," he said and went home. The girl stayed there crying and crying.

After a long time, the girl's two arms fell off, rotting. The girl gets up, walks away from there and ends up in a rich man's garden. She hides daily and lives in the garden at night, feeding on apples. One day the rich man comes out into the garden and sees that one side of many apples on the tree has been eaten. He shows the guard. "Who eats apples like this?" he asks and leaves. While the guard sleeps, the girl comes again and eats. In the morning, the rich man sees that one side of his many apples has been eaten again. He is angry with the guard. The next night he sends his son and says, "You look on one side, the guard on the other, and catch them at all costs."

Once in the night, the son hears something cracking and goes to see who it is. "Who's it?" he shouts.

From that side, "I am the child of a man like you," **(575)** it sounds, "but I am hungry and even naked; I have no time [i.e. way] to show myself to anyone; I am ashamed; bring me a suit of clothes, let me get dressed, then I will appear before you and tell you my fate."

He brought clothes, and she dressed up. She was very beautiful. The younker fell in love with her. The lad [says], "I will marry you."

"No," says the girl, "my father had decreed before his death that I should not marry, but I am weak, I am not fit to marry you."

"You know," says the lad, "I will give you to an old woman, you will sleep there, my father will see you in the morning, and then, if he forgives your sin, you go," he says, and he took her to an old woman's house. The old woman fed her, watered her and put her to bed. And the boy went to his father and told him who had eaten the apples, and he also said, "If that girl came to me, I would marry her."

His father says, "Are you mad, marrying a girl of unknown origin and race, who walks here and there at night, is not suitable for you," but the boy says, "No, I will marry her anyway."

"Last of all, a noose around your neck, but do what you like," says the father.

So the boy goes to the old woman and says, "I'll give you many things; persuade this girl to marry me."

The woman says to the girl, "If it doesn't happen like this, it happens like that; you don't have arms,[24] who will take care of you? This is a rich man's son, marry him; your father is dead and gone, and you must not obey his orders," so she persuaded her. The girl was inclined to the word; [the lad] with a great wedding feast married her, then they lived on together. This lad has gone somewhere [on business]. His wife **(576)** gave birth to a beautiful male child. His father wrote a paper[25] to his son, "Come home on such and such a day; you have a beautiful son; we will give him a name." The lad was [just] staying in the house of the girl's brother. This paper has been brought and handed over. The paper first went to the wife of the son[26] of the [dead] rich man. She opens it, reads it and finds out that her sister-in-law is alive, so he wrote [another paper saying], "Your wife has given birth to a child. It is not a human if you say it is a human; it is not an animal if you say it is an animal; what shall we do with it? Shall we keep it [alive] until you come back, or shall we destroy it?"

The second rich man's son [the younker][27] came, the woman gave him the letter, and the lad[28] read it. As if she knew nothing, the woman asked what they were writing.

---

24 Actually, hands.

25 Actually, a letter.

26 I.e. the girl's brother

27 The guest, the recipient of the letter.

28 He is no longer a lad, but the storyteller still refers to him as a "Jiget", or young lad.

"Nothing," says the lad, "a shop of ours burnt down, nothing else." So the lad wrote to his father, "By the time I get home, destroy my wife and child."

His father has received the paper, but he is sorry to kill this. "Take your child and leave here, so wrote your husband," he says, and, tying the child to her waist, they sent them away from the house. The woman went away.

Those days being sweltering, the woman got thirsty and went down into a well, but the child began to choke. Meanwhile, a voice in her ear said, "Reach out your hand and take the child!" So the woman reached out her hand, and her hand was as it had been before; she could pick up the child.

Slowly, she made her way to her hometown, her brother's house, and begged to be taken in as a guest. They don't recognise her. Her brother and her husband tell her, "If you tell us a tale, we will take you in."

"I don't know any tales," she says, "but I'll tell you what happened to me."

Her sister-in-law **(577)** has felt something and says, "Come on, what can a beggar like this say, don't let such an unclean person into the house."

"No," says her brother, "she can't be refused; it's a pity for her; she has a child; let her stay a while; let her lie down somewhere on the side."

The guest's child started running around. The girl's brother takes a liking to the child and says, "Give him to us; you are weak, and it will be hard for you to carry him, I will take care of him and bring him up."

"No," says the woman, "it's my treasure; when I'm bored, I'll amuse myself with it." They had fun with the child and forgot what they had asked the woman to do. As they finished their coffee and tea and were about to leave, they remembered, "Oh, we wanted to ask her what happened to her," they said, and went back to their seats.

Her sister-in-law got angry again, "Mind your own business," she said, "you don't need to know what happened to her."

"Our business is none of your business," they say [to her]. "Tell [your story]!"

"I'll only tell it," says the woman, "if no one interferes." She began to tell the story of her father's will. After she had told all, they understood. Her brother immediately cuts off his wife's head. The other rich man's son also took his wife and child home, and they lived their lives achieving their desires.

32 A Tale of Seven Girls

(583) Once upon a time, there was an old man and an old woman, and they had seven daughters, a dog and a cat. The husband and wife were going to visit. On their departure, they say to their daughters, "When we are gone, be careful not to beat this cat; if you hit it you will be destroyed," they said, and they went away. One day after they left, the cat peed on the window, and one of the girls hit it. The cat went and peed on the fire and put the fire out. They lived without fire for two days. There is no fire in this place. One of the girls went out into the courtyard, and as she looked this way and that way, smoke was seen in a distant place. She goes to get fire. As she gets there, she sees that an old mother lives there, greets her and goes in. "My child, you've come at the right time," she says, "let me lie down on your knees and massage my head." She does it. They talk to each other. The girl says, "Old mother, come to live with us; our mother and father are not at home, you have no one, we will have fun." A little later, the old woman gives her some fire, a little ash in one hand, and says, "You go on spreading this ash, I will go to your house following the ash."

The girl went away, came home and said to her sisters, "I have found some fire and an old mother." They also had a dog. The old [woman] came, but the dog wouldn't let her come near. The old [woman] turned around and went away, the girls were waiting for her, but she didn't come, they went to bed and fell asleep.

It was morning, two girls went to the old [woman] and asked her, "Why didn't you come last night?" **(584)**

The old woman said, "I went, but your dog wouldn't let me into the yard. Break the dog's two legs so it cannot walk, and then I will go." The girls return and break the dog's legs. In the evening, the old woman came again. The dog, even without legs, did not let her into the court again. In the morning, the girls went again and asked. The old mother says, "I will not come to you until you kill your dog and bury him." The girls came [back], killed, and buried the dog. In the evening, the old mother arrived; they ate, drank and went to bed.

The oldest girl was lying next to the old mother. In the morning, they woke up and saw that one girl was missing; the old woman had eaten her. The girls asked about their sister. The mother says, "She didn't go to bed like you when you fell asleep; she went on a pilgrimage." It was night again; they went to bed. In the night, he ate two of them again. In the morning, they asked again, and the old woman said what she had said before. In the evening, they went to bed again, and the old woman started eating one of the girls again.

The youngest girl wakes up and sees the old woman eating something. "Old mother, what are you eating?" she asks her.

"My husband left me a few things," she says.

The girl says nothing but guesses what the old woman is eating. The old woman has fallen asleep. This girl woke her sisters; they took a whetstone, a comb, and a mirror and escaped. In the morning, the old woman wakes up, sees no girls, turns a mortar into a horse, sits on it, turns a mallet into a kamchi*, takes it in her hands and starts chasing the girls.

The girls were going; the old [woman] was going. The old [woman] was approaching the girls. As soon as the girls saw her, they threw the comb behind their backs. **(585)** The comb has become a thick forest, and the old woman sat down[29] and made a whetstone of it; spitting one [from the spit] she made an axe and began to cut the forest.

29 I.e., to cut out.

When little was left, a fox came and said, "Old mother, you are obviously tired, go, now I will cut it." The old woman gave her axe to the fox, and she lay down to sleep. The fox cut a little, left the axe in the meadow, and ran away. The old woman woke up and saw that the fox had gone. Then she took her axe and began to cut. A little later, another fox came and asked for the axe.

"No," says the old [woman], "one of you had already come to cut; he had deceived me and gone."

"No," says the fox, "it was a red fox, I'm black, I'm not fooling you," but he says [so] in vain.

"You're all the same," said the old mother, and she cut off the fox's head and threw it away, and she cutting and cutting finished the forest. When she got near the girls, the whetstone was thrown down, and a big rock was made. The old woman went up, but she fell back down. Much time has passed.

After a few days, however, with a great struggle, she got up and almost caught up with the girls. The girls threw the mirror to the ground; it became a big sea. The girls on that side of it, the old woman on this side.

The old woman asks the girls, "How you got to the other side?"

The girls answer, "We tied a stone to our feet, got on a shell and came here."

The old woman also ties big rocks on her two feet, sits down in a shell, immediately dives into the sea, and drowns. The girls got free. They go, they go, came to a forest, ate wild apples, then went up a tree and sat there.

(586) The Padishah and his Viziers went just into that forest that day. They sat down under the tree on which the girls were sitting. One of the girls had to pee and peed from there; it fell like rain on the heads of the Padishah and the Viziers. They thought it was raining. One of them said, "No, it can't be raining in such fine weather; there are people in the tree." So they sent one of the soldiers up the tree, and he saw three girls. The soldier took the news, and they made the girls come down. The Padishah fell in love with them, married all three, and there was his wife, so there were four.

One day the Padishah was about to go for an excursion to another town, and he called his wives to him and said, "I'm going to such and such a place; what will you prepare for my return?" The older woman promises various dishes, the second woman promises trousers made of fox skin, the third promises gloves made of lice skin, and the fourth says, "I will give birth to a child with a golden head." The Padishah has gone. As his return approached, each began to make the things he had promised. Finally, the fourth woman gave birth to a child with a golden head. But the three women plotted together enviously. They stole the child and took him to the forest. They put a puppy in its place. In the woods, a witch found the child and took it to her house and cared for it.

When the Padishah returned, the women welcomed him and took out their prepared things. When he asked about his fourth wife, the women said, "Your wife, who wanted to make a child with a golden head, **(587)** had given birth to a puppy."

"Take her and her child quickly into the woods and leave her there; keep out of my sight," says the Padishah. They took her away.

A long time passed. One day, the puppy was walking in the woods and saw the golden-headed child; he said [in human language], "Come on, I will take you to your mother."

"No," says the child, "my mother is here."

"No," says the puppy, "she is not your mother," and he takes him to his mother, telling him everything.

But the witch has heard this, and she catches them up and says, "My child, give her to me."

They are quarrelling.

At last, the witch finds out that the child belongs to this woman, and she gives it to her mother and says, "Take as much gold as you like from my house here and here." So they went and took a lot of gold and built a grand palace.

One day, one of the Viziers of the Padishah passed by saw the palace and recognised the woman. The Vizier entered as a guest, and the woman told him everything that had happened to her, and the Vizier told it to the Padishah. The Padishah said nothing to anyone, went to see her and saw that she was indeed his wife. He brought her. He imposed various punishments on the three women and had them killed, and he lived with the fourth.

33 The Tale of Lord Five Grape Bunches

(594) At the beginning of time, in the month of the Last Judgement, [the frogs suddenly grew wings to fly, the sea fish hired (a cart) to migrate, gentleman-uncle's foxhound rolled up his trouser leg to cross the Black Sea when my mother was in the cradle, and my father was in the lap, and I was not yet in the world, I was a footman at my father's wedding][30] there was a gentleman called [Lord] Five Grape Bunches, he had five bunches of grapes. As he had no other wealth, he lived every year by selling these grapes. Lord Five Grape Bunches, as soon as his grapes ripened, he put them in his hut. Before he could eat it himself, a fox would grow into the habit of eating grapes. One day it goes and eats the grapes, and Lord Five Grape Bunches starved. The lord hunts the fox for a few days to catch it, but he can't catch it, until finally, after a long hunt, he does.

The fox says, "My lord brother, let me go; you may need my help one day."

"You are a fox," says the lord; "what can you do for me?"

The fox says, "Whatever you need, I will do."

"If so, in this and this place, there is a Padishah, I have fallen in love with the daughter of the Padishah, I cannot afford to marry her; she will marry a rich man. I have no wealth; if you take this girl, I will have no other wish," says the lord.

30 It is a fragment of a *tekerleme**.

The fox says, "My lord brother, Five Grape Bunches, know that I will do this for thee," and he has given its word.

Lord Five Grape Bunches says, "If you can do this for me, I will set you free, and so he lets it go."

In this way, six months, **(595)** autumn has passed, the fox walks mountain and valley all smooth; he walks in mud up to his waist; clear weather is pitch dark; wolves and dogs howl together, he himself, who is such a valiant lad turned weak [soft].[31]

Just at the end, it goes to the Padishah's house, up forty steps, next to the stairs there was a chair [the throne], on which he was sitting every day. It goes there, sits on the chair [throne], and the Padishah's Viziers come and see it. They go and tell the Padishah, "My Padishah, on your golden chair [throne], a fox is sitting."

"Go," says the Padishah, "hit it with a stick, and let it go."

The Viziers come to chase it away.

The fox says, "You must not chase me away, I have a word to speak to the Padishah."

They go to the Padishah and tell him.

The Padishah says, "Let it come and tell me, let me see what it has in mind." The fox goes to him and greets him, "Salam, my Padishah."

The Padishah accepts its greeting and says to it, "My younger brother fox, what do you want? Why have you come?"

"My Padishah," says the fox, "there is a gentleman here called Lord Five Grape Bunches, and he could not find a vessel [a bushel] anywhere to weigh the silver in his treasury. So he sent me to you to ask you for your silver measuring bushel."

The Padishah is amazed: what kind of a man can this be, that he has so much money? and he says, "Hey, my younger brother fox, tell me, can't you get this lad for me, as my son-in-law?"

"My Padishah," says the fox, "it is straightforward to make him your son-in-law, but you must accept him as your inner son-in-law; otherwise, he will not marry your daughter."

"Why?" he asks.

"My Padishah, the air of the place where he lives is not very suitable for him, so he thinks he wants to be the inner son-in-law of a Padishah," **(596)** says (he). "If so, brother, do this for me," says he. The fox says, "You may be sure of it, my Padishah."

The Padishah gives it the bushel to measure the silver and sends it away. The fox takes the bushel and goes out into the field. When he takes the bushel from the Padishah, there is a silver coin stuck in one edge. The fox thinks, not knowing what to do, how to deceive the Padishah. He turns the bushel this way, turns it that way and sees the silver. He pulls the silver out of the corner of the bushel, puts it in and goes to the Padishah.

---

31 It is a fragment of a *tekerleme**.

The Padishah sees that there is a silver coin left in the bushel. "My brother fox," he says, "there's a silver coin left in the bushel, don't leave it there, take it out."

"My Padishah," says the fox, "our Lord Five Grape Bunches like this has so much silver scattered along the road; what shall I do with this one coin," and with that, it leaves it there. Then, having amused itself for a few days, it goes again to the Padishah and says to the Viziers, "I want to see the Padishah". They send word to the Padishah, the Padishah comes out, and the fox asks the Padishah for a gold-measuring bushel, and the Padishah gives it to it. The fox takes it, and again, as before, he turns it to and fro and sees that a gold piece is stuck in the corner of the bushel. The fox puts the gold in the bushel and brings it to the Padishah, saying, "Here you are, my Padishah, I have done my work."

The Padishah sees that there is one gold left in the bushel. "My brother fox," he says, "there is one gold left in the bushel."

"My Padishah," says the fox, "what shall I do with it if I take it away, so much gold like this has fallen **(597)** by the wayside."

Then he leaves there and goes to Lord Five Grape Bunches, knocking on his door in the evening. The lord comes to the door and asks who it is. "It's me," says the fox. The lord opens the door, and the fox says salam*, and the lord accepts its salam. "Welcome, brother, I am pleased to see you."

And the fox goes up to the place of honour and sits down, and after inquiring about each other's health and welfare, the lord asks the fox, "What business hast thou come hither?"

The fox says, "By the command of Allah I will give the daughter of the Padishah to thee, if thou wilt consent, and my sin of eating five bunches of grapes once was forgiven thee, and I will repay thee and do thee some good that I said I had done. And the Padishah, when I went into his court to eat hens, took me for a chase to kill me, and thou hast no other goods now but thy five bunches of grapes, now come and I will take the Padishah's daughter for thee."

The lord says, "My fox brother, the Padishah will not give his daughter to a poor man like me," but he sees that the fox [does not budge].

"You just agree to take her!" says it, "the Padishah will give her to you, and I will take her from him."

The gentleman agrees.

The poor fox sets off on foot, arrives at the Padishah's palace, climbs forty steps, and sits in the golden chair. The Viziers come and see a fox sitting on the chair. They go and tell the Padishah. "Call it," says the Padishah, "let it come to me." The fox goes to the Padishah, and the Padishah welcomes it. **(598)**

"Allah be pleased with you, my Padishah," says the fox.

"In what matter have you come, brother?" asks the Padishah.

"My Padishah," says the fox, "I have come to you as our lord's suitor; I ask for your daughter by Allah's command."

The Padishah agrees, and the wedding is arranged.

The fox goes from there and comes to the lord. "The Padishah has accepted you as his son-in-law, and he wants to see you. Now we must go there."

"With this appearance," says the lord, "how can I go to him?"

The fox says, "Let's go; we'll find a way." So they set off, and when they come near the Padishah's palace, the fox says, "In this place in the swamp, roll around nicely, I am going to the Padishah." The gentleman in the marsh by the spring has well been covered in mud.

The fox goes to the Padishah, "My Padishah, as I was bringing your son-in-law, Lord Five Grape Bunches, in a carriage, the horses were frightened by something, and they turned your son-in-law's carriage into the marsh near a spring; hurry, everything was covered with mud and slush, the horses ran away, where they went is unknown. There he is, waiting. Please give me a change of clothes and a three-horse cart to fetch him." The Padishah immediately gives him a three-horse chariot and a set of fine clothes, has the servants load the clothes onto the chariot and sends it. The poor gentleman sees a three-horse cart coming, and they come to the spring. The fox gives the gentleman the clothes there, washes him well in the spring and dresses him. At the same time, he says to him, "Take care that when you go in this dress, you do not look at it too much, as a man who has never seen such a dress."

They board the carriage **(599)** and arrive near the Padishah's palace. The Padishah comes out to receive the son-in-law-child. The Viziers also welcome and lock him up with the Padishah's daughter for the wedding night. The lord makes also the Padishah's daughter eat a bunch of grapes, the wedding feast is held, then the Padishah holds the djez alti*, they come to see the son-in-law. [As if cursed is the day of the one who has not seen him.][32] Lord Five Grape Bunches beats off the dust of his clothes with his finger. The Padishah notices this and says, "Wow, it seems he's never seen such a thing." Hearing him say such a thing, the poor fox comes at once and says to the lord, "Never mind your clothes, or the Padishah will give you to the executioner." [No one loved ignorant, backward people less than the Padishah.][33]

Then the fox quickly goes to the Padishah. "My Padishah," says it, "Your son-in-law does not like the way you have made the clothes for him; he wants another suit of clothes; can't you have another made for him?"

"Very well," says the Padishah, "a suit for my son-in-law."

He has it made. The gentleman takes off those clothes, wears an even nicer suit, and lives like the Padishah himself.

One day the fox, when there was no one in the lord's room, said, "Oh, sir, when you had nothing but five bunches of grapes, [did you think] you would have such a life, no one in this country lives like you. I have made thee the son-in-law of the Padishah, now

---

32 The sentence is crossed out in the original manuscript.

33 Crossed out in the original manuscript.

art thou greater than all, now art thou higher, now art thou more tranquil than all, now how wilt thou return my kindness to you?"

The lord replies, "Oh, brother fox, how can I repay your kindness, when you are dead, I will not bury you in the ground, I will make you a golden cage and hang it on the ceiling above my sleeping place, **(600)** I will never forget you." So one day, the fox will apparently fall and die.

The servants of the lord see the fox dead, and they come and tell the lord. "Tell the workers," says the lord, "to put it in a sack and take it to a secluded place outside the town, throw it out so it doesn't stink here." The servants go and tell the workers. The workers immediately put it in a sack, take it outside the city, and leave it in a ditch in a secluded place. As soon as the workers have gone, the poor fox slowly gets up, follows them, and goes straight to the lord's palace. The lord sits sprawled out. "My lord, my lord-brother," says it, "I know how you keep your word and is this what you would do to me?" The lord does not know what to say. For a while, the fox lived again by the lord's side. One day the fox dies, but he does not believe the fox is dead. He says it is my test again, and he has a golden cage made and hangs it from the ceiling of his resting-place. This gentleman Lord Five Grape Bunches marries the Padishah's daughter, and the fox lies in a golden cage. They have achieved their goal.

34 The Tale of the Thief and the Pickpocket

(608) Once upon a time, there was a thief; once upon a time, there wasn't; he had a mistress. The thief lies at home during the day, and when it gets dark at night, he goes out to steal. He doesn't come home all night, only in the morning. His lover also finds another man for herself, and he is a pickpocket who is at home at night. When morning comes, he leaves. These two were not aware of each other. They live like this for quite some time.

One day, the thief tells his wife, "I'm going to another town; there's nothing left to steal here; cook me something to eat." The wife makes a meat pie. She gets the food and sets off on her journey. The journey is also a few days. He travels for two days and comes to a spring, there are big trees near it, and he lies down at the foot of the trees. The pickpocket also goes to his wife one day and says, "I am going to another town. Prepare some food." The woman cooked the pickpocket the same food she cooked the thief. He takes the food, sets off, and comes straight to where the thief lies. He says salam, then sits down.

"Where are you from?" asks one of them, and he says where he is from.

The thief says, "We both belong to the same village." They sit a little longer and take out their food to eat. The thief sees that the pickpocket's food is the same [from the same place] and says, "This food is my wife's cooking."

The pickpocket says it is his wife's cooking.

They ask each other, "In which part of the village do you live?"

They find out that they both live in the same place. They start to quarrel.

They turn around and go home. The thief says, “My wife,” the pickpocket says, “My wife.”

The woman says to them, “I am the wife of both of you, **(609)** but you did not know about each other.”

“Now, what are we doing?” they ask.

The woman says, “Test yourselves; whichever of you wins will have me.”

The thief has got a piece of silver; he takes the silver and goes to sell it. The pickpocket sees this, goes to his side, steals the silver from his pocket, takes it to his wife and says to her, “Be careful; if the thief comes, don’t say I brought it; say, ‘You left it here.’”

The thief sees that the silver is not in his pocket, I must have left it at home, he says. He comes home and asks the woman. The woman says, “Here, it’s still here.” The thief puts the silver in his pocket and goes off to sell it again. The pickpocket steals it again. He goes a little further and sees that there is no silver. “Oh,” says he, “the pickpocket has stolen it.” And he took it to the woman again. “Be careful,” he says, “don’t tell him I gave it to you.” The thief comes, and the woman takes the silver and gives it to him. The thief takes it and binds it under his knee. The pickpocket sees this too. When the thief walks down the road, the pickpocket comes up to him, takes the thief’s hat and throws it on the roof. The thief asks why he threw his hat up. “I was joking,” he replies. “Take it off and give me my hat. My feet are sore,” he says; “you stand on my shoulder and take off your hat.” The thief stands on the shoulder of the pickpocket, and when he takes off the hat, the pickpocket steals the silver.

The thief takes his hat, goes to the man he wants to sell the silver to, and sees that the silver is missing again. He returns home and says to the pickpocket, “That’s your skill; you’ll see mine tonight.” They sit till evening. At one time in the night, they leave. The thief goes straight to the courtyard of the Padishah; the pickpocket is afraid. **(610)** The thief takes a goose straight from the goose pen, brings it, cuts it, and gives it to the pickpocket, lights a fire, and says, “I will go into the Padishah’s palace, talk to the Padishah a little, and then I will speak to you, and tell you to turn the goose so that it does not burn, and then you will roast the goose on the fire.” The thief goes into the Padishah’s palace. The Padishah is asleep, and his servant is sleeping beside him. He gags the servant’s mouth, ties his arms, grabs him and takes him to another room, and he lies down in the servant’s place.

Soon the Padishah will wake up. “My Padishah, tell me, are you bored, or shall I tell you a story?”

“Tell me,” says the Padishah.

“Once upon a time,” he says, “there was a thief and a pickpocket,” and he tells his story. When he had finished, the thief says, “Turn the goose so that it does not burn; he that came in, let him not go out.”

“‘Turn the goose so that it does not burn.’ What does that mean?” asks the Padishah.

The amusement of the conversation, he tells the Padishah, is that while he was telling the story, the pickpocket in the courtyard was baking the goose.

"In that case, whose is the woman?" says the Padishah. "The thief's craft is greater, and the woman will be the thief's."

So the tale ends; the Padishah [falls] asleep; the thief goes out of the palace; he says to the pickpocket, "Come on, let's go."

They come home.

"Did you hear what the Padishah said?"

"I heard," says the pickpocket.

In the morning, the Padishah wakes up, sees that his servant is not there, and starts to cry out, but the servant does not come. He himself gets up and looks for him and sees him lying in one of the rooms, his mouth and hands tied. He unties him and asks, "What has happened to you?"

"I don't know, my Padishah," he says, "I went to bed with you last night, and when I woke up, I saw myself in this state."

The Padishah finds out there was a thief there last night. He has the bellman proclaim **(611)** in town: "Whoever went into the Padishah's palace last night, whoever he may be, come, the Padishah will call him, no harm will come to him." They go to the Padishah and tell their story one by one.

"The woman belongs to the thief," says the Padishah. He takes out silver and gives them five hundred [coins] each. To the pickpocket, he says, "You go out of this town, don't come back again." And to the thief, he says, "Don't steal anymore!" The woman becomes engaged to the thief, they give her to the thief, and the thief also lives his life contentedly, happily ever after.

35 A Tale of Two Boys

(616) At one time, a man had two sons. They spent one out of every two days wrestling. One day their father got angry with the older one and chased him out of his house. And he goes away unawares, goes to Yalta and starts to work as a porter. At night he lies in a man's café. One day he gets tired of carrying bags tells the café keeper his troubles and asks him for advice. The café keeper was a woman-chaser and said, "I will teach you, how you will earn money with your tool. Here are five paras* for you; buy a cucumber for five paras, then go to the yellow bath. Sit on a tree, put the cucumber in your trousers, and sit opposite the women so that the women coming out of the bath can see you. You know that part yourself."

The boy goes to the bathroom. Five to ten women come and go into the bath, and when they come out, one of the women sees his tool and says to him, "Take this suitcase and let's go, I'll give you some para." So they go and arrive at the woman's house. He puts down her bag and wants to go back. The woman says, "What's the hurry; you must be hungry. Come in and have something to eat." In short, the woman craved this one's tool, and so they settled the matter. After they had got up from the thing, the woman gave him ten pieces of silver and said, "Now go away from here, for it is time for my husband to return home. Tomorrow at ten o'clock you will come," she says. The trade of the woman's husband was running a coffee house. Every morning he went to the coffee house and came back in the evening **(617)**.

The guy goes out of the woman's house and comes to the café. "How much did you earn?" he [the owner of the café] asks him. "Ten pieces of silver," he says, and takes it out and puts it on the table. When he asked how and how it was, the café owner thought and realised it was his wife. He bites his lip but doesn't say a word. Finally, he asks him, "Will you go tomorrow too?"

"Of course, I will," he says, "at ten o'clock."

The next day, he leaves, and the café keeper also goes after him and sees that the door is locked. So he knocks on the door. The woman inside, sensing at once that her husband has come, hides the lad in a large basket, then opens the door and says, "Oh, why have you come untimely?" So he makes up a story and says. He enters, immediately begins to search everywhere, finds nothing, and because he is very thirsty in his confusion, the vinegar made of apple water of ninety-six degrees of concentration [strength] in the window was there. Because it is white, he thinks it is water and drinks it. Immediately, he begins to feel nauseous, then drinks from the water in the wash bowl with which the lad's back was washed, and then goes to his café.

The woman closes the gate after him and pulls the lad out of the basket, they enjoy the pleasures of the bed, and then he gets his ten pieces of silver and comes to the café to tell him how it was. He [the café keeper] asks if he will go again tomorrow. "Of course, I will," he says. So on the third day, he goes to his wife's house, and the café keeper also comes after him. **(618)** The door is locked, and he starts knocking. "Oh, my husband has come," says the woman. On the floor, she scrolled rush mats. She rolled him up in them and set him up against the wall by the door. She goes quickly and opens the door. The café keeper is angry, takes out his knife and begins to rummage about it, first goes to the basket in which he was hiding yesterday and sticks the knife in it, but there is nothing in it. In short, he searches the whole house, unable to find the boy. The woman is getting angry with her husband.

Strange, he thinks to himself and assumes that it might be another woman, and he takes his wife by the windy end [i.e. her bottom] and starts to flatter her, makes coffee, after drinking it, leaves again, goes back to his café and goes about his business. The woman takes the lad out, they do the thing again, and then she gives him another twenty-five roubles. The boy taking it goes for a walk and then comes to the café.

The café keeper gets up, comes to him and asks, "Well, let me see [hear] what you've done."

He tells how the lady's husband has come again and how he's been looking [for him]. He's amazed, [and the lad] takes out the money and shows it to him.

After a while, he asks, "Will you go again?"

"Yes," says the boy. In the morning, he goes out again at that time, and he comes after him, watching. In her yard, there was a water basin. The woman sent him [the boy] into the basin, put a hollowed-out pumpkin on his head, hid him in the water, and then went and opened the door. The husband, now furious, goes in, searches the basket and the rush mats inside, and finds nothing. Angry with himself, **(619)** he concludes that this might have happened in another man's house, calls his wife, and flatters her, then goes out of the house and goes straight to his café. Let us come back to the boy.

She takes it out of the pool, and after their wish is granted again, [he] gets ten roubles and goes to the café again. When the café keeper sees him, he asks him again, "What have you done?"

The boy explains in detail how he hid in the pool.

This [café keeper,] being very angry with the wife, asks [the boy], "If I were to have a gathering at my house, and I were to invite the people of the district, would you tell in honest terms all these things that have happened to you, how you have made money without a single kapek*, would you tell it?"

The boy says, "Yes, I will tell." [The woman was also unaware that the café where (the boy) used to go [to live] is her husband's.][34]

Holding a gathering in his café, this man invites people, invites people from the district, wants to separate his wife, wants to please her, and puts her in charge of the food. The boy at the end of the evening comes in through the back door and is let into his house. They [the guests] sit in one room. After eating, they start asking him questions. He begins to tell how it was.

People start laughing, and the woman with the door open a crack sees trouble coming. What should she do? There was a cat in her house. So, she lets the cat in. As soon as the lad sees the cat, he realises that this is the house he has been to. Suddenly, without people noticing, he says, "What I will finally tell you is that I open my eyes and what I see is that it is morning, [and it seems] I slept on the bath tree. So, it appears that I was only dreaming [it all]. May fortune turn to goodness."

36 A Lack-beard's Tale

(625) Once upon a time, a Padishah had a son. The Padishah bequeathed his wife, "After I am dead, have my son educated until he is twenty-five years old, and after he has finished his education, he shall not receive or return the greeting of blind, lame and lack-beard." Soon after the Padishah dies, the boy enters his twentieth year and completes his studies. In another place, he had an uncle who was a Padishah; he went to visit his uncle.

On the way, he meets a lack-beard, and the lack-beard greets him with salam*. The boy turns his head away. He turns back from his journey, comes home and says to his mother, "Mother, I saw a lack-beard today, going to my uncle's house, he greeted me with

34 Crossed out in the manuscript.

salam, but I didn't accept his salam [greeting], I turned my head away, I couldn't go to my uncle's house because I met a lack-beard."

"Well done, my child," says his mother.

On the second day, suspiciously, he leaves again for his uncle's, and as he goes, he meets that lack-beard again. He says salam to the boy. But, not accepting his salam, he returns home again and says to his mother, "Mother, that lack-beard met me today, I didn't acknowledge his salam again, and I came home."

"You did very well, my son," she says.

The third time he meets the lack-beard again, and he says salam to him. The boy scolds the lack-beard for his beard.

"My child," says the lack-beard, "I am not the lack-beard you suspect; why do you scold me?"

The boy says, "It is my father's will **(626)** that I should not speak to the blind, the lame, or the lack-beard."

Says the lack-beard, "What you said is very correct, but the lack-beard you mentioned is another lack-beard, is not me. Your father was a good friend to me; he always gave me money, never let me bow my head to anyone, and loved me very much."

The lack-beard asks the lad, "My son, where are you going now?

"Here and here," says the boy, "the Padishah is my uncle, I'm going there to visit my uncle."

The lack-beard says, "If so, you are the son of a Padishah, and I am a poor man, take me with you, I will carry your horse, and give it its sack of grain, and be at your service."

The boy gets him a horse, too, puts him on it, and the two of them set off together.

Whether they go a little or a lot, they sit down at a spring one day. The lad takes off his cloak and his sword and rests. The lack-beard puts on his clothes, lays down his own, mounts the lad's horse, comes to the lad and says to him, "I am now the son of the Padishah; you will be the lack-beard; otherwise, I will have your head off."

The boy finishes performing his prayer and begins to beg the lack-beard, "Don't do that, don't do that to me, my lack-beard friend," he says, but the lack-beard is not at all impressed by the lad's plea. The boy can't find a solution, "What can I do, you're the Padishah's son, and if I'm the lack-beard, I'll be the lack-beard," he says.

The lack-beard makes the lad swear that until he dies and rises again, he will tell no one who the lack-beard is or his father's son. The boy claims he will not tell.

(627) There, they immediately mount their horses and set off. One day, they arrive in the country where the lad's uncle is. The Padishah also knew that his brother's son was coming to visit him as a guest. Sitting at the window in the house in the cool of the evening, looking out over the road, suddenly [he sees] his brother's son coming. They went out to meet him that very hour to welcome him with musical instruments, but he saw an ugly lack-beard coming and that the servant behind his brother's son was a thousand times more handsome than he. "Strange," says the Padishah, "that my brother should have such a son, coming here to make me ridiculous. Well, his servant is rather

fit for a Padishah's son. What if that lad were my brother's son, not this ugly, lack-beard one?"

"But," say his Viziers, "my Padishah, five fingers can't all be alike; what shall we do; however ugly, he is your brother's son."

What shall the Padishah do; he resigns.

The lack-beard and the lad spend five or ten days, but the Padishah keeps saying, "Behold, who is fit for a Padishah's son, a lad like a slender sapling is serving the lack-beard." The lack-beard is afraid that he will let him know. How to get rid of this lad? Finding no other way out, he decides to kill the lad, but he can't find a way to kill him.

One day he says to the Padishah, "Uncle, I would like the daughter of the Great Padishah."

"Oh, shameless lack-beard," says the Padishah, "how can you marry her, for each of the lads who have gone to marry the Great Padishah's daughter is worth a hundred lack-beards like you; **(628)** he has made a forty-floor palace out of the heads of such lads; how many hundreds of thousands of lads have fallen prey to his love for her up to now. Not one of them has been able to marry her; perhaps you will marry her. How will you marry her?"

"My servant," says the lack-beard, "is such a lad that he will bring her, but you command him to get her."

"Not my brother's son, but whoever you may be," says the Padishah, "are you not sorry to send such a lad there and have him killed for the sake of such a lack-beard as you are?"

The lack-beard immediately begins to weep and says, "You are my uncle, and because I am lack-beard, you pity my servant more than me. What is that?"

The Viziers also begin to plead, "My Padishah," they say, "send that lad away, let him go, that he may fetch that girl for your brother's son." So the lad mounts his horse and goes. They put a kamchi* in his hand.

The poor boy is going down the road, crying, when he meets an old man who asks him, "Why are you crying, my boy? What sorrow have you?"

The lad replies, "What shall I do? How shall I do it? Who shall weep if I weep not? Now I am going to marry the daughter of the Great Padishah; may I be the one to marry his daughter?"

That old man was prophet Hizir, "My son," he says, "it is hard to take the daughter of the Great Padishah. So he has a forty-floor palace made out of the heads of the lads who went to take her, and he sits on the top of it and gives orders, but I will give you a handful of earth, and when you come near the palace of the Great Padishah you shall scatter half of this earth, and half of it you shall scatter in the courts when you enter through the palace gate," he says. "So the Padishah's daughter **(629)** will come out to walk in the rose garden, you will take her with the cool of the afternoon, but until you are fifty chakirim* [miles] from the Padishah's palace, you will not turn and look back, or else you will perish," he says.

The lad, will be what will be, goes from there and comes near the palace. On all sides of the court was a wall so high that there was no way of getting in. So he scattered half a handful of the earth, and the walls came down, which amazed him. After a little while, the lad went in, and on every side of the Padishah's house stood forty dragons guarding the Padishah's daughter. At this place, he scatters what is left of the earth, so everything inside and outside the palace fall lifeless. From here, the lad in the rose garden puts the Padishah's daughter on his own horse, and with a single clapping spurs to his horse, it covers fifty chakirim at a time.

On the way, the girl asks the lad, "Hey, lad, who are you carrying me like this?"

The lad says, "I take thee to the lack-beard son of the Padishah."

The girl says, "Hey, lad, I'm not going to anybody but you."

But the lad says, "No, I didn't bring you to myself, I gave my word that I'd get you to the lack-beard son of the Padishah, and now I'm taking you there," he says.

After a while, they see the lad bringing the daughter of the Great Padishah. The poor lack-beard doesn't know what to do, "I seem to have no escape from this lad; one day, he will expose my shameful things, he says," and he doesn't know what to do.

The lad brings the girl to his Padishah and says, "Behold, Padishah, I have done your bidding, and he hands her over."

The lack-beard is locked up that night with the girl as a bridegroom, but the girl shoos him away, "Go away, pig **(630)** lack-beard, do not insult me." The woman won't even let the lack-beard near her.

What shall the lack-beard do? He goes straight to the Padishah and says, "You can't do it without cutting this lad's head off."

When he asks, "Why?" [the lack-breard] says, "I cannot live while this lad is alive; behold, she will not let me come near her; she has her eye on him."

"Thou lack-beard fellow," says the Padishah, "thou infidel, can one do such a thing? Behold, he hath done so many things, he hath brought the daughter of that Great Padishah, who hath made a forty-floor palace of men's heads, what more dost thou want?" says he, the lack-beard swine began to weep, "Thou art my uncle, and thou thinkest me less than one of my servants, thou undervalue me to one servant."

The Padishah finds no way out, the lack-beard pig does what he does, and the lad is killed, but the lad does not die. The prophet Hizir prayed for him not to die, only to those it seemed he was dead. So they bring the lad and bury him. And the lack-beard rejoices, "I am free," he says, and he does not know what to do in his joy. When the lad is buried, a dervish comes to the girl that night in her sleep, "Daughter," says he, "go quickly and take the lad out of the grave." The girl goes and takes the lad out of the grave. As he came out of the grave wearing a worker's uniform, he went to the gate of the Padishah and asked, "Isn't a man needed to serve as a worker-servant?" As soon as he sees the lad, the Padishah sees that he looks very much like the handsome lad who was victimised because of the lack-beard. Because he looked so much like the dead lad, the Padishah took a liking to him and hired him as a labourer. And the lack-beard one lived, kept on eating and drinking. "I killed the lad," he said, **(631)** rejoicing. In the meantime, some time had passed. One day, guests come to the Padishah, and to amuse them, it was widespread in the old days to tell tales.

As many tale-tellers were in town, the Padishah collected them. So one day, he made the tale-tellers tell tales to the guests, and the lad also was there serving the arriving and departing guests. Each one tells a tale according to his own knowledge, but the Padishah is very fond of this lad because he resembles the lad who died.

"You lad," says the Padishah, "you seem to be a lad who's seen the world, have you not a fairy tale to tell?"

"Padishah," says the lad, "I'll tell you a tale, but you shut the doors, you let no one out."

"Very well," says the Padishah, and gives the order, and the lad begins to tell the tale, the things that happened to him during his wandering about away from home, from the time his father told him not to speak, not to greet nor receive the greetings a lack-beard, lame and blind, and how the lack-beard told him, "Your father was my friend," but he tells the story in such a way that it is about a certain lack-beard. Then he tells how he received the lack-beard as his servant, then how the lack-beard took the lad's clothes while he was praying, and, attacking him [threatening him], how he offered the lad to be his servant. The lack-beared [meanwhile] is wondering, "My dear, what kind of a thing is this fellow? Has he come with us, or is he that fellow himself?"

The lack-beard's body is burning in fire, but there is no way out. Then he tells how the Padishah receives them and how the Padishah is happy for them. He tells how the Padishah cursed that such a lad should be his servant and that such a lack-beared **(632)** should be his brother's son, and who is fit to be the Padishah's son. Then he tells how the lack-beard did not rest until he had the lad killed, and how he misled the Padishah and sent the lad to the Great Padishah's daughter. The lack-beard was beginning to get the picture. "My stomach is in knots; my belly aches," he says, but there's a hangman by the door, and he won't let anyone out. The lad goes on and on with his tale, but he won't listen to the lack-beard. He also tells how he met the prophet Hizir on the road and how he brought the girl, how the girl didn't even look at the lack-beard, how the lack-beard, again unable to get rid of the lad, asked the Padishah to kill him, how the Padishah scolded the lack-beard, and how the lack-beard, claiming kinship, talked the Padishah into it. One by one he told everything.

The Padishah's eyes rolled like blody pots, all red; he heard the thing. The lad goes on and on about how the lad was killed and buried in the grave, how the girl went and brought him out, how the lad served as a labourer for the Padishah, how the Padishah invited the company and how they came to tell tales. The lad says, "Well, my lack-beard friend, you have made me vouchsafe to speak [of this] only after I have risen from the dead. Well, my lack-beard, I have now come to life after my death, my turn has come, so I have told it." And he saith unto the Padishah, "The lack-beard was my servant, I am thy brother's son, thou art my uncle," and he rises and kisses his hand.

"My dear cousin," says the Padishah, kissing him on the eye and weeping. The lack-beard is tied to a tree, one day his ears are cut off, and the next day his nose is cut off, and he is killed.

The lad, having died and come back to life, meets his uncle, and the lack-beard finds what he deserves. May Allah bring us to our goal. This tale ends like this.

37 The Two Rogues' Tale

(642) There was once a rogue in the Crimea, and there was a rogue in Egypt. The Egyptian rogue was very famous in Egypt. He gave permission to all the rogues in that region to become thieves. The rogue who was the ataman* of the thieves in the Crimea and the thief from Egypt heard about what each other had done and wanted to meet, so they wrote letters [to each other] so that they could meet. The Egyptian rogue has the Crimean rogue invited to the land of Egypt. "So let us go," says the Crimean rogue, let us meet.

One day he gets on the road and starts walking. He walks and walks; one day, in the desert of Egypt a lad sitting in the shade of a tree by a spring.

The Crimean rogue goes up to him and says, "Salam aleykum"*.

The Egyptian thief says to him, "Sit down, lad."

The Crimean rogue sits down.

"Light a cigarette!" says the Egyptian rogue, gives him tobacco, and asks him, "Lad, where has the good luck brought you from?"

"I come from the Crimea," says the Crimean rogue.

"Where will the good luck take you?" asks he.

"I am going to Egypt," says he.

The Egyptian rogue gets suspicious: 'I wonder if this is the Crimean rogue,' says he [to himself], and asks, "To whom are you going?"

"I heard a rumour of a local rogue, I came to see his skill," he says.

"Well then, I am that rogue," says [the other]. They are very glad to see each other.

Then the Crimean rogue says, "Show me one of your skills; let me see."

The Egyptian rogue says, "I will show you a great skill."

On the tree in whose shade they were sitting **(643)** a baby bird was sitting [in the nest]. "Behold," saith the Egyptian, "from the nest of a bird in this tree, wherein the nestling sitteth, I will take out its eggs without its noticing."

The Crimean rogue says, "Go on, do it, let me see!"

The Egyptian rogue climbs up the tree, takes the eggs, and when he starts to put them in his trouser pocket, the Crimean rogue climbs up after him and pulls his trousers off the Egyptian thief without him noticing. The Egyptian rogue comes down and sits in his place, saying, "I have shown a trick. Well, did you like my skill?" he asks the Crimean.

The Crimean rogue says, "It's not a skill to get the egg out of the bird's nest, but to steal the trousers of the man who climbed up to steal it without being noticed, yes. I came here from The Crimea to learn a skill, theft [stealing] from you, [but] you showed me egg stealing. When I was a five-year-old child, I started with petty thefts like this, stealing eggs, you were learning to steal," he says.

The Egyptian thief kisses the Crimean thief's hand and says, "You are the master of your trade. "Oh, ataman," says he, "teach me also this your craft."

And the Crimean thief takes him and brings him to the Crimea.

He teaches him to steal for a long time, and he is no less a thief than the thief in Crimea. After a time, the Crimean rogue was shot in the city of Or*, **(644)** in the nest of thieves, when he went to steal; his companions brought him to the city of Or and buried him. In Crimea, as many rogues, as there are, all mourn the death of their ataman. Then all of them, seeing the Egyptian rogue as a suitable ataman, make him their ataman. The Egyptian rogue became the first [chief] in the place of the Crimean thief. In the Crimea, even today, there are no such famous rogues anywhere like in the city of Or.

38 The Tale of Chora-batir

(648) In the old days, there was Narik. He was a particularly clever, witty man and also a hunter. Narik would go to older people wherever he saw them and listen to their conversation. An old man says, "If one serves at a Khan's door, he will become a man, but if he serves at a rich man's door, he will hit him with a stick and chase him away." Narik listened to these words, went straight to a Khan, and served him, but no exact annual salary was set. There they made him a postman.

One day Narik was delivering the post, and on the way, he stayed at a shepherd's lodge. In the evening, the shepherds gathered, ate, and drank coffee. After their enjoyment was complete, they began to talk about hunting. One of the shepherds said, "Every morning, a white rabbit jumps up and runs away past the sheepfold. We'll show it to Narik and see if he could have it caught by his greyhound." They got up in the morning and went to the sheepfold. The rabbit jumped up again and ran away. The hound was sent after the rabbit. Narik mounted his horse and went after the hare. He chased it, chased it, and finally, he came to a forest, and in that forest was a big cave. The rabbit ran into the cave, and the hound followed it.

Narik looked in; the cave was in a black [dense] fog. "The cave is dark," he says to himself, "the rabbit is white; maybe I can catch it." Narik went into the cave, but the cave was very dark, and the inside was full of pebbles; it was slippery underfoot, and he couldn't walk in it. Slowly, slowly, he turned back, **(649)** picked up a handful of gravel and pocketed it. He came out of the cave, but his mouth and nose were twisted, and he collapsed on the spot and lay unconscious. The shepherds saw that Narik had not returned, and wondered where he had stayed. They sent a shepherd to look for him. He went and found Narik lying unconscious in front of the cave. He took him on his back and carried him to the lodging. They brought word to the Khan. The Khan brought a doctor, and mullahs came, prayed, and made medicine for him. Narik became conscious and came to his senses.

They asked him, "Where were you? What happened?"

Narik began to tell them what had happened to him. "The rabbit went into a dark cave, I went in after him; the cave was full of pebbles inside, that kept rolling out from under my feet, and I fell. I put a handful of pebbles in my pocket, of such pebbles," he said and took them out of his pocket, and they saw that what was in his pocket was all gold.

Now they gathered, went to the cave, and saw that there was a black [thick] fog. They started to pray, they found [dissolved] the spell, the black [dense] fog lifted, they started to empty [from there] the gold. They loaded the gold on forty mules and brought it to the Khan's treasury.

They say to the Khan, "We have profited greatly from this Narik; let us also do good with him," they say.

"Let us first marry him," says the Khan, "but whose daughter he asks for, let us marry him." They themselves each had a daughter. "If he asks for your daughter, you marry her to him; if he asks for my daughter, I will marry her to him." They [also] gave him three villages. It was announced that Narik would marry and that as many daughters as there were, **(650)** would all gather in the rose garden. The girls gathered there; they gave the news to Narik/ "Here, whose daughter you like, you will marry her," they said. Among the girls, he liked the daughter of the Vizier, and they told the Khan. He said, "Blessed be she." Narik was married.

But the son of that Khan also loved this girl. He watched her for six months, and his longing for her struck him in the heart. So he says to an old woman, "Go to Narik's wife, tell her I'll go to her tonight, let her know."

The old woman went to Narik's wife and told her.

Narik came back from the village and said, "I'm going to the bazaar tonight."

"Don't go anywhere tonight," says his wife, "guard your house."

Night fell, and the Khan's son went straight to Narik's house and knocked on the door. The wife opened the door, and the Khan's son came in and sat down. After a while, he started talking. The wife put the coffee on the stove but put the djezve* away from the fire [flame]. He looks and looks; the coffee is not brewing. "I didn't come to you for coffee, he says, get up; let's go to bed."

"What kind of a word is that? What kind of a disgrace is that? You came as the son of a Khan to fornicate with the wife of a [simple] Tatar man. If your Khan father hears this, what will he say? And if my father, the kazi-bey*, hears it, what will he say."

"Light the lamp, I'm leaving." The woman lit the lamp and handed it to him. As he left the door, Narik sat there and immediately cut off his head. He dug a hole, and he was buried.

But the son of this Khan came home once every week. **(650/a)** After the marriage of the Vizier's daughter, he came home once every two weeks; then he did not come at all. The Khan sensed that his child had disappeared. He called his Vizier. "My son is missing, find him," he said.

The Vizier says, "When a man has killed another man, look him in the eye and see." [The people were called together] and [the Vizier] began to look everyone in the eye.

(654) When his son-in-law passed by, he looked into his eyes [and saw] his son-in-law had killed him. He came home to his own house and sent a message to his son-in-law to leave quickly from here.

The two of them prepared two horses and mounted them. In the afternoon they had the horses swum across the river and passed to Etli-kan. After arriving at Etli-kan, [they see] they have nothing. There an okka* of meat costs one gold, and they [the Nariks] have become very poor. Let us come to the Khan.

The Khan and his adviser [the Vizier] were great friends. After the death of the Khan's son, the Khan became estranged from the Vizier. The Vizier asked him: "Why have you had enough of me?"

"So that I would not resent you when your son-in-law killed my only child in this world."

"My son-in-law killed him, but he went to commit adultery with his wife, and that is why he killed him."

"Why didn't you tell me?" says the Khan. "He cut [stabbed] him in one place, I would have cut him in five. Quickly tell him to come back. I gave them three villages as a gift." Then they wrote a letter, their sins were forgiven, and three villages were given to them. The letter was sent.

Narik was coming home from hunting. As he was passing through the village, the postman came and handed him the letter. Narik read the letter; it was great news. He came home, smiled at the door and entered. His wife stood up, "Why are you smiling? Is it news from the village?" **(655)** she asked. "Your [i.e. our!] sins have been forgiven, and they are inviting [us] to the village," he says. The next morning, they got up, and loaded the carriage. They had horses; they put them in the cart and set off. On the way, they met a group of wanderers and befriended them, so they went on together. They were then so poor that they had nothing to eat or drink. If Narik goes and hits a rabbit, they eat; if he misses, they don't eat. Narik got up and went to look for a rabbit. But then his wife was in [childbirth] pains; she saw that a child was coming into the world. An old woman from among the wanderers became a midwife. Narik had a slave; they said to him, "Run, call Narik; here is good news."

"Uncle Narik!" he cried, "You have a son," he said.

Narik had no money; what should he do, how should he do it [reward the bearer of the good news]? "Give your name **(656)** and be free!" he says.

The child was given the name Chora. The slave was freed. "Run and put [the horses] to your car." They put [them to the cart], got on the cart and went to their own village. They saw that their sin was forgiven. Three villages were given to them as a gift. They came out to meet them, kissed each other on the cheeks and eyes and welcomed them.

They lived one year, two years, five or six years. Chora was sent to school, studied for three years, became a very skilful, big boy, and surpassed all children. He never ran away from a child; the children were afraid of him. One day Chora said to the children, "When I move off from my place, raise a flag for me [i.e. in front of me], and I will not hurt you."

"Where," they say, "shall we get a flag?"

"I will find you a flag," he says. There was a calf-herd boy over there; he took one of its calves, cut it off, skinned it, and gave it [i.e. the hide] to the children: "Make it into a flag and hold it up." The children took the flag. Chora did not hurt the children, and the children surrendered to him.

The child who was looking after the calf **(657)** came home crying, "Chora cut off our calf," he said.

"Go and fetch Chora!" says the calf's owner.

They called for Chora, and he came.

"Chora," says the calf's owner, "why did you skin my calf?"

Chora says, "If your calf were[35], his legs would go up your arse."

The calf's master was pleased [with the answer], laughed, and said, "Chora, I will give you a horse," he says.

"God be pleased with thee if thou givest me a present," saith he.

Now they go with the horseman into the field. The horseman is to catch a horse for him and give it to him. They come to the horseman and start looking at horses. The horseman begins to catch the best horses and bring them to Chora. Chora doesn't like any of them, so he starts looking at them himself. Following the horses, he searched for their tracks. Searching and searching, he found a good trail, and on that trail, he found a mangy colt, "Here, I liked this one," **(658)** he says.

The horseman, "The people of our village will laugh," he says, "I will not bring such a horse to the village," he says. Chora said, "If you give me a horse, give me this one; if you don't give it to me, let its legs go up your ass." In the meantime, he found a strap in the stable, tied it around the foal's neck and said that it should be called the Strapped Neck. He brought it home. Everyone was amused at this. "Out of five hundred horses," they say, "he found only one mangy horse for himself." Chora started to look at it and found it the best of five hundred horses. There was another famous horse in a nearby village, and its name was Djel-djetmez*.

Chora heard about this horse and said, "Tell him [his owner] to send me that horse."

They went and told him, "This horse is fit for him [Chora]."

He gave it and sent it to him. At that time there was another horse in another neighbouring village; it was very fast. Chora sent a message to him also to bring and give his horse to him.

They went and told him.

"I will not raise my horse for him," he says.

They told Chora, "He would not give up his horse."

Chora got very angry, sat on the back of the Strapped Neck, and as soon as he moved from his place, fire burst from the top of his head. Then the horse's owner saw that Chora was coming. Immediately he took his horse [called] Sheep-keeper and set out. Chora was

35 You mean, "If you had given birth to it."

arriving. "You wanted the horse; here it is; I was bringing it," says Chora. "You don't bother; I'll take it myself." He took the horse and rode away. Now then, there was another great hero, **(659)** Ahtadji Ali-bey was his name, and besides him, he had forty other pelivans*. This Ali-bey is very angry with Chora. "Scoundrel," he says, "he plays the hero with his horse, I will go and take his horses, if he has already lost his horses, he can do nothing."

One day Ali-bey got on a horse, and the pelivans knew where he was going, "Don't go to Chora-batir; if you still go, let's go together," they said.

Ali-bey says, "I will not go there".

He hid it from them. He took two Arab batirs with him and went straight to the house of Chora. Chora was going to a wedding that day; he was not at home. Ali-bey came between afternoon and evening; **(660)** he saw that Chora was not there. He immediately slaughtered sheep three or four years old and had bal* [bragget] and raki brought for thirst. He kept demanding the Strapped Neck. Chora's father refused to give him the Strapped Neck. The people of the village said, "There is a bigger one than Chora; Chora's mother told his fortune," they said. Chora's mother told them a poem:

Its hair and tail are like silk,
Its running is like that of a Satan[36]
Which of you has a son,
Like Narik's son Chora?

Then Ali-bey went into the stable and grabbed the neck of the Strapped Neck. Chora's sister Kun-kurmez* Mengaru ran up, grabbed the horse's head and refused to give her horse to him, but Ali-bey of Ahtadji said nasty, indecent words to her. She became ashamed and let the horse go; [Ali-bey] hit the head of your Chora's father. [He hit a woman called Arab, causing her great pain, and said very bad words to the woman called your mother.][37] They had many quarrels with your sister [called] Kun-kurmez* Mejnaru; they said bad words to each other. Then Ali-bey took his [Chora's] horse and left. Chora's father mounted his horse and went to give the news to Chora. On the way, they ask him, "Where are you going?" Narik answered them with a stanza:

Oh, where is a wise [knowledgeable man]
Come on, I'll tell him my sorrows
(674) You will hear it too – he said and walked on.

On arriving at the feast, he asked for Chora. Chora-batir was told, "An old man outside is asking for you," they told him. Chora went outside and saw that her father had come. His father, as soon as he saw Chora, said this poem:

36 These are the characteristics of Chora's horse!

37 Crossed out in the manuscript.

Eh, Choram, Choram
When you were squatting on a horse
On your old father's head
You have brought all kinds of trouble.
Chora [also] recited a poem to his father:
Oh, my father, oh, my father
What you said burns me up
What my enemy said
I really like that.

Then he told him how Ali-bey of Ahtaji came and how he took the Strapped Neck:

He came between the afternoon and evening
On Tube-tau he slaughtered three or four-year-old sheep,
He had bal* [bragget] and raki brought for thirst
He kept asking for the Strapped Neck
The poor man who is said to be your father
Was beaten over his head and eyes
To the woman said to be your mother
He said bad words
The woman said to be Arab
(661) [He hit causing great pain.]

"They had many quarrels with you sister Kun-kurmez* Mejnaru. He took the Strapped Neck."

"Where did he go?" asked Chora.

His father said, "He went to the shores of the Blue Wild Plum, to the shores of the Blue Spring." Chora was angry, and as he mounted his horse, its feet sank into the ground up to its ankles. As he moved from his place, flame came from the top of its head. Ali-bey sensed that Chora had moved from his place. Beside him were forty Arab batirs, all of them afraid. One of the batirs said, "We will not fight against Chora; go and pay for what you have done; maybe he will forgive you."

Then, Ali-bey, sitting on the back of the Strapped Neck, came to face Chora, and Chora came to him, and said this verse:

Ej, Ali-bey of Ahtadji, the enemy['s] lord,
At the sight of an enemy, a cowardly lord.

Ali-bey's reply:

I went between the afternoon and evening
I had sheep three or four years old slaughtered on Tube-tau
I had bal* and raki* brought out for thirst
I'll pay that too, I'll make peace.

Chora says, "You beat the poor man who is said to be my father over his head and eyes, you said bad words to the woman said to be my mother, you hit the woman who is said to be Arab, causing great pain, you quarrelled a lot with my sister Kun-kurmez* Mejnaru: **(662)** how you will pay for it."

Ali-bey said, "I understand you will not forgive me, so shoot you."

Chora said, "Your head is great; you are a hero yourself; it is your turn to shoot." Then Chora waited.

Ali-bey shot an arrow, and Chora held out his palm; the arrow hit Chora's palm and fell to the ground.

Chora said, "Three is the man's game, shoot twice more."

Ali-bey shot twice more; he did nothing.

"Now it's my turn," he says, "defend yourself." Chora fired an arrow; Ali-bey's two legs remained on the horse's back, and the part above the waist was taken off by the arrow. Chora led the Strapped Neck into the village. Ali-bey's legs were on the horse's back.

He gathered the villagers together and asked if Chora's mother was wrong or Ali-bey's mother was wrong. He took his arrow in his hand and said, "Tell me quickly, [or] I will take your life with one arrow.

The people began to weep, they went to his father, and they fell at his hands and feet. "Come," they said to him, "reconcile him with us."

"I cannot make him reconcile with you," says he, "you yourselves, go and tell him." **(670)**

Then they began to plead with his mother. "Don't let him destroy these men; he will obey thee."

Chora's mother says, "Come, my child, forgive the sins of those people for my sake," she says, "but in vain."

Chora does not listen to the word. She takes a two-edged [battle] axe in her hand and strikes two blows with the axe on the breast. "If you kill these people, I will [also] curse the [mother's] milk that has been sucked out."

Chora said, "To take the life of people with an arrow is nothing, but to please a mother is a great thing." Chora let the arrow fall from his hand to the ground, "Take the arrow," he said, "go and put it in its place."

The youths then gathered and began to roll the arrow towards the gate with a crowbar. They could not lift it and put it back in place. Then Chora's (sister) Mengaru, laughing loudly, came leaning out of the gate, and picked up the arrow. "A pity for your youth," she said, and, bending down, she grabbed it with one hand and threw it to its place. A good deal of time has gone by likewise; they have spent their time leading warrior's life.

After a while, news came that war was about to start in Kazan. Chora had a friend called Kulumchak-batir. They became friends. Then they wanted to go to war, but Chora's father said, "My child, first I will make you betrothed, then you go."

Chora said, "I will do as you say." Chora was betrothed.

After a few days, they left with Kulumchak, **(671)** travelling a distance of ten and a half days journey; they arrived at a village. In that village, there were forty batirs. Thus said Kulumchak-batir, "Come on, my brother Chora, let us show this village something miraculous. They started, they presented the miracle, but not one of the forty batirs dared to come and ask them who you were."

Then the head of the batirs says, "Whoever goes and asks who they are, I will give him a handful of gold."

There was a beggar in that place. "I will go and ask," he says. He went up to Chora-batir and asked, "Where are you from, great [. . .] **(673)** [tell me, I will not tell," he says][38]

Then Chora said, "I will go to Kazan; if Kazan is not taken by the Cossacks, until I reach Kazan, if the plank [of its houses] is not set on fire, I cannot give Kazan to the Cossacks, hey."

The beggar came back and told this news to the head of the batirs. The head of the batirs called him, welcomed him there, and gave him much attention and hospitality. The chief of the batirs said, "Chora, we have heard that you are of great valour, [. . .] we are grateful to have seen you, but for our sake shoot an arrow, so that we can hear the sound of your arrow."

Chora said, "I will shoot, but take the news to the village; let everyone know that an arrow is being shot; the sound of the arrow will frighten them."

The news was spread in the village, and the people said, "As if we had never heard an arrow sound, here the batirs shoot arrows every day." Chora went to the square in the forenoon time and shot an arrow. Twenty-nine people were frightened to death by the thunder of the arrow, tiles fell from the roofs of houses, and window panes were broken. Two more days were wasted there, and the batirs were waiting for war. Chora said, "We will stay here for another two days; you check the village and come back."

(672) Kulumchak-batir came to the village that night, riding on the back of the Strapped Neck. First, he went to Chora-batir's betrothed and looked through the window. He saw Chora's betrothed playing with a young man in the house, cuddling. Kulumchak was astonished at this and immediately turned back, came straight to Chora, told him the news that a young man was playing with his betrothed in such and such a way. Chora-batir, hearing this news, mounted his horse, turned back, came to their village and went into their house. Chora-batir's sister, Mingero [Mejnaru], jumped into her broth-

38 Crossed out in the manuscript.

er's neck, embraced him, and kissed his cheek and eyes. Then she said, "Brother, I have done you wrong because I have missed you so much. I put on your clothes and went to your fiance; we played cuddling from evening till midnight."

"It is very well," says he, "that you have told me now, for I would have cut off her head."

In the morning he rose and went to battle again and met Kulumchak. They had heard the news of the war, that it would begin on a certain day. It has reached Kulumchak; they say that a frightful [bad] dream means luck [happiness]; mine ended happily. **(675)** Then a paper came; it was written to these people, "Get here very quickly to the war." They set off with Kulumchak-batir. Chora said, "Let us shoot an arrow to let them know we are on our way." Chora shot an arrow, and at the foot of the minaret of the mosque of Kazan country there was a marble stone, and the arrow was bored into that marble stone. Eighty-two batirs were there, they gathered, they went there, they took the arrow, they pulled it, but they could not pull it out. The head of the batirs was Sari-batir; his mother came and saw the arrow, "My son, you cannot pull this arrow out of this stone; it is the arrow of Chora. Put a guard on the road, Chora may be here in a day or so." The war will begin in three days; they are preparing for it. Chora has come; he has pitched a tent on the battlefield. He tied his horse to the tent.

Nobody recognised Chora. Chora said, "You sit by the horses; I'll go and see how things are going." He went straight to the company. They were gathered in a house, all the batirs drinking bal*. Chora entered, said a salam*, and sat down in a narrow little place. No one knew Chora; not even his salam was received. When the bal* was poured, Chora was left out. Then Chora recited a poem to them:

In turn, I see, in turn,
Tomorrow, when your head is in big trouble,
and pull back the head of your horse,
you will get to know your brother.

Sari-batir said, "What kind of beggar is this; pick him up and throw him out." Chora quickly rose from his seat, went out, went straight to his tent and lay down. [The fight] began. From this side they went in front of the Cossacks. They too were preparing for war. Sari-batir's mother saw that Chora had arrived. Quickly, she says, go, Chora has come, have good luck, bring him here. Four or five batirs go and kneel at the feet of Chora. "We didn't recognise you," they say, "don't shame us," so they cajole him, [and Chora says, "They are not to blame because he did not introduce himself"];[39] he rises from his seat and comes straight into the company. All the batirs kissed Chora's face and eyes. "We did not recognise you; why did you not introduce yourself to us?" They

39 Crossed out in the manuscript. It should be written like that: "They are not to blame because I did not introduce myself".

started drinking bal*. A glass of drink was sent to Kulumchak-batir; he drank it too. Sari-batir begins to give gifts to everyone, some were given swords, some arrows, and some shields. To Chora, he gave a [tobacco] box. He puts it in his pocket, goes into his tent and throws it under the horse's feet. "He gave each batir a fighting tool," he says, "and he gave me a tobacco box. I don't smoke; what shall I do with a tobacco box?"

Then, at sunrise, **(677)** the battle began. They also started from this side. Chora, without lifting his head, did not look, offended by the tobacco box. Then the batirs came up and bowed down at his feet, and begged him, "Why art thou offended, Chora," said they, begging him.

"To every batir," says [Chora], "thou hast given a fighting tool; to me, thou hast given a tobacco box. I do not smoke; what shall I do with your box?"

Sari-batir said, "What have you done with the box?"

"It's on the ground, under the horse's feet." Sari-batir jumps up from his seat, takes the box, and opens the top; there's a zilfikar* inside, takes it and puts it in Chora's hand. Chora, seeing the zilfikar, got delighted, took it in his hand and sat on the back of Strapped Neck. When he moved from his place, his horse's feet sank into the ground up to its ankles. For three days and three nights, he fought with the soldiers, encircling Kazan's castle in front and behind it, fighting with blood dripping from his horse's loins.

In seven days, he destroyed [crushed] the [enemy's] troops, and took captive those who remained. The war was over. They rested for a few days. Then they returned to their homeland. The people of the village came to welcome them and began to tell them the news of the war. It was such a big battle that the old people, who were ninety years old, had never heard never seen such a war.

(678) The news came again that they were preparing to start a war again. Chora-batir sent word to Kolumchak and said it had become necessary to go to war again. Let's go, he said, he got ready, took his arrow and mace, sat on the Strapped Neck, said goodbye, and they were on their way. As they were going along the road, the people on the road asked, "Chora-batir, where are you going in such a hurry?"

He said to them, "There is no time for much talk, I will say a poem; if you understand, you will know:"

I'm going to Kazan,
Until I get to Kazan
If the Cossacks don't take Kazan by then
He would not set his board on fire,
I would not give Kazan to the Cossacks.

On the way, a rider comes in front of him; he sees two men walking very gallantly. He asks where they are going. Chora also told him in verse that he was Chora. The rider was coming just to Chora, bringing news of the war. So he said to Chora, "If you don't get there in a day, Kazan will be taken by the Cossacks; get there quickly." As soon as Chora heard this news, he jumped up, and the legs of the Strapped Neck sank into the ground up to its ankles. “I’ll get into the middle of the fight.” Saying [this], he’s run away from there. Unable to prevent the castle from being taken, he jumped down from the castle and fell into the sea. They see that Chora went into the sea, and the war was won by the Cossacks, **(679)** and Kazan was taken.

“Chora will come again,” saying, they don’t believe he’s dead, so they put a guard where he went into the sea. To this day there is still a guard at the place where Chora went into the sea. After the war was over, they told Chora’s mother that Chora had gone into the sea. Chora’s mother said, “Even if he went to sea, there would be no harm. When Chora was born, fire burst out of my mouth; it fell into the water and was not extinguished. I asked the efendi* afterwards what it meant that fire came out of my mouth; it fell into the water and was not extinguished. The efendi said, ‘The son whom you have brought into the world will have a watery destiny; he will fall into the water and will not die.’” So Chora’s mother did not grieve much; she held up her hands and prayed, “May Allah bring us all also together, and let me see my son too. If Allah wills, the time will come, and he will occupy Kazan again.”

39 The Tale of the Smallest Bandit

(700) There was once a poor old man. This old man had three daughters and one son. The old man wanted to send his son to the medrese* when he was seven years old to be taught a lesson. The old man goes to the bazaar and brings back a lot of books. There was a medrese about fifty chakirims* from their village, and he took his child to that medrese. But along the road to the town, there were large forests. They were going along the road when forty-one bandits came out of the forest in front of them. These highwaymen want to kill them, take their horse and cart, they say to each other. By this time, it was very dark at night, and the bandits could not make out who was in the cart, so they went alongside, lit a match, and what they saw was a poor old man and a child in the cart.

They see the old man with the child; some of them say, “Let’s kill them”; some of them say, “Let’s not kill these poor people.”

One of them comes out and says, “If he gives us his child, we won’t kill him; if he doesn’t, we will.”

They say, “Let’s try to ask the old man for the child.”

Then they ask the old man, “Will you give the child to us, if you give the child to us, he will study with us, and we will even give him money,” they say.

The old man is afraid of them, so he gives the child to them. He himself turns back, and when he gets home **(701)** he tells his wife what has happened to him. His wife says, “Allah be blessed at last, if fate wills it, we shall see our child one day.” In this place let us come [back] to the child.

In the meantime, seven or eight years have passed, and the child, living with them in the woods and rock caves, doesn’t even know what literacy is. The forty-first one of the forty bandits was their boss-in-chief, and the child served as a footman under this ataman*. One day the child remembers, "I have lived with them for seven or eight years, and I have never been allowed to go back to see my father and mother. It's impossible for me not to escape now," he thinks. As he is thinking, one by one, the bandits all come home from their business. In the evenings, they gather and talk while eating, "Here, in these woods, we have been robbing for fifteen years, there is no place we have not robbed, there is no man left whom we have not robbed, but in the town near the forest where we live there is a rich man, a man with innumerable wealth and money, and we have never once thought of going to him. It is impossible not to go to him," they talk like that, and the child listens to them talking behind the door. Thus, they talk, they eat, they drink, they fill their bellies, they mount their horses, and they gird their weapons. "However we may, we shall strike at his **(702)** palace tonight; we shall take his money and his goods," saying, they are going away, but their ataman is left at home in his abode.

Now, let these forty robbers go; back at home, the child goes to the ataman and says, "Hey, my ataman brother, I have been walking with you for ten years. If I keep on walking like this, I will not get anything for myself, so if you give me a horse and a sword, I would try myself, maybe I too, I could be a robber like those robbers, I would be good for something," he says.

The ataman says, "Very good," and brings out a scabby colt and a dull sword and gives it to the child. Let the child go after the forty bandits, and let us come after the bandits.

The bandits in the city go to this man's house and ambush him, kill him and the guards, and they are just starting to bag the money buried in the pit when the child arrives there; he has tied his horse to a post near the treasury, and he himself starts shouting "Guard!"

And the bandits [are] thinking, "They attacked us with soldiers." Those who could take some money would take it; those who couldn't ran away from that place without a single coin, and the child, as soon as he noticed that they had run away, went in and started filling the sacks with money. From that place, he loaded the money onto a cart, **(703)** as much as his horse could haul, and from there let him go on his way. The forty bandits went with that run straight to their quarters, and told their story to their ataman, "We had bad luck, and only with difficulty escaped by running away; thanks," they said, "that our lives were preserved intact."

The child, not long afterwards, brought the money in the cart and went to the house where they were staying. "My master, get up, come, help me to unload the money!" he cried.

The ataman comes out, and what does he see: he has gone away once and brought back more treasure and money than they themselves have collected in fifteen years. "Well done, my friend, my brother," he says, "you brought back so much wealth and money for a little child; forty bandits have gone away; they have found nothing to bring," he says. Then they brought in the money, the gold was piled up in one room, the silver in another.

The evening has arrived. In the evening, the forty bandits all came from different directions, went into the house and sat down, then they came to eat, their ataman also ate with them in the same place, and while he was doing so, he spoke to the forty bandits about the money brought by the child: “The money you have brought since you have been robbers, that child went once and brought more; you are not as much as a small child,” he said, scolding his bandits. **(704)**

One of the bandits says, “We were in town when we raided the lord’s treasure chest, we were putting the money in our bags when someone came and scared us, the night was dark too, we couldn’t think what to do, we left it and ran away, it was probably this child who scared us, he put the money on the cart himself and took it. If he’s a lad, today he can go off and bring so much wealth alone, we’ll find out what kind of lad he is,” he says, “if he doesn’t bring it, we’ll chop his head off so he doesn’t drive us all mad and make us all sad.”

The child heard what they said. “So tonight, I will go myself and fetch something, or they will kill me,” he says, and, tying the former blunt sword about his waist, he mounts his mangy colt, and sets out from their lodgings and comes near a village. He could not, however, get hold of anything; at last, he was stealing a horse from somebody’s stable, when they ran after him shouting, “A thief is stealing our horse.” This boy, not knowing what to do, left the horse and ran away. From there he approached another village, taking out his watch he saw that it would soon be midnight; what should he do then; he felt that nothing would be done; if he returned, the bandits would kill him; very slowly, slowly, he returned to his own lodgings; **(705)** he tied his horse carefully to the post in front of the house, he himself went into the kitchen; from there he looked inside the house; all forty-one bandits were asleep as usual. What this should do, he calls out the name of one of the bandits: “My brother batir, I have brought a chest full of treasure; it is very heavy, I cannot lift it and bring it in, come and help me,” he says. As soon as the robber leaves the house, the child draws his blunt sword and strikes the robber, cuts off his head and leaves him there, drawing him aside. Doing the same, he kills all 40 bandits one by one; then it’s the ataman’s turn, and he shouts at the ataman as well as at them. The ataman wakes up from his sleep, “Well done, brother, you’re a [real] bandit now,” he says, and as he steps out he cuts off his head, too; he himself is very tired. He goes to his place, lies down and sleeps; it is dawn, as it should be. With that early morning bedtime, it was only about two hours before noon when he got up, washed his face and arms, went out and looked around. All the robbers lay there frozen in blood. The child thought about what to do, he said. Immediately, **(706)** he took a wooden shovel in his hand and dug a pit like a pit under the threshold of the house and brought the forty-one robbers and put them in that particular pit, and put their blood in the pit, and pressed it down well with earth from above, but he began to think to himself, “If any of their brothers or those close to them come from somewhere and kill me, I must escape from there,” he thought.

The next morning, he rose early, prepared fifty pairs of oxen for fifty wagons, put silver on some of the wagons, gold on some, other valuable things on some, and set out, trusting in God; "I am going to my own native land, where I grew up," he says.

He went on like this for a few days, and then one day he arrived in the country where he was raised. Then he asked where his father and mother were; they showed him. They brought him near a house, "Behold, this is your house," they showed him; he looked at their house. Their house had no window, its door was broken, its walls were on the verge of collapsing, in the yard grew summer plants, grass, in the grass an old man and an old woman; **(707)** poor people were eating grass.

The lad goes to the old men with a greeting selam alejkum*. The poor bezonians accept the boy's greeting but have no idea that it is their child. The boy asks, "My old mother, if you didn't eat this grass, would you have nothing else to eat?"

The old people say, "What shall we do, my child, we have no one to look after us. We have three daughters, but what use would they be to us, they don't serve us like sons; they are no use to themselves; how could they be of any use to us; we had a son, and the bandits took him away; we don't know where he went," they say.

The boy says, "And if you saw your son if he were standing here talking to you like I am, would you know him, would you be able to tell him apart?"

The old men [say], "Yes, we would, my son," they say.

The boy says, "Well, you could not know me either, I am your son".

Then the old men said, "Alas, my child, it is you; welcome, my little child," and they embraced their child; and they wept together as child and a father.

Then he brings the money to the doorstep, spills it on the green grass, and surrounds the area with wagons fencing it off so no one can touch it, saying he is letting the oxen out on the grass to graze. Then, saying that he needs a place to keep that certain money, he finds it in the city **(708)** and brings the first [builder] master, and, measuring it from the threshold, has a fine house built in the courtyard and places the money in it. Then he says to the master, "I will have a palace built; such a palace shall it be; in it shall be stoves of marble stone, it shall have seven storeys, one storey of gold, one storey of silver, one storey of marble, one storey of amber, seven [seventh?] storeys of various precious stones, can you do it?"

The master says, "I will do it, and he takes it on." It will be a palace that the Padishah will envy; if he sees it, it will be ten times better than the Padishah's palace, the head of the masters will do it and finish it, and he gives his word that he will do it in three months.

They are doing two floors of the house. This boy puts [the oxen] in the wagons again and goes to fetch the money that is left. He comes to the place where the money of those certain bandits was; the money left he loads up on to twenty-five pairs of oxen. In the neighbourhood of the place where they used to live [with the bandits], there was a big palace, he had never been in that palace, now he went into that palace, inside the palace there was a big chest, on the top of the chest there was a key, he takes the key, opens the chest, and what he sees, inside the chest there is an even smaller chest, he opens the small chest and sees that there is a beautiful ring inside. As soon as he took the ring in his hand and looked at it, **(709)** he put it in his mouth, and as soon as he took it in, the lad turned into a cat, and when seeing himself in the form of a cat, he was very frightened, and he took the ring out of his mouth, and as soon as he took the ring out, he was again as he had been before. Seeing this, he thought, there must be something in this ring, he said, he put the ring in his pocket, put the treasure [valuables] on it as it should be, and from this place, he is coming back to his country safe and sound.

Let him be coming home with his fortune, and let us come here; five floors of the house have been finished, and two floors have remained. There was a powerful Padishah in this country, who went to visit another Padishah two hundred chakirims* from this country. When he passed the palaces built by the boy, the Padishah's eyes were caught by the palaces. The Padishah had his three-horse chariot stopped, and, gathering the people of this village, he asked them, "Who was building these palaces?"

The people here [say], "There is a poor old uncle here, he has a son, the son has it made," they say.

The Padishah immediately calls the masters together and asks them, "By whose permission are you building these palaces?"

The masters [say], "Padishah, we are workers here; we work for money; the one who has this done is the son of a poor old man," they say; [the Padishah:] "Who gave him permission, for my part, he is not allowed to build such palaces," **(710)** he says. "Don't work, stop it; this palace is nicer than my palaces; as Padishah, I don't allow to build palaces nicer than mine. Who would be the one who builds a palace bigger than mine? Where is that boy, go, call him, let him come here."

The first people say, "My Padishah, that boy has gone away somewhere; he hasn't come back yet; when he comes back, we will send him to you."

And the Padishah said, "As soon as he comes, tell him to come to me," and he goes away from here in anger. The Padishah turned back from his way to visit.

Let the Padishah return in his anger, and we will come back to the place here, where the boy will come after two days, after the Padishah has left, with twenty-five [wagons] loaded with value. He empties that fortune again in a pit in the manner just described. The aforesaid old father tells his son, as he must, that the Padishah has come, and then he goes away angrily, after which he says, "My son, my child, because of this wealth, these palaces, great trouble will obviously come upon us, the Padishah here will not leave us alone, and even you are summoned, 'come to me', he said, and went away."

The son [says] to the old man: "Father, you have nothing to fear from such things, I'll stay here, I'll look after the workers; you go to the Padishah and talk to him," he says.

The old man [says], "My son, if I go there and the Padishah asks, 'Where did your son get this wealth and money', what will I say?"

The boy [says], "Don't you worry about that; **(711)** you tell it like it is true, 'I have a bandit, thief, rogue son, stolen property is his,' tell him."

The old man leaves here, and goes to the Padishah's palace, "Selam aleykum," he says.

The Padishah receives his greeting, "Sit down, old man," he says.

The old man sits down.

"Old man," says the Padishah, "you were a poor old beggar, now you are building better palaces than mine; where did you find so much money, but tell the truth; if you don't tell the truth I will cut your head off."

The old man says, "My Padishah, I have a son, a thief, a real bandit; this is the money he has gathered from theft, not knowing where to give his money; 'I want my palaces to be better than his [i.e. Padishah's]', he said, and he is having them made."

The Padishah says, "You call your son a great thief, so if tonight, when my wife and I are lying on the mattress, he steals my mattress without waking us from our sleep, I will admit he is a good thief, if he does not steal it, I will cut off both your heads tomorrow," he says.

The old man says, "I'll go and tell my son"; he goes out, he comes to his son, he says to his son, "My son, you're making these houses in vain; the Padishah said, 'He has got to do such and such things if he can't do it, I'll cut his head off.'"

The son says, "Father, don't worry about that, to do such a theft is not a [big] thing".

And the Padishah also said, "If he can steal [the mattress], **(712)** I have only one daughter in this world, but I will give my daughter to your son."

The boy wanders here and there during the day, passing the time. After dark, he puts the ring in his pocket, passes by the Padishah's house once more, and what does he see: the Padishah has guards on all sides of the palaces. This boy goes to one side of the palace, takes out the ring, sucks it in, and immediately takes the form of a cat, and at the feet of the guards goes into the Padishah's palace and lies under the bed. It is evening; the Padishah is talking to his wife on the mattress; the Padishah says, "Wife, tonight the poor man's son will steal the bed [mattress] under us; tonight, how shall we do it so that we don't fall asleep if he steals the bed, he will take our daughter and shame me," he says. As they talk like this, [the boy] interrogates them. The Padishah's wife [says], "Come on, this is impossible, all our sides are guarded, to steal a bed [mattress] under a man and escape from so many guards, is it possible?" Then it was two o'clock; "Let's go to sleep," they say, they are falling into a very sweet sleep.

Let them sleep. The boy then takes the ring out of his mouth, and as soon as he has taken it out he is a lad again; he returns to his appearance, and what to do; **(713)** he cannot find a way to take the bed [mattress], then he opens a chest in the house and looks inside, the Padishah's wife had festive clothes inside, he puts them on and becomes like the Padishah's wife, goes to the kitchen, there is a bag of flour, in a bowl he mixes the flour with [a little] water and makes pasta and pours it on the Padishah's mattress, and lays his wife on a cushion in the sleeping chamber. Then he herself goes to the Padishah's side and wakes him up, the Padishah's hand gets stuck in the dough while getting up, and the Padishah asks what this is. "Which of us has soiled ourselves like this?" he says.

The former boy says to the Padishah, "Take off your trousers and shirt and give them to me, I'll wash them and bring them back so that no one will see."

The Padishah takes them off and gives them to him, the boy goes to the door to take them out, but the guards won't let him out; he comes back and says to the Padishah, your guards won't let me out; the Padishah goes out and shouts at them, "Let her out, it's your Padishah-mother, why won't you let her out." The guards let her out.

The boy washes by the well, then sneaks out and escapes. The Padishah waits for an hour or two, no trousers, no shirt, no bed [mattress], then, what a thing is this, did this boy do this to me, saying, he goes up and down searching, then [sees] his wife in the sleeping-cabin is asleep, the Padishah wakes his wife and asks her, **(714)** "What kind of thing is this?"

The wife says, "I cannot know what kind of thing it is," and takes clean trousers and a shirt from the Padishah's chest, the Padishah puts it on and sits down.

It is morning, and he gathers the Viziers and tells the matter as it was; there was a Vizier among the Viziers; this Vizier also wanted to marry the daughter of the Padishah, so this matter pained him, and he said, "My Padishah, it is not proper to give your daughter to a thief."

The Padishah [says], "What shall I do?"

The Vizier says, "We have a grey horse in your stable, placing the throne next to the threshold of your gate, you sit on it, holding the stirrup of that horse, and you will put guards and patrols everywhere, and if he steals that too, we will look for some solution."

Let them say like this: the boy stole the Padishah's pants and bed [mattress] and came home. He sent his father to the Padishah; his father arrived at the Padishah. "What do you want now," he asks.

The Padishah says, "Old man, your son is a skilled thief, but I have a grey horse, I will patrol around it while sitting on my throne holding it [by his stirrup], if he takes it without anyone, not even me, noticing, I will give him my daughter; otherwise, I will cut off both your heads and take the money for myself." The old man comes to his son and tells him the matter, and the son says, "My dad, don't worry about this matter, I'll come back on the grey horse, rejoicing and laughing," he says.

(715) The boy then goes again to the Padishah's palace, taking the previous ring with him, but the palace is surrounded on all sides. The boy, what to do, takes the ring in his mouth, turns into a cat, and again goes under the Padishah's bed and lies down. The Padishah waits outside, saying the boy will come and take the horse; the morning approaches, and he gets sleepy, calls one of the Viziers and says to him, "Sit on my throne and here, take this horse, I will go and sleep a little, give it to no one but me," he says and goes in, undresses as he should and falls asleep. The boy at once puts on the Padishah's clothes and crown and goes out looking like the Padishah, or else he has remembered the name of the Vizier who is staying there instead of the Padishah. He calls out the name of the Vizier, and the Vizier asks what you say, Padishah, the boy, imitating the Padishah himself, says, "Saddle my grey horse and bring it here, I will go and cut off the poor old man's head and his son's, and come back." The Vizier saddles up the horse and brings it to the boy; the boy mounts it, and tells the guards to open the gate; they open it immediately, and the boy, in the manner of the Padishah himself, whirled away.

Early in the morning the Padishah wakes up, gets up and goes to the stables, and what does he see: no horses, no soldiers, no Viziers. He immediately has his Viziers called, "What have you done, I told you not to give the horse to anyone but me."

The Vizier says, "My Padishah, you yourself had the horse saddled, 'I will strike the head of the old man and his son', saying **(716)** you have gone away, what is my fault?"; the Padishah went into the house, and seeing that neither his crown nor his clothes were there, immediately he assembled his Viziers and held a council. "What shall I do, how shall I do it, how shall I get rid of this child?"

Let us come to this side. The horse, as good as it was under the Padishah's hands, the son took care of it, kept it well, put the Padishah's clothes in a nice box and locked it up, and sent the old man to the Padishah again, the old man arrives at the Padishah's, and goes in with a greeting.

"Selam aleykum, my Padishah." The Padishah accepts his greeting, "Well, old man, I have no words for your son's stealing, but let him show one more game [trick?], then I'll give [him] my daughter without a word," he says.

"My Padishah, what game?" asks the old man.

The Padishah says, "When we pass by your palace on our way to the Padishah, which is two hundred chakirim* from your city, if he steals my wife, myself, my coachman, and my three-horse carriage without my knowledge, I will not have a word, I will give you my daughter," he says.

The old man comes to the boy and tells him. The boy was sitting in the window of the palace, watching, when the Padishah's three-horse carriage rushed by the boy's gate like the wind; the boy wondered what to do, but couldn't think of any solution. Immediately, at that place, he rolls up his sleeves and trouser legs and runs after the Padishah's carriage.

Let him run on like that... The Padishah, having passed twenty chakirim* from the place in front of the boy's palace, stops for a rest. **(717)** They sit down for a rest, and the boy catches up with them, running, and passes in front of them, unnoticed by them. On the road inside a large forest, at the edge of the road, were two large trees. The boy had gathered a bunch of green grass from the forest, wrapped himself in it, not looking like a man, and hung himself crosswise on the trees in the road. And the Padishah, driving by in his carriage, saw this hanged thing, made the carriage stop, and asked him, "Are you a man or a genie?"

The boy says, "I am a savage."

The Padishah says, "Can you catch such a boy for me and bring him to me?"

"I will bring him," saith he, "[just] you wait a little while in this place, I will bring him," saith he.

The raspberries on the roadside were just about ripe. The Padishah drives the cart to where the raspberries are; let's rest here, saying, they are sitting. And the boy beside them, “I'll bring you that boy,” he says, and turns around and sits under their wagon. The Padishah's wife gets down to relieve herself.

The Padishah says, “I'll have some raspberries,” and gets down, and the coachman gets down to piss. The son gets on the cart and drives away, comes back to his palace, puts on the clothes in which he escaped from the Padishah, and puts his father and mother on the cart **(718)** and he himself gets on the Padishah's grey horse, [the cart] driven by an attractive cart-driver, they go to fetch the Padishah who is left on foot.

Then they come to the palace, seating the Padishah next to the old man and the Padishah's wife next to the poor old woman. From that time on, forty days and forty nights of wedding and feasting, the Padishah marries his daughter to the son. The boy and the girl grant each other's wish, pick the ripe apricot and eat it. The two wishes become one. May God also bring us to our destination.

Explanatory Glossary

[ ] – Significant and/or obvious gaps in the text are indicated by square brackets. In square brackets are those passages of which original Tatar equivalents Ignác Kúnos has crossed out in the manuscript, but of which publication and translation seemed necessary.

abdas ~ aptas – ritual ablution before performing prayer in Islam
abdast see abdas
aga – lord, elder brother
aka – elder brother, uncle
akcha – money, silver coin
Akmesdjit – Turkish-Tatar name of Simferopol
aptas see abdas
arshin – ell, yard (Turkish measure of the length of about 68 cm)
ashik – sheep's ankle bone, a tool of the favourite nomadic game
ataman – ataman, hetman, (robber) chief
Azrail – the angel of death
bal – bragget, metheglin
batir – epic hero; champion, valiant
batman – batman (a measure of weight varying from 2,5 to 10 kgs
Bay-terek – the name of a magic tree, ‘Lord/Rich Tree’
belesh – turnover pie (a kind of national food of Tatars)
bey– gentleman; (after names) Mr.
bismilla! – In the name of Allah! (Muslim supplication before taking an action.)
bolgach – pirog, a kind of puff pastry
bulamik – pap made of flour
bulgacha – pirog, a kind of puff pastry

chagirim see chakirim
chakirim – versta, verst or Russian mile; length measure, "1067 m; The Tatar word "Chakirim" (shout) suggests that originally this length measure meant the distance for which the shout was heard.
cheburek – pirog, flaky pastry with mutton or other filling fried in fat
chinchi– performer of songs *chin/chyn* (cf. Crm. Tat. *chyng (čıñ)* 'song of improvised character a folkloric genre in verse (see the text "Golden bracelet" at the beginning of the collection).
Chin-Machin – China and the Far East in general.
cholak – crippled in one hand
deer – according to Kúnos' note made in brackets, the Crimean Tatar word *karağa* means 'wild goat'.
Djel-djetmez – (horse name) 'no wind can catch it; winds can't catch up'
djez alti an exhibition of the marriage portion; a gathering and entertainment at the girls' house before the wedding;
djezve – coffee pourer with handle.
efendi – gentleman, learned man
fez – a type of Muslim headdress; an inverted flowerpot-shaped tasselled headdress, usually made of red post;
Gench Amed – Young Amet (Ahmet)
hadji – a Muslim making the pilgrimage to Mecca
helva-seller – vendor of helva (~halva) a popular oriental sweet that can be made at home using hazelnuts, walnuts, cocoa, almonds or pistachios
Hizir(-prophet) – the patron saint of travellers
Horasan lime – a special recipe for a binder that could be used to make a very strong wall
inshalla – if Allah wills (Ar.)
Jew – Hebrew; Here comes up as merchant, peddler or watchmaker
kadi(-bey) – (lord-)judge of Islamic canon law)
Kaf see Mount Kaf
kalakay – a kind of pirog
kamchi – whip, short riding-whip
kazi-bey (~kadi-bey) – lord-judge, see *kadi* and *bey* above
kobete – meat pie
kapek – kopeyka (smallest Russian coin)
Kibla – South (Ar.).
kismet – fate (Ar.)
kumech – brioche, plain cake; bun, muffin
Kun-kurmez – read, "Kün-kürmez" – (beauty like) no other under the sun; epic epithet for Chora's sister (Mengaru, Mejnaru, Mingero), literally, "Does not see the sun"
medrese – school, secondary or post-secondary Mohammedan school
mullah(h) – Muslim priest, teacher

Mount Kaf – fabulous mountains that ring the earth and hold up the sky; home to fantastic creatures
muezzin – a Muslim cleric who calls the faithful to prayer
mulla(h) – Muslim priest, teacher
namaz – prayer and a series of movements performed simultaneously with prayer
Nogay – Kipchak-Turkish people; are looked down upon and often mocked because of their nomadic lifestyle, relative backwardness and peculiar dialect.
okka – (a measure of weight) equals 3 pounds or 1.28 kg
Or – original, Tatar name of Perekop
Padishah– ruler, sovereign
para – money, change, penny
pehlivan – champion, wrestling champion, wrestler, hero
pejlivan see pehlivan
pejliven see pehlivan
pelivan see pehlivan
pound (put ~ pud) – a unit of weight and mass used in Russia until 1924, over 16 kg
raki – aniseed brandy
salam – salutation, short greeting, salute (Ar.)
Salghur – a river in the Crimea.
Salgir – a river in the Crimea
scroll – A fairy-tale element at the beginning of a story, consisting of nonsense (gibberish) and consisting of fixed stylistic elements.
selam alejkum – (greeting), peace be with you!; be welcome (greeting on meeting), meaning peace be upon you (Ar.)
sofra – short-legged table
sufu (sofu) – Muslim mystic (Sufi)
tekerleme – a playful folktale element consisting of nonsense (gibberish) told at the beginning of a story, consisting of fixed stylistic elements
Uf – Uf-Arab see uff
uff – read, "Ooff" 'phew'
Yali-boyu – (toponym) the name of the southern coast of the Crimea
Yendj Osman – Young Osman
zilfikar – a type of medieval sword, according to Islamic tradition the double-ended sword of the Prophet Muhammad and later the Prophet Ali, with magical powers, which, when wielded, can defeat outnumbered enemies.

Glossary

A

a (25) *Interj* (before noun) expresses reproach; oh! so! | ах!; а; о! ведь; же; э!

abab (2) *N Ar.* friend | друг

ababil s. **ebabil**

abadan (2) *Adj P.* comfortable, cozy | благоустроенный, уютный

abala– (2) *V* to crawl, to wiggle | ползать; извиваться

abdas (13), **aptas** *N Ar.* ritual ablution | ритуальное омовение; **abdas al–** to perform an ablution | совершать ритуальное омовение

abdıra– (5) *V* to wince, to recoil (in horror), to be scared (of) | вздрагивать, встрепенуться, испугаться (чего/ кого) → → **aptıra–**

Abdukerim (1) *Pn Ar.* Abdukerim | Абдукерим

Abdurajım (1) *Pn Ar.* Abdurrahim | Абдурраим

Abdurrahman (6), **Abdurraman** *Pn Ar.* Abdurrahman, Abdurraman | Абдуррахман, Абдурраман

Abdurraman s. **Abdurrahman**

Abdurrājim (1) *Pn Ar.* Abdurraim | Абдураим

Abdurrešıd (1) *Pn Ar.* Abdurreshid, Abdulreshit | Абдуррешид, Абдулрешит

Abdussadiχ (2) *Pn Ar.* Abdussadikh | Абдуссадих

Abi (1) *Pn Ar.* Abi | Аби → **aka**

Abibulla (1) *Pn Ar.* Abibulla | Абибулла

Abibullah (1) *Pn Ar.* Abibullah | Абибуллах

ač (9) *Adj* hungry | голодный

ač– (114), **aš– II** *V* to open | открывать; распускаться (о цветах)

ačıχ s. **ačık**

ačık (5) *Adj* open | открыт(ый); **ačık ava** open air | открытая площадка; **appačık, appačıχ** quite open, wide open | совершенно открытый

ačık– (2) *V* to feel hungry, to grow hungry | чувствовать голод, проголодаться; **karnı ačık–** to feel hungry | голодать → **aǯıχ–**

Ačık–teñiz (1) *Tn* Achyk–tengiz (open sea) | Ачык–тенгиз (открытое море, морской простор) → **teñiz**

ačıl– (17) *Vpass* to be opened; to blow | открываться, отворяться; распускаться (о цветах); **sabā ačıl–** to dawn | рассветать

ačılaštır– (1) *Vfact* to render bitter; to sadden; to exasperate | раздражать, опечаливать

ačlıχ (1) *N* hunger, starvation | голод, голодание → **ašlık II**

ačtır– (1) *Vfact* to make sb open sg; to have sg opened | велеть открыть, приказывать открыть

ačulan– (27) *Vpass* to be angry, to grow angry | злиться

ačulandır– (1) *Vfact* to anger, to infuriate | раздразнивать, озлоблять

ačulı (2), **ačulė** *Adj* angry, irritated, evil | сердитый, злой

ačuv (1) *N* anger, malice | зло, злоба; **ačuγundan** because of his (her) anger | из–за его (её) гнева

ačuvan– (1) *V* to be angry | злиться

aǯajb s. **aǯajıb**

https://doi.org/10.1515/9783111442891-004

aǯajıb (2), **aǯajb** *Adj Ar.* amazing, unusual, unique | удивительный, необыкновенный

aǯeb (8) *Adv Ar.* wonder if, wether | интересно было бы знать; разве, неужели, ли, так ли → **aǯeba**

aǯeba (1) *Adv Ar.* wonder if, wether | интересно было бы знать; разве, неужели, ли, так ли → **aǯeb**

aǯele (4), **ǯele** (3) *Adv Ar.* hastily, hasty, urgent(ly) | спешно, спешный, срочно, срочный

Aǯem (3) *N/Adj Ar.* Persian, Iranian | перс, иранец; персидский

aǯi (3) *N Ar.* one who has performed the rites of the pilgrimage at Mecca; pilgrim, hadji | хаджи (человек, совершивший паломничество в Мекку) → **aǯı II, haǯi**

aǯi– s. **aǯı**–

aǯik– (2) *V* to make haste, to be in a hurry; to grow impatient | спешить, торопиться; проявлять нетерпение

aǯit– s. **aǯıt**–

aǯı I (9) *Adj* bitter, sour | горький

aǯı II (4) *N Ar.* one who has performed the rites of the pilgrimage at Mecca; pilgrim, hadji | хаджи (человек, совершивший паломничество в Мекку) → **aǯi, haǯi**

aǯı– (4), **aǯi**– *V* to feel sorrow (for); to grudge (a thing) | жалеть кого/ что

aǯıχ– (1) *N* to feel/grow hungry | проголодать → **ačık**–

aǯın– (1) *V* to feel sorrow; to grieve | жалеть кого, сожалеть; сочувствовать

aǯıt– (2), **aǯit**– *V* to cause pain, to hurt; to make sb feel sorrow | причинять боль (страдания); огорчать, расстраивать

ad (36) *N* name, proper name | имя, название; собственное имя; **ad kuj**– to name, to give a name | дать имя (название), наречь; **adına Kůlbastı kojγannar, adıni Kůlbastı kujγannar** they named him as Kulbasti | его назвали Кюльбасты; **adını Gůlzar kujγannar** they named him as Gulzar | его назвали Гюльзар

ada (1) *N* isle | остров

adam (116), **adem** *N* man, human (being), person | человек, лицо; мужчина; **adam sen de!** do not bother, do not worry about him/her/it!; never mind! | да ладно! подумаешь! да что там! не обращай внимания!; **er adam** man, male | мужчина

add (1) *N Ar.* right; basis | право; правомочие; основа

adem s. **adam**

Ades (1) *Tn* Odessa (city) | (город) Одесса

adet (3) *N Ar.* custom, usage; habit | обычай; привычка, манера

adı s. **ajdı**

adım (3) *N* step (in walking) | шаг; **adım al**– to (take a) step, to tread | шагать, ступать

afat (1) *N Ar.* person of bewitching beauty; beauty; smart appearance | писаная красавица; красота; шикарный вид; прекрасность

aferim (5), **apperėm** (2), **āferim** *Interj P.* bravo! well done! | молодец! браво! → **aferin**

aferin (2), **aferın** *Interj P.* bravo! well done! | молодец! браво! → **aferim**

aferın s. **aferin**

aff (1) *N Ar.* forgiving, forgiveness, pardon | прощение, извинение; **aff it**– to forgive, to pardon | извинять

afijet (3), **afıjat** *N Ar.* health | здоровье; **afijet olson/ olsun** bon apétit!, good health! may it do you good! | на здоровье! приятного аппетита! с лёгким паром!

afıjat s. **afijet**

afser (1) *N R.* officer | офицер

Afstrija s. **Avstrija**

afta (3) *N P.* week | неделя → **hafta**

aɣ (1) *N* net, web | сеть, сетка; **balıχ aɣı** fishing net | (рыболовная) сеть, невод → **au II, čilme**

aɣa (106) *N* (rural) lord, master; elder brother; brother; Mister (used in addressing); *–agha* (occurs as a secondary component of personal names) | хозяин, господин; старший брат; брат (обращение); *–ага* (после имени, второстепенный компонент мужских имен); **Ameša aɣa** Mister Amesha; brother Amesha | Амеша–ага; господин Амеша, брат/дядя Амеша; **Memet aɣa** Mister Memet; brother Memet | Мемет–ага; господин Мемет, брат/дядя Мемет; **tuɣan aɣa** own elder brother | родной старший брат

aɣač (15), **aɣaǯ** (5) *N* tree; wood, wooden | дерево, деревяный

aɣaǯ s. **aɣač**

aɣaj (8) *N* agha; elder brother; (rural) lord, master; Mister, brother, uncle (used in addressing) | ага; старший брат; братец, дядя, дядюшка (обращение) → **aɣa**

aɣar–duvaɣar s. **ak–**

aɣır (5) *Adj* heavy, hard; difficult | тяжелый, трудный; **aɣır hasta, aɣır kasta** seriously ill | тяжело больной; **aɣır–ajaklı** pregnant | беременная → **avur**

aɣır– (1) *V* to ache, to ail, to be unwell | болеть; хворать, поболеть → **avur–**

aɣıt– (1) *V* to cause to flow, to pour, to shed | лить, наливать

aɣız (2) *N* mouth | рот, уста → **avız, avuz I**

aɣla– (62), **aɣle–** (2), **ajle–** (2) *V* to weep, shed tears; to cry | плакать, рыдать, разливаться, кричать

aɣlaš– (2) *V recipr* to weep together; to lament together | плакать, рыдать (вместе с кем–л.), плакаться, жаловаться

aɣlat– (3) *Vfact* to make sb weep/ cry | заставить плакать

aɣle– s. **aɣla–**

aɣu (1) *N* poison | яд, отрава

ah (23) *Interj* oh! alas! | ой!; **ah ček–/ et–** to sigh, to heave a sigh | охать, вздыхать; **ahinen–vahinen umur kečir–** to spend one's life with sighs and laments | проводить свою жизнь со взодохом и воплем; **ah–zār ček–** to sigh, to heave a sigh | охать, вздыхать → **aχ I**

Ahmad s. **Ahmat**

Ahmat (84), **Ahmad** (13), **Āhmad** (1) *Pn Ar.* Ahmat, Ahmad | Ахмат, Ахмад; **Ahmat–batır** Hero Ahmat | Ахмат–богатырь → **Ahmed**

Ahmed (10) *Pn Ar.* Ahmed, Ahmet | Ахмед, Ахмет; **Ahmed–batır** Hero Ahmed | Ахмед–богатырь → **Ahmat**

aht–ı–zar (1) *N* sighs and tears | охание и плач; **aht–ı–zarim kalaǯak** my langour will stay (with sb) | моё охание и томление останется (у кого–л.) → **zār**

ahtaǯı (4) *N* stable hand who deals with castrating and stigmatizing horses | конюх, занимающийся кастрацией и клеймением лошадей

aχ I (44) *Interj* oh! | а! ой! → **ah**

aχ II (2) *Adj* white | белый → **ak I**

aχ– (14) *V* to flow, stream; to be poured out | протекать, течь, литься → **ak–**

aχbarǯılık (1) *N* profession or occupation of a porter | занятие или профессия носильщика; **aχbarʒılık et–** to work as a porter | работать носильщиком

aχča (4) *N* money | деньги → **akča**

aχir s. **aχır II**

aχıl (2) *N* reason, mind, memory | ум, разум, рассудок, память → **akıl**

aχılʒık (1) *N dim* reason, mind, memory; smart | разум, рассудок, памаять → **akılʒık**

aχıllı (1) *Adj* clevet, bright, wise | умный, мудрый → **akıllı**

aχır I (3) *N* stable, shed, barn | стойло, конюшня, хлев

aχır II (5), **aχir** *Adj Ar.* aftermost, the last one, final; end | что–л. последнее; последний, конечный; конец, исход, окончание; **aχir zeman** the end of time | время светопреставления; **zaman aχri** the end of time | время светопреставления

aχjar I (1) *N Ar.* others, strangers; rivals | другие, чужие, соперники

Aχjar II (2) *Tn* Sevastopol (city) | (город) Севастополь

aχmaχ s. **aχmak**

aχmak (3), **aχmaχ** *Adj Ar.* foolish, silly, stupid | глупец; глупый; дурень

aχot (1) *N R.*? hunting?, hunter | охота?, охотник; **aχot–mejzin** hunting muezzin?; hunter–muezzin | охота–муэдзин?; охотник–муэдзин → **mejzin**

aχran (1) *Adv Ar.* last time | (в) последнее время → **aχır II**

aχret (5) *N Ar.* an other world beyond the grave; hereafter; the next world | потусторонний (загробный) мир → **akret**

aχšam (5) *N* evening | вечер → **akšam**

aj I (51) *N* moon, month | луна, месяц

aj II (55) *Interj* oh! (exclamation of surprise) | ай! (воскрик сюрприза/удивления); **aj gidi saña!** oh, you little rascal! | ах ты, шалун!; **ajja aj!** oh! dear! | ой–ой!

aja– (1) *V* to feel sorrow for sb, to commisarete | жалеть кого

ajaχ s. **ajak**

ajak (61), **ajaχ** (2) *N* foot, leg | нога

ajakčı (1) *N* cup–bearer (title), footman | ответственный за сервировку (титуль), лакей

ajakkabė (<**ajakkabe)** (1) *N* shoes; footwear | обувь

ajaklan– (1) *V* to get on one's feet; to set off, spring up | подниматься на ноги; отправляться; вскакивать

ajaklı (2) *Adj* having a foot or leg, footed | имеющий (какие–л.) ноги

ajan–bejan (1) *Adj Ar.* very clear | совершенно ясно, явно

ajat (3) *N Ar.* kitchen, corridor, covered court; | кухня, веранда, крыльцо

ajazlan– (1) *V* to become very cold; to become chilly and clear (weather, night) | сильно охладиться, становиться ясной и морозной (о погоде, ночи)

ajb s. **ajıb**

ajbetli (1) *Adv/Adj* groomed; neat, imposing; well cared–for | ухоженный, хороший, добрый; внушающий уважение, производящий сильное впечатление

ajda (4) *Interj* hurry up! now then! go on! | ну! айда! давай! → **ajde, ajdı**

ajda– (20), **ajde–** (3) *V* to drive on (animals) by shouting | гнать животных выкрикивая

ajdamak (46) *N* bandit, brigand, robber | разбойник, грабитель → **hajdamak**

ajdamaklık (2) *N* brigandage, rapine | бандитизм, разбой, грабёж

ajdat– (3) *V* to make sb drive on (animals) by shouting | заставить гнать (животных) кричая

ajde (14) *Interj* hurry up! now then! go on! | айда! давай! ну пошел! → **ajda, ajdı**

ajde– s. **ajda–**

ajdı (5), **adı** *Interj* hurry up! now then! go on! | айда! давай! ну пошел!; **ajdıñız, ajdınız** hurry up! go on! (plural or polite 2nd person) | айда, давайте! → **ajda, ajde**

ajdın (1) *Adj* lucid, clear, light | ясный, светлый; **kůziñiz ajdın bolson!** Congratulations! | Поздравляю вас!

ajėr– s. **ajır–**

ajγir s. **ajgir**

ajgir [~**ajγir (4)**] (5) *N* stallion, steed, mount | жеребец, конь, скакун → **ajγır**

ajgır s. **ajγır**

ajγır (14), **ajgır** (3) *N* stallion, steed, mount | жеребец, конь, скакун → **ajgir**

ajib s. **ajıb**

ajir– s. **ajır–**

ajirma (1) *N* separating; (here) sorting out, selecting | разделение; (здесь) разборка, отбор, развод; **ajirma jul** road–junction/forking | ответвление пути/дороги

ajıb (10), **ajib** (3), **ajb** *N Ar.* shame, disgrace | стыд, стыдно; **ajb/ajib et–** to behave improperly, to put sb to shame, to behave in an unseemly manner | вести себя не подобающим образом, пристыдить, посрамлять кого–л.

ajın– (2) *V* to become conscious, to come to one's senses | очнуться, прийти в себя

ajır– (10), **ajėr–** *V* to select, sort out, distinguish | разделять; различать; выделять

ajırı s. **ajrı**

ajırıl– (2) *V* to be selected, sorted out; to divorce | отделяться, разлучаться, разводиться; **ajırılmaj dust** inseparable friend | неразлучный друг

ajırt– (1) *Vfact* to have sg selected, get sg sorted out | заставить разделять что–л.; велеть различать; заставить выделить

ajja (1) *Interj* oh! ah! dear me! | ой–ой! ох!; **ajja–aj!** Oh! Dear! | ой–ой!

ajkır– (2) *V* to cry out, shout, scream | вопить, закричать

ajla– (1) *V* to get around; to go, walk round | обходить что–л.

ajlak I (1) *Adv* very, highly | очень, неужто

ajlak II (1) *Adv* vagabond, idler, lazy | праздный, праздношатающийся, ленивый

ajlan– (33) *V* to turn, go round; revolve | поворачиваться, крутиться, вертеться

ajlanajım (6), **ajlenejım** (1) *Interj* my dear! darling! dearest! | ах, ты, мой милый, дорогой; **ajlanajım kůziniñ karasından** I adore your (black) eyes | обожаю твои (чёрные) глаза

ajlandır– (8) *Vfact* to make sg turn, revolve | поворачивать, крутить, вертеть

ajle– s. **aγla–**

ajlenejım s. **ajlanajım**

ajlı (2) *IndefNum* many, much; good deal | много; изрядно; **br ajlı vakitten son** after quite lots of time | спустья довольно много времени

ajnaǯı (1) *Adj P.* trickster, tricky | хитрец, хитрый

ajnık (3) *Adj* sober | трезвый

ajrı (7), **ajırı** *Adj* other, different; another | отдельный, различный; другой;

ajırı tůš– to become separated from each other | быть разлучённым с кем

ajrıl– (22) *Vpass* to part, separate from one another | разделяться, разъединяться, отделяться

ajrılık (6) *N* separation, separate life | отделение, разлука; развод; **ajrılıγa tůš–** to become separated from each other | быть разлучённым с кем

ajrılma (1) *N* separation, divorce, leave | расставание, разлука; развод

ajsa (41), **ajse** (6) *Particle* in that case, if so, then | тогда, в таком случае, а как же, а как иначе → **ejse**

ajse s. **ajsa**

Ajše (12) *Pn fem Ar.* Ayshe | Айше; **Ajše hanım** lady Ayshe | госпожа Айше

ajt– (393), **ejt–** *V* to tell, say; to utter; to make offer of marriage | говорить, произносить, сообщать; сватать чью–л. дочь

ajtır– (=**ajttır–**) (1) *Vfact* to make sb say (sg) or speak | заставить говорить

ajtıš– (9) *V recipr* to talk (to one another) | переговариваться

ajtıu (1) *N* speech, conversation; making offer of marriage | разговор; выговор; беседа; сватовство; **sůz ajtıuγa kel–** to come to make a speech, to come to make a formal proposal of marriage | прийти чтобы произносить речь, сватать

aju (1) *N* bear | медведь

ajva (2) *N* quince (Cydonia vulgaris) | айва

ajvan (7) *N Ar.* (domestic) animal | животное → **hajvan**

ak I (12) *Adj* white | белый → **aχ**

ak II s. **akk**

ak– (31) *V* to flow, stream; to be poured out | протекать, течь, литься; **aγar–duvaγar** flowing–swirling | текущий–клубящийся → **aχ–**

Ak–İdıl (1) *Tn* (here) Volga river; (originally) Belaya (White) river | (здесь) река Волга; (первоначальное значение) Белая (река)

aka (64) *N* elder brother (used also in addressing elder men); | старший брат, обращение к старшему; **Abi aka** brother/uncle Abi | брат/ дядя Аби; **bij–aka** gentleman–uncle | господин–дядя; **Kerim aka** uncle/ brother Kerim | дядя/брат Керим; **Narik–aka** Narik–aka (as pn) brother/uncle Narik (as addressing) | Нарик–ака (имя), брат Нарик (обращение); **Narik–akam** my Narik–aka (as pn), my brother/uncle Narik (as addressing) | мой Нарик–ака (имя), мой брат Нарик (обращение); **Osman–akam** my Osman–aka (as pn), my brother/uncle Osman (as addressing) | мой Осман–ака (имя), мой брат Осман (обращение); **Temr–aka** Temir–aka (pn) | Темир–ака (имя) → **aγa**

akaj (39) *N* man, husband; brother, uncle | мужчина, муж; брат, дядя

akča (62) *N* money | деньги → **aχča**

akčaǯik (1) *N dim* money | денежки

akıl (22) *N* reason, mind, memory | ум, разум, рассудок, память; **akıl ůgret–** to put sb up/wise to sg | научить кого–л. чему–л.; инструктировать; **akıldan nuχsan** crackbrained, simpleminded | скудоумный, слабоумный; **aklı bašina kel–** to come to one's senses, suddenly realize sg, to sober down | прийти в себя, опомниться; взяться за ум, образумиться; **bašından aklı kit–** to lose one's senses, to go out of one's mind | быть ошеломлённым, быть сбитым с толку; **aklına kel–/tůš–** to occur to one's mind, to recur to one's memory | прийти на ум/в голову, вспомнить(ся); **aklından kit–** to go out of one's mind, to forget | выйти из головы, забыть; **aklını al–** to take away one's sanity | лишать кого–нибудь здравомыслия → **aχıl**

akılǯık (1) *N dim* reason, mind, memory; smart | ум, разум, рассудок, память → **aχılǯık**

akıldane (2), **akıldāne** *N* sensible, wise man, (here) councillor or adviser (of the khan); | умница, премудрый; (здесь) советник хана

akıldāne s. **akıldane**

akıllı (8) *Adj* clever, bright, wise | умный, мудрый → **aχıllı**

akılsız (3) *Adj* stupid | глупый

akırtından (1) *Adv* gently, slowly | медленно, тихо

akk (5), **ak II** *N Ar.* the right; justice; equity; share; righteous, almighty (God's epithet) | право, справедливость, доля, часть; справедливый, всевышний (эпитет бога); **akkın rusulı** righteous, almighty God's prophets | профеты справедливого, всевышнего бога

akla– (1) *V* to whitewash | белить

akmak (2) *Inf (mak)* flowing | течение

Ak–mečid s. **Ak–mečit**

Ak–mečit (8), **Ak–mečid** (2), **Ak–mečıd** *Tn* Akmedjid (Simferopol city) | Акмеджид (город Симферополь) → **Ak–meǯid, Ak–mešǯid**

Ak–mečıd s. **Ak–mečit**

Ak–meǯid (6), **Ak–meǯit** *Tn* Akmedjid (Simferopol city) | Акмеджид (город Симферополь) → **Ak–mečit, Ak–mešǯid**

Ak–meǯit s. **Ak–meǯid**

Ak–mešǯid (1) *Tn* Akmedjid (Simferopol city) | Акмеджид (город Симферополь) → **Ak–mečit, Ak–meǯid**

akop (2), **akup** *N R.* dug–out, trench | окоп

akraba (1) *N Ar.* kinfolk, relative | родня, родственник

akret (1) *N Ar.* an other world beyond the grave; hereafter; the next world | потусторонний (загробный) мир; **tůnja akret** the world beyond the grave; till the afterlife | потусторонний (загробный) мир; до потустороннего мира → **aχret**

akrın (1) *Adj/Adv* slow(ly), gentle, gently | тихий, тихо, медленный, медленно; **akrınγına** quietly, gently, slowly | тихо, по–тихоньку, медленно

aksır– (6) *V* to sneeze | чихать; **aksıra–tůksire** continuosly sneezing and coughing | непрерывно чихая и кашляя

akšam (75) *N* evening | вечер; **akšam χajır ajt–** to say good evening to sb | сказать добрый вечер кому–л.; **akšamγaǯıχ** till the evening | до вечера; **akšam(lar) χajır/kajır** good evening! | добрый вечер! → **aχšam**

akšamlık (2) *N* special to evening use; evening dress; (here) supper | предназначанный для вечера, вечерний; (здесь) ужин

aktar– (1) *V* to seek, search | искать, перерыть, перевернуть

Ak–tingiz (1) *Tn* the Mediterranian (Sea) | Средиземное море

akup s. **akop**

al I (49) *Adj* red, rosy | алый, румяный; **al ǯiren** reddish yellow (color of a horse) | рыже–жёлтый (масть лошади); **al**

basma red chintz | красный ситец; **al kelin** bride in red? | невеста в красном платье?; **al kırmızı** pembe | розовый

al II (22) *N Ar.* state, condition; fate | состояние, обстоятельства; **al katır suraš–** to inquire after each other's health | (взаимно) справляться о чьём–л. здоровье, спашивать друг–друга о самочувствии; **bu alinen** thus, in this way | так, таким образом → **hal**

al III *N* front, space in front (of a thing) | пространство (место) перед кем–чем–л.; перёд, передняя часть → **ald**

al– (833) *V* to take, get; to buy, receive; to accept; to seize, grasp | брать, добывать, покупать; получать, принимать; захватывать; **ap (<alıp)** shortened variant of the gerund on *–p* | сокращенный вариант деепричастия глагола на *–п*;
Vaux **aša almaj** (he/she) cannot eat | (он/она) не может есть; **bile almaj** (he/she) cannot know | (он/она) не может знать; **bere almam** I can't give it **bola almaj** he/she/it cannot be/exist | (он/она) не может быть; **bašın čavıp ala** he bashes his head (hits and cuts it off) | отрубает (ему/ей) голову

ala I (1) *Adj* spotted, pied | пестрый, пегий → **alaǯa**

ala II (1) *Adj* brown; hazel (eyes) | карий (карие глаза)

alaǯa (2) *Adj/N* motley, variegated; a kind of (striped) stuff | пегий, вид (полосатого) материала; **alaǯa at** pinto | лошадь пинто пегий

Alaγač (2), **Alaγaš** *Tn* Alaghach (a settlement in the Crimea) | Алагач (поселение в Крыму)

Alaγaš s. **Alaγač**

alaj I (6) *Adv* so, like that, in that manner | так, таким образом

alaj II (1) *Adv* all, the whole; a word put into the text of folk poems | целый, полный; весь; слово вставляемое в текст народных песен → **allaj**

alajkům s. **selam**

alak (2) *N Ar.* destruction, death | гибель, смерть; **alaket–** to perish, to destroy, to kill | уничтожать, губить, убивать; **alak ol–** to perish | быть уничтоженным, гибнуть, погибать, умирать → **elak, halak**

alal s. **elal**

alallaš– s. **elallaš–**

alam s. **alem**

alaman (1) *Adj/Adv* very quick; fast, quickly | очень быстрый; быстро, скоро; **alaman ajak** nimble–footed | быстроногий, легконогий

alamet (3) *N Ar.* sign, mark; symbol | признак, знак

alaša (10) *N* steed, mount, (saddle–)horse | мерин, конь

alat (2) *N Ar.* instrument, tool, device | инструмент, прибор, приспособление

alčaǯıχ (2) *N dim* rather low, very low | низкий, низенький

alčaχ s. **alčak**

alčak (15), **alčaχ** (5) *Adj* low; cowardly | низкий; подлый

alčakla– *V* to become low, to bend down | склонять, сгибать, нагибать

ald (37) *N* front, space in front (of a thing) | пространство (место) перед кем–чем–л.; перёд, передняя часть (чего–л.); **aldı** the front of | перёд кого/чего–л.; **aldına** to the front of | впереди кого/чего–л.; **aldında** in front of | впереди кого/чего–л.;

aldından from the front of | спереди кого/чего–л.

alda– (14) *V* to cheat, seduce | обманывать, соблазнять

aldan– (2) *Vpass* to be deceived, taken in (by) | обманываться, быть соблазненным

aldat– (17) *V* to cheat, deceive | обманывать, соблазнять

aldau (1) *N* deceit, trickery; trap | обман, ловушка; **aldau bol–** to be deceived/cheated | быть/стать обманутым

aldır– (11) *Vfact* to cause or allow to take/fetch/buy | заставить брать, купить

ale (1) *Adv* yet, still | всё ещё, до сих пор → **ele, ele–ele**

alejet–salam (1) *Ar.* may Allah bless him! peace be upon him! | да благословит (его) Аллах

alem (5), **alam** *N Ar.* world, universe | мир, вселенная

alγı (32), **aligi** (6), **aligė**, **aligı** (3), **alıgi**, **alıgı** (2), **alı γı** (2) *Adj* erstwhile, former; before; preceding; aforenamed | прежний, предидущий; названный

Ali (21) *Pn Ar.* Ali | Али; **Ali–bej** Ali–bey | Али–бей

aligi s. **alγı**

aligė s. **alγı**

aligı s. **alγı**

Alim (6), **Alım** *Pn Ar.* Alim | Алим

alıgı s. **alγı**

Alıl (1) *Pn Ar.* Halil | Халиль

alın– (2) *Vpass* to be taken, to be bought, to be seized | быть взятым, быть покупленным, полученным; быть захваченным

alıš– (1) *V* to get used to | привыкать, свыкаться

alıš–beriš s. **alıš–veriš**

alıš–veriš (2), **alıš–beriš** *N* buying and selling, shopping | купля–продажа, торговля; **alıš–veriš et–** to shop, to trade; to do business | заниматься торговлей заниматься перепродажей;

alıštır– (1) *V* to accustom (to), to train | приучать; приручать

alka (4) *N* ring | кольцо → **halka**

Alla s. **Allah**

Allah (44), **Alla** (10) *N/Interj Ar.* Allah, God | Аллах; бог; создатель; **Allah Allah!** Oh, my God! | Ой, Боже мой!

allaj (8) *Interj* a word put into the text of folk poems | слово, вставленное в текст народных песен → **alaj II**

allan– (1) *Vpass* to redden | алеть

allı–gủllủ (1) *Adj* motley, rose–rose, purple | пестрый с преобладанием красного, пурпурный → **al I, gủl**

alma (50) *N* apple | яблоко

Alma–su (1) *Tn* Alma–su (Alma River) | Алма–су (Река Алма)

almak (4) *Inf(mak)* taking (away) | взятие

alsız (1) *Adj* weak, exhausted; unhealthy | истощённый, нездоровый, несостоятельный; слабый → **al II**

alt (5) *N* bottom, underside, lower | низ, низкий; **altında** under something | под чем–нибудь

altı (18) *Num* six | шесть

altıla– (1) *V* to make (it) to be sixfold? | переделать шестикратным?

altılau (1) *N* making (it) to be sixfold | становление шестикратным

altın (148), **altun** *N/Adj* gold, golden | золото; золотой

altın–kulak (1) *Adj* having golden ears | с золотыми ушами → **kulak**

altınǯı I (1) *OrdNum* sixth | шестой

altınǯı II (2) *N* goldsmith | золотых дел мастер

altmıš (5), **atmiš**, **atmıš** *Num* sixty | шестьдесят

altun s. **altın**
Alušta (2) *Tn* Alushta (settlement) | Алушта (поселение)
alva (1) *N Ar.* halvah | халва
ama (55) *Conj Ar.* but, yet, however | но, однако, зато
amajl (2), **amajli** *N Ar.* hamail, amulet | амулет, талисман
amajli s. **amajl**
amam (21) *N Ar.* bath(house) | баня, купальня → **hamam**
amamǯi s. **amamǯı**
amamǯı (2), **amamǯi** *N* poprietor or keeper of a public bath | содержатель бани
aman I (15) *Interj Ar.* oh! mercy! for goodness sake! | ох! ой! боже!; **amanım aman!** oh my, oh! | ой, горе мне!
aman II (63), **eman** (4) *Adv P.* at once, right now, immediately | тотчас, тутже, сразу → **haman**
amanet (4) *N Ar.* deposit, trust | оставление на попечение, хранение, обережение; **Allaɣa amanet!** let (him/her/it) be entrusted to Allah! | да хранит (его/её и т. д.) Аллах!; **Allaɣa amanet bolson!** let (it/him/her) be entrusted to Allah! | пусть будет оставлен на попечение Аллаха; да хранит (его/её/их) Аллах!; **saña amanet olson!** let (him/her/it) be entrusted to you! | да будет оставлен на твоё попечение! → **emanet**
Amat (1) *Pn Ar.* Amat, Ahmat | Амат, Ахмат
ambar (2) *N Ar.* barn, storehouse | амбар, склад
amber (1) *N Ar.* ambergris | амбра (ароматное вещество)
Ameša (5) *Pn Ar./P.* Amesha (Ametsha) | Амеша (Аметша)
amma (5) *Conj* but, yet, still | но, однако, зато → **ama**
ana I (224) *N* mother | мать, матка
ana II (10) *Adv* here is, there is | вон; **ana–mına** look, here | вот вот, вот здесь
anačıχ s. **anačık**
anačık (2), **anačıχ** *N dim* mommy, mum | мамочка, маменька, мамуля
anahtar s. **anaχtar**
anaχtar (15), **anahtar** (6), **anaktar** (2), **anattar** *N Gr.* key | ключ
anaj (23) *N* mommy, mum | мама, матушка
anaktar s. **anaχtar**
analıχ (1) *N* gammer, old woman, | старуха, старушка
anattar s. **anaχtar**
Anatul (1) *Tn* Anatolia | Анатолия
anau (6) *Pron* yonder, that (one) | тамошний, вот!, вон!, вон тот! → **anavı**
anavı (1) *Pron* yonder, that (one), there it is! | тот самый, вон он! вот тот! → **anau**
anǯa (6), **anǯe** *Adv* only, merely; just | только, лишь только
anǯak (2) *Adv* only, merely; just | только, лишь только
anǯe s. **anǯa**
and s. **ant**
anda (44) *Adv* thereat, thither | там; туда
andagi s. **andaki**
andaki (2), **andagi** *Adj* being there, found there | находящийся там
andan (37), **annan** (3) *Adv* from there; after, whereupon | оттуда, после того
ane (47) *N* mother | мать → **ana, anne**
anečık (1) *N dim* mom, dear mother | мамочка → **anneǯık**
anej (2) *N dim* dear mother, mom (dim form of mother) | мамочка → **anaj, annej**

angi (3), **angı** *Adj* which?, which one? | который?, кто именно?; **angımız** which of us | который из нас

anı s. **o**

ankıjt(a)– s. **ankıt**–

ankıt– (3), **ankıjt(a)**– *V* to cause to smell nicely, to give out a (nice) smell | издавать приятный запах, пахнуть, благоухать

anlat– s. **añlat**–

annan s. **andan**

annar s. **u I**

annat– s. **añlat**–

anne (14) *N* mother, mommy | мать, мама → **ane**

anneǯık (1), *N dim* mom, dear mother | мамочка; **anneǯıjim** mom, my dear mother | мамочка → **anečık**

annej (1) *N dim* mommy, mum (dim form of mother) | мама, матушка → **anaj, anne**

ant (6), **and** (3) *N P.* oath | клятва, присяга; **and et**– to take an oath | давать клятву; **ant (~ and) ičtir**– to make swear an oath (someone) | заставить поклясться

antar s. **anter** *N*

anter (4), **antar** (2) *N Ar.* dress, loose robe | платье, женское платье (с рукавами); халат → **anteri**

anteri (11), **enteri** (4), **enterı** *N Ar.* dress, loose robe | платье, женское платье (с рукавами); халат → **anter**

añ (1) *N* mind, consciousness | память, сознание; **añ et**– to call to mind, to remember; to mention | упоминать, вспоминать, восстанавливать чтло–л. в памяти

añ– (2) *V* to call to mind, to remember; to mention | упоминать, вспоминать, восстанавливать чтло–л. в памяти; замечать

aña– s. **añla**–

añat– s. **añlat**–

añla– (21), **aña**– (11), **añna**– (2) *V* to understand, comprehend; to find out; to know (about) | понимать, осознать, узнавать

añlašıl– (1) *Vpass* to be understood | понимать друг друга, договариваться

añlat– (26), **añat**– (13), **anlat**– (11), **annat**– (2) *V* to explain | объяснять, разъяснять

añlatmak (1) *Inf (mak)* explaining | объяснение, разъяснение

añna– s. **añla**– *V*

añnaš– (1) *V recipr* to understand each other; to come to an agreement | понимать друг друга; договариваться

ap s. **al**–

apaj (22) *N* woman, wife | жена, женщина → **apakaj, apšık, eš, hanım, hatın, hatun, χatın, χatun, kadın, katın, karı, šerfe**

apakaj (40) *N* woman, wife | жена, женщина

apalaχ–ǯapalaχ s. **ǯapalak**

apıs (3), **apus** *N Ar.* imprisonment, prison | арест, тюрьма → **hafız**

appačık (2), **appačıχ** *Adj* quite open, quite manifest, very clear | совершенно открытый, ясный → **ačık**

appak (1) *Adj* pure white, all white | совершенно белый, белый–пребелый

apperėm s. **aferim**

apšik s. **apšık**

apšık (4), **apšik** (2) *N* woman, wife | женщина, жена → **apaj, apakaj**

apta s. **apte**

aptas s. **abdas**

apte (3), **apta** *N* elder sister | старшая сестра

aptıra– (3) *V* to wonder at, to be astonished, surprised at; | удивляться; расстеряться; недоумевать → **abdıra–**

apus s. **apıs**

ara (50) *N/Postp* the space between, distance (postp); state of relation (between two persons) | промежуток, между (послелог), отношения между людьми

ara– (7) *V* to see, look (for), to search | искать; **kismet ara–** to look for one's fortune/chance | искать счастье

arab (73), **arap** (3) *N Ar.* Arab, Arabic | араб, арабский

araba (12), **arba** *N* carriage, cart, wagon | арба, телега, повозка; **dủrd tegerčik araba** four–wheeled carriage | четырёхколосная арба/телега

arabačık (1) *N dim* small cart | тележка

arabaǯı (2) *N* driver (of a cart, wagon) | возчик, извозчик, кучер

arabčık (1) *N dim* little Arab | маленький араб

aralaš– (1) *V* to be found in the space between two objects | находиться между двумя предметами/местами; **kỏkke aralaš eken br saraj** sky–high palace | дворец достигающий неба

aralıχ s. **aralık**

aralık (3), **aralıχ** *N* space, gap, interval; side street | шель, промежуток, переулок,

aram (2) *Adj Ar.* forbidden by religion; unlawful, wrong, disgusting, shameful | нечистый, запретный, корыстный; **aram et–** to make unlawful, to forbid the use (of something according to Muslim laws), to make sg unclean | налагать запрет (по мусульманским законам); сделать незаконным; → **haram**

aramla– (1) *V* to rape, violate, to make sb unclean | осквернять, насиловать

aran I (30), **arañ** *N* stable, shad, barn | сарай, стойло

aran II (1) *N* proportion, ratio, rate | пропорция

arančıχ (1) *N dim* stable, shad, barn of small size | небольшой сарай, маленькое стойло

arañ → **aran I**

arap s. **arab**

ard s. **art**

arχa s. **arka**

arχadaš s. **arkadaš**

arık I (5) *Adj* weak, skinny; thin, lean | худой, худошавый, тощий

arık II (1) *N* ditch, trench, canal | арык, канава

Arılgul (1) *Tn* Arilgul (settlement?) | Арильгуль (поселение?)

arka (18), **arχa** *N/Postp* the back; back part, rear; the space behind, beyond | спина; задняя сторона; за, позади, сзади (послелог)

arkadaš (31), **arχadaš** *N* comrade, companion, friend | товарищ, друг

arkalık (1) *N* backrest; log, support, bar | спинка; балка, бревно

arlan– (1) *Vpass* to feel ashamed, to be ashamed | стыдиться

arman (3) *N* threshing | молотьба, обмолот; **arman bas–** to thresh | молотить

armanǯı (1) *N* thresher | молотильщик

armıt s. **armut**

armud s. **armut**

armut (5), **armıt, armud** (3) *N* pear | груша

arnavut (1) *En* Albanian | албанец, албанский

arpa (6) *N* barley | ячмень, ячменный

arslan (2), **aslan** *N* lion | лев, львиный; **arslan jůrekli** heroic, brave, courageous, fearless | мужественный, бесстрашный

aršala (3) *N Ar.* heaven | рай

aršın (3), **aršun** *N* ell, yard (Turkish measure of length of about 68 cm) | аршин (мера длины, около),^* м)

aršun s. **aršın**

art (114), **ard** (7) *N/Postp* back, hinder part, behind, rear; the space behind | спина, задняя сторона; задняя часть чего–л.; пространство находящееся позади (кого–чего–л.); задний; **artına kajt**– to turn back | возвращаться; **artında** behind, after | за, позади; **artından** from behind (of sb/sg) | сзади, за, из–за; **artka** back, backwards, behind, rearwards | назад, обратно; **artta** behind | сзади, позади; **arttan** from behind, in the back | сзади, с задней стороны

art– (1) *V* to increase; to proliferate | возрастать, увеличиваться, размножаться

artıχ s. **artık**

artık (4), **artıχ** *Adv* now, well then; (not) any more, (no) longer | уже; больше; же;

artıl– (1) *Vpass* to become full | наполняться

aru (8) *Adj* good, well | хороший, хорошо

arzuhal (1) *N Ar.* written application | опрошение, челобитная; заявление

as– (19) *V* to hang up; to put (e. g. a pot) on the fireplace | вешать, подвешивать; класть посуду на плиту

asaba (13) *N Ar.* relative, heir; friend, darling | родственник, наследник; дорогой

Asan (14) *Pn Ar.* Asan | Асан → **Hasan**

asil s. **asıl II**

Asim (1) *Pn Ar.* Khasim | Хасим → **Hasim**

asıl I (4) *N Ar.* essence, base, root, origin, base, basis | суть, сущность; происхождение, корень; основа; **aslı nesli** his/her relatives | его/её родственники, родня; **aslı nesli bilinmegen** (his/her) origin is not known | неизвестного происхождения, безродный → **nesil**

asıl II (5), **asil, asl, āsıl** *Adj Ar.* result, produce, gain | следствие, результат чего–л.; вывод, итог; **asil bol**–, **asıl bol**–, **āsıl bol**– **hasil ol**– to be attained, to happen, to occur, to come into existence | получаться (возникать, происходить, образовываться) в результате чего; **asl et**– to produce, to acquire, to gain | образовывать (создавать, получать) в результате чего; **müradlarına asil bol**– to achieve their aim/pupose | достигнуть их цели → **hasil**

asıl– (5) *Vpass* to hang, to be hung up | вешаться, виснуть; быть повешенным

asılı (3) *Adj* hanging, being hung up | повешенный, висячий

askar s. **asker**

askem (1) *Nvoc* my khaseki! (the khan's courtier or favourite wife) | мой хасеки! (ппридворный хана или любезная жена хана)

asker (33), **askar** (12) *N* soldier | солдат; **asker al–/ǯıj**– to enlist, to recruit | набирать в солдаты, вербовать солдат

askerlik (1) *N* soldiership, military service | солдатство, военная служба

asl s. **asıl II**

asla (3) *Adv* never, by no means | вовсе, отнюдь, никогда, никак; ни в коем случае

aslan s. **arslan**

asobni (1) *Adj R.* special, especial | особый, особенный

asra– (5) *V* to feed, nourish; to raise (animals) | выкармливать, выращивать

asret (4), **azret** *N Ar.* longing, yearning, homesickness; suffering from the lack of sb/sg | тоска, томление, вожделение; **azret ček–** to long to see again (person or thing), to miss, to suffer from the lack of | тосковать, томиться, сильно скучать по кому/чему–л. → **hasret**

asretli (1) *Adj Ar.* full of sorrow, full of ardent desire | полный тоски, тоскующий; истосковавшийся, полный страстного ожидания

asretlik (6) *N Ar.* sorrow for or pining after sb | тоска, томление (по кому или чему–л.) → **hasretlik**

ast (13) *N/Postp* space or place beneath a thing; bottom, underside, below, under | низ; (послелог) под; **astına** under, beneath, below | под (что–л.), вниз; **astında** under, beneath, below | под (чем–л.), внизу; **astından** from beneath, from underneath | из–под (чего–л.), снизу

astır– (2) *Vfact* have sb/sg hanged/hung up | велеть повесить

aš (87) *N* food, meal, cooked food | еда, пища, кушанье; **aš u su** food and water | еда и вода (питьё), пища и вода; **aš pišir–** to cook food | готовить пищу; **ašsız kal–** to remain without food | оставаться без еды → **ašajǯak, ašau**

aš u su s. **aš**

aš– I (2) *V* to surpass, exceed | переваливать, переправиться

aš– II s. **ač–**

aša (13) *N/Adv* down, downwards; the lower part; the space below | вниз, низкий, вниз; нижний, нижняя часть

aša– (177) *V* to eat, to feed, to consume | кушать, есть *V;* **aš/ašau aša–** to eat, to feed, to consume | кушать, есть, поесть

ašaγı (2) *N* down, downwards; the lower part; the space below | вниз, нижний; нижняя часть/сторона

ašajǯak (2) *N* food, meal | еда, пища, кушанье → **aš, ašau**

ašat– (7) *V* to feed | кормить

ašau (1) *N* meal, food; eating | пища, еда, кушанье → **aš, ašajǯak**

ašχ s. **ašk**

ašχ– (1) *V* to hurry, rush | спешить, торопиться → **ašık–**

ašik s. **āšik I**

ašıχ s. **āšik I**

ašık s. **āšik I**

ašık– (**ašıγ–**) (2) *V* to hurry, rush | спешить, торопиться → **ašχ–**

ašk (4), **ašχ** *N Ar.* love, passion | любовь

ašla– (1) *V* to inoculate | прививать

ašlama (1) *N* seedling | саженец

ašlan– (1) *Vpass* to be inoculated | быть прививленным

ašlıχ s. **ašlık I**

ašlık I (3), **ašlıχ** *N* cereals, meant for cooking, provisions, foodstuffs | злак, зерно; хлеб; еда, съестные припасы, запасы провизии; **jol ašlık** provisions for the journey | провиант для дороги

ašlık II (1) *N* hunger, starvation | голод, голодание

at (218) *N* horse, stallion | лошадь, конь

at– (98) *V* to cast, throw (away, out, down etc.), to drop; to shoot, fire | бросать, кидать; стрелять; **čilme at**– to cast a net | забросить сети (невод)ь **kůz at**– to set one's eye on sb/sg; to run an eye over | приглядывать; **naz at**– to be coy, to seek to please | быть застенчивым, стремиться понравиться; **tañ atkanı** the first blush of dawn | (утреняя) заря; **tañlar at**– to dawn | рассветать; **top at**– to cannonade | обстреливать из пушек; **tůfenk at–, tüfek at**– to fire a rifle, to shoot | выстрслять из ружья;

Vaux **bašın kesip atkan** 'he has cut his head off' | он отрезал ему/ей голову

ata (46) *N* father, ancestor | отец; предок

ataj (15) *N* father! (addressing) | папа; папочка!

ataman (16) *N* ataman, hetman, robber chief | атаман, вождь разбойников

ataš s. **ateš**

ateš (27), **ataš** (10) *N P.* fire | огонь; **atešge/ateške jan**– to be consumed by fire, to be alight, to catch fire | зажигаться, разгораться (в огне)

atıl– (7), **atl**– (3) *Vpass* to be thrown (out), to dash, to strive | быть выброшенным, бросаться, устремляться

atıš (2) *N* shooting, archery | стрельба (из лука)

atıš– (2) *V recipr* to throw at each other; (here) to fire at ech other | бросать друг в друга; (здесь) перестреливаться

atkar– (1) *V* to conduct, to manage, direct | проводить, устраивать (мероприятие)

atl– s. **atıl**–

atla– (3) *V* to jump, leap over; to step over | прыгать, перепрыгивать

atlama (1) *N* jumping over; steping, crossing (over) | (пере)прыгивание, прыжок, скачок

atlas (4) *N* satin | атлас

atlat– (1) *V* to omit, leave out | обносить

atlı I (2) *N* rider, horseman | всадник, конник

atlı II (10) *Adj* having a horse; horsed, horsedrawn (vehicle) | конный, имеющий коня; **důrt atlı dilǯan** coach drawn by four horses | четвероконный фаэтон/экипаж; **ůč atlı (br) fajton** phaeton drawn by three horses | трёхконный фаэтон

atmıš s. **altmıš**

atna (1) *N P.* week | негделя

attir–ottir s. **attir–uttır**

attir–uttır (2), **attir–ottir** *Onomatop*

au I s. **av**

au II (1) *N* trawl, fishing net | (рыболовная) сеть, невод → **čilme, aγ**

au III (1) *Interj* hi! hallo there! | эй! ну!

au– (1) *V* to fall (down), overturn | накрениться, перевалить

auǯı (11), **auǯu** *N* hunter; fisherman | охотник, ловец

auǯu s. **auǯı**

audarıl– (1) **(audarl–)** *Vpass* to collapse, fall (down) | свалиться, рухнуть; опрокидываться

audarl– s. **audarıl**–

aula– (3) *V* to hunt | охотиться

aulak (1) *N* distance, remoteness | дальность, расстояние

aulan– (2) *Vpass* to hunt; to amuse oneself by hunting | охотиться; развлекаться охотой

auna– s. **avuna**– *V*

av (13), **au** (9) *N* hunt, hunting, chase | охота; **av/γa kit**– to go hunting | идти на охоту; **auγa čık–/čıχ**– to go out hunting | отправляться на охоту

ava (19), **ave** (3) *N Ar.* air; weather; musical tune, melody | воздух; погода; мелодия, напев → **hava**
avala– (3) *V* to take to the air, to fly | взлетать
avalan– (1) *Vpass* to be aired, to take in air | взлетать
ave s. **ava**
aves (5) *Adv Ar.* mood; strong desire (for), inclination, passion | увлечение, охота, страсть; **aves bol–** to be eager, to have aspirations | быть стремящимся; иметь сильную волю; **aves et–** to have a strong desire/interest | (серьёзно) интересоваться чем–л.; иметь сильное желание
aveslen– (1) *Vpass* to desire, to long (for) | увлекаться, томиться, желать
aviǯ s. **avıǯ**
avištır– s. **avuštır–**
avıǯ (17), **aviǯ** *N* handful, the full grasp of the hand | горсть; ладонь; **koš avıǯ** two handfuls of | две горсти чего–л.; **sıñar avıǯ** a handful of | горсть чего–л. → **avuč**
avıǯılık (1) *N* hunt, hunting | охота
avır– s. **avur–**
avız (24) *N* mouth, lips; opening;road–forking | рот, уста; отверстие; ответвление дороги, устье → **aγız, avuz**
avla– (2) *V* to bark, to bay | лаять
avlak (4) *N* distance, remoteness | дальность, расстояние
avlan– (2) *Vpass* to be caught/shot at hunting; to hunt; to amuse oneself by hunting | быть пойманным на охоте; охотиться; развлекаться ототой
Avstrija (4), **Afstrija** *Tn R.* Austria | Австрия → **Avustrija**
avuč (1) *N* handful, the full grasp of the hand | ладонь; горсть → **avıǯ**
avuǯla– (1) *V* to grasp, grip; to take by grasps of the hand | набирать горстями (пригоршнями); загребать руками
avuna– (2), **auna–** *Vpass* to roll, tumble about | валяться, кататься
avur (2) *Adj* heavy, difficult | тяжелый, трудный → **aγır**
avur– (7), **avır–** (2) *V* to ache | болеть → **aγır–**
avurt– (4) *V* to ache, to cause pain | причинять боль кому–л.
avuru (1) *N* ache, pain | боль
Avustrija (4) *Tn R.* Austria | Австрия → **Avstrija**
avuš– (1) *V* to move/go (from one place to another) | перемешаться, переходить (с места на место)
avuštır– (2), **avištır–** *V* to take/carry (over/across), to put sg in another place | перемещать, переносить
avuz I (6) *N* mouth (of river), lips; opening; road–forking | рот, уста; ответвление дороги, устье → **aγız, avız**
avuz II (1) *N Ar.* artificial basin, pond | бассейн, пруд → **havuz**
az (33) *IndefNum* few, a few; not much, little | мало, немного; недостаточно; **azgınačık** very little, very few | немножко, немножечко; **az kete, kůp kete** (in tales) he goes few, he goes much | (в сказках) мало ли идёт, много ли идёт; **az kete, uz kete** (in tales) he goes short, he goes long | (в сказках) коротко ли идёт, долго ли идёт; **az kůjům** my little village | моя маленькая деревня; **br az** a little, a few | немного
az– I (1) *V* to lose weight; to become less | худеть, тощать, уменщиться

az– II (1) *V* to act outrageously, to become mad | выходить из себя, буйствовать
azaǯıχ s. **azaǯık**
azaǯık (3), **azaǯıχ** *IndefNum dim* just a little bit; a few; brief (time) | немножко, немножечко, короткое время → **azgınačık**
azad (8) **azat, āzād (2)** *Adj* free, not enslaved; independent | свободный, волный; **azat/āzād et–** to set free | освободить
Azamet (5) *Pn Ar.* Azamet | Азамет; **Azamet aka** uncle Azamet | дядя Азамет
azap (1), **azab** *N Ar.* pain, torment, torture | мука, страдание; **azap ček–** to agonize, to suffer | мучиться, страдать
azat s. **azad**
azbar (33) *N P.* court, courtyard | двор (для скота)
azčık (2) *Adj dim* just a little bit; a few; brief (time) | немножко, немножечко
azdır– (1) *V* make sb weak, to bring low, to emaciate | расслаблять, отощать
azgınačık (1) *Adv dim* just a little bit; a few, brief time | немножко, немножечко, короткое время → **azaǯık**
azirǯavab (2) *Adj Ar.* quick at repartee, witty | находчивый, сообразительный; **azirǯavab bol–** quick at repartee, witty | находчивый, смышлёный, сообразительный
azirla– s. **azırla–**
azirlan– s. **azırlan–**
aziz (1) *Adj Ar.* dear, beloved; respected, highly esteemed | дорогой, уважаемый; **azizim!** my dear! | дорогой мой!
azık (2) *N* provisions, food | провизия; продукты
azır (1) *Adj Ar.* ready, prepared | готовый, пригиотовленный → **hazır**
azırla– (16), **azirla–** (5) *V* to prepare, make ready; to cook | готовить (еду), подготавливать → **hazırla–**
azırlan– ~ **azėrlan–** (3), **azirlan–** *Vpass* to prepare, to get ready | готовитъся → **hazirlan–, hazırlan–**
azırlık (1) *N* readiness | готовность, подготовленность
azlık (1) *N* paucity, scarcity | меньшинство
Azrail s. **Azrāil**
Azrāil (7), **Azrail** *Pn Ar.* Azrael (name of the angel of death) | Азраил (ангел смерти)
azret s. **asret**
ažda (12) *N P.* dragon, kite | дракон, змей
ā (2) *Interj* ah! ugh! woe! | аха!
āferim s. **aferim**
Āhmad s. **Ahmat**
āl s. **al II**
ālā (1) *Adv Ar.* yet, still | до сих пор, поныне; всё ещё, пока ещё
ālem (4) *N Ar.* world, universe | мир, вселенная
ālim (2) *N Ar.* learned, wise, scholar | умный, мудрый, ученый
āmin (1) *N Ar.* amen | аминь! да будет так!
āšiχ s. **āšik II**
āšik I (10), **ašiχ, ašik, ašıχ, ašık** (2), **āšıχ** *N Ar.* lover, in love | влюблённый, любезный, любовник; **ašik bol–, ašıχ bol–** to fall in love (with) | влюбиться (в кого–л.)
āšik II (4), **āšiχ** *N* knucklebones | таранная кость, бабка, альчик; **āšik ujna–** to play knucklebones | играть в альчики
āšıχ s. **āšik I**
āzād s. **azad**

B

baba (221) *N* father | отец, папа
babaǯık (7) *N dim* daddy | папочка
babaj (4) *N* dad! pap! (addressing) | отец!, папа! (обращение)
babalı (1) *Adv/Adj* having a father | имеющий отца
baǯa (3) *N* chimney | дымовая труба
baǯak (4) *N* foot, leg | нога
badem (5) *N P.* almond | миндаль; **badem–terek** almond–tree | миндальное дерево
baġa (1) *N* frog | лягушка → **baka**
baɣ (13), **bau II** *N P.* vineyard, garden | виноградник, сад
baɣana (4) *N* column, pillar | столб, стойка
baɣanačıχ (1) *N dim* small column, pillar | маленький столб, стойка
baɣča (4), **baɣče** (2) *N P.* garden | сад
baɣčaǯı (1) *N* gardener | садовод
baɣče s. **baɣča**
baɣır (6), **bauɣır** *N* chest, breast; heart, belly, stomach | грудь, внутренности, сердце, желудок; **baɣrın ač–** to open one's shirt; to open one's arms wide | открыть объятия; распростирать руки; **baɣrın ez–** to break one's heart | разрывать его/её сердце; **baɣrına bas–** to clasp/hold in one's arms | прижать кого–л. к (своей) груди → **bavır**
baɣır– (7) *V* to shout, cry out, to howl | кричать, реветь → **bakır–**
baɣırt– (1) *Vfact* to have sb shouted, to make sb cry | велеть кричать
baɣıš (1) *N P.* grant, donation | дар, подношение
baɣıšla– (17) *V* to give gratis, to donate; to forgive, to pardon | дарить, подарить; прощать
baɣıšlan– (3) *Vpass* to be donated; to be forgiven | быть подаренным; быть прощенным
baɣıšlat– (2) *Vfact* to have sb donated; to make sb forgive | велеть/приказать дарить; велеть прощать
baɣla– (20) *V* to tie, fasten, bind | связывать, привязывать → **bajla–, bejle–**
baɣlama (1) *N* folk instrument with three double strings, played with a plectrum | баглама (щипковый инструмент вроде мандолины) → **bajlama**
baɣlat– (1) *V* to have sb tied, fastened | велеть связывать
baɣlı (1) *Adj* bound, chained; tied, fastened | связанный, привязанный → **bajlı**
Baɣ–saraj s. **Bakčasaraj**
baɣrı–bůtůn (1) *N* a sort of watermelone | разновидность дыни
baɣrıš– (1) *V recipr* to cry out together; to yell at each other | кричать друг другу, кричать вместе с другими
Bah–saraj s. **Bakčasaraj**
bahar (5) *N/Adj P.* spring | весна, весенний
bahča (41) *N P.* garden | сад, садовый; **bahča–bahča(dan) kez–** to walk from garden to garden | гулять по садам → **bahče**
bahčalı (1) *Adj* gardened, having a garden | с садом, имеющий сад
bahče (7) *N* garden | сад; **gůlbahče, gůlbakča, kůlbahče** rose–garden | розарий → **baɣča, bahča, bakča**
bahš (2) *N P.* grant, giving, gift; a forgiving | подарок; прощение; **bahšıš bol–** to be given (to be donated); to be forgiven | быть даренным, быть данным; извиняться; **bahš et–** to give (to donate); to forgive | дарить, давать; прощать, извинять → **baχıš**

baχ– (104) *V* to look (at), to take care (of), to look (after) | смотреть, заботиться (о ком/чём) → **bak–**

baχča (5) *N P.* garden | сад → **baγča, baγče, bahče, bahča, bakča**

Baχčasaraj (1) *Tn* Bakhchisaray (city) | (город) Бахчисарай → **Bakčasaraj**

baχır (1) *N* copper, of copper | медь, медный, из меди → **bakır**

baχıš (1) *N P.* grant, giving, gift | подарок → **bahš**

baχšiš s. **baχšıš**

baχšıš (5), **baχšiš, bakšiš, bašıš** *N P.* tip, baksheesh; gift, present | дар, подарок; подачка, чаевые, бакшиш; **bašıš et–** to give (to donate); to forgive | дарить, давать; прощать, извинять → **bahš, baχıš**

baχt (1) *N P.* luck, fortune, destiny; good luck | счастье, судьба; удача

baj (47) *Adj/N* rich (man) | богатый; богач; **Baj–terek** Bay–terek (name of a fabulous fruit–tree) | Бай–терек (имя сказочного плодового/фруктового дерева)

baj– (1) *V* to grow rich | обогатеть

Baj–terek s. **terek**

bajakı (1) *Adj/Adv* quite (a lot), fairly, rather, pretty (much) | изрядное (количество), изрядно; довольно, довольно–таки;

bajaz s. **bejaz**

bajazlan– (1) *Vpass* to become white | становиться белым, белеть

bajazlı (1) *Adj* spotted with white | беловатый, с белизной

Bajdarava (1) *Tn* Baydarova (settlement) | Байдарова (поселение)

bajγuš (2) *N* poor (man), wretched; | бедняга, бедалага

bajıl– (4) *Vpass* to faint, to feel faint | падать в обморок; лишаться чувств (сознания)

bajır (9) *N* hill, slope, ascent | холм, бугор; склон горы, косогор, возвышение

bajla– (25) *V* to tie, fasten, bind; to connect to | связывать, привязывать; **kısχadan bajla–** to cut one's speech short | сокращать (свою) речь; **kušak bajla–** to brace, to girdle, to put on a belt | опоясаться → **baγla–, bejle–**

bajlama (2) *V* folk instrument with three double strings, played with a plectrum | баглама (шипковый инструмент вроде мандолины) → **baγlama**

bajlan– (3) *Vpass* to be tied | быть связанным

bajlattır– (1) *Vfact* have sg tied by sb | велеть/приказать связывать

bajlı (3) *Adj* bound, tied, fastened | связанный, привязанный → **baγlı**

bajrak (6) *N* flag, standard | знамя, флаг

bajram (6), **bejram** (4) *N* festival | праздник; **bajram habarı** news of feast | известие о празднике; **bajram küni** feast–day | праздник

bajtal (12) *N* mare | кобылица

bak– (226) *V* to look (at), to take care (of), to look (after) | смотреть, заботиться; **bakčı** look here! listen! | смотри же! слушай! **Kuj–bakar** Kuy–bakar (name of a horse meaning one, who takes care of sheep) | Куй–бакар (имя лошади, обозначающее лицо, которе ухаживает за овцами);

Vaux **deñip baka** is trying to figure out | пытается выяснить; **tırıšıp baka kačmaγa** is trying hard to escape |

изо всех сил пытается убежать → **baχ–**

baka (6) *N* frog | лягушка → **baġa**

bakaj (3), **bakkaj** *N* knucklebones, ankle-bone of a horse | бабка, козон

bakča (16) *N* garden | сад; **gůlbakča** *N* rose–garden, rosary | розарий → **baγča, baγče, bahča, bahče**

Bakčasaraj (6), **Baγ–saraj, Bah–saraj** (2), **Bakčesaraj** *Tn* Bakhchisaray (city) | (город) Бахчисарай

bakčı s. **bak–**

bakır (1) *N* copper | медь (красная); **bakırlar** copper utensils | медная посуда → **baχır**

bakır– (3) *V* to shout, cry out, to howl | кричать, реветь → **baγır–**

bakkaj s. **bakaj**

bakla (3) *N* bean | бобы

baklalı (1) *Adj* containing bean | фасолевый

bakma (6) *Inf(ma)* looking (at); taking care (of), looking (after) | зрение, ухаживание за кем–л., уход

bakmak (1) *Inf(mak)* looking (at); taking care (of), looking (after) | зрение, ухаживание за кем–л., уход

bakšiš s. **baχšiš**

baktır– (5) *Vfact* to have someone looked (after); | велеть ухаживать, приказать заботиться о ком

baktırt– (1) *Vfact* to have someone looked (after); | велеть ухаживать, приказать заботиться о ком

bal (17) *N* honey; bragget, metheglin; honeyed boza (beverage made of fermented millet) | мед; буза с медом (вид напитка)

bala (307) *N* child | ребенок, дитя; детский; детёныш; **balaγa kal–, balajı kal–** to become pregnant, to be got with child | забеременеть; **er bala** male child, boy | мальчик; **kaz bala(sı)** gosling | гусёнок; **kız bala** daughter | дочь, дочка; **kujan bala(sı)** kit, kitten | зайчонок; **kuš–bala(sı)** young bird, nestling | птенец, птенчик; **pič bala** illegitimate child | незаконнорождённый ребёнок

bala–čaγa (1) *N* household, family, wife and children | детвора, семья, жена и дети, ребятишки → **bala–čala**

bala–čala (1) *N* household, family, wife and children | детвора, семья, жена и дети, ребятишки → **bala–čaγa**

balaban (37) *Adj* big, great, large | большой, крупный

balačıχ s. **balačık**

balačık (5), **balačıχ** *N dim* babe, little child, lad, youngster | деточка, детка; детёныш

balaγuz (1) *N* wax | воск

balah s. **balak**

balak (12), **balah** *N* lower part of the trouser leg | штанина; **balah/balak tire–, balak–pača tire–/sıva–** to roll up one's trouser legs; to prepare to set about some important business (deed) | засучить рукава (штанины); серьёзно взяться за дело → **pača**

balak–pača s. **balak, pača**

balala– (3) *V* to litter, throw; to bring forth young; | плодить (о животных)

balalı (1) *Adj/Adv* having a child; behaving in childish manner | имеющий детей; с детьми; по детскому

balalık (2) *N* childhood, infancy | детство

balčıχ s. **balčık**

balčık (6), **balčıχ** *N* wet clay, mud | глина

baldır (1) *N* (anatomical) calf, shin, shank | икра, голень

baldırγan (1) *N* asafetida | Ферула вонючая, каврак

baldız (1) *N* wife's sister, (here) sister–in–law | свояченица; (здесь) золовка

balıχ s. **balık**

balıχčı s. **balıkčı**

balıχčılıχ (1) *N* fishery, fishing | рыболовство

balık (33), **balıχ** (3) *N* fish | рыба; рыбий

balıkčı (8), **balıχčı** (3) *N* fisherman, fisher | рыбак

balkun (1) *N Fr* balcony | балкон

balkurt (1) *N* bee, wasp | пчела

balla– (2) *V* to smear with honey, to spread honey (on) | намазывать мёдом

balta (15) *N* axe, hatchet; battle axe | топор, секира

bar (458) *Ecp* there is; existent, present; to have, 've got; | есть, имеющийся → **var**

bar– (502) *V* to arrive (at), to go towards, to reach, approach | ходить, доходить, доезжать, пойти, приходить; **suya/suva bar–** to go for some water; to reach water (e.g. river) | идти за водой; доходить до воды (реки) → **var–**

barabar s. **beraber**

baraber s. **beraber**

barak I (1) *N* dog | собака, пёс

barak II (1) *N R.* barrack(s) | барак

bardaχ s. **bardak**

bardak (3), **bardaχ** *N* cup, mug, glass | стакан, бокал, кружка

barım (3) *Adv* at least, for once | по крайней мере, хотя бы

Barın (1) *Tn* Barin (settlement) | Барин (поселение)

barırač (2) *Onomatop* **barırač barılač** (play of words) | (игра слов)

bariš– (1) *V recipr* to make peace (with one another) | мириться

barmačık (1) [< barmak + –čık] *N dim* finger | палец → **barmak I**

barmaχ I s. **barmak I**

barmaχ II s. **barmak II**

barmak I (17), **barmaχ I** (7) *N* finger | палец → **parmak**

barmak II (3), **barmaχ II** *Inf* going (to), arriving (at) | ход, уход, отъезд

barut (1) *N* gunpowder | порох

bas– (23) *V* to tread (upon), press, to enter; to raid, attack suddenly; | давить; ступать; входить, наступать; нападать внезапно

basamak (6) *N* rope ladder; step, stair | верёвочная лестница; ступень; **basamak baši** the top of the stairs | верхняя часть лестницы → **baskıč**

baskıč (1) *N* (rope) ladder; staircase, stairs | ступенька (лестницы) → **basamak**

basma (5) *N/Adj* printed cloth | ситец; ситцевый; **allı–gůllů basma** pink printed chintz/cloth | розовый ситец; **basma kůlmek** chintz shirt | ситцевая рубашка

bastır– (1) *V* to suppress, superimpose, to cover up (e.g. with earth) | прикладывать, накладывать

basturuk (1) *N* prison | тюрма

baš (345) *N* head; chief, leader; beginning, entrance; | голова; начальник; главный; вход; начало; **baš et–** to deal with sb, to gain the upper hand, to prevail (over sb) | сладить/справиться с кем; **baš kuj–** to decide to do something | решаться делать что–л.; **baš ůstůne** with pleasure!; yes (sir)! yes indeed! | с удовольствием! очень хорошо!; есть! слушаюсь!; **bašуa čık–** to oppose, to stand up/against to | выступать против кого–л.; **bašın alıp kit–** to go away (unawares), to leave by oneself; to duck out | уйти незаметно, улизнуть; **bašın(ı) al–** to behead; to mislead; to set about to sg | отрубить чью–л. голову; сбивать с толку; решаться делать что–л.; **bašın ašaa al**– to hang one's head, to bend down one's head | опускать/склонить/понурить голову; **bašını kolla**– to look after, to protect | заботиться о ком, покровительствовать кому; **bašın kutar**– to save one's life | спасать чью–л. жизнь; **bašına ǯet–/jet**– to cause harm/trouble | натворить зло (кому–л.); причинять ущерб (кому–л.); **bašına jazıl**– to be written (pre–destined) by fate | быть написано (судьбой); **bašına kel**– to remember; to happen (to), to befall | приходить в голову; случиться, обрушиться на голову (о несчастье); **bašına tüš**– trouble, misfortune, love etc. happens to sb. | переживать/испытывать горе, несчастье, любовь и тд7; **bašından aklı kit**– to lose one's senses, to go out of one's mind | быть ошеломлённым, быть сбитым с толку; **bašından keč**– to happen, to occur (to) | переживать, испытывать что, приключаться, случаться с кем; **baška čık**– to oppose, to stand up/against to | выступать против кого–л.; **bašta** early, previously, before, in front;| рано, впереди; прежде, в начале; **baštaki** one being ahead/in front; previous | находящийся впереди; предидущий; **baštan čık**– to get out of control, to be led astray, to be corrupted | испортиться, развратиться; сбиться с пути, быть сбитым с толку; **baštan čıkar**– to lead astray, to corrupt, to seduce | совращать, развращать; толкать на дурное; сбивать с пути (с толку); **čal/čāl baš** grey (–haired) head | чалая голова; **češme bašında** at the fountain itself, nearby the fountain | у (самого) источника; **küše bašında** at the corner of the street | на углу улицы; **oǯak bašinda, uǯak bašında** by/beside the fireplace | у (домашнего) очага; возле топки/печки; **tau bašinda** at the peak of the mountain | на вершине горы

bašχa s. **baška**

bašıš s. **baχšıš**

baška (57), **bašχa** (9) *Adj/Postp* other, another, different; except (from), other (than) | другой, иной; кроме, помимо, исключая; **üzinden baška** without oneself | кроме самого себя

baškadan (1) *Adv* again, once more, anew | снова, заново

bašla– (162) *V* to begin, attempt; to set about | начать, приняться; **kurka bašlaj** begins to fear | начинает бояться

bašlan– (11) *Vpass* to be begun | начинаться

bašlı (17) *Adj* headed, heaving a head | имеющий голову; с головой; главный

bašta (14) *Adv* formerly, at the beginning (of) | прежде, сперва, вначале

baštan (6) *Adv* from the beginning | с начала

bašvezir (1) *N* prime (grand) vizier/minister | премьер–министр

bašvezirlik (1) *N* dignity of prime (grand) vizier | достоинство (чин) главного визира (премьер–министра)

bat– (6) *V* to sink, to set (sun etc.) | тонуть, заходить (о солнце)

batak (1) *N* swamp, marsh | болото, топь, трясина

batı (1) *N* sunset; West | закат; запад; **kün batı** sunset; West | закат солнца; запад

batır (48) *N* epic hero; champion; wrestler, mighty man | богатырь, витязь; храбрый

batır– (4), **batėr–** *Vfact* to plunge, to dip, dunk | погружать; макать; утопить

batırlık (3) *N* heroism, act of heroism; courage | героизм, храбрость

batman (2) *N* batman (weight varying from 2,5 to 10 kg) | мера веса (от 2,5 до 10 кг)

Batum (2) *Tn* Batumi (town) | (город) Батуми

bau I (1) *N* tie, bond; string, lace | связь, завязка

bau II s. **baγ**

bauγır s. **baγır**

bavır (1) *N* breast, chest; belly | грудь; брюхо → **baγır**

bazar (27) *N P.* bazaar, marketplace | базар, рынок → **pazar**

Bazarǯıχ (1) *Tn dim* Bazardjik (settlement) | Базарджик (поселение)

bazarlıχ s. **bazarlık**

bazarlık (3), **bazarlıχ** (2) *N* bargain, deal | торг; **bazarlık/bazarlıχ et–** to dicker, to bargain | торговаться (о цене)

bazirkan s. **bazırγan**

bazırγan (2), **bazirkan** *N P.* merchant, peddler | купец, торговец,

bāde (2) *N P.* bowl, cup, glass of wine; wine | кубок; вино

bāri I (2) *Adv P.* at least, for once | по крайней мере, хотя бы, в таком случае

Bāri II (11) *Pn* Baari | Бари

be I (1) *N* letter “b” | буква “б”

be II (3) *Interj* oh!, you! (in vocatives expressing reproach) | эй ты! послушай! ведь ... же!

beg (1) *N* gentleman, Sir; notable, country gentleman; (after names) Mr. | бей, господин (в роли приложения к мужским именам); **Mustā–beg** Musta–beg | Муста–бег; **Ümer–beg** Umer–beg | Умер–бег

begen– (18) *V* to like, admire; to be pleased (with) | одобрять, нравиться; находить по вкусу

begit– (3) *V* to close, lock (up); to fasten, tie (up), to keep down | закрывать, вязать кого–л., схватить

bej (10) *N* gentleman; (after names) Mr. | господин (в роли приложения к мужским именам); **Ali–bej** Ali–bey | Али–бей; **Memet–bej** Memet–bey | Мемет–бей; **Mustafa–bej** Mustafa–bey | Мустафа–бей; **Süjün–bej** Suyun–bey | Сююн–бей; **Šertlan/Šertlen–bej** Shertlan–/Shertlen–bej | Шертлан–/Шертлен–бей; **Usejn–bej** Useyn–bey | Усейн–бей → **beg, bij**

bejan (1) *N/Adj Ar.* declaration, statement; known | изложение, сообщение; известный; **bejan bol–** to become known | становиться известным

bejaz (24), **bajaz** (5), **bijaz** (11) *Adj Ar* white | белый

bejit s. **bejt**

bejıt s. **bejt**

bejle– (6) *V* to tie (up), fasten; to dress a wound | связывать, привязывать; перевязывать рану → **baγla–, bajla–**

bejoγlı (3), **bejuγlı** *N* son of a “bej” (see), richman’s son; younker | сын бия; господский сын(ок), молодой барин → **oγıl, uγıl**

bejram s. **bajram** *N*

bejt (29), **bejit** (9), **bejıt** (5) *N Ar.* couplet, distich | куплет, двустишие, стих, бейт; **bejt ajt**– to recite a verse/poem | читать бейт (стих)

bejuγlı s. **bejoγlı**

bek (93) *Adv* very, much, extremely | весьма, очень → **bik, pek**

bekar s. **bekjar**

bekči (19) *N* watchman, guard | часовой

bekem s. **bekim**

bekėm s. **bekim**

bekim (12), **bekem**, **bekėm** (2), **bekım** (2) *Adv* perhaps, maybe | авось, может быть, возможно *Adv* → **belki, belkim**

Bekir (1) *Pn* Bekir | Бекир

bekjar (6), **bekar** (4) *N Ar.* unmarried, bachelor | холостяк, холостой; **bekjar jiget** unmarried lad, bachelor | холостяк, холостой парень; **bekjar kız** unmarried girl, maiden | девица, незамужняя девушка

bekle– I (33) *V* to wait (for), to guard, to watch (over) | ждать, выживать; хранить, караулить, сторожить

bekle– II (2) *V* to close, lock (up) | закрывать, запирать, замыкать → **bikle–**

beklen– (1) *Vpass* to shut, to be closed, to be locked (up) | закрываться, замыкаться

beklevli? (1) *Adj* closed? | запертый? *Adj;* **kanušnada čal ajyırım bešlevli/ beklevli (?) kaldı** my grey–haired stallion has been locked in the stable | мой чалый жеребец заперт в конюшне → **bekle–**

bekli (2) *Adj?* shut, closed | закрытый

bel (47) *N* waist; passway, pass in a mountain ridge | поясница, талия; проход, перевал; **beli dolaš**– to swing or rock one's hip to and fro | качать талию

bela (3), **belā** *N Ar.* trouble, misfortune | беда, напасть

belālı (1) *Adj* calamitous, troublesome, unfortunate; causing misfortunate | несчастливый, приносящий несчастье

belčik (1) *N dim* waist | поясница, талия

beleš (2) *N* turnover–pie (a kind of national food of Tatars) | большой (круглый) пирог с начинкой (татарское национальное блюдо)

belki (1) *Adv* perhaps, maybe | авось, может быть, возможно → **bekim, belkim**

belkim (1) *Adv* perhaps, maybe | авось, может быть, возможно → **bekim, belki**

belle– (11) *V* to suppose, think, believe | считать, полагать

belli (16) *Adj* obvious, known | известный, известно; **bell‿et– [<belli et–]** to show, to express, to let sg know | обнаруживать, проявлять; давать знать

ben I (357) *Pron pers Sing1* I, me | я; **baña** to/for me | мне → **men**

ben II s. **beñ**

bend (1) *N P.* weir, dam | запруда, плотина

benek (2) *N* small pot, speckle, fleck | пятно, крапинка; **benek–benek** very potted, all in speckles | весь в пятнах

beniadam (2) *N Ar.* man, human being; sons of Adam | человек, сыны Адама

benze– s. **beñze–**

beñ (8), **ben** *N* mole, beauty spot | родинка

beñze– (11), **benze**– (7) *V* to resemble, to look like | быть похожим

ber– (453) *V* to give, to deliver | давать; **sadaka ber**– to give alms | давать

подаяние, давать подачку; **ǯevab ber**– to answer, to reply (to) | отвечать; **izin ber**– to grant permission (to, for), to allow, to permit | давать разрешение, разрешать, позволять; **kız ber**– to give a girl in marriage (to), to marry (a girl) | выдавать (дочь) замуж за кого–л.; **mejil ber**– to consent, to assent (to) | дать согласие на что–л., способствовать к тому, что; **sȯz ber**– to give one's word, to promise | дать слово; **šeftali ber**– to give a kiss | давать поцелуй;
Vaux **(urbaların) baka berme** avoid (looking) staring (at your clothes) | не смотри пристально (на свою одежду); **ǯiberi ber**– to send away quickly, to let go suddenly | быстро отправить, отпустить; **(urbaların tuzın) kaka bere** shakes (dust off the clothes) quickly | быстро отряхивает (пыль с одежды); **sala/salı ber**– just to let go, to set free suddenly/quickly | разрешить идти, идти быстро, неожиданно → **ver**–

beraber (9), **barabar** (5), **baraber** *Adv Ar.* together, accompanying, jointly | вместе, сообща

berber (1) *N It.* barber, hairdresser | парикмахер → **berberǯi**

berberǯi (4) *N* barber, hairdresser | парикмахер → **berber**

bereket (2) *N Ar.* abundance, increase, fruitfulness; blessing, divine gift | изобилие, счастье; дар, благоденствие бога; **bereket bersin!** May God give you blessed increase!; Fortunately, thank heaven! | К счастью, слава богу!

bereketli (2) *Adj* blessed, fruitful; fertile, abundant | благодатный, изобильный, урожайный; **bereketli bol**– may it be plentiful! (said to a person who is gathering in the crops) | хорошего урожая! (пожелание удачи при сборе урожая)

bergi (2) *Adj* the nearest, the nearer one; of this place | находящийся на этой стороне, близлежащий

beri (4) *Postp* since, for (space of time) | от, с, вот уж; **čoktan beri** long (since), for ages | сдавних пор, издавна

beril– (4) *Vpass* to be given | быть данным → **veril**–

beriš (1) *N* giving, delivery | дар, дача; продажа → **alıš–veriš**

berli (6) *Postp* since, for | от, с, вот уж

berlikte (1) *Adv* together, accompanying, jointly | сообща, вместе, совместно

berme (6) *Inf (ma)* the act of giving, delivering | дача, передача, вручение → **beriš**

besle– (7) *V* to feed, nourish; to fatten (animal) | кормить, откармливать

beš (21) *Num* five | пять

Beš–salkım (12) *Pn* Five–bunches of grape | Пять–виноградных гроздей; **Beš–salkım–bij** Besh–salkim–bey (Mr. Five Grape Bunches) | Беш–салкимбий (Бий имеющий пять–виноградных гроздей)

bešik (4) *N* cradle | колыбель, люлька

bešlevli? s. **beklevli** (?)

bet (12) *N* face | лицо

betli (1) *Adj* with (such and such) a face | имеющий какое–л. лицо, с каким–л. лицом; **tabaχ betli** with plate–like face; having round face | имеющий тарелкообразное лицо, круглоликий

bez– (4) *V* to get tired (of), to become disgusted (with) | надоесть, разочароваться, испытывать отвращение к кому–л.

bezdir– (1) *V* to sicken, to disgust | надоедать (кому чем), донять

bezgelek (1) *N* little bustard (Tetrax tetrax) | стрепет (малая дрофа, драхва)

bezil– (1) *V* to get tired (of), to become disgusted (with)? | разочароваться, испытывать отвращение к кому–л.?

biber (8), **bůber** *N* red pepper (Capsicum), pepper (Piper nigrum) | перец (красный или чёрный); **kara biber** black pepper | чёрный перец; **kızıl biber** red pepper | красный перец

bičım (1) *N* shape, form | покрой, сложение; **bičım–bičım** of all sorts, various | различные, разнообразные, всевозможные

bij (36), **bıj** *N* gentleman; (after name) Mr. | господин (в роли приложения к мужским именам); **Beš–salkım–bij** Besh–salkim–bey (Mr. Five Grape Bunches) | Беш–салким–бий (Бий имеющий пять–виноградных гроздей) → **beg, bej**

bijaz s. **bejaz**

bije (4) *N* mare | кобыла

bik (6) *Adv* very, much, extremely | весьма, очень → **bek**

bike (2) *N* (land)lady, mistress; housewife, woman | госпожа; хозяйка; **usta bike** housewife, housekeeper | домашняя хозяйка

bikle– (1) *V* to close, lock (up) | закрывать, запирать, замыкать → **bekle– II**

bil– (160) *V* to know, to be informed, aware; to recognize | знать, узнавать, вспоминать; **menden bil!** I promise! | я обещаю!; *Vaux* to be able to | уметь, мочь; **kajtara bil**– to be able to bring back, to take back | мочь возвратить, мочь принести назад; **kůre bil**– to be able to see | мочь увидеть кого; **sara bil**– to be able to embrace | мочь обнимать кого

bildir– (11) *Vfact* to make know, to communicate | уведомить, известить

bile (4) *Particle* even, alredy | даже

bilejzik s. **bilezik**

bilek (2) *N* wrest | запястье

bilezik (6), **bilejzik** (4) *N* bracelet | браслет

bilin– (2) *Vpass* to be known, to be aware | быть узнанным; выясняться

bilvar s. **bılvar**

bin s. **biñ**

bin– (2) *V* to get up on, to board, mount; to ride (a horse) | садиться верхом; садиться (на какой–л. вид транспорта) → **min**–, **pin**–

binamaz (1) *N P.* one who does not perform the daily worship | (человек) не совершающий намаза

binā (4), **bina** *N Ar.* building | здание

biñ (26), **bin** *Num* thousand | тысяча; **bin bir, biñ bir** thousand one | тысяча один; **jikermi biñ** twenty thousand | двадцать тысяч; **kırχ biñ** forty thousand | сорок тысяч → **miñ**

biñbaškoraj (2), **bıñbaškoraj** *N* eryngo (Eryngium campestre) | перекати–поле

bir (1831), **bėr, bi** (2), **bı, bır** (10), **br** (1224) *Num, Art indef, N* one, once; a, an | один, раз; одинаковый; **bir taha** once again | ещё раз; **bir talaj** a number of, a lot of | довольно много

bir– (22) *V* to give, to deliver | давать → **ber**–, **ver**–

biraz (64), **braz** (43) *IndefNum* a little, some | немного, немножко; несколько, чуть–чуть; **birazdan/ brazdan** in a little while, before long, shortly | немного спустя, вскоре, через некоторое время

birbiri (6), **br–bri** *Pron* each other | один с другим (взаимно)

birde (6), **brde** (27) *Conj* and, plus, (together) with | и, да, вместе

birden (16), **brden** (12), *Adv* at once, suddenly; at a time, in one lot | вдруг, сразу, разом

bircr (28), **brer** (22) *Num* one each, one apiece | по одному *Adv;* **birer–birer, brer–brer** singly, one by one | по одному, один за другим

birge (4), **brge** *Adv* together, at the same time | вместе, одновременно; **onunnen brge** together with him/her | вместе с ним/с ней

biri (131), **bri** (39), **brı** *IndefPron* one of, someone | один из них; кто–нибудь; **bir–birinden** from one another | от друг друга; **br–brlerinen** with one another | с друг–другом,

birinǯi (8), **brėnǯė** *OrdNum* first | первый; **kėrėk birinǯi** forty–first | сорок первый

bir–kač (10) *IndefNum* some, a few | несколько

birlau (1) *N* making sg to be one (single) | быть (сделать) единократным/одинарным?

birle– (1) *N* to make (it) one, to unify? | объединять; сделать единократным/одинарным?

birlik (2), **berlik?** *N* unity, oneness; sameness; union | единство, союз

bismilla (2) *Interj* in the name of Allah | во имя Аллаха

biš (52) *Num* five | пять; **biš jůz** five hundred | пятьсот; **biš un** some, a few, several | несколько, немного → **beš**

bišinǯi (1) *OrdNum* fifth | пятый

bišir– (4) *V* to cook, to bake; to ripen | варить, готовить → **pišir–**

bišlau (1) *N* making (it) fivefold | делать пятикратным *N* увеличение в пять раз? → **bišle–**

bišle– (1) *V* to make (it) fivefold? | увеличать в пять раз?

bit (1) *N* louse | вошь

bit– I (17) *V* to end, to come to an end; to be exhausted finish | кончаться, изнемочь → **pit– I**

bit– II (2) *V* to grow | расти → **pit– II**

bitėr– s. **bitir–**

bitir– (13), **bitėr–** *V* to finish, complete, terminate; to kill | кончать, завершать; губить, уничтожать → **pitir–**

biz (281), **bız, bz** *Pron pers* we | мы

bizau? s. **buzau**

bıčak (2) *N* knife | нож, ножевой → **pıčak**

bıj s. **bij**

bıl– s. **bol– I**

Bılal s. **Sejd Bılal**

bıldır I (2) *Onomatop Adv;* **bıldır–bıldir ak–, bıldır–bıldır ak–** to flow/run babbling/purling | журчать, течь (струёй) → **bůldůr**

bıldır II s. **bıllır**

bıllır (1), **bıldır?** *Adj Ar.* crystal | кристалл, хрусталь; кристальный, хрустальный; **bıllır [bıldır] šiše** crystal bottle | кристальная/хрустальная бутылка

bılvar (4), **bilvar** (3) *N R.* boulevard | бульвар

bıñbaškoraj s. **biñbaškoraj**

bır s. **bir**

bırau (1) *N* auger, gimlet, drill | бурав, сверло, коловорот

bız s. **biz**

bızau (6), **bizau** *N* calf | теленок → **buzau, tana**

bızla– (1) *V* to ice over, to freeze | леденеть

bo (167) *Pron demonst* this | этот; **boɣa** to this | этому/этой; для этого/этой; **bonda** in/at this, here | в/на этом/этой, здесь; **bondan** from/of this | от/из этого/этой, отсюда; **bonı** (of) this | этот/эту/это, (у) этого → **bol II, bu**

boǯaχ s. **boǯak**

boǯak (2), **boǯaχ** *N* corner, nook, region, country | угол, уголок, укромное место; **boǯaχ–boǯaχ** in every hole and corner | по всем углам → **buǯak**

boden s. **boten I**

boɣ– (1) *V* to strangle | душить

boɣaj (1) *Adv* probably, presumably, most probably | вероятно, очевидно, наверно

boɣaz (4) *N* throat, neck | горло, шея → **buɣaz**

boɣdaj (7) *N* wheat | пшеница → **buɣdaj**

boχ s. **bok**

boj (3) *N* height, stature; length, size; edge (of a road), bank (of a river) | рост; длина; вдоль чего–л. → **buj**

boja (1) *N* paint, dye, color | краска

boja– (2) *V* to paint, to dye | красить, окрашивать → **buja**–

bojalı (1) *Adj* painted, dyed | крашенный, окрашенный → **bujalı**

bojdaχ s. **bojdak**

bojdak (4), **bojdaχ** *N* bachelor, celibate, single | холостяк, холостой; **bojdaχ kal**– stay (old) bachelor | оставаться холостяком/неженатым → **bujdak**

bojla– (2), **bujla**– *V* to go (travel) lengthwise (of a road, a coast or a river) | идти (ехать) вдоль чего

bojlo (1) *Adj* tall, high; grown–up | имеющий какой–л. рост, рослый → **bujlı**

bojor– (4) to order to be done, to command, rule (over) | повелевать, пожаловать → **bujur**–

bojun (4) *N* neck, throat | шея, горло; **bojnına al**– to undertake/take on/up | принимать/брать на себя что–либо → **bujın, mojın, mujın**

bok (3), **boχ** *N* shit, excrement | дерьмо, помёт; **boχtan bol**– to be made of shit; (here) to be false | быть сделанным из дерьма, (здесь) быть ложным; **čam boku** shit/turf of pine? | дерьмо/торф сосны?

bokaulı (1) *Adj* being put in chains (irons) | закованный в кандалы

boklu (2) *Adj* shitty, dirty, trashy | дрянной, поганый; **boklu kůt** shitty ass | дерьмовая задница → **buklı**

bol I (9) *Adj* abundant, full | обильный; **bol–bol** abundantly, plentifully | в изобилии, вдоволь, с излишком

bol II (1) *Pron demonst* this | этот → **bo, bu**

bol– I (982) *V* to be, to exist; to become, to happen, to be completed | быть, совершаться, находиться, иметься у кого–чего–л., существовать, становиться; **amanet bol**– to be entrusted (to) | быть оставлен на попечение; **belli bol**– to come to light, to be discovered, to turn out (to be) | обнаруживаться, проявляться; делаться ясным; **br/bir bol**– to become one, join with | собъединять усилия; **bosson [<bolson]** let it

be | пусть будет; **süz bol–** to become the subject of gossip | стать темой пересудов → **ol–, bul– II**

bol– II *V* to find, to meet | находить, достигать → **bul– I**

bolaj (5) *Adv* such, so, in this way | такой, так, таким образом

bolan– (1) *Vpass* to become turbid, to be soiled | пачкаться, становиться мутным → **bulan–**

bolanıχ (1) *Adj* turbid, soiled, muddy | мутный

Bolat (78) *Pn* Bolat (steel) | Болат (булат, сталь); **Temir Bolat** Temir Bolat | Темир Болат

boldır– (1) *V* to get tired | уставать → **boldor–**

boldor– (10), **boldur–** *V* to get tired | уставать → **boldır–**

boldur– s. **boldor–**

bolgač (1) *N* a kind of puff pastry | вид слоёного пирога

bolγala– (1) *V* to immerge, plunge (sg into), to turn (in) | слегка обваливать, валять

bolomsız (3) *Adj* clumsy, silly, doltish | беспомощный; неумелый, глупый

bolon– (1) *Vpass* to be realized, to happen, take place | осуществляться, случаться

bolot (6) cloud | облако → **bulut**

bomboš (1) *Adj* quite empty | пустым пустой, совершенно пустой

bonǯa (1) *IndefNum* this much, so much | столь много, столько → **bunǯa**

bondaj (9) *Adv* such, like this | такой, похож на этот, подобно этому

bondajın (10) *Adv* such, like this | такой, похож на этот, подобно этому → **bundajın**

bor– (2) *V* to twist, wring; to bow one's head | крутить, вертеть; приклонить голову; **bujnın bordı** to bow one's neck, to bow one's head; to show humility | наклонять голову; проявить смирение → **bur–**

bora I (2) *N Gr* storm, tempest | бора → **boran**

bora II (4) *N* camel hengst | верблюд (самец)

boran (5) *N* tempest, storm | буран → **bora I**

borč I (1) *N R.* borsch | борщ; **kapısta borčı** borsch, cabbage–soup | борщ

borč II s. **borǯ**

borǯ (5), **borč** *N* debt | долг → **burǯ**

bork (1) *N* fur cap; fur, pelts | мех, меховый; меховая шапка? → **bürk**

boru (1) *N* pipe, tube | труба

bosaγa (9) *N* threshold, doorstep | порог

bosson s. **bol– I**

bostorγaj (6), **postorγaj** (2) *N* (grey) sparrow (*Passer domesticus*), lark (*Alauda arvensis*) | серый воробей, жаворонок

boš (2) *Adj* empty | пустой

boš–bošına (1) **[<boši–bošına]** *Adv* quite in vain, uselessly | совершенно напрасно, впустую → **bošına**

boša– (1) *V* to set free; to become free | освобождать(ся), стать свободным

bošalt– (4), **bošat–** *V* to empty, to carry out | опоражнивать; разгружать, выгружать

bošına (1) *Adv* in vain, useless, needless | напрасно, впустую, зря; напрасный, тщетный → **boš–bošına**

botalı (1) *Adj* having camel colt; (female camel) with camel colt | (верблюдица) с верблюжонком; **botalı bije** she–camel, female camel with calf | верблюдица с верблюжонком

botem s. **boten II**

boten I (7), **boden** *Adj/Adv Ar.* needless, useless, vain; in vain | напрасный, напрасно; беполезный; in vain | напрасно, зря

boten II (3), **botem** *Adj* other, another, different | другой, иной; **botem jerde** elswhere, somewhere else | на другом месте, не там же

botka (1) *N* flour soup; mush, pap | каша; мучной суп?

boz I (1) *Adj* (ashen) gray, roan | серый; сивый (о масти) → **bus**

boz II (2) *N* ice | лёд, ледяной → **buz**

boz– (3) *V* to demolish, ruin, destroy | портить, повреждать, разорять → **buz–**

bozık (1) *Adj* broken, destroyed, spoiled | испорченный, плохой; **bozık tůš–** to go wrong; to go bad, to get spoiled | испортиться; тухнуть → **buzuk**

bȯjle (5) *Adv* so, thus, in this way | так, такой; таким образом → **bůjle**

börek (1) *N* flaky pastry with thin layers of cheese or other filling | бурек → **bůrek**

br s. **bir**

br–bri s. **birbiri**

braz s. **biraz**

brden s. **birden**

brer s. **birer**

breu (23), **breů** *IndefPron* one of them; one person; someone | один из; кто–либо

breů s. **breu**

brge s. **birge**

bri s. **biri**

brsi s. **biri**

bu (915), **bı** (3), **bů** *Pron* this (one) | этот/эта/это → **bo, bol II**

buǯak (1) *N* corner, nook; region | угол, уголок; укромное место → **boǯak**

budaχ (1) *N* branch, twig | ветка, ветвь → **butaχ, butak, pıtak, putak**

budama (1) *Adj* pruned, being lopped | (от)резанный

budur– (1) *Vfact* to have sb/sg tied tightly; to have sg squeezed together | приказать туго завязывать кого–чего–л., велеть зажимать чего–л.

bugun s. **bůkůn**

buγaχ (1) *Adj* foggy, misty, vaporous | окутанный (покрытый) туманом → **buγak**

buγaj (1) *Adv* probably, presumably, most probably | вероятно, очевидно, наверно

buγak (1) *Adj* foggy, misty, vaporous | окутанный (покрытый) туманом → **buγaχ**

buγaz (3) *N* throat, neck | горло, шея → **boγaz**

buγdaj (2) *N* wheat | пшеница; **misır buγdajı** corn, maize | кукуруза, маис → **boγdaj**

buγıl– (5), **buγul–** (2)*Vpass* to suffocate, to get drowned | быть удушенным; задыхаться; тонуть; захлебнуться;

buγul– s. **buγıl–**

Buχara (2) *Tn* Bukhara (city) | (город) Бухара

buχčalan– (1) *Vpass* to be wrapped up in a bundle; to prepare for a journey | быть завязанным в узел; подготовиться к путешествию

buj (32) *N* height, stature; length, size; edge (of a road), bank (of a river) | рост; длина; вдоль; **bujdan buja** all over, from end to end | целиком, полностью; из конца в конец, во всю длину → **boj**

buja– (3) *V* to paint, to dye | красить, окрашивать → **boja–**

bujalı (2) *Adj* painted, dyed | крашенный, окрашенный → **bojalı**

bujan– (1) *V* to be painted, colored | быть окрашенным

bujdaχ (1) bachelor, celibate | холостяк, холостой → **bojdak, bujdak**

bujdak (1) *N* bachelor, celibate | холостяк, холостой → **bojdak, bujdaχ**

bujık (1) *N* mustache | усы; **kajtan bujık** having a long, thin moustache | имеющий длинные и тонкие усы

bujın (29) *N* neck, throat | шея, горло; **bujnına al–** to undertake/take on/up | принимать/брать на себя что–либо; **bujnına ur–** to meet one's end/destiny; to come off badly | достигать судьбы → **bojun, mojın, mujın**

bujla– s. **bojla–**

bujlı (2) *Adj* tall, high; grown–up | имеющий какой–л. рост, рослый → **bojlo**

bujur– (3) *V* to order to be done, to command, rule (over) | повелевать, пожаловать; **bojroñoz** please! here you are! | пожалуйста! → **bojor–**

bukla– (1) *V* to soil, befoul | пачкать, гадить

buklı (1) *Adj* shitty, dirty, trashy | дрянной, поганый → **boklu**

bukun s. **bủkủn**

bukur (1), **[bủkủr]** *Adj* humpbacked | горбатый

bukủn s. **bủkủn**

bul– I (19) *V* to find, to discover | находить

bul– II (10) *V* to be, to exist; to become, to happen, to be completed | быть, совершаться, находиться, иметься у кого–чего л., существовать, становиться; **buluγı** his being (a), his existence | его бытие, его существование → **bol– I**

bulaj (9) *Adv* so, like this, such, in this way | так, таким образом

bulamık (2) *N* mush, pap made of flour | мучная каша

bulan– (2) *Vpass* to become turbid; to be tumbled (e.g. in flour, dirt etc.) | становиться мутным; мутнеть; быть обвалянным (в муке и т. п.) → **bolan–**

bulašıχ (1) *N* smeared, soiled; (here) aching | грязный; испачканный; (здесь) больной

bulaštır– (1) *V* to smear, to soil | пачкать

buldur (1) *Adj* languid, torpid | вялый

bulgača (1) *Adv* a kind of puff pastry | вид слоёного пирога

bulγala– (1) *V* to immerge, plunge (sg into), to turn (in) | слегка обваливать, валять

bulγavıš (1) *N* wooden spoon, stirrer | деревянная ложка, мешалка

bulu (1) *N/Inf(uw)* being, existing | бытие, существование; **buluγı** his being, his existence | его бытие, его существование → **bul– II**

bulun– (7) *Vpass* to be found, to be present, to be | находиться где–нибудь, иметься, бывать

bulut (9) *N* cloud | облако; **bulut al–** to be covered by a cloud | погрузиться в облако → **bolot**

bunǯa (4), **bunǯe** *IndefNum* this much, so much | столь много, столько → **bonǯa**

bunǯe s. **bunǯa**

bundajın (1) *Adj* such, like this | такой, похож на этот, подобно этому → **bondajın**

bur– (3) *V* to twist, wring; to bow one's head | крутить, вертеть; приклонить голову; **bašın bur–** to bow one's head; to show humility | наклонять голову; проявить смирение; **karnı bura** one's stomach is seized with cramp | чувствует судорожные боли в желудке → **bor–**

burada (1) *Adv* here | здесь

buradan (4) *Adv* from here, hence | отсюда

buralı (1) *Adj* local, native of this place | здешний, местный (житель)

burǯ (3) *N* debt | долг → **borǯ**

burǯlan– (1) *Vpass* to get into debt, to become indebted | задолжать

burda (1) *Adv* here | здесь

burgu (1) *N* drill, auger, gimlet; corkscrew | бурав, сверло

burıl– (1) *Vpass* to twist oneself about; to writhe | поворачиваться, извиваться → **burul–**

burma (4) *N* screwed, twisted, spiral | закрученный

burmaχ (2) *Inf* twisting, screwing | крутить, завинчивание

burul– (2) *V past* to twist oneself about; to writhe | поворачиваться, извиваться → **burıl–**

burun (10) *N/Postp* nose | нос; **bašlamazdan burun** before it starts | до начала, перед тем как начался

burungu (2) *Adj* earlier, former, preceding; ancient | прежний, предшествующий; древний прежний, предшествующий; предыдущий; древний → **burunki**

burunki *Adj* previous, former, earlier, preceding | прежний, предидущий → **burungu**

bus (1) *Adj* (ashen) grey, roan | серый; сивый (о масти) → **boz I**

bustan (13) *N P.* vegetable garden; (here) melon bed, (water–)melon–field(s); melon, watermelon | огород; дынник, бакча; дыня, арбуз

bustanǯi s. **bustanǯı**

bustanǯı (6), **bustanǯi** *N* vegetable gardener; gardener raising/growing (water–)melons | огородник; бахчевод, бахчевник

but (1) *N* thigh | ляжка, бедро

butaχ (1) *N* branch, twig | ветка, ветвь → **budaχ, butak, pıtak, putak**

butak (1) *N* branch, twig | ветка, ветвь → **budaχ, butaχ, pıtak, putak**

buz (12) *N* ice | лед, ледяной → **boz II**

buz– (8) *V* to demolish, ruin, destroy | портить, повреждать, разорять → **boz–**

buzau (2) *N* calf | теленок → **bızau, tana I**

buzauǯu (1) *N* calfherd | телятник, пастух телят

buzaula– (1) *V* to calve | телеться, отеляться

buzaulat– (1) *Vfact* to have [the cow] calved | велеть теляться

buzaulatma (1) *Inf (ma)* having [the cow] calved? | веленье об отелянии?

buzil– s. **buzul–**

buzıl– s. **buzul–**

buzla– (1) *V* to be covered with ice; to freeze | леденеть, замерзать

buzlı (1) *Adj* iced, freezed | со льдом, мерзлый

buzuk (1) *Adj* broken, destroyed, spoiled | испорченный → **bozık**

buzul– (4), **buzıl–**, **buzil–** *Vpass* to become spoiled, tainted; break off (magic) | портиться, прекращаться (о волшебстве)

bů s. **bu**

bůber (1) *N* pepper, red pepper | перец (красный/чёрный) → **biber**

bůǯek (1) *N* bug | насекомое

bůjı– s. **bůjů–**

bůjle (58) *Adv* so, thus, in this way | так, таким образом → **bőjle**

bůjuk s. **bůjůk**

bůjů– (6), **bůjı–** *V* to grow (up); to boast, put oneself up | расти, возноситься; становиться взрослым

bůjůk (42), **bůjuk** *Adj* great, large, big; grown–up, elder | большой, крупный; могучий; взрослый, старший

bůjůt– (1) *V* to enlarge; to bring up (child) | возвышать; увеличивать пастить, воспитывать

bůklen– (1) *Vpass* to be bent, crooked, folded | быть сложенным; быть загнутым

bůkrejt– (1) *V* to bend, to fold, spin | гнуть, сгибать

bůkůl– (3) *Vpass* to bend, bow; to fold | быть загнутым; **beli bůkůlgen** humped back | сутулык, сутуловатый

bůkůn (29), **bugun, bukun** (2), **bukůn** (3) *Adv* today | сегодня

bůl– (1) *V* to divide, distribute | делить

bůlbůl (23), **bulbul** *N* nightingale (*Luscinia megarhyncos*) | соловей; соловьиный

bůldůr (1) *Onomatop Adv* **bůldůr–bůldůr aχ–** to flow rolling along | течь сплошным потоком → **bıldır I**

bůldůr– (2) *Vfact* to have sg cut (off, into pieces); to get sg broken (into pieces) | делить

bůlkůl– (1) *Vpass* to shake, tremble; to shiver | дрожать, стрястись

bůlůn– (1) *Vpass* to be divided, separated; (here) to splinter, shatter | делиться; (здесь) раскалываться

bůr– (1) *V* to do laying, crinkle, plait (on clothes) | делать складки, сборки

bůrčůklen– (1) *Vpass* to bud, burgeon | пускать почки, почковаться

Bůrʒůmek (1) *Tn* Burdjumek (settlement?) | Бюрджюмек (поселение?) *Tn;* **Bůrʒůmek kırında** on the field of Burdjumek (settlement?) | на луге (поселения?) Бюрджюмек

bůrek (3) *N* flaky pastry with thin layers of cheese or other filling | пирог, пирожок; **ʒıllı ʒımšaχ bůrek** warm soft burek (see there); (here in figurative sense) hot soft vulva | тёпло–мягкий бурек; (здесь в переносном смысле) тепло–мягкое влагалище → **börek**

bůri (2) *N* wolf | волк

bůrk (1) *N* fur cap | меховая шапка → **bork (**bőrk)

bůrlen– (1) *Vpass* to bloom; to bud, burgeon | распускаться (о почках), пускать почки, почковаться

bůrme *N* pleat, fold | складка

bůrmeli (1) *Adj* pleated | со складками, со сборками

bůtůn (11), **půtin** *Adj* all, whole | весь; все; целый; **bůtůn alem/ālem/ʒihan** the whole world | весь мир

bz s. **biz**

Č

čabala– (1) *V* to strive, struggle | стараться, прилагать усилия; **iše čabala–** to struggle with a task | стараться/трудиться над заданием

čabdırıl– (1) *Vpass* to slam, dash (into, on, against sg) | ударяться, бросаться

čabık (2), **čapık** *Adj/Adv* quick | быстрый, скорый

čač (7), **čaš** (2) *N* hair | волосы → **sač, šač**

čač– (1) *V* to scatter, to sprinkle | сеять, сыпать → **sač–**

čačıl– (1) *Vpass* to sprinkle, spread; to fall (down) | сыпаться, рассеиваться; упасть

čadır (5) *N* tent | шатёр, палатка

čaγ (5) *N* age, period, time | возраст, период жизни; **balalık čaγı** childhood | детский возраст, детство

čaγa (1) *N* child | младенец, ребенок; **bala čaγa** household (all the members of a family) | семья, домашние, детвора, весь дом → **bala**

čaγana (3) *N* a kind of music instrument; row, hullabaloo | род музыкального инструмента; музыка (вообще); буча; **čalγı čaγana** musical instruments; music; row | инструменты; музыка; шум, какофония

čaγarak (1) *N* chimney; flue | дымоходная труба

čaγır– (38) *V* to call | звать; приглашать → **čakır–**

čaγırt– (6) *Vfact* to have sb called | позвать (пригласить) кого–л. через кого–л.

čaγla– (1) *V* to warble (nightingale); to burble, purl | щёлкать, щекотать (о соловье); журчать, шуметь (о воде); **bůlbůl čaγlaγan** the nightingale warbled | соловей щёлкал, щекотал

čaj (4) *N* tea | чай

čaj– (1) *V* to crush; to melt (in liquid), to dissolve | толочь, мять; растворять

čajχa– (1) *V* to rinse, wash | полоскать, мыть → **šajχan–**

čajır (4) *N* meadow, (alpine) pasture | чаир, луг (горный)

čajnık (1) *N R.* tea pot, tea kettle | чайник

čakirim s. **čakrım**

čakıl (5) *N* pebble, gravel | галька

čakıltaš (1) *N* pebble, gravel | галька

čakır– (30) *V* to call | звать; приглашать → **čaγır–**

čakırǯı (3) *N/Pn* falconer | сокольник, сокольничий; **Čakırǯı** Chakirdji | Чакырджы

čakırım s. **čakrım**

čakırt– (8) *Vfact* to send for sy; to have sy called | позвать (пригласить) через кого–л.

čakrim s. **čakrım**

čakrım (12), **čakirim, čakrim, čakırım** (5) *N* verst (Turkic measure of length of about 1 km) | верста (мера длины, около ! км)

čal (34), **čāl** *Adj* grey(–haired), white; greybeard, old man | чалый; **čal at** grey horse | серый конь; **čal/čāl baš** grey(–haired) head | чалая голова; **čal sakallė** having grey beard | с серой (седой) бородой

čal– I (35) *V* to play (on) an instrument; to steal; to strike, to cut off | играть, исполнять; красть, похищать; бить, косить, резать; **baγlama čal–** to play baghlama | играть на багламе; **čalγı čaγana čal–**, **čalγı čal–** to play (on) an instrument | играть на инструменте; **davul čal–** to beat the drum | бить (ударять) в барабан; **dāre čal–** to play the tambourine | играть на тамбурине; **kemane/kemanče čal–** to play the violine | играть на скрипке; **kurban čal–** to kill (an animal) as a sacrifice | резать (барана) в жертву, заклать жертву; **zurna čal–** to play the zurna | играть на зурне

čal– II (1) *V* to steal | воровать, красть

čalaš (11) *N* felt covered round tent; hut made of brushwood | шалаш

čaldır– (7) *Vfact* to make sb play an instrument | велеть играть (на инструменте), заставить играть (напр. музыканта); **davul čaldır**– to make sb beat the drum | велеть бить в барабан; **dāre čaldır**– to make sb play the tambourine | велеть играть на тамбурине; **kemane dāre čaldır–, kemen dāre čaldır**– to make sb play the violine and tambourine | велеть играть на скрипке и тамбурине; **kurban čaldır**– to make sb kill (an animal) as a sacrifice | велеть резать (животного) в жертву, заставить заклать жертву; **zurna čaldır**– to make sb play the zurna | велеть играть на зурне

čalgi s. **čalγı**

čalgı s. **čalγı**

čalγı (7), **čalgi** (2), **čalgı** *N* music; playing an instrument | музыка; игра на музыкалном инструменте; **čalγı čaγana čal**– to play musical instruments, music plays | играть на музыкалном инструменте, звучат музыка и музыкальные инструменты; **čalγı čaγananen karšıla**– to meet sb with playing musical instruments; to meet sb with music/row | встречать кого–л. с музыкой и шумом; встречать кого–л. с звучающими инструментами

čalı (3) *N* bush, shrub | куст

čalın– (1) *Vpass* to be played (an instrument) | исполняться (на инструменте); **saz čalın**– to be played on *saz* | играют/играли на сазе

čalıš– (1) *V* to work, to try, strive | работать, стараться

čalka (1) *Adv* on one's back, backward | навзничь; **čalka–čalka tůšep (tůšėp) jat**– to lie on one's back | лечь на спину, лечь навзничь

čalka– (1) *V* to agitate, to shake, to toss about | взбалтывать, встряхивать

čallan– *Vpass* to turn grey | поседеть

čalma (1) *N* a kind of turban | чалма (сорт тюрбана); **čalma čırma**– to wrap around a turban | окутывать/обматывать чалму (тюрбан)

čalman (5) *N* (military) coat | шинель (солдатский)

čam (1) *N* pine *(Pinus)* | сосна; **čam boku** shit/turf of pine? | дерьмо/торф сосны?

čaman (1) *Adj* languid, weak, feeble; lazy | вялый, ленивый; **čaman at** lazy horse | ленивая лошадь

čamašır (2) *N* laundry | бельё; **čamašır tep**– to wash laundry by kicking/treading on | мыть бельё топтанием

čamır (3), **čamur** *N* mud, slush; mortar | грязь, слякоть

čamur s. **čamır**

čan s. **čañ**

čana (2) *N* sled, sleighs | сани, санки; **at čana** horse–sleigh | конные сани

čanaχ s. **čanak**

čanak (2), **čanaχ** *N* earthenware pot; tureen | чашка, миска, посуда; **bal čanaχ** honey pot | медовая миска

čañ (4), **čan** *N* (large) bell; gong | колокол, звонок; **čan kak**– to ring (hit) the bell | звонить в звонок

čañlı (1) *Adj* furnished with bells or gongs | оснащённый звонками

čap– (12) *V* to run, trot; to chase; tu hurtle; to beat | бежать, гнать (лошадь), мчаться; бить, ударять; **atı čap**– to let the horse run | пустить лошадь вскачь

čapa (2) *N* mattock, hoe | цапка, мотыга

čapala– (1) *V* to hoe | мотыжить

čapan (1) *N* outer gown or robe with long skirts and sleeves; robe of honor, caftan | верхний халат с длинными подолами и рукавами; почётная мантия

čapık s. **čabık**

čapıš– (2) *V recipr* to run/trot together, to make a concerted rush, to chase sg/sb together with sb | бежать вместе с другими, гнать (лошадь) вместе с другими, мчаться вместе с кем–л.; бить/ударять друг друга

čapkala– (1) *V* to rush about, to run to and fro | носиться туда сюда

čapkalan– (1) *Vpass* to rush about, to run to and fro | носиться туда и обратно

čapkıl– (2) *V* to dash into/against sb, to fall on sb | бросаться, кидаться (на кого–л.)

čaptır– (4) *Vfact* to have sg cut, to have (sb's head) chopped off | велеть/приказать отрубить (кому–л. голову)

čapuχ (1) *Adj* quick, fast, swift; quickly, speedily | быстрый, быстро → **čabık**

čara s. **čare**

čardaχ s. **čardak**

čardak (3), **čardaχ** *N P.* bower, trellis; portico; mansion | чердак; дворец

čare (38), **čara** (20), **čāre** (3) *N P.* way, means, device | средство, выход; **čare bul**– to find means (to) | находить выход; **čara tap**– to find means (to) | находить выход;

čarıχ (2) s. **čarık**

čarık (5), **čarıχ** (3) *N* rawhide sandals | чарыки; постолы (род крестьянской самодельной обуви из сыромятной кожи) → **čaruk**

čaršı (11), **čarši** (11), **čaršu** *N* the bazaars, market quarter/place | рынок

čaruχ s. **čaruk**

čaruk (11), **čaruχ** *N* rawhide sandals | чарыки; постолы (род крестьянской самодельной обуви из сыромятной кожи) → **čarık**

čaš s. **čač**

čašt (1) *N R.* happiness, felicity | счастье

čatal (3) *N* fork; forking, road–junction | рогатина, развилина; **čatal čık**– to ramify, to branch off | ответвляться, разветвляться;

čatla– (2) *V* to crack, split | трескаться

čatlat– (1) *Vfact* to track, fissure | надбивать, надламывать

čatmıš (1) *Past Participle* joined (eyebrows) | (брови) соединённые в концах

čauš (1) *N* senior (in rank); foreman, headman | старший (над рабочими), инспектирующий

čav– (18) *V* to pluck, to break off | отрывать, срывать

čāre s. **čare**

čebůrek (4) *N R.* flaky pastry with mutton or other filling fried in fat | чебурек → **čiberek, čůberek**

čeč– (2) *V* to untie, unfasten, unbutton | развязывать, расстёгивать

čeček (9) *N* flower, blossom | цветок; **bir top čeček** a bunch of flowers | букет цветов → **čicek, šeček, šešek**

čečeklen– (1) *Vpass* to flower, to come into flower | цвести, расцветать

čegerli (1) *Adj* bushy | кустистый; **čegerli oba** bushy hill | кустистый холм

ček– (43) *V* to pull, draw; to strech (out); to mesure | тянуть, взвешивать; **ah ček**– to heave a sigh | ахать; **ah zār ček**– to sob one's heart out | горько (безутешно) плакать; **asret/asretlik ček**– to be homesick | тосковать по родине; **azab ček**– to suffer torment | мучиться, страдать; **hasret/**

hasretlik ček– to be homesick | тосковать по родине; **χorloχ ček**– to be/feel ashamed (of) | стыдиться чего–л.; **šer ček**– to endure/suffer wikedness of sb | страдать от зла/злодеяния кого–л.; **uk ček**– to stretch bow | натягать лука

čekil– (11) *Vpass* to withdraw, draw back; to get away | быть тянутым, взвешиваться; **ǯerbi čekil**– recruiting/conscription is arranged, to be enlisted, to be declared fit for military service | быть призванным в солдаты; быть призванным на военную службу

čekin– (1) *Vpass* to be bashful, shy; to scruple | стыдиться, робеть, смущаться

čekiš– (2) *V recipr* to suffer pain/tortures; to be dying | мучиться, умирать; **ǯan čekiš**– to suffer pain/tortures; to be in the throes of death | мучиться, страдать; быть в агонии, агонизировать

čekmen (2) *N* outer gown or robe with long skirts and sleeves; caftan | чекмень, кафтан

čektir– (1) *Vfact* to have sg ground by sb | велеть/заказывать молочь

čelebi (2) *N* man of refinement; well–bred, educated; gentleman | благовоспитанный, образованный; господин

čelek (4) *N* bucket (made of wood) | бадья

čelik (1) *N* steel | сталь

čeltek (2) *N* grating, bars | решето

čengel (1) *N* hook | кручок

čepče s. **čipče**

Čerkes (6), **Čerkez** *En/Pn* Cherkess | черкес

Čerkez s. **Čerkes**

čert– (2) *V* to strike togeteher, to touch | стукнуть, тронуть; щёлкать; **pıski čert**– to fillip/flick with the finger | щёлкать, дать щелчок; **şırpı čert**– strike a chip/mach | чиркнуть палочкой/спичкой

češhane (1) *N P.* rifles | (нарезное) охотничье ружьё

češin– (1) *Vpass* to undress (oneself) | раздеваться

češit (3) *N* kind, sort | вид; **češit tụrlụ musibet** different variety of trouble/misfortune | разновидные сорта беды и несчастья

češme (31) *N* fountain with spout | источник, фонтан; **češme bašı** fountain–head, surroundings of the fountain | фотанная головка (сам источник и его окружность)

čet (24) *N* the far/remote brink/verge of sg | край, берег, обочина, граница

čevir– (3) *V* to turn, rotate | переворачивать; **evirep čevirep (čevirėp)** turning round and round | крутить, вертеть, рассматривать (со всех сторон); тщательно проверять

čeviz s. **ǯeviz**

čevre (7) *N* surroundings, circumference, environs, periphery | окружность, окрестность

čevril– (1) *Vpass* to turn round | переворачиваться, перейти, превратиться

čez– (8) *V* to untie, unfasten, unbutton; to solve; to free | развязать, распутать, расстегнуть, решить

čezil– (2) *Vpass* to be solved, to be unfastened; to disappear | развязываться, решаться

čiberek (1) *N* flaky pastry with mutton or other filling fried in fat | чебурек → **čebụrek, čụberek**

čibuk s. **čıbuk**

čiček (6) *N* flower, blossom | цветок → **čeček**

čift (5), **čipt** (2) *N/Adj P.* couple, pair; (pair of) yoke/draught–animal (s); plough | пара; чёт; рабочая скотина; пахота; **babamnıñ čift ủgủzi tegilsıñ** you are not my father's pair of yoke/draught–oxes | ты не пара (рабочих) волов моего отца; **čift ủgủz** yoke/draught–oxes | (рабочие) волы → **čifte**

čifte (5) *Adj* double, paired | парный, двойной → **čift**

čiɣara s. **čıɣara**

čik– s. **čık**–

čil (1) *N* hazel grouse/hen (Tetrastes bonasia); partridge | рябчик, куропатка пустынная

čille (1) *N* suffering | мучение, страдание

čilma (1) *N* hook, trip (with the foot) | подножка, зацеп (ногой); **čilma at**– to trip, to cast the net up | подставить ногу, дать подожку; закидывать сети (невод) → **čilme**

čilme (1) *N* trawl, fishing net, web | (рыболовная) сеть, невод; **čilme tašla**– to leave/throw the net; to cast the net up | закидывать сети (невод) → **aɣ, au II, čilma**

čimen (5), **čủmen** *N* meadow | луг, лужайка

čimendifer (1) *N Fr.* railway, train | железная дорога

čimenlik (3) *N* meadow; grassy place | луг, лужайка; занятая лугами

Čin–mačin (4), **Čin–mačın** (2) *Tn* common name of China and the Far–East | Китай и страны Дальнего Востока

Čin–mačın s. **Čin–mačin**

čingene (1) *N* Gipsy | цыган

čini (3) *N P.* tureen; glazed tile, porcelain | миска; фаянс

čiñ–čiñ (1) *Onomatop* ? | ; **čiñ–čiñ čın (čini) amam** ringing resounding bath | сильно звучающая баня (хамам)

čipče (8), **čepče, čıpče** (2) *N* chiken | цыплёнок

čipčik (1) *N* sparrow | воробей

čippej (4) *N* young (of birds), nestling | птенец, птенчик; **čippejim!** my dear! honey! dearest! my darling! (gentle addressing) | мой милый! моя детка! (ласкательное обращение) → **čủppej**

čipt s. **čift**

čiput s. **čuput**

čirikkin s. **čirkin**

čirkin (6), **čirikkin** *Adj* ugly, unsightly | некрасивый, дурной

čit (2) *N* the far/remote brink/verge of sg | край, берег, обочина, граница

čiv (2) *N* nail | гвоздь, нагель

čivilde– (1) *V* to chirp | чирикать

čizme (4), **čızma, čızme** *N* boots | сапоги

čıbın (4) *N* fly | муха

čıbırt– (1) *Vfact* to align, to line up | велеть тянуться?

čıbuχ s. **čıbuk**

čıbuk (7), **čibuk**, **čıbuχ** *N* twig, rod | чубук, прут, розга → **šıbuk**

čıda– (4) *V* to tolerate; to endure, stand | терпеть, выносить

čıfut s. **čufut**

čıyar– (3) *V* to take out, to remove, take off | снимать, выводить, выкладывать

čıɣara (2), **čiɣara** *N Sp.* cigarette | папироса, сигарета; **čiɣara/šıɣara jak**– to light a cigarette | закурить сигарету → **šıɣara**

čıyart– (3), **šıyart**– *V* to have sg taken out, to have sg removed, have sg taken off

| приказать снимать, велеть выводить, приказать выкладывать

čıχ– (124), **čiχ**– *V* to go/come out, to set out | выходить, уходить, отправляться → **čık**–

čıχar– (52) *V* to take out, to pull/push out; to take off | выводить, вытягивать; снимать (одежду) → **čıkar**–

čıχart– (1) *Vfact* to have sg taken out, to have sg pulled/pushed out; to have sg taken off | приказать выводить, велеть вытягивать; приказывать снимать (одежду) → **čıkart**–

Čıχmıš *Tn* Chikhmish | Чыхмыш

čık– (229), **čik**– (13), **šık**– *V* to go/come out, to set out | выходить, уходить, отправляться → **čıχ**–

čıkar– (49), **čikar**– *V* to take out, to pull/push out; to take off | выводить, вытягивать; снимать (одежду) → **čıχar**–

čıkart– (1) *Vfact* to have sg taken out, to he sg pulled/pushed out; to have sg taken off | приказать выводить, велеть вытягивать; приказывать снимать (одежду) → **čıχart**–

čıl (2) *Adj* freckled, speckled | пёстрый, пятнистый; **čıl–čıl barak** very pied dog | очень пёстрая собака

čın (1) *N/Adj* China, Chinese | Китай, китайский

čınčı *N* performer of songs chin/chyn (cf. CrmTat. chyng (čıñ) 'song of improvised character; traditional form of folk poetry' | исполнитель чина/чына (см. Кртат. чынъ "чин" – песня импровизационного характера; традиционный вид народно–поэтического творчества)

čınčılik (1) *N* art of chinchis, performing songs chin/chyn | мастерство чинчи/чынчы, исполнение песен чин/чын → **čınčı**

čıñaš– **[<čıñlaš–]** (1) *V recipr* to perform songs chin/chyn together, competing | петь/исполнять чины/чыны вместе, соревнуясь друг с другом → **čınčı**

čıpče s. **čipče**

čıpit s. **čuput**

čıpıt s. **čuput**

čıplak (3) *Adj* naked | голый; **čırlak–čıplak** half–naked | полуголый, полураздетый

čıput s. **čuput**

čıraχ (1) *N* candle | свеча, лучина → **čırak I**

čıraj (2) *N* facial expression, look, | выражение лица; вид

čırak I (3) *N* candle | свеча, лучина; **čırak jak**– to light a lamp | зажигать лампу → **čıraχ**

čırak II (1) *N* apprentice, servant | слуга, подмастерье

čırlak–čıplak s. **čıplak**

čırma– (1) *V* to wind, wrap around | обвивать, окутывать, завернуть; **čalma čırma**– to wrap around the turban | обматывать чалму (турбан)

čız– (1) *V* to solve, explain; to decipher | развязать, распутать → **čez**–

čızma s. **čizme**

čızme s. **čizme**

čoban (45) *N* shepherd, herdsman | чабан

čočamaj (1) *N* lark, skylark (Alauda arvensis), crested lark (Galerita cristata), | жаворонок; **čočamaj torγaj** skylark | жаворонок

čočka (1) *N* pig, swine | свинья

čoǯuk (3) *N* child | ребенок

čoχ (30) *Adj/Adv, IndefNum* much, great (in amount), long (in duration), too much, very | много, очень, долго → **čok**

čoχčalan– (2) *Vpass* to become numerous | размножиться

čoχčı– (4) *V* to pick/scratch about (for food) | пощипывать

čoχmar (1) *N* butt, club, cudgel | дубинка, дубина, палица → **čokmar**

čoχraχ (2) *N* wellspring | родник, ключ, источник → **čokrak**

čoχu– (1) *V* to peck | клевать

čojun (2) *N* cast iron | чугун

čok (43) *Adj/Adv, IndefNum* much, great (in amount), long (in duration), too much, very | много, очень, весьма, долго → **čoχ, šok**

čokčalan– (1) *Vpass* to become numerous | становиться больше, размножаться

čokır (2), **čukır** *N* pit, hollow | яма → **čukur**

čokıš [čỏkıš?] (1) *N* hammer, gavel | молоток

čokmar (16) *N* butt, club, cudgel | дубинка, дубина, палица → **čoχmar**

čokraχ s. **čokrak**

čokrak (12), **čokraχ** (4) *N* wellspring | родник, ключ, источник → **čoχraχ**

čolak (10) *N* crippled in one hand | человек с изувеченной рукой

Čongar (1) *Tn* Chongar (settlement) | Чонгар (поселение)

čonkaj– (1) *V* to crouch down, to sit down | приседать

čontoχ s. **čontuχ**

čontuχ (2), **čontoχ** *N* short, curt, lean | куцый, короткий

čoñtaj (2) *N* little leather bag | кожаный мешочек

čora (117) *N* child (born out of intermarriage); servant; slave | ребёнок от смешанного брака, раб; слуга; **Čora–batır** Chora–batir | Чора–батыр

čorab s. **čorap**

čorap (3), **čorab** *N* stocking, sock, hose | чулок, чулки

čorba (5) *N* soup | суп → **čurba, šorba**

čorbaǯı (1) *N* master, owner | хозяин → **šorbaǯı**

čoš– (2) *V* to enjoy; to become enthusiastic (about sg) | наслаждаться, разойтись

čỏl (9), **šỏl** *N* steppe, meadow | степь, поле → **čủl, šỏl**

čỏplủk (2) *N* rubbish, litter | мусор

čöple– (1) *V* to pick/scratch about (for food) | пощипывать

čubar (2) *Adj* spotty, motley | пёстрый, чубарый

čubır– s. **čubur**–

čubuk s. **čıbuk**

čubur– (5), **čubır**– *V* to scatter, to get spilt | высыпаться

čuburt– (1) *Vfact* to arrange in a row | выстраивать, строить в ряды → **čıbırt**–

čufut (3), **čıfut** *En Ar.* Hebrew, Jew | еврей, иудей → **čuput, ǯuput**

čukır s. **čokır**

čukur (3) *N* pit, hollow | ямя

čuput (15), **čiput, čıpit, čıpıt** (3), **čıput** (4) *N Ar.* Hebrew, Jew | еврей, иудей → **čufut, ǯuput**

čurba (1) *N* soup | суп → **čorba, šorba**

čuval (16) *N* sack, bag | мешок

čủberek (2) *N* flaky pastry with mutton or other filling fried in fat | чебурек → **čebủrek, čiberek**

čủj (4) *N* nail | гвоздь

čủk– (2) *V* to crouch down, to sit down | оседать, опускаться

čủkel– (1) *V* to crouch down, to sit down | оседать, опускаться; приседать на корточки

čůl (11) *N* steppe, meadow | степь, поле → **čȯl, šȯl**

čůmen s. **cimen**

čůnki (2) *Conj P.* because, where as | потому что, так как, ибо, поскольку

čůp (4) *N* chip, splinter | лучина, щепка

čůppij (8) *N P.* young (of birds), nestling | птенец, птенчик; **čůppijim, čůppijım** my dear! honey! dearest! my darling! (gentle addressing) | мой милый! моя детка! (ласкательное обращение) → **čippej**

čůr– s. **ǯůr**–

čůri– s. **čůrů**–

čůrů– (4), **čůri**– *V* to rot, to go bad | гнить, испортиться

čůrůk (1) *Adj* rotten | гнилой, испорченный

čůz– (1) *V* to solve, explain; to decipher; to losen, unbind | развязать, распутать → **čez**–

ǯ

ǯadı (~**ǯadi**) (14), **ǯadı χatın (ǯadi χatın)** *N* witch, hag | ведьма → **obur**

ǯaγ– I (1) *V* to rain, to snow | идти, падать (о дожде, снеге) → **ǯav– I, jaγ–, jau–, jav**–

ǯaγ– II s. **ǯak**–

ǯaγın (2) *Adj/Adv* near (in place or time) | близкий, недавний; → **ǯak ın, jakın**

ǯaγmır (1) *N* rain | дождь → **jaγmır, jaγmur**

ǯahil (1) *Adj Ar.* stupid person, uneducated | невежда, неграмотный

ǯaj– (4) to spread, to scatter | расстилать, раскатывать → **jaj**–

ǯajan (2) *N/Adj* on foot; pedestrian | пешком, пеший → **ǯajau, jajan**

ǯajau (1) *N/Adj* on foot; pedestrian | пешком; пеший → **ǯajan, jajan**

ǯajau–ǯalpı (1) *Adv* barefooted and travelling on foot | на своих двоих (буквально пешком и босиком)

ǯajıl– (3) *Vpass* to spread (out) | стлаться; разливаться *Vpass* → **jajıl**–

ǯak (3) *N Postp* side, flank; edge; bank | сторона; бок, берег; послелог в значении около, у, при → **jaχ, jak**

ǯak– (4), **ǯaγ– II** *V* to burn, to set on fire | жечь, зажигать → **jaχ–, jak**–

ǯaka *N* collar, edge | воротник, край

ǯaka → **jaka**

ǯakıl– (1) *V* to pour one's woes | жаловаться

ǯakın (2) *Adj/Adv* near (in place or time) | близкий; ближний → **ǯaγın, jakın**

ǯaklav (2) *N* brink/verge of sg; (here) doorpost | край, обочина; (здесь) косяк → **jakla**–

ǯal (3) *N* repose, rest, ease, relaxation | отдых, передышка, чуток отдохнуть

ǯala (1) *N* gossip, slander, calumny | клкевета, оговор, поклёп

ǯala– (16) *V* to lick with the tongue | лизать → **jala**–

ǯalan (2) *N/Adj* lie, falsehood, false | ложь, неправда; ложный, фальшивый; **ǯalan tůnja** the false world; the transitory life | ложный мир; тленный/непостоянный мир → **jalan**

ǯalanajak (1) *Adv* barefoot | босиком

ǯalda– (3) *V* to swim, float | плавать → **jalda**–

ǯaldat– (1) *Vfact* to float; to wave | заставить плавать (купаться), развевать что–л. → **jaldat**–

ǯalγa– (1) *V* to tie, connect; (here) to get tied to one another; to meet sb | прибавлять, соединять; привязывать; (здесь) связываться, встречаться (с кем–л.)

ǯalın (1) *N* mane, hair | грива, шерсть

ǯallat (1) *N Ar.* executioner, hangman | палач

ǯalma– (2) *V* to eat ravenously | жрать to eat ravenously | жрать; **ǯalmap ǯut–** to gulp down, to devour | проглатывать, жадно проглотить что–л. → **jalma–**

ǯalpak (2) *Adj* flat and wide; smooth, even | плоский → **jalpak**

ǯalpı (1) *Adj* broad; spreading out | широкий, плоский → **ǯajau ǯalpı**

ǯaltıra– (1) *V* to shine, to glitter | сверкать, блестеть → **jaldızla–, jıltıra–**

ǯalvar– (1) *V* to beg | умолять → **jalvar–**

ǯam (1) *N P.* glass, window pane | стекло

ǯaman (1) *Adj* bad, skinny | плохой, худой, скверный → **jaman**

ǯami (9) *N Ar.* mosque | мечеть → **ǯāmi**

ǯan I (78) *N P.* soul, life; dear, lovable | душа, жизнь; дорогой, любимый; **ǯan dust** intimate friend | душевный друг; **ǯanım** my dear | душа моя; **ǯanım kůzim** my dear one! apple of my eyes! | мой дорогой!, душа моя!; **(bek) ǯandan seu–/sůj–** to love with all one's heart | душевно любить, любить от всей души; **ǯanı sıχıl–, ǯanı sıkıl–** to be bord, to feel dull | скучать, томиться

ǯan II (2) *N/Postp* side | сторона, бок; (послелог) → **jan**

ǯan– (2) *V* to burn; to suffer; to complain (of sb/sg) | гореть, страдать, досадовать, жаловаться кому–л. на что–л. → **jan–**

ǯanabat (1) *N/Adj Ar.* gunk, pagan; cursed, damned, hateful | дрянь, погань, проклятый, ненавистный → **ǯenabet**

ǯanaj (43) *N* rogue, tramper, naughty (kid) | жулик, проказник, бродяга, шалун

ǯanǯık (1) *N dim.* little soul; (my) dearest; little darling | душечка (сокольник); миленький

ǯangız (7), **ǯanγız** (2) *Adj/Adv* alone, by oneself | один, одиноко → **jangız, jañgız**

ǯanγız s. **ǯangız**

ǯanij (2) *N* (my) dearest; little darling | мой милый, дорогой

ǯanık (2) *N* burned; suffering of love | ожог; страдающий от любви → **janık**

ǯanlı (1) *Adj* living, lively | живой, оживлённый

ǯap– I (2), **ǯav– II** *V* to do, to make; to build | делать, строить; **kjamet ǯap–** to dwel, to stay | обитать, жить, проживать

ǯap– II (2) *V* to cover (up/over), to wrap up | прикрывать; закрывать → **jap– II**

ǯapalaχ s. **ǯapalak**

ǯapalak (4), **ǯapalaχ** (3) *N* snowflake (s) | хлопья; **apalaχ–ǯapalaχ kar jav–** to snow/fell in large flakes/blotches? | хлопьями падает снег

Ǯaper (1) *Pn* Djaper (Djafer) | Джапер (Джафер)

ǯapma (1) *N* sheet of peat, plate of turf | торфяная плита

ǯar I (2) *N* coulee; steep slope (of mountain) | овраг, крутой берег

ǯar II (1) *V P.* darling, lover | дорогой, любовник → **jar**

ǯar– (1) *V* to split, rend | изрезывать, раскалывать → **jar–**

ǯara (1) *N* wound, injury | рана, ранение → **jara**

ǯara– (1) *V* to be befit, to be suited for, to be of use | годиться, быть годным, угодить → **jara**–

ǯaramazlık (1) *N* uselessness, naughtiness | негодность, шаловливость → **jaramazlık**

ǯaraštır– (2) *Vfact* to adorn, decorate; to honour (sb with sg) | украшать; почтить → **jaraštır**–

ǯardım (2) *N* help, aid | помощь; **kol ǯardım** helping hand (s) | помогающая рука → **jardım**

ǯarık (2) *Adj* bright | яркий; **ǯap–ǯarık** fully cracked | полностью разрезанный → **jarık**

ǯarıl– (1) *Vpass* to rend, tear, split (up) | расколоться

ǯarlı (28) *Adj/N* poor | бедный, неимущий

ǯarlı–pakir s. **pakir**

ǯartı (1) *Adj* old, dilapidated | старинный, устарелый, ветхий

ǯartı–ǯurtu (3) *Adj* old and tattered things, castoffs | старый и изношенный → **ǯartı**

ǯasa– (1) *V* to do, to make; to build, to arrange | делать, строить, устраивать, налаживать → **jasa**–

ǯaš (1) *Adj/N* young, youngman | молодой, молодой человек → **jaš I**

ǯaša– (2) *V* to live | жить → **jaša**–

ǯat– (22) *V* to lie down, to go to bed | лежать, ложиться → **jat**–

ǯatak *N* bed, couch, mattress; nest, den | постель, кровать; гнездо, стан → **jatak**

ǯatkız– *Vfact* to lay something down; to put to bed | уложить (в постель) → **jatkız**–

ǯau (2) *N* enemy | враг

ǯaulıχ s. **ǯaulık**

ǯaulık (6), **ǯaulıχ** (2) *N* napkin, head scarf, towel | платок → **ǯauluk, jaulıχ, jaulık, jauluk**

ǯauluk (2) *N* napkin, head scarf, towel | платок → **ǯaulık, jaulıχ, jaulık, jauluk**

ǯaur (1) *N* upper part of a person's back; back | спина

ǯav– **I** (6) *V* to rain, to snow | идти (о дожде, снеге) → **ǯaγ– I, jaγ–, jav**–

ǯav– **II** (1) *V* to do, make, build | делать, создавать, строить

ǯavab (15), **ǯevab** *N Ar.* answer | ответ; **azir–ǯavab kiši** quick at repartee, witty man | находчивый, смышлёный, сообразительный человек; **ǯavab ajt**– to answer | отвечать

ǯavaš (2) *Adj/Adv* slow(y) | медленный, медленно → **javaš**

ǯavun (4) *N* rain | дождь → **javun**

ǯavur (1) *N* upper part of a person's back, back | спина

ǯaz (2) *N* summer | лето → **jaz**

ǯāhil (2) *Adj Ar.* stupid person, uneducated | невежда, неграмотный

ǯāhillik (1) *N* ignorance, folly | невежественность, глупость

ǯāmi (16), **ǯami** *N Ar.* mosque | мечеть

ǯebelek (2) *N* fluff of (common) dandelion | пушок (одуванчика); **ǯel ǯebelek** with lightning speed; hurriedly; in great haste | (по)спешно

ǯedi seven | семь → **jedi, jidi**

ǯefā (1) *N Ar.* suffering, pain | мучение, страдание

ǯehennem (2) *N Ar.* hell, Gehenna | геена, ад

ǯejiz (1) *N Ar.* trousseau | приданое

ǯek– (5) *V* to hitch (a team), to harness, to put (the horses) to a carriage | запрягать, впрягать → **jek**–

ǯel (9) *N* wind | ветер; **ǯel ǯebelek** with lightning speed; hurriedly; in great haste | (по)спешно *N;* **Ǯel–ǯetmez** Djel–djetmez (the wind won't catch it), the name of the hero's horse | Джел–джетмез (ветер не догонит), кличка коня героя → **jel**

Ǯel–ǯetmez s. **ǯel**

ǯele s. **aǯele**

ǯelemnik (1) *N* pleasant, cooling breeze | приятный, прохладный ветерок

ǯellat (1) *N Ar.* executioner | палач; **ǯellat ettir–** to have sb executed | велеть/приказывать казнить кого–л.

ǯellen– (1) *Vpass* to break wind, poop | пускать ветры

ǯem (2) *N* fodder; grain | корм, фураж → **jem**

Ǯemajı (1) *Pn fem Ar.* Djemayi | Джемайи

ǯemāt (7), **ǯamāt** *N Ar.* company, crowd; people | общество, компания; люди, народ

ǯemiden (2) *N R.* suitcase, valise | чемодан

ǯemijet (5), **ǯemjet** (2) *N Ar.* gathering, assembly, society | община; общество

ǯemiš (2) *N* fruit, fruits; crop | фрукты; урожай → **jemiš**

ǯemjet s. **ǯemijet**

ǯenabet (1) *Adj Ar.* cursed, damned; disgusting | проклятый, ненавистный → **ǯanabat**

ǯenaze (1) *N Ar.* funeral | похороны

ǯenge (5) *N* the elder brother's wife, sister–in–law | жена старшего брата, невестка

ǯenk (9) *N Ar.* battle, combat; war | сражение, битва; война

ǯennet (22) *N Ar.* heaven, paradise | рай; **sekiz ǯennet kapısı** the door of eight heavens/paradise | дверь восьми небес/раев

ǯep (29) *N* pocket | карман

ǯer (37) *N* the earth; ground; place | земля, место → **ǯir, jer, jir**

ǯerbi (2) *N Ar.* a causing to come; call | привлечение, призыв

ǯes s. **ǯez I**

ǯešil (1) *Adj* green | зелёный → **ješil**

ǯet– (17) *V* to suffice, to reach sg | хватать, достигать (чего–л.), догонять; **bašına ǯet–** to cause harm/trouble | натворить зло (кому–л.); причинить зло (кому–л.); **Ǯel–ǯetmez** Djel–djetmez (horsename meaning 'wind won't catch it up') | Джел–джетмез (кличка коня со значением 'ветер его не догонит') → **jet–**

ǯeter (1) *Adj/Interj* enough | достаточный, достаточно → **jeter**

ǯevahir (2), **ǯevair** *N Ar.* jewel, gem; precious things | драгоценные камени; **ǯevahir dašı** precious stone | драгоценный камень

ǯevair s. **ǯevahir**

ǯeviz (3), **čeviz, ǯiviz** *N* walnut | грецкий орех

ǯez I (5) *N/Adj* bronze, brass | латунь, жёлтая медь; **firenkli jez** Europaean copper | европейская медь; **sarı ǯez** brass | жёлтая медь → **jez**

ǯez II (3) *N* trousseau | приданое; **ǯez altı** exhibition of marriage portion | выставка приданого невесты

ǯeza (5), **ǯezā** *N Ar.* punishment, fine | наказание, штраф; **ǯezalar et–** to punish | наказывать, оштрафовать

ǯezā s. **ǯeza**

ǯezbe s. **ǯezve**

ǯezve (2), **ǯezbe** *N Ar.* pot with long handle (for making Turkish coffee) | кофей-

ник (турка) с длинной ручкой (для приготовления кофе алатурка)
ǯėr (1) *N* song | песня, сказ → **jır**
ǯiber– (67), **ǯiver**– *V* to set free, to send | отпускать, посылать;
Vaux **urıp ǯiberir** quickly hits/beats | быстро/внезапно ударяет; **jazıp ǯibersin** let her quickly write | пустъ она быстро напишет → **jiber**–
ǯiger (2) *N* soul, heart | душа, сердце; **ǯigerimin kůšesi** my heart's delight | радость моего сердца
ǯigerle– (1) *V* to saddle | седлать (коня)
ǯiget (139), **ǯıget** (5), **ǯigėt** (1), **ǯigit** (7), **ǯigıt** (6) *N* fellow, guy, lad | парень, молодец → **jiget, jiket**
ǯigėt s. **ǯiget**
ǯigit s. **ǯiget**
ǯigıt s. **ǯiget**
ǯihan (4) *N P.* world, universe | мир, вселенная
ǯij– s. **ǯıj**–
ǯijištir– (1) *Vfact* to have sg collected, to have sg gathered (together) | велеть собираться, велеть скопляться?
ǯijlek (3) *N* raspberry | клубника
ǯil *N* year | год → **ǯıl, jıl**
ǯilan snake, serpent | змея → **jılan**
ǯilki s. **ǯılkı**
ǯilt–ǯılt (1) *Adv* sparkling; shining brightly | сверкая, очень ярко; искрясь; **ǯilt–ǯılt et**– to flicker | мигать
ǯilve (3) *N* grace, coquetry, coquettish airs | прелесть, очарование, кокетство; **ǯilve–naz** coquetry | кокетничание, кокетство; **ǯılve [ǯilve] sat**– to play the coquette | кокетничать
ǯilveli (1) *Adj* delightful, charming, coquettish | прелестный, очаровательный, кокетливый
ǯimiš (8) *N* fruit, fruits | фрукты
ǯin (19), **ǯins II** (4) **ǯıns** *N* djinn, genie, demon, evil spirit | джинн, дух, бес; **ins–misin, ǯins–misin?, ıns misiñ, ǯıns misiñ?** are you a human being or a djinn? | человек ли, демон ли ты? → **ins**
ǯinli (1) *Adj* haunted; possessed by a demon | заколдованный
ǯins I (1) *N* sort, type, kind | род, вид, сорт
ǯins II s. **ǯin**
ǯir (13) *N* the earth; ground; place | земля, место, окраина → **ǯer, jer, jir**
ǯiren (4), **ǯıjren** *Adj* sorrel, bay (horse), | рыжий, игреневый; **ǯiren kaška alaša** a bay horse with blaze on its forehead | игреневый конь с белой отметина на лбу; **ǯiren sač** red hair | рыжие волосы
ǯit– (1) *V* to suffice, to reach sg | хватать, достигать (чего–л.) → **ǯet**–, **jet**–
ǯivan (4), **ǯıvan** *N P.* young man, handsome young man | юноша; красивый (пригожий) парень
ǯiver– s. **ǯiber**–
ǯiviz s. **ǯeviz**
ǯıget s. **ǯiget**
ǯıgıl– s. **ǯıγıl**–
ǯıγıl– (6), **ǯıgıl**– *Vpass* to collapse, to be gathered; to collect or mass together | валиться, обрушиваться; собираться → **ǯıjıl**–, **jıkıl**–
ǯıj– (8), **ǯij**– *V* to collect in a heap, to pile up, to accumulate | собирать, наваливать, накапливать
ǯıjıl– (10) *Vpass* to collect or mass together; to be concentrated; to be gathered | скопляться, собираться
ǯıjım (1) *N* gathering; heap, pile | собрание, скопление, сбор

ǯıjın (6) *N* heap, pile, crowd | сборище, толпа

ǯık– (4) *V* to ruin | разрушать

ǯıkıl– (7) *Vpass* to collapse, to fall down | валиться, обрушиваться → **jıkıl–**

ǯıl (12) *N* year | год → **jıl**

ǯıla– (34) *V* to weep, cry; to complain; to grieve | плакать, горевать; **ǯılaj kal–** to be softly crying | поплакать, всхлипывать

ǯılama (9) *N* weeping, crying; lamentation | плач

ǯılan (30) *N* snake, serpent | змея → **jılan**

ǯılaš– (3) *V recipr* to weep together; to lamment together | плакать, рыдать (вместе с кем–л.)

ǯılat– (1) *Vfact* to cause sb to weep | доводить до слёз; заставить плакать

ǯılduz (1) *N* star | звезда → **jıldız, julduz**

ǯılγa (3) *N* valley, coulee, glen | долина, овраг, лощина

ǯılıš– (3) *V recipr* to gather together, to come together | собираться (все вместе)

ǯılkı (47), **ǯilkı, ǯılki** *N* horse; herd of brood mares and stallions with their colts loose in the fields | лошадь, табун → **jılkı**

ǯılkıǯi s. **ǯılkıǯı**

ǯılkıǯı (4), **ǯılkıǯi** *N* horseherd | табунщик → **jılχıǯ**

ǯıllı (2) *Adj* hot, heat | тёплый

ǯılpın–ǯılpın (1) *Adv* glimmering, flickering | мигая, мигающий, мерцающий

ǯımırtka (5) *N* egg | яйцо → **jımırta**

ǯımšaχ (2) *Adj* soft, mild | мягкий → **jımšak**

ǯın s. **ǯin**

ǯıns s. **ǯin**

ǯıput s. **ǯuput**

ǯırık (1) *N* torn, slit trench | трещина, щель, расщелина

ǯıvan s. **ǯivan**

ǯıvır– (2) *V* run | бежать → **ǯuvur–**

ǯıvuk (1) *Adj* near | близкий, близко

ǯıvukla– s. **ǯuvukla–**

ǯıvuru (1) *N* running | бег, пробег

ǯoj– (1) *V* to lose | терять, потерять

ǯojt– (1) *V* to lose | терять, потерять → **ǯujt–**

ǯojėl– s. **ǯojıl–**

ǯojıl– (9), **ǯojėl–, ǯojul–** (2) *Vpass* to disappear, to get lost | исчезать, теряться → **ǯujıl–**

ǯojuk (4) *N* lost | потерянный, пропавший

ǯojul– s. **ǯojıl–**

ǯok (14) *NegEcp* non–existent, not present; not to have; no | не имеется; нет → **jok**

ǯokı (1) *N* sleep, nap; sleepiness | сон, сонливость → **joku**

ǯokla– (4) *V* to sleep | спать → **jokla–**

ǯokluk (1) *N* absence, non–existence | отсутствие, недостаток

ǯol (12) *N* road | дорога; **ǯolγa čık–** to set out on a journey or a way | отправиться в путь → **jol**

ǯorga s. **ǯorγa**

ǯorγa (7), **ǯorga** *N* ambler, pacer | рысца, иноходец; **ǯorγa kız** nimble/lively girl | проворная/поворотливая девушка

ǯorok (1) *N* row, line; order | порядок, уклад

ǯuγur– (2) *V* to run | бежать

ǯuj– (7) *V* to lose, perish | терять, потерять, губнуть

ǯujdur– (1) *V* to annihilate, destroy | уничтожать, губить

ǯujıl– (4), **ǯujul–** *Vpass* to disappear, to get lost | исчезать, теряться → **ǯojıl–**

ǯujt– (1) *V* to lose | терять, потерять → **ǯojt**–
ǯujul– s. **ǯujıl**–
ǯukla– (13) *V* to sleep | спать → **jukla**–
ǯuku (2) *N* sleep, nap; sleepiness | сон, сонливость → **juku**
ǯul (6) *N* road, way | путь, дорога → **jul**
ǯum– (8) *V* to shot, to close (eye) | закрывать (глаза) → **jum**–
ǯuma (3) *N Ar.* friday | пятница
ǯumaertesi (2), **ǯumerte**, **ǯumertesı** *N* saturday | суббота
ǯumalak (1) *Adj* round, globular | круглый, шарообразный
ǯumertc s. **ǯumaertesi**
ǯumertesı s. **ǯumaertesi**
ǯumle (2) *Adj Ar.* all | весь, всё
ǯuput (3), **ǯıput** *En Ar.* Hebrew, Jew | еврей → **čufut, čuput**
ǯura– (2) *V* to interpret, to explain | относить (чему–л.); **tůš ǯura**– to explain/interpret a dream | разгадывать/растолковывать сон
ǯurt (15) *N* native country, home, habitation | родная страна, родина, дом → **jurt**
ǯurt– (2) *V* to trot | бежать трусцой; переть, переться
ǯuš– (1) *V* to enjoy; to be enthusiastic (about sg) | наслаждаться, разходиться, восторгаться
ǯušan (1) *N* wormwood of Judea (Artemisia of Judaica) | полынь
ǯut– (2) *V* to swallow, to gulp down | глотать, поглощать → **jut**–
ǯuv– (11), **ǯu**– (3), **ǯū**– *V* to (give a) wash | мыть, умывать (кого–л.); **čamašır/kır ǯuv**– to wash laundry | мыть бельё → **juv**–
ǯuva (3) *N* nest | гнездо → **juva**
ǯuvındır– (1) *Vfact* to have sb/sg washed | мыть, умывать (кого–л.) → **juvundır**–
ǯuvuk (7) *Adj* near, close | близкий, недалёкий
ǯuvukla– (3), **ǯıvukla**– *V* to draw near, to approach | приближаться
ǯuvun– (1) *V* to wash (oneself) | мыться, умываться, купаться → **juvun**–
ǯuvunduk (1) *N* laundry; dirty | бельё; грязь
ǯuvur– (7), **ǯuvar**–? *V* run | бежать → **ǯıvır**–
ǯū– s. **ǯuv**–
ǯůbbe (3) *N Ar.* robe (worn by imams, judges, barristers and professors with full sleeves and long skirts) | джуббе, мантия
ǯůgen (2) *N* rein, bridle | повод, узда, вожжи → **ǯügen**
ǯůjrůk (1) *Adj* (horse) that goes (runs) fast, fleet | быстроходный (о коне)
ǯůk (1) *N* load, burden | груз, вьюк → **ǯük**
ǯůmle (11), **ǯůmne** *Adj* all | весь, вся, всё; **ǯůmle alem** the whole world | весь мир
ǯůmne s. **ǯůmle**
ǯůr– (41), **čůr**– (1) *V* to go in any way or direction, to move; to go into service, to work somewhere | ходить, ехать; идти в прислуги, работать где–нибудь/у кого–нибудь → **ǯür**–, **jůr**–
ǯůrek (1) *N* heart | сердце → **jůrek**
ǯůriš s. **ǯůrůš**
ǯůrset– (2) *Vfact* to lead (sb by the hand) | водить (кого–л. за руку)
ǯůršen (1) *Adj/Adv* fast, swiftly, brisk | быстрый; резвый; быстро
ǯůrůš (2), **ǯůriš** *N* gait, way of walking | ходьба, походка, аллюр (вид походки лошади) → **jůrůš**
ǯůtkůr– (2) *V* to cough, to clear one's throat | кашлять, крякать; откашливаться
ǯůz (1) *N* face, cheek | лицо → **jůz I**

ǯůzůk (6) *N* ring | кольцо → **jůzůk, jüzük**

ǯügen (2) *N* rein, bridle | повод, узда, вожжи → **ǯůgen**

ǯük (1) *N* load, burden | груз, вьюк → **ǯůk**

ǯür– (20) to walk, to move; to shoulder; to go into service, to work somewhere/for sb | ходить; идти в прислуги, работать где–нибудь / у кого–нибудь; **ajdep ǯür–** to drive on (animals) by shouting | гнать, подгонять (животных криками); → **ǯůr–, jůr–**

ǯüz (1) *Num* a hundred | сто → **jůz II**

D

da (102) *Conj/Particle* too, also; and | тоже, же, также; и, да и → **ta**

dabır–dubur (2) *Onomatop* imitates the sounds of footsteps | подражает звукам шагов

dadaj (12) *N* elder sister | старшая сестра, тётя

daγ (6) *N* mountain; forest, wood | гора; лес → **tau**

daγıt– (1) *V* to scatter, disperse | разбрасывать, рассеивать, распускать → **taγıt–**

daha (20), **daa** (4), **dā** (8) *Adv* more, further; still | ещё, пока ещё; **bir/br dā** once more | ещё раз → **taha**

dajim (1) *Adv Ar.* always, lasting, continuously | постоянно, всегда

dajım (1) *Postp* time, turn; again, once more | (послелог) раз; **er dajım** always | всегда

dajın (32) *Postp* like | как, подобно; **balalari dajın** like their children | как его/её/их дети; **kaj dajın?** what sort? what kind of? | какой? какого рода?; **men dajın** like me | как я, подобно мне; **ne dajı?, nedajın?** what sort? | какого рода?; **ondajın** like him/her | как он/она, подобно ему/ей; **sen dajın** like you | как ты, подобно тебе; **undajın** like him/her | как он/она, подобно ему/ей → **nedajın, tajın**

dajre (1) *N Ar.* tambourine | бубен → **dāre**

dak– (1) to put on, to wear, to give (a name) | нацеплять; надевать; давать имя → **tak–**

dal (4) *N* branch, twig; willow | ветвь, ветка → **tal II**

dal– (1) *V* to dive, plunge (into) | окунаться, нырять; **jukuγa dal–** to fall asleep | погрузиться в сон → **tal– II**

daldır– (1) *Vfact* to submerge (in), to plunge, dip (into) | окунать, погружать

dalfes (3) hubcap, fez without a turban | колпак, турецкая феска → **talfes**

dallan– (3) *Vpass* to branch out | ветвиться, разветвляться, давать побеги

dam (5) roof, terrace; roof garden; house | крыша, кровля; дом → **tam I**

dam– (1) *V* to drip | капать → **tam–**

damsız (1) *Adj* tasteless, flavourless | не вкусный, безвкусный → **tamsız**

dane (9), **dāne** (2), **dene** (3) *N P.* grain, seed; piece; one's darling | штука (при счете); зерно, семя; дорогой/дорогая → **tane**

dangalak (2), **danγalaχ** *N* blockhead, stupid; butt, club | дубина

danγalaχ s. **dangalak**

dar–důnja (2), **dar–i dünja** *N Ar./P.* this (transitory, ephemeral) world | этот мир, тленный (бренный) мир → **důnja, dünja**

dara (2) *N* tare | тара

darab (1) *N Ar.* blow, hit, fight | удар, драка

daraɣač (2) *N* gallows | виселица

darı (1) *N* millet (Panicum) | просо

darıl– (3) *Vpass* to become cross, angry (with) | сердиться; бранить

daš (30) *N* stone | камень → **taš I**

dašlı (1) *Adj* stony, rocky | каменистый

dat (2) *N* taste, flavor | вкус, наслаждение; **dat–lezzet** taste–flavour | вкус–смак → **tat I, lezzet**

dava s. **dāva**

davaǯı (1) *N* (law) plaintiff, claimant | истец, жалобщик

davalı (1) *Adj* (law) defendant; (law) in disput | спорный, оспариваемый; скандальный, претенциозный

davam (1) *N Ar.* continuation | продолжение; **davam et–** to last, to go on | длиться, продолжаться

davıl s. **davul**

davılǯı (2) *N* drummer | барабанщик

davıš s. **davuš**

davran– (1) *V* to bestir oneself | действовать (каким–л. образом); поступать

davul (3), **davıl** *N Ar.* drum | давул, большой барабан

davuš (6), **davıš** *N* sound, voice | голос, звук

davušla– (1) *V* to say something; to sound | издавать звук, заговорить

dāne s. **dane**

dāneǯik (1) *N dim.* granule, little grain | штучка, зёрнышко

dāre (5) *N Ar.* tambourine | бубен, тамбурин; **kemane dāre čaldır–** to make sb play the violine and tambourine | велеть играть на скрипке и тамбурине → **dajre**

dāva (7), **dava** *N Ar.* lawsuit, process (at law) trial; quarel | (судебный) процесс, (судебное) дело, спор, тяжба; **dāva darab jok** no disputing and litigation | нет спора и тяжбы; **dāva kes–** to pass sentence/judg(e)ment on | присуждать, приговаривать; **dāva kůr–** to settle the case (of) | предъявлять иск, подавать жалобу

dāvalaš– (1) *V* to quarel, to wrangle; to litigate | ссориться, спорить, судиться

de (114) *Particle* too, also; and | тоже, же, также; и, то, да и → **te**

de– (809) *V* to say (to), to tell | говорить, сказать; **ne demek?** what does it mean? | что значит это? → **di–, dı–, te–, ti–**

deg– (1) *V* to touch, to reach | трогать, касаться

degil (2) *Neg Cop* (final) not, (isolated) no; (initial, or in anticipation of verb) not only, let alone | не (именное отрицание), нет; **degil–mi?** is not (so)? isn't it? .... did you? ..., didn't you? etc. | не так ли? → **tegil**

degin (1) *Postp* until | до (послелог)

degirmen (4), **dermen** *N* mill | мельница → **tegirmen**

dej– (3) *V* to touch, to hit, reach, attain | трогать, касаться

deje (4) *Postp* saying; thinking that, hoping that, fearing that | говоря что, думая что, полагая что; для того, чтобы; так называемый, по имени → **de–**

deju (63) *N P.* demon, devil; giant | див, злой дух, демон; гигант → **deu, diju**

dek (2) *Postp* until | до (послелог); **ajaɣadek** (from head) to foot | (с головы) до ног; **tizinedek** to his/her knees | до его/ее колена

deli (19) *Adj/N* fool, idiot | дурак

deli–divāne (1) *N* crazy (about love) | обезумевший (от любви); **deli–divāne bol–** to be crazy (about love), to go crazy (over love) | обезуметь (от любви)

delik (4) *N* hole, gap, orifice, opening | дыра, отверстие, брешь; **delik–tešik** full of holes | весь продырявленный, в дырах; изношенный, ветхий (об одежде и т. п.)

delikanlı (3) *N* youth, young man | парень, юноша, молодец

delir– (1) *V* to go mad, become insane | сойти с ума

dellal (1) town–crier, broker, middleman | глашатай, маклер, посредник → **tellal**

demek (2) *Particle* so, thus, therefore, in this case | значит

demet (1) *N* bunch, bouquet, bundle | связка, пучок, букет

demir (1) *N* iron | железо → **temir**

demirli (1) *Adj* containing iron; anchored | имеющий (содержащий) железо, (стоящий) на якоре; **Demirli Asan** Demirli Asan | Дэмирли Асан

dene s. **dane**

deniš– s. **deñiš–**

denk (2) *Adj* similar, equal, match; dear, pair | подобный, сходный; пара, партнёр

deñ– (1) *V* to folow pursue; to watch (for sb) | испытывать, исследовать → **deñe–**

deñe– (1) *V* to folow pursue; to watch (for sb) | испытывать, исследовать → **deñ–**

deñiš– (7), **deniš–** *V* to change, to be replaced; | меняться (товарами), изменяться; купить–продавать

deñiz (2), **deñız** *N* sea | море → **teñiz**

dere (24) *N* valley, stream, rivulet | горная долина, балка; ручей → **tere**

dereli (1) *Adj* valleyman, belonging to valley | обитатель (горной) долины; принадлежащий к долине

dere–tůpe (2), **tere–tůpe** *N* up hill and down dale | везде, где попало (буквально в гору и под гору); **dere–tůpe tůs ket–**, **tere–tůpe tům–tůz ket–** to keep on going quite smoothly/incessantly up hill and down dale | идти по дороге со спусками и подъёмами совершенно ровно → **dere, tůpe**

derja (9), **terja** (3) *N P.* a large body of water, river | большая полноводная река

derman (1) *N P.* remedy, medicine, cure | лекарство, лечебное средство

dermen s. **degirmen**

dert (~derd) (30) *N P.* pain, disease; trouble, sorrow, suffering | недуг, хвор; страдание; **derde tůš–** to be chagrined | быть огорчённым; **dertler ač–** to confine one's troubles (to) | рассказать о свём горе → **tert**

dertli (3) *Adj* pained, sorrowful, troubled, worried | страдающий, переживающий

derviš (9) *N P.* dervish | дервиш

destan (1) *N P.* epic, epic poem | дастан, эпический сказ; **destan bol–** to become object of a gossip, to win fame | становиться предметом толков; приобретать известность

deste (1) *N P.* bunch, bouquet, packet | связка, пачка, кипа, букет → **teste I**

destır (1) *N P.* permission, leave | позволение, разрешение

deu (21), **dev** *N P.* demon, devil; giant | див, демон, исполин, великан, гигант → **deju, diju**

dev s. **deu**

deve (4) *N* camel | верблюд → **tůje**

di– (1126), **dı–** (11) *V* to say, to tell; to mention, to think etc. | говорить, сказать; полагать, думать и тд. → **de–, te–, ti–**

dilber (13) *N/Adj P.* beautiful woman; beloved; heart–captivating; attractive | красавица, возлюбленная; покорительница сердца; красивый, привлекательный → **dủlber, tilber**

dilʒan (1) *N Fr.* phaeton | дилижанс; **dủrt atlı dilʒan** four–horse phaeton | четырёхконный дилижанс

dindin (2) *Onomatop* sound of a spindel | звук шпинделя; **dindin barıj, dindin kelij** (sounding) dindin is going, (sounding) dindin is coming (spindel) | (звуча) диндин отходит, (звуча) диндин приходит (это шпиндель) → **ủreke**

dinle– s. **diñle–**

dinne– s. **diñle–**

diñe– (6) *V* to listen to, to hear; to pay attention to, to obey | слушать, внимать; слушаться → **diñle–, tiñle–**

diñle– (5), **dinle–, dinne–** *V* to listen to, to hear; to pay attention to, to obey | слушать, внимать; слушаться → **diñe–, tiñle–**

dip (2), **dıp** *N* bottom, foot, lowest part | дно, низ; низшая часть (чего–л.) → **tip**

direk (3) *N* column, pillar | столб; колонна → **tirek**

divar s. **dıvar**

divāne (2) *Adj P.* foolish, insane, crazy | безумный, сумасшедший; **deli–divāne** crazy (about love) | обезумлён (от любви) → **deli**

dıjdırmaj s. **dujdurmaj**

dıvar (4), **divar** *N P.* wall | стена → **duvar**

dolaf (1) *N Ar.* cupboard, wardrobe | стенной шкаф; ниша → **tolap**

dolan– (2), **dulan–** *V* to wind (around a thing), to go around in a circle | обходить, ходить (вокруг чего) → **tolan–**

dolaš– (8), *V (recipr)* to go around, to wander | бродить, обходить, побывать всюду; **beli dolaš–** to swing or rock one's hip to and fro | качать талию → **dulaš–, tolaš–, tulaš–**

dolaštır– (1) *Vfact* to take out for a walk, to walk sb | водить (гулять), прогуливать (кого)

dolma (1) *N* cabbage roll | голубцы

domoz (1) *N* pig, swine | свинья → **tomoz**

dompaja (1) *Adv* goggle–eyed | с выпучивающимися глазами; **kủzleri dompaja** goggling his/her eyes | выпучивая его/её глаза → **opajsa**

dompal s. **romalama dompal**

don (1) *N* pair of drawers | кальсоны → **ton, tun**

dora (6) *Adj/Adv/Postp* straight(ly); directly; honest; truthful(y) | прямой, прямо; в направлении; правда, правдивый, честный → **dorı, doru, torı, toro, toru**

dorı (2), **dori** *N /Adv/Adj/Postp* straight(ly); direct(ly); honest; truthful(ly) | прямой, прямо; в направлении; правда, правдивый, честный → **dora, doru, torı, toro, toru**

doru (5) *N/Adv/Adj/Postp* true, right; toward, towards | верный, правильный; прямо; (послелог) по направлению к, мрямо к; к, под, около (о времени) → **dora, dorı, torı, toro, toru**

dost (2) *N P.* friend, friendly | друг → **dust**

dỏrdủnʒi (1) *OrdNum* fourth | четвёртый

dỏrt (4) **dort** *Num* four | четыре → **dủrt**

dỏrteu (1) *Num Adj* four | четверо

Drama (2) *Tn* Drama (river) | Драма (река)

dua (12), **duā, duha** *N Ar.* prayer | молитва, моление; **duha kıl–** to pray | молиться → **duva**

dubur s. **dabır–dubur**

dugıl (1) *Neg Cop* (final) not, (isolated) no; (initial, or in no anticipation of verb) not only, let alone | не (именное отрицание), нет → **tugıl, tügel**

duγra– (1) *V* to cut into slices and pieces | нарезать на мелкие куски → **tuγra–, toγra–**

duha s. **dua**

duj– I (6) *V* to hear, to learn (of); to feel | чуять, чувствовать → **tuj– I**

duj– II (1) *V* to become filled and satiated | насыщаться, наедаться → **toj– I, tuj– II**

dujdur– (2), **dıjdır–** *Vfact* to make sb feel sg, make sb conscious of sg | дать почувствовать кому–л. что–л.; извещать кого–л. о чём–л.

dujdurmaj (4), **dıjdırmaj, dủdủrmaj** *Adv* unnoticed | незаметно

dulab (3), **dulap** *N Ar.* cupboard, wardrobe | стенной шкаф; ниша → **dolaf**

dulaf s. **dolaf**

dulan– s. **dolan–**

dulap s. **dulab**

dulaš– (3), *V (recipr)* to go around, to wander | бродить, обходить, побывать всюду; **sačı beli dulaš–** to swing or rock one's hip and hair to and fro | качать талию и волосы → **dolaš–, tolaš–, tulaš–**

duman (6) *N* fog, mist | туман → **tuman**

dunja s. **dủnja**

durba (1) *N* bag, little sack | торба, небольшой мешок, сума → **torba**

durt– (1) to prod, to goad; to butt, to rub, to massage | бодать, чесать; массировать; **šamar durt–** to give a slap on the face or a box on the ear | дать пощёчину → **tủrt–**

dust (27) *N P.* friend | друг; **ajrılmaǯaχ dust** inseparable friend (s) | неразлучные друзья; **ǯan dust** intimate friend | душевный друг → **dost**

dustlık s. **dustluk**

dustluk (3), **dustlık** *N* friendship | дружба

dut (1) *N P.* mulberry *(Morus)* | шелковица

dut– (3) to hold, to hire, to engage | держать, хватать; ловить; нанимать ; **sủzin dut–** to keep one's word, to take the advice of, to obey | слушаться чего–л. совета, слушаться кого → **tot–, tut–**

duva (2)*N Ar.* prayer | молитва, моление → **dua**

duvaǯı (1) *N* one who prays | молящийся → **dua**

duvak (1) *N* vail | вуаль

duvak– (aγar–duvaγar?) s. **ak–**

duvar (8) *N* wall | стена

duvarǯı (1) *N* bricklayer, stonemason | каменщик

dủbủr–dủbủr (1) *Onomatop*; **dủbủr–dủbủr dủrteủ** four of them throbbing–rumbling | четверо стучая–громыхая

dủg– (1) *V* to hammer, forge | ковать, колотить; **temir dủg–** to beat iron | ковать железо

dủgme (1) *N* button | пуговица

dủgủn (10) *N* wedding feast; rich (meal) repast | свадьба; празднество; **dủgủn et–** to rejoice, to hold a wedding | справлять свадьбу → **tủgủn**

dủkan (2) **dükjan** *N Ar.* shop | лавка, магазин → **tủkan**

dủlber (11), **dülber** (2) *N/Adj P.* beautiful woman; beloved; heart–captivating; attractive | красавица, возлюбленная; покарительница сердца; красивый, привлекательный → **dilber, tůlber**

důlberlik (3) *N* beauty | красота → **dülberlik, tůlberlik**

důldůl (12) *N Ar.* fabulous horse with wings; magic steed | сказочный, крылатый конь

důn (3) *Adv* yesterday | вчера → **tůn**

důn– (4) *V* to circle, to turn round/back | вращаться, возвращаться (домой) → **tůn– I**

důnen (3) *N* foal at its age of 4–5 years | жеребенок по 4–5–ому году

důnja (17), **dunja** (7) *N Ar.* world, universe | мир, свет → **dünja, tůnja**

důnki (1) *Adj* yesterday's | вчерашний

důp (1) bottom, foot, stock | дно, низ → **tůp, tüp**

důrdane (1) *Adj P.* precious, valuable, high estimated | ценный, дорогой, высоко оцениваемый

důrdinʒi (3) *OrdNum* fourth | четвертый → **důrdůnʒi, důrtinʒi**

důrdůnʒi (1) *OrdNum* fourth | четвертый → **důrdinʒi, důrtinʒi**

důrt (důrd) (32), **durt, dürt** *Num* four | четыре → **dȯrt**

důrteu (1) *Num* four | четверо

důrtinʒi (5) *OrdNum* fourth | четвертый → **důrdinʒi, důrdůnʒi**

důrtlau (1) *N* making (it) to be fourfold? | становление четырёхкратным

důrtle– (1) *V* to make (it) to be fourfold? | делать (чего–л.) четырёхкратным

důš (1) *N* dream, vision | сон, сновидение → **tiš II, tůš, tüš**

důšman (7), **düšman, důšmen** (3) *N P.* enemy, foe | враг, неприятель → **tůšman**

důšmen s. **důšman**

dův– (3) *V* to beat; to thrash | колотить, бить → **düv**–

důzemnik (1) smoothness, flatness | глядь → **tůzemnik**

důzenlik (1) *N* order, peace | порядок, устройство

düfek rifle, gun, shotgun | ружьё, винтовка → **tifek, tůfek, tüfek**

dükjan s. **důkan**

dülber s. **důlber**

dünja (2) *N Ar.* world, universe | мир, свет; **dar–i dünja** this world, transitory world | этот мир, тленный (бренный) мир → **důnja, tůnja**

dürt s. **důrt**

düv– (1) *V* to hammer, forge | колотить, бить → **dův**–

E

e– (174), **ė**– *V* to be (substantive verb) | быть (вспомогательный глагол); **edi** was | был(а/о); **ediler** they were | они были; **eken** (it/there) was | был, было; **ekende** when (it/there) was | когда (это) было/происходило → **i**–

ebabil (3), **ababil** *N Ar.* goatsocker, whippoorwill *(Caprimulgus europaeus)* | горная ласточка, обыкновенный козодой

ebe (1) *N* old woman; midwife | старуха, повивальная бабка, акушерка

ebet (2) *Particle* yes, of course, certainly | да, конечно, разумеется, несомненно → **evet, hebet**

ečki (31) *N* goat | коза

ečkičik (1) *N dim* kid | козленочек

eʒel (12) *N Ar.* the appointed hour of death; death, fate, destiny | смертный час; смерть, судьба; **eʒel šerbetin ič**– to drink off the sherbet (sweet fruit drink) of deth | выпить шербет смерти; **eǯeline ǯet–** to cause one's death | причинять смерть кого–л.; **eǯelsiz ül–** to die prematurely | преждевременно умереть

edije (1) *N Ar.* gift, present | дар, подарок

efendi (12), **ependi** (3) *N Gr.* gentleman; master | господин; хозяин, владелец

efkjar (1) *N Ar.* sadness, sorrow, ennui, sad thoughts | грусть, тоска, грустные думы

Eftade (1) *Pn fem Ar.* Eftade | Эфтаде

Eftadeǯik (2) *Pn fem, dim* Eftadedjik | Эфтадеджик

eg– *V* to bend, to bow | склонить, нагнуть

eger I (2) *N* saddle | седло, лука; **eger kašı** (saddle–)bow | лука (седла) → **jiger**

eger II (57) *Conj* if, whether | если → **ejer**

egerek (2) *N* sheepfold, shady refuge for (sheep) flocks | овчарня

egerle– (1) *V* to saddle | седлать (коня)

egerlen– (1) *Vpass* to be saddled | быть осёдланным (о коне)

egil– (3), **ejgil–** *Vpass* to bend, to bow | наклоняться

eglen– (4) *Vpass* to have a good time, to enjoy oneself; to stay, to tarry (swhere) | развлекаться; задерживаться, задержаться (где–л.) на некоторое время → **ejlen– I, ilen–**

egme (2) *N* bending, bowing | склон, нагиб; приседание,

egri (1) *Adj* crooked, bent, curved | кривой; косой

eh *Interj* ah! | ах!

ej (60) *Interj* hi! hallo there! | эй! слушай! → **aj**

ejer (1) *Conj* if, whether | если → **eger II**

ejgil– s. **egil–**

eji (4) *Adj* good, well | хорошо, благой → **ijgi**

ejiǯe (1) *Adv* pretty good, rather well; thoroughly, completely, carefully | как следует, хорошенько, основательно

ejil– (3) *Vpass* to bend, to bow (down) | наклоняться

ejilik (2), **ejlik** *N* goodness, favor, charity, benefaction | довро; благо, благодеяние → **ijgilik, ijilik**

ejle– (8) *V* to make, to do | делать, совершать; **imdad ejle–** to help, to come to the rescue (of) | помогать, оказывать помощь → **ile–**

ejlen– I (4) *Vpass* to have a good time, to enjoy oneself | развлекаться, забавляться, весело проводить время → **eglen–, ilen–**

ejlen– II (2) *Vpass* to stay, to tarry (somewhere) | задерживаться где–нибудь

ejlen– III (1) *Vpass* to swing around, to turn round | поворачиваться, оборачиваться; **ejlenip bak–** to look round | оглядываться → **ajlan–**

ejlenǯe (2) *N* amusement, entertainment | развлечение; затея, забава

ejlendir– (1) *Vfact* to entertain | развлекать

ejlik s. **ejilik**

ejrılγan s. **erılγan**

ejse (1) *Conj* as for, if; however | если, хотя, же → **ajsa**

ejt– s. **ajt–**

ejvah (6) *Interj* alas! | увы!

Ejvaz (1) *Pn fem* Eyvaz | Эйваз

ek– (5) *V* to sow | сеять

ekeje (1) *N Ar.* story, tale | рассказ, сказка → **ikaje, ikeje**

ekeu (2) *Num/Adj* two, two of | двое, по двое → **ikeu**

eki (197), **ekė** (2), **ekı** (8) **ėki** (2) *Num* two | два; **kırk eki** forty–two | сорок два → **iki**

ekilau (1) *N* making two, (re)doubling | удвоение

ekim (6) *N Ar.* physician, doctor of medicine | врач

ekin (1) *N* crop, growing grain | посев; **ekin korıǯısi** ranger | полевой сторож

ekinči s. **ekinǯi**

ekinǯi (31), **ekinči** *OrdNum* second | второй; **ekinǯu/ekinǯi u̇j** second house (cook house) | второй/другой (нижний) дом, кухонный домик

ekindi (6) *N* time of the afternoon prayer | время (молитвы) между полуднем и заходном солнца

ekišer (3), **ekšer** *Num/Adj* two each, two at a time; in twos | по двое

ekjet (8) *N* (fairy) tale | сказка → **hekjet**

ekle– (1) (<**ekile**–) *V* to make (it) two | удваивать что–л.

eksil– (1) *Vpass* to decrease, to grow less | уменьшаться; убавляться

ekšer s. **ekišer**

ektir– (1) *Vfact* to bend, to bow | наклонять, сгибать; **mojın ektir**– to have sb submitted to, to have sb humiliated one'self (before) | подчинять кого–л.

el I (60) *N* hand | кисть руки; **el bas**– to swear | присягать, клясться; **elden ket–/kit**– to be lost, to pass out of ome's hand | уплыть из рук; пропасть; **eline al**– to take in one's hand | взять в (его/её) руки; **eline keč**– to be caught, to come into one's possession | попасть в руки, перейти в чью–л. собственность; **elinde kal**– fall into someone's hands | попасть в чьи–то руки

el II (13) *N* people, stranger, alien | народ, люди; другие, чужие люди; **el χızı** bride, woman, wife | невеста, женщина, супруга, жена

ela (7) *Adj* hazel (eyes) | карий; **ela kaš** hazel eyebrows | карие бровьи; **ela ku̇z** hazel eye (s) | карие глаза → **ala II**

elak (2) *N Ar.* destruction, death | гибель, смерть; **elak et**– to perish, to destroy, to kill | уничтожать, губить, убивать → **alak, halak**

elal (2), **alal** *N Ar.* canonically lawful, legitimate | дозволенный (разрешённый) шариатом, на счастье, законный, оправданный (религией); **alal bolson, elal bolsun** let it be given to him freely; let all claim be given up to him | да будет (ему) во благо, на счастье!; бог с ним, не жалко!; **elal et**– to give up a legitimate claim to another; not to begrudge (something given) | прощать что кому; освобождать кого от какого–л. обязательства (по своей доброй воле)

elallaš– (5), **alallaš**– (3) *V recipr* to take leave (of), to bid farewell (to) | проститься, попрощаться → **halallaš**–

elamdulilla (1) *N Ar.* Thank God! Glory be to God! | слава Аллаху! хвала Аллаху!

elālem (1) *N Ar.* all the world, (strange) people | все люди, чужие люди

elbet (9), **elbette** (3) *Particle Ar.* certainly, surely | несомненно, конечно, безусловно, разумеется

elbette s. **elbet**

elbise (1) *N Ar.* clothes, dress | верхняя одежда, платье

elči (2) *N* ambassador, envoy | посол; посредник

ele (1) *Adv* yet, still | всё ещё, до сих пор → **ale**

ele– I (4) *V* to sift, to sieve | сеять, просеивать

ele– II s. **ejle–**

ele–ele (1) *Particle* now then! well! at last! | ну и ну!, вот это да!, наконец–то

elek (5) *N* fine sieve | сито; **un elegi** sieve for flour | сито муки

elif (6) *N Ar.* name of the first letter of the Arabic alphabet | элиф (название первой буквы арабского алфавита)

elifbe (1) *N Ar.* (the Arabic) alphabet | азбука, (арабский) алфавит

elki (1) *Adv* former, earlier, above mentioned | предидущий, прежний; вышесказанный

elle– (1) *V* to handle, to touch with the hand | касаться руками, трогать, щупать

elli (15) *Num* fifty | пятьдесят

elli–belli (1) *Onomatop;* **elli–belli et–** to put/send to sleep by singing lullaby | усыпить ребёнка колыбельной песней

elmas (1) *N Ar.* diamond | алмаз, бриллиант

elvaǯı (1) *N* maker or seller of helva | приготовляющий или продающий халву → **helvaǯı**

em (40) *Conj* and also, as well as | и, также → **hem**

em– (7) *V* to suck | сосать

eman s. **aman II**

emanet (4) *N Ar.* a trust, anything entrusted to another for safekeeping | оставление на попечение; вещь, оставленная на хранение; **emanet et–** to deposit, to entrust | оставлять на попечение, сдавать на хранение → **amanet**

emček (2), **emčik** *N* teat, nipple | сосок, грудь (женская)

emčik s. **emček**

emǯe (14) *N* paternal uncle (familiarly used in addressing an older man) | дядя (тоже обращение к старшим мужчинам)

emdi (1) *Adv* now, at present, the present time | теперь, сейчас, только что

emen (7) *Adv P.* at once, instantly, right now | тотчас, тутже, сразу → **hemen, aman II**

Emine (24) *Pn fem Ar.* Emine | Эмине

emir ~ **emr** (39) *N Ar.* order, command | приказ; **Emir** Emir (male name) | Эмир (мужское имя); **emr ejle–, emr et–** to order, to command | приказывать

Emir s. **emir**

emız– (1) *V* to suckle, to give suck | кормить грудью

emr s. **emir**

emzirt– (1) *Vfact* to make sb suckle? | приказать кормить грудью

en I (1) *N* width, breadth | ширина

en II (12) *Particle* most | самый, больше всего, наиболее → **eñ**

en– (9) *V* to come down, to fall down, to descend | сходить; спускаться, опускаться → **in–**

ende s. **endi**

endečik (1) **[<endek–čik]** *N dim Ar.* (small) ditch, trench | канава, ров

endi (38), **ende** (13), **endė** (4), **ėndė** *Adv* now, at present, the present time | теперь, сейчас, только что → **inde. indi**

enekej (1), *N dim* (little) younger brother | младший брат → **eni**

engin (1) *Adj* open, wide, vast; (here) the open sea, offing | широкий; (здесь) открытое море, морские просторы

eni (2), **enne [ėnnė?]** *N* younger brother | младший брат → **ini**

eniš (6) *N* (downward) slope, incline | спуск, откос; **eniš eniš** entirely downhill | весь под уклон; **eniše tůš–** to go downhill | идти под уклон

enište (5) *N* sister's or aunt's husband | дядя (муж сестры матери)

enkas (1) *Adv Ar.* feigned, pretended | нарочно, намеренно, умышленно

enkastan (5), **enkazdan, enkasten** *Adv* feigned, pretended | нарочно, намеренно, умышленно

enkazdan s. **enkastan**

enne s. **eni**

entel– (1) *V* to rush upon, to hurl oneself (upon) | набрасываться на кого–л.

enten (1) *Adv* henceforth, from now on | с этого времени, отныне; **enten son** from now on | после этого, отныне

enteri s. **anteri**

entešeli s. **menteševli** *Onomatop*

eñ (6) *Particle* most | наи–; самый, больше всего, наиболее; **eñ bůjůk** the biggest; the eldest (brother) | самый большой; старейший → **en**

eñri–bůyri (1) *Adj/Adv* bent and crooked, contorted, twisted | искривлённый, кривой; криво и косо

ep I (10) *Adv* all, the whole, wholly | всё; всё ещё; **ep bir** all the same | всё равно; **epimiz/epmiz** all of us | все мы → **hep**

ep II (10) *N* way, means, device | путь, способ; **ebin(i) tap–** to find a way to do something | найти способ сделать что–нибудь

epdeš (1) *N* comrade, companion, friend | товарищ, друг; спутник → **ipdeš**

ependi s. **efendi**

eppi (1) (< ***epij**) *Adv* pretty well, a good many, a good deal of | довольно–таки много, порядочно

epsi (32) *Pron* all of it, all of them | все (они)

er I (29) *N* male; (brave) man, manly man; husband | муж; мужчина; **er adam** manly man | мужественный человек; **er bala** male child | (мужской) ребёнок; **erge ber–** to marry off (to) | выдать замуж

er II (42) *Adj P.* every, each | каждый, всякий; **er br** each, every single | каждый, всякий; **er dajım** always | всегда, постоянно; **er kim** whoever | всякий, кто бы ни; **er kůn** every day | каждый день, ежедневно; **erbiri** each one, every one (of) | каждый ис них, каждый в отдельности; **erbirisi** each one, every one (of) | каждый ис них, каждый в отдельности → **her**

er– (2) *V* to attain, reach | достигать, доходить, добираться → **ir–**

Erǯeb (1) *Pn Ar.* Erdjeb (Eredjeb) | Эрджеб (Эреджеб)

erek (3), **erik** *Adj/N* far, distant; distance, remotness | далёкий; **erekke** far away | далеко

eren (1) *N* saint man, patron saint | святой человек, покровитель

ereze (1) *N* hinge of two loops | дверной крюк, задвижка, запор

erif (10), **erıf** *N Ar.* fellow (always derogatory), lad; man | парень; тип, субъект; человек → **erip**

eriflik (2) *N* state of bravado/bravery, prowess | молодцеватость; мужество, бравада; **eriflik sat–** to pass oneself off as a lad | выдаваться за парня

erik I (2) *N* plum | слива, сливовый
erik II s. **erek**
erip (3) *N* fellow (always derogatory), lad;; man | парень; тип, субъект; человек → **erif**
eriš (5) *Adj* obstinate (person), pig–headed | непослушный
erišleš– (1) *V recipr* to attack each other; to make a fool of each other | разыгрывать друг друга
erištir– (1) *Vfact* to get (through) to, to forward (sg to sb) | помогать достигнуть
erıf s. **erif**
erılyan (3), **ejrılyan** *N P.* lilac (Syringa vulgaris) | сирень
erkek (3) *N* man, male | мужчина; самец; **erkek bala** male child | (мужской) ребёнок *N;* **erkek urbası** man's clothes, suit | мужское платье, мужской костюм
erkeklik (1) *N* masculinity, manliness, courage | мужество
erken (1) *Adv* early | рано
erkes (15) *Pron P.* everybody, everyone | все (люди); каждый, всякий → **herkes**
erkez s. **erkes**
erte (17) *Adv* early | рано; **pazar–erte** Monday | понедельник
erten (6) *Adv* next day, next morning | следующий день, следующее утро; **ertenge** the next, the following (morning, day etc.) | на следующее утро, на следующий день
ertesi (12), **ertesı** *Adv* the next, the following (day, week etc.) | следующий (о дне, неделе и тд7); **ertesi kůn, ertesı kůnı** next day | на следующий день
ertesı s. **ertesi**
Erzurum (3) *Tn* Erzurum (city) | (город) Эрзурум
es (21) *N* reason, mind(s), brain, sense, consciousness | сознание, рассудок; ум, разум; **esi kel**– to come to one's senses, suddenly realize sg, to sober down | прийти в себя, опомниться; взяться за ум, образумиться; **esi kit**– to go out of one's mind, to lose one's senses | ошеломиться, сбиться с толку; **esi bašına kel**– to come to one's senses, suddenly realize sg, to sober down; to think the better of it | прийти в себя, опомниться; взяться за ум, образумиться; **esi/esı bašından ket**– to slip one's mind, to lose one's senses; to weaken, to get tired | растеряться, не знать что делать; обессилеть, устать; **esini bašına topla**– to come to one's senses, to collect oneself; to gather one's wits together, to think and consider well | взяться за ум; опомниться; **eske tůš**– to come into one's mind, to be remembered | прийти на ум, вспомнить(ся)
es– (7) *V* to blow | веять
esap (1) *N Ar.* computation, calculation | исчисление, вычисление, счёт
esen (2) *Adj* healthy | здоровый; **sau esen** alive and in good health | жив и здоров
esenleš– (1) *V* to bid each other farewell | (по)прощаться
eserke– (1) *V* to underrate, undervalue; to hold sb of lower rank | недооценивать; не признавать чью–л. ценность
esir– (1) *V* to be drunk | стать пьяным → **isr**–
esirke– (1) *V* to drink | опьянеть

esirt– (1) *Vfact* to make sb drunk | поить допьяна

eskele s. **iskele**

eski (15) *Adj* old, ancient; former, worn out | старый, прошлый, выношенный

Eski Kırım (1) *Tn* Eski Kyrym (town) | Эски Кырым (город)

esle– (3) *V* to observe, notice | замечать; подметить

esli (2) *Adj* aged, advanced in years; staid, experienced | пожилой, опытный; **esli akıl** sharp wit | острый ум

Esma (1) *Pn fem Ar.* Essma | Эсма

esmeri (2) *Adj Ar.* dark skinned, swarthy; brown | смуглый, коричневый

essiz (2), **essız** *Adv/Adj* in a faint, unconscious | бессознательный

essizlen– (1) *Vpass* to (fall into a) swoon, faint | падать в обморок

essız s. **essiz**

Estergum s. **Estirgum**

Estirgum (3), **Estergum** *Tn Hung.* Estergom (Esztergom, a town in Hungary) | Эстергом (город в Венгрии) → **İstirgum**

eš (3) *N* one of a pair, mate, fellow; husband; wife | пара, партнёр, один из супругов; **eš–dust** friends and acquaintances | друзья и знакомые

ešeǯik (1) *N dim* burro, little donkey | ослик

ešek (4) *N* donkey | осёл

ešχ (1) *N Ar.* goodness, prosperity; love | благо, благополучие, любовь; **ešχ bolsın** well done! bravo! | браво, молодец → **ašk, ašχ**

ešil– (3) *Vpass* to crumble, moulder; to be scattered/dispersed | обвалиться, осыпаться

ešit– (29), **ešıt–** *V* to hear | слышать → **išit–**

ešitil– (1) *Vpass* to be heard | быть услышанным, сьышаться → **išitil–**

et (33) *N* meat | мясо

et– (365) *V* to do, to make | делать, совершать, производить; **ǯilt–ǯilt et–** to glimmer, to flicker | мигать; сверкать, брезжить(ся); **elāl et–** to give up a legitimate claim to another; not to begrudge (something given) | прощать что кому; освобождать кого от какого–л. обязательства (по своей доброй воле); **fedā et–** to sacrifice | пожертвовать, принести в жертву; **kauga et–**, **kauγa et–** to quarrel (with), to fight | ругаться, ссориться с кем; драться с кем; **kotkodaχ et–** to cluck, to cackle | кудахтать; **kůz et–** to set one's eye on sb/sg; to run an eye over | приглядывать; **naz et–** to feign reluctance | жеманиться, церемониться; **ne et–** [>n‿et– ~ net–] what to do | что делать; **pešman et–** to repent | заставить раскаяться, заставить кого пожалеть о том, чтоь **pikir et–** to occur to one's mind | размышлять, думать, обдумывать; **rahat et–** to be at ease; to make oneself comfortable, to rest | успокаиваться, отдыхать, наслаждаться покоем; **sabr et–** to be patient, to endure | терпеть, выносить, выдерживать; **salam et–** to send one's compliments or respects | приветствовать кого, передавать привет (поклон); **sejran et–** to go for a trip, to make an excursion | делать экскурсию, гулять; **šak et–** to flap (wings), to clap | щёлкать крыльями; **šap et–** to smack, to make smacking sound | чмокать; **šapır–šupur et–** to plash, to gurgle | плескаться, журчать; **šůkůr et–** to thank, to praise (God), to be thankful for | выражать благодарность, благодарить (Аллаха); **taχ–taχ et–** to knock repeatedly, to strike with noise | стучать, хлопать; **tın–tın et–** to ring, to resound, to peel | сильно звучать; **toj/tuj et–** to have a rich meal/repast; to feast | устраивать пир(шество); устраивать свадебный пир; **traš et–** to shave, to cut | бриться; **ůmůt et–** to hope, to expect; to be hopeful | надеяться, уповать; ждать → **it– II**

etek I (2) *N* skirt (of a dress); sole(s) | юбка; пола, подошва, подножие; **etek [tolısı] para** heaps/lots/loads of money | много/множество/навалом денег

etek II (1) *N* shoe | башмаки

etli I (2) *Adj* fleshy, meaty; cooked with pieces of meat | мясистый, мясной, с мясом

Etli II (2) *Pn* Etli (khan) | Этли (хан); **Etli kān** Etli kan | Этли кан (хан)

etmek (1) *N* bread | хлеб → **ůtmek**

etraf (2) *N/Postp Ar.* sides, surroundings | округа, окружение

ettir– (5) *Vfact* to have sg made | заставить кого–л. делать что–л.

euel s. **evel**

euelki (11) *Adj* the former; the (year, month, week) before last | прежний, старый, предыдущий → **eveliki, evelki, evvelki**

eulendir– (1) *Vfact* to marry off | женить, выдать замуж

evel (14), **euel** *Postp Ar.* earlier, before, first | раньше, прежде, до

eveligij s. **eveliki**

eveliki (2), **eveligij** *Adj* the former; the (year, month, week) before last | раньше, прежние времена (год, месяц, неделя) → **evelki, evvelki, euelki**

evelki (8) *Adj* the former; the (year, month, week) before last | прежний, старый, предыдущий, прежние времена (год, месяц, неделя) → **eveliki, evvelki, euelki**

evet (5) *Particle* yes, certainly | да, конечно → **ebet, hebet**

evir– (1) *V* to turn | крутить, вертеть → **čevir–**

evlad (4) *N Ar.* child, son; children | дитя, сын; дети

evladlik (1) *N* adopted child, foster child | приёмный ребёнок; **evladlik et–** to adopt a child | адоптировать, усыновлять

evlija (1) *N Ar.* (Muslim) saint; saintlike person | святой, угодник

evvel (1) *N Ar.* earlier, before, first | раньше, прежде, до

evvelden (1) *Adv* beforehand, previously | раньше, заблаговременно; прежде (до)

evvelki (1) *Adj* the former; the (year, month, week) before last | прежний, старый, предыдущий → **burungu, burunki, eveliki, evelki, euelki**

ez– (13) *V* to crush, to pound, bruise, to smash | давитъ, мять, разминать; **bayır ez–** to cause intense pain (to) | причинить большие страдания; **balak ez–** to beat trouser legs (with beetle) | сбивать штанины; колотить штанины вальком → **jez–**

ezan (3), **ezen** *N Ar.* (islam) a call to prayer; time of morning prayer | (рел.) азан, призыв муэдзина к молитве; время утренней молитвы; **ezan/**

ezen uku– to call to prayer ("azan") | прокричать (оглашать) призыв к молитве (4азан4)

ezdir– (1) *Vfact* to make sg crushed/smashed | велеть (заставить) давить, мять, разминать

ezel (2) *N Ar.* past eternity, time without beginning | вечность, давность; **ezelden** from eternity | с давних пор

ezen s. **ezan**

ezil– (1) *Vpass* to be crushed | быть раздавленным

eždeṙ (5) *N P.* dragon | дракон, змей → **eždeṙha**

eždeṙha (6) *N P.* dragon | дракон, змей → **eždeṙ**

ē–ē (2) *Interj* eh–eh | эй–эй

ėndė s. **endi**

F

Fadime (2) *Pn fem Ar.* Fadime (Fatime) | Фадиме (Фатиме) → **Fatime**

Fadmeǯik (4) *Pn fem, dim Ar.* (little) Fadime (Fatime) | (маленькая) Фадиме (Фатиме)

faχir (12) *N/Adj Ar.* poor, poor fellow | бедный, несчастный; бедняк → **fakir, pakir**

fajda (7), **fajde** *N Ar.* profit, use | польза, выгода

fajdasız (1) *Adj* useless, vain | бесполезный, невыгодный

fajde s. **fajda**

fajtın s. **fajton**

fajton (20), **fajtın** *N Fr.* phaeton | фаэтон

fajtonǯi (1) *N* driver of phaeton | извозчик

fakir (7), *N/Adj Ar.* poor, poor fellow | бедный, несчастный; бедняк → **faχir, pakir**

falan (6) *Adj/N/Pron Ar.* so and so, and so on; such–and–such; and such like, approximately | такой–то; имярек; и так далее; примерно → **felan, filan, pılan**

fallan– (3) *Vpass* to tell somone's fortune | предсказывать чью–л. судьбу

farfur (1) *N P.* porcelain | фарфор, фарфоровый

Fatime (1) *Pn fem Ar.* Fatime | Фатиме → **Fadime**

Fatma (2) *Pn fem Ar.* Fatma | Фатма

Fazile (7) *Pn fem Ar.* Fazile | Фазиле

fedā (3), **fejda, fida** *N Ar.* sacrifice | жертва, быть принесённым в жертву; **fejda/fida et–** to sacrify | пожертвовать, принести в жертву; **fedā bol–, fejda ol–** to be sacrified | быть принесённым в жертву

fejda s. **fedā**

felan (13) *N/Adj/Pron Ar.* so and so, and so on; such–and–such; and such like, approximately | некто, такой–то, имярек; и так далее; примерно; **felan jerge** to such–and–such a place | на такое–то место → **falan, filan**

felek (1) *N Ar.* fate, destiny | судьба, рок, фортуна

fenālik (1) *N Ar.* badness | зло, плохое

fener (5) *N Gr.* lantern, (street)lamp | фонарь; фары

ferde (3) *N P.* curtain | занавес, занавеска; **ferdege kir–** to go behind the curtain to spend the bridal night | заходить за занавес (чтобы проводить брачную ночь)

fereǯe (17) *N Ar.* dustcoat formerly worn by women when they went out | ферадже (верхняя женская накидка, одетая женщинами когда они вышли из дома)

ferik (2) *N* chick, chicken; (here) young woman | молодая курица, молодой петушок; (здесь) молодая женщина
ferište (2) *N P.* angel; (here) miracle worked through the agency of a saint | ангел; (здесь) чудо (совершённое святым)
fes (6), **pes** *N Ar.* fez (a felt headdress in the shape of a short cylindrical peakless hat) | феска → **dalfes**
fesligen (1) *N Gr.* sweet basil (*Ocimun basalicum*) | базилик огородный
fetjebil (1) *N P.* volunteer soldier | добровольческий солдат
fida s. **fedā**
fidan (5) *N Gr.* sapling, young tree | саженец, росток; **fidan bujlı** slender | стройный, статный → **pidan**
fikırlı (1) *Adj* clever, sharp–witted | умственный, разумный
fil (2) *N Ar.* elephant | слон
filan (5), **fılan** (3) *N/Adj/Pron Ar.* so and so, and so on; such–and–such; and such like, approximately | такой–то; имярек; и так далее; примерно → **falan, felan**
filǯan (6), **finǯān** *N* (coffee/tea) cup | чашка (кофейная/чайная)
filiz (2) *N Gr.* young shoot, bud, tendril | побег, росток; **filiz bujlı** slender | стройный/изящный как росток → **pilis**
finǯān s. **filǯan**
findık s. **fındık**
fino (1) *N It.* pet dog, lap dog | комнатная собака, болонка; **fino kŭpek** lap dog | комнатная собака
firak (1) *N Ar.* separation, grief at separation | расставание, разлука, горе от разлуки
firenk (2) *N P.* European; (here) syphilis | европеец, иностранец; (здесь) сифилис
firenkli (1) *Adj* European, foreigner | европеец, иностранец; **firenkli jez** european brass/copper/bronze | европейская медь/латунь
firılda– (1) *V* to spin around, to circle | крутиться, кружить
fitil (1) *N Ar.* wick | фитиль; шнур
fılan s. **filan**
fındıχ → **fındık**
fındık (12), **fındıχ** (3), **findık** *N Ar.* hazelnut, peanut | фундук → **fundik, funduk**
fırla– (1) *V* to fly out | вылетать, взлетать (о птице)
fırt (3) *Adv* **fırt fırt** continually (moving in and out) | непрерывно, постоянно, всё время (о движении)
fırtına (1) *N It.* storm, tempest | буря, шторм
fısılda– (1) *V* to whisper, sizzle | шипеть
fıstık (1) *N Ar.* pistachio nut | фисташка
fokišli (1) *Adj R.* magic, bewitching | волшебный, жонглёрский; чернокнижный
fokusčı (1) *N R.* magician, wizard, conjurer | фокусник
fukara (2) *Adj/N Ar.* poor, pauper | бедный, неимущий (человек) → **pukara**
fundik (3) *N Ar.* hazelnut, peanut | фундук → **fındık, funduk**
funduk (4) *N Ar.* hazelnut, peanut | фундук → **fındık, fundik**
furınǯı s. **furunǯi**
furun (6) *N Gr.* oven; bakery | печь → **purun**
furunǯi (2), **furınǯı** *N* baker | пекарь

G

gaχ (1) *Onomatop;* **gaχ et–** to belch, to eruct | срыгивать

gajb (7), **γajb** (2) *N Ar.* loss, lack, absence | потера; отсутствие; **gajb ol–** to be lost, to disappear from sight | быть потеранным, теряться, пропадать

gajet s. **gājet**

gajrı (4) *Postp Ar.* other than, besides, apart from | кроме, помимо

galava (3), **γalava** *N R.* head, chief; (here) (army) oficer | голова

galiba (1) *Particle Ar.* probably, presumably | вероятно, кажется, видимо, наверно

Galitsa (3) *Tn* Galitsa (historic region in Eastern Europe) | Галица (область в Восточной Евтопе) → **Galitsja**

Galitsaja s. **Galitsja**

Galitsja (2), **Galitsaja** *Tn* Galitsya (historic region in Eastern Europe) | Галиция (область в Восточной Евтопе) → **Galitsa**

gam (5) *N Ar.* grief; gloom | горе, печаль

garib (5) *Adj Ar.* poor, needy, miserable; stranger, person away from home | бедный, несчастный; живущий на чужбине

garla– (1) *Onomatop V* to bubble (water) | (звукоподражательный глаголь) (вода) бурчит

gastinča (1) *N R.* inn, hostelry | гостиница

gazel (1) *N Ar.* lyric poem of a certain pattern | газель – лирическоее или вакхическое стихотворение

Gazi s. **Gāzi**

gājet (2), **gajet** *Adv Ar.* very, extremely, greatly | очень, крайне, чрезвычайно, весьма

Gāzi (2), **Gazi** *Pn Ar.* Ghazi; ghazi – one who fights on behalf of Islam; hero | Гази (имя собственное); гази, борец (за веру), победитель (неверных); герой

geč– (6), **ječ–** *V* to pass by, to pass through, to pass (over) | проходить, переходить, переправляться, проникать, миновать → **keč–**

geče s. **geǯe**

geǯe (6), **geče** (2) *N* night | ночь → **keǯe, keče, kiǯe**

gel– (8) *V* to come (from, to), to arrive | прибывать, приходить → **kel–**

gemi (9), **jemi** (2) *N* ship, boat | корабль → **kemi**

genč (5), **jenǯ** *N/Adj* young (man) | молодой (человек); **Genč–Amed** Gench–Amed | Генч–Амед; **Jenǯ Osman** Yendj Osman | Йендж Осман → **kenč**

gene (21) *Adv* again, yet, but yet | снова, опять → **gine, kene**

geniš (1) *Adj* wide, large | широкий

gerčekten (2) *Adv* in fact, actually | действительно, по–настоящему → **kerčekten**

gerdan (2) *N P.* neck, throat; front of the neck, double chin | шея (спереди), двойной подбородок → **gerdane**

gerdane (1) *N P.* necklace, neckband | ожерелье, колье → **gerdan**

gerek (2) necessary, needed | нужно, необходимо, надо; нужный, необходимый → **kerek**

geri s. **keri** *Adv*

Gerič (7), **Geriǯ** (4) *Tn* Kerch (a town in the east of the Crimea) | Керч (город в восточном Крыму)

Geriǯ s. **Gerič**

getir– (1) *V* to bring (to) | приносить, привозить; приводить → **ketir–**

gez– (7) *V* to walk about, to stroll | гулять, прогуливаться → **kez–**

gezin– (1) *Vpass* to wander about without any definite purpose | прогуливаться → **kezin–**

gezme (6) *Inf(ma)/N* walk (ing), stroll, pleasure trip | гулянье, прогуливание

gibik (2) *Postp* like; as; (after predicate) almost, nearly | как, подобно, будто → **kibi, kibik, kůbůk**

gidi (3) *N* rascall, good–for–nothing, shameless | бесстыдник, мерзавец, подлец; **aj gidi saña** hey, you, rascall! | ах, ты, мерзавец!

gine (3) *Adv* again, yet, but yet | снова, опять → **gene, kene**

girman (4), **gırman** *En* German | герман

git– (9) *V* to go (away), to leave, depart | уходить; отправляться, идти → **ket–**

gizlen– (3) *Vpass* to hide oneself | скрываться, спрятаться

gizli (3), **gizlı** *Adj* hidden, secret | скрытый, тайный, секретный → **jizli, kizli**

gizlı s. **gizli**

gonǯe (1) *N P.* flower bud | бутон

gora (1) *N R.* mountain | гора; **Karpat gorası** Karpat mountains | Карпатские горы

gȯt (2) *N* bottom, backside, ars | зад, задница; **gȯtini je–** to lick one's ars | лизать задницу кому–л. → **kȯt**

gȯzgů (1) *N* mirror | зеркало

gradosluχ (1) *Adj R.* of (...) degrees | (...) градусный

guberna (1) *N R.* government | губерния

gubernatir s. **gubernatır**

gubernatır (4), **gubernatir** *N R.* governor | губернатор

guna (2) *N P.* sin, guilt | грех

gurbetlik (5), **kurbetlik** (2) *N Ar.* foreign travel, exile | скитание на чужбине

Gurgur (1) *Pn* Gurgur | Гургур

gurz I (6), **gůrs** *N P.* iron club, mace used in battle | железная палица, булава

gurz II (1) *En* Georgian | грузин; **Gurzın jašı** Georgian youth/youngsters | грузинские юноши

Gurzuf (1) *Tn* Gurzuf (or Hurzuf, a settlement in Yalta Municipality, Crimea) | Гурзуф (или Хурзуф, поселение в Ялтинском муниципалитете Крыма)

guvardije (1) *N R.* watchman | гвардия

guzel s. **gůzel**

gůččůk (2) *Adj* small, little, young | маленький, младший → **kůčůk**

gůdůrde– (4) *V* to thunder, rumble | греметь, громыхать; **kůk gůdůrde–** to thunder (sky) | гром (небо) гремет

gůgen (1) *N* blackthorn, hawthorn (Prunus spinosa), wild plum | тёрн (плод), терновый → **kůgen**

gůgům (4), **gůgůn** *N Gr.* copper jug with handle, spout and lid | кувшин (медный)

gůgůn s. **gůgům**

gůjůn (1) *N* heart, soul | сердце, душа

gůl (62) *N P.* rose, flower | роза, розовый, цветок; **gůl ač–/ačıl–** to bloom/open (roses/flowers) | распуститься (о розах/цветах) → **kůl II**

gůlbahča (3), **gůlbahče** *N* rose–garden, rosary | розарий → **gůl, bahča, bahče, gůlbakča, kůlbahče**

gůlbahče s. **gůlbahča**

gůlbakča (1) *N* rose–garden, rosary | розарий → **gůl, bakča, gůlbahča, kůlbahče**

gůlbešeker (1) *N P.* rose jam | засахаренные лепестки розы; розовое варенье
gůllů (1) *Adj* flowered, full of roses | украшенный розами/цветами → **allı–gůllů**
gůlšen (2) *N P.* rose garden | розарий
Gůlzar (1) *Pn fem P.* Gulzar (flower garden; rose garden) | Гюльзар (цветник; цветущая)
gůn (3) *N* day | день → **kůn**
gůneš (1) *N* sun | солнце → **kůneš, kůñeš**
gůr (1) *N P.* tomb, grave | могила
gůr– (1) *V* to see | видсть → **kőr–, kůr–, kür–**
gůrdůrde– (1) *V* to curse, swear; to make a noise | ругаться; шуметь; орать
gůrle– (1) *V* to thump; to beat; to thunder | гулеть, рокотать, греметь; **jůrek gůrle–** to throb/flutter (heart) | биться/колотиться сильно (сердце)
gůrs s. **gurz I**
gůrůlde– (1) *V* to coo | ворковать; гудеть → **kůrůlte–**
gůrůldi (1) *N* a humming, a buzzing noise; uproar | гул, рокот → **kůrůlti**
gůrůn– (1) *Vpass* to be seen; to look (as) | показаться, выглядеть → **kůrůn–**
gůrz s. **gurz I**
gůtir– (1) *V* to take away; to carry away | уносить; отвозить
gůzel (74), **guzel** (2) *Adj* beautiful; good, nice | красивый; → **güzel, kůzel**
gůzellik (7) *N* beauty | красота → **küzellik**
gůzerlik (2) *N* harmal (Peganum harmala) | гармала
güzel (4) *Adj* beautiful; good, nice | красивый; **dunja güzeli** a ravishing beauty | писаная красавица, королева красоты → **gůzel, kůzel**

Γ([1])

γajb s. **gajb**
γajrı (15), **gajri, γajri** (6) *Postp Ar.* other than, besides, apart from | кроме, помимо; **benden γajri** apart from me, besides me | кроме меня
γalava s. **galava**
γaliba (1) *Adv Ar.* probably, presumably | вероятно, кажется, видимо, наверно
γarib (8), **γarip** *Adj Ar.* poor, needy, miserable; stranger, person away from home | бедный, несчастный, живущий на чужбине
γarip s. **γarib**

H

ha (1) *Interj* well now; what about ...? oh, I see! oh, yes | ну–ка! да, кстати! → **hā**
habar s. **haber**
haber (4), **habar** *N Ar.* news, message | весть, известие → **χabar I, χaber, kabar, kaber I**
hač (4) *N Ar.* pilgrimage to Mecca | хадж (паломничесство в Мекку)
haǯi (14), **haǯı** (2) *N Ar.* one who has performed the rites of the pilgrimage at Mecca; pilgrim, hadji | хаджи (человек, совершивший паломничество в Мекку) → **aǯi, aǯı II**
haǯı s. **haǯi**
haǯlik (1) *N Ar.* pilgrimage (to Mecca) | паломничество (в Мекку)
hade s. **hajdı**
hafız (2), **hafuz** *N Ar.* prison | тюрма
hafta *N P.* week | неделя → **afta**
hafuz s. **hafız**
hajdamak (1) *N* cattle–rustler, marauder | разбойник → **ajdamak**

hajde s. **hajdı**

hajdı (4), **hade, hajde** *Interj* hurry up, now then, go on | ну, айда, давай → **ajda**

hajėrli s. **hajırlı**

hajır I (2), **hajr** *Particle* no | нет

hajır II (1) *N Ar.* goodness, boon, behoof, happiness | добро, благо, польза, выгода; счастье; **akšamlar hajır** good evening | добрый вечер

hajırlı (5), **hajerli (hajėrli)** *Adj* good, happy, lucky | добрый, счастливый; **hajırlı bolson** let sb be lucky/happy/successful; let's hope for the best! | пусть будет благополучным/удачным; дай бог, чтобы всё обошлось (наилучшим образом)! **hajırlı kůn** good/lucky day | добрый, счастливый день → **kajırlı**

hajr s. **hajır I**

hajret (1) *N Ar.* amazement, astonishment, surprise | удивление, изумление, недоумение

hajvan (3) *N Ar.* (domestic) animal | животное → **ajvan**

hal (25) *N Ar.* state, condition; fate | состояние, обстоятельства; **halıñ niǯe?** how do you feel? | как ты себя чувствуешь? какое у тебя здоровье? → **al II**

halak (1) *N Ar.* destruction, death | гибель, смерть; **halak et–** to perish, to destroy, to kill | уничтожать, губить, убивать → **alak, elak**

halallaš– (1) *V recipr* mutually to give or forgive all that has been or may have been unjustly taken or done; to take leave | прощать друг другу, освобождать друг друга (от какого–л. обязательства, долга и т. п.) прощаться

halk (11), **hāk** *N Ar.* folk, people, crowd | народ, люди; толпа → **χalk, kalk**

halka (1) *N Ar.* ring | кольцо; окружение → **alka**

hamam (19) *N Ar.* bath | баня → **amam**

haman (1) *Adv P.* at once, instantly, right now | тотчас → **aman II**

hamur (1) *N* dough | тесто → **kamır**

hana (4), **hane I** *IntPron/Adv* where? where is it? well? you know! | ну–ка, где → **kana**

hančer (1) *N Ar.* short curved dagger, khanjar | кинжал

hane I s. **hana**

hane II s. **hāne**

hanım (16) *N* lady, woman, wife | госпожа, дама, жена

haram (1) *Adj Ar.* forbidden by religion, unlawful, impure, unclean | нечистый, запретный → **aram**

Hasan (5) *Pn Ar.* Hasan | Хасан → **Asan**

hasil (1) *N Ar.* result, produce, gain | следствие, результат чего–л.; вывод, итог; **hasil ol–** to be attained, to happen, to occur, to come into existence | получаться (возникать, происходить, образовываться) в результате чего → **asıl II**

Hasim (12), **Hasım** *Pn Ar.* Hasim | Хасим → **Asim**

hasir (1) *N Ar.* rush mat | циповка, мат → **χasır**

Hasım s. **Hasim**

hasret (4) *N Ar.* a longing, yearning, homesickness | тоска, тоскующий; **hasret ol–** to long to see again sb/sg; to yearn (for) | тосковать, сильно кучать по кому–чему → **asret**

hasretlik (2) *N Ar.* longing, yearning, craving for sb/sg | тоска (по кому или чему–л.); **hasretlik ček–** to long to see again (person or thing), to suffer from the lack of sb/sg | тосковать, томиться, сильно кучать по кому–чему → **asretlik**

hasta (2) *N P.* ill, sick | больной → **χasta, kasta**

hastalan– (1) *Vpass* to become ill | болеть, заболеть, захворать

hašer (1) *N Ar.* Day of Resurrection, Judgement Day | день воскресения из мертвых, судный день

hata (1) *N Ar.* mistake, error, sin | ошибка, погрешность

hatın (28) *N* lady, woman, wife | женщина, жена → **hatun, χatın, χatun, kadın, katın**

hatta (1) *Particle Ar.* even, so much, so that | даже, и при этом, вдобавок

hatun (13) *N* lady, woman, wife | женщина, жена → **hatın, χatın, χatun, kadın, katın**

hava (1) *N Ar.* air; weather; musical tune, melody | воздух; погода; мелодия, напев → **ava**

havuz (2) *N Ar.* artificial basin; pond | бассейн, пруд → **avuz II**

hazirlan– (1) *Vpass* to prepare, to get ready | готовитъся → **azirlan–**

hazır (3) *Adj Ar.* ready, prepared | готовый, приготовленный → **azır**

hazırla– (4) *V* to prepare, make ready | готовить, приготовлять → **azırla–**

hazne (5) *N Ar.* treasure, treasury | казна; сокровищница

hā (1) *Interj* oh, I see! oh, yes | ну–ка! да, кстати!→ **ha**

hāk s. **halk**

hālā (3), **hala** *Adv Ar.* yet, still | всё ещё; пока ещё

hāne (3), **hane II** *N P.* house | дом; **kajve–hāne** coffee shop | кофейня*;* **mafuz–/māfuz–hane** prison | тюрма

hebet (2) *Particle* yes, of course, certainly | да, конечно, разумеется, несомненно → **ebet, evet**

heč s. **hič**

hekjet (1) *N Ar.* (fairy) tale | сказка → **ekjet**

helvaǯı (1) *N* maker or seller of helva | приготовляющий или продающий халву → **elvaǯı**

hem (19) *Conj* both ... and, and also, as well as | ещё также, тоже, и ... и → **em**

hemen (9) *Adv P.* at once, instantly | тотчас, немедленно → **aman II, emen**

hep (15) *N/Adv P.* all, the whole; always | всё, всё ещё

hepsi (3), **hepsı** *Pron* all of it, all of them | всё, все они

her (18) *Adv P.* every, each | каждый, всякий → **er II**

herif (1) *N Ar.* fellow (always derogatory), lad; man, male, husband | человек, парень; тип, субъект → **erif**

herkes (2) *Pron P.* everybody, everyone | все (люди); каждый, всякий → **erkes**

heva (2) *N Ar.* love | любовь; **hevalar bolsun** let be happiness/love | пусть будет счастье/любовь

hič (9), **heč, hiš** *Adv P.* nothing; (in negative sentence) no, none whatever, none at all, never | ничего; (в отрицательных предложениях употребляется для усиления отрицания) нисколько, совершенно, совсем, вовсе → **ič II**

hisap (1) *N Ar.* account; consideration, supposition | вычисление, счёт; предположение, расчёт; **hisap et–** to suppose, to assume | предпологать, считать

hisse (1) *N Ar.* lesson | урок, назидание
hiš s. **hič**
Hizir (6) *Pn Ar.* Khizir (legendary person who attained immortality by drinking from the water of life) | Хызыр (легендарное лицо, которое стало бессмертным выпив из воды вечной жизни) → **Kidir–nebi**
hizmet (2) *N Ar.* service, duty; work | служение, прислуживание; работа → **kizmet**
hizmetči (9), **hizmetčı** *N* servant | слуга, работник → **kizmetči**
hizmetčı s. **hizmetči**
hınzır s. **χınzır**
hırsız (4), **hirsiz, hırsiz** *N* thief | вор → **χırsız, kırsız**
hırsızlık (1) *N* theft | воровство → **χırsızlık, kırsızlık**
hoǯa (15) *N P.* teacher; hodja (Muslim teacher); husband | учитель, (мусульманский) преподаватель; муж → **χoǯa, oǯa**
hopla– (1) *V* to caress with the hand, to stroke | ласкать → **upla–**
horan (1) *N* hedge, hedgerow, railing | ограда, изгород
horas s. **horaz**
horaz (7), **horas** *N P.* cock, rooster | петух
hoš (2) *Adj P.* pleasant, nice; good | хороший, приятный → **oš**
hošbeš (1) *N P.* greeting | приветствие
hu (1) *Interj/Pron Ar.* hey, there!; (mystical orders) greeting formula; He (i. e. Allah) | эй! послушай–ка! (возглас, которым подзывают); мусульманское приветствие, которым обмениваются дервиши; Он (Аллах); **jā–hū!** Hey, see here! Look at me! (expresses impatience); O, God! | эй, ты! послушай! ну, ну и что; господи!
huda s. **χuda**
Hurasan (1) *Tn P.* Khorassan | Хорасан
hůkům (1) *N Ar.* adjudgement, sentence, decision, decree | приговор, решение; распоряжение; **hůkům et–** to rule, to dominate; to sentence, to judge | властвовать, господствовать; выносить решение, присуждать
hůrmet (1) *N Ar.* respect | уважение; **hůrmet et–** to respect, to honor | чтить, почитать, уважать → **ůrmet**

X (χ)

χabar I (56), **χaber** (2) *N Ar.* news, message | весть, новость; сообщение; **χabar ber–, χabar bir–** to announce (to), to inform (someone of something) | сообщить, информировать, довести до сведения; **χabar et–** to send news, to send a message | дать знать, сообщить, в известность поставить → **haber, kabar, kaber I**
χaber s. **χabar I**
χač– (2) *V* to run away, to escape | бежать, убежать → **kač–**
χadar (6) *Postp* until, till, to | до, столько; **bo χadar, bu χadar** this much | такое–что количество, столько → **kadar**
χadir (~ **χadri**) (1) *N Ar.* worth, value; price, cost | ценность, цена → **kadir I**
χafas (1) *N Ar.* kage, coop | клетка → **kafas, kafes**
χajalı (1) *Adj* rocky (place) | скалистое (место) → **kajalı**
χajır (7), **χajr** (2) *N Ar.* goodness, boon, behoof, happiness | добро, благо, польза, выгода; счастье; **akšam**

χajır ajt– to say good evening | сказать добрый вечер; **χajır et–** to do good to, to do sb a good turn | совершать благое (доброе) дело, делать добро кому; **χajırγa bolson** let sb be lucky/happy/successful; let's hope for the best! | пусть будет благополучным/удачным; дай бог, чтобы всё обошлось (наилучшим образом)! **χajr bol–** to be of use, to be lucky, to have a good turn | быть полезным/благополучным; **χajrın kůr–** to enjoy the advantage or profits (of) | видеть добро (пользу) от кого–чего; извлекать пользу (выгоду) → **hajır II, kajır**

χajırlaš– (1) *V recipr* to agree (on sg) | соглашаться

χajırsız (1) *Adj* good for nothing, useless; unlucky (fellow) | не делающий добра; злополучный

χajnata (1) *N* father–in–law | тесть, свёкор → **kajnata**

χajr s. **χajır**

χajt– (4) *V* to turn back, to return | возвращаться → **kajt–**

χal– (8) *V* to remain, stay; as an auxiliary verb expressing perfect or continuous acting | оставаться, останавливаться; в роли вспомогательного глагола указывает на совершенность или длительность действия → **kal–**

χalk (5) *N Ar.* folk, people, crowd | народ, люди; толпа → **halk, kalk**

χan (5) *N* khan, sovereign, ruler | хан, правитель → **kān**

χanet (1) *N* wing | крыло → **kanat**

χanım (26) *N* lady, wife, mistress (of a household) | госпожа (дома), жена, дама

χannan (1) *IntPron* from where | откуда

χantar (1) *N Ar.* steelyard for weighing | весы

χapısta (1) *N R.* cabbage | капуста → **kapısta**

χapo (2), **χapu** *N* door | дверь → **kapı, kapo, kapu**

χapu s. **χapo**

χar (1) *N* snow | снег → **kar**

χar‿biber s. **karabiber**

χara (2) *Adj* black | чёрный; **χar‿biber (<χarabiber)** black pepper | чёрный перец → **kara**

χara– (1) *V* to look, watch | глядеть, смотреть; видеть → **kara–**

χaranfil (1) *N Ar.* garden pink, carnation (*Dianthus plumarius*) | гвоздика → **karanfil, karanpil**

χarar (1) *N* habit, wont; character; decision, determination | привычка, характер; решение → **karar**

χardaš (2) *N* brother or sister; (in addressing) (my) friend, my dear | брат или сестра; (при обращении) мой друг, милый мой → **kardaš**

χarγa (4) *N* crow *(Corvus)* | ворона, грач → **karγa**

χarılγač (1) *N* swallow | ласточка → **karılγač**

χarši (1) *N/Postp* opposite, opposite side or direction; (as a postposition) against, contrary (to), towards, a little before | то, что находится напротив; (в функции послелога) против, напротив, перед, к → **karšı**

χart (1) *N/Adj* oldman, greybeard | старый, старик → **kart**

χas (9) *Adj Ar.* special, particular, private | присущий, свойственный

χasab (1) *N Ar.* butcher | мясник → **kasab**

χasavetlen– (1) *Vpass Ar.* to be anxious (about), to become sorrowful (at, over) | беспокоиться; печалиться о чём

χasır (3) *N Ar.* rush mat, matting | циновка → **hasır**

χasta (6) *N P.* sick, ill | больной *N P.* → **hasta, kasta**

χastalan– (1) *Vpass* to become ill | болеть, заболеть

χastalık (1) *N* sickness, illness | болезнь, заболевание

χatı (4) *Adj* hard, stiff | твёрдый, жёсткий → **katı**

χatın (97) *N* lady, woman, wife | женщина, дама, госпожа; жена → **hatın, hatun, χatun**

χatınlık (3) *N* womanhood; ladyship; (here) capacity of wife, as a wife | женский пол; женщины; (здесь) качество жены; **χatınlıkka al–** to marry, to wed | брать в жёны кого–л.

χatır (10) *N Ar.* respect, memory; thought, one's feelings | уважение, память; **χal χatır surıš–** to inquire after each other's health; to inquire after one another | расспрашивать друг друга (о жизни, об остоятельствах); **χatır kal–** to feel hurt, or offended, to take offense (at) | обижаться, огорчаться, сердиться на кого; **χatırı/ χatrı χošla–** to please (sb) | уважить кого, доставлять удовольствие, радость кому–л.; **χatırı tůš–** to lose interest in sg, not to be in the mood any more | отпадать охота у кого; **χatırından čık–** to forget, to slip one's memory | вылететь (выскочиыь) из головы, ускользнуть из памяти; **χatırında kal–** to be remembered | сохраниться в памяти, запоминаться кому,ь **χatırı kal–** to feel hurt, or offended, to take offense (at) | обижаться, огорчаться, сердиться на кого → **katır**

χatırǯık (1) *N dim;* **χatırǯıγı kal–** to feel hurt or offended, to take offense (at) | обижаться, огорчаться, сердиться на кого → **χatır, katır**

χatun (16) *N* lady, woman, wife | женщина, дама, госпожа; жена → **hatın, hatun, χatın**

χavun (1) *N* melon (*Cucumis melo*) | дыня → **kavun**

χazan– (1) *V* to win, to gain, to earn | зарабатывать, выигрывать, завоевать → **kazan–**

χazıl– (1) *Vpass* to be dug, to be excavated | рыться, пассив от копать → **kazıl–**

χijar s. **χıjar**

χizmet (3) *N Ar.* service, work | служение, работа → **hizmet, χızmet, kizmet**

χıbırdan– (1) *Vpass* to keep on moving or quivering of itself | шевелиться, приходить в движение → **kıbırdan–**

χıčır– (3) *V* to howl, roar; to cry (out), shout | кричать, выкрикивать → **kıčır–**

χıčıtkan (1) *N* (stinging) nettle (*Urtica urens*) | крапива (жгучая); **χıčıtkan ot** stinging grass | жгучая трава, крапива → **kıǯıtkan**

χıdir– s. **χıdır–**

χıdır– (14), **χıdir–, χıdr–** *V* to seek, to look (for), to search | искать, разыскать → **kıdır–**

χıdr– s. **χıdır–**

χıj– (1) *V* to perform (a ceremony) | осуществлять, справлять; **nikah χıj–** to perform the ceremony of marriage; to marry | заключать брак, вступать в брак → **kıj– II**

χıjar (6), **χijar** *N* cucumber (*Cucumis sativus*) | огурец; **χijardaj** like a cucumber | как огурец; **langa–**

χıjarı a kind of cucumber; slicing cucumber? | сорт огурца

χılıč s. **χılıč** (2)

χılıǯ s. **χılıč**

χınzir s. **χınzır**

χınzır (3), **hınzır, χınzir** *N Ar.* pig | свинья; боров → **čočka, domoz**

χır (1) *N* countryside, uncultivated land | поле, степь, луг → **kır I**

χırmıska (1) *N* ant | муравей → **kırmıska**

χırmızı (2) *Adj.* red | красный → **kırmızı**

χırsız (3), **hırsız** *N* thief | вор, жулик → **kırsız**

χırsızla– (1) *V* to steal | воровать → **χırsla–, kırsla–**

χırsızlık (1) *N* theft | воровство → **hırsızlık, kırsızlık**

χırsla– (1) *V* to steal | воровать → **χırsızla–, kırsla–**

χısmet (1) *N Ar.* chance, destiny, luck, fortune | удача, судьба, участь, счастье → **kismet**

χız (15) *N* girl, daughter | девочка, девушка; дочь, дочка → **kız**

χızdır– (1) *V* to light, to set light to | зажигать → **kızdır–**

χoǯa (1) *N* teacher; hodja (Muslim teacher); husband | учитель, (мусульманский) преподаватель; муж → **oǯa, hoǯa**

χoχla– (1) *V* to smell | нюхать → **kukla–**

χoj– (4) *V* to put, place, set | класть, положить → **χuj–, koj–, kuj– I**

χolumla– (1) *V* to foal | жеребиться → **kolumla–, kolunla–**

χomsar (1) *N R.* officer | комиссар

χonuš– (1) *V* to speak | говорить → **konuš–**

χopar– (2) *V* to break off | отломить, обломать → **kopar–**

χora (2) *N* (wattle) fence; | плетень; изгородь; → **kora**

χoraz (1) *N P* cock | петух → **koraz**

χorχ– (2) *V* to be afraid, to fear | бояться, пугаться → **korχ–, kork–, kurχ–, kurk–**

χorloχ (1) *N* meanness, anguish, torment | мука, мучение, страдание; **χorloχ ček**– to suffer (tortures), to struggle | страдать, мучиться

χoš (10) *Adj/Adv* pleasant, agreeable, good | приятный; привлекательный; **χoš keldige/keldikke kel**– to come to welcome sb | выходить встречать/приветствовать коро–л. (напр. гостей); выходить чтобы сказать добро пожаловать; **χoš safa kel**– welcome! | добро пожаловать! → **hoš, koš III, kuš II**

χošla– (1) *V* to like, to delight, to rejoice | любить, уважать; радовать, веселить

χubare (1) *N* dome, cupola | свод, купол

χuda (5) *N* matchmaker; marriage broker; father–in–law | сват; тесть, свёкор → **kuda**

χudaj (1) *N P.* God; My Lord! | Бог; Боже мой! → **kudaj**

χudalıχ (1) *N* proposal (of marriage), matchmaker's activity/work | сватовство; действие/дело свата → **kudalık**

χuj– (1) *V* to put, place, set | класть, положить → **χoj–, koj–, kuj– I**

χujın (3) *N* bosom | пазуха → **kujın**

χum (1) *N* sand | песок → **kum I**

χunšu (1) *N* neighbor | сосед, соседний → **koñšu, kuñšu**

χupar– *V* to break off | отломить, обломать → **χopar–**

χurma (1) *N P.* date | финик, финиковый

χutur– (1) *V* to go mad; to be infected with rabies | беситься → **kutur–**

İ (i)

i! (1) *Interj* hi! hallo there! | эй! ну! гей!
i– (246) *V/Vaux* (predicative verb) to be | быть (вспомогательный глагол); **idė, idi** he/she/it was | был/а/о; **idiler** they were | они были; **iken** he/she/it was | был/а/о; **ise** if, although; however; as for | а, если, хотя, же
İbrahim (2) *Pn Ar.* Ibrahim | Ибрахим
ibret (1) *N Ar.* example, lesson | поучительный пример, урок
ič I (190) *N* the inside of anything; interior | внутренная часть; **ičinde** in (it), inside (it), within (it) | внутри чего–л.; **ičinden** from within, from inside | из, от чего–л., изнутри; **ičine** into, inwards | вовнутрь
ič II (35) *N/Particle* nothing; (in negative sentence) no, none whatever, none at all, never | ничего; (в отрицательных предложениях употребляется для усиления отрицания) нисколько, совершенно, совсем, вовсе, ни за что; **ič br** nothing, nothing at all | ничего, совершенно ничего; **ič kimse** nobody, no one | никто, ни один человек; **ič kılınmaz** there is nothing to be done, there is nothing you can do | ничего невозможно сделать → **hič**
ič– (85), **iš–** (3) *V* to drink; smoke | пить; курить; **tůtůn ič–/iš–** to smoke (tobacco) | курить
ičeǯek (3) *N* beverage, drink | напиток; питьё
ičer– (2) *Vfact* to give to drink, to water | поить
ičeri (54) *N* inside, inner part | внутрь; **ičerge, ıčerige** to the inside | вовнутрь
ičil– (4) *Vpass* to be drunk | быть выпитым; питься
ičin (44), **ičın (5), išın, (–čin (2), –čın (4))** *Postp* for, on account of; by reason of; in order to | для, ради, из–за; чтобы → **ičün**
ičir– (1) *Vfact* to give to drink | поить
ičın s. **ičin**
ički (1) *N* (alcoholic) drink | (спиртный) напиток
ičme (7) *Inf(ma)* drinking | питьё
ičtir– (1) *Vfact* to make sb drink sg | поить
ičůn s. **ičün**
ičün (1), **(~–čun, ~–čůn (3), ~–čün (4))** *Postp* for, on account of; by reason of; in order to | для, ради, из–за; чтобы → **ičin**
iǯare (1) *N Ar.* rent | арендная плата
iǯra (1) *N Ar.* performance | исполнение
İdıl s. **Ak–İdıl**
iger (1) *N* saddle | седло, лука → **eger I, jiger**
igėt s. **jiget**
ijbe (3) *N* saddle–bag | перемётная сумка, кошель, котомка
ijde (2), **ıjde** *N* oleaster, wild olive *(Elaeagnus angustifolia)* | лох, джида
ije (1) *N* master, possessor, owner, patron | хозяин, владелец, обладатель
ijgi (2) *Adj* good, well | хороший, благой → **eji**
ijgilik (2) *N* goodness, kindness; good side | добро, доброта → **ejilik**
ijlik s. **ijilik**
ijilik (~ijlik) (1) *N* goodness, kindness; good side | добро, доброта → **ejilik, ijgilik**
ijne (2) *N* needle, pin | игла

ikaje (2) *N Ar.* (short) story; story–telling; tale | рассказ, повествование; сказка → **ekeje, ikeje**

ike (ikė) (20) *Num* two | два → **eki**

ikeje (8) *N Ar.* (short) story; story–telling; tale | рассказ, сказка; повествование; → **ekeje, ikaje**

iken s. **i–**

ikešti (4), **ikišti** *V* was; were | был(а), было, были

ikeu (1) *Num* two, two of | двое, по двое → **ekeu**

iki s. **eki**

ikišti s. **ikešti**

il– (3) *V* to hang up; to suspend | вешать (одежду)

ilaǯ (3), **laǯ** (2) *N Ar.* medicine, drug | лекарство; **laǯ et–** to medicate, to cure | лечить

ilaǯla– (4) *V* to medicate, cure | лечить

İlah (1) *N Ar.* God, Allah | бог, Алла, Рабби; **İlahim!** My God! | Боже мой! → **Allah**

İlāhi (1) *Interj Ar.* Oh God! My god! | обращение к богу; **jā–ılāhi** Oh, my God! | боже мой! ах, боже мой!

ile (~–le) (9) *Postp* with, together with; and | с, вместе с, и, с помощью

ile– (1) *V* to make, to do | делать, совершать → **ejle–**

ilen– (2) *V* to have a good time, to enjoy oneself | развлекаться, забавляться, весело проводить время → **ejlen– I**

ilevli (1) *Adj* hanging, suspended | висячий, подвешенный

ilim (3) *N Ar.* science, knowledge | наука, знания

ille (6) *Adv* without fail, by all means, particularly | непременно, обязательно

ilmon (2) *N R.* lemon | лимон

ilvan (3) *N Ar.* decoration, adornment; grace; jewelry | укращение

imam (2) *N Ar.* imam, prayer leader | имам (духовный глава мусульман)

iman (1) *N Ar.* faith, belief | вера

imdad (2) *N Ar.* help, aid | помощь; содействие; **imdad eyle–** to help sb | помогать кому–л.

İmdal (2) *Tn* Imdal (river) | Имдал (река) → **İndal?**

imdi (1) *Adv* now, at present | теперь, сейчас → **inde, indi**

imkjan (1) *N Ar.* possibility | возможность

imtam (<imtan) (1) *N Ar.* examination, test, trial | экзамен

in– (5) *V* to come down, to descend | сходить; спускаться, опускаться

inan– (4) *V* to believe | верить, поверить → **ınan–**

inǯe (24) *Adj* thin, fine | тонкий, изящный; **inǯe belli** slim waisted | девушка с тонкой талией; **inǯe–maraz** tuberculosis? | туберкулёз?

inǯi (5) *N* pearl | жемчуг; бисер

İnǯil (2) *N Ar.* the New Testament | Евангелие

inǯir (1) *N P.* fig | инжир, винная ягода

inǯit– *V* to hurt, to injure | причинять боль, обижать

İndal (1) *Tn* Indal (river?) | Индал (река?) → **İmdal?**

inde (9) *Adv* now, at present; the present time | теперь, в это время → **imdi, indi**

inde– (4) *V* to call out (to sb), to say sg | крикнуть, позвать

indi (3) *Adv* now, at present; the present time | теперь, в это время → **imdi, inde**

ine (3), **ıne** *N* needle, pin | игла, иголка → **ijne**

inge (2) *N* a woman's sister–in–law | жена старшего брата, невеста

ini (6), **ınnı** (3) *N* younger brother | младший брат → **eni**

inne– (1) *V* to sound, to call out | подать голос

ins, **ıns** *N Ar.* human (being), a member of human race | род человеческий, человек; **ins misin, ǯins misin?; ıns misiñ, ǯıns misiñ?** are you a human being or a djinn? | человек ли, демон ли ты?

insan (16) *N Ar.* human being, mankind | человек

inšallah (3) *Interj Ar.* if Allah (God) wills; I hope so; all being well | если будет угодно Аллаху; дай–то бод; надеюсь, что

ip (7) *N* string, rope; cord | шнур, верёвка → **jip**

ipdeš (2) *N* fellow taveler; comrade, companion, friend | спутник, товарищ

ipek (1) *N* silk | шёлк → **jipek, jůfek**

iplik (1) *N* thread, yarn, sewing cotton | нитка, нить

ir– (7) *V* to reach | достигнуть; **muratına ir**– to attain one's desire | добиться цели, достигнуть желаемого → **er**–

irgat (11), **irɣat** *N Gr.* farm hand, farm–labourer, servant | батрак; **irgat jůr** to serve/work as a day–laborer/farm hand/servant | работать батраком, служить слугой у кого–л. → **ırgat, ırɣat**

irɣat s. **irgat**

iri– (2) *V* to melt | таять → **jiri**–

iršat (1) *N Ar.* directing, guiding, teaching the right way | наставление на истинный путь, указание правильного пути; **iršat–name** book of moral instructions, code, law–book | книга (моральных) наставлений

isar (2) *N Ar.* stone–wall | забор (из камня), каменный забор

ise s. **i**–

İsfahan (3) *Tn* Isfahan (town) | (город) Исфахан

isim (1) *N Ar.* name | имя

isirtkič *N* drink; beverage(s) | напиток, питьё

iskadron (1) *N R.* cavalry regiment | эскадрон

iskele (8), **eskele, sikele** *N It.* landing place, harbour, port | пристань, причал, мол; **iskele bašı** entrance to the harbour | вход в порт

islam (1) *N Ar.* Islam; Muslim | ислам

ispan (2) *N* sleeping compartment? | (небольшое) спальное помещение?

isporlašma (1) *N R* dispute; quarelling, fray | спор, ссора, брань

isr– **(isru)** *V* to get drunk | стать пьяным; пьянеть; **isrėp jür– (isrėp jürmek)** to be in a state of drunkenness | пьянствовать → **esir**–

isrek *Adj* drunk(en) | пьяный, хмельной; **isrek baš** drunken (person) | пьяный (человек)

İstambol s. **İstambul**

İstambul (8), **İstambol** (3), **Stambul** *Tn* Istanbul (city) | (город) Стамбул

istarica (1) *N R.* senior (in rank), head, chief, leader | старица

iste– (119) *V* to want, to wish for, to ask for | хотеть, желать, просить; **kız iste**– to make a formal proposal of marriage | сватать(ся), просить руки (девушки)ь **murad iste**– to purpose?, to wish determinedly | желать, хотеть

istet– (1) *Vfact* to make sb ask (sb) to/for, to send for sb | заставить просить;

попросить/велеть позвать (к себе) кого–л.

istikam s. **istikan**

istikan (2), **[~istikam!]** *N R.* glass | стакан; **istikam** my glass | мой стакан; **istikañ** your glass | твой стакан;

İstirgum (1) *Tn Hung.* Estergom (Esztergom, a Hungarian town) | Эстергом (венгерский город) → **Estirgum**

iš I (129), **ıš** (2) *N* work; service; business, job | работа, дело, занятие; **iš kür–** to work, to do a job | работать, заниматься делом; **iše bak–** to attend to business, to occupy oneself with work | заниматься (своим) делом; **iše čabala–** to do one's best at work | прилагать усилия в работе; **iše kit–** to go to work | идти на работу

iš II s. **ič II**

iš– s. **ič–**

išχola (2) *N R.* school | школа

išit– (19) *V* to hear | слышать → **ešit–**

išitil– (2) *V* to be heard | быть услышанным, сьышаться → **ešitil–**

išın s. **ičin**

išle– (23) *V* to work, to labor, to perform, to do | работать, делать; **nakıš/nañıš išle–** to embroider | вышивать; **nešle– [=ne išle–?]** what to do | что делать?; **nešlijim [=ne išlijim?]** what shall I do? | что мне делать?; **nišle– [=ne išle–?]** what to do | что делать?; **nišlijik? [=ne išlijik?]** what shall we do? | что нам делать?

išlet– (3) *Vfact* to make sb work, (here) to make sb embroider (with sg) | заставить работать; (здесь) расшивать, вышивать

išlettir– (1) *Vfact* to make sb work, (here) to make sb embroider (with sg) | заставить работать, (здесь) заставить вышивать

išmar (1) *N Arm.* sign, nod, wink | знак (жест, подмигивание); **išmar et–** to make signs; to wink | делать знаки (подмигивать)

ištan (13), **ıštan** *N R.* baggy trousers; shalwar | брюки, штаны

ište (42), **šte** (28) *Particle* look; here; there; now; | вот, итак, такь именно, как раз

it (7) *N* dog | собака

it– I (3) *V* to push | толкать, отталкивать

it– II (11) *V* to do, to make | делать, совершать, производить; **aff it–** to forgive | извинять → **et–**

iz (4) *N* trace, track | след; **iz et–** to leave traces (behind) | оставлять следы → **jiz**

izgara (1) *N Gr.* gridiron; grate | решётка; рашпер; жаровня

izin (22) *N* permission, consent, permit | разрешение, позволение; **izin al–** to obtain permission | получить разрешение, получить согласие; **izin ber–** to permit, to grant permission | разрешать, дать разрешение

izle– (4) *V* to track by the footprints; to trace; to follow after | выслеживать, преследовать

İzmir (1) *Tn* Izmir (a city, Turkey) | Измир (турецкий город)

İzmit *Tn* Izmit (a town, Turkey) | Измит (турецкий город)

I (ı)

ıč I s. **ič I**

ıč II s. **ič II**

ıjde s. **ijde**

ınan– (6) *V* to believe | верить, поверить → **inan**–

ınnı s. **ini**

ıns s. **ins**

ıraχ (1) *Adj* far off (in time or space), distant, remote | далёкий, отдалённый

ırgat s. **ırγat**

ırgatlık (1), *N* slavery, servitude | рабство, доля прислуги

ırγat (28), **ırgat** (17) *N Gr.* farm hand, farm–labourer, servant | батрак, слуга; **ırgat jůr**–/**ǯür**– to serve/work as a day–laborer/farm hand/servant | работать батраком, служить слугой у кого–л.; **ırgat tut**– to engage a farm hand/servant etc. | нанимать батрака/слугу и тд. → **irgat**

ırık (1) **(~irık)** *N* uneven, rough, rugged (surface, place) | ухабистый

ırılda– (2) *V* to snarl, to snore continously, to growl | рычать

ırmaχ (2) *N* river | река

ısıǯaχ (1) *Adj* hot, heat | горячий, жаркий

ıslan– (1) *Vpass* to get wet, to be wetted | промокнуть

ıslat– (1) *Vfact* to wet, to soak (in water) | мочить

ıšıklık (2) *N* courtyard | двор

J

ja (14) either, or; oh! | а, или, либо, разве; ой, ох → **jā**

jaban (2) *N/Adj P.* foreign land/country; wild, bleak land | чужой; дикая пустынная местность

jafraχ s. **jafrak**

jafrak (4) *N* leaf | лист → **japrak**

jaγ (7) *N* fat, oil | жир, масло, сало

jaγ– (6) *V* to rain, to snow | идти (о дожде, снеге) → **jau**–, **jav**–, **ǯaγ–I**, **ǯav– I**

jaγdır– (1) *V* to sprinkle, spread, to pour (out/forth) | засыпать, забрасывать

jaγla– (6), **jaχla**– *V* to oil, to grease | намазывать масло, смазывать маслом

jaγlan– (1) *Vpass* to be oiled, to be greased | смаызваться

jaγlık (1) *N* napkin, handkerchief | платок → **ǯaulıχ**, **ǯauluk**, **jaγlık**, **jaulıχ**, **jauluk**

jaγmır s. **jaγmur**

jaγmur (12), **jaγmır** *N* rain | дождь; **jaγmur aldı** caught him/her the rain | его/её захватил дождь → **ǯaγmır**

jaχ (3) *N/Postp* side, flank; edge; bank | сторона; бок, берег?; послелог в значении около, у, при → **ǯak**, **jak**

jaχ– (8) *V* to burn, to set on fire | жечь, зажигать → **ǯak**–, **jak**–

jaχla– s. **jaγla**–

jaχši (35), **jaχšė**, **jahši** *Adj* good, well; pretty, allright, excellent; agreeable, | хорошо, ладно, согласен; хороший, благой; **jaχši et**– to improve, to cure, to restore (sb) to health | исправлять; вылечить; → **jaχšı**, **jakši**, **jakšı**

jaχšı (42), **jahšı** *Adj* good, well; pretty, allright, excellent; agreeable, | хорошо, ладно, согласен; хороший, благой; **jaχšı etėp** well, thoroughly | хорошо, основательно → **jaχši**, **jakši**, **jakšı**

jaχšılık (1) *N* goodness, kind(li)ness, charity | добро, благо, благородство

jaχtır– (1) *Vfact* to have sg set on fire, to have sg burnt | приказывать сжечь, велеть зажигать → **jaktır**–

jaj (2) *N* bow (of an archer), bow for playing a stringed instrument | лук, смычок; **jaj ček**– pull the bow | натянуть лук,

jaj– (13) *V* to spread, to scatter | расстилать, раскатывать → **ǯaj–**
jajan (4) *Adv* on foot | пешком → **ǯajan, ǯajau**
jajıl– (1) *Vpass* to spread (out) | стлаться; разливаться → **ǯajıl–**
jajla (11) *N* high plateau, summer camping ground, pasture | яйла (летнее горное пастбище); нагорье; плоскогорье
jajma (1) *N* something spread out as covering; cover, blanket | подстилка; **jajma katın** corpulent/portly woman | жирная/развесистая женщина
jak (84) *N Postp* side, flank; edge; bank | сторона; бок, берег; послелог в значении около, у, при; **jaγına** to his/her side | к его/её стороне; **jaγından** from his/her side | от (кого–чего–л.) стороны; **karšı jak** opposite shor | противолежащий берег; **teñizniñ jaγı** seaside, sea board | морской берег; **u jakka bo/bů jakka** here and there | туда и сюда → **ǯak, jaχ**
jak– (27) *V* to set on fire; to light, to burn | жечь, зажигать → **ǯak–, jaχ–**
jaka (6) *N* collar, edge | воротник, край → **ǯaka**
jakın (17) *Adj/Adv* near (in place or time) | близкий; ближний → **ǯakın, ǯaγın**
jakınla– (1) *V* to draw near, to approach | приближаться, подходить
jakınlaš– (2) *V recipr* to draw near, to approach | приближаться, подходить
jakıš– (1) *V recipr* to be suitable or becoming, to be proper | подходить, быть к лицу
jakla– (1) *V* to border, to edge sg | стоять/находиться на краю; стоять/находиться по сторонам чего–л.
jaklaš– (1) *V recipr* to draw near, to approach | приближаться, сближаться
jakši (1) *Adj* good, well; pretty, allright, excellent; agreeable, | хорошо, ладно, согласен; хороший, благой → **jaχšı, jaχši, jakšı**
jakšı (2) *Adj* good, well; pretty, allright, excellent; agreeable, | хорошо, ладно, согласен; хороший, благой → **jaχšı, jaχši, jakši**
jaktır– (2) *Vfact* to have sg set on fire, to have sg burnt | приказывать сжечь, велеть зажигать → **jaχtır–**
jala– (8) *V* to lick with the tongue | лизать → **ǯala–**
jalan (7) *N/Adj* lie, falsehood; false | ложь, неправда, враньё; лживый, ложный; **jalan sůjle–** to lie | лгать; **jalan tůnja** the false world; the transitory life | ложный мир; тленный/непостоянный мир → **ǯalan**
jalan–ajak (1) *Adv* barefoot | босиком
jalda– (2) *V* to swim | плавать; **jaldaj–jaldaj** swimming | плывя → **ǯalda–**
jaldat– (3) *Vfact* to make sb/sg swim; to wave; to float | заставить плавать, развевать что–л. → **ǯaldat–**
jaldız (1) *N* star | звезда
jaldızla– (1) *V* to gild; to make sg shine/glitter, to beautify | золотить; придавать внешний блеск; украшать
jalγız (1) *Adv* alone, by oneself | один, одиноко → **jalınγız**
jalı (13) *N* bank, shore | берег; **jalı jalı (kez–)** (to walk) along the bank/shore | (гулять) вдоль берега
Jalı–bujı (6) *Tn* Yali–boyu (southern coastline of the Crimea) | Яли–бою (южное побережье Крыма)

jalınyız (2) *Adv/Adj* alone, by oneself | один, одиноко → **jalγız**

jalkar– (1) *V* to implore; to beg piteously or whiningly | умолять, просить

jalma– (1) *V* to eat ravenously | жрать; **jalmap jut–** to gulp down, to devour | проглатывать, жадно проглотить что–л. → **ǯalma–**

jalpaχ s. **jalpak**

jalpak (2), **jalpaχ** *Adj* flat and wide | плоский → **ǯalpak**

Jalta (6) *Tn* Yalta (town) | (город) Ялта

jalvar– (18) *V* to beg, to implore | умолять, просить, уговаривать → **ǯalvar–**

jaman (11) *Adj* bad, violent, cruel, strong | плохой, худой, скверный → **ǯaman**

jamanla– (2), **jamanna–** *V* to disparage, to malign, to besmirch | хаять, охаивать, чернить, порочить, хулить

jamanna– s. **jamanla–**

jamau (1) *N* patch | заплата, латка

jamaulı (1) *Adj* patched | латанный

jan (97) *N/Postp* side, flank | бок; послелог в значении около, у, при; **bir jana** toward, aside, on one side | в староу, на бок; **jana–jana** side by side | около друг–друга; **janıma** to my side | ко мне, к моей стороне; **janına** to his/her side | к его/её стороне; **janında** at his/her side; by his/her side | у его/её стороны → **ǯan II**

jan– (50) *V* to burn; to feel grieved or sorry for; to suffer, to be angry | гореть, страдать, досадовать, жаловать; **ateše jan–** to be on fire; to suffer a loss; to go through a great misfortune | гореть в огне; попасть в беду; нести утрату/потери → **ǯan–**

jan–kesiǯi (21), **jan–keseǯi** (12), **jan–keseǯı, jan–kesıǯi** *N* pickpocket | вор–карманник

jana (6) *Adj* new | новый

janadan s. **jañadan**

janaχ s. **janak**

janak (37), **janaχ** (8) *N* cheek, face | щека

janaklı (3) *Adj* with ... cheek, having ... face | имеющий какие–л. щёки, с какими–л. щеками

janaš– (8) *V* to draw near, approach | приближаться

janaša (2) *Adj* adjoining, near, close | рядом, по соседству

janǯu (1) *N* coachman, carman | ямщик

jandır– (3) *V* to cause to burn, to burn | зажигать, воспламенять, поджигать

jangın (1) *N* fire, conflagration | пожар

jangız (14) *Adv* alone, by oneself | один, одиноко

janıχ s. **janık**

janık (6), **janıχ** (2) *N* burned; pathetic, passionate (lover) | горелый, переживание; **janık al–** to be seized with burning love | горячо/страстно влюбиться в кого–л. → **ǯanık**

jantık (1) *N* (a kind of) flaky pastry with mutton or other filling fried in fat | разновидность чебуреков

jañadan (2), **janadan** *Adv* over again from the beginning, afresh, anew | снова, сначала, опять, сызнова

jañgız (1) *Adj/Adv* alone, by oneself | один, одиноко → **ǯangız, jalıñız**

jañı (7) *Adj* new, raw | новый, сырой

jañıl– (4) *Vpass* to err, to make a mistake, to go wrong | ошибаться, заблуждаться

jañıš (3) *Adj* mistake, wrong, incorrect | ошибка, ошибочный

jañıšlık (1) *N* mistake, error, blunder | ошибка, неправильность

jañlıš (1) *Adj* incorrect, faulty, wrong; mistake | ошибка; плохой, ошибочный

jap– I (82) *V* to do, make, build sg | делать, создавать, строить; **nap– (< ne jap–)?** what to do? | что делать? → **ǯap– I**

jap– II (1) *V* to cover, to veil; to shut (door, window) | покрывать что чем; закрывать → **ǯap– II**

japı (2) *N* building, construction | здание, строение, сооружение

japıχ (1) *N* closed | закрытый

japıl– (5) *Vpass* to be made | готовиться, делаться

japıš– (1) *V recipr* to stick, adhere; to stick to one (as a bore); | приклеиваться; приставать; прицепляться

japıštır– (1) *Vfact* to glue on; | клеить; приклеивать

japračık (4), **japraǯık** [< japrak + –čık/–ǯık] *N dim* small leaf | листочек

japraǯık s. **japračık**

japraχ s. **japrak**

japrak (12), **japraχ** (2) *N* leaf | лист; **japrak ač–** to leaf | покрываться листьями → **jafrak**

japraklan– (1) *Vpass* to come into leaf | покрываться листьями

japtėr– s. **japtır–**

japtır– (20), **japtėr–** *Vfact* to have sg done/made | заставить сделать

jar (189) *N P.* friend, lover, one's beloved | любимый, возлюбленный; **jar tut–** keeping a mistress| содержать любовницу → **ǯar II**

jar– (3) *V* to split, to cut through | колоть; рассекать → **ǯar–**

jar‿keǯe s. **keǯe**

jara (9) *N* wound, injury | рана, ранение → **ǯara**

jara– (15) *V* to be befit, to be suited for, to be of use | годиться, быть годным, угодить → **ǯara–**

jaradan (3) *N* the Creator (God) | создатель, творец (о боге)

jaral– (1) *Vpass* to split, to be splitted; to break one's heart; to be injured | треснуть; обрываться; надрываться

jaralan– (5) *Vpass* to be wounded | получить ранение; быть раненым; пораниться

jaralı (4) *Adj* wounded | раненный

jaramaj (3) *Adj* naughty, rude, improper, nasty | плохой, дурной

jaramaz (5), **jaramas** *Adv* useless, good–for–nothing | озорник, сорванец, негодный

jaramazlık (1) *N* uselessness, naughtiness, rudeness, bad behaviour | озорство, злодейство → **ǯaramazlık**

jaraš– (15) *V recipr* to be suitable, to go well with | подходить, идти кому–л., быть к лицу

jarašık (3) *N* pleasing appearance; elegance; ornament | украшение чего–л.

jaraštır– (1) *Vfact* to make suit; to adorn | украшать, разукрашать → **ǯaraštır–**

jarat– (3) *V* to give existence to; to create | созидать, творить

jardım (6) *N* help, aid | помощь, содействие; **jardım et–** to help, to succour | помогать, оказывать помощь; **kol jardımda tabıl–** to give manual help/aid/assistance to | оказывать ручную (физическую) помощь → **ǯardım**

jarı (27) *N* half | половина, пол–, полу–; **jarı keče** midnight | полночь; **jar ‿ keǯe (jarı keǯe)** midnight | полночь

jarıʒık (1) *N dim* little darling, little sweetheart | маленькая любимая, душенька

jarıχ s. **jarık I**

jarıχlı *Adj* bright, brilliant | блестящий;

jarık I (18), **jarıχ** (7) *N* light (of one's eyes) | луч, свет (очей) → **ʒarık**

jarık II *Adj* split, cracked | разрезанный, расколотый

jarıklandır– (1) *Vfact* to make sg bright or luminous, to make sg illuminated | светить, освещать, озарять

jarım (3) *N* half | половина

jarın (24) *Adv* tomorrow | завтра

jarma (2) *N* billet (of firewood), log | полено; **jarma terek** billet–wood, cut–tree? | колотое дерево

jas (4) *N* mourning (for the dead) | траур; **jas tut**– to mourn, to be in mourning (for) | оплакивать, носить траур

jasa– (37) *V* to do, to make; to build, to arrange | делать, строить, устраивать, налаживать → **ʒasa**–

jasal– (1) *Vpass* to be built, constructed, arranged | делаться, строиться, производиться, быть сделанным

jasat– (21) *Vfact* to have sg made/done; to order sg from sb | заставить сделать (что–л.); допустить сделать; заказывать

jasattır– (5) *Vfact* to have sg made/ done; to order sg from sb | заставить сделать (что–л.); допустить сделать; заказывать

jasıl– (1) *Vpass* to be done/made | делаться, быть сделанным → **jasal**–

jaslan– (1) *Vpass* to lean against something; to support oneself | опираться, прислоняться

jasman (1) *N* woodden utensil for carrying water; flask | деревяный сосуд для перевозки воды

jastıχ s. **jastık**

jastık (9), **jastıχ** (2) *N* pillow, bolster; cushion | подушка

jaš I (99) *Adj/N* young (man); age, years (of a person) | юноша, молодой человек; молодой; возраст; **jašına kel**– to be ... year(s) old | исполниться ... год/лет; **jaš jiget** young lad | юноша, молодой парень; **jaš kız** young girl | молодая девочка, девушка → **ʒaš**

jaš II (8), **ješ** (1) *N* tears | слеза

jaša– (34) *V* to live, exist | жить, существовать; обитать; **kůp jašasın!** long live! | да здравствует! → **ʒaša**–

jaščık (1) *N dim* little young (man) | молоденький (парень) → **jaš I**

jaščıklı (1) *Adj* very young, youthful | молоденький

jašır– (2) *V* to hide, to conceal | скрывать; **til jašır**– to keep silent, not to speak | неметь, не говорить

jašlı (3) *Adj* suffused with tears | слезливый → **jaš II**

jašlıχ s. **jašlık**

jašlık (4), **jašlıχ** (3) *N* youth, youthfulness | молодость; **jedi jašlıχ** seven–year–old | семилетний

jat– (151) *V* to lie down, to go to bed, to sleep with someone; to get laid; as an auxiliary verb expresses continuous action | лежать, ложиться (спать); спать с кем–л.; (в роли вспомогательного глагола выражает длительность действия); **čalka tůšep jat**– to lie on one's back | лечь на спину, лечь навзничь; **jatar vakit** bed–time | время сна;

Vaux **bola jatkan** sg happening now/ at present | сейчас происходящее (событие, действие); **jata toryan** lying (at present) | лежащий (в этот момент/сейчас); **keli jat**– to be coming | приходить; **keli‿jatkanda [< kele jatkanda]** on the way (from); being coming; (when) was coming | приходя; приходивши; **keti‿jatkanda [< kete jatkanda]** on the way to; being going (to/towards), (when) was goimg (to/towards) | уходя; уходивши; **uzata jatχan** was reaching something out to someone | протянул что–то кому–то; **jasatıp jat**– to be having palaces made | заказывать построить дворцы → **ʒat**

jatak (11) *N* bed, couch, mattress; nest, den | постель, кровать; гнездо, стан → **ʒatak**

jatıl– (2) *Vpass* to be lain (passive of jat–) | ложиться; **jatılası zamanı** (it's) time to go to bed, bed–time | пора ложиться спать, пора сна

jatır– (3) *Vfact* to lay something down; to put to bed | класть, укладывать; ложить (в постель)

jatıš (1) *N* lying; resting, relaxing | лежание, отдых, успокоение; затихание

jatkız– (1) *Vfact* to lay something down; to put to bed | уложить (в постель) → **ʒatkız**–

jatu (1) *Int(uw) Adv* lying | лежание

jau– (4) *V* to rain, to snow | идти, падать (о дожде, снеге) → **ʒav– I, jaγ–, jav–**

jaulıχ s. **jaulık**

jaulık (16), **jaulıχ** *N* napkin, handkerchief | платок → **ʒaulıχ, ʒauluk, jaγlık, jauluk**

jauluk (8) *N* napkin, handkerchief | платок → **ʒaulıχ, ʒauluk, jaγlık, jaulık**

jav– (9) *V* to rain; to snow | идти /о дожде, о снеге/; **jaγmır/jaγmur jav**– to rain | идти – о дожде; **javun jav**– to rain | идти – о дожде; **kar jav**– to snow | идти – о снеге → **ʒav– I, jaγ–, jau–**

javaš (7) *Adj* slow(ly), gentle; gently; careful(ly) | осторожный, медленный; осторожно, медленно → **ʒavaš**

javaščiktan s. **javaščıktan**

javaščıktan (3), **javaščiktan** *Adv* slowly, carefully | медленно, потихоньку

javaštan (3) *Adv* slowly, carefully | потихоньку; незаметно; тихо

javun (2) *N* rain | дождь → **ʒavun**

javur (1) *N Ar.* giaour, unbeliever, non–Muslim, heathen | гяур, неверный, не мусульманин; язычник

jaz (15) *N* summer | лето → **ʒaz**

jaz– I (46) *V* to write | писать; **suret jaz**– to draw, paint, to take one's picture/portrait photo | (на)рисовать, (на)писать картину/портрет кого–л., (с)фотографировать

jaz– II (2) *Vaux* (an event) nearly happened, (a deed) almost done | (в роли вспомогательного глагола) чуть было не; **bola jazdı** to be about (to be), to be going (to be) | почти, чуть не, чуть не сбылось, чуть было не (случилось); **kùre jazγan** he was about to see | чуть было не видел; **òle jaza** he is about to die | чуть не умер; **ùle jazγan** she was about to die | чуть было не умерла

jazı (11) *N* writing; letter | письмо, писание; **kara jazı** bad luck, misfortune | несчастная судьба, несчастье → **jazu**

jazıʒı (1) *N* scribe | писец, писарь

jazıχ s. **jazık**

jazık (10), **jazıχ** *N* a pity; a shame | сожаление; стыд

jazıl– (10) *Vpass* to be written, to be decreed by fate | быть написанным (судьбой), быть назначенным судьбой; **bašına jazıl–** one's fate is to | быть суждено кому; на роду быть написано кому

jazılı (4) *Adj* written; decreed by fate | письменный, написанный/назначенный (судьбой)

jazmaχ (1) *Inf (mak)* writing | писание

jazu (1) *N* writing; letter | письмо, писание → **jazı**

jā (29) *Conj* either, or; oh! here there! | a, или, либо, разве; ой, ох → **ja**

jāki (1) *Conj* or, else, otherwise | или

je– (17) *V* to eat | есть, кушать

ječ– s. **geč–**

jedek (1) *N* reserve, spare | резерв, запас

jedi (31), **jedė** (3) *Num* seven | семь → **ǯedi, jidi**

jedilau (1) *N* making it seven/sevenfolded | поднятие на семь

jedile– (1) **(~jedle–)** *V* to make it seven | поднять на семь

jedinǯi (1) *OrdNum* seventh | седьмой

jedle– s. **jedile–**

jek– (4) *V* to harness, to put (the horses) to a carriage | запрягать, впрягать; **at jek–** to put a horse (to a carriage) | запрягать лошадьь **saban jek–** to put a horse to a plow | запрягать лошадь в плуг → **ǯek–**

jekirmi (1) *DefNum* twenty | двадцать; **jekirmi tokoz** twenty–nine | двадцать девять

jektir– (2) *Vfact* to make sb harness (a horse to a wagon) | велеть запрячь

jel (8) *N* wind | ветерь → **ǯel**

jelken (3) *N* sail | парус; **jelken ač–** to hoist sail | поднимать паруса

jem (2) *N* fodder | корм, фураж → **ǯem**

jemek (6) *N* food, meal | еда, пища

Jemen (3) *Tn* Yemen | Йемен

jemi s. **gemi**

jemin (1) *N* oath | клятва, присяга; **jemin ettir–** to make sb swear, to put sb on oath | приводить к присяге

jemiš (2) *N* fruit | фрукт, плод → **ǯemiš**

jen (1) *N* sleeve, cuff | рукав

jen– s. **jeñ–**

jenǯ s. **genč**

jene (104) *Adv* again; yet, but yet, still | опять, ещё → **jine**

jeñ– (4), **jen– (**1) *V* to overcome, to win | побежать, одолевать

jeñejden s. **jeñiden**

jeñiden (2), **jeñejden** *Adv* over again from the beginning, afresh, anew | снова, опять

jeñišė (1) *N* cotoneaster? | кизильник?; **tuzlık [tůzlık?] jeñišė** flatland cotoneaster? | равнинный кизильник

jer (198) *N* ground, place | земля, место; **bir jerde** in one place | на одном месте; **bu jerde** in this place | на этом месте; **felan jerge** to so and so place | на такое–то место, к такому–то месту; **jerine** to his/her place; instead of him/her | вместо кого–чего–л.; **jerine kuj**– to replace by sg; to mistake sb for sb; to mix sb up with sb | перепутывать; принимать кого–л. за кого–л. → **ǯer, ǯir, jir**

jerčik (1) *N dim* dim.–hypochoristic form of "jer" | уменьшительно–ласкательная форма слова "йер" → **jer**

jerleštir– (1) *V* to place, to put in a certain place | помещать

jerli (1) *Adj* local; native (resident) | местный; уроженець **kaj jerlisiz?** where are you from? | из каких мест вы? откуда вы родом?

jesir (3) *N Ar.* captive, slave | пленный, раб

ješel s. **ješil**

ješil (16), **ješel** *Adj* green | зелёный → **ʒešil**

ješillen– (1) *Vpass* to become green, to be freshened | зеленеть, становиться зелёным

jet– (26) *V* to reach, to arrive; to suffice, to be enough | догонять, доставать; доходить; хватать, быть достаточным; **artına jet–** catch up with sb | догонять → **ʒet–, ʒit–**

jeter (3) *Adj* enough | достаточно → **ʒeter**

jetim (1) *N* orphan | сирота

jetiš– (19), **jetıš–** *V* to reach, to arrive; to suffice, to be enough | достичь, дойти, догонять; хватать, быть достаточным

jetištir– (1) *Vfact* to rase, grow, to breed; to educate; to deliver | производить, разводить; воспитывать; доставлять

jetmiš (1) *Num* seventy | семьдесят; **jetmiš toχoz** seventy–nine | семьдесат девать

jez *Adj/N* bronze, brass | латунь, жёлтая медь → **ʒez I**

jez– (1) *V* to crush, to suppress; (here) to beetle (wash) | давитъ, мять, разминать → **ez–**

jezid (1) *N Ar.* fiend; base; monster | изверг

jiber– (2) *V* to let go, to let loose; to send (away) | отпускать, посылать → **ʒiber–**

jibrišim (1) *N P.* silk thread; silk | шёлковые нитки; шёлковая ткань

jiger (1) *N* saddle | седло → **eger I**

jigerle– (2) *V* to saddle | оседлать → **ʒigerle–, egerle–**

jigerlet– (1) *Vfact* to make sb saddle (a horse) | приказать кому–л. оседлать (коня)

jiget (175), **igėt** (1), **jigėt** (10), **jıget** *N* young man, fine manly youngster; hero; brave, stouthearted | парень, молодой человек; молодец, храбрец; → **ʒiget, jiket**

jigetlik (1) *N* courage, heroism | смелость, удаль, отвага

jigėt s. **jiget**

jigil– (1) *Vpass* to be harnessed | быть запряжённым

jigirim (9), **jigirem**, **jıgerem**, **jıgırım**, **jıγırım** *Num* twenty | двадцать → **jikirem**

jikerem s. **jikirem**

jikerme (7), **jıkerme** *Num* twenty | двадцать; **jikerme jüz** twenty hundred | двадцать сот (две тысячи) → **jigirim, jikirem**

jiket (20) *N* young man, fine manly youngster; hero; brave, stouthearted | парень, молодой человек; молодец, храбрец; → **jigėt, jiget,**

jikirem (5), **jikerem** twenty | двадцать → **jigirim**

jilduz s. **jılduz**

jimirtka s. **jımırtka**

jine (15), **jıne** (1) again; yet, but yet, still | опять, ещё → **jene**

jip (7), **jıp** *N* rope, string; thread | нитка, нить, верёвка → **ip**

jipek (8), **jıpek** (2) *N* silk | шёлк → **ipek, jüfek**

jir (40) *N* ground, place | земля, место → **ʒer, ʒir, jer**

jiri– (1) *V* to melt | таять → **iri–**

jiz (2), **jız** *N* trace, track | след → **iz**

jizle– (1) *V* to hide, conceal | скрывать, прятать

jizli (1) *Adj/Adv* hidden, secret(ly) | скрытый, тайный, скрыто, секретно → **gizli, kizli**

jıgerem s. **jigirim**

jıgırım s. **jigirim**
jıγ– (2) *V* to pile up, to collect in a heap | собирать, наваливать
jıγırım s. **jigirim**
jıγlıš– **(<jıγılıš–)** (1) *V recipr* to crowd together | собираться, скопляться
jıχıl– (6) *Vpass* to pull down, to ruin | свалиться, падать → **jıkıl**–
jık– (14) *V* to pull down, to ruin | валить, рушить
jıket s. **jiget**
jıkıl– (9) *Vpass* to collapse, to fall down | свалиться, падать; **attan jıkıl**– fall down from the horse | падать с лошади → **jıχıl**–
jıl (26) *N* year | год; **bı‿jıl [< bu jıl]** this year | в этом году; **bı‿jılkı [< bu jılki]** this year's | текущего года → **ǯil, ǯıl**
jılan (10) *N* snake, serpent | змея → **ǯilan, ǯılan**
jıldız (11), **jıldiz** *N* star | звезда; **tügerek jıldız** round star | круглая звезда; **sabā jıldızı** morning star | утренняя звезда → **ǯılduz, jılduz, julduz**
jılduz (10), **jilduz** *N* star | звезда → **ǯılduz, jıldız, julduz**
jılγı s. **jılkı**
jılχı s. **jılkı**
jılχıǯi s. **jılχıǯı**
jılχıǯı (2), **jılχıǯi** *N* horseherd | табунщик → **ǯılkıǯı, jılkıǯı**
jılkı (3), **jılγı, jılχı** *N* herd of brood mares and stallions with their colts loose in the fields | табун → **ǯılkı**
jılkıǯi s. **jılkıǯı**
jılkıǯı (13), **jılkıǯi** *N* horseherd | табунщик → **ǯılkıǯı, jılχıǯı**
jıllık (2) *N* one year old; a year's sallary | ежегодник; годовая плата
jıltıra– (2) *V* to shine, sparkle | сверкать, блестеть → **ǯaltıra**–
jımırta (3), **jımorta** *N* egg | яйцо → **ǯımırtka, jumurta**
jımırtka (14), *N* egg | яйцо → **jımırta, jumırtka, jumurta**
jımorta s. **jımırta**
jımšak (1) *Adj* soft, mild | мягкий
jıp s. **jip**
jıpek s. **jipek**
jır (1) *N* song | песня, сказ → **ǯėr**
jırak (1) *Adj* distant, far off | далёкий, отдалённый
jırla– (1) *V* to sing | петь, напевать
jız s. **jiz**
joχ (53) *NegEcp* not to be; no; there is (are) not | нет; не имеется → **jok, juχ, juk**
joχari s. **jokarı**
joχla– (1) *V* to control, to visit | навещать, проверять → **jokla–, juχla–, jukla– II**
joχo (4) *N* sleep, sleepiness | сон, сонливость → **ǯokı, ǯuku, joko, joku, juχu**
joχsem (1) *Particle* if not, otherwise; or; or else | или, разве, что ли
jok (181) *NegEcp* not to be; no; there is (are) not | нет; не имеется; **jok et**– to destroy | уничтожать → **joχ, juk, juχ**
jokarı (2), **joχari** *Adj/Adv* high, upper, top; upward(s) | верх, верхняя часть; верхний; наверху, вверх → **jukarı**
jokla– (3) *V* to control, to visit | навещать, проверять → **joχla–, juχla–, jukla– II**
joko (2) *N* sleep, sleepiness | сон, сонливость → **ǯokı, ǯuku, joχo, joku, juχu**
joksa (3) *Conj* if not, otherwise; or; or else | или, разве, что ли → **juksa, juχsa**
joku (3) *N* sleep, sleepiness | сон, сонливость → **ǯokı, ǯuku, joχo, joko, juχu**
jol (23) *N* road, way | дорога, путь; **jolγa čık**– to set out on a journey or a way | отправиться в путь → **ǯol, ǯul, jul**

jolaχ (2) *N/Adj* stripe, band; striped, streaked | полоса; грядка; полосатый

joldaš (1) *N* fellow traveler, comrade | попутчик, спутник → **juldaš**

jolla– (2) *V* to send, to dispatch, to send off | посылать, отправлять → **julla–**

jonga (1) *N* shavings | стружка; щепка, щепа

jonγaǯıχ (1) *N dim* small shavings | стружка; щепка, щепа

jorkan (1) *N* quilt | одеяло → **jurγan**

josma (3) *N* pretty and attractive (woman) | прелестная (о женщине) → **jusma**

ju– (4) *V* to wash | мыть → **ǯuv–, juv–**

juχ (4) *NegEcp* not to be; no; there is (are) not | нет; не имеется → **jox, jok, juk**

juχla– (4) *V* to control, to visit | навещать, проверять → **joχla–**

juχsa (4) *Conj* if not, otherwise; or; or else | или, разве, что ли → **joksa, juksa,**

juχu (12) *N* sleep, sleepiness | сон, сонливость → **ǯokı, ǯuku, joχo, joko, juku**

juk (25) *NegEcp* not to be; no; there is (are) not | нет; не имеется → **joχ, jok, juχ**

jukarı (4) *Adv* high, upper, top; upward (s) | вверх, верхняя частъ → **joχarı, jokarı, uγarı**

jukla– I (34), **juχla–** *V* to sleep | спать → **ǯukla–, ujukla–**

jukla– II (1) *V* to control, to visit | навещать, проверять → **joχla–, jokla–, juχla–**

juksa (3) *Conj* if not, otherwise; or; or else | или, разве, что ли → **joksa, juχsa**

juktur– (1) **[< ujuktur–]** *Vfact* to put/lull to sleep; to rock to sleep | усыплять; убаюкивать

juku (17) *N* sleep, sleepiness; dream | сон, сонливость; **jukusın ač–** to wake up | просыпаться → **ǯokı, ǯuku, joχo, joko, juχu**

jul (162) *N* road, way | дорога, путь; **julγa tůš–** to set out on a journey or a way | отправиться в путь → **ǯul, jol**

julčı s. **julǯı**

julǯı (3), **julčı** *N* traveler, passenger | путник

juldaš (1) *N* fellow traveler, comrade | спутник, товарищ → **joldaš**

julduz (3) *N* star | звезда → **jıldız**

julla– (26) *V* to send, to dispatch, to send off | посылать, отправлять → **jolla–**

jullan– (1) *Vpass* to depart, to start out (to) | отправляться (в путь)

julluχ (1) *N* provisions for a journey | дорожные припасы

jum– (3) *V* to shut, to close (eye) | закрывать (глаза); **kůz jumup ačχanǯe** in the twinkling of an eye | не успеть (глазом) мигнуть → **ǯum–**

jumarlaš– (1) *V recipr* to tumble about embracing each–other | катиться, валяться обнимая друг–друга → **jumurlaš–**

jumırtka (1) *N* egg | яйцо → **jımırta, jımırtka, jumurta**

jumruk (4) *N* fist; blow with the fist | кулак; удар кулаком

jumurlaš– (2) *V* to tumble about embracing each–other | катиться, валяться, обнимая друг–друга → **jumarlaš–**

jumurta (1) *N* egg | яйцо → **jımırta, jımorta, jumırtka**

jun– s. **juvun–**

jur– s. **jůr–**

jurγan (3) *N* quilt | одеяло → **jorkan**

jurγunnık (2) *N* weariness, fatigue | усталость, утомляемость

jurıl– (4) *Vpass* to get tired | утомляться

jurt (5) *N* yurt; home, habitation | юрта; жилище; родина; **babañ jurtına!** to your father's home/yurt! (when

swearing) | в дом/юрту твоего отца! (ругательство) → **ǯurt**
jusma (1) *N* pretty and attractive (woman) | прелестная (о женщине) → **josma**
jut– (5) *V* to swallow, to gulp down | глотать, поглащать → **ǯut**–
juv– (7) *V* to wash | мыть → **ǯuv**–
juva (3) *N* nest; hole | гнездо; нора → **ǯuva**
juvarlan– (4) *Vpass* to revolve, to turn round; to tumble down (from sg) | скатиться, свалиться, упасть
juvarlat– (1) *Vfact* to cause to rotate, to make round | катить
juvun– (2), **jun**– *Vpass* to wash oneself, to have a bath | мыться, купаться → **ǯuvun**–
juvundır– (1) *Vfact* to (give a) wash (to) | мыть, умывать кого–л. → **ǯuvındır**–
jůfek (1) *N* silk | шёлк → **ipek, jipek**
jůkle– (5) *V* to load | вьючить, грузить, нагружать
jůklen– (4) *Vpass* to be loaded (up), to be charged | нагружаться, навьючиваться
jůn (1) *N* wool, woolen | шерсть, шерстяной
jůnli s. **jůnlů**
jůnlů (2), **jůnli** *Adj* woolen | с шерстью, волосатый
jůr– (31), **jur**–, **jür**– *V* to walk, to move; to shoulder; to go into service | ходить; двигаться, идти в прислуги; идти в прислуги, работать где–нибудь/у кого–нибудь; **ırgat jůr**– to serve/work as a day–laborer/farm hand/servant | работать батраком, служить слугой у кого–л.
Vaux **tůšůnep jůrgen** he was thinking (for long) | (долго) думал; **esirep jůreler** they are getting drunk | они пьянствуют → **ǯůr–, ǯür–**
jůreǯik (2) **[<jůrek-čik]** *N dim* little heart | маленькое сердце
jůrek (31) *N* heart | сердце; **jůrekten sůj**– to love with all one's heart | душевно любить, любить от всего сердца → **ǯůrek**
jůreklen– (1) *Vpass* to take heart, to become bold | набраться храбрости
jůrekli (2) *Adj* brave, bold | смелый, храбрый
jůren– s. **ůren**–
jůret– (2), **jüret**– *V* to teach | учить, обучать → **ůgret**–
jůrů– (8) *V* to walk, to move | ходить
jůrůš (1) *N* gait, march | походка → **ǯůrůš**
jůver– (43) *V* to send, to dispatch; to leave (in), to let go; to throw (away)/loose | отправлять, посылать; пустить; отбрасывать
jůz I (34) *N* face | лицо → **ǯůz**
jůz II (10) *N* hundred | сто; **biš jůz** five hundred | пятьсот → **ǯüz, jüz**
jůz– (6) *V* to break, to tear, to split | ломать, разрывать, раздирать → **ůz**–
jůzůk (101) *N* ring | кольцо → **ǯůzůk, jüzük**
jůzůkči (1) *N* maker or seller of rings; jeweller/goldsmith | делающий или продающий колца
jůzům (20) *N* grape | виноград; **jůzům kůtůgi** vine–trunk | виноградный чурбан; **jůzům salkımı** a bunch of grapes | виноградная кисть
jür– s. **jůr**–
jüret– s. **jůret**–
jüz (2) *Num* hundred | сто → **jůz II**
jüzük (9) *N* ring | кольцо → **ǯůzůk, jůzůk**

K

kab–kara s. **kap–kara**

kabab (2) *N Ar.* roast, broiled flesh or meat | кебап; **šiš–kabab** meat roasted on a spit or skewers | кебап (жареное на углях мясо)

kabahat (13), **kabāt** *N Ar.* fault, offense; guilt, sin | ошибка, грех

kabaχ s. **kabak I**

kabak I (4), **kabaχ** *N* pub, tavern | кабак

kabak II (4) *N* pumpkin *(Cucurbita pepo)* | тыква

kabakčı (3) *N* tavern keeper, barkeep | кабатчик, трактирщик → **kabak I**

kabar (6) news, message | весть, новость; сообщение; **kabar ber–** to announce (to), to inform (someone of something) | сообщить, информировать, довести до сведения; **kabar ǯiber–** to send news, to send a message | посылать новости, посылать сообщение → **haber, χabar I, kaber I**

kabāt s. **kabahat**

kaber I (7) *N Ar.* news, message | весть, новость; сообщение → **haber, χabar I, kabar**

kaber II s. **kabir**

kabir (13), **kaber** (1), **kabr** *N* grave, tomb | могила; склеп

kabık s. **kabuk**

kabırγa (1) *N* rib; side–wall of a building | ребро; боковая стенка

kabr s. **kabir**

kabuk (6), **kabık** (1) *N* outer covering of anything, rind, peel, skin etc. | кора, скорлупа, кожура

kabul (3) *N Ar.* acceptance, reception | принятие, приём

kaburčak (2) *N* shell | раковина, ракушка

kač (73) *IntPron* how many? how much? | солько?; **bir/br kač** (a) few, several | несколько

kač– (97), **kaš–** (4) *V* to run away, to escape | бежать, убежать → **χač–**

kačan (6) *IntPron* when | когда

kačır– (2) *V* to make or let escape | дать возможность убежать; упустить

kačıš (1) *N* running (away), fleeing away | убегание, бегство

kada (4) *N* younger brother/sister | младший брат, младшая сестра

kadaj (7) *N* younger brother/sister; my dear, darling (as addressing) | младший брат, младшая сестра; мой дорогой

kadar (76) *Postp* until, till, to | до, столько; **ne kadar, ni kadar** how much/many | сколько; **šo kadar** this much | такое–что количество, столько; **u‿γadar [<u kadar]** that much | в такой степени, настолько, так → **χadar**

kadarčık (1) *Postp dim* until, till, to | до, столько → **kadar**

kade (10) *N Ar.* drinking glass, tumbler, goblet | рюмка, стопка, чарка, чашка

kademsız (1) *Adj Ar.* unlucky | несчастный

kadi s. **kādi**

kadife (1) *N Ar.* velvet | бархат

kadir I (1) *N Ar.* worth, value, rank | ценность, цена → **χadir**

Kadir II (1) *Pn Ar.* Kadir | Кадир (Къадир)

kadı s. **kādi**

kadın (3) *N* woman, wife | женщина, жена → **hatın, hatun, χatın, χatun, katın**

Kaf (2) *Tn* Caucasus | Кавказ; **Kaf tavı** Caucasus mountains | Кавказские горы

kafas s. **kafes**

kafes (6), **kafas** (9) *N Ar.* kage, coop | клетка → **χafas**

kaftan (3), **kaptan** (1) *N Ar.* caftan | кафтан

kaγad s. **kahat**

kaγid s. **kaγıt**

kaγıt (3), **kagıt, kaγid [kaγıd?]** (2) *N P.* paper, of paper | бумага, бумажный → **kahat, kijat**

kahad s. **kahat**

kahat (22), **kaγad** (1), **kahad** *N P.* paper, of paper | бумага, бумажный; **kahat ujna–** to play cards | играть в карты → **kaγıt, kijat**

kahpolu s. **kahpe**

kahpe (4), **kappe [< kahpe]** *Adj/N Ar.* harlot, prostitute | распутная женщина, проститутка; **kahpe ulan, kahpolu [< kahp‿olu < kahpe olu], kahpuli [< kahp‿uli < kahpe uli]** son of a bitch; rascal, bastard | сукин сын, ублюдок, продувная бестия, плут, мерзавец, подлец;

kaj (12) *IntPron* what, what kind of; which | какой, который; **ka‿jakka [<kaj jakka]** where to, to which direction | куда, в какую сторону; **kaj dajın** what? what kind of? | какой

kaja (35) *N* rock | скала

kajalı (1) *Adj* rocky (place) | скалистое (место) → **χajalı**

kajb (1) *N Ar.* loss | потеря, отсутствие; **kajb et–** to lose, (here) to kill, destroy | потерять, (здесь) убивать, уничтожать

Kajbulla (2) *Pn Ar.* Kaybulla | Къайбулла

kajda (36) *IntPron* where, in which direction | куда, где; **kajdadır** wherever, no matter where | где–бы ни; где–бы, куда–бы

kajdan (13) *IntPron* from where, whence | откуда

kajgana (1) *N* scrambled eggs, omlet | яичница → **kajγanaχ**

kajγanaχ (1) *N* scrambled eggs, omlet | яичница → **kajgana**

kajγır– (8) *V* to be anxious (about sg/for sb), to worry (about) | беспокоиться, переживать, думать (о ком–л.)

kajγu (1) *N* regret, sorrow; care; worry | печаль, забота; грусть

kajıχ (3) *N* boat, rowboat | лодка, ладья → **kajık**

kajıχčı (1) *N* boatman | лодочник

kajık (5) *N* boat, rowboat | лодка, ладья → **kajıχ**

kajın–ata s. **kajnata**

kajır (1) *N* goodness, boon, behoof, happiness | добро, благо, польза, выгода; счастье → **χajır**

kajır– (1) *V* to sprain (joint); to twist; to turn (out) | загнуть, выворачивать; **sač kajır–** to make curly (hair) | завивать (волосы)

kajıra– s. **kajra–**

kajırlı (1) *Adj* good, happy, lucky | добрый, счастливый; **kajırlı olson** let sb be lucky/happy/successful | пусть будет благополучным/удачным → **hajırlı**

kajıš (2) *N* strap; thong; belt | ремень

kajl (3), **kāil** *N Ar.* consenting; agreeing | соглашащюийся

kajlı (5) *N Ar.* consenting; agreeing | соглашащюийся; **kajlı bol–** to consent, to agree to | соглашаться с кем–чем → **kajl**

kajmaχ s. **kajmak**

kajmak (3), **kajmaχ** *N* cream, clotted cream | пенка, сливки, сметана

kajna– (7) *V* to boil | кипеть

kajnana (10) *N* mother–in–law | тёша, свекровь

kajnaš– (1) *V recipr* to swarm; to mill | кишеть

kajnat– (3) *Vfact* to boil | кипятить

kajnata (2), **kajınata** *N* father–in–law | тесть, свёкор → **χajnata**

kajra– (2), **kajıra–** *V* to sharpen, whet, grind | точить, острить; **pıčak kajra–** to sharpen a knife | точить нож

kajraχ s. **kajrak**

kajrak (2), **kajraχ** *N* whetstone, grindstone | точило, наждак

kajsı (8) *IntPron* which, whichever | который

kajt– (164) *V* to turn back, to return | возвращаться → **χajt–**

kajtan (2) *N* coton or silk cord; braid | шнур(ок), басон, нашивка, тесьма

kajtar– (21) *V* to get out of (doing), to let return, to recall; to refuse | вернуть, возвращать, отклонять; отказаться

kajtıl– (1) *Vpass* to turn (round/back) | вернуться, возвращаться

kajtmaχ (1) *Inf (mak)* turning (round/back) | возвращение

kajva s. **kajve**

kajve (47), **kajva** (1) *N* coffee | кофе; **kajve tukmayı** mallet used for pounding coffee in a mortar | пест применяемый для толчения кофе в ступе; **kajve–hāne** coffee–shop | кофейня, кафе; **kajvege otor–** sit down in a coffee–shop | зайти в кафе

kajveǯi (8) *N* keeper of a coffee shop | содержатель кофейни; продающий кофе

kajveǯilik (1) *N* coffee trade and service | занятие содержателя кофейни

kak (1) *Onomatop* sound of cackling | звук кудахтанья; **kak et–** to cackle/caw (crow) | кудахтать/каркать (ворона)

kak– (13) *V* to push, to drive a nail in, to shake | стучать, забивать, трясти; **bend kak–** to build/erect a dam | сооружать плотину

kakıl– (1) *Vpass* to be knocked (at) | стучаться, забиваться

kakılda– (1) *V* to cackle, to prattle | кудахтать, квохтать → **kak**

kal– (231) *V* to remain, stay; as an auxiliary verb expressing perfect or continuous acting | оставаться, останавливаться; в роли вспомогательного глагола указывает на совершенность или длительность действия; **χatırı/katrı kal–** to feel hurt, or offended, to take offense (at) | обижаться, огорчаться, сердиться на кого; **χatırında kal–** to be remebered | сохраниться в памяти, запоминаться кому; **mārum kal–** to remain deprived of; to be disappointed | быть лишённым, терять кого–что; разочаровываться; **sauluknan kalıñ!** good–bye! | до свидания!; **tul kal–** to be widowed | овдоветь
Vaux **aγlap kal–** to burst into tears | расюьакаться, заливаться слезами; **čıka kaldı** suddenly appeared in front (of him) | вдруг появился (перед ним); **ǯılaj kal–** to cry for a long time | долго плакать; **karap kal–** to see (long) | смотреть (длительно); **šašıp kala** is left astonished | остается пораженным; → **χal–**

kala (15) fortress, castle | крепость → **kale**

kalaj (18) *Int/Adv* how, in what way | как, каким образом; **kalaj etip** how, somehow | как, кое–как; **kalaj da bolsa** in someway or other | так или иначе

kalajlı (1) *Adj* tinned | лужёный, покрытый оловом
kalajsız (1) *Adj* untinned | нелужёный
kalak (3) *N* shovel; oar; paddle | лопата, совок
kalakaj (5) *N* cake or loaf commonly sweetened; kind of bun | слоёный пирог, сдобный пирог
kalam (8) pencil, (reed) pen | ручка, перо; **kalamnen ček–** to draw (eyebrows) with a pencil **kara kalam [kaš]** long narrow black eyebrows | тонкие длинные чёрные брови → **kalem**
kalat (1) *N Ar.* cloak, gown | халат
kalau (1) *N* stone–wall | забор, кладка
kalb (15), **kalib** (1), **kalip** (1), **kalıb** (2), **kalıp** (1) *N Ar.* heart; soul | сердце
kaldır– (21) *V* to keep/leave (in a certain state); to remove | оставлять, снять
kale (2) *N* fortress, castle | крепость → **kala**
kalem (7) *N* pencil, (reed) pen | ручка, перо; **kalem(daj) kaš** long narrow eyebrows (looking as if they were drawn by pencil) | тонкие (словно нарисованные пером) брови; **kara kalem(daj) [kaš]** long narrow black eyebrows (eyebrows like a black pencil) | тонкие длинные чёрные брови (брови похожие на чёрный карандаш) → **kalam**
kalemtraš (2) *N* pencil sharpener, knife for cutting reed pens | нож для очинки тростниковых ручек; перочинный нож; точилка (для карандашей)
kalχ– s. **kalk–**
kalib s. **kalb**
kalip s. **kalb**
kalıb s. **kalb**
kalın (1) *Adj* thick, stout | толстый
kalıklı (1) ~ **kalikli** *Adj* backward, retrograde (person), left; defective, incomplete | отсталый, оставшийся (от чего–л.); недостаток
kalk (31) *N Ar.* folk, people | народ, люди; **Kırım kalkı** people (folk) of the Crimea | народ Крыма, Крымское население → **halk, χalk**
kalk– (7), **kalχ–** (1) *V* to get up, stand up, to become erect | вставать, подниматься, вставать на ноги
kalkan (1) *N* shield, buckler | щит
kalpačık (1) *N dim* (small) fur cap | шапка, папаха → **kalpak**
kalpaχ s. **kalpak**
kalpak (19), **kalpaχ** (2) *N* fur cap | шапка, папаха
kaltıra– (5) *V* to tremble, shiver, quiver | дрожать, трястись, трепетать
kaltırama (1) *N* trembling, shivering | дрожание, трепет
kama (18) *N* dagger, poniard | кинжал
kamala– (1) *V* to surround, encircle | окружать
kamanče (3) *N* kind of violin | кеманча → **kemanče**
kamander (1) *N R.* commander | командир
kamaš– (2) *V* to be dazzled | ослепляться ярким светом
kamči s. **kamčı**
kamčı (32), **kamči** (4) *N* knout | кнут
kamčıla– (1) *V* to quilt, to lambaste, whip | стегать, хлестать
kamır (3) *N* dough | тесто → **hamur**
kamıš (1) *N* bulrush, reed | камыш; тростник
kamsa (2) *N* anchovy (*Engraulis encrasicolus*) | анчоус, хамса
kan I (28) *N* khan, ruler, sovereign | хан, правитель

kan II (27) *N* blood | кровь; **kanүa buja–** to paint with blood | красить кровью
kan III (1) *N* house; café, inn | дом, кофейня, гостиница → **hāne**
kan– (1) *V* to be persuaded, to believe; to quench one's thirst, to eat one's fill | убедиться, поверить; утолять (жажду, голод), насыщаться
kana (6) *Adv* where? where is it?; after all, well really | а?; ну–ка; где, где же? → **hana**
kanarja (4) *N* moskito | комар
kanat (23) *N* wing | крыло; **iki kanat kapı** two–leaved door | двустворчатая дверь → **χanet**
kanat– (1) *V* to run one's blod out | пустить кровь
kanatlan– (4) *Vpass* to grow wings, to take wing (and fly away) | окрылиться
kanǯa (1) *N* dog–collar | ошейник
kanda (13) *IntAdv* where, anywhere; where to | где; куда; **kanda kitesin** where are you going to | куда ты идёшь
kandır– (3) *V* to persuade sb to do sg, to get sb to do sg | убеждать, уговаривать
kanetčık (1) *N dim* small wing | маленькое крыло
kanfet (1) *N R.* sweets | конфета
kanı (1) *Adj Ar.* convinced, satisfied, relieved | довольствующийся, удовлетворяющийся чем–л.; **kanı bol–** to be sure of sg, to take for certain | быть уверенным в чём–л.
kanlı (1) *Adj* bloody | окровавленный
kanušna (1) *N R.* stable | конюшня
kap– (12) *V* to snatch, seize, catch | хватать ртом
kap–χara s. **kap–kara**
kap–kara (3), **kab–kara, kap–χara** *Adj* entirely black, pitch–black | совершенно черный, чёрный–пречёрный
kapa– (6) *V* to shut, close (door, eyes, book), to cover up | закрывать, прятать, устилать; **kijeu kapa–** to lock up the bridegroom (together with the bride into the same room) | запирать жениха (рядом с невестой в ту же комнату)
kapak (4) *N* lid, cover | крышка
kapal– (1) *Vpass* to shut, close of itself (door, window etc.) | закрываться
kapamak (1) *Inf(mak)* closing, shutting, locking | закрытие
kapat– (4) *V* to close, shut; cover | закрывать, запирать, замыкать; устилать
kapčık (2) *N dim* small sack | футляр, чехол
kapek (5) *N R.* kopeyka (smallest Russian coin) | копейка
kapı (127) *N* door | дверь; **bojalı kapı** painted door | крашеная дверь; **ǯennet kapısı** the gate of paradise | райские ворота → **χapo, χapu, kapo, kapu**
kapıl–kupul (2) *Adj* having shell | имеющий оболочку; **kapıl–kupul baġa, kapıl–kupul baka** terrapin, tortoise | черепаха → **kaplı–kuplu**
kapım (1) *N* quid, slab | кусок
kapısta (2) *N R.* cabbage (*Brassica oleracea/Caulis*) | капуста; **kapısta borčı** borsch, cabbage–soup | щи, борщ; **kapısta kočanı** cabbage–stalk, heart of cabbage | кочерыжка → **χapısta, kapusta**
kapıš– (1) *V recipr* to snatch (for), to pick a quarrel with (sb) | схвататься
kapkač (3) *N* stealing by snatching | похищение; **kapkač ket–** to catch, to grab | схватывать, ловить, поймать
kapkan (2) *N* trap | капкан, ловушка

kapla– (3) *V* to cover over, to overlay (with) | покрывать, облицовывать

kaplı–kuplu (2), **kapli–kuplu** *Adj* having (a) shell | имеющий оболочку; **kaplı–/kapli kuplu baka** terrapin, tortoise | черепаха → **kapıl–kupul**

kappe s. **kahpe**

kaptan s. **kaftan**

kaptır– (3) *Vfact* to make sb snatch sg, to get sb to seize sg, catch | дать прикусить, приказать взять

kapu (25) door | дверь → **χapu, kapı**

kapuska s. **kapusta**

kapusta (4), **kapuska** *N R.* cabbage (*Brassica oleracea/Caulis*) | капуста → **kapısta**

kar (16) *N* snow | снег → **χar**

kara (70) *Adj* black | чёрный; **kara biber** black pepper | чёрный перец; **kủzin karası** pupil, apple of eye | зрачок, зеница (ока) → **χara**

kara– (60) *V* to look, watch | глядеть, смотреть; видеть → **χara–**

Kara–kủz (1) *Pn* Kara–kuz (i. e. black–eyed) | Кара–кузь (т. е. чёрно–окий)

Kara–teñiz (4), **Kara–teñız** *Tn* Black Sea | Чёрное море

Kara–teñız s. **Kara–teñiz**

karabiber (3), **χar‿biber [<χarabiber]** *N* black pepper | чёрный перец → **biber**

karaʒa (2) *N* roe deer (*Capreolus capreolus*); (here) wilde goat | косуля (род небольших оленей); олень; (здесь) горный (альпийский) козёл

karaγač (1) *N* elm (*Ulmus*) | вяз, ильм

karakol (2) *N* patrol, guard | патруль

karakuš (3) *N* eagle (*Aquilla*) | ягнятник, степной орёл

karala– (4) *V* to blacken, to scribble | чернить, писать начерно

karaldı (2) *N* court, courtyard | усадьба, двор

Karalez (1) *Tn* Karalez (a village in the region of Simferopol) | Каралез (деревня в области города Симферополь)

karaman (12) *N* hero | герой

karamanlık (2), **karamannık** *N* heroism | героизм

karamannık s. **karamanlık**

karampıl s. **karanpil**

karanfil (18) *N Ar.* garden pink, carnation (*Dianthus plumarius*) | гвоздика → **χaranfil, karanpil**

karanγı (6), **karangė** *Adj* darkness, blackness | темнота, мрак

karanlıχ (1) *N* darkness, blackness | темнота, мрак → **karannık, karañık**

karannık (2) *N* darkness, blackness | темнота, мрак → **karanlıχ, karañık**

karanpil (8), **karampıl, karanpıl** *N Ar.* garden pink, carnation (*Dianthus plumarius*) | гвоздика → **χaranfil, karanfil**

karañik s. **karañık**

karañı (1) *N* darkness, blackness | темнота, тьма

karañıχ s. **karañık**

karañık (15), **karañik** (6), **karañıχ** (2) *N* darkness, blackness | темнота, мрак → **karanlıχ, karannık**

karañıklan– (1) *Vpass* to grow dark, to become black | темнеть

karar (6) *N* decision, resolution | решение, резолюция; **karar ber–** to decide (to) | решать, принимать решение; **karardan keč–** to give up, to desist (from), not to keep one's decision/promise/word | отступаться, отказываться от чего–л.; **kararına kel–** to grow up | возрастать, вырастать → **χarar**

Karasu (7) *Tn* Karasu (town), now Karasu Bazar | (город) Карасу, ныне Карасубазар

karaul s. **karavul**

karavana (1) *N* copper dish used as a mess tin by soldiers and sailors | бачок (бак) для пищи

karavıl s. **karavul**

karavul (10), **karaul, karavıl** *N* guard | сторож

karčγa s. **karčıγa**

karčıγa (3), **karčγa** *N* facade, frontage | фасад

kardaš (37) *N* brother or sister; (in addressing) (my) friend, my dear | брат или сестра; (при обращении) мой друг, милая моя → **χardaš**

karγa (9) *N* crow (*Corvus*) | ворона, грач

karγa– (1) *V* to curse, to damn | проклинать

karı (19) *N* woman, wife | женщина, жена; **karı koǯa** wife and husband, couple | женщина и муж, супруги

karılγač (1) *N* swallow (*Hirundo*) | ласточка → **χarılγač, karlıγač**

karın (6) *N* abdomen, stomach | живот, утроба, брюхо; **karnı bura** one's stomach is seized with cramp | чувствует судорожные боли в желудке

karınǯa (4) *N* ant (*Formica*) | муравей

karıš (5) *N* span (space) | пядь

karıš– (2) *V* to be mingled, to be mixed | смешиваться

karıštır– (12) *V* to mix, stir | мешать, смешивать

karlı (3) *Adj* snowy, covered with snow | снежный

karlıγač (3) *N* swallow | ласточка → **χarılγač, karılγač**

karmala– (1) *V* to palpate, snoop around | ощупывать; шарить

Karpat (6) *Tn* Karpat (mountains) | Карпат(ские горы); **Karpat gorası** Karpat mountains | Карпатские горы; **Karpat(nıñ) tauları** Karpat mountains | Карпатские горы

karpuz (13), **karpus** *N* watermelon (*Citrullus vulgaris*) | арбуз

Karson (2) *Tn* Herson (a city in the south of Ukraine) | Херсон (город на юге Украины)

karšila– s. **karšıla–**

karšı (58), **karši** (3) *Postp* opposite, opposite side or direction; (as a postposition) against, contrary (to), towards, a little before | то, что находится напротив; (в функции послелога) против, напротив, перед, к; **karši karšija** face to face | лицом к лицу, друг против друга; **karšıda/ karšında** opposite the; on the opposite side of | напротив (кого–чего–л.) → **χarši**

karšıla– (13), **karšila–, karšla–** *V* to go to meet, to welcome | встречать

karšılık (1) *N* return, recompense, answer | противодействие, ответ, отплата

karšla– s. **karšıla–**

kart (149) *N* oldman, greybeard | старый, старик → **χart**

kartaj– (2) *V* to grow old | стареть

kartanaj (1) *N* old woman, gammer | бабушка, баба

kartčık (1) *N dim* old woman, gammer | бабушка, старуха

kartep s. **kartop**

kartij (4), **kartıj** *N vocat* grandmother, old woman; little mother, mammy | бабушка, мамочка, старуха

kartıj s. **kartij**

kartlık (1) *N* old age | старость

kartop (2), **kartep** *N R.* potato | картофель

karu s. **karuv**

karuɣ (2) *N* strength, power; (here) help, goodness, charity | сила; (здесь) помощь; благотворительность; **karuɣı/karuɣu kajtar–** to return, to recompense, to compensate (for) | возмещать, отплачивать → **karuv**

karuv (2), **karu[v]** *N* strength, power; (here) goodness, heartiness, kindness, favour | сила; (здесь) помощь; благотворительность; **karu[v]/karuvu kajtar–** to return, to recompense, to compensate (for) goodness, help | возмещать, отплачивать за доброту, за помощь → **karuɣ**

kasab (6), **kasap** *N Ar.* butcher | мясник → **χasab**

Kasab–ulı *Pn* Kasab–uli (butcher's son) | Касаб–улы (сын мясника)

kasabčı (1) *N* butcher | мясник

kasap s. **kasab**

kasavet (2), **kasvat** *N Ar.* restlessness, anxiety, sadness, bitterness | беспокойство, тоска, печаль; **kasvat et–** to be anxious (about), to be sorroful (at, over) | беспокоиться, печалиться

kasavetlen– (1) *Vpass* to be worried, to grieve | беспокоиться, печалиться

kasta (6) *N* sick (person), ill | больной; **aɣır kasta** seriously ill | тяжелобольной; **kasta bol–** to become ill | заболевать → **hasta, χasta**

kastana (1) *N* chestnut | каштан(овый)

kasvat s. **kasavet**

kaš I (32) *N* eyebrow | бровь; **ela kaš** hazel eyebrows | карие бровьи; **kalam kašlı** having long narrow eyebrows | с тонкими длинными бровьями; **kalem kaš, kalam kaš** long narrow eyebrows | тонкие длинные бровьи; **kara kašlı** having black eyebrows | с чёрными бровьями; **kaš(lar) bas–** to knit the eyebrows | хмурить брови; **sụrmeli kaš** eyebrow tinged with antimony | бровь подведённая сурьмой

kaš II (2) *N* pommel of a saddle | лука седла; **eger kašı** (saddle–)bow | лука (седла)

kaš– s. **kač–**

kašaɣala– (1) *V* to curry, to groom | чистить скребницей

kaščık (1) *N dim* eyebrow | бровь

kašɣır s. **kaškır**

kašı– (1) *V* to scratch (with the nails) | чесать

kašıχ s. **kašık**

kašık (5), **kašik, kašıχ** *N* spoon | ложка

kaška (1) *N* blaze on the forehead of an animal; having a blaze on the foehead | белая отметина на лбу животного, звёздочка

kaškır (5), **kašɣır** *N* wolf | волк

kašlı (2) *Adj* having (black) eyebrows | имеющий какие–л. брови; с чёрными бровями; **kara kašlı** having black eyebrow | имеющая чёрные брови, с чёрными бровями

kat I (33) *N* layer, fold; floor | слой, пласт, етаж; **br kat urba** one set of clothes | один комплект одежды; **kat kat** in layers, layer on layer | слоями, рядами

kat II (56) *N/Postp* side, flank; (as a postp.) beside, at, to | сторона, бок; (в роли послелога) около, у, к, при; **katına** beside him/her | к его/её стороне; к нему/к ней

kat– I (2) *V* to add (to), to join (to) | прибавлять

kat– II (48) *V* (not knowing) what to do? how is one supposed to act? | (не знать) что делать; (не соображать) как поступать; **katıp** ~ **katėp** how, in what way | как, каким образом → **et–, net–**

katala– (1) *V* to revolve, turn round; to move | катиться, вертеться; возиться

katife (4) *N Ar.* velvet | бархат → **kadife**

katı (2) *Adj* hard, stiff | твёрдый, жёсткий → **χatı**

katı–kutu (1) *N* odds and ends, unk; one thing and another | барахло; шурум–бурум, всякая всячина

katık (3) *N* saur milk, yoghurt | катык, кислое молоко

katıl– (1) *Vpass* to join, to be completed | присоединяться, быть добавленным

katın (28) *N* woman, wife | женщина, жена → **hatın, hatun, χanım, χatın, χatun, kadın, katın**

katır (6) *N Ar.* memory, respect | память, уважение, почтение; **al–katır** health and welfare | жизнь и здоровье; **al–katır suraš–** to inquire after each other's/one another's health | расспрашивать друг друга (о жизни, об обстоятельствах); **katırım ičin** for my sake | ради меня, для моего блага; **katırı/katrı kal–** to feel hurt, or offended, to take offense (at) | обижаться, сердиться (на кого)→ **χatır**

katlama (6) *N* meal made from flour; cheesecake, flatbread, pitabread | мучное блюдо; лепёшка (жареная), ватрушка, слоёное тесто с сыром

katlan– (1) *Vpass* to be stacked, to be folded | складываться

katran (1) *N Ar.* tar | дёготь

kattı (1) *Adj* hard, stiff | твёрдый; сильный

katun (1) *N* woman | женщина → **χatun**

kauga s. **kauγa**

kaugalaš– (1) *V recipr* to quarrel, argue, to fight | ссориться, скандалить; драться

kauγa (6), **kauga** (4) *N* quarrel, row; fight | скандал; брань; схватка; **kauγa et–** to quarrel (with), to fight | ругаться, ссориться с кем; драться с кем; **kaugaja ket–** to go to war | идти на войну → **kavγa**

kaurıl– s. **kavrıl–**

kavak (3) *N* poplar (Populus) | тополь

kaval (4) *N* shepherd's pipe, flageolet | пастуший рожок, свирель

kave s. **kāve**

kavga s. **kavγa**

kavγa (4), **kavga** (2) *N* quarrel, row; fight | скандал; брань; схватка → **kauγa**

kavγalaš– (1) *V recipr* to quarrel, argue, to fight | ссориться, скандалить; драться

kavil (6), **kavıl** (3) *N Ar.* agreement, accord | слово; договор, пари; **kavil jap–** to agree (on sg), to make an arrangement with sb | договариваться, прийти к соглашению

kavıl s. **kavil**

kavın s. **kavun**

kavır– (1) *V* to fry, to roast | жарить

kavıš– (4) *V recipr* to meat, to come together | соединяться; встречаться; **kůn kavıškan** the sun went down | солнце зашло → **kavuš–**

kavrıl– (2), **kaurıl–** *N pass* to be roasted, to roast | жариться

kavuklu (6) *Adj* wearing a quilted turban | носящий на голове кавук (турбан)

kavun (8), **kavın** (1) *N* melon (*Cucumis melo*) | дыня; **sarı kavun** melon | дыня → **χavun**

kavur– (4) *V* to fry, to roast | жарить

kavuš– (3) *V recipr* to meet, to come together, to obtain | соединяться; встречаться, приобретать → **kavuš**–

kavuštır– s. **kavuštur**–

kavuštur– (3), **kavuštır**– *Vfact* to unite, reunite | соединять

kaz (35) *N* goose (Anser) | гусь; **kaz balası** gosling | гусёнок; **kaz kůmesi** goose house | птичник для гусей

kaz– (6) *V* to dig, to trench | копать, рыть

kaza (1) *N Ar.* accident, misfortune | несчастный случай, авария; **kazaүa kuj**– to cause harm/trouble | причинить зло, натворить зло кому–л.

kazačıχ (1) *N dim* accident, misfortune | несчастный случай, авария

kazak (7) *N/En R.* Cossack | казак

kazan I (32) *N* cauldron | кастрюля, котёл

Kazan II (14) *Tn* Kazan (city) | (город) Казань;

kazan– (13) *V* to win, to gain, to earn | зарабатывать, выигрывать, завоевать → **χazan**–

kazanıl– (1) *Vpass* to be earned | быть заработанным

kazarma (15) *N R.* barrack (s) | казарма

kazālı (1) *Adj* dangerous, unsafe, causing accidents | опасный, влекущий за собой несчастный случай

kazdır– (1) *Vfact* to have sg dug | велеть копать, заставить рыть

kazi (1) *N Ar.* kadi (judge of Islamic canon law) | кади(й), судья (мусульманский) → **kādi**

kazık (7) *N* pale, stake | кол

kazıl– (2) *Vpass* to be dug, to be excavated | быть выкопанным → **χazıl**–

kazma (3) *N* pickaxe, mattock | заступ, кирка, мотыга; **kazma kůrek** pickaxe | киркомотыга

kaznadar (1) *N Ar.* treasurer | казначей → **kazneǯi**

kazne (5) *N Ar.* treasure | казна

kazneǯi (1) *N Ar.* treasurer | казначей → **kaznadar**

kādi (33), **kadi** (12), **kadı** (2), **kādı** (1) *N Ar.* kadi (judge of Islamic canon law) | кади(й), судья (мусульманский); **kad akam** my brother/uncle kadi | мой брат/дядя кади → **kazi**

kādı s. **kādi**

kāil s. **kajl**

kāın (1) *N Ar.* treacherous, traitor | изменник, предатель; **kāıñ čektim** I suffered from your infidelity/faithlessness | я страдал от твоей неверности

kān (48) *N* khan, sovkereign, ruler | хан, правитель → **χan**

kār (1) *N Ar.* deep sorrow, distress | горе, скорбь, печаль; **kārindan** to his/her shame | от его/её стыда

kātib (1) *N Ar.* scribe, clerk; writer | писарь

kāve (8), **kave** (4) *N Ar.* coffee; coffeeshop | кофе, кафейня → **kajve**

ke s. **ki**

ke– s. **kel**–

keben (3), **kebėn** *N* stack (of corn), rick | стог, омёт

kebete s. **kȯbete**

kebėn s. **keben**

keč (16) *Adv* late, delayed | поздно, поздний; запоздалый; **keč bol**– to be late | опаздывать; **keč kel**– to come late | прийти поздно; **kečke kal**– to be late | опаздывать; **kečkeǯek** until late, for long, a long time/while | до позднего (времени), долго, подолгу

keč– (108), **keš**– *V* to pass by, to pass through, to pass (over) | проходить; переходить; переправляться; проникать; миновать; **karardan keč–** to give up, to desist (from), not to keep one's decision/promise/word | отступаться, отказываться от чего–л.; **u̇zinden keč–** to get enraged, to lose tember/patience | выходить из себя, возмущаться; **vakıt keč–** to pass, to go by (time) | проходить (время); **vāzkeč–** to renounce, to give up (a claim), to cease from | отказываться, оставлять → **geč–**

keče (12) *N* night | ночь; **jarı keče** midnight | полночь → **geǯe, keǯe, kiǯe**

keče–ku̇ndu̇z s. **keǯe–ku̇ndu̇z**

kečeǯik (1) *N dim* night | ночь → **keǯeǯik**

keči (1) *N* goat | козёл; **kijık keči** 'wild goat' | дикий козёл

kečik– (2) *V* to be late, to be delayed | опоздать, опаздывать

kečil– (1) *Vpass* to be passed (by/over) | быть пройденным; **vaz kečil–** to be renounced, to be given up (of a claim) | отказываться

kečin– (4) *Vpass* to live (on), to get along (with) | жить, существовать; проживать; кормиться

kečir– (13) *V* to pass (something through something), to cause to pass, to accompany (someone) to a place | проводить (что–л.); проводить (через что–л.); **u̇mu̇r kečir–** to spend life | проводить жизнь; **vaχıt kečir–** to spend (time) | проводить (время)

keǯe (85) *N* night | ночь, ночной; **jarı keǯe; jar‿keǯe** midnight | полночь; **keǯe–sajın** every night | по ночам, каждую ночь → **keče, kiǯe**

keǯe–ku̇ndu̇z (2) *Adv* day and night, continously | день и ночь

keǯečik s. **keǯeǯik**

keǯeǯik (3), **keǯečik** *N dim* night | ночь → **kečeǯik**

kedaj (1) *N* keday (folk poet and singer) | кедай (народный поэт–песенник)

kef s. **kejf**

Kefe (6) *Tn* Caffa, Kaffa (town); now Feodosia/Theodosiya | (город) Каффа, сегодня Феодосия

kefen s. **kefin**

kefergı (1) *N* lock of hair covering one's forehead | чёлка, прядь падающая на лоб; локон

kefin (6), **kefen** *N Ar.* shroud, winding sheet | саван

kejf (7), **kef** *N Ar.* pleasure, delight, joy | удовольствие, наслаждение; **kejf su̇r–** to enjoy, to savor, to have a good time (with) | кейфовать, развлекаться, веселиться; **tu̇šek kefi jap–** to make the joy of bed | делать наслаждение постели

kekel (2) *N* bangs; taft of hair; lock on the top of the head | хохол, чуб

kel– (1056), **ke–** (2), **kē–** *V* to come (from, to), to arrive | прибывать, приходить; **du̇njaya kel–** to be born | появиться на свет, родиться; **julı kel–** to be one's turn | подходить его/её очередь; **kegen/kēgen** (one) was coming/arriving; (one's) coming/arrival | (кто–л.) пришёл/прибывал; приход/ прибыттие кого–л.; **kegenǯek** until (one) is coming/arriving | пока (кто–л.) придет/прибывает; **tez/tezden kel–** to come quickly/soon | быстро приходить; *Vaux* **alıp kel–** to bring sg (to) | приноцить что–нибудь; **hajvanları ajdap kelgenner** the animals were driven there | животные были загнаны туда; **jetip kel–** to reach, arrive | приезжать, достигать; **kajtıp kel–** to turn back, return | вовращаться; **suvarıp kel–** to water (oxen) regularly | регулярно поливать (волов) → **gel–, kil–**

kelam (3) *N Ar.* word, a saying, speech | слово, речь, беседа; **kelam et–** to talk, to exchange a few words with sb | беседовать с кем–л., изрекать, говорить, переговариваться

kele (9) *N* mortar | ступа

keletke (1) *N* granary, garner | амбар, житница

kelin (25), **kelın** *N* bride, daughter–in–law | невеста; **al kelin** bride in red? | невеста в красном платье?; **kelin–kuman** bride's ewer? | кувшин невесты?

kelinček (7), **kelinčik** (2), **kelinčık** (2), **kelinǯek** *N dim* young wife | молодуха

kelinǯek s. **kelinček**

keliš– (1) *V recipr* to meet, to see each other | встречаться; **rast keliš–** to come across, to meet by chance | встречаться, сталкиваться с кем (неожиданно)

kemanče (3), **kamanče** *N P.* a kind of violin | кеманча → **kemane, kemen**

kemane (7), **kemāne** (2) *N P.* violin with five strings | скрипка; **kemane dāre čaldır–** to make sb play the violine and tambourine | велеть играть на скрипке и тамбурине → **kemanče, kemen**

kemāne s. **kemane**

kemen (1) *N P.* violin with five strings | скрипка; **kemen dāre čaldır–** to make sb play the violine and tambourine | велеть играть на скрипке и тамбурине → **kemanče, kemane**

kemer (1) *N P.* belt, girdle | пояс; марка, свод, дуга

kemi (10), **kemė** *N* ship, boat | судно, корабль → **gemi**

kemik (2) *N* bone | кость

kenar (19), **kėnar** *N P.* edge, border | край, окраина; обшивка

kenarsız (3), **kenarsiz, kenersız** *Adj* not having an edge, border or margin; endless, unlimited | бесконечный, бескрайний, безграничный; **kenarsız kůlmek** unhemmed shirt | рувашка без обтачки, не подшитая рубашка

kenǯal (2) *N Ar./P.* short curved dagger, khanjar | кинжал

kenǯe (1) *Adj* youngest (brother or sister) | младший (брат или сестра)

kendi (28) *Pron pers* oneself; self, own | сам, свой

kene (41) *Adv* again | снова, опять → **gene**

kenef (3) *N Ar.* toilet | уборная, туалет

kenersız s. **kenarsız**

keñ (2) *Adj* broad | широкий

keñešlik (1) *N* wideness | простор, приволье

keramet s. **kiremit**

kerbar (2) *N P.* amber, yellow amber | янтарь

kerček (3) *N* real, true, reality, truth | правда; истина

kerčeklen– (1) *Vpass* to come true | осуществляться

kerčekten (6) *Adv* in fact, actually | действительно; поистине, по–настоящему → **gerčekten**

kere (14), **kerre** (12) *N* time, times | раз; **bı‿kerre** once, just | один раз

kerek (81), **kırek** *Adv/Adj* necessary, needed | нужно, необходимо, надо; ныжный, необходимый; **kerek bol–** to be necessary/needed | необходимо, быть нужным; **kerekė tarzı**

as is due, properly | как следует, надлежащим образом → **gerek**

kerek– (12) *V* to be necessary, needed, required | пригодиться, понадобиться; **kerekmij** not necessary/ needed | не надо, не нужно

keremit s. **kiremit**

kergef (3) *N Ar.* embroidery frame | пяльцы

keri (2), **geri** *Adv* back, backword, behind; rear | назад, обратно; **gerije tün**– to return back | возвращаться обратно; **keri χajt**– to return back | возвращаться обратно

Kerim (1) *Pn* Kerim | Керим; **Kerim aka** brother/uncle Kerim | брат/дядя Керим

kerre s. **kere**

kes (1) *N* (last) will, testament | завещание; **kes et**– to bequeath, to enjoin | завещать, поручать; **kes etu** (last) will, testament | завещание, завет

kes– (78) *V* to cut (through, down or off), | резать, стричь; **dava kes**– to pass sentence/judg(e)ment on | присуждать, приговаривать

kese (2) *N* purse, small bag | кошелёк

kesen–kes (1) *Adv* decisively, categorically | наотрез, твёрдо, решительно, окончательно

kesertki (1) *N* ordinary lizard (Ophisops elegans) | ящерица

kesik (1) *Adj* cut (through or off); (here) short(ened) | отрезанный, усеченный; (здесь) короткий

kesil– (8) *Vpass* to be cut; to end; to come to an end | резаться, кончаться

keskinlik (1) *N* sharpness, clearness; concentration | острота, резкость; крепость

kesmedaš (1) *N* hewn stone, ashlar | пилённый камень

kestir– (1) *Vfact* to cause to cut; (here) to fall down (smog) | приказывать резать; (здесь) спуститься (о тумане)

kestıne (1) *N Gr.* chestnut *(Castanea dentata)* | каштан

ket– (386), **kėt**– (2) *V* to go (away), to leave, depart | уходить; отправляться, идти; ; **kaugaja ket**– to go to war | идти на войну
Vaux **alıp ket**– ~ **ap–ket**– to take away, to carry away | убирать, уносить; **bitep ketken** it's gone, disappeared | исчез, пропал; **čıkıp ket**– to go out | выйти; **gajb bolıp kete** disappears | исчезает; **jakši bolıp kete** has got better | исправлен; **oγorlap ket**– to steal sg | украсть что–нибудь; **patlap kit**– to explode, break | разорваться, сломаться; **učıp kit**– to fly away | улететь; → **git–**, **kit**–

keten (12) *N/Adj Ar.* flax (Linum usitassimum), flaxen, linen cloth | лен, льняной

ketėr– s. **ketir**–

ketir– (196), **ketėr**– (3), **ketır**– *V* to bring (to) | приносить, привозить; приводить → **getir**–

ketirt– (1) *Vfact* to get in, to call for, to send for | заставить приносить/ привозить

ketiš (2) *N* going away, departure | уход, отъезд

ketır– s. **ketir**–

kettig (1) **[ketlik?]** *N* running (functioning, ticking?) (clock) | ход (часов), тиканье

kevde (5) *N* body, trunk, stem | тело, туловище, торс

kez– (23) *V* to walk about, to stroll | гулять, прогуливаться; **tamdan tama kez**– to stroll, to tramp; to live a debauched life (about women) |

таскаться, блудить (о женщинах) → **gez–**

kezdir– (2) *Vfact* to show around, to take through | (вы)водить (гулять), прогуливать

kezin– (1) *Vpass* to wander about without any definite purpose | прогуливаться → **gezin–**

Kezleu s. **Kezlev**

Kezlev (4), **Kezleu** *Tn* Kezlev (town), today Yevpatoria | (город) Кезлев, сегодня Евпатория

kezmek (1) *Inf(mak)* walking about, strolling | гулянье, прогуливание

kē– s. **kel–**

kėrėk s. **kırk**

kėz s. **kız**

kėzdėr– s. **kızdır–**

ki (62), **ke, ku** *Conj* who, which, that | что, который, чтобы, так как

ki– s. **kil–**

kibi (7) *Postp* like; as; (after predicat) almost, nearly, somewhat | как, подобно, будто (послелог) → **kibik, kůbůk**

kibik (69) *Postp* like; as; (after predicat) almost, nearly, somewhat | как, подобно, будто (послелог) → **gibik, kibi, kůbůk**

kibla (5), **kible** (2) *N Ar.* south, southern | юг, южный; **kibla jak** south, southern side; direction of Kaaba in Mecca | юг, южная сторона; направление к Каабе (Мекка); **kiblada** in the south | на юге; **kibleden es–** to have a south–wind blowing, to blow from the south (wind) | дуть с юга (ветер)

kible s. **kibla**

kibrit (1) *N Ar.* match | спичка

kičiǯek (3), **kičıǯik** *Adj dim* small, little; young, younger | малый, молодой, младший

kičik (2), **kičık** *Adj* small, little; young, younger | малый, молодой, младший

kičıǯik s. **kičiǯek**

kičık s. **kičik**

kičırım–kačırum (1) *Adj* winding, meandering | извилистый

kičkene (~kič´kene) (15) *Adj/N* little, young, younger | мелкий, маленький → **kiškene**

kičkir– s. **kıčkır–**

kiǯe (1) *N* night | ночь, ночной → **keǯe**

kidir– I s. **kijdir–**

kidir– II s. **kıdır–**

Kidir–nebi (3), **Kıdir–nebi** *Pn Ar.* Kidir–nebi (prophet Kidir) | Кидир–неби (пророк Кидир) → **Hizir**

kidır– s. **kıdır–**

kigenek (1) *N* tassel?, rag, clout | кисть; тряпка?

kij– I (35), **ki–, kıj–** (2) *N* to put on, to wear | одевать, надевать

kij– II (1) *V* to cut; to chop up fine | рубить (крошить, резать) на мелкие кусочки

kijafet (3), **kijapet** *N Ar.* general appearance and dress (of a person), dress, costume | внешний вид, костюм, одежда

kijamet (12), **kijāmet** (2) *N Ar.* great disaster; last judgement, doomsday | невообразимая суматоха, конец света; **kijamet kop–, kijāmet kup–** all hell to break loose; to erupt (about confusion, combustion) | подняться (о страшном переполохе)

kijapet s. **kijafet**

kijat (1) *N P.* paper, of paper | бумага, бумажный → **kaɣıt, kahat**

kijdir– (5), **kidir–, kıjdır–, kėjdir–** *Vfact* to dress, to clothe (someone) | одевать, заставить надеть что

kijeu (28) *N* bridegroom, son–in–law | жених; зять; **ič–kijeu** inside bridegroom (bridegroom living together with the bride's family) | внутренний жених (жених, живущий у семьи невесты) → **kůjeu**

kijik (4) *Adj* untamed, wild | дикий; **kijik alma** crab(–apple) | дикое яблоко; **kijık keči** 'wild goat' | дикий козёл

kijin– (11), **kıjın–** *Vpass* to dress oneself | одеваться

kijiz (7), **kijız** (2), **kıjız** *N* felt | войлок, кошма

kijım (1), **[~kijim?]** *N* clothing, dress | наряд, одежда

kijız s. **kijiz**

kijmetlė s. **kijmetli**

kijmetli (5), **kijmetlė** (2) *Adj* valuable, precious | дорогой, ценный

kil– (6), **ki–** *V* to come (from/to), to arrive | прибывать, приходить, приезжать; **kigenǯek** until (one) is coming/ arriving | пока (кто–л.) придет/прибывает → **gel–, kel–**

kiliǯ s. **kılıǯ**

kilid (1) *N P.* lock, latch, padlock | замок

kilim (2) *N* woven matting, rug without a pile | ковёр

kiltlen– **[<kilitlen–]** (1) *Vpass* to be locked, latched | закрываться на замок, запираться

kılıǯ s. **kılıǯ**

kilse [<**kilise]** (2) *N Gr.* church | церковь

kim (96), **kım** (3), **kum** (9) *Pron int* who | кто

kimi (6) *Pron indef* who, some, certain | иной, некоторый → **kimisi**

kimisi (7), **kimse II [<kimisė]** (3), **kimisı** *Pron indef* who, some | кто–то → **kimi, kůmůsi**

kimisı s. **kimisi**

kimse I (37), **kümse** *IndefPron* someone, somebody; anyone, anybody | кто–нибудь, кто–то, некто; **ič kimse** nobody | никто

kimse II s. **kimisi**

kinalı (3) *Adj* tinged with henna | окрашенный хной → **kına**

kir– (165) *V* to come in, enter, go (in, into), to get (into) | входить, въезжать; **kapıdan kir–** to enter the gate | входить через дверь/ворота; **kirıp ket–** to go in/inside, to enter | входить, заходитьь **kujnına kir–** to go into one's bossom | прильнуть (прижаться) к груди кого–л.; **mezara/mezarγa kir–** to go into tomb; to be burried; to die | войти в могилу; быть похороненным; умереть

Vaux **barıp kir–** to go in/inside, to enter | войти, зайти; **kelip kir–** to come in/inside, to enter | войти, зайти

kira s. **kıra**

kiramet s. **kiremit**

kiraz (3), **kıraz** *N Gr.* cherry (*Prunus cerasus*) | черешня → **kirez**

kire (1) *Adv* back, backward; behind | назад, обратно

kirek (1) *Adv* necessary, needed | нужный, необходимый

kiremet s. **kiremit**

kiremit (6), **keramet**, **keremit**, **kiramet**, **kiremet** *N Gr.* roof tile | черепица

kirez (4) *N Gr.* cherry (*Prunus cerasus*) | черешня *N Gr.* → **kiraz**

kirgij s. **kırγıj**

kiril– (1) *Vpass* to get in/into | пассив от глагола кир– 4входить4

Kirim (6) *Tn* Crimea | Крым → **Kırım**

kiriš– (1) *V* to meddle, interfere; to attempt | вмешиваться

kirlen– (1) *Vpass* to become dirty or soiled | стать грязным

kirpi (2) *N* hedgehog | ёж

kirpik (2) *N* eyelash | ресница

kirset– (11) *V* to make sb go (in, into), to let sb in, to send in; to put, thrust in | заставить войти, вводить, всовывать; впускать; присылать

kirsiz s. **kırsız**

kirsız s. **kırsız**

kirtle– (2) *V* to lock, latch | закрывать (на замок), запирать

kirtli (2) *Adj* locked, latched | закрыт, запертый замком

kise (3) *N* pocket, purse; small bag | карман, мешочек

kiset– (1) **[<kijset–]** *V* to dress, to clothe (someone) | надевать, одевать

kismet (14) *N Ar.* chance, destiny, luck, fortune | удача, судьба, участь, счастье; **kismet ara–** to look for chance | искать счастье; **kismet bolsa** if he/she has chance | если суждено; **kismet bolsun/bolson!** good luck! | (по)желаю удачи, успеха → **χısmet**

kišej (1) *N* groin | пах

kiši (68), **kši** (6) *N* man | человек

kiškene (6), **kškene [<kiškene]** *Adj, N* little, young, younger | мелкий, маленький, малыш → **kičkene**

kišne– (5) *V* to neigh, to whinny | ржать

kit– (452), **kıt–** (2) *V* to go (away), to leave, depart | уходить; отправляться, идти; **al–kitejim [<alıp kitejim]** I'll take it away, let me take it away | я пусть отнесу (его//её), дай–ка мне отнести (его/её); **al–kit– [<alıp kit–]** to take (it) away, to carry away | относить; отвозить; **elden kit/kıt–** to be lost, to pass out of one's hand | уплыть из рук; пропасть; **patlap kit–** to break one's heart over sg | разрываться (о сердце); **razvet kit–** to go to scout | идти на разведку, разведывать; **šejid kit–** to die in the war as a hero | погибнуть смертью героя, погибнуть на войне; **zaja kit–** to be lost, to be wasted (about one's life, time); to become useless | быть потерянным, потеряться; становиться лишним/бесполезным → **git–, ket–**

kitab s. **kitap**

kitap (9), **kitab** *N Ar.* book | книга

kiz s. **kız**

kizart– s. **kızart–**

kizli (3) *Adj* hidden, secret | скрытый, тайный, тайно, секретно → **gizli**

kizmet (9), **kızmet** *N Ar.* service, work | служение, работа; **kizmet et–** to serve, to render service, to work | служить, работать → **hizmet, χizmet, χızmet**

kizmetči (6) *N* servant, worker | служащий; слуга, работник → **hizmetči**

kıbırda– (1) *V* to move slightly and quickly; to stir | шевелиться, шататься

kıbırdan– (1) *Vpass* to keep on moving or quivering of itself | шевелиться, шататься → **χıbırdan–**

kıčır– (16), **kıčr–** (2) *V* to howl, roar; to cry (out), shout; to utter a loud scream | кричать, выкрикивать → **χıčır–**

kıčkır– (9), **kičkir–** (2) *V* to utter a loud scream; to hiccup; to sob | кричать, выкрикивать

kıčkırt– (2), **kıškırt–** *Vfact* to make sb cry out | заставить выкрикнуть

kıčr– s. **kıčır–**

kıʒıtkan (1) *N* (stinging) nettle (Urtica urens) | крапива (жгучая); **kıʒıtkan otu** stinging grass, nettle | жгучая трава, крапива → **χıčıtkan**

kıdir– s. **kıdır–** |

Kıdir–nebi s. **Kidir–nebi**

Kıdirlez (1) *Pn Ar.* Hıdırlez (Day of Hızır–İlyas, May 6, feast of the arrival of summer) | Хидреллез; день Хызыр–Ильяса, первый день лета (6–ого мая); **Kıdirlezniñ nanesi** peppermint of Kidirlez? | мята Кыдирлеза?

kıdır– (40), **kidir–** (4), **kidır–** (4), **kıdir–** (4), **kıdr–** (3) *V* to seek, to look (for), to search | искать, разыскать

kıdırın– (1), **kıdrın–** *Vpass* to look for, to search, to try to find sb/sg | приискивать, долго искать (всюду/везде)

kıdr– s. **kıdır–**

kıdrın– s. **kıdırın–**

kıj– I (2) *V* to perform (a ceremony) | осуществлять, справлять; **nikah kıj–** to perform the ceremony of marriage; to marry | заключать брак, вступать в брак → **χıj–**

kıj– II s. **kij– I**

kıjalma– s. **kıjama–**

kıjama– (2), **kıjalma–[!]** *V* to spare; to have pity on; to (be)grudge sb sg | проявлять жалость; жалеть дать что–л. кому–л.; щадить → **kıjma–**

kıjdır– (3) *Vfact* to make sb do sg | заставить сделать; велеть совершить (обряд); **nikā kıjdır–** to make sb perform the ceremony of marriage; to have sb married | заставить кого заключить брак, приказать кому вступить в брак

kıjgız– (1), **[~kijgiz–]** *V* to dress, to clothe (someone) | одевать, заставить надеть что

kıjγınnık (1) *N* trouble; slaughter | забота; резня, кровавая расправа

kıjik (2) *N* deer *(Cervus)*, stag, hart | олень

kıjık (1) *N* corner, edge; leer | угол, край; искоса; **kůz kıjıγınnen bak–** to look out of the corner of one's eye (at); to watch with a sideways or oblique glance | смотреть/видеть краем (краешком) глаза, украдкой; смотреть искоса

kıjıkla– (2) *V* to fold | складывать

kıjıl– (1) *Vpass* to be put/placed | быть положенным → **kıj– I, kujıl–, kujul–**

kıjın (1) *Adj* difficult, hard | трудный

kıjıš– (1) *V* to twist, to get out of shape | покоситься, гнуться

kıjma– (3) *V* to spare; to have pity on; to (be)grudge sb sg | проявлять жалость; щадить; жалеть дать что–л. кому–л.; → **kıjama–**

kıjna– (8) *V* to torment, torture | мучить; истязать

kıl (10) *N* hair, bristle | волос; щетина

kıl– (24) *V* to do, make, act, commit | делать, действовать, исполнять; **duva kıl–** to pray | молиться; **namaz kıl–** to perform the namaz, to pray | совершать намаз, молиться

kılǯāz (1) *N dim* hair, bristle | волос, щетина

kıldır– (2) *Vfact* to have sg done, to get sb make sg | заставить сделать; **nıkā kıldır–** to make sb perform the ceremony of marriage; to have sb married | заставить кого заключить брак, приказать кому вступить в брак

kılıblı (1) *Adj* having a case, cover | в обёртке, в чехле, с чехлом, в футляре → **kılıp**

kılıč s. **kılıǯ**

kılıǯ (29), **kiliǯ** (3), **kilıǯ** (2), **kılıč** *N* sword; saber | сабля, меч → **χılıč**

kılın– (1) *Vpass* to be done | делаться, совершаться; **ič kılınmaz** there is nothing to be done, there is nothing you can do | ничего невозможно сделать

kılıp (1) *N Ar.* case, cover | оболочка, кожа, чехол, футляр

kıllı (1) *Adj* hairy, bristly | волосатый

kım s. **kim**

kımıldaš– (1) *Vrecipr* to move slightly, to stir (together with others) | шевелиться, двигаться (вместе с другими)

kın (1) *N* sheath, scabbard | ножны, стручок

kına (1) *N Ar.* henna | хна

kır I (2) *N* countryside, uncultivated land; field; hill | поле, степь, равнина, луг; холм → **χır**

kır II (2) *N* dirty linen, laundry | грязное бельё

kır– I (10) *V* to break, crush | ломать, разбивать

kır– II (1) *V* to clean, make, tidy up | убирать, скрести (чистить)

kıra (5), **kira** *N* rent, hire | аренда; **kıra tot/tut**– to rent | арендовать

kıravat s. **kravat**

kıraz s. **kiraz**

kırdır– (1) *Vfact* to make sb break sg, to have sg broken/crushed | приказать сломать, заставить разбить

kırγıj (6), **kirgij** (2) *N* Eurasian sparrowhawk (*Accipiter nisus*), hawk | ястреб–перепелятник

kırχ s. **kırk**

kırık I (1) *Adj* broken, crushed | разбитый, сломанный

kırık II s. **kırk**

kırıl– (3) *Vpass* to be broken, crushed | сломаться, разбиться

Kırım (35) *Tn* Crimea | Крым → **Kirim**

kırk (151), **kėrėk, kırχ** (14), **kırık** (24) *Num* forty | сорок; **kırχ biñ** forty thousand | сорок тысяч; **kėrėk birinǯi** forty–first | сорок первый; **kırk eki** forty–two | сорок два; **kırk segiz** forty–eight | сорок восемь

kırk– (1) *V* to clip, shear | стричь

kırkılda– (1) *Onomatop V* to bubble | (звукоподражание) булькать

kırkınǯı (2) *OrdNum* fortieth | сороковой

kırmizi s. **kırmızı**

kırmıska (2) *N* ant | муравей → **χırmıska**

kırmızi s. **kırmızı**

kırmızı (9), **kırmizi, kırmızi** (2) *Adj* red | красный; **kip–kırmizi** quite red, very red, bright red | красный–прекрасный, совершенно (весь) красный, яро–красный → **χırmızı**

kırpık (1) *N* eyelash | ресница

kırsız (51), **kirsiz** (2), **kirsız** (3) thief | вор → **hırsız, χırsız**

kırsızlık (10) *N* theft | воровство → **hırsızlık, χırsızlık**

kırsla– (16) *V* to steal | воровать → **χırsla**–

kıs– (2) *V* to squeeze, tighten | защемлять; жать, сжимать; **umız bašın kıs**– to shrink, to pull one's neck between one's shoulders | съёжиться, вобрать голову в плечи

kısχa (4), **kisχa** *Adj* short, brief | короткий, краткий; **kısχa laf** in short, the short of it is | вкратце, короче говоря, не вдаваясь в подробности → **kıska**

kısıl– (1) *Vpass* to be squeezed, tightened | тискаться

kısır (1) *Adj* barren, unproductive | яловый, бесплодный

kıska (5) *Adj* short, brief | короткий, краткий → **kısχa**

kıskan– *V* to be jealous (of), to envy | ревновать, завидовать

kıstır– (1) *Vfact* to have sg squeezed, tightened | прищемить; вложить

kıstırıl– (2) *Vpass* to be squeezed, tightened | тискаться, жаться

kıš (4) *N* winter | зима; **kıšta** in winter | зимой

kıškırt– s. **kıčkırt–**

kıt– s. **kit–**

kıt–kıt (1) *Onomatop* sound of peeping, chirping | звук чирикания, щебетания

kız (740), **kėz** (2), **kiz** (89), **kz** (2) *N* girl, daughter | девочка, девушка; дочь, дочка; **kız al–** to marry a girl | взять девушку в жёны, жениться на невушке → **χız**

kızart– (3) *V* to roast | жарить

kızarttır– (1) *Vfact* to have sg roasted | заставить поджарить

kızčıχ s. **kızčık**

kızčık (4), **kızčıχ** *N dim* little girl | девочка, девчонка → **kızǯık**

kızǯāz (2) *N* poor (unlucky) litle girl | бедная девочка, бедняжка, девчонка

kızǯıχ s. **kızǯık**

kızǯık (3), **kızǯıχ** (2) *N dim* little girl | девочка, девчонка → **kızčık**

kızdır– (6), **kėzdėr–**, **kızdir–** *Vfact* to light, to set light to | зажигать → **χızdır–**

kızıl (5) *Adj* red, ruddy | красный

kızkardaš (8) *N* sister | сестра

kızlı (1) *Adj* hidden, secret | скрытый, тайный, секретный

kızmet s. **kizmet**

kėrėk s. **kırk**

kjafir (1) *N Ar.* infidel, unbeliever; non–Muslim | гяур, немусульманин, негодник; негодяй

kjamet (1) *N Ar.* residence, staying at a place, living | место жительства, пребывание в каком–либо месте, проживание; **kjamet ǯap–** to dwell, to stay

kjar (1) *N P.* profit, gain | выгога

kjā (2) *Conj P.* at one time; at times, now and then | иногда ... иногда, то ... то; **kjā ... kjā** at times, now and then | иногда ... иногда, то ... то

kjāmil (2) *Adj Ar.* perfect, complete; educated, sage | совершенный; эрудированный, знающий

Kjeva (2) *Tn* Kyiv (city) | (город) Киев

koba (19) *N* cave | пещера

kobete s. **kȯbete**

kobuk (1) ~**kȯbu̇k** *N* foam | пена

koč [~kȯč] (2), *N* migration; nomad | переезд, кочевка; **koč kalklar¬** nomad folk(s), nomads | кочевой народ, кочевники; **koč má¿ir** nomad transmigrator, nomad emigrant | кочевник–переселенец → **kȯč–**

koč– s. **kȯč–**

kočan (1) *N* corn cob; heart (of a vegetable) | кочерыга; **kapusta kočanı** cabbage–stalk/stump, heart of cabbage | кочерыжка

kočχar s. **kočkar**

kočkar (5), **kočχar** (2) *N* ram | баран

koǯa (17) *N* husband, master, teacher | муж, хозяин, учитель

Koǯa Salası ~ Koǯa Salaši? (1) *Tn* Kodja Sala (a village in the Crimea) | (деревня в Крыму) Коджа Сала (Саласы)

koǯak s. **kuǯak**

kodalak (1) *N* ox–cart with two solid wooden wheels | двухколёсная повозка, арба (с колёс из сплошных деревянных дисков)

kodor– s. **kudur–**

koget (2) *N* grass, herb | трава

koj I (38) *N* sheep | баран, овца → **kuj**

koj II s. **kȯj**

koj– (7) *V* to put, place, set | класть, положить; **adına ... koj–** to give the name

... (to sb), to name as ... | давать имя (название); наречь, называть; **adını koj–** to give the name ... (to sb), to name as ... | давать имя (название) кому–л.; *Vaux* **azırlap kuj–** to prepare, arrange sg | приготовить, подготовить что–нибудь; **biklep kuj–** to lock sb up somewhere | запереть (кого куда–нибудь) → **χoj–**, **χuj–**, **kuj– I**

kojan (14) *N* hare (Lepus europaeus) | кролик, заяц → **kujan**

kojroχ s. **kojrok**

kojrok (23), **kojroχ** (3) *N* tail | хвост

kojun (3) *N* sheep | баран, овца → **kujun**

kokı– (4), *V* to smell, to have a smell | пахнуть → **kukı–**

koko s. **koku**

kokok (1) *N* cuckoo *(Cuculus canorus)* | кукушка; **kokok bakırıj** the cuckoo is shouting/crying cuckoo | кукушка кукует

koku (5), **koko** *N* smell | запах → **kukı**, **kuku**

kol (19) *N* arm; sleeve | рука, рукав → **kul**

kol–tava (1) *N* frying pan with handle | сковорода с ручкой

kolaj (5) *Adj* easy | лёгкий; **kolajnen** easily, fairly easily | (довольно) легко, без труда → **kulaj**

kolčaχ (1) *N* glove(s) | перчатка

kolla– (2) *V* to keep under observation; to look after, to protect | наблюдать, следить; **bašıñı kolla!** look after yourself! | следи за собой! береги себя!

kolmek (2) *N* shirt, chemise, smock | рубашка; сорочка → **kůlmek**

koltuk (2) *N* armpit | подмышка

Kolumčak s. **Kulumčak**

kolumla– (3) *V* to foal | жеребиться → **χolumla–**, **kolunla–**

kolunla– (1) *V* to foal | жеребиться → **χolumla–**, **kolumla–**

kon– (1) *V* to settle/make a night's halt | останавливаться (на ночлег, кочевье), опускаться, садиться → **kun–**

kona (4) *N* pastry board, kitchen board | доска для теста

konak (5) *N* halting place, inn; guest, temporary visitor | ночлег; гость → **kunak**

konakbaj (1) *N* owner, landlord | владелец, хозяин дома → **kunakbaj**

konan (1) *N* a two– or three–year–old foal | двух–трёхгодовалый жеребёнок → **kunan**

konuk (13) *N* guest | гость; **konuk bol–** to be a guest; to put up, stay | быть гостьем у кого–л., оставаться → **kunuk**

konuš– (2) *V* to speak | говорить → **χonuš–**, **kunıš–**

konušma (1) *N* speech, discussion | разговор

koñšu (3) *N* neighbor | сосед, соседний

kop s. **kȯp**

kop– (7) *V* to break in two, to break off | отламываться, отскакивать; **kijamet kop–** all hell to break loose?; to erupt (about confusion, combustion) | подняться (о страшном переполохе) → **kup–**

kopaj– (2) *V* to become haughty, proud; to boast | кичиться

kopar– (2) *V* to break off | отломить, обломать; **kijamet kopar–** to make a great fuss; to shout and scold terribly | поднять сильный шум (крик), переполох, поднять скандал; орать → **χopar–**, **kupar–**

kopır (6), **kopir** (2) *N* bridge | мост → **kůpůr**

kor I (2) *N* glowing embers, live coal | жар, уголь

kor II (1) *Adj* poor, unlucky; contemptible | горемыка; низкий, презренный; **kor bak–** to look down (upon), to treat as of no account | пренебрегать, презирать, относиться с презрением

kor– I *V* to set up, to establish, to organize | строить, ставить, сооружать, организовать → **kur–**

kor– II s. **kȯr–**

kora (10) *N* (wattle) fence; hurdle; paddock, court | ограда, изгородь; плетень; загон, (скотный) двор; **aj korası** halo/ring/circle round the moon | венец вокруг луны → **χora**

koraj (3) *N* withered (tree, grass etc.), weed, rank | бурьян; **biñbaškoraj, bıñbaškoraj** eryngo (*Eryngium campestre*) | перекати–поле

korala– (1) *V* to fence (in/off) | огораживать

koraz (2) *N* cock | петух → **χoraz, kuraz**

korıǯı (16), **korıǯi** (6) *N* watchman, guard | сторож; **ekin korıǯısi** ranger | полевой сторож

kork– (11) *V* to be afraid, to fear | бояться, пугаться → **χorχ–, korχ–, kurχ–, kurk–**

korkot– s. **korkut–**

korkut– (3), **korkot–** (2) *V* to frighten, to alarm, to cause to be anxious | пугать

korlık (1) *N P.* anguish, torment, trouble | мука, мучение, страдание

koro *Adj* dry, dried | сухой; сушёный → **kuru**

korot– (2) *V* to dessicate, to dry | сушить, высушивать → **kurut–**

kortka (1) *N* old woman, gammer | старуха, старушка → **kurtka**

koster– s. **kȯster–**

koš I (3) *N* a pair | пара; парный; **koš avıǯ** two handfuls of | две горсти

koš II (2) *N* paddock (for sheep), wattle enclosure; herdsman's hut | загон (для овец); хата/хижина пастуха

koš III (10) pleasant, agreeable, good | приятный; привлекательный; **koš keldikke kel–** to come to say welcome to sb | прийти чтобы сказать добро пожаловать кому–л.; **koš geldiñiz/keldiñiz** welcome (polite) | добро пожаловать (вежливая форма); **koš keldige čık–** to go to meet, to welcome, to go to greet sb | выходить встречать кого–л.; **koš keldiñ** welcome! | добро пожаловать! → **hoš, χoš III, kuš II**

koš– (4) *V* to run, to go in haste | бегать; скакать на перегонки

košıl– (1) *Vpass* to join | присоединяться; прибавляться

kotak (1) *N* penis | мужской член

koter– s. **kȯter–**

kotır s. **kotor**

kotkodaχ (1) *Onomatop N* clucking | кудахтание; **kotkodaχ et–** to cluck, to cackle | кудахтать

koto (8) *N* box | коробка, ящик; **tu̇tu̇n kotosı** snuffbox, tobacco–jar | табакерка; коробка для табака → **kutu**

kotor (5), **kotır** (1) *Adj* mangy | паршивый

kov– (2) *V* to chase | гнать, преследовать

kozγa– (1) *V* to excite, agitate; to alarm; to acerbate | возбуждать, тревожить; раздражать

kozla– [<kozula–] (1) *V* to yean | окотиться, ягниться

kozo s. **kozu**

kozu (8), **kozo** (4), **kȯzi[!]** *N* lamb | ягнёнок, барашек → **kuzu**

kȯbek–taš (1) *N* central massage slab in a Turkish bath | (каменная) лавка (в турецкой бане)

kȯbete (4), **kebete**, **kobete**, **kůbete** *N* meat pie | кобете (пирог с мясной начинкой

kȯč– (4), **koč**– *V* to migrate; to move off; to migrate seasonally | переезжать, кочевать, трогаться → **koč, kůč**–

kȯj (12), **koj** (3) *N* village | деревня, село → **kůj**

kȯk (5) *N* sky | небо → **kůk I**

kȯkrek (2) *N* chest, breast; bosom | грудь; пазуха → **kůkrek**

kȯl (2) *N* lake | озеро → **kůl III**

kȯletke (1) *N* shadow, shade | тень

kȯmuš (1) *N/Adj* silver | серебро → **kůmůš**

kȯndelen (1) *Adv* athwart | поперёк; **kȯndelenden** pretended, feigned | притворно, деланно

kȯp (10), **kop** (3) *IndefNum* much, many; long | много, очень, долго

kȯr– (26), **kor**– (3), **kör**– (4) *V* to look | видеть → **gůr**–, **kůr**–

kȯrešmek (1) *Inf recipr* wrestling; fight | борьба

kȯrset– (2) *V* to show | показывать, предъявлять → **kůrset**–

kȯrůn– (5) *Vpass* to show oneself (to), to appear, to be seen, to seem | показаться, появляться; выглядеть

kȯrůš– (1) *V recipr* to see each other, to meet | увидеться, встречаться → **kůrůš**–

kȯse (72), **köse** (2) *N* lack–beard, with little or no beard | безбородный

kȯster– (5), **koster**– (2) *V* to show, to exhibit, to present | показывать, предъявлять; оказывать

kȯše (2), **köše** *N* corner | угол → **kůše**

kȯt (1) *N* bottom, buttocks | зад, задница; задний проход → **gȯt**

kȯter– (2), **koter**– *V* to rouse, raise, to lift | поднимать, повышать → **kůter**–

kȯtekle– (1) *V* to give a beating | избить → **kůtekle**–

kȯtůr– (2), **kötür**– *V* to take away, to carry away | уносить → **kůtůr**–

kȯz (7), **köz** (1) *V* eye | глаз → **kůz II**

köse s. **kȯse**

köše s. **kȯše**

Krako (1) *Tn* Krakow (city) | (город) Краков

kravat (2), **kıravat** *N* bed | кровать

Krım s. **Kırım**

kškene s. **kiškene]**

ku s. **ki**

kubernat (1) *N R.* governor | губернатор

kučak (2) *N* breast, embrace; lap | объятие, охапка → **kuǯak**

kučakla– (4) *V* to embrace, to take in one's arms | обнимать → **kuǯakla**–

kučaklaš– (1) *V* to embrace one another | обниматься → **kuǯaklaš**–

kuǯaχ s. **kuǯak**

kuǯak (8), **koǯak** (1), **kuǯaχ** (2) *N* breast, embrace; lap | объятие, охапка; **kuǯaχlara sıγ**– to fit into one's arms | вмещаться в объятия кого; иметь место в объатиях у кого–л.; → **kučak**

kuǯakla– (3) *V* to embrace, to take in one's arms | обнимать → **kučakla**–

kuǯaklaš– (1) *V recipr* to embrace one another | обниматься → **kučaklaš**–

kuǯur (2) *Adj* funny, weird | странный

kuǯurlan– (1) *Vpass* to find something peculiar/strange | удивляться

kuǯurlı (1) *Adj* curious, strange | странный

kuda (1) *N* matchmaker; marriage broker; father–in–law | сват; тесть, свёкор;

kudaka, kuda aka uncle matchmaker | дядя сват → **χuda**

kudača (1) *N* sister–in–law; matchmaker | сваха по отношению к детям другой свахи

kudaj (8) *N P.* God; My Lord! | Бог; Боже мой! → **Allah, İlah, χudaj, mevla, Subhan, tañrı**

kudalık (1) *N* proposal (of marriage); matchmaker's activity/work | сватовство; действие/дело свата→ **χudalıχ**

kudret (1) *N Ar.* power, might | мощ, могущество

kudur– (3), **kodor**– (1) *V* to seek, to look (for), to search | искать, разыскать

kugum (1), (~**kugům?)** *N* jug, pitcher | кувшин

kuj (11) *N* sheep | баран, овца → **koj I**

kuj– I (124) *V* to put, place, set | класть, положить; **ad kuj**– to give a name to sb | давать имя кому–л.; **baš kuj**– to decide to do sg, to make up one's mind to do sg | решаться; **hafizγa kuj**– to commit to one's memory | запечатлевать в памяти; **jurt kuj**– to build a house | строить дом; **kůz kuj**– to set one's eye on sb/sg; to run an eye over | приглядывать; **sarhoš kuj**– to make sb drunk | напоить кого–л. допьяна → **χoj–, χuj–, koj–**

kuj– II s. **kůj–**

Kuj–bakar (1) *Pn* Kuy–bakar (name of a horse meaning one, who takes care of sheep) | Куй–бакар, кличка коня обозначающая лицо, которе ухаживает за овцами);

kujan (7) *N* hare (Lepus europaeus) | кролик, заяц → **kojan**

kujaš (1) *N* sun | солнце

kujdur– (1) *N fact* to have sb put/placed (sg) to | заставить класть

kujı s. **kuju**

kujıl– (7) *Vpass* to be put/placed | быть положенным → **kıj– I, kıjıl–, kuj– I, kujul–**

kujın (25) *N* bosom | пазуха; **kujnı ač**– to open his/her heart (to sb) | открывать (своё) сердце (кому–нибудь); **kujnına al**– to embrace, to take on one's lap | обнимать, заключать в объятия кого–нибудь; **kujnına kir**– to snuggle, to go to bed with someone | обниматься, ложиться с кем–нибудь; **kujnına sarıl**– to be clasped to one's bosom, to embrace sb | упасть на грудь (к) кому–нибудь, обнимать кого–нибудь; **kujnına tůš**– to fall on someone's bosom | упасть на грудь (к) кому–нибудь

kujruk (3) *N* tail | хвост → **kojrok**

kuju (22), **kujı** (5) *N* well | колодец

kujul– (2) *Vpass* to be put/placed | класть, быть положенным → **kıjıl–, kujıl–**

kujumǯı (1) *N* goldsmith, jeweler | ювелир, золотых дел мастер

kujun (1) *N* sheep | баран, овца → **kojun**

kukı *N* smell | запах → **koku, kuku**

kukı– *V* to smell, to have a smell | пахнуть → **kokı–**

kukıš (1) *N* cock | петух

kukla– (4) *V* to smell | нюхать → **χoχla–**

kuklan– (1) *Vpass* to smell, to have a smell | пахнуть

kuku– (7) *V* to smell, to have a smell; to stink | пахнуть

kukuk (2) *N* cuckoo *(Cuculus canorus)* | кукушка

kukulda– (1) *V* to creak, to rustle | скрипеть, скрежетать

kul (90) *N* arm | рука

kulač (1) *N* fathom | маховая сажень

kulaǯık (1) *N dim* little ear | маленькое ухо → **kulak**

kulaj (7) *Adj* easy, easy way to do something | лёгкий, удобный; **kulajnen** easily, fairly easily | (довольно) легко, без труда → **kolaj**
kulak (21) *N* ear | ухо
kulamıta (3) *N* penis | мужской член
kuletke (1) *N* shadow, shade | тень
Kulımčak–batır s. **Kulumčak**
kuluk s. **kủlủk**
Kulumčak (10), **Kolumčak** *Pn* Kulumchak | Кулумчак; **Kulımčak/Kulumčak–batır** Hero Kulymchak/Kulumchak | Кулымчак/Кулумчак–богатырь
kum I (6) *N* sand | песок → **χum**
kum II s. **kim**
kumač s. **kumaš**
kuman (5) *N* water ewer with handle and long spout, used for ablutions, etc.; ewer, teapot, kettle | кувшин (для омовения); чайник; **kelin–kuman** bride–ewer | кувшин невсты; **legen kuman** basin and water ewer (used for ablutions) | таз и кувшин (для умывания или омовения) → **legen**
kumančıχ (1) *N dim* small water ewer with handle and long spout, used for ablutions, | кувшин (для омовения)
kumar (1) *N* gambling | азартная (карточная) игра
kumaš (3) *N* cloth, tissue, material | кумач, ткань, материя
kumek [kủmek?] (1) *N* pit (fall), hole | яма, погреб
kun s. **kủn**
kun– (22) *V* to settle/make a night's halt | останавливаться (на ночлег, кочевье), опускаться, садиться → **kon–**
kunak (13) *N* halting place, inn; guest, temporary visitor | ночлег; гость; **konak et–** to give a feast, to entertain sb in one's house | принимать (угощать) гостей → **konak**
kunakbaj (4) *N* owner, landlord | владелец, хозяин дома → **konakbaj**
kunaklık (2) a day's journey | однодневный переход; **kunaklık aulak** distance of a day's journey | расстояние однодневного перехода; **kunaklık jer** (territory) distance of a day's journey | (место) расстояние однодневного перехода
kunan (3) *N* a two– or three–year–old foal | двух–трёхгодовалый жеребёнок → **konan**
kunıš– (1) *V* to speak | говорить → **χonuš–, konuš–**
kunuk (1) *N* guest | гость → **konuk**
kuñšu s. **koñšu**
kup– (4) *V* to break in two, to break off | отламываться, отскакивать; **kijāmet kup–** all hell to break loose; to erupt (about confusion, combustion) | подняться (о страшном переполохе) → **kop–**
kupar– (1) *V* to break off | отломить, обломать → **kopar–**
kupır [kủpır?] (3) *N* bridge | мост → **kopır, kủpri, kủpủr**
kuplu s. **kaplı–kuplu**
kur– (3) *V* to set up, to establish, to organize | строить, ставить, сооружать, организовать → **kor– I**
kuraz *N* cock | петух → **χoraz, koraz**
kurban (7) *N Ar.* sacrifice; victim | жертва; жертвенное животное; **kurban bajram** (the Muslim) festival of sacrifice | курбан–байрам (праздник жертвоприношения); **kurban bol–** to sacrifice oneself (for) | пожертвовать себя, стать жертвой; **kurban čal–** to kill as a sacrifice | заклать жертву; **kurban**

čaldır– to have (an animal) killed as a sacrifice | велеть заклать жертву; **kurban ket**– to fall a victim to | стать жертвой; **kurban olayım!** please! I beg you! | прошу! умоляю!

kurbetlik s. **gurbetlik**

kurd s. **kurt I**

kurχ– s. **kurk**–

kurk– (45), **kurχ**– (5) *V* to be afraid, to fear | бояться, пугаться; **kurχa kurχa/ kurka kurka** timorously, timidly, fearfully | боязливо, пугливо, робко → **χorχ–, korχ–, kork–**

kurku (11) *N* fear, fright, anxiety | страх, боязнь, испуг

kurkunıšlı (1) *Adj* terrible, terrifying | страшный

kurkuz– (1) *V* to frighten, alarm; to cause to be anxious | пугать

kurma (2) *N P.* date | финик

kurna (1) *N Ar.* basin of a bath or fountain | бассейн (для купанья или фонтана)

Kurpatkin (1) *Pn R.* Kurpatkin | Курпаткин

kursaǯık (2), **kusaǯak** *N dim* small stomach/belly | желудочек, животик

kursak (14) *N* stomach, belly, rumen; bosom | живот, брюхо; пазуха

kursi (6), **kursı** *N Ar.* small table, bedside table; throne | табуретка, столик; трон, престол

kuršun (4) *N* lead; bullet | свинец, пуля

kurt I (~kurd) (1) *N* wolf | волк

kurt II (1) *N* worm | червь

kurtar– (1) *V* to save, rescue | спасать → **kutar**–

kurtıl– s. **kurtul**–

kurtka (22) *N* old woman, gammer | старушка → **kortka**

kurtul– (2), **kurtıl**– *Vpass* to escape, to get free; to be saved | спасаться; вырваться → **kutul**–

kuru (2) *Adj* dry, dried | сухой; сушёный → **koro**

kuru– (2) *V* to dry, to become dry | сохнуть, высыхать

kurul– (5) *Vpass* to be built, to be put on | строиться, быть построенным; **samavar kurulγan** the samovar was put on | восседали самовар, поставили самовар

kurulu (2) *Adj* established, set up; being put on | поставленный, построенный; **kurulu samavar** samovar being put on | поставленный самовар

kurut– (7) *Vfact* to dessicate, to dry | сушить, высушивать → **korot**–

kus– (2) *V* to throw up, vomit | рвать, блевать; **kobuk kus**– to vomit foam, to foam | запениться

kusaǯak s. **kursaǯık**

kusku (1) *N* (pointed) crowbar; iron wedge | лом

kusur (1) *N Ar.* fault, defect | изъян; недостаток

kuš I (66) *N* bird | птица; **ababil kušu** common swift (*Apus apus*); European nightjar (*Caprimulgus europaeus*), common goatsucker | горная ласточка, обыкновенный козодой, чёрный стриж; **kara kuš** steppe eagle (*Aquilla nipalensis*) | ягнятник, степной орёл; **kuš bala/balası** nestling, young bird | птенец; **šāhin-kušı** hawk (*Accipiter nisus*), falcon | сокол, лячин; **tutı kušu** parrot (*Ales imitatrix*) | попугай; **uma kušı** bird of paradise, bird of chance | гамаюн, райская птица; птица счастья

kuš II (1) *Adj/Adv* pleasant, agreeable, good | приятный; привлекательный; **kuš keldiñ** welcome! | добро пожаловать! → **hoš, χoš, koš III**

kuš– (1) *V* to embrace; to take in one's arms | обнимать, брать в охапку

kušak (9) *N* belt, girdle | пояс

kušan– (6) *Vpass* to gird oneself, to put on a sash; to dress up, to make oneself smart | одеваться, опоясаться, наряжаться

kušlık (1) *N* forenoon; lunch | время между утром и полуднем; обед; **kušlık vaktı** forenoon time | время между утром и полуднем

kušu (7) *N* (horse) race, racing | бег, состязание в беге; скачки

kutar– (20) *V* to save, rescue | спасать → **kurtar–**

kutıl– s. **kutul–**

kutır– s. **kutur–**

kutlı (1) *Adj* lucky, fortunate; happy | счастливый, радостный

kutlulık (1) *N* happiness, joy, gladness | счастье, блаженство

kutu (4) *N* box | коробка, ящик; **tůtůn kutusı** snuffbox, tobacco–jar | табакерка; коробка для табака → **koto**

kutul– (26), **kutıl– (~kutl–)** (3) *Vpass* to escape, to get free; to be saved | спасаться; вырваться → **kurtul–**

kutur– (2), **kutır**– *V* to go mad; to be infected with rabies | беситься → **χutur–**

kuturt– (1) *Vfact* to drive mad/crazy | сводить с ума; доводить до сумасшествия

kuv– (24), **ku**– (6), **kū**– (3) *V* to drive away | гнать, прогонять

kuval– s. **kuvala–**

kuvala– (7), **kuval**– (1) *V* to pursue, to run after, to chase | гоняться, преследовать

kuvan– (17) *V* to be glad, to be pleased | радоваться, испытывать радость; **kuvanγanından** for/with his/her joy | от радости

kuvanč (1) *N* joy, pleasure, delight | радость

kuvet (1) *N Ar.* strength, power, force | сила, мощь, могущество

kuvıš (1) *N* hollow, cavity, cave | дупло

kuzu (1) *N* lamb | ягнёнок, барашек → **kozu**

kuzulı (1) *Adj* big with young (sheep), accompanied by a lamb | с ягнёнком; суягный

kū– s. **kuv–**

kůb s. **kůp I**

kůbůk (7) *Postp* like; as; (after predicate) almost, nearly, somewhat | (послелог) какъ подобно, словно, в роде → **gibik, kibi, kibik**

kůč I (1) *N* strength, power, force | сила, мощь

kůč II (1) *Adj* difficult, heavy; hard | тяжёлый, трудный

kůč– (10) *V* to migrate; to move off; to migrate seasonally | переезжать, кочевать, трогаться → **kȯč–**

kůčelek (4) *N* pup, puppy | щенок

kůčli (1) *Adj* difficult, strong, powerful | сильный

kůčůʒik (5), **kůčůʒık**, **kůčůʒůk** (2) *Adj dim.* little, small | маленький, младший

kůčůʒık s. **kůčůʒik**

kůčůʒůk s. **kůčůʒik**

kůčůk (26) *Adj* small, little, young | маленький, младший; **kůčůgůnde** at his/her little age | в маленьком возрасте; **kůčůkten** from his/her little age, from his/her childhood | с маленького возраста → **gůččůk**

kůgen (3) *N* blackthorn, hawthorn (*Prunus spinosa*), wild plum | тёрн (плод), терновый; **Kůk–kůgen** (toponym) Blue–wild plum | Синий тёрн → **gůgen**

kůgerčın (4), **kůgerčin** *N* pigeon, dove | голубь; **kůk kůgerčin** stock pigeon/dove (Columba oenas) | клинтух → **kůjerʒin, kůveršın**

kůj (75) *N* village | село, деревня → **kój**

kůj– (7), **kuj–** [=**kůj**–!] (2) *V* to burn, to be on fire | гореть

kůjerʒin *N* pigeon, dove | голубь → **kůgerčın, kůveršın**

kůjeu (~kůjev) (6), *N* bridegroom | жених

kůjıs (1) *N* chest, breast, bosom | грудь, грудная клетка → **kůkůs**

kůk I (18) *N* sky | небо → **kók**

kůk II (15) *Adj* blue, grey | голубой, серый

Kůk–čokrak (1) *Tn* Kuk–chokrak (hydronym meaning blu(ish) wellspring) | Кюк–чокрак (гидроним обозначающий голубой источник)

Kůk–kůgen (1) *Tn* Kuk–kugen (hydronym meaning blu(ish) sloe/hawthorn/plum) | Кюк–кюген (гидроним обозначающий (голубой) тёрн; терновую ягоду → **kůgen**

kůkče (1) *Adj* bluish–grey | голубовато–серый

kůkrek *N* chest, breast; bosom | грудь; пазуха → **kókrek**

kůkůs (3) *N* chest, breast, bosom | грудь → **kůjıs**

kůl I (26) *N* ashes | зола, пепель; **kůl–karıštıran** stirring ashes | смешивающий пепель

kůl II (13) *N* rose (Rosa), flower, blossom | роза, цветок → **gůl**

kůl III (1) *N* lake | озеро → **kól**

kůl– (35) *V* to smile | смеяться

kůlbahče (1) *N* rose–garden | розовый сад → **kůl II, bahče; gůlbahče**

Kůlbastı (20), **Kulbasti** *Pn Ar.* Kulbasty (name of the hero of the tale meaning: grilled thin flank) | Кюльбасты (имя сказочного героя; значение имени: мясо, жаренное на рашпере

kůldir– (1) *Vfact* to cause to laugh, to amuse | смешить

kůleš– s. **kůlůš–**

kůletke (3) *N* shadow | тень

kůlge (2) *N* shadow | тень

kůliš– s. **kůlůš–**

Kůl–karıštıran (2) *Pn* Kul–karўshtўran (name of the hero of the tale meaning: mixer (stirring) of ashes) | Кюл–карыштыран (имя сказочного героя; значение имени: мешающий золу)

kůlmek (24), **külmek** (1) *N* shirt, chemise, smock | рубашка; сорочка; **kenarsız kůlmek** skirt without trimming | рубашка без обшивки; **keten kůlmek** flaxen/linen shirt | лен, льняная рубашка → **kolmek**

kůlte (2) *N* bunch, bundle, sheaf | сноп, связка

kůlůk I (4), **kuluk** (1) *N* brood hen | клуша, наседка

kůlůk II (1) *N* pick, mace | кирка, булава; **kůlůk balta** pole–axe | секира

kůlůš– (5), **kůleš– [kůlėš–]**, **kůliš–** *V recipr* to laugh together | смеяться вметс с кем

kům– (25) *V* to bury, to dig into the ground | хоронить, закапывать в землю

kůmeč (3) *N* brioche, plain cake; bun, muffin | калач, булка; **majlı kůmeč** buttered roll | сдобная булка, маслиный калач

kůmenli (1) *Adj* angry | сердитый

kůmes (8) *N* henhouse, chicken coop | курятник, птичник; **kaz kůmesi** goose house | птичник для гусей

kůmů s. **kimi**

kůmůl– (1) *Vpass* to be burried | быть похороненным (погребённым)

kůmůr (6) *N* coal | уголь

kůmůsi (3), *Pron* who, some | кто–то → **kimisi**

kůmůš (50) *N/Adj* silver | серебро; **kůmůšten/kümüšten** (made) of silver | (сделан) из серебра → **kȯmuš**

kůmůšlik (2) *Adj* containing silver | содержащий серебро; **jůz kůmůšlik** worth of hundred silver coins | ценностью в сто серебра

kůn (346), **kun** (2), **kün** (5) *N* day | день; **bajram kůni** festive day | праздник; **bůkůn** today | сегодня; **ǯuma kůni** friday | пятница; **er kůn** every day | каждый день; **kůn tuɣanda** at sunrise | на рассвете; **kůnde pišken jemek** food cooked in the sun | пища обожжённая, поспетая на солнце; **kůnge pišken aš** food cooked in the sun | пища обожжённая/поспетая на солнце; **pazar kůni** sunday | воскресенье; **salı kůni** tuesday | вторник → **gůn**

kůnde (2) *Adv* daily, per/every day | дненной, ежедневно, каждый день → **gůnde**

kůndůz (14), **kůnduz** *N/Adv* day (time), by day | день, днём; **keče–/keǯe–kůndůz** day and night | день и ночь, днём и ночью

kůneš (1) *N* sun | солнце → **gůneš, kůñeš**

kůnın (1) *Adv* on the day of; by day | в день; днём; **degen kůnın alırmen** I'll take it (back) on the fixed day | возьму (обратно) в назначенный день

Kůn–kůrmez (1) *Pn fem* Kun–kurmez | Кюн–кюрмез

kůnlůk (1) *Adj* daily, of the same day; so many days old | дневной, суточнуй; –дневный, на ... дней

kůnne– (1) *V* to be jealous (of), to envy | ревновать, завидовать

kůñeš (2) *N* sun | солнце → **gůneš, kůneš**

kůñůl (10) *N* soul, heart | душа, сердце

kůp I (2), **kůb** *N* large earthenware jar with two ears | большой глиняный кувшин с двумя ушками; чаша

kůp II (25) *IndefNum* a lot, much, many | много; **az kete, kůp kete** (in tales) he goes few, he goes much | (в сказках) мало ли идёт, много ли идёт

kůpe (1) *N* earring | серьга, серёжка

kůpek (23) *N* dog | собака; **fino kůpek** lap–dog | комнатная собачка

kůpri (4) *N* bridge | мост **kůbri (kůpri) bašı** bridgehead | предмостье → **kopır, kůpůr**

kůpůr (9), **kupır [kůpır]** (3) *N* bridge | мост → **kopır, kůpri**

kůr (3) *Adj* blind | слепой; **kůr–topal** blind–lame (=all together) | слепые–хромые (значит: все вместе)

kůr– (246) *V* see | видеть; **dāva kůr**– to settle the case (of) | предъявлять иск, подавать жалобу; **iš kůr**– to work | работать; **kůrči!** do look (at it)! | смотри же!; **kůre jaz**– to come near to seeing sb/sg | чуть не увидеть; **mevanet kůr**– to get help from sb | получить помощь от кого–л.; **rahat kůr**– to have peace/quiet | обретать (душевное) спокойствие, видеть (знать) покоя/удобства; **tůš kůr**– to have a dream | видеть сон; **ůrmet kůr**– to have honour, to be respected | пользоваться уважением, быть уважаемым → **gůr–, kȯr–, kör–**

kůre (2) *Postp* according to, as to, in respect of | согласно (чему–л.), по, соответственно (чему–л.); **bilgenine kůre** according to his/her knowledge | согласно его/её знанию; **ůziñe kůre** accordig to yourself | по–твоему самому

kůre– (1) *V* to shovel (away), to clear away, to clean up | грести лопатой; оттребать, убирать

kůrek (4) *N* shovel; shoulder–blade | совковая лопата; лопатка

kůreš (3) *N* wrestling, fight | борьба, схватка

kůreš– (3) *V* to wrest, to fight | бороться, сражаться

kůrešme (1) *Inf(ma)* wrestling, fight | борьба, схватка

kůrėš– s. **kůrůš**–

kůrilti s. **kůrůlti**

kůrin– s. **kůrůn**–

kůriš– s. **kůrůš**–

kůrın– s. **kůrůn**–

kůrk (2) *N* fur, fur coat | мех, меховая шуба

kůrmek (1) *Inf (mak)* seeing | видение

kůrpe (1) *N* bran | крупа; **arpa kůrpe** barley–bran | ячневая крупа

kůrset– (2) *V* to show | показывать, предъявлять → **kȯrset**–

kůrůl– (1) *Vpass* to be seen | виднеться, видеться

kůrůlte– (1) *V* to make a loud noise; to thunder | шуметь, грохотать, греметь; **kůk kůrůlte**– to make a thundering noise (sky) | греметь → **gůrůlde**–

kůrůlti (3), **kůrilti** *N* noise, uproar | гул, рокот → **gůrůldi**

kůrůn– (21), **kůrin**– (8), **kůrın**– *Vpass* to be seen, to show oneself | виднеться, видеться; показывать себя → **gůrůn**–

kůrůš– (12), **kůrėš**–, **kůriš**– (6) *V recipr* to see each other, to meet | увидеться, встречаться → **kȯrůš**–

kůrůšme (2) *Inf(ma) recipr* seeing each other, meeting someone | видеть друг друга; встречаться

kůrůšmek (1) *Inf(mak) recipr* seeing each other, meeting | встреча, свидание

kůrůštir– (1) *Vfact* to arrange a meeting (for) | устраивать свидание (встречу)

kůs– (2) *V* to tire (of), to get bored (with) | надоесть, наскучить

kůster– (32) *V* to show, to exibit | показывать; предъявлять; оказывать

kůše (8) *N* corner | угол; **kůše baši** corner of the street | угол улицы; → **kȯše**

kůšeli (2) *Adj* cornered, angled | имеющий какой–л. угол; –угольный

kůšk (3) *N P.* summerhouse; pavilion | летний дворец; павильон

kůt (7) *N* bottom, buttocks, backside | зад, задница; **boklu kůt** shitty buttocks/arse | дерьмовая задница

kůt– (2) *V* to wait (for), to await, to expect | ждать, ожидать

kůtek (2) *N* stick, blow, cudgeling | палка, драка

kůtekle– (2) *V* to cudgel | избить → **kȯtekle**–

kůten (2) *N* bottom, buttocks, backside | зад, задница; **kůteni koltuyına al**– to take arse under one's arm? | задницу подмышкой взять?

kůter– (19), **küter**– *V* to rouse, raise, to lift | поднимать, повышать → **kȯter**–

kůtůk (4), **kutuk** (2) *N* tree trunk, balk; log | пень, чурбак, колода, куст

kůtůr– (2) *V* to take away, to carry away | относить, отвозить, увазить → **kȯtůr–**

kůveršın pigeon, dove (*Columba livia*) | голубь → **kůgerčın, kůjerʒin**

kůz I (3) *N* automn | осень

kůz II (135), **küz** (1) *N* eye | глаз; **Kara–kůz** Kara–kuz (Black–eye, a name) | Кара–кюз (Чёрный глаз, имя); **kůz jašı** tear, teardrop | слеза; **kůz jašıñı kurutma!** may your tears not dry up! | чтобы не иссякали твои слёзы!, да не иссякнут твои слёзы!; **kůz et–** to set one's eye on sb/sg; to run an eye over | приглядывать; **kůz jumup ačχanʒe** in the twinkling of an eye | не успеть (глазом) мигнуть; **paranı[n] da kůzi siʒak** the eye of money is also attracktive | глаз денег тоже привлекателен → **kȯz**

kůzajdın (1) *N* good news, congratulation (because of a good news) | радостная весть; поздравление с радостной вестью

kůzʒik [kůzʒık] (1) *N dim* the apple of one's eyes? | свет жизни, свет очей → **kůz II**

kůzel (1) beautiful; good, nice | красивый, хороший → **gůzel, güzel**

kůzgů (7) *N* looking–glas | зеркало

kůzle– (1) *V* to watch, to take aim at | целиться, метить; **jımırta kůzle–** to test/screen eggs by light | тестировать/проверять яйца просвечиванием

kůzli (6), **kůzlı** (2) *Adj* having (such and such) eyes | имеющий какие–л. глаза; **ela kůzli** having hazel eyes | с карими глазами; **kůmůr kůzlı** having coal–black eyes | с угольно–чёрными глазами

kůzlı s. **kůzli**

kümse s. **kimse I**

küzellik (1) *N* beauty | красота → **gůzellik**

kz s. **kız**

L

laʒ s. **ilaʒ**

laf (10) *N P.* word, talk | слово, речь, разговор; **laf et–/ejt–** to talk, to chat | разговаривать, болтать → **lap**

layım (3) *N Ar.* underground tunnel | подземный ход

laj (1) *N* silt, clay | ил, грязь

lajik s. **lājik**

lajılahi (1) *Interj Ar.* oh, my God! | ах господи! боже мой!

lajk s. **lājik**

laki s. **lakij**

lakij (8), **laki**, **lakıj** *N R.* servant | лакей, слуга

lakin s. **lākin**

lakın s. **lākin**

lale s. **lāle**

lampa (4) *N R.* lamp | лампа

langa s. **χıjar**

lap (19) *N P.* word, talk | слово; речь, разговор → **laf**

lap–lap (1) *Onomatop* the sound of the pat of the (bare) feet | шлёп–шлёп (подражание шлёпанию ног)

lazim (1) *Adj Ar.* necessary, needed | необходимый; надо, нужно

lājik (13), **lajik** (2), **lajık, lajk** (3), **lājik** (5), **lājıχ, lājık** *Adj Ar.* worthy, suitable; fitting, deserving, appropriate | подходящий, достойный; **baña/kendime lājık** it is appropriate/fitting to me | подходящий мне; **lajik kȯr–**

to seem worthy or suitable, to find fitting | видеть/рассуждать достойным/подходящим

lājıχ s. **lājik**

lājık s. **lājik**

lākin (83), **lakin** (33), **lākın** (3) *Conj Ar.* but, still, yet, however | но, однако

lāle (3), **lale** *N P.* tulip | тюльпан

lānet (1) *N Ar.* malediction; damnation | проклятие; проклятый

lāzim (2) *Adj Ar.* necessary | необходимый, нужный; надо, нужно

legen (5), **lejen** *N P.* large bowl, basin | лохань, таз; **legen kuman** basin and water ewer (used for ablutions) | таз и кувшин (для умывания или омовения) → **kuman**

lejen s. **legen**

lejla I (1) *N Ar.* night, very dark night | ночь; очень тёмная ночь

Lejla II (1) *Pn Ar.* Leyla | Лейла

lemse (1) *N/Adj* German; (here) Austrian | немец(кий); (здесь) австриец

leppek (3) *Interj Ar.* here!, present!; at your service! | чем могу служить? (готов) к вашим услугам! слушаюсь!

leš (4) *N P.* carcass | падаль

lezzet (1) *N Ar.* taste, flavor | вкус; **dat lezzet** taste–flavour | вкус–смак → **dat, tat I, nezet**

lika–lika (2) *Onomatop* **(kujan) kelij lika–lika** (rabbit) is coming "lika–lika" (special way of rabbit's moving) | приходит "лика–лика" (об особенности движения/хода зайца)

limon (1) *N Gr.* lemon | лимон → **ilmon**

Livov (2) *Tn R.* Lvov (city) | (город) Львов

lokata (1) *N* gratis, for nothing | на даровщину, даром, за чужой счёт

Lumančıχ (1) *Pn* Lumanchikh | Луманчих

M

ma (6) *Interj* see! behold! here (you are)! | вот тебе! нате!

Maǯar I (1) *En* Hungary | Венгрия → **Maǯarja**

maǯar II (2) *N* cart, coach of Hungarian style | телега (венгерской разновидности); **tůje maǯarı** camel carriage | верблюжая повозка

Maǯarja (2) *Tn* Hungary | Венгрия → **Maǯar**

mafuz–hane s. **māfuz–hane**

maɣara (12), **maɣra** *N* cave, cavern | грот

maɣra s. **maɣara**

maɣraš– (2) *V recipr* to bleat together | блеять (вместе, совместно)

mahal (2), **māl** *N Ar.* time | время

mahale s. **mahalle**

mahalle (3), **mahale** *N Ar.* quarter, district | квартал, район, территория

mahbetlik (1) *N Ar.* love, friendship | любовь, дружба

mahkeme (1) *N Ar.* law court | суд

mahmur (2), **māmur** *Adj Ar.* sleepy (languid) eye; soft, tender (look) | томный, сонный, мечтательный

mahšer (2) *N Ar.* (place of) the last judgement; great crowd, great confusion | место, где будут собраны все люди в день страшного суда; столпотворение

maχtan– s. **maktan–**

mail (1), **majil** *Adj Ar.* leaning, inclined; tending towards | наклонившийся; склонный; симпатизирующий (кому–чему); **mail bol–** to be inclined to, to be tending towards; to fall in love with | быть склонным, склоняться к чему; влюбиться в кого → **mejil**

maj (5) *N* fat, oil, grease | масло, жир

majdan (4) *N Ar.* open space; public square; ground | площадь, плац; **majdanγa čık–/čıχ–** to come out into the open; to be revealed, to be found out | выходить на ружу, обнаруживаться, показываться

majil s. **mail**

majilsiz (1) *Adj* not inclined; not tending to | не склонный

majla– *V* to oil, to butter, to grease | намазать (маслом), смазывать

majle s. **māle**

majlı (3) *Adj* fat; greasy, oily | масляный

makas (5) *N Ar.* scissors; shears | ножницы

makbul (1) *Adj Ar.* acceptable, liked | допустимый, принятый

maktan– (6), **maχtan–** *Vpass* to boast, brag | хвастаться, хвалиться

maktavlı (1) *Adj* famous, renowned | хвалёный

mal (72) *N Ar.* property, wealth; an animal owned (cattle, horses etc.) | товар, имущество; скотина, скот (о крупном рогатом скоте); **malγa bola [da]** (even if) it costs livestock) | (даже если) это стоит скота

mal–mǘlk (1) *N Ar.* goods, property | всё имущество; пожитки

malım [malim] (1) *N Ar.* teacher | учитель, преподаватель

Maltabar (2) *Pn* Maltabar (i. e. he will be rich; he will get riches) | Малтабар (т. е. найдет богатства)

mamırlık (1) *N* heaviness after a drunken sleep; pleasant sleepy feeling | томность

mandal (1) *N Ar.* latch; bolt | защёлка, задвижка

mangal (1) *N Ar.* brazier; roasting pan (eastern) | мангъал; жаровня

Manγuš (1) *Pn fem* Mangush | Мангуш

mani (9), **mane**, **manı**, **mānı** *N Ar.* a form of Turkish folk song | мане (вид народнопоэтического творчества)

maniʒi (3) *N* composer and singer of mani | слагатель и исполнителъ мани

mantar (1) *N Gr.* mushroom | гриб

mañlaj (2), **mañnaj** *N* forehead | лоб

mañnaj s. **mañlaj**

mapus (1) *N Ar.* prison | тюрма → **māfuz–hane**

marafat s. **marafet**

marafet (3), **marafat** *N Ar.* skill, talent, skilled trade | умение, мастерство, способность, искусство → **mārefet**

maraz (1) *N Ar.* disease, illness | болезнь; **inʒe–maraz** tuberculosis | туберкулёз; **inʒe–maraz it–** to make sb sick | сделать кого больным

mārebe (1) *N Ar.* war, battle, fight | бой, схатка

marhametsiz (1) *Adj Ar.* merciless, pitiless | без милосердия

martini (2) *N* Martini(–Henry) single–shot rifle | Мартини(–Генри) однозарядная винтовка

Maruška (1) *Pn fem R.* Marushka | Марушка

masal (41) *N Ar.* (fairy) tale, story | сказка; **masal masal maseki (**beginning formula of tales) | (зачин сказок)

masalʒı (2) *N* story teller | сказочник

maseki s. **masal**

masχara s. **maskara**

masχaralık (4), **maskaralık** *N* buffoonery; shame, dishonour | позор, бесстыдство; **masχaralıkka al–** to make fun of sb, to ridicule | надсмеиваться/издеваться над кем–л.

maskara (7), **masχara** (2) *N Ar.* buffoon, laughingstock, ridiculous | предмет насмешек, опозоренный; **maskara et–** to make a laughingstock of sb |

сделать посмешищем; **maskara bol–** to be ridiculed, to be made a laughingstock of | сделаться посмешищем, быть осмеянным

maskaralık s. **masχaralık**

maša (1) *N P.* tongs, pincers | шипцы для камина, мангала

mašina (7), **mašına** (4) *N R.* machine; (here) railway carriage | машина; (здесь) железнодорожный вагон

mat (6) *N P.* (chess) checkmate | шах и мат; **mat et–** to defeat | победить

mator (1) *Adj* goodlooking, handsome | приятной наружности

matu (10), **mātu** *Adj Ar.* silly | глупец; **matu jerine koj–** to take sb for fool | считать кого–нибудь глупцом

mazar s. **mezar**

māǯir (1) *N Ar.* emigrant, fugitive | переселенец; эмигрант, беженец

māfuz–hane (3), **mafuz–hane** (1) *N Ar.* prison | тюрма → **mapus**

mākul (1) *Adj Ar.* reasonable, wise, rational | понятный, разумный, рациональный

māl s. **mahal**

māle (7), **majle** (1), **mālle** (1) *N Ar.* quarter, district, ward | квартал, район

mālle s. **māle**

māmur s. **mahmur**

mānā (1) *N Ar.* meaning, sense | смысл, значение

māni (7), **manı** (2) *N Ar.* a form of Turkish folk song | мане (вид народнопоэтического творчества);

mārefet (1) *N Ar.* skill, talent, skilled trade | мастерство; искусство; фокус, проделка → **marafet**

mārim s. **mārum**

mārum (3), **mārim** (1) *Adj Ar.* deprived, destitute | лишённый, не имеющий /кого– чего–л.; **[–dan] mārim bol–** to be deprived of, to be disappointed in/with | быть лишённым, лишаться чего, терять кого–что; **[–dan] mārim kal–** to remain deprived of | быть лишённым, лишаться чего, терять кого–что

mātu s. **matu**

māze (5) *N Fr.* (large) store, shop | магазин, лавка

mečid s. **mečit**

mečit, (Ak–)**mečit** (4), (Ak–)**mečid** (2), (Ak–) **mečıd** *N Ar.* (small) mosque | мечеть; **Ak–mečid, Ak–mečıd, Ak–mečit** Akmedjid 'White mosque' | Акмеджид 'Белая мечеть' → **meǯid**

meǯid, (Ak–)**meǯid** (5), (Ak–)**meǯit** (1) *N Ar.* small mosque | мечеть; **Ak–meǯid** Akmedjid 'White mosque' | Акмеджид 'Белая мечеть' → **mečit**

meǯit s. **meǯid**

meǯnun (1) *N Ar.* mad, insane; | обезумевший от любви

meded (1) *N/Interj Ar.* aid; help! | помощь, помогите!

medrese (9) *N Ar.* medresse, Muslim theological school; high school | медресе (духовное училище)

meger (2) *Conj P.* but, however; perhaps | а; но; так как; оказывается

Meχail (1) *Pn R.* Mekhail | Михаил

mejdan (5), **mejden** (3) *N Ar.* open space, public square; arena, ground | площадь, плац; **mejdana čıkar–** to bring to light, to reveal | выводить на чистую воду, разоблачать; **mejdana kel–** to turn out, to reveal | раскрываться, обнаруживаться

mejden s. **mejdan**

mejil (3), **mejl** (2) *N Ar.* leaning, inclined; tending towards | наклон, наклонившийся; склонный; симпатизирующий (кому–чему); **mejil/**

mejl ber– to be inclined to, to be tending towards; to fall in love with | увлечься, полюбить кого–л. → **mail**

mejl s. **mejil**

Mejnaru (3) *Pn fem* Meynaru | Мейнару

mejva (5) *N P.* fruit | плод, фрукт

mejvelik (1) *N* fruit garden | фруктовый сад

mejzer s. **mezar**

mejzerlik (1) *N* graveyard, cemetery | кладбище

mejzin (52), **mejzen**, **mejzın** (2) *N Ar.* muezzin, one who calls Muslims to prayer | муэдзин; **aχot–mejzin** hunter–muezzin? | охотник–муэдзин?

mejzın s. **mejzin**

mekjen (2) *N Ar.* place, site; residence | место, местопребывание

mekteb (1) *N Ar.* school | школа

mektup s. **mektůp**

mektůp (24), **mektub** (7), **mektüb** (4), **mektüp** (3) *N Ar.* letter | письмо

mektüb s. **mektůp**

mektüp s. **mektůp**

melek (4) *N Ar.* angel | ангел

melepše s. **melevše**

melevše (6), **melepše** *N P.* violet (*Viola tricolor*) | фиалка; **bir top melevše** a bunch of violet | один букет фиалки; **mor melepše** purple violet | лиловая фиалка

melhem (2) *N Ar.* ointment; healing, cure | мазь, лекарство

melmer s. **mermer**

melte (1) *N* (candle) wick; | плошка, светильник, фитиль; **čırayın meltesi** candle ick | фитиль лучины

meme (26) *N* teat; breast | материнская грудь; сосок

Memet (4), **Memmet** (2) *Pn Ar.* Memet | Мемет

Memiš (2) *Pn dim hypoch* Memish | Мемиш

memleket (9) *N Ar.* country | страна

Memmet s. **Memet**

men (793) *Pron pers Sing1* I | я; **maña** to/for me | мне, для меня; **men dajın** like me | как я; **menė** me | меня; **meni, menim** my, of me | мой, у меня; **meni, menı** me | меня → **ben I**

menteševli (1) *Adj P.* hinged; furnished with hinges; bolted | с металлической петлей, имеющий петельный кручок; шарнирный, коленчатый, подвижный; с угловым железом; **enteševli–menteševli** hinged–hinged(?) | шармирный–шарнирный(?)

Meñaru (3), **Miñero** *Pn fem* Mengaru | Менгару; **Kůn–kůrmez Meñaru** Kun–kurmez Mengaru | Кюн–кюрмез Менгару → **Mejnaru**

merǯan (1) *N Ar.* coral | коралл

merdāne (1) *Adj P.* brave, manly, heroic | храбрый, доблестный

merdimen (4) *N P.* ladder; stairs, staircase | лестница

merekeb (4), **merekep** (1) *N Ar.* ink | чернила

merekep s. **merekeb**

merka (15) *N R.* bushel, measuring vessel | мерка

mermer (8), **melmer** (3) *N/Adj Ar.* marble | мрамор, мраморный; **mermer direk** marble pillar | мраморный столб; **mermer taši** marble stone | мраморный камень

mešavre [<mešavere] (1) *N Ar.* mutual consultation, council | обсуждение, совет; **mešavre et–** to consult (sb about sg), to discuss (sg with sb) | совещаться о чём–л., советоваться с кем–л.

meš̌zid, (Ak–)**meš̌zid** (1) *N Ar.* small mosque | мечеть; **Ak–meš̌zid** Akmeshdjid 'White mosque' | Акмещджид 'Белая мечеть' → **mečit, meǯid**

meše (2) *N P.* oak (Quercus) | дуб

mešrebe s. **mešrepe**

mešrepe (4), **mešrebe** (3) *N Ar.* jar, jug, mug | кружка

metej s. **meti**

meti (3), **metej** *N* barrel, cask | бочка; **ǯavunduk metilerı** dirty water barrels | бочки с грязной водой; **su meti** water–barrel | водовозная бочка

metinni (2) *Adj Ar.* firm, strong, solid; trustworthy | стойкий, непоколебимый

mevanet (2) *N Ar.* help, aid; provisions, food | помощь; провизия; **mevanet kůr–** to get help from sb | получить помощ от кого–л.

mevla (9) *N Ar.* God, creator | бог; **Mevlam** my Lord | Господь мой

mezar (15), **mazar** (2), **mejzer** (1), **mezer** (6) *N Ar.* tomb | могила

mezer s. **mezar**

mezerlik (4) *N* cemetery, graveyard | кладбище; **mezerlik taší** grave stone | надгробный камень

mezlič (2) *N Ar.* assembly, council | собрание

mi (121), **ım** (5), **mı** (14), **mo** (3), **mu** (6), **um** (2) *IntParticle* | (частица вопросительная)

millet (2) *N Ar.* nation, people | народ

min (40) *Pron pers* I | я; **minde** at me, on me | у меня, на мне; **mine** me | меня; **minem** my, of me | мой, у меня; **mini** me | меня

min– (95), **mın–** *V* to get up on, to board, mount | садиться, влезать → **bin–, miñ–, pin–**

mina s. **mına**

minara s. **minare**

minare (14), **minara** (2), **minâre** (2), **mınare** (2) *N Ar.* minaret | минарет

minauvi (1) *Particle* look! here are! here! | вот это(т)! → **mınavu**

minder (3) *N* cushion (to sit on); mattress | подушка, матрас

mindir– (4) *Vfact* to seat sb on sg | подсадить

minset– (3) *Vfact* to seat sb on sg | подсадить

miñ (1) *Num* thousand | тысяча → **biñ**

miñ– (1) *V* to get up on, to board, mount | садиться, влезать → **bin–, min–, pin–**

Miñero s. **Meñaru**

mis (6), **misk** *N Ar.* musk | мускус

misil (1) *N Ar.* similar, equal; equal amount | подобие, равное количество; **misli jok** it is matchless | бесподобный, несравненный

Misır s. **Mısır**

misk s. **mis**

mı s. **mi**

mıχtar (3) *N Ar.* amount | количество

mıjıχ (1) *N* mustache | ус; **sırma mıjıχ** silvery moustache | серебряные усы

mıjız (2) *N* horn | рог

mık (2) *N* nail | гвоздь

mına (13), **mina** (2), **miña** (2), **mıña** *Particle* look! here are! behold! | вот это(т)!

mınau (8), **minau** *Pron* this (one), of this place | (вот) это, этот, здешний

mınavu (1) *Particle* look! here are! here! | вот это(т)!s. → **minauvi**

mında (32), **minda** (4) *Adv* here, hither | здесь, сюда

mından (5), **mınnan** *Adv* from here; of this | отсюда; из этого

mındır (1) *N R.* uniform | мундир

mınnan s. **mından**
mıña s. **mına**
Mısır (19), **Misır** (2) *Tn Ar.* Egypt | Египт; **misır buɣdajı** maize, corn | кукуруза, маис
mıšik s. **mıšık**
mıšıχ s. **mıšık**
mıšık (38), **mıšik** (2), **mıšıχ** (5), **mušuk** (1) *N* cat (*Felis domesticus*) | кот, кошка
moɣaj– (2) *V* to become sorrowful, to be grieved; to be offended | огорчаться, грустить, печалиться; обижаться
mohtač (1) *N Ar.* in want, needy | нуждающийся; **mohtač bol**– to be in need/want of sg | нуждаться в чём–л.
mojın (9) *N* neck | шея; **mojın ektir**– to have sb submitted to, to have sb humiliated one'self (before) | подчинять кого–л. **mojnına al**– to undertake, to take on, to take upon onedelf | брать на себя, объязываться → **bojun**
mojnuz (1) *N* horn | рог, панты
molča (6) *N* bath | баня
molla (22) *N Ar.* mullah, muslim priest and teacher; educated man | мулла; мусульманский учитель и священник; грамотный человек → **mulla**
mollalık (1) *N* the position of mullah; knowledge, education | должность муллы; знание, образованность
monda (1) *Adv* here, hither | здесь, сюда → **munda**
mondor (1) *Adj* black; black–eyed | чёрный; чёрноглазий; **mondor kůz** black–eyed (girl) | черноглазая (девушка)
mor (6) *Adj* violet, purple | фиолетовый, лиловый
mort (2) *Adj* frail, fragile; flagging, weak | хрупкий, ломкий; слабый
Moskov (1) *Tn* Moscow (city) | (город) Москва
msapir s. **musapir**
mu I (35) *Pron demonst* this | это, сё → **bu**
mu II s. **mi**
mubarek (5) *Adj Ar.* blessed; good, beautiful | благословенный; чудной
muɣaj– (1) *V* to become sorrowful, to be grieved; to be offended | впадать в уныние, обижаться, огорчаться
muɣırdık (1) *N* covered wagon | фургон
muhabat (1) *N Ar.* love, friendship, friendly talk | любовь, дружба; дружеская беседа; **muhabat et**– to have a friendly chat | беседовать/болтать с кем по–дружески
muhabbetlik (1) *N Ar.* love, friendship | любовь, дружба
Muhammed (1) *Pn Ar.* Mukhammed | Мухаммед
mujın (9) *N* neck | шея; **mujnına sar/sarıl**– to fall on somebody's neck | бросаться кому–л. на шею → **bojun, bujın, mojın**
mulet s. **mūlet**
mulla (28) *N Ar.* mullah, muslim priest and teacher; educated man | мулла; мусульманский учитель и священник; грамотный человек → **molla**
mum (2) *N P.* candle; wax | воск
mumin (1) *N Ar.* believer (in Islam); modest, peaceable | верующий (мусульманин); скромный, смирный
muna (20) *Particle* look! here are! behold! | вот это(т)!
Munar (1) *Tn* Munar | Мунар; **Munar taui** Munar mountain | гора Мунар
munasib s. **munasip**
munasip (2), **munasib** *Adj* fit, suitable, proper | подходящий, соответствующий, достойный; **munasib/**

munasip kůr– to think proper, to approve (of) | находить (считать) подходящим/приличным

munʒaχ (3), **munʒak** *N* necklace, string of pearls | бусы

munʒak s. **munʒaχ**

munda (41) *Adv* here, hither | здесь, сюда → **monda**

munı s. **bu**

murad s. **murat**

murat (28), **murad** (18), **můrad** (2), **můrat**, **mürad**, **mürat** *N Ar.* intention, aim, desire | намерение, цель, стремление; **muradına kavuš–** to achieve/obtain his/her aim/purpose | добиться цели, достигнуть желаемого; **muradina/muradına er–**, **můratına/müratına ir–** to attain/reach his/her aim/purpose | добиться цели, достигнуть желаемого; **müradlarına asil bol–** to achieve their aim/purpose | достигнуть их цели

murın (2) *N* nose | нос → **burun**

musafir (14), **msapir**, **musapir** (8) *N Ar.* guest, visitor, stranger | гость, посетитель; **musapir et–** to have as a guest | принимать как гостя кого; оставлять гостить кого

musafirlik (6), **msapirlik** (2), **müsafirlik**, **müsapirlik** *N* visit | посещение, визит

musapir s. **musafir**

musibet (3) *N/Adj Ar.* calamity, evil; illomened, calamitous | несчастье, беда; злосчастный

Mustafa (2) *Pn Ar.* Mustafa | Мустафа; **Mustafa–bej** Mustafa–bey | Мустафа–бей

Mustā (1) *Pn Ar.* Musta | Муста; **Mustā–beg** Musta–beg | Муста–бег

mušavere (5), **můšavere** *N Ar.* mutual consultation | совещание, совет; **mušavere/můšavere et–** to confer with sb on sg; to consult sb sbout sg | совещаться скем–л. о чём–л.

mušuk s. **mıšık**

mūlet (7), **mulet** *N Ar.* fixed term of respite; delay | отсрочка, срок

můbārek (6) *Adj Ar.* blessed | благословенный; **můbarek olsun!** good luck with it!; congratulations! may it be blessed | на счастье! (доброе пожелание); пусть будет благословенным!

můftů (1) *N Ar.* mufti; official expounder of Muhammedan law | муфтий

můlk (4), **mülk** *N Ar.* possession, property | собственность, имущество

můmkin (1) *Adj Ar.* possible, potential | можно, возможный

můrad s. **murat**

můrat s. **murat**

můšavere s. **mušavere**

můškůl (10) *Adj/N* difficult, hard | трудный

můškůllen– (1) *Vpass* to become difficult; (here) to be sorow/sad | становиться трудным; осложняться; (здесь) печалиться

můžde (3), **mužde**, **müžde** *N P.* good news, joyful tidings; present given to bearer of good news | радостная весть; награда за добрую весть

můždeʒi (2), **müždeʒi** *N* messenger carrying joyful news | вестник приносящий благую (добрую) весть; человек сообщающий радостную новость; **müždeʒi ber–** to send a carrier of joyful news | отсылать вестника, сообщающего радостную новость

mülk s. **mulk**
mürad s. **murat**
mürat s. **murat**
müsafirlik s. **musafirlik**
müsapirlik s. **musafirlik**

N

načar (1) *Adj P.* forced by necessity, helpless, in distress | вынужденный, безвыходный; **načar kal–** to get into a hopeless situation | попасть в безвыходное/безнадёжное положение

nafile (1) *Adv Ar.* useless, in vain | напрасный, тщетный

naγıš (2), **nañıš** *N Ar.* embroidery | вышивка, вышивание; **naγıš tik–** to embroider | вышивать; **nañıš išle–** to embroider | вышивать → **nakıš**

najt– [< **n‿ajt–** < **ne ajt–]** (3) *V* (not knowing) what to say | (не знать) что сказать → **ajt–, nasa–, nejle–, net–, nišle–**

nakirdi s. **nakırdı**

nakırdı (10), **nakirdi** (3), **nakirtı, nakırdi, nakırtı** (2) *N* talk, words | разговор, речь; **nakırdı/nakırtı et–** to talk, to gossip | разговаривать, сплетничать

nakırtı s. **nakırdı**

nakıš (2) *N* embroidery | вышивка, вышивание; **nakıš išle–** to embroider | вышивать → **naγıš**

nal (9) *N Ar.* horseshoe | подкова; **nal at [<nallı at]** iron–shod horse; shoed horse | подкованный/кованный конь

nalband (5), **nalbat [<nalbant]** (1) *N* blacksmith | кузнец (подковывающий лошадей)

nalbat s. **nalband**

naliva–krugom (1) *V R.* left about face! | налево кругом!

nam (4), **nām** (2) *N P.* name | имя

namaz (26) *N Ar.* prayer | намаз (совершаемая пять раз в день обязательная молитва у мусульман); **namaz kıl–** to perform the namaz, to pray | совершать намаз, молиться

name (1) *N P.* letter | письмо

namus (1) *N Ar.* honor, honesty | честь, достоинство

nane s. **nāne**

nañıš s. **naγıš**

nap– [< n‿ap– < ne jap–] (1) *V* (not knowing) what to do; how is one supposed to act? | (не знать) что делать; (не соображать) как поступать → **jap– I, kat– II, nasa–, nejle–, net–, nišle–**

nar (8) *N P.* pomegranate (*Punica granatum*) | гранат

nargoz (8) *N P.* the Persian New Year's Day | (персидский) новый год (по лунному календарю), весенний праздник

Narik (~ Narik–aka) (31), **Narık** (4) *Pn* Narik, Chora–batyr's father | Нарик, отец Чора–батыра

Narık s. **Narik**

nas s. **nasıl**

nasa– [< **n‿asa–** < **ne jasa–**] (10) (not knowing) what to do? how is one supposed to act? | (не знать) что делать; (не соображать) как поступать → **jasa–, kat– II, nejle–, net–, nišle–**

nasil s. **nasıl**

nasıl (105), **nas** (6), **nasil** (20), **nasl** [< **nasıl**] (17) *IntPron/Adv* how? what sort? | как? каким образом?; **er nasıl** anyway, in any case; what (so)ever, whatever kind of | как бы (то) ни было; какой бы (то) ни был → **naslı**

nasl s. **nasıl**

naslı (8) *IntPron/Adv* how? what sort? | как? каким образом? → **nasıl**

naz (6), **nāz** *N P.* whims, coquetry | изящество, кокетство; **ǯilve–nāz** coquetry | кокетничание, кокетство; **naz at–** to be coy, to seek to please | быть застенчивым, стремиться понравиться

nazar (8) *N Ar.* a looking, glancing at thing; the evil eye | взор, взгляд; дурной глаз, сглаз; **kůz nazar** the evil eye | дурной глаз; **nazar almasın/dejmesin/tijmesin** let the evil eye not touch | пусть взор/сглаз не коснется; **nazarγa uγrat–** to be subjected to the effect of evil eye | вподвергаться действию сглаза

Nazife (7) *Pn fem Ar.* Nazife | Назифе

nazlan– (6) *Vpass* to behave coquettishly, to be coy; to behave in an affected manner | нежиться, капризничать, кокетничать

nazlı (17) *Adj* coquettish, coy; spoiled; kind, charming | изнеженный, кокетливый; милый

nālin (1) *N Ar.* clogs, sabot | деревянные сандалии

nām s. **nam**

nāne (7), **nane** *N Ar.* peppermint (Mentha) | мята; мятный

ne (372), **nė** (2) *IntPron* what? how? which? | что? как? какой? который?; **ne ... ne** neither ... nor | ни ..., ни; **nedaj** what sort? | какой?; **nedajı?** what sort? | какой?; **nege** why? what for? | для чего? с какой целью?; **neje** why? what for? | для чего? с какой целью?; **nejse [< ne ise]** whatever it is | что же, ладно, словом; что бы то ни было → **ni**

nebet (4) *N Ar.* watch (of a sentry); turn (of duty etc.) | караул, дежурство; очередь (стража/дежурства) → **nůbet**

nebetči (4), **nebetčı** (3) *N* sentry, watchman | постовой, караульный

nebetčı s. **nebetči**

nečin [< ne ičin] (18), **nečın [< ne ičın]** (3), **nejčin [< ne ičin]**, **nejčın [< ne ičın]** *IntPron* why, for what | почему, зачем → **ničin**

nečın s. **nečin**

neǯe (4) *IntPron* how many? how much? how? | сколько?; как?, каким образом?; **neǯe–neǯe** very many | довольно много, очень много → **niǯe**

nedaj (3), **nedajı** *IntPron* how? what sort? | какой, как → **nedajın**

nedajı s. **nedaj**

nedajın (3) *IntPron* how? what sort? | какой, как → **dajın, nedaj**

Nefa (1) *Pn Ar.* Nefa | Нефа; **Nefa usta** master Nefa | мастер Нефа

nefes (2) *N Ar.* breath, a breathing | дыхание

nefis (1) *N Ar.* passion, intense emotion/wish | страсть, сильное желание

nejčin s. **nečin**

nejčın s. **nečin**

nejet (1) *N Ar.* intention, purpose | намерение; **nejet et–** to resolve, to intend | намереваться, задумать

nejle– [< n‿ejle– < ne ejle–] (3) *V* (not knowing) what to do? how is one supposed to act? | (не знать) что делать; (не соображать) как поступать; **nejlegen** what did he/she do? | что он(а) делал(а); **nejlejim?** what shall I do? | что мне делать? → **ejle–, kat– II, nasa-, net–, nišle–**

nejse s. **ne**
nene (2) *N* mother, mom | мать
nere (8) *IntPron* what place? what part? | какое место? какая часть?; **nerede** where? wherever? | где?; **nereden** from where? whence? | откуда?; **nerege** to what place? whither? | куда?
nereli (2) *Adj/ IntPron* where from? of what place? | откуда (родом)? из каких мест?
nesil (1) *N Ar.* generation, descendants | поколение, потомство; **aslı nesli bilinmegen** (his/her) origin is not known | безродный, неизвестного происхождения
nešle– s. **nišle–**
net– **[< n‿et– < ne et–]** (1) *V* (not knowing) what to do? how is one supposed to act? | (не знать) что делать; (не соображать) как поступать → **et–, kat– II, nejle–, nišle–**
nezet (1) *N Ar.* taste, flavor | вкус → **lezzet**
nezzetli (1) *Adj* tasty, delightful | вкусный; аппетитный → **nezet**
nė s. **ne**
ni (31), **nı** (1) *IntPron* what? how? which? | что? как?; **ni ... ni** neither ... nor | ни ..., ни; **nijse [< ne ise]** whatever it is | что же, ладно, словом; что бы то ни было; **nise [< n‿ise < ne ise]** whatever it is | что же, ладно, словом; что бы то ни было → **ne**
niče (1) *IndefNum* how many? how much? | сколько?
ničin [< n‿ičin < ne ičin] (23), **ničın [< n‿ičın < ne ičın]** (4) *IntPron* why? for what? | почему?, для чего?, зачем? → **nečin**
ničın s. **ničin**
nižе (3) *IntPron* how many? how much? how? | сколько?; как?, каким образом? **halıñ nižе?** how are you feeling? | как ты себя чувствуешь? → **nežе**
nijet (1) *N Ar.* aim, intenton; wish | цель, намерение; **nijet et–** to resolve, to intend | вознамериться, намереваться
nijse s. **ni**
nika s. **nikā**
nikah s. **nikā**
nikā (8), **nika, nikah, nıkah, nıkā** (3) *N* engagement, marriage | брак, бракосочетание, брачный договор; **nikah χıj–** to perform the ceremony of (engagement) marriage; to marry | заключать брак, вступать в брак; **nikā kıl–** to become engaged to, to marry | заключать брак, обручаться
nikāsız (1) *Adj* unmarried | находящийся в незаконном браке
nine (1) *N* grandmother | бабушка
ninni (1) *N* lullaby | колыбельная (песня)
nise s. **ni**
nišan I (5) *N P.* sign, mark | знак, метка
nišan II (3) *N P.* engagement | обручение; **nišan ber–** to engage (one's son or daughter) | обручать кого–нибудь с кем–нибудь
nišanla– (4) *V* to engage (one's son or daughter) | обручать дочь или сына у кого–л.
nišanlı (13) *N* engaged to be married | помолвленный
nišle– [< n‿išle– < ne išle–] (16) *V* (not knowing) what to do? how is one supposed to act? | (не знать) что делать; (не соображать) как поступать; **nišlejim? [< ne išlejim?]** what shall I do? → **išle, kat– II, nasa-, nejle–, net–**

nı s. **ni**
nıkā s. **nikā**
nıktap (3) *Adv* strongly, tightly; hard | крепко, туго, уплотняв?
Nogaj–Ahmad (1) *Pn* Noghai Ahmad | Ногай Ахмад → **Ahmat**
noγaj (32), **nogaj** (1) *En* Noghai | ногай
nokta (1) *N* halter (of horse) | недоуздок; **nokta sal–** to put halter (on a horse) | надевать недоуздок
nol– **[< ņ ̦ol– < ne ol–]** (4) *V* (not knowing) what to happen?, what to be? | (не знать) что случиться? что происходить? → **ol–**
nuχsan (1) *N Ar.* deficiency; (here) simple(–minded) | недостаток; (здесь) глупый
nuktalan– (3) *Vpass* to be dotted; to be finished/decided | отмечаться точками; кончаться; решаться
numaj (2) *Adj* abundant, plentiful | обильный
nur (2) *N Ar.* light, sparkle, radiance | свет, сияние, луч
nůbet (2), **nübet** *N Ar.* turn (of duty etc.) | черёдь (дежурства); черёд → **nebet**
nübet s. **nůbet**

O

o (137) *Pron pers* (s)he, it; that | он(а), оно; тот; **andan** from him/her | от его/неё; **anı** him/her | его/её; **anıñ** his/her, of him/her | его/её; **aña** to him/her, for him/her | ему/ей, для него/неё; **oγa** to him/her, for him/her | ему/ей, для него/неё; **olar** they | они; **oni** him/her | его/её; **onı** him/her | его/её; **onınčün** [<**onın‿čün** < **onın ičün**] for him/her | для него/неё, из–за него/неё; **onıñ** his/her, of him/her | его/её; **onu** him/her | его/её; **onunnen** with him/her | с ним/ней; **oña** to him/her, for him/her | ему/ей → **u**
oba (14) *N* hill, heap, mound | холм, бугор; большое количество → **uba**
obala– (1) *V* to collect in a heap, to pile up | собирать в кучу, сгребать в кучу
obir **[~ȯbir]** (4), **obır** (2) *Adj* the other one; other | другой, тот; последующий; **obir/obır tůnja** the other world | потусторонний мир
obiri (2), **öbürü** *Pron pers* the other one; other | другой, тот; **obırısi kůn** the day after tomorrow | послезавтра
obirleri (2), **ȯbırlerı** *Pron pers* the other ones; others | другие, остальные, те
obur (15) ~ **obur–ǯadı (obur–ǯadi)** *N* hag, witch, sorceress | ведьма, колдунья → **ǯadı**
obur–ǯadı s. **obur**
oǯa (5) *N P.* teacher; hodja (Muslim teacher); husband | учитель, (мусульманский) преподаватель; муж → **hoǯa, χoǯa**
oǯaχ s. **oǯak**
oǯak (6), **oǯaχ** *N* fireplace, oven | очаг; **oǯak baši** a place near the fireplace | место у (домашнего) очага; место возле топки/печки; → **uǯak**
oda (11) *N* room | комната
odaman (14) *N R.* head shepherd; officer | старший чабан; офицер; **Gurgur odaman** officer Gurgur | офицер Гургур
odon (3), **oton** *N* firewood, log | дрова
ofak (2) *Adj* small, little | мелкий, маленький s. **ufak**
oγlan (2), **olan** *N* boy; youth; lad | мальчик, парень
oγlı s. **oγul**
oγlo s. **oγul**
oγlu s. **oγul**

oɣorla– (1) *V* to steal | украсть

oɣur (1) *N* (peace of good) luck, fortune | счастье; **oɣrın olson** good luck to you! good journey! | желаю тебе счастья! в добрый час! счастливого пути! → **uɣur**

oɣul [~ **oɣli** (5), ~ **oɣlı** (4), ~ **oɣlo**, ~ **oɣlu**, ~ **olu (=ōlu < oɣlu)]** *N* son; boy | мальчик, сын; **kahp‿olu [< kahpe olu]** son of a bitch | отпетый мошенник, плут → **uɣıl**

oh (2) *Interj* oh | ой, ох

oχ (6) *Interj* oh | ой, ох

oχot– (1) *Vfact* to teach | учить, обучать → **okot–, uχut–, ukıt–, ukut–**

oχšu– (1) *V* to retch, to feel sick | испытывать рвотный позыв

oχu– (2) *V* to read; to learn, to study | читать; учиться → **oko–, ukı–**

oj I (1) *N* thought, intention | мысль, намерение

oj II (1) *Interj R.* oh! | ой!

ojla– (2) *V* to think | думать

ojlan– (15) *Vpass* to think | думать

ojle s. **ȯjle**

ojuk (1) *N* hollow, grooved, hollowed out | выемка, выбоина; выдолбленный

ojun (2) *N* game, play; trick | игра, шутка → **ujun**

ok (5) *N* arrow | стрела → **uk**

okı– s. **oko–**

oklau (1) *N* rolling spin | скалка → **uklau**

oko (1) *N* reading; learning, studying | чтение; учение; изучение

oko– (5), **okı**– *V* to read; to learn, to study | читать; учиться → **oχu–, ukı–**

okomaχ (1) *Inf (mak)* reading, studying | чтение

okot– (2), *Vfact* to teach | учить, обучать → **oχot–, uχut–, ukıt–, ukut–**

okotma (1) *N* teaching, tutoring | обучение

ol– (151) *V* to be, to exist | быть, существовать; **amanet ol**– to be entrusted (to) | быть оставлен на попечение (кого–л.) → **bol– I, ul–**

ola (3) *Conj* but, still, yet | только, лишь; но, однако

olaʒa (1) *Adj* all (existing), each (of the), every | всё, все, каждый

olan s. **oɣlan**

olen s. **ȯlen**

olok (1) *N* trough, drinking basin (at) a fountain | колода, жёлоб, жёлоб водопойной колоды; водослив → **uluk**

omer s. **ömer**

omur s. **ȯmu̇r**

on (9) *Num* ten | десять; **on beš** fifteen | пятнадцать; **on bir** eleven | одиннадцать → **un I**

onot– (3) *V* to forget | забывать → **unut–, unıt–**

oñ (2) *Adj* right (–hand side) | правый; **kjā oñında kjā sulında** at one time on his/her right, at one time on his/her left; whether on his/her right, or on his/her left | иногда на право от него/неё, иногда налево от него/неё; либо на право от него/неё, либо налево от него/неё

oñ– (5) *V* to do well, to succeed | преуспевать; везти (в чём–л.) → **uñ–**

oñolmaz (1) *Adj* incurable | неизлечимый

oñuz (1) *N* shoulder | плечо → **umız**

opajsa (1) **Ašadan kelij opajsa, ku̇zleri dompaja – baka.** (She) is comming from below leaping, (her) eyes goggling – it is a frog (riddle). | (Она) идёт снизу прыгая (её) глаза выпучивая – это лягушка (загадка).

Or (8) *Tn* Or (town), now Perekop | (город) Ор, сегодня Перекоп → **Ur**

or– I (1) *V* to chop; to reap, harvest | рубить; жать (серпом)

or– II (1) *V* to hit, to beat, to shoot, to throw; to kill | бить, ударять; убить; стрелять; бросать; втыкать; **šamar or–** to give a slap on the face | давать пощёчину → **ur–**

ora (2) *N* that place | то место

ora– (3) *V* to spin, to fold in | мотать, заворачивать → **ura–**

orada (2) *Adv* there, at that place | там → **urada**

orak (1) *N* harvest; sickle, reaping hook | жатва; серп → **urak**

oral– (1) *Vpass* to be spinned, to be folded | обвиваться, заворачиваться

oraza (1) *N Ar.* fasting, fast | ураза, воздержание от пищи (по предписанию шариата), пост

orın (~**ornı**) (5), **oron** *N* place | место → **urun***

orloχ (1) *N* seed | семя, семена

orman (1) *N* forest, wood | лес

oro I (1) *N* pit (fall), hole; | яма (для хранения зерна) → **uru I**

oro II (1) *N* thief | вор

oron s. **orın**

orta (4) *N* middle, centre | середина → **urta**

ortalık (2) *N* one's immediate surroundings | округа, вокруг

ortanǯa (3) *Adj* middle, middle child (of three) | средний (о детях) → **ortanǯı, urtanǯa, urtanǯı**

ortanǯı (5) *Adj* middle, middle child (of three) | средний (о детях) → **ortanǯa, urtanǯa, urtanǯı**

Osman *Pn Ar.* Osman | Осман; **Osman–aka** brother/uncle Osman | брат/дядя Осман

oš (1) *Adj/Adv* nice, kind | приятный, красивый → **hoš**

oša– (7) *V* to resemble, to look like | быть похожим, походить

ot I (2) *N* grass | трава; **χıčıtkan ot** stinging nettle | жгучая трава, крапива; **kıǯıtkan otu** stinging nettle | жгучая трава, крапива → **ut I**

ot II (25) *N* fire | огонь; **ot tutıš–** to take/catch fire | зажигаться, разгораться

otaj– (2) *V* to overgrow with grass | зарасти травой

otır– (4), **otėr–** (1) *V* to sit (down); to dwell, to live; to stay | садиться; сидеть; жить, проживать → **otor–, otur–, utır–**

otla– (1) *V* to graze | пастись → **utla–**

otlat– (1) *Vfact* to pasture (cattle) | пасти

oton s. **odon**

otor– (53) to sit (down); to dwell, to live; to stay | садиться; сидеть; жить, проживать
Vaux **ujnašıp otor–** to play (lengthily), fiddle | играть (длительно), развлекаться → **otır–, otur–, utır–**

otort– (2) *Vfact* to seat, to place, to let have a seat | сажать, садить

otoz (10), **ottoz** *Num* thirty | тридцать; **otoz beš** thirty–five | тридцать пять; **otoz toχoz/tokoz** thirty–nine | тридцать девять → **utuz**

ottir *Onomatop*; **attir ottir** (play of words) | (игра слов)

ottor– s. **ȯtůr–**

otun (1) *N* firewood, log | дрова → **odon**

otur– (7) to sit (down); to dwell, to live; to stay | садиться; сидеть; жить, проживать → **otır–, otor–, utır–**

oturt– (1) *Vfact* to seat, to place, to let have a seat | сажать, садить

ozγar– (1) *V* to see off, to show the way | провожать кого

Ȯ

ȯbırlerı s. **obirleri**

ȯd (1) *N* gall, bile | желчь; **ȯdů patla**– to be frightened to death | насмерть напугаться

ȯg (8) *Postp* front, space in front (of a thing) | перед чем–л. (в роли послелога); **kapı ȯgi** place in front of the door | место перед дверью; **ȯgi** fore–part of sg | передняя часть чего; **ȯgine** before, in front of | перед кем/чем–л.; **ȯgůnde** in front of sg | перед кем/чем–л.; впереди кого/чего–л.; **ȯgůndeki** sg in front of sg | впереди кого/чего–л.

ȯgůz (11), **ȯgüz** *N* ox | вол → **ȯkůz, ůgůz**

ȯgüz s. **ȯgůz**

ȯjle (2), **ojle** *Adj/Adv* so, in that manner, such, like that | такой, так, таким образом

ȯksek (10) *Adj* high | высокий

ȯkuz (1) [~**ȯkůz**] *N* ox | вол → **ȯgůz, ůgůz**

ȯl– (3) *V* to die | умирать; **ȯle jaza** he/she was about to die | чуть было не умер(ла)

ȯlen (4), **olen** (2) *N* grass | трава → **ůlen**

ȯlů (1) *N* dead | мёртвый, мертвец → **ůlů**

ȯmer (3); **omer** (2) *N Ar.* life | жизнь; **omer et**– to live together with sb | жить вместе с кем–л. → **ȯmůr, ömür, ůmůr**

ȯmůr (9), **omur** (5) *N Ar.* life | жизнь; **omur/ȯmůr et**– to live together with sb | жить вместе с кем–л. → **ȯmer, ömür, ůmůr**

ȯñ (1) *Postp* front, space in front (of a thing) | перед чем–л.; **kapı ȯñinden keč**– to pass in front of the gate | проходить перед воротами

ȯr– (1) *V* to plait; to knit | плести; вязать

ȯreke (1) *N* spinning wheel; winding wheel | стойка, на которой закрепляется шерсть для прадения; прялка → **ůreke**

ȯrken (1) *N* stem, stalk | стебель

ȯt– (5), **ot**– *V* to pass (by, through, over) | проникать, миновать, проходить → **öt–, ůt– I**

ȯtkȯz– (1) *Vfact* to pass (something through something) | (дать) проводить (время)

ȯtȯr– s. **ȯtůr**–

ȯttir– s. **ȯttȯr**–

ȯttȯr– (10), **ȯttir**–, **öttȯr**–, **öttör**– (3) *V* to kill, murder | убивать

ȯtůr– (16), **ȯtȯr**– (4), **ötir**– *V* to kill, murder | убивать

ȯz (47) *Pron pers* self; own, genuine, real; | сам, свой, родной → **öz**

Ö

öbürü (1) *Pron* the other one; other | другой, тот → **obiri**

ölčettir– (1) *Vfact* to have sg measured | велеть измерить

ömer (1) *N Ar.* life | жизнь; **ömer et**– to live togerther with sb | жить вместе с кем–л. → **ȯmůr, ömür, ůmůr**

ömür (1) *N Ar.* life | жизнь → **ȯmer, ȯmůr, ömer, ůmůr**

ös– (2) *V* to grow up | вырастать → **ȯs–, ůs**–

öst (1) *N* upper or outs ide surface, top of anything, upper (most) | верхняя часть чего–л.; верх чего–л.; вершина → **ůst**

öt– (1) *V* to pass (by, through, over) | проникать, миновать, проходить → **ȯt–, ůt– I**

ötir– s. **ȯtůr**–

öttȯr– s. **ȯttȯr**–
öttör– s. **ȯttȯr**–
öz (1) *Pron pers* self; own, genuine, real; | сам, свой, родной → **ȯz**

P

pača (1) *N* shank, trouser leg | нога, штанина брюк → **balak**
padiša (328), **padišah** (4), **padıša** (4), **patiša** (1) *N P.* sovereign, ruler | падишах, царь → **padša**
padišah s. **padiša**
padıša s. **padiša**
padša (250), **patša** (85) *Adj P.* sovereign, ruler | падиша, царь → **padiša**
padvor (1) *N R.* heel of a shoe or boot | каблук
paχıl (1) *Adj* envious | завистливый; **paχıl šejtan** envious devil | завистливый дьявол, чёрт
paχılla– (2) *V* to begrudge, envy | завидовать
paj (7) *N* share, lot, portion | пай, часть, доля
pajda (4) *N Ar.* profit, use, advantage | польза, прок, выгода, толк; **pajda et**– to be useful, to help быть полезным, приносить пользу кому–л.; идти кому–л. на пользу, помогать; **pajda(si) tij**– to come in useful | быть полезным, помогать → **fajda**
pajdasız (1) *Adj* useless, vain | бесполезный → **fajdasız**
pajgambar s. **pajγamber**
pajγamber (2), **pajgambar** *N Ar.* prophet | пророк
pajsız (1) *Adj* endless, unendless, unlimited | бесконечный, безграничный; **učsız pajsız** vast, endless | безграничный, бескрайний
pajton (8), **pajtun** (1) *N Fr.* phaeton | фаэтон → **fajton**
pajtun s. **pajton**
pakir (4) *N/Adj Ar.* poor, poor fellow | бедный, несчастный; бедняк; **pakir–ǯarlı, ǯarlı–pakir** poor–destitute | бедный–неимущий → **ǯarlı, faχir, fakir**
pakir–ǯarlı s. **pakir**
paklava (1) *N Ar.* sweet pastry generally cut into diamond– shaped pieces | пахлава
pakle– (1) *V* to clean; to gobble up | подъедать всё дочиста
paldır–gủldủr (1) *Onomatop* noisy, loud (sound of running water) | шумно; с грохотом (шум текущей воды)
palı (2), **pāli** *Adj P.* expensive | дорогой; **pāliγa sat**– to sell at a high price | продавать дорого
pamador (2), **pamodor** *N R.* tomato | помидор
pamodor s. **pamador**
pampuli (1) *Adj* boastful, boasting | хвастливый, дутый; **pampuli kuraz** boastful dandy | хвастливый дэнди
pamuk (2) *N P.* cotton | хлопок; хлопчатник
pan (1) *Adj* false | фальшивый, подельный (о деньгах)
papı (3) *N* duck | утка
papıč s. **papuč**
papuč (4), **papıč** (1) *N P.* slipper | тапочки, туфли
papur (1) *N Fr.* steamer, steamship | пароход → **paraχod**
para I (36) *N* money | деньги; **turasız para** money without script (coat of arms) | деньги без тугры/герба
para II s. **pare**
paraχod (15), **paraχot** *N R.* steamer, steamship | пароход → **papur**

paraχot s. **paraχod**

parala– s. **parla– I**

parča (15) *N* small piece, fragment | осколок, обломок

parčala– (1) *V* to break into pieces, to cut into parts | ломать вдребезги (на части)

parčalan– (1) *Vpass* to be broken into pieces | ломаться на части, разбиваться вдребезги (на мелкие куски)

parčalat– (1) *Vfact* to make sb break sg | приказывать ломать

pare (7), **para** *N P.* small piece, fragment | штука, кусок

park (1) *N Ar.* difference, discrimination | разница, различие, отличие; **park et**– to discriminate, to treat differently | различать, отличать

parla– I (6) *V* to shine, to gleam, to glitter | блестеть, сиять, сверкать

parla– II (16), **parala**– (3) *V* to break into pieces, to cut into parts | ломать, разламывать, колоть, рубить

parlan– (1) *Vpass* to shine, to gleam, to glitter | сверкаться, сияться

parmak (12) *N* finger; bar, rail | палец; (металлический) прут

Partatur (2), **Patartur** *Tn* Port–Artur | Порт–Артур

pas (1) *N* rust, tarnish | ржавчина

pasta (2) *N It.* watery boiled rice | каша

paša (5) *N* pasha | паша

pašali (1) *Adj* (door) with bolt(–lock)? | (дверь) с задвижкой?

Patartur s. **Partatur**

patilǯan s. **patılǯan**

patiša s. **padiša**

patılǯan (2), **patilǯan** *N* eggplant (*Solanum melongena*) | баклажан

patırda– (1) *V* to make a continuous knocking noise | звукоподражательный глагол передающий звук падающих с дерева плодов

patla– (9) *V* to burst (out); to explode; | взрываться, выходить из себя; **ödü patla**– to be badly frightened, to be frightened to death | сильно испугаться, насмерть напугаться; **patlap kitken jüregi** his/her heart was broken | у него/неё надломилось сердце

patlat– (2) *Vfact* to hit; to slap | шлёпать, хлопнуть; **at patlat**– to beat the horse | шлёпать лошадь

patrun (1) *N R.* cartridge | патрон, заряд

patša s. **padša**

patšalık (1) *N* sovereignty, empire, state | царство, государство

pazar I (5) *N P.* the bazaars, market | базар, рынок → **bazar**

pazar II (1) *N P.* Sunday | воскресенье; **pazar küni** on Sunday | в воскресенье

pazarerte (1) *N* Monday | понедельник

pazvan (3) *N P.* big armlet | большой браслет

pāli s. **palı**

peč (1) *N R.* stove | печь

pečatlan– (1) *Vpass R.* to be stamped with a seal | заштемплёвываться, опечатываться

pehlivan s. **pelivan**

pejliven (2) *N P.* champion, wrestler | чемпион, борец → **pelivan**

pek (8) *Adv* very, much; hard, strong | весьма, очень, сильно

pelaket (1) *N Ar.* disaster, catastrophe | бедствие, катастрофа

pelevanlik (1) *N* wrestling | борьба → **pelivan**

pelivan (8), **pehlivan** *N P.* champion, wrestler | чемпион, борец → **pelivan**

penčere s. **penʒere**
penʒere (37), **penčere** (6), **perʒere** *N P.* window | окно
perčem (7), **perčen**, **perčım** *N* bangs, tuft of hair, lock on the top of the head | чуб
perčen s. **perčem**
perčım s. **perčem**
perʒere s. **penʒere**
perde (3) *N P.* curtain | занавес
peri (12), **perı** *N P.* fairy, good djinn | пери, фея, дух
perı s. **peri**
peršembe (2) *N P.* thursday | четверг
pes s. **fes**
peš (7) *N/Postp P.* footmark, trace; space behind, back | след; позади, сзади кого–чего, за кем–чем (в функции послелога); **pešine tůš**– to follow sb, to run after sb | следовать за кем–то → **piš**
pešin (1) *Adj/Adv P.* paid inadvance, ready (money) | уплачиваемый вперёд, наличные (о деньгах)
peškir (3) *N P.* towel | полотенце
pešman (2) *N P.* regretful, sorry | кающийся, сожалеющий; **pešman bol**– to repent | каяться, раскаиваться; **pešman et**– to make sb repent sg | заставить раскаяться → **pišman**
peštamal s. **peštemel**
peštemel* (4), **peštmel** (3), **peštamal** *N P.* large bath towel; apron | полотенце, передник
peštmel s. **peštemel***
Peter (1) *Tn R.* Peter (Petersburg) | Питер (Санктпетербург)
pezevenk (1) *N P.* pimp, procurer | подлец
pič (1), **pič bala** *N P.* illegitimate child, bastard | ублюдок, выродок, незаконнорожденный ребёнок
pič– (2), **piš**– **II**, **pıč**– (2) *V* to cut | косить
pičaχ s. **pıčak**
pičak s. **pıčak**
pičen (6) *N* hay | сено
pički s. **pıčkı**
pidan (1) *N Gr.* sapling, young tree | саженец, росток; **pilis–pidan** slender–sapling/young tree | статный/стройный росток → **fidan**
pijade (2), **pıjāde** *N/Adv P.* pedestrian, foot soldier; on foot | пехота, пехотинец; пешком
pikir (5) *N Ar.* thought, mind | мысль; мнение; **pikir et**– to come to one's mind | (при)помнить, мыслить
pil (1) *N Ar.* elephant | слон
pilan (1) *Adv Ar.* so and so, and so on | такой–то
pilatir (1) *Onomatop N*; **Bir kiškene pilatir, ʒümne kalkın ʒılatır – ijne.** A shining/glittering tiny one makes all people cry – it's a needle (riddle). | Тоненькая да блестящая но каждого человекла доведёт до слёз – это игла (загадка).
pilau (3), **pılau** *N P.* pilaf (boiled rice prepared with butter, meat, fat, etc.) | плов; **pirinč–pilau** rice–pilaf | рис–плов
pilčan s. **pilʒan**
pilʒan (7), **pilčan** *N Ar.* coffee cup, tea cup | чашка
pilen (8), **plen** *N R.* captive; prisoner (of war) | плен; военнопленник; **pilen al**– to take prisoner (here: prisoner of war) | взять (захватить) в плен; (здесь речь идёт о военнопленных)
pilennik (2), **pilennık** *N R.* captive; prisoner (of war) | пленник; ; военнопленник; **pilennige al**– to take prisoner (here: prisoner of war) | взять (захва-

тить) в плен (дословно: в пленник); (здесь речь идёт о военнопленных)

pilennık s. **pilennik**

pilis (1) *N* young shoot, bud, tendril | побег, росток; **pilis–pidandaj** like a slender/slim like a young shoot | стройный как росток → **filiz**

pin– (5) *V* to get up on, to climb (up), to scramble | карабкаться, взбираться (с трудом); **pindij pindij ketken, pindi pindi pindi ketken** scrambling scrambling was going (away) (it's a hedgehog in a riddle) | (с трудом) карабкаясь проходил (мимо) (это ёж в загадке) → **bin–**

pirinč (2), **pirinǯ** *N P.* rise (*Oryza sativa*) | рис

pirinǯ s. **pirinč**

pirılda– s. **pırılda–**

piš (2) *N/Postp P.* footmark, trace; space behind, back | след; позади, сзади кого–чего, за кем–чем; **pišine tůš–** to follow, to run after sb | следовать по пятам, идти вслед за кем, преследовать кого → **peš**

piš– I (19) *V* to be cooked or baked; to ripen | вариться; созревать; **piše kel–** to grow ripe (grape) | поспеть, созреть (виноград)

piš– II s. **pič–**

pišir– (48), **pišır–** (3), **pışir–**, **pšir–** (2) *V* to cook, to bake; to ripen | варить, выпекать, готовить → **bišir–**

piširt– (2) *Vfact* to make sb cook or bake | заставить варить, жарить

pišır– s. **pišir–**

pišken (5) *Adj* well cooked, well baked; ripe | сваренный; зрелый, спелый

piškir– (2) *V* to gush out, to spurt out, | бить струёй, извергать

pišman (1) *N P.* regretful, sorry | кающийся, сожалеющий; **pišman bol–** to repent, to regret | каяться, раскаиваться → **pešman**

pit– I (11) *V* to end, come to an end, to be all gone | кончаться → **bit– I**

pit– II (2) *V* to grow, to yield | расти → **bit– II**

pitir– (15), **ptir–** (5) *V* to finish, to complete; to kill | окончать, завершать; убивать → **bitir–**

pitlen– (1) *Vpass* to clear oneself of lice | очищать себя от вшей

piva (3) *N R.* beer | пиво

pıč– s. **pič–**

pıčaχ s. **pıčak**

pıčak (14), **pičaχ**, **pičak** (3), **pıčaχ** (2) *N* knife | нож, ножевой → **bıčak**

pıčakla– (1) *V* to stab, to knife | ударить (ранить) ножом

pıčkı (3), **pički** (2), *N* saw, bucksaw | пила, двуручная пила; лучковая пила

pıjāde s. **pijade**

pılan (2) *Adj Ar.* so and so, and so on; and such like, approximately | такой–то; имярек; и так далее; примерно → **falan, felan, filan**

pılau s. **pilau**

pılǯan s. **pilǯan**

pın (2) *Onomatop N* → **pın–**

pın– (1) *Onomatop V*; **Pın pın pınmaz, jirge tůsse sınmaz – japraχ.** [...], if it falls down to earth, it won't be broken – it's a leaf (riddle) | (...), если упадёт на землю не разобьётся – это лист (загадка)

pırılda– (7), **pirılda–** (3) *V* to move around hurriedly and anxiously; to be scattered | быстро кружиться; рассыпаться

pıski (1) *N* fillip, rap | щелчок; **pıski čert–** fillip/flick with the finger | дать щелчок

pıta– (2) *V* to lop | обрезать ветки

pıtak (2) *N* branch, twig | ветка, ветвь → **budaχ, butak, putak**

pojız (2) *N R.* train | поезд

polk (2) *N R.* regiment | полк; **ulaniski polk** regiment of uhlans | уланский полк

polkovnik (1) *N R.* colonel | полковник

portokal (7) *N Fr.* orange | апельсин

post (1) *N P.* skin, leather | шкура, кожа; **postın ser–** to lay out his/her skin (somewhere), meaning spending unpolitely too long time somewhere | расстилать его/её шкуру (где-нибудь), что обозначает неприлично долгое пребывание где-нибудь

postaʒı (1) *N* postman, mailman | почтальон → **poštaʒı**

postorγaj s. **bostorγaj**

pošı s. **pošu**

pošta (1) *N R.* post office | почта

poštaʒı (1) *N* postman, mailman | почтальон → **postaʒı**

pošu (4), **pošı** *N* shawl, homespun woolen cloth | платок → **pušı**

Pravlen (1) *Tn* Pravlen (settlement?) | Правлен (поселение?)

pristav (1) *N R.* recruiting (person) | пристав

prıkaz (1) *N R.* order | приказ

pšir– s. **pišir–**

puχara s. **pukara**

pukara (6), **puχara** *N/Adj Ar.* poor | бедняк, бедный, несчастный (человек) → **fukara**

pul (4) *N P.* scale (of a fish), small coin, plates for decoration | чешуя; блёстки, металлические пластинки (для украшения)

purun (3) *N Gr.* bakery | пекарня, булочная → **furun**

pus– (4) *V* to crouch (down) | присесть (опуститься на корточки); **pusıp pusıp kit–** to move carefully and silently, to go stealthily | осторожно, тихо идти, прокрадываться

pušı (1) *N* shawl, homespun woolen cloth | платок → **pošu**

pušt (1) *N P.* rascal, bastard, brat | подлец, негодяй; обманщик

put (3) *N R.* pound (16.38 kg) | пуд (16,38 кг)

putak (2) *N* branch, twig | ветка, ветвь → **budaχ, butaχ, butak, pıtak**

pu̇sku̇l (5) *N* tassel, tuft | кисть, кисточка, бахрома

pu̇su̇r (2), **püsür** *Adj Gr.* filth, rubbish; dirty, shitty | нечистоплотный, грязный

pu̇tin s. **bu̇tu̇n**

püsür s. **pu̇su̇r**

R

raf (8) *N P.* shelf | полка

rahat (13) *N Ar.* ease, quiet, rest, comfort | покой, спокойствие; **rahat bol–** to be quiet, satisfied, cosy, happy | быть спокойным; быть в довольстве; быть удобным (комфортабельным); быть счастливым

rahatla– (1) *V* to become comfortable, to feel relieved | успокаиваться; чувствовать облегчение

rahatlan– (5), **rātlan–** (2) *Vpass* to rest, to take one's ease, to calm oneself | успокаиваться, быть спокойным; отдыхать

rahatsız (1) *Adj* disturbing, troubling, uncomfortable, unquiet | беспокойный, лишённый покоя; неудобный;

rahatsız et– to disturb, to trouble, to annoy | беспокоить, мешать

raχi s. **raki**

raki (9), **raχi**, **rakı** (3) *N Ar.* raki, arrack | ракы; водка

rakı s. **raki**

Ramazan (3) *N Ar.* Ramazan | месяц поста

raskel– s. **rastkel**–

raskil– s. **rastkel**–

rastkel– (31), **raskel**–, **raskil**– (2) (< P. *rast* + *kel/kil*–) *V* to meet by chance, to come across | встречаться, сталкиваться с кем (неожиданно)

rastkeliš– (1) *V recipr* to meet (each other) by chance, to come across with someone | встречаться, сталкиваться с кем (неожиданно)

razvet (1) *N R.* discovery, a spotting | разведка

rābet (1) *N Ar.* popularity, attention, desire | популярность, внимание, желание, склонность

rāmet (6) *N Ar.* compassion, forgiveness and grace, mercy | милосердие, милость, прощение; **rāmet ejle**–/**ile**– to have mercy on, to show compassion | жалеть; соболезновать; **rāmet ol**– to die | умереть

rātlık (2) *N* ease, comfort, quiet | покой, спокойствие → **rahat**

rāzi (28) *Adj Ar.* satisfied, contented, agreed | согласный, удовлетворённый; **Allah rāzi bolsın** may Allah be pleased/satisfied (with you) | пусть будет Аллах довольным (с вами/с тобой); **rāzi bol**– to be contented | быть довольным; соглашаться, удовлетворяться; **rāzi rāzi aγla**– to cry bitterly; to sob one's heart out | горько/безутешно плакать; плакать навзрыд

rāziliχ (1) *N* agreement, compliance | согласие, реазрешение

reǯa (2) *N Ar.* request | просьба; **reǯa et**– to request (something from someone) | просить кого о чём

reket (7), **rekjet** (2) *N Ar.* a kneeling in worship | рекят (элемент намаза)

rekjet s. **reket**

renk (4) *N P.* colour | цвет, краска

romalama dompal (1) *N* name of a toy or children's game | название детской игры/игрушки; **romalama dompal jap**– to play "romalama dompal" (?) | играть в "ромалама домпал"

ruba (8) *N* clothes, clothing, dress | платье, одежда

ruble (8) *N R.* rouble | рубль

ruhset s. **ruχset**

ruχset (8), **ruhset**, **růhset**, **ruχsat**, **růχset**, **uruχset** *N Ar.* permission, license | разрешение, позволение; **ruχset**/**ruhset ber**– to grant permision (to, for), to permit sg | давать разрешение, позволять; **ruχset iste**– to ask for permission | спрашиваться; просить разрешения; **ruχset tile**– to ask for permission | спрашиваться; просить разрешения; **uruχset ber**– to grant permision (to, for), to permit sg | давать разрешение, позволять → **rukset**

rukset (1) *N Ar.* permission, license | разрешение, позволение → **ruχset**

rusul (1) *N Ar.* prophets | посланники божие, пророки

S

saat s. **sāt**

saba s. **sabā**

sabah s. **sabā**

sabaχ (1) *N* stalk | стебель
sabak I (1) *N Ar.* lesson, class | урок
sabak II s. **sabā**
saban (15) *N* plough | плуг; **saj saban** light wooden plough | лёгкий деревянный плуг
sabanʒı (1) *N* plowman, plower | пахарь
sabā (74), **saba** (5), **sabah** (7), **sabak II** *N Ar.* morning | утро
sabın s. **sabun**
sabır (6), **sabur** (2), **sabr** (3) *N Ar.* patience, mettle | терпение, выдержка; **sabr et**– to be patient, to endure терпеть, выносить, переносить
sabr s. **sabır**
sabun (2), **sabın** *N Ar.* soap | мыло
sabunla– (3), **sapınla**– **(spınla–)**, **sapunla**– *V* to soap, to wash with soap | мылить, мыть с мылом
sabur s. **sabır**
sač (7) *N* hair | волосы → **čač, šač**
sač– (7), **saš**– *V* to scatter, to sprinkle | сеять, сыпать → **čač**–
sačaχ (3) *N* fringe | бахрома, басон → **sačak II**
sačak I (1) *N* eaves of a house, cornice | выступ крыши, карниз, стреха
sačak II (2) *N* fringe | бахрома, басон → **sačaχ**
sačım (2) *Adj* scattered | разбросанный; рассыпанный; **tůkům–sačım** scattered far and wide | разбросано вразбивку; **tůkům–sačım bol**– to be scattered far and wide | рассеяться/рассыпаться → **tůkům**
sačmala– (1) *V* to talk or act in an unreasonable manner | путаться; говорить глупости
saʒajak (1) *N* trivet | таган, треножник (для котла)
sadaka (2) *N Ar.* alms, charity | милостыня, подаяние
safa s. **sefa**
safalı s. **sefalı**
saγ I (6) *Adj* alive, sound in body, healthy | живой, здоровый; **saγ ve salamet/salāmet!** (be) safe and sound! have a good journey! (as a greeting and a wish) | (будь) живым и здоровым! счастливое путешествие! (приветствие и желание); **saγ–salam** safe and sound | жив–здоров, цел и невредим; **saγnen salamet** safe and sound | живой и здоровый → **sau**
saγ II (3) *Adj* right, right–hand | правый
saγ– (3) *V* to milk | доить → **sav–, sıv–**
saγan s. **sahan**
saγat s. **sāt**
saγın (1) *N* mirage | мираж
saγın– (5) *V* to yearn, to pine for | скучать, тосковать
saγır (1) *N* deaf (person) | глухой
sahab (1) *N Ar.* owner, patron, possessor, master | хозяин, владелец → **sāhip**
sahan (3), **saγan** *N Ar.* copper food dish, dish of food | сахан, медное блюдо
sahat s. **sāt**
sahib s. **sāhip**
sahıb s. **sāhip**
saχat s. **sahat**
saχın– s. **sakın**–
saχla– (1) *V* to hide, to keep secret | прятать, скрывать
saχlan– (5) *Vpass* to hide oneself | прятаться, скрываться → **saklan**–
saj (1) *Adj* shallow | мелкий, неглубокий; **saj saban** light wooden plough | лёгкий деревянный плуг
saj– (8) *V* to count; to respect | считать; уважать
sajdır– (1) *Vfact* to have sg counted | велеть считатъ
sajil (2), **sajıl** *N Ar.* beggar | нищий, попрошайка

sajıl s. **sajil**

sajıl– (3) *Vpass* to be counted, to be reckoned; to be esteemed | подвергаться счёту; считаться, рассматриваться; **sajılmaz** uncountable, countless | неисчисляемый, бесчисленное колечиствоо

sajın (12) *Postp* dependent on, as | по мере того как, с каждым разом, по ...; **keǯe–sajın** every night | по ночам; **sủjgen–sajın** as he/she was loved | по мере того, как его/её любили

sajla– (4) *V* to count | подсчитывать, сосчитывать; **sajlap** far and wide, everywhere | вдоль и поперёк, везде и повсюду; **Kırım sajlap** everywhere in the Crimea | везде и повсюду в Крыму

sajlam s. **selam**

sajlan– (3) *Vpass* to be sorted out; to be enlisted | быть выбранным; быть призванным на военную службу

sajlap s. **sajla**–

sajran s. **sejran**

sak (1) *Adj* cautious, watchful, careful | осторожный, бдительный; **sak ol**– to be cautious | быть осторожным → **sakt**

sakal (10) *N* beard | борода

sakallė s. **sakallı**

sakalli s. **sakallı**

sakallı (14), **sakallė**, **sakalli** (2) *Adj* bearded | бородатый; бородач

sakča (1) *Adv* carefully | осторожно

sakči s. **sakčı**

sakčı (8), **sakči** *N* watchman, guard | сторож, караульщик; **sakči tur**– to be on the watch, to keep guard (over sg) | стоять на посту/на страже

sakın– (12), **saχın**–, **sakin**– *Vpass* to take care of oneself by flinching, standing out of the way, etc.; to be cautious in word and deed | беречься; остерегаться; избегать

sakla– (15) *V* to hide; to preserve, shield from danger, to keep, store, save for future use | охранять; скрывать; откладывать; **Allah senı saklasın!** may Allah protect you! | побереги Аллах!

saklan– (8) *Vpass* to hide (oneself); to preserve oneself from sb/sg | прятаться; скрываться; храниться → **saχlan**–

saklı (1) *Adj* hidden, put aside, preserved | сохраняемый, спрятанный; **saklıγa kijgen urba** holiday attire | праздничное платье, одежда для праздника

sakt (1) *Adj* watchful, careful | бдительный, осторожный, чуткий, бережный; **sakt bol**– to be careful | быть осторожным/бдительным → **sak**

sal (1) *N* raft | плот, паром

sal– (28) *V* to put (on, down), to build, to beat; to throw | класть; строить, бить; бросать; **sala/salı ber**– just to let go, to set free suddenly/quickly | разрешить идти, идти быстро, неожиданно; **ủj sal**– to build a house | строить дом

sala (1) *N* village | деревня; **Koǯa Salası** Kodja Salasi (Kodja's village in the Crimea) | Коджа Саласы (деревня Коджи в Крыму)

salam (80), **sajlam**, **selam** (13) *N Ar.* salutation, greeting, salute | доброе пожелание, приветствие, привет; **sajlam ber**– to greet sb | приветствовать; **salam alejkum/alejkủm!** peace be with you! | мир вам! здравствуй(те)!; **salam/selam al**– to acknowledge a salute oranother's bow | отвечать на приветствие; **salam ber–** to greet sb | приветствовать; **salam bolson** greetings (to) | привет; **salam et–** to greet sb | приветствовать кого, передавать привет (поклон); **salam sůle–** to send sb's compliments | передавать привет (поклон); **salum alejkum** peace be with you! | мир вам! здравствуй(те)!; **selam alejkum/alejküm!** peace be with you! | мир вам! здравствуй(те)!; **selam bir–** to greet sb | приветствовать

salamet (7), **salāmet**, **selamet** *N Ar.* peace, safety, health; liberation, happiness | мир; согласие; спасение; безопасность; здоровье; благополучие; **saγ ve salāmet** safe and sound | живой и здоровый; **sauluχ salamet** safe and sound | живой и здоровый ; **saunen salamet, sau ve salamet** safe and sound | живой и здоровый;

saldat (13) *N R.* soldier | солдат

saldır– (3) *Vfact* to have sg built; to have sg put up | велеть построить, заставить поставить

Salγır s. **Salγur**

Salγur (2), **Salγır** *Tn* (a river in the Crimea) Salghur | (река) Салгур в Крыму

sali (1) *Adj Ar.* righteous, pious; glorious | благочестивый, праведный

salı (5) *N* Tuesday | вторник; **salı kůni** (on) Tuesday | (во) вторник

salın– (1) *Vpass* to settle on, to be spread over | налетать, покрывать, устилать

salınǯak (3) *N* seesaw | качели

salındır– (1) *Vfact* to stoop. to tilt; to swing | наклонять, нагибать, свешивать

salkin s. **salkın**

salkım I (6) *N* hanging bunch of grapes | гроздь, кисть винограда; **ůzům salkımı** cluster/bunch of grapes | виноградная гроздь, кисточка винограда

salkım II s. **salkın**

Salkım–bij (1) *Pn* Salkim–biy (Mr. Grape Bunch) | Салкым–бий (Господин гроздь/кисть винограда) → **salkım I**

salkın (3), **salkin**, **salkım** *Adj* cool, chilly | прохладный; **ekindi salkininnen** with the cool air of the time of the afternoon prayer | прохладным ветером времени полудненной молитвы **salkım** [< **salkın**] **jel** cool air | прохладный ветер

salkınnık (1) *N* coolness | холод, прохлада

salla– (7) *V* to swing or rock to and fro | качать, колебать, развевать что–л.

sallan– (9) *Vpass* to be about to fall; to swing, to sway | колебаться, шататься

salpırat– (1) *Vfact* to hang, to cause to hang | опускать, свешивать

salum–alejkum s. **selam**

samavar (9) *N R.* samovar | самовар; **kurulu samavar** samovar being put on | поставленный самовар; **samavar kurulγan** the samovar was put on | восседали самовар, поставили самовар

samsun (3) *N* cembra pine | кедр?

san (1) *N* esteem, reputation | слава, известность, уважение; **sanγa kat–** to grade sb according to one's reputation/rank etc. | причислять/оценивать кого по репутации/рангу

san– (19) *V* to seem, to suppose, to think | считать, думать, предполагать

sanč– (1) *V* to set up in the earth (lance, etc.) | вонзать, втыкать (в землю)
sandal (2) *N Gr.* rowboat | лодка
sandėk s. **sandık**
sandiχ s. **sandık**
sandik s. **sandık**
sandıγ s. **sandık**
sandıχ (14), **sandiχ** (3), *N* pantry, chest, box | кладовая; сундук, ящик → **sandık**
sandık (~**sandıγ**) (36), **sandėk**, **sandik** (6), *N Ar.* pantry, chest, box | кладовая; сундук, ящик → **sandıχ**
sanki s. **sankı**
sankı (4), **sanki** *Adv* as if, as though, supposing that | будто, словно, как будто
sankım (3) *Adv* as if, as though, supposing that | будто, словно, как будто
santırač (1) *Adj Ar.* checkered, cross–lined | клетчатый, (ткань) в клетку; **santırač jaulık** cross–lined napkin | клетчатый платок
sap–sau s. **sau**
sapa s. **sefa**
sapınla– s. **sabunla–**
sapla– (1) *V* to thrust into; to pierce | вонзать, втыкать
sapunla– s. **sabunla–**
sar– (30) *V* to wind, to wrap; to embrace | обёртывать, окрушать, охватить, объять; **bujnına sar–** to wrap around one's neck | обнимать шею у кого; **sarık sar–** to wind, to wind up a turban | онаматывать тюрбан
sara– (1) *V* to turn yellow | желтеть
saraj (135) *N P.* palace | дворец
Saraj–murun (1) *Tn* Saray–murun (Sarayburnu, i.e. Cape Palace), a quarter of Istanbul city | Сарай–мурун (Сарайбурну, т. е. дворцовый мыс), знаменитый район Стамбула
saral– (2) *Vpass* to turn yellow | желтеть
saran (1) *Adj* miser, mean | скопидом, скряга
sarar– (4) *V* to turn yellow | желтеть
sarf (1) *N Ar.* a spending, expense, outlay | расходование; трата; **sarf et–** to spend | расходовать
sarhoš (8), **saroš** (3), **sarroš**, **saruš**, *Adj P.* drunken | пьяный; **sarhoš bol–**, **saroš ol–** to get drunk | стать пьяным, опьянеть; **sarhoš kuj–** to make drunk | сделать пьяным, напоить → **sarχoš**
sarhošlan– (1) *Vpass* to get drunk | стать пьяным
sarχoš (2) *Adj P.* drunken | пьяный → **sarhoš**
Sari–batır s. **Sarı–batır**
sarı (19) *Adj* yellow | жёлтый; **sarı su** jaundice | жёлтуха
Sarı–batır (6), **Sari–batır** (2) *Pn* Sari–batir | Сары–батыр → **batır**
sarıχlı (1) *Adj* turbaned, wearing a turban | носящий чалму
sarık (2) *N* turban | чалма, тюрбан; **sarık sar–** to wind, to wind up a turban | онаматывать тюрбан
sarıl– (36) *Vpass* to embrace sb | обнимать, бросаться кому–л. на шею; **bujnına sarıl–** to be wrapped around one's neck | броситься на шею (в объятия) кому; **kujnına sarıl–** to be clasped/folded in one's arms | броситься на грудь (в объятия) кому
sarımsaχ (1) *N* garlic (Allium sativum) | чеснок
sark– (4) *V* to hang down, to let oneself down | свешиваться, опускаться
sarkıt– (1) *Vfact* to cause to hang, to bring down upon | опускать, свешивать
sarma–burma (1) *N* a sort of shotgun (with breakable barrel) | разновид-

ность охотничьего ружья (с переломным механизмом); **düfegim sarma–burma** I have a shotgun with breakable barrel | моё ружьё с переломным механизмом

sarmašıχ (1) *N* common ivy, hedera (*Hedera helix*) | вьюнок

saroš s. **sarhoš**

sarroš s. **sarhoš**

saruš s. **sarhoš**

sası– (2) *V* to stink, to smell badly | дурно пахнуть

Sastopol (1) *Tn* Sevastopol (city) | (город) Севастополь

saš– s. **sač–**

sat– (44) *V* to sell; to boast; to pass oneself off as | продавать; хвастаться; выдавать себя за кого–л.; **eriflik sat–** to pass oneself off as a lad | выдаваться за парня; **ǯılve [ǯilve] sat–** to play the coquette | кокетничать; **pālıγa sat–** to sell expensively | продавать дорого; **šeftāli sat–** to sell peck | продавать поцелуй

satıl– (6) *Vpass* to be sold | быть проданным, продаваться

satın (2): **satın al–** *V* to buy | купить

sattır– (1) *Vfact* to make sb sell sg | заставить продать что кому

sau (17) *Adj* alive, sound in body, healthy | живой, здоровый; **sap–sau** very healthy; strong | очень здоровый; абсолютно здоровый; сильный; **sau bol!** good–bye! farewell! | будь здоров! прощай!; **sau esen boloñız** be in good health! | живите в здоровье! здравствуйте!; **sau ve salamet** safe and sound | живой и здоровый; **saunen salamet** safe and sound | живой и здоровый → **saγ I**

sauklaš– (3) *V recipr* to say good–bye to each other, to take leave of | прощаться, раставаться с кем → **saulaš–**

saulaš– (1) *V recipr* to greet or salute each other | здороваться, приветствовать друг друга → **sauklaš–**

saulık (5) *N* health | здоровье

saulıklaš– (8), **sauluklaš–** (2) *V recipr* to say good–by to each other, to take leave of | прощаться, раставаться с кем

sauluχ s. **sauluk**

sauluk (7) *N* health | здоровье; **sauluχ salamet** safe and sound | живой и здоровый; **sauluknan kalıñ!** good–bye! farewell! (polite) | будьте здоровы! прощайте!; **sauluknen kal!** good–bye! farewell! (informal) | будь здоров! прощай!

sauluklaš– s. **saulıklaš–**

saurıl– (1) *Vpass* to be scattered, to be ventillated | быть развеянным; рассеиваться; разбрасываться → **savrul–**

sav– (1) *V* to milk | доить → **saγ–, sıv–**

savab (1) *N Ar.* merit acquired by a good action; reward | благодеяние, награда, воздаяние

savatlı (1) *Adj Ar.* engraving in black on silver; niello | украшенный узорами с чернью, черненый (о серебре)

savır– s. **savur–**

savrul– (1) *Vpass* to be scattered, to be ventillated | быть развеянным; рассеиваться; разбрасываться → **saurıl–**

savul– (1) *Vpass* to become sound, healthy, to get well again | вылечиваться, выздоравливать

savult– (1) *V* to heal, to cure, to restore (sb) to health | вылечивать, излечивать, исцелять

savur– (2), **savır–** *V* to throw into the air, to blow violently | веять; брызгать, пускать (дым); **sıyarčıyın savur–** to blow the smoke of his cigarette | выдувать дым своей сигареты

savuš (1) *N* struggle, fight, battle, war | битва, схватка, война

savut (1) *N* tableware, vessel, pot | посуда, сосуд

saz I (2) *N P.* musical (folk)instrument, any string instrument played by plucking | щипковый народный музыкальный инструмент; **saz čal–** to play saz | играть на сазе

saz II (8) *N* reeds, rushes; marsh, morass | тростник; топь, болото

sazlık (1) *N* place covered with rushes, reed bed, marshy place | болотистое место, заросли камыша, топи, трясина

sāb s. **sāhip**

sāde (1) *Adj P.* mere, simple(-minded), simplehearted; (here) only, solely | простой, простодушный, простосердечный; (здесь) только, исключительно

Sāfije (8) *Pn fem* Safiye | Сафие

sāhi (1) *Adj Ar.* true, real | верный, истинный

sāhip (23), **sahib** (4), **sahıb**, **sābi (=sāhibi)** (10), **sābı (<sāhibı)** (4), **sāhib** (4) *N Ar.* owner, possessor | хозяин, владелец; **sābı** his/her/its owner | его/её хозяин → **sahab**

sāt (63), **sat** (2), **saat** (9), **saүat**, **sahat** (9), **saχat** *N Ar.* watch, clock; hour | часы; час; **o saat, o‿ssaat** at that hour, that time, immediately | в тот час, сейчас; **šo sāt, šo‿ssāt** at this hour, immediately | в этот час, сейчас; **šu sāt, šu‿ssāt** at this hour, immediately | в этот час, сейчас

sebeb (4), **sebep** (3) *N Ar.* cause, reason | причина, повод

Sebihe (6) *Pn fem Ar.* Sebihe | Себихе

sečil– (1) *Vpass* to be chosen | быть выбранным

sefa (19), **safa** (6), **safā**, **sapa**, **sefā** (3), **sepa** *N Ar.* enjoyment, delight, pleasure | удовольствие, наслаждение; **χo‿ssafa [<χoš safa] keldiñiz!** welcome! I am pleased to see you! | добро пожаловать! я; **safa sůr–** to lead a life of pleasure; to enjoy oneself | проводить жизнь в удовольствиях, наслаждаться жизнью; блаженствовать; **sapale [<sapa ile] keldiñ!** welcome! I am pleased to see you! | добро пожаловать!; **sefa ettir–** to have sb pampered/amused by sb | велеть развлекать кого; **sefada kal–** to lead a life of pleasure; to enjoy oneself | проводить жизнь в удовольствиях, наслаждаться жизнью; блаженствовать; **sefanen** with pleasure | с радостью; **sefā keldiñ!** welcome! I am pleased to see you! | доб[1]ро пожаловать!

sefalı (3), **safalı** *Adj* that gives pleasure or enjoyment, pleasant | приятный, веселый, довольный, беззаботный

sefā s. **sefa**

sefer (5) *N Ar.* time, occurrence | раз

sehar (1) *N Ar.* time just before dawn; early morning | рассвет, заря; **sabā sehar vaktında** in the morning, at dawn | утром, на рассвете/на заре

Sejd Ahmat (8), **Sejd Amad**, **Sejd Amat** (5) *Pn* Seyd Akhmat | Сейид Ахмат, Сейдамет, Сейтамет

Sejd Amad s. **Sejd Ahmat**

Sejd Amat s. **Sejd Ahmat**

Sejd Bılal (1) *Pn* Seyd Bilal | Сейтбилял

Sejd Ůmer (1) *Pn* Seyd Umer | Сейид Умер

sejir (6), **sejır**, **sejr** (3), **sijir** *N Ar.* observation, onlooking; spectacle | наблюдение, зрелище; **sejr et–** to look, to observe | смотреть, наблюдать

sejirlen– (1) *Vpass* to be astonished/amazed at sg | удивляться

sejır s. **sejir**

sejr s. **sejir**

sejran (3), **sajran**, **seran** *N P.* observation, contemplation | наблюдение, зрение; **sejran et–** to catch sight of, to behold | зреть, созерцать

sekir– (2), **sıkır–** (**~sikir–**) *V* to jump, to gallop | прыгать, скакать

sekiz (8), **segiz** (2), **sekėz** (2) *Num* eight | восемь; **kırk segiz** forty–eight | сорок восемь; **sekiz ǯennet** eight heavens | восемь небес/раев ; **un sekiz** eighteen | восемьнадцать → **sikez**

seksen (4) *Num* eighty | восемьдесят

sel (8) *N* torrent, flood | сель, поток; бурный паводок, стремительное течение

sela (1) *N Ar.* a chant from the minarets on occasions of the Friday namaz; a call to a funeral service and similar sacred occasions | призыв к совершению молитвы (пред намазом в пятницу или для отпевения покойника)

selam s. **salam**

selamet s. **salamet**

selbi (6) *N P.* cypress (*Cupressus sempervirens*) | кипарис вечнозелёный; **selbi–terek** cypress–tree | дерево кипарис

sele (10) *N Ar.* flattish wicker basket | широкая корзина

semir– (1) *V* to grow fat | толстеть, тучнеть

semiz (1) *Adj* fat, fleshy, overweight | жирный, полный, тучный, сочный

sen (864), **señ** (2) *Pron pers* you *(singular)* | ты; **saγa** to/for you | тебе, к тебе, для тебя; **sana/saña** to/for you | тебе, для тебя; **sende** at/with/by you | у тебя; **sendeki** sg which is at/with you | что–нибудь находящийся у тебя; **senden** from you | от тебя; **seni** you | тебя; **seniñ/seni/senıñ** of you, your | твой, у тебя

sene (2) *N* year | год

senelik (2) *Adj* for a year; of a year | годовой

sep– (13) *V* to sprinkle, to scatter | брызгать; сыпать; сеять → **serp–**

sepa s. **sefa**

sepet (2) *N P.* basket | корзина

sepıl– (1) *Vpass* to be sprinkled | сыпаться

ser– (3) *V* to spread out on the ground, to spread over | расстилать, уложить

seran s. **sejran**

seratlık (4), **serātlık**, **sėratlık** *N* wandering about away from home | скитание на жужбине

serātlık s. **seratlık**

seren (2) *Adj* cool, cool weather or air | свежий, прохладный

sergi (2) *N* shop–front; temporary display of goods; (here) small street shop | уличная палатка, ларёк

sernik (1) *N R.* kibrit | спичка → **šėrpė**

serp– (2) *V* to sprinkle, to scatter | распылать (воду), брызгать; рассеивать → **sep–**

sersem (2) *Adj P.* stupified, foolish, silly | одуревший; ошалевший

sersemčik (2) *N dim* stupified, foolish, silly | глупенький

ses (18) *N* sound | голос; **ses čıχar**– to sound, to speak, to squeak | подать голос, откликнуться

sesken– (1) *V* to wake up (with a start), to be startled | встрепенуться, вздрогнуть

sesle– (1) *V* to say sg, to speak, to utter | сказать, вымолвить

seu– (9) *V* to love, to like, to pet | любить, ласкать → **sev–, sủj–**

seuda (1) *V* love, passion, intense longing; strong wish or desire | любовь, страсть → **sevda**

sev– (23) *V* to love, to like, to pet | любить, ласкать → **seu–, sủj–**

sevda (3), **sevdā** *N* love, passion, intense longing; strong wish or desire | любовь, страсть; **sevda ček**– to be deeply in love | сильно увлечься, испытывать сильную любовь → **seuda, sủjda**

sevdalık (2) *N* love, passion, intense longing; strong wish or desire | любовь, страсть → **sủjdalik**

sevdā s. **sevda**

sevgili (3) *Adj* beloved, dear, darling; lovable | любимый; возлюбленный → **sủjgili**

sevin– (1) *N pass* to be glad, to be pleased | радоваться

seviš– (4) *V recipr* to love or caress one another; to make love | любить друг друга; совершать половой акт → **sủjıš–**

sez– (1) *V* to perceive, to feel | чувствовать, смекать, предчувствовать

si s. **sij**

si– I s. **sı–**

si– II s. **sij–**

siǯak I (1) *Adj* hot, heat | горячий, жаркий

siǯak II? s. **sijah**

siǯim s. **sıǯım**

sigiz s. **sikez**

sigizlau (1) *N* making sg to be eight; increasing sg to eight | делать что–л. восьмикратным; становление восьмикратным → **sikezle–**

sij (5), **si** (2) *N* hospitable reception/ entertainment; treat | угощение; **si et–, sijlar et**– to receive/entertain sb in one's house | угощать кого–л.; **sij ủrmetler kủr**– to experience hospitable reception and respect | испытывать гостеприимное угощение и уважение

sij– (14), **si**– (2), **sıj**– (3) *V* to piss, to pee | мочиться

sija (3), **sijā, sıja** *Adj* black | чёрный; **sim–sija** jet–black | черным–черно

sijah (2), **siǯak [=sijak?]** *Adj P.* black | чёрный

sijak? s. **sijah**

sijdir– (2) *Vfact* to have sb pissed, to make sb pee | заставить мочиться

sijir s. **sejir**

sijırčık s. **sıjırčık**

sijla– s. **sıjla–**

sijrek (1) *Adj* rare | редкий

sikele s. **iskele**

sikez (5), **sigiz** (1) *Num* eight | восемь; **un sikez** eighteen | восемьнадцать; → **sekiz**

sikezle– (1) *V* to make it to be eight | делать что–л. восьмикратным → **sigizlau**

siklet (1) *N Ar.* heaviness, uneasiness; sadness, gloom | тяжесть; тягость, тоска; **sikletge kal**– to find oneself in heavy, uneasy or dangerous situation, to be exposed to haviness or danger | попасть в трудное или опасное положение, оказаться в трудном, опасном положении

sil– (4) *V* to wipe; to hit | вытереть; стереть; ударить

sila I s. **silā**

sila II s. **sıla II**

silā (5), **sila, sıla** (2) *N Ar.* weapon, arm | оружие

silāsız (1) *Adj* unarmed | безоружный, невооружённый

sileǯek (1) *N* large bath towel | полотенце

silik– (1) *V* to shake | трясти

silken– s. **silkin–**

silkin– (2), **silken–** *Vpass* to shake oneself | трястись

silte– (2), **sılt–** *V* to swing, to flourish, to wave | размахивать

sin (17) *Pron pers* you *(sing.)* | ты; **sinden** from you (sing.) | от тебя; **sine** you | тебя; **sini** you (sing.) | тебя; **sinin/siniñ** of you, your | твой; **sinnen** with you | с тобой → **sen**

sin– s. **siñ–**

sini I (3), **sını** *N P.* round metal tray | круглый поднос

sini II (~**sını**) (1) *N P.* bossom, breast. chest | грудь, лоно; **sınıne al–** to clasp/fold in one's arms | прижать кого–л. к (своей) груди

siñ– (3), **sin–** *V* to crouch down (to hide oneself); to be hidden | впитываться; прятаться

sirik s. **sırık**

sirke (1) *N* vinegar | (винный) уксус

siz (117), **sız** *Pron pers* you *(plural, or polite for singular)* | вы

sı– (7), **si–** *V* to fit into | вмещаться → **sıγ–**

sıč– (2) *V* to shit, to defecate | испражняться; гадить

sıčan (11) *N* mouse | мышь → **sıčkan**

sıčavul (2) *N* rat | крыса

sıčkak (2) *Adj* shitty | испражненный

sıčkan (17) *N* mouse | мышь → **sıčan**

sıčra– (3), **sıčrı–?** *V* to cavort, to jump | прыгать

sıčrı–? s. **sıčra–**

sıǯım (5), **siǯim, sıžım, šıǯım** *N* rope, cord | верёвка, шнур, бечёвка

sıγ– (9) *V* to fit into | вмешаться → **sı–**

sıγarčık (2) *N dim* sigarette | сигарета

sıγır (7) *N* cattle, cow | крупный рогатый скот; корова

sıγırčık (2), **sıγırǯıχ** *N dim* starling (*Sturnus vulgaris*) | обыкновенный скворец → **sıjırčık**

sıγırǯıχ s. **sıγırǯık**

sıχ– (1) *V* to squeeze, to tighten, to put pressure on | жать, сжимать, сдавливать → **sık–**

sıχıl– *Vpass* to be squeezed | сжиматься, стесняться; **ǯanı sıχıl–** to be bord, to feel dull | скучать, томиться → **sıkıl–**

sıχındı (1) *N* discomfort; need; difficulty | стеснённость → **sıkıntı**

sıχtırmač (2) *N* a kind of tightly fitting trousers | род узких штанов

sıj– s. **sij–**

sıja s. **sija**

sıjır– (2), **sıjir–** *V* to flay, to skin; to tear, to peel off; to scrape off | снимать кожу, облупить; содрать

sıjırčık (1), **sijırčık** *N dim* starling (*Sturnus vulgaris*) | обыкновенный скворец → **sıγırčık**

sıjla– (5), **sijla–** *V* to treat with marks of respect and distinction (guest), to entertain sb in one's house | уважительно угощать кого–л.; **sij sıjla–** to host, to give a dinner in someone's honour | оказывая почести угощать кого–л.

sık– (2) *V* to squeeze, to tighten, to put pressure on | жать, сжимать, сдавливать → **sıχ–**

sıkıl– (5) *Vpass* to be squeezed | сжиматься, стесняться; **ʒanı sıkıl–** to be bord, to feel dull | скучать, томиться → **sıχıl–**

sıkıntı (1) *N* discomfort; need; difficulty | стеснённость → **sıχındı**

sıkır– s. **sekir–**

sıklan– (1) *Vpass* to begrudge, envy; to want, to wish sb | завидовать; желать, хотеть

sıla I s. **silā**

sıla II (7), **sila II** *N Ar.* visit to one's native place and relatives | посещение родни, родины; **silaha [silaja!] bar–** to go to visit one's relatives | пойти посещать родни; **sılaja kel–** to come to visit one's relatives | приходить посещать родни

sıla– (1) *V* to whitewash; to paint, to moisten | (вы)белеть; намочить; смазывать; **sılau sıla–** to whitewash, to paint | белить известью, красить

sılak (1) *Adj* wet | мокрый

sılau *N* whitewashing, whitening, painting | побелка (известью), обмазка

sın– (1) *V* to be broken, to be broken into pieces | ломаться, разбиваться

sını s. **sini I**

sıñar (3) *Adj* a single thing; half (of sg), odd number; | один из пары; парное к ...; **sıñar avıʒ** a handful (of) | горсть (чего–л.)

sıñe– (1) *V* to try, to test; to attempt | испытывать, пробовать

sıñır (1) *N* frontier, border, limit | граница; рубеж; предел

sıpa– (2) *V* to stroke, to caress | гладить

sıpırtkı (1) *N* broom, brush | метла, веник → **sůpůrge**

sıptırıl– (2) *Vpass* to slip out of one's hands, to escape, to run away | ускользнуть, вырваться; убегать

sır (1) *N Ar.* secret | тайна, секрет

sıra (26), **sira** (3) *N* row, order; turn | ряд; порядок; **sirajlan** in rows, in turn, in order | по очереди; рядами

sırık (~**sirik?)** (1) *N* pole, stick | шест, жердь

sırlı (~**sırli?**) (1) *Adj* glazed | глазурованный

sırma (5) *N* lace or embroidery of silver or silver gilt thread; golden (hair); silver with a small admixture of gold | серебряная или золотая нить; **sırma čač** shiny golden blond hair | блестящие золотистые светлые волосы; **sırma–kulak** ear decorated with "sirma" | ухо украшенное "сырмой"

sırmalı (2) *Adj* embroidered with gold or silver thread | шитый золотом или серебром; **sırmalı–kulak** having ears decorated with "sirma" | имеющий уши украшенные "сырмой"; **sırmalı sač** like shiny golden blond hair | как блестящие золотистые светлые волосы

sırt I (11) *N* back | спина

sırt II (2) *N* North, northern | север, северный

sıt– (1) *V* to beat up (eggs) | бить (яйца)

sıv– (1) *V* to milk | доить; **sıɣır sıvar** (she) milks a cow | (она) доит корову → **saɣ–, sav–**

sıva– (1) *V* to plaster, to stucco | штукатурить

sız– (1) *V* to leak; to ooze | чертить; сочиться

sızɣır– (1) *V* to whistle; to hiss (snake) | свистеть, шипеть (о змее)

sıžım s. **sıʒım**

slaučan (1) *N* common earthworm (*Lumbricus terrestris*) | земляной червь

s'o s. **šo**

soba (1) *N* stove, hothouse | печь, печка

sofra (2) *N* dining table; wooden or metal tray serving as a table | обеденный стол; накрытый (сервированный) стол → **sufra**

soɣan (2) *N* onion | лук → **suɣan**

soɣıš (1) *N* fight, struggle, battle, war | драка, битва, война

soχta s. **suχta**

soj s. **suj**

soj– (2) *V* to peel, to flay; (here) to slaughter (kill by cutting one's throat) | чистить кожура, обдирать; резать; бить (скот); (здесь) убить перерезывая шею у кого/чего → **suj–**

sok– (1) *V* to strike, to hit, to beat | втыкать, всовывать; бить → **suk– II**

sokak (1) *N* street | улица → **sukak**

sokır (4) *N* blind | слепой → **sokor**

sokor (17) *N* blind | слепой → **sokır**

sol (2) *Adj* left | левый

sol– (7) *V* to languish, to wilt, to fade | вянуть, блёкнуть → **sul–**

solakaj (1) *Adj* left–handed | левша

soldır– (~**soldir–**) (3), **soldor–** *Vfact* to wither, to parch | дать увянуть, обесцвечивать; **al janaklın soldor–** to make your ruddy–cheeked (face) pale | дать увянуть (щеку) твоей румяно–лицевой (девушки) → **suldır–**

soldor– s. **soldır–**

son (170) *N/Adj/Postp* end, last, latter, final; later than | конец; потом, после; **tap sonı/sonu** just at the end | как раз в конце → **soñ, sun**

sonra (14) *Adv/Postp* hereafter, then, afterwards, later (than) | потом, немножко погодя → **soñra**

soñ (120) *N/Adj/Postp* end, last, latter, final; later than | конец, потом, после; **soñında, soñunda** at/in the end (of) | в конце чего; наконец; **soñki** the last one | последний; **tap soñı** just at the end | как раз в конце; **ůjleden soñ** in the afternoon | после обеда → **son, sun**

soñra (6) *Adv/Postp* hereafter, then, afterwards, later (than) | потом, немножко погодя, позже → **sonra**

sor– (3) *V* to ask, to inquire (about) | спрашивать, осведомляться о чём–л. → **sora–, sur–, sura–**

sora– (21) *V* to ask, to inquire (about) | спрашивать → **sor–, sur–, sura–**

sozol– (1) *V* to spread out, to stretch out | растягиваться; тянуться; удлиняться

sȯz (6) *N* word, speech; promise, agreement | слово; **sȯz ber–** to give one's word, to promise | давать слово; обещать; **sȯz br [bir] et–** to agree on (sg) | договариваться (о); **sȯzům jok** I have no objections; well and good | я ничего не скажу (против); не возражаю; всё в порядке; **sȯzům kalmaj** I will not have a word (against), I will have no objections | у меня ни слова не будет (против); не буду возражать → **sůz**

spınla– s. **sabunla–**

Stambul s. **İstambul**

stavaj (4) *V R.* get up, stand up (wake up) | вставай

su (133) *N* water | вода; **akan su** flowing water | текущая вода; **su baši** spring; water–side; shore | источник; берег (реки, озера); **su četinde** at the edge of water (i.e. river, lake, sea) | на краю/берегу воды (т. е. реки, озера и тд7); **suɣa, suja, suva** (in)to the water | в воду, к воде → **suv**

Subhan (1) *Interj Ar.* God (Allah) | Бог (Аллах)

suč (1) *N* fault, guilt, sin | вина, ошибка, проступок

sufa (1) *N Ar.* divan, sofa | софа, диван

sufra (5) *N Ar.* dining table; wooden or metal tray serving as a table | обеденный стол, накрытый (сервированный) стол → **sofra**

sufu (8) *N Ar.* godly, mystic, Sufi; devotional | набожный, благочестивый, праведный

suγan (1) *N* onion | лук → **soγan**

suγıš (24), **suγuš** (3) *N* struggle, fight, war | битва, война

suγuš s. **suγıš**

suγuš– (1) *V recipr* to struggle, to fight with each other | сражаться, воевать, драться

suχ– s. **suk– I**

suχta (7), **soχta** (1) *N Ar.* Muslim theological student | софта (ученик медресе), студент

suj (7) *N* race, ancestors; descendants; sort, kind, species | род, родня; сорт; **suj sup** (2) (all) relatives; ancestors and descendants | родственники, родичи, родня

suj– I (38) *V* to peel, to flay; (here) to slaughter (kill by cutting one's throat) | чистить кожура, обдирать; резать; бить (скот); (здесь) убить перерезывая шею у кого/чего), бить (скот) → **soj–**

suj– II s. **sůj–**

sujdigi s. **sůjdigi**

sujdır– (4) *Vfact* to have sb slaughtered | заставить резать/бить

sujıl– (2), **sujul–** *Vpass* to be flayed/ slaughtered | быть резанным; быть очишенным от кожи

suj sup s. **suj**

sujul– s. **sujıl–**

suk– I (5) *V* to thrust into | втыкать, всовывать → **suv– I**

suk– II (4) *Inf* to hit, to beat, to strike | бить, ударять → **sok–, suv II**

sukak (2) *N* street | улица → **sokak**

sukıl– (1) *Vpass* to push, ti invade sg, enter by force, to intrude | влезать; втискиваться; вмешиваться

suklan– (1) *V* to be jelous (of sb), to be envious (of) | завидовать, ревновать

sul (4) *Adj* left | левый → **sol**

sul– (1) *V* to languish, to wilt, to fade | вянуть, блёкнуть → **sol–**

suldır– (~ suldir) (1) *Vfact* to wither, to parch | дать увянуть, обесцвечивать; **al janaγın suldır– (~suldir)** to make one's ruddy cheek pale | дать увянуть её румяное лицо → **soldır–**

Sulejman (1) *Pn Ar.* Suleyman | Сулейман

sultan (2) *N Ar.* sultan, ruler, sovereign | султан; монарх; государь

sumarla– (4) *V* to order, to book | заказывать

sun (8) *N/Adj/Postp* end, last, latter, final; later than | конец; потом, после; **ketkenden sun** after he/she went away | после того как ушёл; **tap sunında** just at/in the end (of) | в конце концов; наконец → **son, soñ**

sun– (1) *V* to languish, to wilt, to fade | вянуть; блёкнуть → **sol–, sul–**

suñ (1) *N/Adj/Postp* end, last, latter, final; later than | конец, потом, после; **tap suñı** just at/in the end | как раз в конце → **son, soñ, sun**

sup s. **suj**

supa (1) *N* thick stick, club, cudgel | дубина

sur– (63) *V* to task, to inquire (about) | спрашивать, осведомляться о чём–л. → **sor–, sora–, sura–**

sura– (105) *V* to task, to iquire (about) | спрашивать, осведомляться о чём–л. → **sor–, sora–, sur–**

suraš– (1) *V recipr* to task one another, to inquire | расспрашивать друг друга

suret (11) *N Ar.* form, shape; picture | рисунок; портрет; **suret jaz–** to draw a picture/portret | рисовать картину/рисунок

surıš– (3) *V recipr* to task one another, to inquire | расспрашивать друг друга; **hal χatır surıš–** to inquire after each other's health and circumstances | распрашивать друг друга (о жизни, об остоятельствах)

susa– (5) *V* to be thirsty, to thirst | испытывать жажду; хотеть пить

susum (3) *N* thirst; beverage | жажда; напиток

suv (18) *N* water | вода → **su**

suv– I (2) *V* to thrust into | втыкать, всовывать → **suk– I**

suv– II (1) *V* to strike, to hit | бить, ударять → **sok–, suk– II**

suval (2) *N Ar.* question | вопрос

suvar– (5) *V* to water | поливать, орошать; подмывать; **čarši suvar–** to water the bazaars | поливать, подмывать базар

suviš s. **suvıš**

suvıril– (~ suvırl–) (1) *V* to buckle to, to set about doing sg | засучиваться

suvıš (~suviš) (3), **suvuš** *N* war, battle | война

suvlık (1) *N* watery, damp (place) | водный, водное место

suvu– (2) *V* to become estranged/alienated (from sb) | разлюбить; остыть

suvuk (2) *Adj* cold, frost; chilly and clear | холод; стужа; холодный, студёный

suvuš s. **suvıš**

sů– s. **sůj–**

sůg– (2) *V* to curse, to swear | ругать, бранить

sůj– (80), **suj–** (2), **sů–** *V* to love | любить → **sev–, seu–, süj–**

sůjda (1) *N* love, passion | любовь, страсть → **sevda**

sůjdalik (2), **sůjdalıχ** *N* love, passion | любовь, страсть → **sevdalık**

sůjdalıχ s. **sůjdalik**

sůjdigi (15), **sujdigi** (1) *N* beloved, darling | возлюбленный → **sůj–**

sůjgi (1) *N* love, affection; compassion | любовь; нежность

sůjgili (1) *N* beloved, dear | любимый, дорогой → **sevgili**

sůjgilik (1) *N* love, passion | любовь, страсть

sůjın– s. **sůjůn–**

sůjıš– (1) *V recipr* to love each other; to make love | любить друг друга; совершать пловой акт → **seviš–**

sůjken– (2) *V* to lean (against, on) | опираться, прислоняться; тереться?; **ǯaklavına sůjkenip** leaning against the doorpost | прислоняясь к косяку

sůjle– (28), **sůle–** (15) *V* to speak, to say, to tell; to sing | говорить, сказать, рассказывать; петь; **jalan sůjle–** to lie | лгать; **salam sůjle–** to send one's compliments | передавать приветствие

sůjlen– (6), **sůlen–** (1) *Vpass* to be told | быть сказанным; говориться

sůjri (1) *Adj* sharp, pointed | острый; остроконечный; **sůjri bašlı erenner** dervishes with cone caps | дервиши с конусообразными колпаками

Sůjůn bej (3) *Pn* Suyun–bey | Сююн бей

sůjůn– (2), **sůjın–** *Vpass* to be glad, to be pleased | (об)радоваться

sůk– (3) *V* to unstitch, to unbraid | распарывать, расплетать

sủkủl– (1) *Vpass* to come unstitched, to tear | распарываться, расплетаться

sủkủn– (1) *Vpass* to be cursed, to be sworn | быть обруганным, быть оскарбленным (словами)

sủle– s. **sủjle–**

sủlen– s. **sủjlen–**

sủleš– (1) *V recipr* to speak to one another, to converse or consult with one another | беседовать, переговариваться

sủn– (7) *V* to be extinguished; to go out (fire) | гаснуть

sủndủr– (6) *Vfact* to extinguish | гасить, тушить

sủngi (1) *N* bayonet | штык

sủpủr– (2) *V* to sweep | мести, подметать

sủpủrge (1) *N* broom | метла, веник → **sıpırtkı**

sủr– (18), *V* to drive away, to advance, to spend (life, time), to go on, to push on, to pass (time) | изгонять, ссылать, двигать перед собой; продолжаться, длиться, проводить; **ǯủz sủr–** to pay humble respect, to throw oneself down before sb | преклоняться перед кем–л., падать ниц; **saban sủr–** to plough | пахать → **sür–**

sủriš– (1) *V recipr* to go (come) by crowds; to proceed (to a place) together with others | идти массой, идти толпою

sủrme (1) *N* antimony | сурьма (краска)

sủrmeli (1) *Adj* tinged with antimony | подведённый сурьмой (о глазах)

sủrt– (1) *Vfact* to wipe (sg clean), to wipe away (tears) | вытирать, протирать, утирать (напр. слёзы)

sủrủ (4) *N* herd, flock | стадо

sủrủn– (3) *Vpass* to rub on; to drag oneself along the ground; to come to a sudden stop | тереться; спотыкаться; волочиться

sủt (12) *N* milk | молоко

sủv– (**~suv–**) (1) *V* to love, to like, to pet | любить, ласкать → **sủj–**

sủz (43) *N* word, speech, talk, promise | слово; **sủz ajt–** to make/deliver a speech | произносить речь; **sủz bol–** to become the subject of gossip | стать темой пересудов; **sủzi kesil–** to be interrupted, to be disturbed, to be cut short | запинаться в речи, перебиваться, застревать в горле (слова); **sủzin tut–** to keep one's word, to take the advice of, to obey | слушаться чего–л. совета, слушаться кого; **sủzinden čık–** to disregard the advice of (someone) | ослушаться → **sỏz**

sủz– (2) *V* to strain, to filter; to pour out (drink), | цедить; выливать, разливать (напиток); **šerabı sủz–** to draw off wine, to bottle wine | разливать вино по бутылкам

sủzıl– (**~sủzl–**) (1), *Vpass* to filter/ooze through | цедиться

süj– (1) *V* to love | любить to love | любить → **sev–, seu–, sủj–**

sümük (1) *N* mucus (ot the nose) | сопли, слизь

sür– (2) *V* to drive away, to advance, to spend (life, time), to go on, to push on, to pass (time) | изгонять, ссылать, двигать перед собой; продолжаться, длиться, проводить; **sürip (~sürıp) bar–** to march (in); to enter | двигаться, следовать; входить → **sủr–**

Š

šač (18) *N* hair | волосы → **sač**

šafran (1) *N* kitchen cabinet, commode | кухонный шкаф

šaɣird (1) *N P.* pupil; apprentice | ученик, подмастерье

šahan (1) *N P.* hawk, falcon | сокол, лячин → **šāhin**

Šahislam (2) *Pn P.–Ar.* Shahislam | Шахислам → **Šehislam**

šaχa s. **šaka**

šaχılda– s. **šakılda–**

šaj (46) *Adv* such, so, in this way/manner | так, таким образом; **šaj etip/itep (itėp)** so, thus, in this way | так, таким образом

šajχan– (1) *Vpass* to sway about; to rock/sway | полоскаться, встряхиваться; колыхаться → **čajχa–**

šajlı (4) *Adj* good, nice, beautiful, fine | хороший, миловидный

šak (1) *Onomatop N*; **šak et–** to flap (wings), to clap | щёлкать крыльями

šaka (3), **šaχa** *N* joke, fun | шутка; **šaχa et–** to joke, to make a fun | шутить

Šakaǯı (1) *Pn* Shakadji (i. e. joker, kidder), byname | Шакаджи (т. е. шутник), прозвище

šakı– (1) *V* to sing loudly (nightingale, canary) | (громко) петь (о певчих птицах), заливаться

šakılda– (3), **šaχılda–** *Onomatop V* to burble, to murmur (water) | журчать, шуметь (о воде)

šakırla– (1) *Onomatop V* to rattle, to make a loud noise | греметь, стучать, хлопать

šal (10) *N* shawl | шаль; **topozlı šal** shawl decorated with pearls | шаль с бусинками

šaltıra– (1) *Onomatop V* to clatter, to make a cracking sound | дребезжать, стучать, трещать

šaluar s. **šalvar**

šalvar (4), **šaluar** ~ **šaular** *N* shalvar, baggy trousers | шаровары, штаны

šam–alaǯı (2) *N* striped silk and cotton material for dresses (first made in Damascus); dress made of it | камчатная ткань; пёстрый ситец первоначално сделан в Дамаске; одежда сщитая из такой ткани

šamama (2) *N P./Ar.* a sort of pear; muskmelon | сорт груш; сорт мелкой душистой дыни

šamar (6) *N* a slap on the face; palm (of the hand) | пощёчина; оплеуха; **šamar dart–/dùrt–** to give a slap on the face | давать пощёчину; **šamar or–** to give a slap on the face | давать пощёчину

šamata (1) *N Ar.* great noise, uproar | большой шум

šamdan (1) *N P.* candlestick | подсвечник

šan (1) *N Ar.* fame, renown; dignity, honour | слава, доброе имя, честь

šan– (1) *V* to hope, to trust (in), to rely (on) | надеяться, доверять → **ušan–**

šandır–šındır (1) *Onomatop N* sound of shuffling | звук волочения ног

Šangeri (6) *Pn* Shangeri (khan) | Шангери (хан)

šanγırdat– (2) *Vfact* to make sg rattle/clatter; to make a noise by crashing sg | разбивать со звоном, треском (посуду, стекло)

šap (1) *N* smack, smacking | чмок, чмоканье; **šap et–** to make smackling sound | чмокнуть; **šap etip ùp–** to give a smacking kiss | давать звонкий поцелуй

šapır–šupur (1) *Onomatop N* sound of bubbling water | звук клокотающей воды; **šapir šupur et**– to wimple | булькать, журчать; **šapir šupur su** bubbling, boiling water | клокотающая, кипящая вода

šapka (2) *N R.* hat, fur cap | шапка → **šapki**

šapki (3), **šapkı** *N R.* hat, fur cap (Russian plural) | шапки → **šapka**

šapkı s. **šapki**

šarab (5), **šerab** (2) *N P.* wine | вино

šart (6) *Onomatop N* crack, crash | звукоподражание треска; **šart et**– to crackle | трещать, щёлкать

šaš– (31) *V* to wonder, to be surprised; to miss | удивляться, опешить; **julını šaš**– to lose his/her way, to take another direction | заблуждаться; сбиваться с дороги; **jurɣunnıktan šaš**– to get exhausted by weariness | истощаться от усталости

šašχın s. **šaškın**

šašır– (2) *V* to be confused (about something), | ошарашить, изумить; **julını šašır**– to lose one's way, to go bad | заблуждаться; сбиваться с (правильной) дороги

šaškın (2), **šašχın** *Adj* simple (–minded), loafer; stupid | растерявшийся; зевака, разиня

šašmala– (2) *V* to be confused, to lose one's head | растеряться, опешить

šatırda– (4) *V* to make a noise, to jangle, to clink | издавать шум, треск

šatırdı (1) *N* noise, crack | шум, треск

šatır–güdür **(~šatır–gudur)** (1) *Onomatop Adv* with crack and thump | с шумом и грохотом

šatır–šatır (1) *Onomatop N* big noise, crack | треск, шум

šatlat– (1) *Vfact* to delight, to gladden | веселить, обрадовать

šatlık (1) *N* joy, happiness | веселье, радость

šaule s. **šavle**

šavle (8), **šaule** (5) *N* beam, light, brightness | луч; **aj šaulesi** moonlight, moonbeam | свет/луч луны, лунный свет; **šavle bas**– to flood with light/brightness | озарять светом, блеском

šāhin (3) *N P.* hawk, falcon | сокол, лячин; **šāhin kušı** falcon/hawk [bird] | (птица) соколь/лячин → **šahan**

šebetij (1) *N* amusement, entertainment | развлечение

šeček (1) *N* flower | цветок → **čeček**

šeddet (1) *N Ar.* hardness, sverity, strength | сила, резкость, строгость

šeftali (10), **šeftalı**, **šefteli** (3) *N P.* peach, (fiurative) kiss | персик; (перен.) поцелуй; **šeftali ber**– to give a kiss | давать (дарить) поцелуй

šeftalı s. **šeftali**

šefteli s. **šeftali**

šeher (63), **šehir** (13) *N P.* city, town | город

šehir s. **šeher**

Šehislam (1) *Pn Ar.–P.* Shehislam | Шехислам → **Šahislam**

šehit (~šehid) (1) *N Ar.* one who has died while in military service; hero | герой, погибший за правое дело; **šehid bol**– to die in a battle (for islam) | быть убитым на войне; погибнуть смертью героя (о мусульманине)

šehr (2) *N Ar.* month | месяц

šej (70) *N* thing, deed, fact, matter | вешь, дело; предмет → **šij**

šejid s. **šejit**

šejit (4), **šejid** (1) *N* one who has died while in military service; hero | погибший за правое дело, герой; **šejid kit–** to die in a battle (for islam) | умереть мученической смертью; погибнуть смертью героя (о мусульманине); **šejit bol–** to die in a battle (for islam) | быть убитым на войне; погибнуть смертью героя (о мусульманине)

šejtan (4) *N Ar.* satan, devil | сатана, чёрт, дьявол; **šejtanya uj–** to let oneself be led astray; to yield to temptation | податься соблазну (искушению) → **šejten**

šejten (23) *N Ar.* satan, devil | сатана, чёрт, дьявол → **šejtan**

šeker (10) *N P.* sugar, sweet | сахар; сладкий

Šeker–ulan (1) *Pn* Sheker–ulan | Шекер–улан → **ulan**

šekerli (2) *Adj* sweetened with sugar, containing sugar, sugared | с сахаром, сахарный; сладкий

šemšer (1) *N P.* boxwood (*Buxus sempervirens*) | самшит

šen (2) *Adj P.* joyous, cheerful, glad | весёлый, радостный

šer (1) *N Ar.* evil, wickedness, harm; enmity | зло, вред, вражда; **šer cek–** to endure/suffer wikedness of sb | страдать от зла/злодеяния кого–л.

šerap s. **šarab**

šerbet (2) *N Ar.* sweet fruit drink | шербет; сладкий ароматизированный напиток; **eǯel šerbetin ič–** to drink up the sherbet of the hour of death; to die | выпить шербет смертного часа; умереть

šeren (1) *N* haystack, shock of sheaves | скирд, копна

šerfe (1) *N Ar.* woman, wife | женщина, жена

šerget (1) *N Ar.* companion, partner | спутник, напарник

Šerif (1) *Pn Ar.* Sherif | Шериф

šerjat (1) *N Ar.* (Muslim) canonical law, law code | шариат

Šertlan → **Šertlen**

Šertlen (4) *Pn;* **Šertlan–bej** (2), **Šertlen–bej** Shertlen–bey | Шертлен–бей

šešek (1) *N* flower | цветок → **čeček**

šeuk (1) *N Ar.* light, sunlight | свет, огонь

šėrpė (1) *N* match | спичка → **sernik**

ši s. **šij**

šij (68), **ši** (6), **šıj** (8) *N* thing, deed, fact, matter | вещь, дело, предмет → **šej**

šijčik (1) *N dim* little thing | маленькая вешь

šimde s. **šimdi**

šimdi (59), **šimde**, **šımdi** *Adv* now, at present | теперь, сейчас; **šimdijeǯek** until this time | до сих пор, до сего времени → **šindi**

šindė s. **šindi**

šindi (57), **šinde**, **šindė** (8), **šindı** (2), **šındė**, **šındi** *Adv* now, at present | теперь, сейчас; **šindijeǯek** until this time | до сих пор, до сего времени → **šimdi**

šindiki (1) *Adj* the present, the actual (state, time etc.) | теперешний, нынешний; настоящий

šipaniski s. **šipanski**

šipanski (2), **šipaniski** *Adj R.* Spanish (breed of sheep) | испанский; испанская порода овец

šiš (1) *N* broach, skewer | вертел; **šiš–kabab** meat roasted on a spit or skewers | кебап (жареное на углях мясо)

šiš– (2) *V* to swell, to become inflated | пухнуть, опухать

šiše (11) *N P.* bottle | бутылка; **penčere šišesi** window–glass | оконное стекло

šišman (2) *Adj* fat, obese | полный, толстый

šıbırtkı (3), **šėbirtkı, šıpırtkı** *N* knout; whip | плеть, кнут, нагайка

šıbuk (1) *N* stick, twig | палка; прут; розга → **čıbuχ, čıbuk**

šıǯım s. **sıǯım**

šıγart– s. **čıγart–**

šıγara (1) *N Sp.* cigarette | папироса, сигарета → **čıγara**

šıχar– (1) *V* to take out, to bring out | выводить, вытягивать

šıj s. **šij**

šılab s. **šılap?**

šılap? (1), ~ **šılab?** *N* gallop | галоп, карьер; **šılab‿aγγır, šılap aγγır** (a horse breed?) | (порода лошадей?)

šımdi s. **šimdi**

šındė s. **šindi**

šınkıj–šınkıj (1) *Adv* whimpering | хныкая

šıpırındı (1) *Onomatop N* the sound of shuffling | звук шарканья

šıpırt– (2) *Vfact* to beat, to lash | сечь, хлестать

šıpırtkı s. **šıbırtkı**

šır (1) *Particle* all, mere, pure, bare | (усилительная частица) сплошь; сплошной; совершенно; **šır χara** jet–black | черным–черно

šıra (1) *N P.* juice | сок

šırılda– (2) *Onomatop V* to flow in a stream with a babbling sound | журчать

šırpı (1) *N* chip, clipping, dry twigs; match–stick | обрезанные ветви деревьев; **šırpı čert–** strike a chip/mach | чиркнуть палочкой/спичкой

šırpıntı (1) *N* palpitation, trembling | трепетание

šo (143) *Pron demonst* this, that | этот, тот; **šo kadar** so much | столько; **šo saat** this time; immediately | в этот час; сразу; **šo sāt, šo‿ssāt** at this hour, this time, immediately | в этот час, в это время, сразу, сейчас; **šonın‿čün [<šonın ičün]** because of this one | из–за этого

šoγon (3) *Adj* ugly, rude, unsuitable | дурной, грубый, неприличный; **šoγon sůz** indecent/rough word | неприличное, дурное слово

šok (2) *Adj/Adv* much, great (in amount), long (in duration), too much, very | много, очень, весьма, долго → **čoχ, čok**

šol (1) *Pron* this, that | этот, тот → **šo**

šolaj (6) *Adv* so, in this way | так, таким образом → **šulaj**

šondaj (5) *Adv* such, so, in this way | такой, так, таким образом → **šundaji**

šondajın (8), **šondajin (**1) *Adv* such, so, in this way | так, такой

šorba (1) *N* soup | суп, похлёбка → **čorba, čurba**

šorbaǯı (2) *N* master, owner | хозяин → **čorbaǯı**

šort–šort (1) *Onomatop N* noise of snipping | (подражение слабому треску)

šȯl (1) steppe, meadow | степь, поле → **čȯl, čůl**

šu (142), **šů** (1), **šü** (6) *Pron demonst* this, that | это, тот, сё; **šuna buna** hither and thither, to and fro | туда и сюда; взад и вперёд; **šunda** here, in this place | здесь; на этом месте; **šundan**

from here, out of this place | отсюда, от этого места; **šu sāt, šu‿ssāt** [< **šu sāt**] at this hour, immediately | в этот час, сейчас; **šů‿jerde** [< **šu jerde**] in this place; at once, instantly | тотчас, немедленно; **šü‿jerde [< šu jerde]** in this place; at once, instantly | тотчас, немедленно; **šü‿jirden [< šu jirden]** from this place; at once, instantly | тотчас, немедленно

šufa (1) *N Ar.* recovery, cure, remedy for a disease | исцеление, выздоровление; средство от болезни

šulaj (9) *Adv* such, so, in this way/manner | так, таким образом → **šolaj**

šunda s. **šu**

šundan s. **šu**

šundaji (šundaj) (1) *Adv* so, such, in this way/manner | так, такой, таким образом → **šondaj**

šundi (1) [~**šundij** < šundıj < sundaj] *Adv* so, such, in this way/manner | так, такой, таким образом

šuvultı (1) *N* rustle, sigh (ing) | шум

šů s. **šu**

šůjle (2) *Adv* so, such, just, in this way/manner | так, таким образом

šůkůr (4), **šukůr**, **šükür** *N Ar.* gratitude, thanksgiving | благодарение, слава; **šůkůr et–** to thank | поблагодарить

šü s. **šu**

T

ta (11) *Conj/Particle* too, also; and | тоже, же, также; и, да и → **da**

tabaχ s. **tabak I**

tabak I (9), **tabaχ** (3) *N* plate, dish | тарелка, блюдо; **tabaχ betli** with plate–like face, dish; having round face | имеющий тарелкообразное лицо, круглоликий

tabak II s. **tabaka**

tabaka (1), **tabak** *N Sp.* tobacco box, cigarette box | коробка для табака; табакерка; портсигар

taban (2) *N* sole (of a foot or shoe) | подошва

tabijat (1) *N Ar.* nature, habit | натура, характер

tabıl– (2) *Vpass* to be found | находиться

tabıldır– (1) *Vpass, fact* to fulfil, to finish | выполнять, исполнять; завершать

tabın (5), **tabun** (2) *N* herd of brood mares and stallions with their colts loose in the fields | табун

tabır (1) *N* battalion, troop, camp | батальон, отряд, лагерь; **tabır asker** a battalion of soldiers | батальон солдат

tabjatlı (~tabijatlı) (1) *Adj* having such and such a nature, decent person; (here) stubborn, headstrong | порядочный, благородный; нравственный; (здесь) упрямый человек

tabun s. **tabın**

taǯ (2) *N P.* crown | корона

taǯlı (1) *Adj* crowned | имеющий корону; с короной

tayın– (3) *V* to shoulder, to undertake | брать на плечо (ружьё); взваливать (себе) на плечо; накинуть (себе) на плечи

tayıt– (3) *V* to scatter, to disperse | разгонять; рассеивать; рассыпать → **dayıt–**

taha (60), **tā** (2) *Adv* more, further, still | ещё, пока ещё, всё ещё; **bir taha** once again | ещё раз → **daha, tan**

taht (4), **taχt** *N P.* (sovereign's) throne | престол, трон

tahta s. **taχta**

tahtalan– s. **taχtalan**–
taχ taχ (1) *Onomatop N* the sound of thud, patter, clack | звук треска; **taχ taχ et**– to knock repeatedly, to strike with noise | стучать, хлопать
taχat s. **takat**
taχt s. **taht**
taχta (11), **tahta** (3) *N* board, plank | доска → **takta**
taχtalan– (3), **tahtalan**– *Vpass* to be planked, to be boarded | забивать досками, обивать досками
taj (13) *N* foal, colt | жеребёнок
taja– s. **taj**–
taj– (1) (~**taja**–) *V* to sideslip, to glide? | скользить? *V;* **taktadan tajaγan** slipped down on a board/plank? | скользив (или скользил) по доске?
tajaχ s. **tajak**
tajak (13), **tajaχ** (3) *N* stick | палка
tajan– (23) *Vpass* to lean (against, on), to be based on, to resist; to endure | опираться; прислоняться; противостоять; выдержать; **tajanmaz ol**– to be unbearable | стать/быть невыносимым
tajandır– (1) *Vpass, fact* to lean, to repose against/on | опереть, прислонить
tajau (~tajav) (1) *N* brace, support | опора
tajav s. **tajau**
tajγalat– (1) *Vfact* to make sg slippery | делать скользким
tajın (3) *Postp* like, almost, nearly, somewhat | как, подобно, будто; **kůmes tajın** like a hen house | подобный курятнику; **zıp tajın** like pitch | как дёготь, подобный смоле → **dajın**
tajmak (1) *Inf* gliding, sliding | скользание
tak– (10), **taχ**– *V* to put on, to wear, to give (a name) | нацеплять; надевать; давать имя; **čarık tak**– to put on rawhide sandals | надевать чарыки → **dak**–
takat (6), **taχat** *N Ar.* strength, power | сила, мощь
takatsızlık (2) *N* exhaustion, weakness | бессилие, слабость
takdir (2), **taktir** *N Ar.* appreciation, approval | одобрение, оценка; **takdir bol**– to agree, to estimate, to value | оправдывать; **taktirge turul**– to be appreciated, to be estimated, to be valued | оцениваться
takıl– (1) *Vpass* to be affixed? | цепляться?; **su takıl(a) kač**– going cleaving waves/waters (ship) | идти рассекая волны/воду (пароход)
takıldı (1) *N* clatter of hoofs | конский топот; звук копыт
takım (7) *N* a set, suit (of clothes), set of tools; toop | комплект, набор, снаряжение, команда; **bir takım** some, certain, a set of | некоторое количество, целый ряд; **takım ur**– to harness (a horse) | надевать сбрую ña лошадь)
takın– (2) *Vpass* to wear, to gird sg on | опоясаться оружиями; надевать на себя что
takıštır– (1) *V* to gather (together), to prepare (for), to make ready for | собрать, приготовить
takje (1) *N Ar.* skullcap, (high) fur cap | ермолка, тюбетейка, колпак
takta (6) *N* board, plank | доцка → **taχta**
taktaǯık (1) *N dim* small board, small plank | дощечка
taktir s. **takdir**

taktır– (1) *V* to have sg affixed, to make sb attach sg (to) | заставить прикреплять, заставить нацепить

tal I (5) *N* willow (*Salix*) | ива, верба; **tal-terek** willow–wood | ива

tal II (10) *N* branch, twig | ветвь, ветка → **dal**

tal– I (1) *V* to get tired, to weaken | утомляться, ослабеть

tal– II (1) *V* to dive, plunge (into) → **dal–**

tala– (2) *V* to bite (into), to catch | кусать(ся), грабить; **ilim tala**– to pick up science/knowledge | вкусить в науку/знание

talaj (15) *IndefNum* a number/lot/bunch of; a part/half of | довольно много, скопище; **br talaj vakit** lots of time, longwhile | много времени; долгое время; **br talajdan son** long after | спустя долгое время

talak (1) *N* spleen, milt | селезёнка

talalajlam (10) *Interj/Onomatop* **talalajlam talalajlam** (line without meaning) | (стих без значения)

talaš– (13) *V recipr* to quarrel violently, to struggle; to suffer pain | грызться, ссориться; страдать; терпеть боль; **bar talaš!** do suffer tortures! | измучивайся!

talǯı (1) *N* diver | ныряльщик, водолаз

talda (4) *N* shade, semi–darkness | тень

taldır– (3) *Vfact* to tire (out), to exhaust | утомлять

taleke (2) *N R.* (farm–)wagon, dray | телега

talerka (2) *N R.* plate | талерка

talfes (3) *N* hubcap, fez without a turban | колпак, турецкая феска → **dalfes**

talfesli (2) *Adj* wearing a fez without a turban; wearing a hubcap | носящий колпак/турецкую феску

talı (6) *Adj Ar.* destiny, fate | судьба, рок, участь; **talım jok** I have no luck | я не счастливый, мне не везёт

tallal s. **tellal**

tam I (9) *N* roof, terrace; roof garden; house | крыша, кровля; дом; **tamdan tama kez**– to stroll, to tramp; to live a debauched life (about women) | таскаться, блудить (о женщинах) → **dam**

tam II (1) *Onomatop N* **tam et**– to clap, to plop | удариться, хлопать

tam– (2) *V* to drip | капать → **dam–**

tamam (18) *Adj Ar.* complete, correct, exact | целый, полный; **tamam bol**– to be completed/finished/ended | заканчиваться, завершаться, кончаться; **tamamnen** comletely, entirely | вполне, полностью

tamaša (2) *N Ar.* spectacle, performance; amusement | зрелище, представление; развлечение

tamčı (1) *N* drop | капля

tamčıla– (1) *V* to drip | вливать по капле, капать, сочить

tamγa (4) *N* brand, stigma; (here) cicatrice, scar | тамга, клеймо, метка; (здесь) рубец; **tamγa ur**– to stamp, to brand | ставить печать (метку, клеймо), клеймить (животное)

tamla– (6), **tamna**– (1) *V* to drip | капать

tamna– s. **tamla–**

tamsız (1) *Adj* tasteless, flavourless | невкусный, безвкусный

tamur (1) *N* root; kinsfolk | корень; родня

tan s. **tane**

tana I (12) *N* calf | телёнок → **buzau, bızau**

tana II s. **tane**

tanau (1) *N* nostril | ноздря

tandır (1) *N* oven, stove; bottom of stove | печь, дно очага

tane (~tan) (39), **tana** (2), **tāne** (4), **tene** (21) *N P.* grain, seed; piece; one's darling | штука (при счете); зерно, семя; любимый/любимая; **bir/bır tan [< tane]** once more/again | ещё (раз), один раз → **dane**

tanı– (37) *V* to know, to be acquainted with | знать, узнавать; признавать

tanıš– (3) *V recipr* to become acquainted | знакомиться

tanıt– (1) *Vfact* to acquaint | знакомить, познакомить

tañ (9), **tan** *N* dawn | рссвет, утреняя заря; **tañ atkanda** at dawn | на рассвете; **tañ atkanı** the first blush of dawn | (утреняя) заря; **tañ/tañlar at**– to dawn | рассветать;

tañmaj (1) *N* palate | нёбо

tañrı (2) *N* God | бог

tap (14) *Particle/Conj* complete, exact, exactly like; down to, up to | аж, даже; так что, вплоть до; **tap sonı/soñı/sonu/suñı/sunında** just in the end | в самом конце, наконец

tap– (115), **tav**– *V* to find, to guess, to find out; to bear | находить, угадать, родить; **borǯγa tap**– to find (get!) as a loan | находить (приобретать!) взаймы; **ulančıχ tap**– to give birth to a little boy | родить мальчишку

tapıl– (5) *Vpass* to be found, to be present, to exist | быть найденным; иметься

tapmaǯa (1) *N* enigma, riddle | загадка

tapta– (1) *V R?* to trample | топтать

taptır– (1) *Vfact* to make sb bear; to assist a childbirth | велеть/приказывать родить (ребёнка)

tar (4) *Adj* narrow | узкий, тесный → **dar**

tara– (5) *V* to comb | расчёсывать

taraf (29) *N* side, edge | сторона; направление

taraχ s. **tarak**

tarak (12), **taraχ** (2) *N* comb | гребёнка, расчёска

taral– (2) *Vpass* to be scattered | рассеиваться, разбредаться, разбегаться

taran balıγı (2) *N* roach (Rutilus rutilus Heckeli) | тарань

taraš (15) *N* hound, foxhound | дворняжка

tarı (10) *N* millet (*Panicum miliaceum*) | просо → **darı**

tarıl– (9) *Vpass* to become cross/angry (with); to be offended | сердиться; обижаться; разозлиться → **darıl**–

Tarkan (2) *Tn* Tarkan | Таркан

tarla (7) *N* field, plough–land | поле, пашня

tars: **tarz at**– to catch? | поймать?

tart– (10) *V* to pull, to draw; to mill; to hit strongly | тянуть, взмешивать, молоть; ударить сильно; **bašına tart**– to hit/knock on one's head | бить по голове; **kůzine tart**– to hit/knock on one's eye | бить по глазу

tartıl– (2) *Vpass* to draw back, to move off/away | пятиться; отступать; удаляться

tartıš– (2) *V recipr* to struggle/argue with sb | тягаться с кем–л.; ругаться с кем–л.

tarttır– (2) *Vfact* to make sb draw sg; (here) to have sb bound to sg | натягивать; (здесь) велеть привязать кого–л. к чему–л.

tarz (42) *Postp Ar.* mode, manner, style, | образ, способ, форма, вид, манера; **ajındıγı tarzı** as soon as (s)he recovered/regained his/her consciousness; as soon as (s)he recovered/gathered his/her strength

| ка только он(а) пришёл/пришла в себя/в сознание; **kůrgeni tarzı** as (soon as)/when (s)he saw sg/sb | как (только) он(а) увидел(а); **ůzımız tarze** in our own way; according to our manner | по нашему (способу)

tas (4) *N Ar.* metal drinking pot, chalice, bowl, mug | таз, сосуд; чаша, миска

tasma (4) *N* collar (of a dog, horse etc.), strap | тесьма, ошейник, ремешок

tasmalı (2) *Adj* having a collar with strap, strapped | имеющий тесьму с ремешком

Tasmalı–bujın (13), **Tasmalı–bujin** the name of Chora–batir's steed meaning "leashed neck" | Тасмалы–буйын (т. е. шея с ошейником), кличка скакуна Чора–батыра → **bujın**

taš I (19) *N* stone | камень → **daš**

taš II (1) *N* outside, outer appearance; the outside, the space outside | внешняя сторона; пространство вне чего–л.

taš– (1) *V* to overflow, to run over, to boil over | выходить из берегов, выливаться через край; переливать (кипящей жидкости через край)

taši– (5) *V* to carry, to transport (from one place to another); to wear (a clothe) | носить (одежду), перевозить, таскать

tašla– (73) *V* to leave, to quit; to let; to give birth to a child | бросать, покидать; родить; **pič bala** to give birth to an illegitimate child | родить незаконного ребёнка

tašlat– (1) *Vfact* to make sb leave (sb or sg) | заставить (велеть) бросить

tat I (1) *N* taste, flavor | вкус

tat II (2) *En* Tat | тат

tat– (1) *V* to taste, to try the taste of; to grow into the habit of | пробовать (на вкус); пристрашаться к чему–л.

tata (1) *N* elder sister; sister–in–law | (старшая) сестра; свояченица

tatar (11) *N/En* Tatar | татар, татарин

tatlı (16) *Adj* sweet | сладкий; **tatlı su** fresh–water, drinking water | пресная вода, питьевая вода; **tatlı sůz** kind/amicable word | любезное слово, дружелюбное слово; **tatlı til** kind/amicable language/conversation | дружеский язык/разговор, любезная беседа

tau (96), **tav** (2) *N* mountain; forest | гора; лес; **kara tau** thick forest | густой лес; **Karpat(nıñ) tauları** Karpat mountains | Карпатские горы; **tau baši** the peak of the mountain | вершина горы

tauk s. **tavuk**

tauriski (1) *Adj* (of) Taurida/Tavrida | таврический; **tauriski guberna** Government of Taurida | Таврическая губерния

taušan (2) *N* rabbit | заяц

tav s. **tau**

tava (4) *N* frying pan | сковорода

tavan (7) *N* ceiling; attic | потолок; чердак

tavrıl– (1) *Vpass* to become haughty, to be self important, to put on airs | возноситься; воображать; чваниться

tavuk (12), **tauk** (2) *N* hen | курица; **kara tavuk** eurasian blackbird (*Turdus menula*) | чёрный дрозд

tavusıl– (1) *Vpass* to become thin and weak, to be exhausted | ослабевать; иссякать

taz (72) *Adj* bald | безволосый, плешивый

tazeden (3) *Adv* anew, again, once more | снова, опять

tazelen– (1) *Vpass* be refreshed | освежиться, обновляться

tazerle– s. **tazirle–**

tazirle– (3), **tazerle–** *V* to tell off, to chide | отчитывать; бранить, выругать; укорять кого–то за что–то

tazı (10), **tazi** (2) *N P.* borzoi, greyhound | борзая (собака)

tā I s. **taha**

tā II (2) *Adv* even until, even as, so that; complete, perfect | аж, даже; да; так что, вплоть до → **tap**

tālim (3) *N Ar.* training, tutoring, practice, exercise | учение, обучение, упражнение; **tālime čik–, tālimge čıχ–** to go out to exercise/drill ground | выходить на учебный участок; выходить на учебное поле

tāne s. **tane**

tārif (1) *N Ar.* explanation, description | объяснение; описание; **tārif et–** to describe, to explain | описыватьъ объяснять,

tāze (1) *Adj P.* fresh, new; other | свежий, новый; другой

te s. **de**

te– (1) *V* to say (to), to tell | говорить, сказать → **de–**

tedarik (1) *N Ar.* preparation(s) | заготовка

Tefide (10) *Pn fem Ar.* Tefide | Тефиде

tegerčik (1) *N dim* small wheel (of vehicle) | (маленькое) колесо; **dửrt tegerčik araba** four–wheeled vehicle/carriage | четырёхколёсная повозка

tegil (22), **tegıl** (5) *Neg Cop* (final) not, (isolated) no; (initial, or in anticipation of verb) not only, let alone | не (именное отрицание)

tegin (1) *Adj* silent, calmer | тихий, спокойный

tegirmen (5) *N* mill | мельница → **degirmen**

tegirmenči (1) *N* miller | мельник → **tekirmenǯi**

tegıl s. **tegil**

tek (12) *Adj* a single thing, unique, alone; only (once), merely | один, одинокий; единственный

teke (3) *N* he–goat | козёл

tekerleme (1) *N* tongue twister, rigmaroll, tinkle? | присказка, прибаутка

tekirmenǯi (1) *N* miller | мельник → **tegirmenči, termenči, tirmenči**

teklif (1) *N Ar.* proposal, offer | предложение; **teklif et–** to propose, to offer | предлагать, рекомендать

tekne (1) *N* trough | корыто

tel (10) *N* (here) hairspring; string | (здесь) волосок, волосинка; повод, струна

telef (2), **telep** *N Ar.* ruination, death | гибель, смерть; уничтожение; **telef bol–** to be destroyed, to be ruined, to be killed, to die | уничтожаться, разрушаться, быть убитым, умирать; **telep et–** to destroy, to ruin, to kill | уничтожать, разрушать, губить, убивать

telep s. **telef**

tellal (2), **tallal** *N Ar.* town–crier, broker, middleman | глашатай, маклер, посредник → **dellal**

telle– (1) *V* to string (a musical instrument); to adorn with golden wire or thread | натягивать струны на какой нибудь инструмент; украшать с золотой или серебряной нитью; **kemāneji telle–** to string a violine | натягивать струны на скрипку

tellen– (2) *Vpass* to be stringed (about a musical instrument); to be adorned with golden wire or thread | быть натянутым со струнами (о каком нибудь струнном инструменте); украшаться, убираться золотой или серебряной нитью

telli (1) *Adj* decorated with golden wire or thread | (наряд) с золотой канителью; **telli χoraz** pied/multi–coloured cock | пёстрый петух

telmir– (2), **tilmir–** *V* to wish, to set one's eye on sb/sg | желать, жаждать

tembel (1) *Adj P.* lazy | ленивый

tembele– (2) *V* to give instructions; (here) to call attention (of sb to sg) | давать наставления, (здесь) предупреждать кого–нибудь

temel (2) *N* basics, justice; reality | основа, правда; действительность; **ev temeli** foundation of a house | фундамент дома

temėz (1) *Adj* clean | чистый

temir (14), **temır**, **temr** *N* iron | железо; **Temir Bolat** *Pn* Temir Bolat | Темир Болат; *Pn* **Temir–/Temr–aka** Temir–/Temr–aka | Темир–/Темр–ака → **demir**

Temir Bolat s. **temir**

Temir–aka s. **temir**

temirǯi (1) *N* blacksmith | кузнец

temır s. **temir**

Temr–aka s. **temir**

ten I (5) *N P.* body | тело

ten II (1) *Adj* equal, match | равний, одинаковый; **ten kel–** to balance, to equal | приходить в равновесие, сравняться с кем–л.

tene s. **tane**

tenha (1) *Adj P.* shady, shaded; out–of–the–way; quiet | уединённый, укромный, тихий; пустынный

tentek (4) *N/Adj* silly, stupid; oaf | глупец, дурный, дурень

tentir– (1) *V* to tremble; to be about to fall, (way of walking of old people) | делать движения свойственные престарелым людям

teñiz (54), **teñız** (22) *N* sea | море; **aćık teñiz** open sea | открытое море; **Kara–teñiz/teñız** *Tn* Black Sea | Чёрное море; **teñiz bujı** edge/bank of the sea | берег моря, по берегу моря → **deñiz, tiñiz**

teñız s. **teñiz**

tep– (4) *V* to kick, to spurn, to step, to tread (on, down) | лягать, брыкать, пинать

tepe (3) *N* peak, crown of the head | вершина, макушка

tepele– (2) *V* to knock on the head, to kill the enemy | бить, избивать; уничтожать, истреблять (противника) → **tȯpele–**

tepsi (1) *N* saucer, salver, small tray | блюдце, поднос

teptir– (1) *Vfact* to make sb kick sb/sg | велеть ударять ногой; заставить вскочить?

ter (1) *N* perspiration, swet | пот, испарина

teraze (2), **terezi** *N P.* pare of scale | весы

terbet– (2) *V* to rock, to swing | качать, колебать

tere (3) *N* valley, brook, (dial.) creek | горная долина, (диал.) ручей → **dere**

tere–tüpe s. **dere–tüpe**

terek (85) *N* tree | дерево; **alma–terek** apple–tree | яблоня; **badem–terek** almond–tree | миндалевое дерево; **Baj–terek** Bay–terek (name of a fabulous fruit–tree) | Бай–терек (имя сказочного плодового/фруктового дерева); **kiraz–terek** cherry–tree | черешня; **kůl–terek** rose–bush | розовое дерево, розовый куст; **meše–terek** oak–tree | дуб; **tal–terek** willow–wood | ива → **tirek**

teren (3) *Adj* deep; dark (colour) | глубокий, темный (о цвете); **terenden aχ et**– to have a deep sigh | вздохнуть глубоко

terezi s. **teraze**

teri (17) *N* lether, hide | кожа

terja s. **derja**

terle– (5) *V* to sweat, to perspire | потеть

termenči (2) *N* miller | мелъник → **tegirmenči, tekirmenǯi, tirmenči**

ters (2) *Adj* rude, not kind, not pleasant | нелюбезный, грубый; **ters jer** uninhabited place | заброшенное место

tert (~terd) (1) pain, disease; trouble, sorrow, suffering | недуг, хвор, страдание → **dert**

terzi (2) *N P.* tailor, sewers | портной

teslim (2) *N Ar.* surrender | сдача, капитуляция; **teslim bol**– to surrender, give oneself up | сдаваться, капитулировать

teste I *N P.* bunch, bouquet | связка, букет, пачка, кипа → **deste**

teste II s. **testi**

testi (3), **teste** *N P.* earthenware jug, pitcher | кувшин, сосуд; **su testisi** water–pitcher /monkey | кувшин для воды

teš– (2) *V* to lance, to overpunch | подырявливать, пробивать; проходить, пробираться

tešiχ s. **tešik**

tešik (6), **tešiχ**, **tešık** (2) *N* pierced, burst open; hole | подырявленный, проколотый, дыра; **delik–tešik** full of holes; the worse for wear, worn out (clothe) | весь продырявленный, в дырах; изношенный, ветхий (об одежде и т. п.)

tešikli (2) *Adj* holed, with holes (in it) | дырявый

tešil– (1) *Vpass* to become holed | быть подырявленным (проколотым), пробиваться

tešık s. **tešik**

teumetlen– (8), **teümetlen**– *Vpass* to hang in doubt, to suspect | сомневаться, подозревать

teümetlen– s. **teumetlen**–

tevekkel (1) *N Ar.* carefree, resigned, who leaves things to chance/fate | беспечный, надеющийся на авось; не напрасно, не зря

Tevide (2) *Pn fem Ar.* Tevide [< Tevhide] | Тэвиде [< Тэвхиде]

tez (24), **tiz** *Adj/Adv P.* quick(ly), speedy, swift(y) | быстрый, скорый; быстро, скоро

tezden (3) *Adv* quickly, swiftly | скоро, быстро

teze (9) *N* maternal aunt | тётка (со стороны матери)

tezja (3) *N P.* (weaver's) loom, workbench | (ткацкий) станок, верстак

ti– **I** s. **tij**–

ti– **II** (1) *V* to say (to), to tell | сказать **ti**–→ **te**–

tiber– (1) *V* to give a push, to shove | толкнуть

tifek (1) *N* rifle, gun, shotgun | ружьё, винтовка → **důfek, tůfek, tüfek**

tigenek (2) *N* thorn, pine | колючка, шип

tij– (17), **ti**– (2), **tıj**– *V* to touch, to reach; to drop in (to sb) | трогать, касаться, задевать; заходить; **nazar tijmesin!** may the evil eye not reach it! | пусть не сглазят его/её!

tije s. **tůje**

tije– (8) *V* to load | нагружать, погружать

tik– (4) *V* to sew | шить; **nayıš tik**– to embroider | вышивать

tiken (3) *N* thorn, pine | колючка, шип

tikenek (1) *N* thorn, pine | колючка, шип → **tigenek**

tikenlik (1) *N* place with many thorns and brambles | место, поросшее колючками; кустарник

tikil– (3) *Vpass* to be sewn | шиться

tikle– (2) *V* to set up, to erect | поставить стоймя, водружать

tiktir– (2) *Vfact* to have sg sewn | отдавать в пошивку что-л.; шить себе что-л.

til (31) *N* tongue, language | язык, речь; **ālemniñ tili** the common talk, gossip | язык сплетни; сплетня; **tile destan bol**– to become the object of talks | стать предметом разговора, быть у каждого на языке; **tilge kel**– to start speaking | заговаривать

til– (2) *V* to bite, to tear apart, (here) to tear in pieces, to cut into slices | укусить; разорвать; разрезать на куски

tilber s. **dilber**

tile– (48) *V* to wish, to desire, to want to | желать, хотеть

tilek (5) *N* wish, request, demand | желание, запрос, потребность

tileme (1) *Inf (ma)* wish, appetency | желание

tilen– (5) *Vpass* to beg | клянчить

tilenǯi (2) *N* beggar | нищий, попрошайка

tilim (2) *N* chum, slice, strip | ломоть, (отрезанный) кусок

tilki (3) *N* fox | лиса

tilmir– s. **telmir**–

tilsim s. **tılsım**

tilsiz (1) *Adj* dumb, taciturn | немой, молчаливый

tima– (1) *V* to roll into a ball; (here) to doze rolling into a ball | съёживаться; свёртываться клубком; (здесь) дремать свёртываясь

Timer Bolat (2) *Pn* Timer Bolat | Тимер Болат

tingiz s. **tiñiz**

tinne– s. **tiñle**–

tiñ– (1) *V* to listen to; to pay attention to, to obey | слушать, внимать; слушаться → **tiñe**–, **tiñle**–, **tıñ**–, **tıñla**–

tiñe– (3) *V* to listen to; to pay attention to, to obey | слушать, внимать; слушаться → **tiñ**–, **tiñle**–, **tıñ**–, **tıñla**–

tiñiz (1), **tingiz** (2) *N* sea | море → **deñiz**, **teñiz**

tiñle– (3), **tinne**–, **tiññe**– *V* to listen to; to pay attention to, to obey | слушать, внимать; слушаться

tiñleme (1) *N* listening to, paying attention to | слушание, внимание

tiññe– s. **tiñle**–

tip (3) *N* bottom, foot, lowest part | дно; низ → **dip**

tire– (3) *V* to roll up (sleeve), to tuck up | засучивать (рукава)

tirek (4) *N* tree | дерево → **terek**

tirel– (3) *V* to resist, to be stubborn | быть подпёртым, сопротивляться; упрямиться

tireviš (1) *N* support, prop | балка, упор

tiril– (7), **tırıl**– *Vpass* to return to life, to be revived | оживать, воскресать

tirmenči (4) *N* miller | мельник → **tegirmenči**, **tekirmenǯi**, **termenči**

tiš I (6), **tıš** (2) *N* tooth | зуб

tiš II (4) *N* dream | сон, сновидение; **tiš kůr**– to have a dream | видеть сон → **důš**, **tůš**, **tüš**

tišle– (6) *V* to bite | кусать, откусывать

tiz I (21) *N* knee | колено; **tizinedek** to his/her knee | до его/её колена; **tiz čůk**– to kneel down | опуститься на колени

tiz II s. **tez**

tiz– (3) *V* to line up, to arrange in a row | выстраивать(ся) в ряд; **inʒi tiz**– to string pearls | нанизывать бисер

tizgin (1) *N* rein, bridle | повод, узда, вожжи

tizil– (1) *Vpass* to be arranged in a line or row | выстраиваться в ряд; нанизываться

tıγır– (1) *V* to revolve, to turn round; to roll along | катиться

tıγırčıklan– (1) *Vpass* to revolve, to turn round; to roll along | катиться, вертеться

tıγırla– (2) *V* to roll down; (tears) trickle from one's eyes | катиться; **kůz jaši tıγırlaj** tears trickle from his/her eyes | слёзы катятся из его/её глаз

tıχ– (1) *V* to put in, to shove in | совать

tıj– (1) *V* to shut off, to block up; to hold back (one's tears) | запрудить, сдержать (слёзы)

tıklı (1) *Adj;* **tıklı jerge saban** for [...?] soil plough | для (...?) земли плуг

tılsım (10), **tilsim**, **tilsım** *N* witchcraft, magic; magic word | волшебство, чары; **jazılı tilsım** written talisman | письменный талисман; **tilsim et**– to cast a spell upon sg/sb, to enchant | зачаровывать, заколдовывать

tınıš (1) *N* breather, rest; a little time for resting | вздох, передышка

tınıšlık (1) *N* composure, quiet | спокойствие, мир

tıñ– (2) *V* to listen to; to pay attention to, to obey | слушать, внимать → **tiñ**–, **tiñe**–, **tiñle**–, **tıñla**–

tıñat– s. **tıñlat**–

tıñla– (4), **tıñna**– *V* to listen to; to pay attention to, to obey | слушать, внимать → **tiñ, tiñe**–, **tiñle**–, **tıñ**–

tıñlat– (2), **tıñat**– *Vfact* to make sb listen (to) | дать/заставить (вы)слушать

tıñna– s. **tıñla**–

tıñtıñ (1) *Onomatop N*; **tıñtıñ et**– to tremble | трепетать, стрястись

tıril– s. **tiril**–

tırılt– (1) *Vfact* to revive | оживлять; **anasın tırıltgen** revived his own mother | оживлял свою мать

tırıš– (3) *V* to take pains | стараться

tırkel– (1) *V pass* to be tied behind one another, to hang | прицепляться, сцепляться

tırmaš– (2) *V* to clamber (up) | карабкаться, взбираться

tırnaχ s. **tırnak**

tırnak (7), **tırnaχ** (2) *N* fingernail, talon; hoof | ноготь, коготь; копыто; **at tırnaγı** hoof | копыто

tırnakʒık (1) *N dim* small fingernail, small hoof | ноготок, коготок

tırpın–tırpın (3) *Adv* (to move) scuffling | (идти) шаркая; **tırpın–tırpın et**– to shuffle | шаркать

tırpıntı (1) *N Onomatop.* scuffling | шарканье: волочение ног

tıš (7) *N* the outside, the space outside | внешняя (наружная) сторона чего–л.; вне

tıšarı (25), **tišarı** (7), **tšarı** (8) *N/Adv* the outside, the space outside | пространство вне чего–л., снаружи, на дворе; **tišarγa** out, outside, towards the outside | наружу, на открытое место; **tıšarda** outside | вовне, снаружи; **tıšarγa, tšarγa** out, outside, towards the outside | наружу, на открытое место

toban I (7) *N* straw | солома
toban II (2) *N* millet–flour | пшённая (просяная) мука
tobuk (1) *N* heel; ankle | пятка, каблук, лодыжка, щиколотка
tobulɣu (1) *N* meadowsweet | таволга
tofraχ (2) *N* earth, soil, ground | земля; почва → **toprak, tufrak, tuprak**
toɣ– (2) *V* to be born | родиться
toɣaj (6) *Ndat* meadow, pasture | луг
toɣajlık (2) *N* meadowland, pasture | сплошные луга
toɣra– (1) *V* to cut into slices or pieces; to carve | нарезать
toχmaχ (1) *N* mallet, hammer | колотушка, молот → **tokmak, tukmak**
toχo– (1) *V* to weave | ткать → **tuku–**
toχoz (7) *Num* nine | девять; **jetmiš toχoz** seventy–nine | семьдесят девять; **otoz toχoz** thirty–nine | тридцать девять → **tokoz, tuχız**
toχozlau (1) *N* making it nine, increasing sg to nine | становление девитикратным → **tokozla–**
toχsan (2) *Num* ninety | девяносто → **tuχsan**
toχta– (5) *V* to stop | остановиться → **toqta–, tuχta–**
toχu– (1) *V* to weave | ткать
toj (29) *N* wedding, (wedding) feast | (свадьебный) пир, свадьба; **toj et–** to have a rich meal/repast; to feast | устраивать пир(шество), устраивать свадебный пир → **tuj**
toj– I (13) *V* to become filled and satiated | насыщаться, наедаться → **tuj– II**
toj– II (2) *V* to feel, to perceive; to hear | чуять, чувствовать; слыхать → **duj–, tuj– I**
toj–du̇gu̇n (4), **toj–tu̇gu̇n** (2) *N* wedding feast | празднества свадьбы; **toj–du̇gu̇n et–** to rejoice, to hold a wedding | справлять свадьбу; пировать; пиршествовать → **tuj–du̇gu̇n**
toj–tu̇gu̇n s. **toj–du̇gu̇n**
tojan– (1) *V* to become filled and satiated | насыщаться
tojdor– (1) *Vfact* to fill (up), to satiate | насыщать, накормить
tojmaχ (1) *Inf* satiety | сытость, насыщение → **tujmak**
tok– (1) *V* to strike, to hit, to knock | бить, побить, колотить
tokal (6) *Adj* curved, crooked; blunt; with a tail cut off | загнутый, закруглённый; куцый; тупой; **tokal kılıǯ** blunt sword | тупой меч; **tos–tokal šač** very curly/frizzy hair | очень кудрявые/вьющиеся волосы
toklu (2), **tokul (!)** *N* yearling lamb | годовалый барашек
tokmak (1) *N* mallet, hammer | колотушка, молот → **toχmaχ, tukmak**
tokos s. **tokoz**
tokoz (12), **tokos** (1) *Num* nine | девять; **jekirmi tokoz** twenty–nine | двадцать девять; → **toχoz, tuχız**
tokozla– (1) *V* to make it nine, to increase to nine | делать что–л. девитикратным → **toχozlau**
tokta– (9) *V* to stop, to break off, to stand | остановиться, прекращать → **toχta–, tuχta–**
toktat– (1) *Vfact* to make sb stop | остановить
toktattır– (1) *Vfact* to have sb stopped | заставить остановить; велеть остановить
tokul[!] s. **toklu**

tol– (13) *V* to become full; to expire, to come to (an) end | наполняться, быть выполненным → **tul–**

tolan– (1) *V* to go around in a circle, to wander around | обходить, кружить → **dolan–**

tolap (2) *N Ar.* cupboard, wardrobe | стенной шкаф, ниша → **dolaf**

tolaš– (4) *V* to go around, to ramble, to wander | бродить, разгуливать; обходить (весь город) → **dolaš–**

toldır– (10), **toldor–** (4), **toldur–** (1) *V* to fill (up), to satiate | наполнять; насыщать, накормить

toldor– s. **toldır–**

toldur– s. **toldır–**

Toliman (1) *Pn* Toliman | Толиман

tolı (1), *Adj* full | полный, изобилующий → **tolo, tolu, tulı, tulu**

tolkun (1) *N* wave | волна

tolo (7) *Adj* full | полный, изобилующий → **tolı, tolu, tulı, tulu**

tolu (16) *Adj* full | полный, изобилующий → **tolı, tolo, tulı, tulu**

tomalačık (2) *Adj dim* fleshy, plump; stubby | круглый, полный, толстый

tomalak (2) *Adj* round (as a ball), plump | круглый; **tomalak baldır** roundish, muscular calf | круглая, мускулистая икра

tomoz (1) *N* pig, swine | свинья → **domoz**

ton (1) *N* fur (coat); clothe | шуба, тулуп; одежда → **don, tun**

toñ– (1) *V* to freeze | мёрзнуть

toñka (1) *Adj* dug | тупой

toñoz (1) *N* swine, pig | свинья → **domoz, χınzır**

top I (8) *N* bunch, bouquet, packet | связка, пачка, кипа; букет

top II (1) *Adj/N* ball, ball–shaped, round | шар(ик); мяч; круглый

topal (26) *Adj* lame | хромой; **topal–mopal** laming–limping | хромой–прихрамывающий

topallık (1) *N* lameness | хромота

topčık (1) *N dim* small bunch, small bouquet | маленький букет, пучок

topla– (13) *V* to collect, to gather | собирать → **tupla–**

toplan– (1) *Vpass* to be gathered | собираться → **tuplan–**

toplaš– (1) *V recipr* to gather together, to come together | собираться (все вместе) → **tuplaš–**

topoz (1) *N* globule; (a single) pearl | шарик/бусинка → **tupuz**

topozlı (1) *Adj* having globules; decorated with small balls | имеющий шарики; украшенный шариками (бусами) → **tupuzlı**

toprak (3) *N* earth, soil, ground | земля; почва → **tofraχ, tufrak, tuprak**

tor s. **tȯr***

tor– (28) *V* to stand (up), to get up; | вставать, стоять;
Vaux expresses duration of acting | вспомогательный глагол, выражающий длительность действия; **bile tor–** to know, to be aware of sg | знать иметь сведения; **ǯuklap tor–** to sleep for a long time | спать долгое время; **dep tor–** to keep saying | продолжать говорить; **kıjnap tor–** to torture sb for a long time | мучить кого долгое время; **kůrůnip tor–** to be seen | видняться → **tur–**

tora (16) *Adj/Adv* straight(ly); directly; honest; truthful(y) | прямой, прямо; в направлении чего–л.; правда, правдивый, честный → **dora, dorı, doru, torı, toro, toru**

torat– (1) *V* to direct; to aim (sg at sg), to point (gun at sg) | направлять; прицеливаться

torba (1) *N* bag, little sack | торба, небольшой мешок, сума → **durba, turba**

torγaj (3) *N* house sparrow (*Passer domesticus*) | воробей → **bostorγaj, turγaj**

torχ (4) *N R.* tenders, deal, business | торг; **torχ et–** to announce, to put up for sale? | оглашать, опубликовывать, продавать с торгов

torı (1) *Adj* straight(ly); directly; honest; truthful(y) | прямой, прямо; в направлении чего–л.; правда, правдивый, честный → **dora, dorı, doru, toro, toru**

toro s. **toru**

toru (4) *Adj* straight(ly); directly; honest; truthful(y) | прямой, прямо; в направлении; правда, правдивый, честный → **dora, dorı, doru, torı**

tos–tokal (5) *Adj* very curved/crooked; very blunt | ohen; очень загнутый/закруглённый; очень тупой; **tos–tokal sač** very curly/frizzy hair | очень кудрявые/вьющиеся волосы → **tokal**

tostoγan (2) *N* wooden cup/jug/jar/bowl | деревянная чашка

toš I (1) *Adj/Adv* (place) opposite to, facing | напротив, противоположный; **toš kajtar–** to turn back | обращать назад

toš II [tȯš] (1) *N* bossom, breast; caracas | грудь, грудинка, передняя часть туши; **hınzırıñ toši II [tȯši]** pig–sided | грудь свиньи

tot– (7) *V* to hold, to hire, to engage; to obey | держать, хватать; ловить; нанимать; слушаться → **dut–, tut–**

totaj (7) *N* elder sister; sisterinlaw | (старшая) сестра; свояченица; **totaj kız** elder daughter | старшая дочь → **tutaj**

totır– (2) *Vfact* to fill (up), to satiate | наполнять; насыщать, накормить → **tojdor–, totor–, tottor II**

totol– (1) *Vpass* to come to a sudden stop; to be caught; to lose (one's breath) | быть пойманным/захваченным (дыхание)

totor– (4) *Vfact* to fill (up), to satiate | наполнять; насыщать, накормить; **kursak totor–** to stodge/stuff the stomach | наполнять желудок → **tojdor–, totır–, tottor– II, tudır–, tuttır–, tutur–**

tottor– I (4) *Vfact* to have sb hold/seize sg | дать (в руки), заставить держать что–л. → **tuttur–**

tottor– II (3) *Vfact* to fill (up), to satiate | наполнять; насыщать, накормить → **tojdor–, totır–, totor–, tuttır–, tutur–**

tovar (1) *N* sheep or goats; cattle | скот, овцы, козлы; скотина; **tovar deñiš–** to change/sell cattle | меняться (скотом), продавать скот → **tuvar**

toz (2) *N* dust | пыль

tȯben (1) *N/ Adv* down, downwards, the lower part | ниц, вниз

tȯgerek (2) *Adj* round | круглый, круг

tȯk– (4) *V* to pour; to shed (tear) | лить → **tök–, tůk–**

tȯpele– (1) *V* to kill, to destroy, to annihilate | убивать, уничтожать → **tepele–**

tȯr* ~ **tor?**(1) *N* pride place, place for the respectful guest | почётное место, красный угол; **torγa** [(sic!] [= **tȯrge?]** **čik–** to go up to the pride place | восходить на почётное место → **tůr**

tȯše– (1) *V* to spread (over), to lay down (carpet); to cover, to wrap | стлать, расстилать (ковёр)

tȯšek (2) *N* bed, mattress | постель, матрац → **tösek**

tök– (2) *V* to pour; to shed (tear) | лить; **su tök**– to urinate, to piss; to spill water | писать, мочиться; наливать воду → **tȯk–, tůk–**

töle– (4) *V* to pay | платить, оплачивать → **tůle–**

tölemek (1) *Inf(mak)* paying | платёж, (рас)плата

töre (10) *N* master, owner, mister; gentleman | господин, человек знатного происхождения

töšek (2) *N* bed, mattress | постель, матрац → **tȯšek, tůšek**

traš (2) *N P.* shaving | бритьё; **traš et**– to shave | бриться; **traš ettir**– to make sb shave | велеть побриться

tšarɣa s. **tıšarı**

tuč (1) *N* bronze; (here) steel | бронза; (здесь) сталь

tudak (1) *N* lip | губа, губы

tudır– (2) *Vfact* to fill (up), to fulfil; to load (up) | наполнять, заполнять, выполнять; нагружать

tudurt– (1) *Vfact* to make sb fill (up), load (up) sg | приказывать нагружать

tufrak (7) *N* earth, soil, ground | земля; почва → **tofraχ, toprak, tuprak**

tugel s. **tůgel**

tugıl (3) *Neg Cop* (final) not, (isolated) no; (initial, or in no anticipation of verb) not only, let alone | не (именное отрицание); нет → **dugıl, tůgůl**

tugul s. **tůgůl**

tuɣ– (32), **tu**– (4), **tuχ**– (1) *V* to be born; to rise (sun, moon) | рождаться; восходить (солнце, луна); **aj tuɣaj** the moon is being born, the moon is rising | луна рождается, луна восходит; **kůn tuχaj** the sun is being born, the sun is rising | солнце рождается, солнце восходит

tuɣan (9), **tūɣan** (2) *N* blood (brother); relative | родной (брат); родственник, родич

tuɣanǯık (1) *N dim* little brother; (here) (my) dear little child | (миленькая) дитя моя; братец

tuɣı (1) *N* eye?, (a) grape? | глаз?, ягода винограда?

tuɣır– (7), **tuɣr**– *V* to give birth to, to bear (children) | родить, рожать → **tuɣur–**

tuɣma (2) *Inf(ma)* birth, born | рождение; **anadan tuɣma kůzleri** his/her eyes from birth? | его/её глаза как будто врождённые?

tuɣra– (1) *V* to cut into slices | разрезать на куски, дробить → **duɣra–, toɣra–**

tuɣrı s. **tuɣru**

tuɣru (2), **tuɣrı** *Adj/Adv* straight(ly); directly; honest; truthful(y) | прямо, прямой, верный

tuɣrul– (1) *Vpass* to be directed (to, towards), to go/march towards | направляться (к чему/кому–л.), держать путь (к чему/кому–л.)

tuɣu (1) *Inf (uw)* birth, born; rise | рождение; восход (солнца/луны); **kůn tuɣu** sunrise | восход солнца

tuɣur– (6) *V* to give birth to, to bear (children) | родить, рожать → **tuɣır–**

tuχ– s. **tuɣ–**

tuχız (1) *Num* nine | девять; **tuχsan tuχız** ninety–nine | девяносто девять → **tokoz**

tuχsan (5) *Num* ninety | девяносто; **tuχsan tuχız** ninety–nine | девяносто девять → **toχsan**

tuχta– (2) *V* to stop, to stand | остановиться → **toχta–, tokta–**

tuj (24); **tuj kur–** to hold a wedding feast | справлять свадьбу → **toj**

tuj– I (50) *V* to feel, to perceive; to hear | чуять, чувствовать; слыхать → **duj–, toj– II**

tuj– II (8) *V* to become filled and satiated | насыщаться, наедаться → **toj– I, duj– II**

tuj–dügün (5) *N* wedding feast | свадьбы, празднества → **toj–dügün**

tujaχlı (1) *Adj* hoofed | копытный; с копытом → **tujak**

tujak (1) *N* hoof | копыто

tujdır– s. **tujdur–**

tujdur– (7), **tujdır–** (1) *Vfact* to notify, to make known, to announce | уведомлять, оповещать

tujıl– (1) *Vpass* to get satiated | насыщаться, наедаться

tujır– (1) *Vfact* to notify, to make known, to announce | сообщать, извещать

tujmaχ s. **tujmak**

tujmak (1), **tujmaχ** *Inf* satiety | сытость, насыщение → **tojmaχ**

tujrım (2) *Adj* crooked | кривой?; **tujrım–tujrım** tortuous, twisted | кривой, искривлённый, с извилинами

tujul– (1) *Vpass* to be feeled, to be heared | чувствоваться; слышаться

tukmak (2) *N* mallet, hammer | колотушка, молот; **kajve tukmaγı** coffee–mallet | кофейная колотушка → **toχmaχ, tokmak**

tukta– (2) *V* to stop, to stand | остановиться → **toχta–, tokta–**

tuku– (3) *V* to weave | ткать → **toχo–**

tukum (2) *N* gender, sort, kind; tribe | род, порода, вид; племя

tul (6) *Adj* widow | вдова, вдовец; **tul χatın** widow | вдовая женщина; **tul kal–** to get widowed | овдоветь

tul– (6) *V* to become full, to fill; to be fulfilled | наполняться, быть выполненным → **tol–**

tulaš– (2) *V* to go round about, to wander | бродить, обходить, побывать всюду → **dolaš–**

tuldır– (~tuldir–) (8), **tultır–** (2), **tuldur** (2) *Vfact* to fill (up), to satiate | наполнять; насыщать, накормить

tuldur– s. **tuldır–**

tulı (4) *Adj* full | полный, изобилующий → **tolı, tolo, tulu**

tulu (1) *Adj* full | полный, изобилующий → **tolı, tolo, tulı**

tultır– s. **tuldır–**

tulup (6) *N* skin made into a bag, sack; hide | мешок; шкура, содранная целиком;

tumalak (1) *Adj* round | круглый

tuman (12) *N* fog, mist | туман, дым; **küzleri tuman** one has a humid gaze | его/её взгляд влажный

tun (1) *N* fur (coat); clothe | шуба, тулуп; одежда → **don, ton**

Tuna s. **Tunaj**

Tunaj (2), **Tuna** (1) *Tn* Danube (river) | (река) Дунай

tunat– (1) *V* to equip, to deck out, to ornament | оборудовать, обставлять

tunja s. **tünja**

tupla– (3) *V* to collect, to gather | собирать → **topla–**

tuplan– (1) *Vpass* to be gathered, to come together | собираться → **toplan–**

tuplaš– (4) *V recipr* to gather together | собираться → **toplaš–**

tuplat– (1) *Vfact* to bring together | заставить собрать

tuprak (1) *N* earth, soil, ground | земля; почва → **tofraχ, toprak, tufrak**
tupuk (1) *N* heel, ankle | пятка, каблук
tupuz (1) *N* ball, globe | шарик → **topoz**
tupuzlı (1) *Adj* having globules; decorated with small balls | имеющий шарики; украшенный шариками (бусами); **tupuzlı šal** shawl decorated with pearls | шаль с бусинками → **topozlı**
tur– (195) *V* to stand (up), to get up; | вставать, подниматься, стоять; **ajakka tur–** to stand up, to rise to one's feet | вставать, подниматься; **ǯel turγanda** when the wind is blowing | когда ветер дует; **julında tur–** to block one's way | загородить/преградить кому–л. путь; **tek tur–** to remain alone | оставаться один(оким); **turša turša!, turšı turšı!** do stand up! | ну–ка, вставай же!
Vaux expresses duration of acting | вспомогательный глагол, выражающий длительность действия; **ǯava tur–** it rains/snows a lot/for long | долго (длительно) идет (дождь/снег); **dep tur–** to keep saying | продолжать говорить; **ojlanıp tur–** to keep thinking | продолжать думать; **utıra tur–** to sit about, lounge | посиживать → **tor–**
tura (2) *N* the sultan's monogram; (here) blazon, coat–of–arms | тугра, герб; **turalı** (script) with a monogram | (письмо) с гербом, монограмом; **turasız** (script) not having a mongram | (письмо) не имеющее монограмма/герба; **turasız para** money without script (coat of arms) | деньги без тугры/герба
turak (2) *N* dwelling place | местонахождение; местожительство
turba (2) *N* bag, little sack | торба, небольшой мешок, сума → **durba, torba**
turγaj (2) *N* house sparrow (*Passer domesticus*) | воробей → **bostorγaj, torγaj**
turγuz– (1) *Vfact* to make sb stand/stop | заставить встать
turlu s. **ṫurlů**
turna (3) *N* crane (*Megalornis grus*) | журавль, журавлиный
tursat– (3) *Vfact* to make sb stand/stop; to wake up | заставить встать; разбудить
turšu (5) *N* pickle | маринованные (солёные) овощи (фрукты)
turuba (1) *N R* pipe | труба
turul– (1) *Vpass* to settle down, to acquiescence (in) | успокаиваться; примиряться с чем–л.
tut– (78) *V* to hold, to hire, to engage | держать, хватать; ловить; нанимать; **nefes tut–** to hold one's breath | задержать дыхание; **oraza tut–** to keep fast | поститься → **dut–, tot–**
tutaj (2) *N* elder sister; sisterinlaw | (старшая) сестра; свояченица → **totaj**
tutar (1) *N* sum; (here) debt | сумма; (здесь) долг
tuti (1) *N* parrot (*Ales imitatrix*), cockatoo (*Cacatua*) | какаду, попугай; **tuti kušu** parrot | попугай
tutıl– (2) *Vpass* to be caught; to get confused | быть пойманным; (здесь) прийти в смущение
tutıš (1) *N* bout, clash; wresling | схватка, борьба
tutıš– (8) *V recipr* to clutch (each other), to cling to, to pick a quarrel with | схватиться, сцепиться
tutluk (1) *N* getting stuck (of the sun); taking place (of the eclipse of the sun)

| застой, приостановка, задержка; (здесь) происшествие затмения; **kůn tutluktı** the Sun was eclipsed (by the Moon); the eclipse of the sun took place | (Луна) затмевала Солнце; произошло солнечное затмение

tutma (4) *Inf (ma)* seizure, capture; rent | захват, поимка; наём

tuttır– (1) *Vfact* to fill (up), to satiate | наполнять; насыщать, накормить → **tojdor–, totır–, totor–, tutur–**

tuttur– (1) *Vfact* to have sb hold/seize sg | дать (в руки), заставить держать что–л.; велеть схватить кого–л./чего–л. → **tottor– I**

tutul– (1) *V* to be caught | быть пойманным

tutur– (1) *Vfact* to fill (up), to satiate | наполнять; насыщать, накормить → **tojdor–, totır–, totor–**

tutuš– (3) *V recipr* to clutch (each other), to cling to, to pick a quarrel with sb | схватиться, сцепиться

tuvar (2) *N* bullock, ox, draught animal | вол, рабочий скот; **tuvar deñiš–** to change/sell cattle | меняться (скотом), продавать скот → **tovar**

tuvar– (1) *V* to set free; outspan, unyoke | освобождать; распрягать

tuvarči s. **tuvarčı**

tuvarčı (17), **tuvarči** (3), **tuvarǯı** *N* herdsman, cattle–herd, cowman | скотник, скотница

tuvarǯı s. **tuvarčı**

tuvıš (1) *N* birth; (here) sunup, sunrise; East | восход (солнца); восток; **kůn tuvıš** sunrise; East | восход (солнца); восток

tuz (7) *N* salt | соль

tuzlı (2) *Adj* salted | солёный

tuzlık [tůzlık?] (1) *N* plain, flatness | ровность, равнина; **tuzlık [tůzlık?] jeñišė** flat land Cotoneaster? | кизильник равнины?

tūγan (1) *Adj* blood, native, birth– | родной; кровный родственник

tůbe I s. **tůpe**

tůbe II (1) *N* promises, oath | зарок; **tůb‿et– [< tůbe et–]** to promise repeatedly | давать обещания

Tůbetau (2) *Tn* Tubetau | Тюбетау

tůfek (16), **tůfenk** *N* rifle, gun, shotgun | ружьё, винтовка → **důfek, tifek, tüfek**

tůfenk s. **tůfek**

tůgel (8), **tugel** (3) *Neg Cop* (final) not, (isolated) no; (initial, or in no anticipation of verb) not only, let alone | не (именное отрицание); нет → **dugıl, tugıl, tůgůl**

tůgerek (4) *Adj* round | круглый

tůgůl (16), **tugul** (3) *Neg Cop* (final) not, (isolated) no; (initial, or in no anticipation of verb) not only, let alone | не (именное отрицание); нет → **dugıl, tugıl, tůgel**

tůgůn (11) *N* wedding feast; rich (meal) repast | свадьба, празднество; ликование

tůj I (8) *N* millet | пшено; **tůj pasta** millet–pap/mush | пшённая крупа/каша; **tůj šorba** millet–soup | пшённый суп

tůj II (1) *N* hair, bristle | щетина, ворс

tůjdůr– (1) *Vfact* to make sb smash sg, to make sb crack sg | велеть или приказать толочь, размельчать

tůje (14), **tije** (2) *N* camel | верблюд → **deve**

tůjeǯi (2) *N* camel driver (owner) | погонщик верблюдов, владелец верблюдов

tůjlen– (1) *Vpass* to become cross/angry?, to be offended? | рассердиться?

tůjme (5) *N* button, bud | пуговица; **tůjme/tůkme čez–, tůjme čůz–** a knob unbutton | расстегнуть пуговицу

tůjůk (1) *N* ?: **toldı jůregim tůjůk** my heart broke | был убитым; моё сердце наполнилось горем

tůk (5) *N* hair, bristle | волосок, щетина, ворс

tůk– (8) *V* to pour, to shed, to spill | лить, сыпать; **kůz jašı tůk–** to shed tears | лить слёзы; **ter tůk–** to sweat heavily | сильно вспотеть, обливаться/покрываться потом → **tȯk–**

tůkan (8), **tůken** (3), **tůkjan** *N Ar.* shop | лавка, магазин → **důkan, tükjen**

tůkanči s. **tůkanǯı**

tůkanǯı (4), **tůkanči, tůkančı, tůkenči** *N* shopkeeper | лавочник, торговец

tůken s. **tůkan**

tůken– (3) *V* to end, to come to an end; to run low/short, to run out | кончаться, иссякать

tůkenči s. **tůkanǯı**

tůkil– (3) *Vpass* to be poured out | вылиться, пролиться

tůkjan s. **tůkan**

tůkme (1) *N* button, bud | пуговица

tůksır– (1) *V* to sneeze | чихать; **aksıra–tůksıre** continuosly sneezing and coughing | непрерывно чихая и кашляя

tůktir– (1) *Vfact* to have sg poured/overturned | (понудительный от) лить, опрокидывать

tůkůl– (8) *Vpass* to be poured out | выливаться, проливаться; опадать (литься) → **tůkil–**

tůkům (2) *N* pouring out, scattering; falling (down) | отливка, разливка; падение; **tůkům–sačım bol–** to be scattered far and wide | рассеяться/рассыпаться → **sačım**

tůkůr– (5) *V* to spit | плевать

tůkůrůk (2) *N* spittle | плевок

tůlber (4) *N Adj P.* beautiful woman; beloved; heart–captivating; attractive | красавица, возлюбленная; покарительница сердца; красивый, привлекательный s. **dilber, tilber, důlber**

tůlberlik *N* beauty | красота → **důlberlik, dülberlik**

tůle– (2) *V* to pay | платить, оплачивать → **töle–**

tůleu (1) *Inf(uw)* payment | платёж, расплата?

tůlki (62), **tůlkı** (2), **tůlků** *N* fox | лиса

tůlkı s. **tůlki**

tůlků s. **tůlki**

tům–tůz (2) *Adj* quite even, quite smooth | совершенно ровный (гладкий); **tere–tůpe tům–tůz ket–** to keep on going quite smoothly, incessantly up hill and down dale | идти по дороге со спусками и подъёмами совершенно ровно → **tůz**

tůn (2) *Adv* yesterday | вчера; **tůn keǯe(de)** yesterday night | вчера ночью → **důn**

tůn– I (3) *V* to circle, to turn round/back | вращаться, возвращаться (домой); **kerije tůn–** to turn back | возвращаться → **důn–**

tůn– II (1) *V* to be closing in (evening), to be falling (night), to be getting dark | наступать (о вечере), вечереть, темнеть; **kůndůz tůngen** day–time turned to night | день превратился в ночь

tůnja (43) **tunja** (6) *N Ar.* world, universe | мир, свет; **alaj tůnja** the whole world | весь мир, вселенная; **ǯalan tůnja**

the false world; the transitory life | ложный мир; тленный/непостоянный мир; **tůnjaɣa kel**– to come into the world6 to be born | появиться на свет, родиться → **důnja, dünja**

tůnteril– (1) *V* to overturn, to pan and tilt | опрокидывать, переворачивать

tůp (64), *N/Postp* bottom, foot, stock | дно, низ → **důp, tüp**

tůpe (21), **tůbe** (5) *N* hill, apex, peak, upper; crown of the head | холм, вершина, верх, темя; верхняя часть чего–л.; **kůl tůpe** rose–hill/ rose mound | розовый холм

Tůpesi–ačıχ s. **Tůpesi–ačık**

Tůpesi–ačık (2), **Tůpesi–ačıχ** *Tn* Tupesi–achik (~ Open–Top) the name of a village | Тюпеси–ачык (Деревня с открытым верхом?) → **ačık**

tůr (1) *N* pride place, place for te respectful guest | почётное место, красный угол → **tȯr**

tůr– (1) *V* to fold, to roll up | закатывать; свёртывать трубочкой

tůrbe (2) *N Ar.* tomb, grave; mausoleum | усыпальница, гробница; мавзолей

tůrků (9) *N* folk song | тюркю (народная песня); **tůrků sůjle**– to sing a folksong | петь народную песню

tůrkůčik (1) *N dim* folk song | народная песенка тюркю

tůrlen– (1) *Vpass* to be on the turn, to vary | меняться, изменяться

tůrli–tůrli s. **tůrlů**

tůrlů (36), **turlu** *Adj* sort, kind, variety, various | разный, различный; **tůrli–tůrli, tůrlů–tůrlů** of all sorts, various | различные, разнообразные → **türli**

tůrlů–tůrlů s. **tůrlů**

tůrt– (5) *V* to prod, to goad; to butt, to rub, to massage | бодать, чесать; массировать

tůrttůr– (3), **tůttůr**– *Vfact* to get sb to massage, to make sb scatche | заставить массировать; велеть чесать; заставить скоблить

tůs s. **tůz**

tůš (21) (4) *N* dream | сон, сновидение; **tůš kůr**– to have a dream | видеть сон → **důš, tiš II, tüš**

tůš– (145) *V* to fall down, to come down, to descend, | падать, слезать; **ajrılıɣa tůš**– to get separated/isolated from sb | разлучаться с кем–л.; **aklına tůš**– to occur to one's mind, to recur to one's memory | прийти на ум/в голову, вспомнить(ся); **attan tůš**– dismount from horse | сходить с лошади; **derde tůš**– to get into trouble | попасть в беду; **eline tůš**– to get/fall into sb's hands | попадать кому–л. в руки; **eniše tůš**– to set foot on slope | ступать на склон; **esine tůš**– to come into one's mind, to be remembered | прийти на ум, вспомниться; **χasta tůš**– to fall/get ill | заболевать; **χatır(lar)ı tůš**– to lose interest in sg, not to be in the mood any more | отпадать охота у кого; **julɣa tůš**– to set out on a journey or a way | выходить/отправиться (в путь); **jullarɣa tůš**– to set out in search of someone | отправляться по многим местам (чтобы разыскивать кого–чего–л.); **kujnına tůš**– to be clasped/folded in one's arms | упасть на грудь (к) кому–л.; **kůneš tůš**– to set, to go down (sun) | садиться/заходить (о солнце);

pešine tůš– to follow in sb's track, to run after | следовать по пятам, идти вслед за кем, бросаться в погоню за кем–л.; **plene tůš**– to be taken prisoner | попасть в плен

tůšek (9) *N* bed, mattress | постель, матрац → **tȯšek, tӧšek**

tůšel– (3) *Vpass* to be spread | быть устланным

tůšin– s. **tůsůn**–

tůšman (12), **tůšmen** (7) *N P.* enemy, foe | враг, противник → **tüšmen**

tůšmen s. **tůšman**

tůšůn– (28), **tůšin**– (3) *V* to think (out) | мыслить, думать

tůšůr– (12), **tůšır**– *Vfact* to drop, to throw off | ронять; сбрасывать → **tüšür**–

tůte– (1) *V* to give out smoke, to fume | дымиться, коптить

tůttůr– s. **tůrttůr**–

tůtůn (21) *N* tobacco (*Nicotiana tabacum*) | табак; **tůtůn ǯak**– to have a smoke, to light a cigarette | закурить; **tůtůn ič–/iš**– to smoke (tobacco) | курить; **tůtůn kotosı/kutusı** snuffbox, tobacco–jar | табакерка; коробка для табака

tůz (2), **tůs** *Adj* smooth, even | плоский, гладкий, ровный; **dere tůpe tůs ket**– to keep on going quite smoothly/ incessantly up hill and down dale | идти по дороге со спусками и подъёмами совершенно ровно

tůzel– (1) *V* to improve and reach a right and proper condition | исправляться, налаживаться

tůzemnik (2), **důzemnik** *N* smoothness, flatness | гладь

tůzle– (1) *V* to smooth, to flatten, to level | ровнять, сглаживать

tüfek (1) *N* rifle, gun, shotgun | ружьё, винтовка → **důfek, düfek, tifek, tůfek**

tükjen (1) *N Ar.* shop | длавка, магазин → **důkan, tůkan**

tüp (~tüb) (3) *N, Postp* bottom, foot, stock | дно, низ → **důp, tůp**

türli (3) *Adj* sort, kind, variety, various | разный, различный → **tůrlů**

tüš (1) *N* dream | сон, сновидение; **tüš kör**– to have a dream | видеть сон → **důš, tiš II, tůš**

tüš– (6) *V* to fall down, to come down, to descend, | падать, слезать → **tůš**–

tüšmen (4) *N* enemy, foe | враг, неприятель → **tůšman**

tüšür– (1) *Vfact* to drop, to throw off | ронять; сбрасывать → **tůšůr**–

U

u I (442) *Pron pers* he/she/it; that; such | он (она, оно), такой (такая, такое); тот; **annar** they | они; **ular** they | они; **unda** at him/her; (over) there | у него/ неё; там; **undan** from there; from him/her | оттуда, с того места; от него/неё; **unın‿čın (< unın ičın)** for him/her; therefore, so, for this reason, because of it | для него/неё, потому; из–за (э)того; **unnan** from there | оттуда, с того места; **unnar** they | они; **uña** for/to him/her | ему/ей, для него/неё; **ů‿jerden [< u jerden]** from that place, on the spot | оттуда, с того места, тут же, сейчас же

u II (1) *Conj Ar.* and; also, too | и, а также; **aš u su** food and water | еда и вода (питьё), пища и вода

uba (10) *N* hill, heap, mound | холм, бугор; большое количество → **oba**
ubir s. **ủbir**
uč s. **ủč**
uč– (30), **uš**– (2) *V* to fly | летать
učkundur s. **ủčkủndủr**
učma (3) *Inf(ma)* flight, flying | лёт, полёт
učmak (1), **ủčmak** (!) *Inf(mak)* to fly, flying | летать, лёт
učsız (1) *Adj* endless, boundless | бесконечный
učun (1) *Adj* veritable, authentic, true, right | настоящий, подлинный
učundan (1) *Adv* in fact, actually | действительно, поистине; по–настоящему
učur– (6), **očır**– (1), **učır**– (2) *Vfact* to let fly, to throw | дать улететь; упустить
uӡ (15) *N* end, tip, point | конец, оконечность; остриё; **uӡdan uӡa** from one end to the other | от одного конца до другого; **uӡında** at its end/point | на конце, в конце (чего–л.); **uӡından** from its end/point | от его конца
uӡaχ s. **uӡak**
uӡak (2), **uӡaχ** *N* fireplace, oven | очаг; **uӡak baši** a place near the fireplace | место у (домашнего) очага; место возле топки/печки → **oӡak**
uӡuz (1) *Adj* cheap | дешевый
uda (10) *N* room; (here) new and better house | комната; (здесь) новый и лучший дом
Udaman (1) *Pn* Udaman | Удаман
uf (9) *Interj* uff | уф, ой
ufaӡık (1) [< ufak + –čık] *Adj dim* very small, weeny | очень мелкий, крошечный
ufak (2) *Adj* small, little | мелкий, маленький, небольшой; **ufak jaš** young days, youth | юный возраст, юность → **ofak**
ufakla– (3) *V* to make smaller, to break up and crumble | мелчать, крошить
uɣarı (1) *Adv* high, upper, top; upward(s) | верх, вверх, верхняя частъ → **joχarı, jokarı, jukarı**
uɣıl (~uɣlı) (5), *N* son | сын → **oɣul, uɣul**
uɣra– (6) *V* to drop in (to sb) | заходить ненадолго, заглянуть
uɣrat– (1) *V* to be subjected to difficult position; to put sb in a tight corner | подвергать чему–л. плохому; ввергать в беду
uɣul (3) *N* son | сын → **oɣul, uɣıl**
uɣulsız *Adj* not having a son | не имеющий сына
uɣur (9) *N* (a peace of good) luck, fortune | счастье; **uɣrın olson, uɣrıñ bolson** good luck to you! good journey! | желаю тебе счастья! в добрый час! счастливого пути!; **uɣurlar bolson/olson** good luck! good journey! | желаю счастья! в добрый час! счастливого пути! → **oɣur**
uχut– (2) *Vfact* to teach; (here) to make sb read | учить, обучать; (здесь) заставить читать → **oχot**–, **okot**–, **ukıt**–, **ukut**–
uj– (2) *V* to suit, to fit, to submit to, to join | слушаться, повиноваться
uja (1) *N* nest | гнездо
ujal–? (1) *Vpass* to be ashamed | стыдиться
ujan– (17) *V* to wake up, to awake | проснуться
ujanıχ (2) *Adj* vigilant, wakeful | бодрствующий; бдительный, неусыпный
ujat– (9) *V* to awake up | (раз)будить

ujdır– (2) *V* to ensconce, to put in order | прилаживать дела, устраивать дела
ujdur– (1) *Vfact* to have sg carved | заставить выдолбить
ujın s. **ujun**
ujınʒı (1) *N* player, dancer | игрок, танцор → **ujunčı**
ujken s. **u̇jken**
ujku (1) *N* sleep | сон → **juku**
ujlan– (6) *V* to think | думать
ujlanmak (1) *Inf(mak)* thinking | мышление, думать
ujmaχ s. **ujmak**
ujmak (4), **ujmaχ** (2) *N* thimble | напёрсток
ujmalı (1) *Adj* chisel(l)ed, carved | резной
ujna– (35) *V* to play, to dance; to move to and fro | играть, танцевать; ходить взад и вперёд; **kahat ujna**– to play cards | играть в карты; **kumar ujna**– to play games of chance | играть в азартные игры
ujnak (5) *Adj* playful, frisky | игривый, резвый
ujnaš (2) *N* lover | любовник, любовница
ujnaš– (2) *V recipr* to play one with another | играть (вместе)
ujnat– (3) *Vfact* to cause to play; to cause to move | подвигать, пошевелить чем–нибудь; **at ujnat**– to make the horse dance | заставить лошадь танцевать
uju– (2) *V* to sleep | спать
ujukla– (1) *V* to sleep | спать → **ʒukla**–, **jokla**–, **jukla**–
ujul– (1) *Vpass* to be scooped out; to be carved | быть выдолбленным
ujun (~ujn) (5) *N* game, play; trick, joke | игра, шутка; **ujnı** his/her game/trick | его/её игра/шутка → **ojun**
ujunčı (1) *N* player, dancer | игрок, танцор → **ujınʒı**
uk (40) *N* arrow | стрела; **uk ček**– to stretch arrow (bow!) | натягать стрелу (лук!)
ukı– (12), **ukė**–(2), **uki**– (3) *V* to read, to learn/study | читать, учиться → **oko**–, **uku**–
ukıt– (1) *Vfact* to teach, to educate | учить, обучать
ukka (1) *N* oke/okka/oka (measure of weight, equals 3 pounds or 1.28 kg) | окка (мера веса, равная 3 футам, 1,28 кг)
uklau (2), **uklav** *N* rolling pin | скалка → **oklau**
uku (1) *Inf(uw)/N* reading, learning | чтение
uku– (11) *V* to read, to learn/study | читать, учиться
ukujan (2) *N* one who is reading/learning; (here) educated, literate one, who can write and read | читающий; (здесь) образованный, грамотный человек; умеющий читать и писать
ukumet (1) *N Ar.* government, state; (here) local authority | правление; (здесь) власть, ведомство
ukut– (1) *Vfact* to teach; (here) to make sb read | учить, обучать; (здесь) заставить читать → **okot**–, **uχut**–, **ukıt**–
ul (196) *N* son | сын
ul– (4) *V* to be, to exist, to become to | быть, существовать; стать → **bol– I**, **ol**–
ulaχ s. **ulak**
ulak (4), **ulaχ** (2) *N* kid | козлёнок
ulan (267) *N* boy, lad | мальчик, парень
ulančıχ (2) *N dim* little boy, lad | мальчонка, мальчуган → **ulanʒık**
ulanʒıχ s. **ulanʒık**

ulanǯık (3), **ulanǯıχ** (2) *N dim* little boy, lad | мальчонка, мальчуган → **ulančıχ**

ulaniski (2) *Adj R.* variant of Russian adjective "ulanskij" meaning 'belonging to uhlan(s) (member(s) of the light cavalry armed with lance); of cavalry | разновидность русского имени прилагательного уланский (улан – член лёгкой кавалерии вооружённый пиками); кавалерийский

ulu (2) *Adj* great, large, big | великий, большой

uluk (4) *N* trough, drinking basin (at) a fountain | колода, жёлоб, жёлоб водопойной колоды; водослив → **olok**

uluš– (2) *V recipr* to howl together with one another | выть в один голос

uma (1) *N P.* mythical bird, bird of paradise; happiness, | Умай (гамаюн, райская птица); счастье; **uma kuši/kušu** bird of paradise | райская птица

umač (3) *N* a special sort of soup made of flour | затируха (суп); **umač čorbası** a sort of noodle soup; grated noodle soup | затируха (суп)

umalaχ (1) *N* ball, globule | шар, шарик

Umer–oǯa (1) *Pn Ar.* Umer–odja | Умер–оджа

umız (9) *N* shoulder; shoulder–joint | плечо, плечевое сплетение

umızbaš (3) *N* point of the shoulder | плечевое сплетение; **umızbaš kis**– to hold the neck between the shoulders | поджимать шею между плеч

umud s. **ůmůt**

umur s. **ůmůr**

umut s. **ůmůt**

un I (53) *Num* ten | десять; **beš/biš un** a few, several | несколько; **un beš/biš** fifteen | пятнадцать; **un dort** fourteen | четырнадцать; **un eki** twelve | двенадцать; **un sekiz/sikez** eighteen | восемнадцать → **on**

un II (4) *N* flour | мука

unaǯı (1) *Adv* till, until, up to that time | до того (времени)

unda s. **u I**

undan s. **u I**

unıt– (3) **(~unt–)** *V* to forget | забывать → **onot**–, **unut**–

unna– (1) *V* to powder with flour | посыпать мукой

unnan s. **u I**

unut– (10) *V* to forget | забывать → **onot**–, **unıt**–

uñ– (2) *V* to do well, to succeed | преуспевать; везти (в чём–л.) → **oñ**–

uñız (4) **(~uñzı)** *N* shoulder | плечо; **uñzına ur**– to carry on one's shoulder, to shoulder | накинуть/набросить на чьи–л. плечи → **uñuz**

uñuz (1) *N* shoulder | плечо → **uñız**

upla– (1) *V* to fondle, to flatter (sb) | ласкать → **hopla**–

upůpuš (1) *N* hoopoe (*Upupa epops*) | удод

Ur (1) *Tn* Or (town), now Perekop | (город) Ор, сегодня Перекоп → **Or**

ur– (95) *V* to hit, to beat, to shoot, to throw; to kill | бить, ударять; убить; стрелять; бросать; втыкать; **kuršun ur**– to fire a gun | выстрелить, выпустить пулю; **tajaχ ur**– to give a beating | побить, поколотить; **takım ur**– to harness (a horse) | надевать сбрую ña лошадь); **tamγa ur**– to stamp/brand | ставить печать (метку, клеймо), клеймить (животное); **ur tut** bargaining hard | торгуясь твёрдо → **or– II**

ura– (2) *V* to spin, to fold in | мотать, заворачивать → **ora**–

urada (4), **urda** (1) *Adv* (over) there, at that place | там

urak (1) *N* sickle; harvest(ing), reaping | серп; жатва

urba (41), **uruba** (1) *N* clothe, battle–clothes/war–clothes | одежда, боевая одежда

urda s. **urada**

urdur– (1) *Vfact* to make sb hit sb; to have sb beaten | позволить бить, заставить ударить

urıl– (6) *Vpass* to beat oneself; to be beaten by sb | быть подбитым, биться, ударяться

urın* (~urnı) (9) *N* place | место, местность → **orın**

urman (2) *N* forest, wood | лес → **orman**

urpi–čurpı (1) *Onomatop;* **urpi–čurpı** (play of words) | (игра слов)

uršen (1) *Adj P.* ful of gaiety, entertaining | весёлый, развлекательный; **uršen ǯir** place of amusement | место развлечений

urta (8) *N/Adj* middle, midst | середина, средний; **urtada** in the middle | на середине

urtanǯa (2) *Adj* middle, middle child (of three) | средний (по отошению к двум другим) → **ortanǯa, ortanǯı, urtanǯı**

urtanǯı (1) *Adj/N* middle, middle child (of three) | средний (по отошению к двум другим) → **ortanǯa, ortanǯı, urtanǯa**

uru I (3) *N* pit (fall), hole; | яма (для хранения зерна) → **oro I**

uruba s. **urba**

uruγ s. **uruv**

uruχset s. **ruχset**

urul– (7) *Vpass* to be beaten; to be bemused; to fall in love (with) | быть подбитым; быть растрелянным; влюбиться в кого–л. → **vurul–**

urus (3) *N En R.* Russian | русский; **urusnıñ saldatı** Russian soldier | русский солдат

uruš– (1) *V recipr* to fight with one another, to strike each other | сражаться, биться

uruv (2), **uruγ** *N* gender, genus | род; порода

us– (1) *V* to grow (up) | вырастать; расти

Usejn–bej (2) *Pn* Useyn–bey | Усеин–бей

Uskut (1) *Tn* Uskut | Ускут; **Uskutıñ tauları** Uskut mountains | горы Ускут

uslan– (2) *V* to smarten up, to wise up; to change one's mind | образумиться, одуматься

usta (24) *N P.* master (of a trade or craft), skilled, clever | мастер, умелец; **Nefa usta** master Nefa | мастер Нэфа; **usta bike** housewife, housekeeper | домашняя хозяйка

usta–talǯı (2) *N* master diver | мастер–водолаз

uš– s. **uč–**

ušak (9) *N* boy, youth, child; small; male servant | подросток, парень; малый; слуга

ušan– (1) *Vpass* to believe, to hope; to trust in sb | надеяться, доверять

ušat– (1) *Vfact* to liken (to), to compare (with) | находить похожим

ut I (12) *N* grass | трава → **ot I**

ut II (3) *N* fire | огонь → **ot II**

uta (19) *N* room | комната

utan– (11) *Vpass* to be ashamed | стыдиться

utır– (82), ~ **uter–** (2) ~ **utėr–** (3) ~ **utir–** (5) *V* to sit (down); to dwell, to live; to stay | садиться; сидеть; жить, проживать; *Vaux* **beklep utır–** to be waiting for sb/sg | ожидать кого/чего–нибудь → **utur–**

utırt– (4) *Vfact* to seat, to place; to stab into | сажать, усаживать; вонзать

utla– (1) *V* to graze | пастись → **otla–**

utur– (16) *V* to sit (down); to dwell, to live; to stay | садиться; сидеть; жить, проживать → **utır–**

uturɣuč (1) *N* small bench, small chair | скамеечка, стульчик

uv– (2) *V* to rumple, to chafe, to rub | мять, тереть

uva I (1) *N* hill, mound | холм, сопка → **uba, oba**

uva II (1) *N* dale, valley | долина

uz (1) *Adv* distant, remote, far off/away | далеко; **az kete, uz kete** (in tales) he goes short, he goes long | (в сказках) коротко ли идёт, долго ли идёт

uz– (6) *V* outride, outrun, catch up with, to go/come before | обгонять, перегонять

uzaχ s. **uzak**

uzak (15), **uzaχ** (2) *Adj* distant, remote, far off/away | далёкий, далеко, долгий

uzan– (2) *Vpass* to spread, to stretch oneself out, to be extended or prolonged | простираться, тянуться; возвышаться

uzat– (14) *Vfact* to extend, to prolong; to show the way out, to see sb out | растянуть, протянуть; провожать; **el uzat–** to lay hand on; to stretch out the hand (to) | затрагивать (рукой) что–л.; протягивать руку помощи

uzdır– (1) *Vfact* to let go (off), to send sb off | отправлять, отпускать

uzɣar– (4) *V* to see off | провожать, напутствовать кого–л.

uzun (19) *Adj* long, lengthy, durable | длинный, долгий, длительный

U̇

u̇ s. **u**

u̇bir (3), **ubir** (1) *Adj* the other (one), different | тот, другой

u̇biri (1) *Pron* the other (one), that one | (тот) другой

u̇č (122), **uč** (9), **u̇š** (2), *Num* three | три; **u̇č–biš** a few, several, a number of | несколько; **u̇čkeče[k]?**, **u̇čkeǯek** until three; until the third attempt | до трёх; до третьей попытки → **üč**

u̇čeu (1) *Num* three, of three | трое

u̇čınči s. **u̇ču̇nǯi**

u̇čınǯi s. **u̇ču̇nǯi**

u̇čku̇ndu̇r (5), **učkundur** (1) *N* beet | свекла

u̇čle– (1) *V* to make three | доводить до трёх; складывать втрое

u̇čleš– (1) *V recipr* to amount to three | утраиваться

u̇čunǯı s. **u̇ču̇nǯi**

u̇ču̇nče s. **u̇ču̇nǯi**

u̇ču̇nǯi (20), **učunče** (1), **u̇čınči** (1), **u̇čınǯi** (2), **u̇čunǯı** (1), **u̇ču̇nǯı** (3) *OrdNum* third | третий, третичный

u̇ču̇nǯı s. **u̇ču̇nǯi**

u̇fu̇r– (1) *V* to blow (into or upon) | дуть

u̇g (39) *N Postp* front, space in front (of a thing), headmost | передняя часть чего–л., перёд; передний; **u̇gimden** from before me | передо мной; **u̇ginde** in front of him/her; before; in the presence (of) | перед ним/ней; впереди него/неё; **u̇gine** to his/her/its front? | перед ним/ней; **u̇ginden/u̇gu̇nden** from his/her/its front | перед ним/ней

u̇giz s. **u̇gu̇z**

u̇gren– (2) *Vpass* to learn | учиться *pass* → **üren–**

u̇gret– (7) *V* to teach, to educate | учить, обучать → **ju̇ret**–

u̇gu̇z (26), **u̇giz** (2), **u̇gız** (2) *N* bullock, ox | вол; **čift u̇gu̇z/u̇gız** yoked ox/draught–ox | рабочий/гужевой вол

u̇j (259), **uj** (1) *N* house, home | дом; **u̇j arkalıyı** balk of a house | балка дома; **u̇jde** at home | дома; **u̇j sal**– to build a house | строить дом → **üj**

u̇jken (10), **ujken** (2), **u̇ken** (3) *Adj* elder, eldest, senior | старший

u̇jle I (32), **u̇jli II** (1) *Adv* so, such, in that manner, like that | так, такой; **u̇jlemeli bu̇jlemeli** in some way or other? (play of words) | (игра слов) таким или иным образом? → **bu̇jle**

u̇jle II (4) *N* noon, afternoon | полдень; **u̇jleden soñ** (in the) afternoon | (в) полдень

u̇jlen– (18) *Vpass* to marry, to get married | жениться; выходить замуж

u̇jlendir– (5) *Vfact* to marry off, to give in marriage (to) | женить; выдать замуж

u̇jli I (1) *Adj* married (man/woman) | женатый; замужняя

u̇jli II s. **u̇jle I**

u̇jrıl– s. **u̇jru̇l**–

u̇jru̇l– (4), **u̇jrıl**– (1) *Vpass* to go round, to spin, to cycle back | вертеться; закручиваться, закружиться; возвращаться

u̇ken s. **u̇jken**

u̇ksek (9) *Adj* high, tall | высокий

u̇ksu̇r– (1) *V* to cough, to have a cough | кашлять

u̇ksu̇z (4) *N* orphan | сирота; **u̇ksu̇z kal**– to become an orphan | осиротеть

u̇ku̇n– (1) *V* to long for, to yearn for, to be sorry for, to regret | грустить (о ком, о чем) пожалеть, сожалеть

u̇ku̇z (1) *N* ox | вол → **ȯgu̇z, ȯkuz, u̇gu̇z**

u̇l– (101) *V* to die, to decease, to perish | умереть, кончаться; дохнуть → **ül**–

u̇lč– (1) *V* to measure | мерить, измерять

u̇lče– (4) *V* to measure of capacity (for grain) | мерить, измерять

u̇ldu̇r– (8) *V* to kill | убивать → **u̇tir**–, **u̇tu̇r**–, **ȯtu̇r**–, **ȯttȯr**–

u̇le (9) *N* noon | полдень; **u̇le ašı** midday meal | обед → **u̇jle II**

u̇len (3) *N* grass | трава → **ȯlen**

u̇li (4) *N* dead | мёртвый, мертвец → **ȯlu̇, u̇lu̇**

u̇lke (7) *N* country | страна

u̇lu̇ (17) *N* dead (man), corpse | мёртвый, мертвец → **ȯlu̇, u̇li**

u̇lu̇m (15) *N* death | смерть

u̇maj (1) *N* mythical bird, bird of paradise; happiness, | Умай (гамаюн, райская птица приносящая счастье); счастье → **uma**

U̇mer–beg (2) *Pn* Umer–beg | Умер–бег

u̇mu̇r (16), **umur** (2) *N Ar.* life | жизнь; **u̇mu̇r kečir**– to spend life | проводить жизнь → **ȯmu̇r**

u̇mu̇t (4), **umud, umut** *N* hope | надежда; **umut et**– to hope | надеяться

u̇mu̇tlen– (2) *Vpass* to be hopeful, to rise one's hopes | питать надежду, надеяться

u̇ner (14) *N* skill, ability, art, talent | умение, мастерство, способность, искусство → **üner**

u̇nerli (2) *Adj* skillful, clever, talented | искусный, умелый

u̇p– (20) *V* to kiss | целовать; **ku̇zinden u̇p**– to kiss the eyes (of a younger person) | целовать глаза (у младших) → **üp**–

u̇pke (1) *N* gall, bile; friendly reproach | жёлч; гнев, злоба; дружеский упрёк, укор;

ủpkele– (1) *V* to reproach, to reprove | укорять, упрекать (по–дружески)

ủpter– s. **ủptir**–

ủptir– (6), **ủpter**–, **ủptủr**– *Vfact* to let sb kiss sb; to make sb kiss sb | разрешить поцеловать; заставить поцеловать

ủptủr– s. **ủptir**–

ủr– **I** (8) *V* to plait, to knit | плести, вязать; **χora ủr**– to weave a fence | плести плетень

ủr– **II** (1) *V* to bark | лаять

ủreke (1) *N* spinning wheel; winding wheel | стойка, на которой закрепляется шерсть для прядения; прялка → **ỏreke**

ủren– (2) *V* to learn | учиться → **ủgren**–

ủrk– (1) *V* to be frightened | испугаться

ủrmet (1) *N Ar.* honour, regard, respect | уважение, почтение, почёт; **ủrmet kủr**– to be respected, to have respect | пользоваться уважением → **hủrmet**

ủrnek (1) *N* example, sample | образец, пример

ủrtủk (1) *Adj* covered, wrapped | закрытый, прикрытый

ủrtủklen– (1) *Vpass* to cover oneself, to be covered | покрываться, укрываться

ủs– (7) *V* to grow up | вырастать → **ỏs**–, **ös**–

ủst (105) *N/Postp* upper or outs ide surface, top of anything, upper (most) | верхняя часть чего–л.; верх чего–л.; вершина; **ủstủnde** on top of it; on, over | поверх (чего–л.), на (чём–л.), над (чем–л.)ь **ủstủne** on top of; on, over | наверх, поверх, сверху, на кого–что–л. → **öst, üst**

ủstủl (1) *N R.* table | стол

ủšlau (1) *Adv* making three; triple, triad | доводить до трёх; складывать втрое

ủt (4) *N* gall, bile | желчь; **ủdi patla**–, **ủti patla**– to be frightened to death | сильно испугаться, насмерть напугаться

ủt– I (6) *V* to pass (by, through, over) | проникать, миновать, проходить → **ỏt**–, **öt**–

ủt– **II** (1) *V* to sing | петь

ủtir– (35) *V* to kill, to unnihilate, to destroy; to waste (time) | убивать, уничтожать, убивать время, попусту тратить время → **ủtủr**–, **ỏtủr**–, **ỏttỏr**–

ủtirt– (3) *Vfact* to make sb kill sb, to have sb killed | велеть убивать

ủtkủr (1) *Adj* sharp/keen (witted), quick | резкий, острый; проникновенный

ủtmek (22) *N* bread | хлеб

ủttir– (1) *V* to spend | проводить (время, жизнь); **ủmủr ủttir**– to pass away/spend life | провести жизнь (каким–л. образом)

ủttủr– **I** s. **ủtủr**–

ủttủr– **II** (1) *Vfact* to make sb sing | заставить петь

ủtủr– (30), **ủttủr**– **I** (1) *V* to kill | убивать

ủtủrt– (7) *Vfact* to have sb killed | приказать убить, заставить убить кого–л.

ủz (221) *Pron pers* own, self | сам

ủz– (6) *V* to break, to tear, to split | ломать, разрыватъ раздирать; **jủregi ủz**– to break one's life/heart | быть убитым горем; → **jủz**–

ủzdủr– (1) *Vfact* to make sb break sg, to have sg torn | велеть отрывать, срывать

ủzek (2) *N* pith, core, hardcore, kern | сердцевина, ядро, сердечник; **ủzegi tal**– to get exhauste, to be tired out | утомляться, изматываться, изнуряться

ủzen (9) *N* river | река

u̇zengi (1) *N* stapedess, stirrup | стремя
u̇zere (1) *Postp* (after infinitive) on condition that, for the pupose of; (as postp.) on, upon, according to | (с инфинитивом) с тем, чтобы; для того, чтобы; (послелог) сообразно с (чем–л.); по, как; **ku̇pri u̇zere** to the bridge, on the bridge | на мост, на мосту
u̇zerinde (1) *Postp* on, over, about (him, her, it) | на; в; над; о/об, по
u̇zerine (4) *Postp* onto, over, about (him, her, it) | на (кого, что–л.)
u̇zu̇m (1) *N* grape | виноград → **ju̇zu̇m**

Ü

üč (4) *Num* three | три → **u̇č**
üj (3) *N* house, home | дом → **u̇j**
ül– (1) *V* to die, to decease, to perish | умереть, кончаться; дохнуть → **u̇l–**
üner (5) *N* skill, ability, art, talent | умение, мастерство, способность, искусство → **u̇ner**
üp– (1) *V* to kiss | целовать → **u̇p–**
üst (1) *N/Postp* upper or outs ide surface, top of anything, upper (most) | верхняя часть чего–л.; верх чего–л.; вершина → **u̇st**

V

vah (8), **vaχ** *Interj* oh! alas!; sorrow; sighing | ах! жаль! увы!; **aχ dij vaχ tij** sighs | вздохнет; **ahinen–vahinen umur kečir–** to spend one's life with sighs and laments | проводить свою жизнь со взодохом и воплем; **ah–vah et–** to sigh and lament | ахать и охать; вздыхать, причитывать; → **ah**
vaχ s. **vah**
vaχit (~ **vaχtı vaχti**) (37), **vaχıt** (5) *N Ar.* time, it's time, period | время, пора; **er vaχit** all the time | всегда; **vaχtinen** in the past, once, at one time, at its proper time | вовремя, когда–то, некогда, в своё время → **vakit**
vaχıt s. **vaχit**
vaχıtsız (2) *Adj/Adv* untimely | безвременный, несвоевременный; безвременно, несвоевременно
vaj (47) *Interj* (interjection expressing surprise or regret) oh! alas! alack! | ах!, ай!, увы!, как жаль! → **aj II**
vajna (14) *N R.* war | война
vajvajla– (1) *V* to wail, to yammer, to lament | ахать, охать; вопить
vakisle– (1) *V R.* to polish up, to shine | ваксить
vaket s. **vakit**
vakėt s. **vakit**
vakit (~ **vakte,** ~ **vaktı,** ~ **vakti**) (129), **vaket** (2), **vakėt** (1) **vakıt** (11) *N Ar.* time, it's time, period | время, пора; **br talaj vakit** lots of time, longwhile | много времени; долгое время; **er vakit** all the time | всегда; **her vakıt** all the time | всегда; **sehar vaktında** at the time just before dawn, early morning | рано утром, на заре → **vaχit**
vakıt s. **vakit**
vakıtsız (1) *Adv/Adj* at the wrong time, untimely | безвременно, безвременный; преждевременно, преждевременный
vakzāl (1) *N R.* railway station, train station | вокзал
var (1) *Ecp* there is; existent, present; to have, 've got; | есть, имеется → **bar**
var– (2) *V* to go, to arrive | идти; прийти → **bar–**

Varna (1) *Tn* Varna | Варна
Varvara (2) *Pn fem R.* Varvara | Варвара
vasijet (5), **vasjat** (3) *N Ar.* last will and testament | завещание; **vasjat et–** to bequeath | делать завещание
vasjat s. **vasijet**
vatan (1) *N Ar.* homeland, motherland | родина
vazgeč– (1), **vāzgec–** *V* to give up, to cease from, to abandon | отречься; прекратить → **vazkeč–**
vazkeč– (5) **vāzkeč–** (2) *V* to give up, to cease from, to abandon | отречься; прекратить → **vazgeč–**
vazkečil– (3), **vāzkečil–** *Vpass* to swerve (from a course), to change one's mind | отступать, отказываться (от своего слова, обещания)
vāde (1) *N Ar.* promise, fixed term or date | обещание; **vāde kıl–** to promise | обещать
vāzgeč– s. **vazgeč–**
vāzkeč– s. **vazkeč–**
vāzkečil– s. **vazkečil–**
ve (22) *Conj* and, also, too | и, также
Veli (5) *Pn Ar.* Veli | Вели
Veliš (4) *Pn dim Ar.* Velish | Велиш; **Velıš–aka** Velish–aka (brother/uncle Velish) | Велиш–ака (брат/дядя Велиш)
Velıš s. **Veliš**
ver– (7) *V* to give, to deliver | давать → **ber–**
veril– (1) *Vpass* to be given | даваться→ **beril–**
veriš s. **alıš–veriš**
vers (1) *N R.* verst (an obsolate Russian unit of length, equals to 1.0668 km) | верста (устарелая русская единица измерения расстояния, равная 1066 метрам)
vezir (86) *N Ar.* vizier | визирь
viran (1) *Adj P.* ruined, devastated | разоренный, опустошенный
vire (1) *Adv It.* continuosly | беспрерывно, беспрестанно
viz (1) *Onomatop N* whizz, wirr | жужжание; (kuršun) **viznen kel–** to come by whizzing (bullet) | прийти жужжанием (пуля)
vızır (2) *Onomatop N* (a humming and buzzing sound); to be trembling/shaking | (о быстром и легко совершающемся действии); дрожать, трястись; **vızır–vızır** swinging very much | сильно колыхающийся; **vızır–vızır memeler** swinging breasts | колыхающиеся груди
vuǯud (1) *N Ar.* body, trunk | тело, туловище, корпус
vur– (7) *V* to beat, to shoot, to throw; to kill | бить, ударять; убить; стрелять; бросать; втыкать; **bujnı vur–** cut someone's throat | перерезать кому–нибудь горло → **ur–**
vurul– (1) *Vpass* to beat oneself, to be beaten | быть подбитым, биться, ударяться → **urul–**
vuslet (1) *N Ar.* meeting; union (with one's beloved) | встреча; соединение (возлюбленных)

Z

zabit (2) *N Ar.* soldier; officer | солдат; офицер
zaif (2), **zāif** *Adj Ar.* weak, effete, sapless | слабый, истощенный
zaja (1) *N/Adj Ar.* lost, gone; perished; in vain | потеря, утрата; зря; **zaja kit–** to get lost | потеряться
zalim (2) *Adj Ar.* cruel, pitiless | жестокий, безжалостный

zaman (77), **zeman** (4) *N Ar.* time; period, epoch | время, период; эпоха; **aχır zaman** the end of time | время светопреставления; **jatılası zamanı** time to go to bed | время отхода ко сну; пора ложиться; **sarılası zamanı** time to embrace | время объятия; пора обниматься; **zaman aχrı** the end of time | время светопреставления

zambak (1) *N Ar.* lily (*Lilium*) | лилия

zampara (1) *N P.* womanizer, rake | волокита, бабник

zamparaǯı (1) *N* petticoat–chaser, womanizer, rake | юбочник, бабник

zampıl–zumpul (1) *Onomatop* **zampıl–zumpul ujna–** to frisk/thrash/toss about | вертеться; ёрзать; метаться

zan (2) *N Ar.* opinion, belief, conviction | мнение; убеждение; **zanγa bar–** to form an opinion, to think, to convince oneself of | думать; убеждаться

zanāt (4), **zenat** *N Ar.* skill, ability, art, talent | умение, мастерсайлан–тво, способность, искусство

zapas (1) *N R.* reserve | запас; **zapas jašlar** reserve youngmen | юноши (запасные) резервисты

zar I (1) *Adv P.* as if, as though | разве

zar II s. **zār**

zarar (4) *N Ar.* harm, loss, waste | вред, урон; **zarar et–** to do harm to sb/sg | причинять кому–/чему–нибудь вреда

zarf (1) *N Ar.* saucer, zarf | чашечная подставка

zauk–safa (1) *N Ar.* pleasure and enjoyment | удовольствие и наслаждение → **zevk, sefa**

zauklan– s. **zeuklen–**

zavalli s. **zavallı**

zavallı (8), **zavalı** (1), **zavalli** (3) *Adj* miserable, unlucky | бедный; несчастный

zāif s. **zaif**

zālim (4) *Adj Ar.* cruel, pitiless | жестокий, безжалостный

zāmet (1) *N Ar.* work, trouble, difficulty, ado | труд, беспокойствие, хлопоты, забота

zār (2), **zar** *N P.* crying, weeping, to sob, to groan | плач, рыдание, стон; **ah zār** bitter tears | охание и стон, горький плач; **zār ček–** to shed bitter tears | стонать, горько плакать

zāri (2) *Adv P.* bitterly | горько, печально; **zāri zāri aγla–** to weep bitterly | горько (безутешно) плакать

zāt (1) *N Ar.* person, individual | особь, особа, личность, лицо

zāten (1) *Adv Ar.* essentially, in any case | по сути; вообще; собственно говоря; именно

zeher (13), **zehir** *N P.* poison, bane | яд; отрава

zehir s. **zeher**

zejtın (1) *N Ar.* olive | маслина, олива

zekjat (2), **zekjet** *N Ar.* alms (prescribed by Islam, generally one fortieth of one's income) | закат (ежегодная милостыня, которая равняется одной сорокой части годового дохода); **zekjat olsun** let it be alms, charity | пусть будет милостыня

zekjet s. **zekjat**

zeman s. **zaman**

zemane (1) *N Ar.* the present time | современность; **zemanede** at this time, now | в это/настоящее/текущее время; теперь

zembil (5) *N Ar.* basket (woven of rushes or palm–leaves) | корзина; зембиль, плетёная кошёлка

zenat s. **zanāt**

zengin (4) *N/Adj P.* rich, wealthy; wealthy man | богатый

zeuk (2) *N Ar.* pleasure, enjoyment, fun | радость, удовольствие, наслаждение

zeuklen– (2), **zauklan–** *Vpass* to amuse oneself, to enjoy | развлекаться, наслаждаться

Zevaj (2) *Pn Ar.* Zevay | Зевай

Zibide (1) *Pn fem. P.* Zibide | Зибиде

zijade (6), **zijāde** *Adj Ar.* more, much, too much | больше; более; слишком

zijan (1) *N P.* loss, damage, harm | ущерб, вред, убыток; **zijan et–** to suffer loss or damage | терпеть убыток/ущерб

zijāde s. **zijade**

zijāfet (1) *N Ar.* feast, banquet | угощение, пиршество

zil (3) *N P.* bell ring, ringer | звонок

Zilfikar s. **Zulfikar**

zilif (1) *N P.* side lock of hair, lovelock | локон, завиток → **zůlůf**

zinalık (2) *N Ar.* adultery, fornication, immorality | прелюбодеяние, блуд

zinǯir (3), **zınǯer** *N P.* chain | цепь

zingar (1) *Adj P.* blue | голубой

zip (4), **zıp**, **zıpt** *N Ar.* pitch, tar | смола, дёготь, вар

zıldan (1) *N P.* prison, dungeon | тюрма

zınǯer s. **zinǯir**

zıp s. **zip**

Zıpır (1) *Tn* Siberia | Сибирь

zıpt s. **zip** *N*

zor (1) *Adj/N P.* hard, difficult; strength, force; difficulty; suffering | тяжёлый, трудный, сильный; сила, трудность; страдание; **zornen** by force, with difficulty | силой, насильно; с трудом → **zur**

zorla– (2) *V* to force, to compel | принуждать, заставлять → **zurla–**

Zulfikar (3), **Zilfikar** (2) *N Ar.* Zulfikar | Зульфикар

zulum (3) *N Ar.* compression, violence | притеснение; насилие

zumpa (1) *N P.* pestle, mallet | пест, пестик (ступки); колотушка

zur (8) *Adj/N P.* hard, difficult; strength, force; difficulty; suffering | тяжёлый, трудный, сильный, мощный; сила, трудность; страдание; **zurɣa kal–** to be put in a difficult position | попасть в затруднительное положение; **zurnen** by force, with difficulty | силой, насильно; с трудом → **zor**

zurla– (6) *V* to force, to compel | принуждать, заставлять → **zorla–**

zurlan– (1) *Vpass* to be forced, to be compelled | быть принуждённым

zurna (2) *N P.* a primitive double–reed instrument played with a drum (davul) in folk music | зурна (простой деревяный духовой музыкальный инструмент с двойной тростью обычно выступает в народной музыке вместе с давулом)

zůlf s. **zůlůf**

zůlůf (~**zůlf**) (2) *N P.* side lock of hair, lovelock | локон, завиток → **zilif**

zůmbůl (3) *N Ar.* hyacinth (*Hyacinthus orientalis*) | гиацинт восточный

www.ingramcontent.com/pod-product-compliance
Lightning Source LLC
Jackson TN
JSHW051912211025
92956JS00004B/36
9783111441979